Plantation Forestry in the Tropics

Plantation Forestry in the Tropics

The Role, Silviculture, and Use of Planted Forests for Industrial, Social, Environmental, and Agroforestry Purposes

Julian Evans
John W. Turnbull

THIRD EDITION

OXFORD
UNIVERSITY PRESS

OXFORD
UNIVERSITY PRESS

Great Clarendon Street, Oxford OX2 6DP

Oxford University Press is a department of the University of Oxford.
It furthers the University's objective of excellence in research, scholarship,
and education by publishing worldwide in

Oxford New York

Auckland Bangkok Buenos Aires Cape Town Chennai
Dar es Salaam Delhi Hong Kong Istanbul Karachi Kolkata
Kuala Lumpur Madrid Melbourne Mexico City Mumbai Nairobi
São Paulo Shanghai Taipei Tokyo Toronto

Oxford is a registered trade mark of Oxford University Press
in the UK and in certain other countries

Published in the United States
by Oxford University Press Inc., New York

© Julian Evans, 1982, 1992
© Julian Evans and John W. Turnbull, 2004

First edition published 1982 (hbk), 1984 (pbk)
Reprinted 1986
Second edition published 1992
First reprinted 1992
Reprinted (with corrections) 1996, 1999

A catalogue record for this title is available from the British Library

Library of Congress Cataloging in Publication Data

(Data available)

ISBN 0 19 852994 5 (hbk)
ISBN 0 19 850947 2 (pbk)

10 9 8 7 6 5 4 3 2 1

Typeset by Newgen Imaging Systems (P) Ltd., Chennai
Printed in Great Britain
on acid-free paper by Antony Rowe Ltd, Chippenham, Wiltshire

To

All who benefit from trees, but especially
those whose livelihoods depend on them

Preface

It is 12 years since the first author (JE) wrote the preface to the second edition and 22 years since the first edition of this book appeared. Much has changed. In particular, there is the expectation now that 'planted forests'—the emerging term to embrace tree planting in all its forms—will, in our lifetimes, become the principal wood fibre resource in the world. Moreover tree-planting and creation of new forests may play a key role in mitigating climate change as carbon is stored, at least for a time. New technologies are also playing their role, new understanding of full involvement of people—stakeholder participation, and new appreciation of ecological imperatives and what sustainability really means, all make a new edition essential.

For this new edition, with its greater complexities and capturing of advances in plantation forestry, John Turnbull has played a huge part. While we have kept the structure and approach of the previous editions, which appeared to be so much appreciated, every page of every chapter has been revised and updated and two new chapters—clonal forestry and ecological restoration—have been added. And, owing to my (JE) wife's untimely death in early 2002, John took on an even greater role by taking under his wing much of the revision.

As well as updating the text, the book is enriched by drawing upon visits to many tropical countries not before included, most notably Bangladesh, Bolivia, Burkina Faso, China, Ecuador, Mali, Nepal, and Vietnam. John's long association with the Center for International Forestry Research (Indonesia) and Australian Centre for International Agriculture Research and Julian's continuing visits to Ethiopia and Swaziland, travels on behalf of the Christian relief and development agency, Tearfund, and his key role with the two UNFF intersessional meetings on planted forests—Chile (1999) and New Zealand (2003), complement each other and have greatly added to our experience as we have revised this book.

The aim of the book remains the same as before, namely, to give an account of plantation forestry in the tropics and subtropics for the student in developing countries and for all who intend to work there. It is not a manual of practice as such, rather it is an overview that sets plantation silviculture in the wider context of the development processes and its social, environmental, and ecological impacts. Plantation forestry is more than simply growing trees efficiently. It is this wider embrace that we tackle here and earnestly hope it will continue to be of use to many across our world as we struggle with poverty, with environmental degradation, and with the injustice of corruption and unfair trading, which all seem to focus on or become magnified for those who live in what we still call the third world.

This last remark betrays Julian's particular Christian convictions, but both of us deeply desire to see the well-being and livelihoods of all in the tropics enhanced and less fragile. We know the particular role that well-conceived planted forests can play and hope this new edition will advance this aim.

Julian Evans
John W. Turnbull
October 2003

Acknowledgements

Many colleagues and friends have provided advice, comments, and contributions as we have worked on this third edition. We have endeavoured to acknowledge them below.

Where material from previous editions of this book is again used no specific acknowledgement is repeated here.

All of the following have assisted us, and we are most grateful: Y.Ali Hussin, J. Ball, M. Baxter, F. Blakeway, K. Bubb, N. Byron, E. Campinhos Jr., C. Cossalter, H.R. Crompton, L.Cunningham, B. Dell, J.C. Doran, K.G. Eldridge, J.L. Gava, J.L.M. Gonçalves, A.R. Griffin, B.V. Gunn, C.E. Harwood, Ryde James, A. Kenny, P. Lal, D. Lamb, I. Last, N.E. Marcar, J.N. Marien, A. Mitchell, A.R. Morris, D.G. Nikles, P. Saenger, J. Simpson, M.J. Stewart, Xu Daping, N. Yalimaitoga

In Australia, special thanks to Dr Glen Kile, former Chief, CSIRO Forestry and Forest Products for his support and to his staff Kron Aken, who assisted with photographs, and Ailsa George, for assistance with literature searches.

In the United Kingdom thanks are due to Imperial College London, where Julian works part-time as Professor of Tropical Forestry, for permitting some of his revision work to be included in his duties.

Last but not least, John's wife, Jennifer, provided tremendous support and assistance editorially.

Julian also acknowledges his late wife who supported his work on earlier editions of this book and who was so pleased that this time round he had a co-author. How apposite that was to prove to be, since Margaret died unexpectedly in January 2002, only God knew.

Specific acknowledgements for the third edition

The following have kindly granted the right to reproduce material in this new edition.
E. Campinhos Jr., Fig. 11.6(a)
J. Carle (UNFAO) for data on plantation areas in Chapter 3.
Mr C. Cossalter for Figs 11.1(b) and 22.2.
Dr K.G. Eldridge for Fig. 11.1(a)
Dr C. Harwood for Fig. 11.6(b)
Dr D.G.Nikles for Fig. 11.4
Dr P. Lal for Figs 11.8 and 11.9
Mr M. Baxter for Figs 11.7(a), (b), and (c).
Mr J.L. Gava and Dr J.L.M. Gonçalves, for Figs 12.6(a) and (b).
Prof. P. Saenger for Fig. 23.9(a) and (b)

CSIRO Publications and Dr B.Dell for Table 14.3

SAPPI (Usutu) and Dr. A. Morris for Fig. 24.6

Contents

PART I

Introduction

CHAPTER 1

Introduction

Plantation forestry in the tropics is a large subject. The aim of Chapter 1 is to indicate the scope of this title by defining and describing briefly for the purpose of this book, the tropics, the tropical environment, forest plantations, and tropical tree species.

Most forest plantations are distinct from rain forest or savanna because their orderliness and uniformity show they are artificial. But more careful definition of 'plantation' or 'planted forest' is needed since much tree-planting is not done in regular blocks. Many trees are planted within existing forest, that is, enrichment planting, or as single rows, that is, linear planting. Though both involve tree-planting, they often do not constitute conventional plantations. Moreover, older stands of planted forest may take on many features of natural forest, which further blurs definitions. In addition, tree-planting in association with food crops and livestock management (agroforestry) in the tropics is increasing. Trees in most agroforestry systems are not intended to become closed canopy forest but nevertheless they represent a growing stock of planted trees of value to society.

The tropics

The 'tropics' can be easily defined in geographical terms. They are that part of the world lying between latitudes 23° 27' north and south of the equator. But this is of limited usefulness since land-uses or plantation forestry typical of tropical regions is not bound by strict latitudinal limits. The point is seen in the suggestion that the tropics are 'where bananas grow'! Moreover, the characteristic warmth of tropical regions and the associated vegetation types may be found outside the geographical tropics and, conversely, on the mountains of equatorial Ecuador, New Guinea, or Uganda typically temperate conditions prevail. Nevertheless, it is because of many similarities rather than divergences in the tropical environment, including social and economic conditions, that allow plantation forestry in the tropics to be considered as a whole.

Though the geographical definition of the tropics is not of much value for delimiting land-use, some indication is needed of the regions considered within the embrace of this book. South of the equator these extend to at least 27°S to include, for example, the plantations in southeast Queensland (Australia), those of KwaZulu Natal and Swaziland in southern Africa, and afforestation in the state of São Paulo, Brazil. A similar latitude north of the equator includes the Caribbean, most of Mexico, the Gangetic plain of India and irrigated plantations of Pakistan, the Chinese fir (*Cunninghamia lanceolata*) plantations of southern China and Taiwan, and the arid zone plantings in the African Sahel.

Within this broad belt lies some 40% of the world's land surface (rather more than 5000 million ha), almost half of its population, but comparatively little of the world's economic prosperity. Though most nations in this region are classified as 'developing countries', there is great potential for increased output in food and plantation forestry.

As background to the important developments taking place in the tropics, particularly in land-use practices including tree-planting, four broad aspects of the tropical environment are considered in outline.

Climate

Climate is the most important influence on vegetation and it is climate that distinguishes the tropics from other parts of the world. But the distinction is not so much in magnitude, such as great heat, but in constancy and uniformity. Except in more subtropical parts, seasons only occur because of variation in rainfall, not because of variation in temperature, which causes summers and winters in temperate lands. Excellent general accounts of the tropical climate will be found in Jackson (1977) and Nieuwolt (1977) but while most texts provide long-term averages for climatic factors we should stress that these do not always correspond closely to actual conditions that may be experienced in the short term at a particular location.

The productivity of trees in the tropics is affected by many environmental factors but temperature, incident radiant energy and rainfall are the main climatic factors. Understanding the ecophysiological basis for productivity of tropical tree plantations and how various environmental factors interact with the physiology of the species is important for developing and managing plantations (Gholz and Lima 1997).

Temperature
The tropics have been defined using average air temperature, for example, all regions within the 20°C isotherm, or where the coldest month has a mean temperature of at least 18°C. More meaningful, however, is to delimit the tropics as where mean monthly temperature variation, between the average of the three coldest and three warmest months, is less than 5°C. This accommodates the much lower absolute temperatures experienced in tropical highlands and emphasizes the uniformity of temperature throughout the year, which is the main reason why many temperate crops will not grow.

Lack of variation in mean monthly temperature does not necessarily apply to temperatures during a single day. In cloudy equatorial regions temperatures may vary only a little with a daily range of 7 or 8°C, but in arid, often cloudless regions such as the Sahel, a range of 25 or even 30°C is not uncommon. Constancy of temperature within 1 day is not a characteristic of the tropics, as occasionally damaging frosts to coffee and other crops readily confirm.

A useful though simplified generalization is that temperature extremes increase away from the equator: towards the subtropics both daily and seasonal temperature variation gradually become more pronounced.

Solar radiation and daylength
In the tropics daily solar radiation levels average about twice that of temperate regions (Landsberg 1961) and are not so variable owing to less annual variation in the angle of the sun's rays. Differences in cloudiness are the chief cause of any variation. The higher radiation levels are because the sun's rays strike the surface at or near to perpendicular. Thus a unit of the sun's energy is intercepted by a smaller surface and the thickness of atmosphere through which the rays pass is less.

This higher radiation, when combined with a long sometimes continuous growing season, produces very high yields since more energy is available for photosynthesis. One of the most productive forest plantations in the world was *Eucalyptus deglupta* in Papua New Guinea (PNG) located just 5°S of the equator growing on deep river alluvium and experiencing a year-round rainfall (Fig. 1.1(a)). At 3 years of age mean height was 24 m and stand volume 288 m^3 ha^{-1} thus achieving > 90 m^3 ha^{-1} y^{-1} increment (Eldridge *et al.* 1993). Over many thousands of hectares of clonal plantations of *E. grandis*, *E. urophylla* and their hybrids at Aracruz in the moist coastal plain of eastern Brazil

Figure 1.1 (a) Highly productive plantation of native *Eucalyptus deglupta*, age 7 years. The tree in the centre is 38.3 m tall and 39.5 cm diameter breast height, Papua New Guinea. (b) Highly productive plot of exotic *Eucalyptus grandis* standing at over 1000 m^3 ha^{-1} at 30 years of age located 28°S but in the 'tropical' environment of coastal KwaZulu-Natal, South Africa. (This stand is the one featured on the book's cover.)

(19°S) the mean annual increment with a 7-year rotation is 45 m^3 ha^{-1} (Campinhos 1999).

Daylength (photoperiod) in the tropics is almost constant at about 12 h. Even at 25° latitudes, variation during the year is only 3 h, from a minimum of about 10.5 h to a maximum of 13.5 h. Although total radiation decreases at higher latitudes there may be some compensation in the increased daylength in summer. Eucalypts are daylength neutral and grow whenever temperature and moisture levels are adequate. In contrast the growth of most poplars (*Populus* spp.) is constrained when they become dormant in response to short daylength in the tropics and commercial plantations are rarely found below 27° latitude.

Rainfall
In the tropics rainfall is the most important climatic influence on agriculture and forestry. The amount and distribution of rainfall are the main criteria used to classify tropical climates and to delimit seasons. Annual rainfall amounts vary from zero to more than 10 000 mm. Moisture is not considered seriously limiting at any time of the year for only about 28% of arable land in the tropics. For 42% of arable land moisture is limiting for 4–7 months, and for the remaining 30% for more than 8 months. In plantation forestry this great variation affects choice of species, rates of growth, quality of wood produced, timing of operations, and even the purpose for growing trees.

Climates of tropical and subtropical regions can be broadly classified into four main types based on the amount and distribution of rainfall:

1. *Humid*: high rainfall (over 1800 mm) with only short dry periods and humid for over 9 months of the year. Rainfall exceeds potential evapotranspiration in most months. About one-quarter of the tropics are included, mostly in equatorial regions, for example, part of Amazonia, Congo Basin, much of Southeast Asia, and many Pacific Islands. Rain forest is the characteristic climax vegetation.
2. *Monsoonal wet/dry*: high rainfall (1000–1800 mm) but characterized by well-defined wet and dry seasons and a humid period of 4–9 months. Potential evapotranspiration generally exceeds rainfall for 4–6 months of the year. Almost half of the tropics fall into this category. This climatic type occurs in eastern Indonesia, northern Thailand, parts of Myanmar, southern and eastern India, much of central and western Africa (south of the Sahel), north-eastern Brazil and Venezuela. Climax vegetation may be savanna, deciduous, or semi-deciduous woodland.
3. *Subtropical humid and subhumid*: humid areas have high rainfall (1000–1800 mm) with only short

dry periods and are humid for over 9 months of the year. Subhumid areas have moderate rainfall (600–1500 mm) with less humid conditions in most of the year. Some of the more important industrial plantations have been developed in this climatic type, for example, in eastern Australia, southern Brazil, Paraguay, eastern South Africa, Zimbabwe, parts of northern India, northern Vietnam, and parts of southern China. Climax vegetation may be evergreen or semi-evergreen woodland and forest, with rain forest in more humid areas.
4. *Arid and semiarid*: low, seasonal, and erratic rainfall usually in 1–4 months of the year with long dry periods. Arid areas have less than 200 mm y^{-1}. Many great deserts border the tropics, vegetation is sparse or absent, and food cannot be grown without irrigation. Semiarid areas have 200–600 mm annual rainfall concentrated in up to 4 months. About one-sixth of the tropics are included, for example, the African Sahel, parts of south Asia, much of tropical Australia, and areas of northeast Brazil. Vegetation is sparse and is often dominated by *Acacia* scrub. Establishment of plantations occurs to only a very limited extent, tree-planting is primarily for protection (soil stabilization and erosion control, sand-dune fixation), for shelter and shade, and for building poles, fencing materials, and fodder.

As well as rainfall distribution, intensity is important. Heavy thunderstorms are frequent and it is not uncommon for 200–400 mm of rain to fall in a few hours. Such heavy rain is often erosive because of large droplet size and because much water runs off the surface rather than soaking in. Not only is less moisture retained by the soil but more surface erosion occurs.

One extremely important aspect of rainfall in the tropics is its unreliability both in total amount from year to year and in onset and duration of the wet season especially in more arid areas. This is largely determined by the relative movement north and south during the year of the Inter-Tropical Convergence Zone (ITCZ). Failure to reach higher latitudes means failure of adequate rains. A tragic illustration was the prolonged drought affecting the African Sahel, and especially Ethiopia and Sudan, from 1982 to 1985 and again in 1990–91 when many people and domestic animals perished. However, even relatively small changes and irregularities can damage plants. In Fiji very low rainfall in 1977 led to many deaths of newly planted *Pinus caribaea*. Periodic prolonged winter droughts in Swaziland have killed *E. grandis* and *P. taeda* and much reduced growth of *P. patula* even though the total annual rainfall was about average (Evans 1978; Morris 1993a).

Other forms of precipitation such as mist and hail are locally important. Mist is a common phenomenon in tropical highlands and may augment plant moisture supply. Species of tropical pines differ markedly in the way water droplets condense on their needles, which appears related to the mistiness of their natural habitat (Lamb 1973) and some eucalypts, for example, *E. urophylla*, develop a 'drip tip' on their leaves, apparently as a means of shedding excess moisture. Violent hailstorms often occur in areas of broken topography, particularly in more subtropical regions, and can severely damage some tree species owing to fungal invasion of scarred bark (Evans 1978).

Wind
Wind is usually only locally important in the tropical climate when high wind speeds of great destructiveness develop in cyclones, thunderstorms, and sandstorms, or when salt-laden near to the shore. However, one general exception is the desicating harmattan wind that blows off the Sahara affecting all West African countries.

Vegetation and soils

Vegetation
The diversity of vegetation, ranging from species-rich rain forest to barren desert, provides enormous variety in the tropics. Climatic variation, particularly rainfall, is the dominant cause for this great diversity. The coverage of the main vegetation types are approximately: savanna 42%, rain forest 30%, semi-deciduous and deciduous woodland 15%, desert shrubs, grasses, or no vegetation 13% (Sanchez 1976). These figures are not static; the present trends are decline in forest and increase in grassland and desert.

For the purposes of plantation forestry four points are noted:

1. Not only is vegetation an important indicator of climate, but also within one area, variation in plant species and luxuriance of growth usually indicate local differences in soil and site conditions.
2. The presence of many species and seemingly profuse growth, as in most rain forest ecosystems, do not necessarily indicate high soil fertility. Many of the soils under rain forests are infertile; the forest subsists on efficient reuse (cycling) of nutrients through rapid decay and re-incorporation of litter—dead leaves, fallen fruits, branches, etc. By contrast many poorly vegetated steppe and semi-desert soils have quite high fertility.
3. Probably some 80% of all plant species in the world are found in the tropics. This is a genetic resource barely examined yet which is already diminishing as tropical forest is cleared for

'development' or by shifting cultivation. Not only does the value of many species await discovery but, equally important, many species already known to man are underused (National Academy of Sciences 1975). Agriculture development tends to narrow the variety of food crops cultivated. Today about twenty crops, out of the 1000 or more known to man, produce 95% of the world's food. Similarly in forestry more than half of all plantations in the tropics are eucalypts or pines, yet many untried species of plantation potential must exist in the rich tropical flora.
4. The type and density of vegetative cover directly affect plantation development because of the need to prepare a site for planting.

Soils
Tropical soils vary in their properties at least as much as temperate ones. Parent materials climate, relief, surface vegetation, and other organisms interact over time to produce numerous kinds of soils. There are many national soil classifications. The main systems used internationally are 'Soil Taxonomy' (Soil Survey Staff 1999) and the FAO/ UNESCO terminology now updated as the World Reference Base for Soil Classification (WRB 1998). Soil names in this book follow the Soil Taxonomy system. The great diversity of soils in the tropics is indicated by the recognition of 1 million families, 200 great groups and 11 orders (Eswaran *et al.* 1992). Five soil orders, Oxisols, Ultisols, Aridisols, Alfisols, and Entisols, extend over 75% of the land area of the tropics. Maps showing the distribution of these main soil types in the tropics are shown in Lal (1997). A comprehensive world soil map is available (FAO 1988*a*) and can be obtained in digitized form. Regional soil and terrain databases are also available (e.g. FAO/UNEP/ISRIC 1999).

Most tropical soils available for commercial tree plantations have low natural fertility. Those with high levels of fertility are limited in extent and nearly all already support densely populated centres of agriculture, for example, the volcanic soils of East Africa and Indonesia, the Vertisols of the Sudan and India, and alluvial soils (Inceptisols) of the river basins in Asia. Oxisols and Ultisols and Alfisols, predominant soils of the humid tropics, are highly acidic and deficient in P, K, Ca, and Mg but have generally satisfactory physical characteristics. In the subhumid and semi-arid tropics Alfisols, Aridisols, and Vertisols predominate and these soils have reasonably high fertility but poor physical structure and so are very susceptible to compaction and erosion (Lal 1997).

There now is greater awareness of how to manage tropical soils (e.g. Sanchez 1976; Young 1989;

Eswaran *et al*. 1992; Van Wambeke 1992). Many soils of equatorial lowlands and in monsoonal climates are infertile and are unable to support annual cropping, as practised in temperate countries, without regular addition of fertilizers or manures. For example, only 7% of soils in the Amazon basin are free of major limitations for agriculture, and about 55% of soils in tropical America are too poor for farming, although they are capable of supporting forests (Wadsworth 1997). In many tropical countries human-induced degradation of soils of various types, such as salinity, and water or wind erosion, is a serious problem that is endangering agricultural sustainability and environmental stability. The extent of degraded areas has been assessed and a database and world map produced (Oldeman *et al*. 1991).

Protection of the soil by a vegetative cover is an important reason for afforestation and those concerned with plantation development have to recognize the diversity of tropical soils and adjust species and management practices accordingly (e.g. Gonçalves *et al*. 1997).

Land-use and farming practices

Land-use is not static and the potential for change in the tropics is very great. Clearing of forest to provide higher returns has a long history. Over the centuries agriculture has expanded and much forest has disappeared. In recent years the rate of conversion has been particularly high in the tropics and in 1990–2000 almost 1% (14.2 million ha) of tropical forest was cleared annually (FAO 2001*a*).

It has generally been accepted that population growth drives deforestation through the demand for more agricultural land for food production although many governments have cleared forests to stimulate economic development and to provide employment for rural populations. Most people living in the tropics work on the land, for example, in India and south China more than 80% do so, and although the rural population has been declining in some countries due to industrialization and urbanization, the area of arable land per capita in many countries is decreasing. In some countries improvements in agricultural productivity has been greater than needed to keep pace with population growth, but elsewhere much prime agricultural land has become degraded due to inappropriate farming methods.

Although the direct link to population growth and shifting cultivation seems less valid than previously thought, the demand for agricultural land remains a major factor leading to deforestation (FAO 2001*b*). These facts emphasize that there will be continuing pressure on tropical forests to provide land for agriculture and that agricultural practices need to change, especially on marginal lands, as do illegal logging and other corrupt practices that are widespread.

Over 30 years ago Kellogg and Orvedal (1969) pointed out that most of the world's unused but potentially arable land was in the tropics, and potential exists in the tropics for increased output of food and timber from improvements in land-use. Indeed, as the Brundtland report makes clear (Brundtland 1987), two of the three main tropical regions, Latin America and south and Southeast Asia, have seen steadily increasing food production; only in sub-Saharan Africa has it declined.

Shifting cultivation, which is by far the most extensive farming system, wholly depends on natural cycles to sustain it and is often in upland hilly areas. Similarly, nomadic herding in arid regions is dictated by the natural availability of grazing and forage. Numerous attempts to increase agricultural production on such marginal lands have failed and have resulted in degradation through soil erosion or salinization. It is now clear that appropriate farming systems must be developed and implemented on these fragile tropical lands. But, responsible land-use depends on land tenure security, and problems of rights to land and ownership are possibly the single greatest obstacle to sustainable development of tropical lands.

Problems of land-use and changes in farming practices can be illustrated with an example from Asia. Southeast Asia is characterized by a high population growth rate and a low area of arable land per capita, a combination that has led to migration and dense settlement of upland areas. Most agriculture is on marginal uplands many of which are part of about 180 million ha of erodible, acidic Ultisols. Traditional shifting cultivation has been largely replaced by more intensive permanent agriculture with annual cropping and soil losses from erosion are typically 2–4 cm year^{-1} (200–400 t ha^{-1} year^{-1}) (Fujisaka 1991). Annual cropping in these uplands carries a higher risk of soil degradation than systems with a tree or shrub component and it has been recommended that agroforestry technologies will improve the long-term productivity and sustainability of agriculture, and reduce off-site damage due to flooding and siltation of dams etc. (Blair and Lefroy 1991). In some areas local agroforestry systems evolved from traditional shifting cultivation practices. These can be extended and new technologies introduced. However, the adoption of these 'improved' technologies will depend on socioeconomic and policy changes that ensure security of land tenure, provision of credit and/or subsidies, access to markets and market information, appropriate pricing and taxation regulations, etc.

Economic, social, and developmental status

Nearly all countries in the tropics are described as 'developing' and have a low Gross National Product (GNP) per person. Small family incomes mean that most people in the tropics are poor. Many countries in Africa have a GNP below US$500 per capita per year, and the huge populations of China and India had GNPs of US$688 and US$392, respectively in 1997 (FAO 2001a). Within a country there can be very unequal distribution of wealth and many rural populations practising subsistence agriculture in unfavourable environments experience severe poverty.

There is considerable variation in population growth and economic development among countries in the tropics. Between 1995 and 2000 the global average annual population increase was 2.4% compared with 1.4% in Asia and the Middle East, and 1.5% in South America. Asian countries such as China, Malaysia, and Vietnam have relatively high growth of GNP whereas growth is slow in most African countries.

Production is often based on a few agricultural and mining commodities, many of which are exported wholly or partly unprocessed (coffee, copra, timber, copper, tin, etc.) and exposed to wide fluctuations in price. Developed countries frequently restrict access of agricultural products from developing countries or impose quotas and/or tariffs to protect their own farmers. The incentive to manufacture goods is also dampened by trade barriers restricting access to markets of developed countries. Of equal concern is that much development has involved producing commodities, often using advanced technology, exclusively to satisfy demand in developed countries rather than improving directly the welfare of the nation itself. Though such projects, when successful, earn foreign exchange, produce tax revenue, and raise living standards of those associated with them, they have frequently diverted attention away from more fundamental, sustainable development, often better achieved through small, technologically simple improvements.

On top of this dismal picture is often political instability, arising from past constraints of colonial rule, tribal rivalries, and unfulfilled aspirations following independence, that leads to frequent and often violent change in government. In Africa, since 1945, at least 50 wars have been fought involving over half of the continent's nations of which most sought to overthrow the ruling regime or arose because of tribal, religious, or minority grievances.

Such instability, compounded by wastage of resources in warfare, militates against progress and development. Weak governance in the forest sector is common in many countries. This allows forest crime and corruption such that illegal operations proliferated. There is increasing awareness of the immense costs associated with corruption and illegal activities and the disincentives to international investments (FAO 2001b). Political instability and weak governance of the forestry sector especially affect tree-planting and plantation forestry since development takes many years and investment is committed for a long time.

Defining a plantation

Simply the act of planting trees defines the subject, but there is a continuum of types of plantation and the interface between some planted forests and natural forests is indistinct. Most people have their own idea of what constitutes a 'plantation' but agreeing on a precise definition of this and associated terms such as 'afforestation', 'reforestation', and 'deforestation' is not easy. A visit to the Internet website of Lund (2000) provides a plethora of definitions from many countries and illustrates the complexity of the problem but efforts are being made to harmonize forest-related definitions (e.g. FAO 2003).

Planted forests may be classified in several ways, for example, by their species composition, by their scale, by their complexity, or by the purpose for which they are planted (Evans 1999a). Although planted forests are frequently managed intensively for wood production they can also be managed less intensively for conservation, protection, or other socioeconomic purposes. Distinction between different types of planted forests are important for global assessments and international communication. For definitions relevant to resource assessment see FAO (2001a) and Carle and Holmgren (2003). For a discussion of definitions relevant to climate change and carbon sequestration see Noble *et al.* (2000).

FAO (2001a) defines plantation as: 'Forest stands established by planting or/and seeding in the process of afforestation or reforestation. They are either:

• of introduced species (all planted stands), or
• intensively managed stands of indigenous species, which meet all the following criteria: one or two species at plantation, even age class, regular spacing.'

While this definition covers most situations there remains the question of the meaning of 'intensively managed'. It was intended to exclude stands established as plantations but are now considered seminatural because they have not been managed intensively for a significant period. The FAO definition was not intended to replace existing national classifications because national inventories, terms and definitions have specific purposes of relevance to each for each country (FAO 2001*a*). There are many types of plantations ranging from short-rotation industrial plantations through to 'close-to-nature' plantations that vary in intensity of management and other management practices according to whether the objectives are to maximize wood production, to maximize environmental values, or some combination of production and conservation objectives. Close-to-nature plantations are complex production systems using more than one species that may be uneven-aged, and several management practices, such as a mixture of coppice and standards, to provide a range of products and environmental services (Kanowski 1997). Lack of agreement on interpretations of the definition of 'plantation' or the now widely used term 'planted forest' causes problems in comparing forest resource statistics from different sources.

The boundary between planted and natural forests can be imprecise. The first consideration is one of origin. Between the extremes of afforestation and unaided natural regeneration of indigenous forest, there is a range of forest conditions where intervention occurs to a greater or lesser extent in regeneration. Four forest types can be identified according to their origin.

1. Afforestation is the act or process of creating forest land where it 'historically' did not exist (Lund 2000). Others have specified the time factor as 'where there has been no forest for at least 50 years' (Evans 1992) or 'which previously did not carry forest within living memory' (FAO 2001*a*). Afforestation of grasslands falls into this category (Fig. 1.2) and planting to stabilize sand-dunes, etc.
2. Reforestation is the act or process of changing previously deforested lands back to forest land

(Lund 2000). A distinction can be made on the basis of whether the previous crop is replaced by the same or a different crop. An example of the latter is where rain forest is logged, cleared, and then part replanted with a single tree species, for example, *Acacia mangium* or *Paraserianthes falcataria* (Fig. 1.3). The former is less common but *Araucaria* plantations in PNG and some *Triplochiton* plantations in West Africa are examples of this forest type since the previous forest was often dominated by the same species as used in the plantations (Fig. 1.4). Replanting is the re-establishment of planted trees, either because afforestation or reforestation failed, or the tree crop was felled and regenerated (FAO 2001*a*).
3. Forests established by natural regeneration with deliberate silvicultural intervention and manipulation.
4. Forests, which have regenerated naturally without human assistance, for example, most natural forests in the tropics.

This book, plantations are the forest types in classes 1 to 2 above, that is, artificial regeneration is the basic criterion.

It is common to differentiate further between 'industrial' plantations', which are established totally or partly to produce wood for industry, mainly sawlogs, veneer logs, pulpwood, and mining timbers; and 'non-industrial' plantations established for one or more of the following objectives: fuelwood, wood for charcoal, wood for domestic consumption, non-wood products, and soil protection. It is not always possible to differentiate rigidly between these types of plantations.

In addition to their origin, several other characteristics tend to be associated with plantations. These include:

• Well-managed plantations usually have higher yields of useful wood than natural forests. Many commercial plantations in the tropics have an

Figure 1.2 Afforestation of grassland with *Pinus patula* on the Viphya plateau, Malawi.

Figure 1.3 Reforestation with *Paraserianthes falcataria* on recently cut-over rainforest in southern Philippines.

Figure 1.4 Reforestation using the same species (*Araucaria* spp.) that naturally dominate the lower montane forest near Bulolo in PNG.

annual growth rate of 10–30 m³ ha⁻¹ compared with 1–5 m³ ha⁻¹ for natural forests.
• Plantations produce wood quickly and of a more uniform size and quality than from natural forests. This facilitates harvesting, transport, and conversion.
• Plantations can be located wherever infrastructure and suitable land are available, and near to population centres or wood processing units thereby making them more easily accessible and reducing transport costs.

Traditionally, trees have been planted for landscape enhancement, prevention of soil erosion, providing shelter against heat and wind, and in agroforestry systems. They may also be carbon sinks to help mitigate global warming.

There are billions of trees planted in cities on farms and along roads not included in the FAO definitions of 'plantations' and 'forests'. These 'trees outside the forest' are not included in plantation statistics but nevertheless make a substantial contribution to the environment and provide significant social and economic benefits. In Kenya trees on farms produced almost 10 million m³ of wood in 2000 and in some densely populated areas contributed 18–51% of total household incomes (FAO 2001*a*). Reference is made to such plantings throughout this book.

Shape

Plantations are usually of regular shape with fixed and clearly defined boundaries. Most industrial plantations are established in large blocks, although retention of areas of natural forest among the blocks for conservation purposes has become more common, for example, at Aracruz Florestal in Brazil (Campinhos *et al.* 1993) and is an essential requirement for almost any type of certification. Planting small blocks, belts, and strips of trees, and often single lines, is an integral part of much rural development. The 'four around' planting in China involved planting strips of trees along roads, railways, waterways, and farms, and small block planting (<10 ha) of trees for industrial purposes is common in outgrower schemes in several countries.

Stocking

This refers to the number of usable trees per unit area, and the very words 'forest' or 'plantation' imply that land is reasonably well wooded. Essentially, a satisfactory plantation is where the planted trees are reasonably uniformly distributed and are using most of the productive capacity of the site. Optimal spacing of trees can vary considerably depending on the type of plantation, site productivity, and the effect of individual stem size on its value. It is the usual practice to plant clonal plantations or plantations using improved seedling stock at wider spacing as thinning of poorly formed trees is not necessary. In intensively managed industrial plantations growers aim to have a stocking of over 90% and 600–1500 trees ha^{-1}. In the Congo, clonal eucalypts are planted at 800 trees ha^{-1}, and in both the Philippines and PNG *E. deglupta* is planted at 4×4 m spacing, giving a stocking of only 625 trees ha^{-1}. Maximum annual increment may occur at moderately high densities (2500–3000 trees ha^{-1}) and where there is a use for very small sized poles this may be a suitable spacing. Farmers in Ethiopia often plant their small woodlots at even higher densities than this as they can progressively thin out the trees and have a variety of uses for the thinnings. In agroforestry systems the spacing between trees and rows must be based on a consideration of the effects of shade and root competition of adjacent crops.

Naturalization

Plantations of exotic (introduced) species are obviously artificial and could not occur on the site naturally. But if, subsequently, the species becomes well adapted to its new environment, sets seed freely, and can be easily regenerated naturally it is said to be 'naturalized'. However, even though naturally regenerated, such forest is sometimes arbitrarily defined as still being 'manmade' for a long period after the original introduction.

Forestry and agricultural tree crops

There is no clear distinction between agricultural and forestry tree crops. Rubber plantations (*Hevea brasiliensis*), grown for the latex which exudes from cut bark, have historically not been classed as forest plantations whereas plantations of *Acacia mearnsii*, grown primarily for bark tannins, usually are. But, recent advances in wood technologies have seen rubber, coconut palm (*Cocos nucifera*), and African oil palm (*Elais guineensis*) becoming increasingly important as a source of wood products. These species account for nearly 28 million ha of plantations, mainly in Southeast Asia (FAO 2001*a*). Such crops are not considered in this book.

Agroforestry is simply the mixing or blending of tree growing with food crops or livestock. It is defined and described in Chapter 20. Sometimes terms such as 'tree plantations' or 'tree crop forestry' are used to describe tree crops cultivated for non-timber products with 'forest plantations' used for conventional wood production. In China, tree crop forestry is considered to be forestry rather than agriculture and a wide range of species are being grown to produce oils, fruits, medicines, etc. (He and Hu 1991).

Mixed regeneration systems

Where enrichment planting supplements existing forest, the forest is normally classified as man-made if the planted trees ultimately form more than half the final crop (Fig. 1.5).

Figure 1.5 Plantation of mahogany (*Swietenia macrophylla*) in Fiji derived from enrichment line planting.

Forest restoration and rehabilitation plantations

In recent years there have been attempts to accelerate the recovery of degraded forests and deforested lands to restore productivity, biodiversity, and other values. Even large-scale industrial plantations have the potential to be used as successional catalysts to assist natural regeneration of trees and shrubs by modifying the microclimate of their understorey, increasing soil organic matter etc. (Lamb 1998). Developing plantations that can produce valuable forest products and at the same time accelerate regeneration of species-rich forest ecosystems is a challenge that is now being addressed (Parrotta *et al.* 1997b). Examples are given in Chapter 22.

Tropical plantation species

Clearly, any species widely planted within the limits of the 'tropics' described earlier is included in this heading. However, a great many species have been tried in trial plots, and some further definition is required. In this book the major emphasis will be on species which fall in one of two categories.

1. Species occurring naturally within the tropics, for example, *A. mangium, E. urophylla, P. caribaea, Swietenia macrophylla,* and *Tectona grandis.*

2. Species planted widely and successfully in the tropics, but often in the cooler highland regions, the natural distribution of which lies largely or wholly outside the tropics. Three examples of major importance are *E. grandis, Grevillea robusta,* and *P. elliottii.*

Species such as *E. globulus, P. radiata, P. taeda,* and *Populus deltoides* will be largely excluded from consideration. Though they have been tried in many parts of the tropics their natural range is wholly outside the tropics and, with some exceptions, for example, *E. globulus* in the highlands of China, Ethiopia, India, and Peru, they have not been a success even though all are important plantation species in the cooler subtropics and/or warm temperate regions.

CHAPTER 2

Why plantations?

Interest in plantations and tree-planting in the tropics continues to increase rapidly. The area of plantation increased over thirteen times between 1965 and 2000, and the rate of planting in the 1990s was double that of the 1980s (Table 3.2). Why should this be so? Why should small countries such as Fiji, Swaziland, and Zimbabwe undertake plantation programmes covering tens of thousands of hectares, and why have millions of hectares being planted in China, India, and Brazil? In India, for example, not only are government and parastatal bodies involved but some 10 000 non-government organizations (NGOs) are concerned to promote tree-planting of one kind or another. This interest in plantation forestry is understandable in countries lacking natural forests where rural populations are heavily dependent on planting, but many countries with large natural reserves of timber are also implementing major projects. For example, Brazil, with one of the largest remaining areas of natural forest in the world, the Amazon rain forest, in 1966 embarked on a planting programme, which reached an annual planting rate of nearly half a million hectares during the late 1970s and early 1980s (Fig. 6.1). Indonesia, with the world's third largest tract of tropical forests and a forest cover of about 100 million ha, embarked on a massive plantation programme starting in the mid-1980s and by 1997 the government had allocated permits to investors seeking to establish a total of 4.5 million ha of plantations (Barr 2001).

In seeking to answer the question of why we need plantations, it must be emphasized that overall there is no immediate worldwide shortage of timber. In 2000 the estimated area of natural forest and forest plantations in the world was 3869 million ha, about 30% of the earth's surface (FAO 2001a). The area of plantations was 187 million ha. The total amount of wood in the world's forests is estimated at 386 500 million m^3.

Several studies have addressed global supply and demand for wood. Sedjo and Lyon (1996) estimated that average annual demand for industrial roundwood would increase from 1700 million m^3 in 1995 to about 2300 million m^3 in 2045. Sohngen *et al.* (1997) analysed regional supply from existing forest resources and the potential role of plantations and FAO (1998) has estimated consumption, production, and trade in forest products globally to 2010 (Table 2.1). By 2010, Asia, driven by increasing population growth and economic development, Europe and the former Soviet Union are expected to increase their share global industrial roundwood consumption while the North and Central American region is expected to decrease its share (FAO 1998). Consumption of pulpwood for wood-based panels and paper is expected to approximately double to 1330 million m^3 in 2045. In 1996, 70% of industrial roundwood was used in developed countries (FAO 1999a).

Solberg (1996) reviewed recent projections for fuelwood, which accounts for over 50% of global wood harvests. Although rising incomes and urbanization may result in less consumption of fuelwood in some developing countries, consumption is expected to rise from 1900 million m^3 in 2000 to 2200 million m^3 in 2010 (FAO 1998). Asia is the largest consumer of fuelwood (49% of the total in 1994) followed by Africa 25%, and South America 12%. Asia's consumption is likely to rise while Africa and South America should maintain their shares at the existing level. In 1996 developing countries used 90% of all fuelwood and charcoal, but wood fuels are also becoming more attractive than fossil fuels as a modern, renewable source of energy in some developed countries for economic and environmental reasons (FAO 1999).

Table 2.1 Current and future forecast global forest production/consumption by products, 1996 and 2010

Product (million m³)	Production/consumption		Growth 1996–2000 (%)
	1996	2010	
Industrial roundwood	1490	1872	26
Sawn wood	430	501	17
Wood-based panels	149	180	20
Pulp	179	208	16
Paper and paperboard	248	394	39
Fuelwood and charcoal	1860	2210	27

Source: Based on FAO (1998, 1999a).

With these consumption levels and a very modest growth rate, the existing 3700 million ha of natural forest could supply world wood needs for many years to come. If present and future demand for wood will not itself cause an overall shortage why then are plantations being established and tree-planting so widely encouraged? Indeed, expansion of such programmes is one of the principal strategies of the National Forest Action Programme (NFAP), formerly the Tropical Forestry Action Plan (FAO 1985a). Many reasons can be put forward, although from country to country the most important factor will not always be the same.

With changes in the type and quantity of industrial roundwood required it is expected that plantations will largely satisfy future increases in global wood supply. Also, natural forests have provided wood more cheaply than plantations but the price advantage is becoming less due to improvements in productivity of intensively managed plantations, decreasing availability of wood from natural forests and increasing costs of transportation of logs from these forests. Annual industrial wood supply from plantations is expected to increase from 624 million m^3 to 1043 million m^3 from 2000 to 2040, a rise from 35% to 46% of the total supply (ABARE-Jaakko Pöyry 1999). Some recent reports argue that 'well before 2050 virtually all wood and fibre products will be derived from managed, planted forests' Dyck (2003).

Factors favouring plantation development

In the tropics, more than anywhere else, there are silvicultural, economic, social, and environmental benefits, which make tree-planting and plantation forestry attractive when compared with natural forests. Flexibility in site location to reduce the distance wood has to be transported to processing facilities, planting of gentle terrain to facilitate mechanized harvesting, and the opportunities for intensive management of selected species to produce more wood per unit area, all reduce costs. These and other factors discussed below favour plantation development.

Past and continuing destruction of natural forest

For a long time natural forest has been exploited, cleared, suffered man-made damage, and has gradually declined in extent. In some countries loss of forest has gone on steadily over thousands of years, in others it is a recent occurrence. But only in the last

150–200 years has net destruction of forest taken place in almost every country and, in recent years, the rate of resource depletion has increased sharply in most tropical countries while forest cover is stable or even increasing in many temperate areas. Globally the annual net rate of deforestation (the balance of loss of natural forest and the gain in forest area through afforestation and natural expansion of forest) has decreased slightly from 1980–90 (13 million ha) to 1990–2000 (9.4 million ha) but it remains high in the tropics with an annual loss of 14.2 million ha converted to other land-uses and a further 1 million ha converted to plantations (FAO 2001a). In Africa, the average loss of forest cover for the 10-year, period 1990–2000 was 7.8% and for countries such as Côte d'Ivoire (31%), Nigeria (37%) Land Sierra Leone (29%) it was very high. Deforestation rates were less in Asia and the South America. In countries with extensive tropical rain forests the 10-year loss was 13 million ha (12%) in Indonesia and 23 million ha (4%) in Brazil (FAO 2001a).

Examples of decline in forest cover abound. In the last 200 years Haiti has changed from being the richest French colony to the poorest country in the western hemisphere largely because nearly all the once extensive forest and most of the topsoil are gone. In 1995 only 1% of forest cover remained. In Ethiopia, forest cover has declined from 40% to just 2.9% since 1850. Between 1965 and 1985 the percentage forest cover in Thailand dropped from 55% to 29% although logging has now been suspended. In Peninsular Malaysia extant rain forest is now largely confined to parks and reserves. In West Africa, Côte d'Ivoire and Nigeria lost more than 5% of forest each year during the 1970s and are still are losing 3–4% annually. Similar reports can be cited for many other tropical countries.

Puri (1960), in *Indian forest ecology*, devoted a chapter to the ways man has influenced the forests of India. He concluded that 'the activities of man, on the whole, acted against the normal development and succession of forest vegetation'. In historic times India was extensively covered with dense forest and now (2000) it is 21%. In the days of Alexander the Great (350 BC) the whole of north Punjab was forest; now most of this large area is dry and treeless. The decline continued between 1951 and 1980 with 4.3 million ha of forest converted to other uses, such as agricultural crops, river valley projects, and establishment of industry. However, since the 1980 forest conservation act this loss of forest cover has been greatly reduced and was only 38 000 ha between 1990 and 2000 (FAO 2001a).

Deforestation and/or forest degradation are sometimes referred to as 'forest decline' (Contreras-Hermosilla 2000). Although it is difficult to provide precise definition of these terms, forest decline can be taken to include losses of forest productivity in terms of wood and non-wood products and environmental services. It directly threatens the livelihoods of those millions of forest dependent peoples in the tropics and is a source of great conflict between governments, private companies, and rural populations.

Is forest decline harmful? The answer depends on an assessment of the positive and negative impacts on the economy, environment, and other socio-economic factors. Some land currently under forest could be better used for agriculture and governments may clear forests to generate wealth for infrastructure development and other government services such as health and education. However much deforestation and forest degradation provides profits for a few individuals or companies must be considered socially and environmentally undesirable as well as economically unsatisfactory for the country. This 'inappropriate deforestation' is a significant problem that needs to be addressed (Kaimowitz *et al.* 1998).

Logging companies and farmers, with and without government approval, are usually the main direct agents of forest decline. In the Brazilian Amazon, Mexico, and Central America clearing to establish large cattle ranches has been a major cause of forest loss, elsewhere in South America millions of hectares have been cleared for soybean cultivation, and in Southeast Asia vast areas of rain forests have been cut and burned to establish oil palm and other agricultural tree crops (Kaimowitz 2000). Urban and industrial expansion, including mining and hydroelectric schemes, shifting cultivation (Fig. 2.1), and fuelwood gathering (Figs 2.2, 2.3, and 2.4) also contribute to forest decline. But natural causes, such as typhoons, pests and diseases and fires, cause severe degradation of forests. Devastating fires in Indonesia in 1983 and 1996–97 burned millions of hectares on the island of Borneo and left many forests severely degraded (Mori 2000).

These direct causes of deforestation are usually the result of more fundamental forces dictated by macroeconomic policies and the distribution of

(a)  (b)

Figure 2.1 Clearance of rain forest by shifting cultivators in the Philippines (a), and Sabah, Malaysia (b).

Figure 2.2 Firewood gathered for cooking and heating in the Sahelian zone of Chad. (Photo P W Turner.)

(a)
(b)

Figure 2.3 (a) Pats of cowdung for sale as fuel in Pakistan. Two kilograms of dung is sold at the same price as 1 kg of flour. (b) Wood fencing in Senegal.

Figure 2.4 Party of young women returning after a day gathering wood in the bush (southern Ethiopia).

economic and political power. Much literature has been devoted to examining why deforestation occurs and who is responsible, for example, Repetto (1990), Brown and Pearce (1994), and WCFSD (1997). Kaimowitz and Angelsen (1998) reviewed 150 models of tropical deforestation and analysed the complexities and interactions of the causes. They suggested higher prices for agricultural and forest products, road construction, subsidies for agriculture in forested areas, and land tenure policies that encourage land speculation contribute significantly to deforestation. Commodity booms sometimes triggered by currency devaluations and trade liberalization may also play a part. Contreras-Hermosilla (2000)

critically reviewed the underlying causes and identified:

1. Market failures
   - unpriced forest goods and services;
   - monopolies and monopolistic forces.
2. Mistaken policy interventions
   - wrong incentives;
   - regulatory mechanisms;
   - government investment.
3. Governance weaknesses
   - concentration of landownership;
   - weak or non-existent ownership and land tenure arrangement;
   - illegal activities and corruption.

4. Broader socio-economic and political causes
- population growth and density;
- economic growth;
- distribution of economic and political power;
- 'excessive' consumption;
- toxification;
- global warming;
- war.

Deforestation and forest degradation are a complex phenomenon with many socio-economic, political, and cultural causes. Pressures to exploit and clear natural forest have not stopped and several factors suggest that natural forests, especially in the tropics, will continue to decline and have less capacity to provide wood for industrial and other uses. It follows that if the predicted increases in consumption of wood products are correct more wood will have to be provided from plantations.

Problems of access to existing forest

1. A resources survey (FAO 2001*a*) indicates that the main limit to accessibility of wood from tropical forests is remoteness from transportation infrastructure. This is especially the case in the Amazonian region. This alone makes these forests, at least temporarily, inaccessible as a timber resource. Subtropical forests are relatively accessible to roads, rivers, etc.
2. Natural forests conserve biological diversity, store carbon for the mitigation of global climate change, protect fragile ecosystems in mountains and dry areas reducing soil erosion and providing clean water, and provide recreational and employment opportunities. These environmental and social services of forests were highlighted at the UNCED conference in Rio de Janeiro and in subsequent international treaties. They are receiving increasing emphasis and large areas are being legally protected from logging, for example, in China, India, and Thailand (Durst *et al.* 2001). Protected areas are the major limiting factor for access to forests in some parts of the tropics. For example, in Asia accessibility is mainly restricted by protected areas and altitude limits, and to a lesser extent by remoteness (FAO 2001*a*).
3. Moves towards more sustainable harvesting practices through reduced impact logging may decrease harvesting intensities in natural forest and have an impact on supply potential. A shift from harvesting primary to secondary forest may also result in lower volumes (FAO 1999*a*).
4. Some areas of natural forest cannot at present be logged because of physical limitations such as swampy ground and steep, mountainous terrain. In Papua New Guinea (PNG), of 38 million ha of forest, 21 million ha are classified as unexploitable for these reasons. Also, it is on many such inaccessible sites that forest cover is most needed to prevent soil erosion.
5. Past exploitation of forest has frequently left remaining areas located far from the chief wood users. The extensive eucalypt plantations around Belo Horizonte in Brazil are for charcoal production for smelting iron and steel because all accessible forest within 500 km has long since been cut. In parts of Africa scarcity of natural woodland results in women having to trek many kilometres each day to collect wood (Fig. 2.4). A major advantage of most industrial and social forestry plantations is that the trees are grown and can be used near where people live.
6. Much tropical forest does not contain high volumes per hectare of currently utilizable timber. And, even if all trees are felled for pulpwood or charcoal, yields from much forest in the Amazon, Southeast Asia, and many parts of Africa, only average about $100 \text{ m}^3 \text{ ha}^{-1}$. This low stocking makes timber harvesting unprofitable in more remote areas.

Unsatisfactory natural regeneration and failures of management

There is a vast area of natural tropical forests, mainly rain forests, but there has been a near universal failure to manage it satisfactorily and it is estimated that as little as 5% is being purposely managed for the sustainable production of wood (Armitage 1998). However, it should be recognized that tribal peoples have often enriched the forests they inhabit by planting. In Indonesia, for example, local communities have developed several million hectares of 'agroforests' comprised of dipterocarp trees mixed with fruit trees, rubber, and other perennial crops (Foresta and Michon 1993).

Most systems of tropical forest management have sustained yield as a guiding principle and there is potential for these forests to be managed sustainably (FAO 1989*a*, 1989*b*). There is also no lack of scientific methods for managing many types of tropical forests, for example, dipterocarp forests in Asia (Appanah 1998, 2001), and much research continues to refine silvicultural techniques (e.g. Bertault and Kadir 1998). There are major efforts to provide criteria and indicators (e.g. Prabhu *et al.* 1999), management guidelines (e.g. Armitage 1998), and codes of practice for sustainable tropical forest management. But the resource continues to decline in quality and quantity. Many economic, social, and institutional reasons, as well forest management and silvicultural factors, have been identified as contributing to the decline (e.g. Wyatt-Smith 1987; Panayotou and Ashton 1992).

Compared with the relative simplicity of plantation work, the silvicultural expertise and effort needed to encourage natural regeneration of desired timber species are considerable. Whether the reasons are administrative, such as careful timing of fellings, leaving seed trees, ensuring timely weeding, cleaning, and release cuttings, etc., or simply that little is known about tropical ecology, it remains the general case that natural regeneration of desired species is difficult to achieve. Fox (1976) noted: 'Success is more likely with forests tending to single dominance or when the desirable crop species grow rapidly in response to light'—two features typical of plantations.

One way of overcoming inadequate natural regeneration, while largely retaining the natural forest structure and cover, is to supplement it by planting lines or groups of the desired species so-called 'enrichment planting' (Figs 1.5, 12.16 and Chapter 12). Such mixed regeneration has been tried throughout the tropics, mostly in rain forests (see Dawkins 1958). Poore (1989) also cites several examples but concludes that almost all failed to proceed beyond the project scale. Overall the success of enrichment planting has been variable and so its efficacy has been questioned and its promotion and application have declined. Major problems of applying enrichment planting on a large scale in production forests are the difficulty of planting supervision, and the high cost and timeliness required to weed seedlings and release them from regrowth.

Another treatment applied for a time in a number of countries was the practice of improvement cuttings to favour growth of selected trees. In Gabon, 120 000 ha of *Aucoumea klaineana* forest was treated in this way only to be abandoned in 1962 in favour of plantations of this species. Bertault *et al.* (1993) outline some of the reasons why in the 1960s much investment in natural forest management in Africa and elsewhere switched to plantation establishment.

Changes are occurring rapidly in how forests are managed, by whom and for what purposes. Forest management in some countries, for example, India's successful joint forest management initiative (Saxena 2001), has been devolved to local communities and there is a greater awareness of the many benefits natural forests can provide in addition to wood supplies. In this environment it would be a mistake to assume that plantations are the sole answer to future wood supplies. Research must continue into the complex ecology of mixed tropical forest and also address the many social, political, and economic problems that make management of these forests so

difficult. If some of these problems can be solved it may be possible to take advantage of the fact that the difference between the costs of timber production in many plantations and natural forest, despite the low productivity of the latter, may not be as great as is commonly believed (Leslie 1987).

Land availability

Availability of land for planting is, and will continue to be, a key factor determining where plantations are developed. In recent years major tree-planting enterprises have been developed in the southern hemisphere as a result of comparative advantage over northern areas based on the availability of large areas of surplus agricultural land in parts of Asia, South America, and Oceania (ABARE – Jaakko Pöyry 1999).

Many tropical countries have low population densities and large areas of virtually unused land. Much land, marginal for agriculture, is potentially available for forest plantations. In Brazil, for example, most of the population lives in the industrial south and southeast and a vast area of its central region is very thinly populated. Over 100 million ha of it is 'cerrado', land previously cleared and roughly grazed but now covered with scrub and almost wholly unused (Allan 1979). Nearly all cerrado is suitable for planting.

In general, land shortage is unlikely to be a problem in the foreseeable future. To place the requirement in perspective, in 1995 the total area of plantations in the tropics was about 55 million ha, about 1% of the total land area. Over the tropics as a whole it is estimated that there are 2077 million ha of degraded land, of which 758 million ha have a theoretical potential for forest replenishment (Grainger 1988*b*). Nilsson and Schopfhauser (1995) estimate about 345 million ha available for plantations and agroforestry but Trexler *et al.* (1993) cited in Dubas and Bhatia (1996) suggest it would make 'economic sense' to convert only 67 million ha to plantations.

Despite the apparent ready availability of land, there are often social and economic factors that will restrict the actual area available for plantations in a particular location. The main land problem is that of tenure. It may be difficult for companies to obtain ownership or lease of large areas of land without conflicts with local communities who dispute land and tree tenure arrangements. Difficulties in obtaining large areas of land is a major reason why companies in the private sector have shown increasing interest in forming partnerships with communities and individuals to grow industrial wood on their land. Most large corporations producing pulpwood in the tropics now either operate an outgrower scheme or provide extension services to small-holders to grow trees for them. These partnership arrangements have in most cases provided a socially acceptable solution to the vexed question of acquiring land for plantation development. They respond to political pressures for local control and are part of the reason why plantations in large blocks are decreasing and forest farms are growing in importance (Mayers 2000). Some of these issues are discussed in more detail in Chapter 5.

High productivity

Apart from rapid growth, which is the notable feature of many tropical plantations, all plantations when successfully established have several other management advantages, which lead to high productivity per unit area.

1. Stands usually consist of one species selected to produce wood that meets the needs of the intended user or market. The wood produced is relatively uniform and, especially if clones are used, the quality can be closely matched with the need of the processor. In industrial plantations unmarketable species are not planted.
2. Planting ensures full stocking on a site and fullest use of its potential. This is not only efficient land-use but the high volumes per unit area which result lead to cheaper harvesting costs.
3. Initial spacing of trees, thinning regime and rotation length can be manipulated to produce the desired mix of tree sizes for poles, pulpwood, sawn timber, veneer, etc.
4. A careful pre-planting survey can help match species with site conditions for optimum growth and can identify all land not suitable for planting.

Figure 2.5 Fast growth of plantations in the tropics. (a) *Eucalyptus grandis* age 3 years at Aracruz Florestal, SE Brazil. (b) *Pinus caribaea* var. *hondurensis* age 5.5 years at Jari, Amazon, Brazil. (c) *Acacia mangium* age 3 years, Sabah Softwoods, Sabah, E. Malaysia. (d) *Tectona grandis* age 10 years in PNG.

Table 2.2 Average growth rates attained in some tropical plantations

Plantation development	Species	Mean annual increment ($m^3 ha^{-1} yr^{-1}$)	Rotation (years)
Usutu Forest, Swaziland	*Pinus patula*	19	15–17
Viphya Pulpwood Project, Malawi	*P. patula*	18	16
Fiji Pine Ltd., Fiji	*P. caribaea*	15–20	17–20
Jari Celulose SA, Brazil	*P. caribaea*	20	16
ECO-SA, Congo[a]	*E.* hybrids	20–25	7
Aracruz Celulose SA, Brazil	*E. grandis* and *E.* hybrids	45	7
Shiselweni Forestry, Swaziland	*E. grandis*	18	9
PICOP[b] Resources Inc., Philippines	*Paraserianthes falcataria*	28	10
Sabah Softwoods, Malaysia	*A. mangium*	20–30	7–8
PT Musi Hutan Persada, Indonesia	*A. mangium*	29	6
Jari Celulose SA, Brazil	*Gmelina arborea*	20	10
Department of Primary Industries Queensland Forestry.	*Araucaria cunninghamii*	15	40–50
Seaqaqa plantations, Fiji	*Swietenia macrophylla*	14	30

[a] Eucalyptus du Congo Société Anonyme.
[b] Paper Industries Corporation of the Philippines.

Table 2.3 Growth rates of managed forest and plantations

	Yield ($m^3 ha^{-1} year^{-1}$)	Rotation (years)
Canada average	1.0	—
Siberia (Russia)	1.0–1.4	—
Sweden average	3.3	60–100
US average	2.6	—
UK average (conifers)	12	40–65
New Zealand pines	18–30	20–40
South African pines	10–25	20–35
Subtropical eucalypts	5–30	8–25
Teak plantations	4–18	40–80
Tropical hardwood plantations	25–45	8–20
Tropical pines	15–45	8–30
Tropical eucalypts	up to 70	7–20
Tropical high forest (managed)	0.5–7	—
Southeast Asia dipterocarp forest (managed)	up to 17	—

Source: Modified from Wood (1975).

Only suitable species and suitable land need ever be planted.

These advantages apply to both industrial plantations and tree-planting in social forestry, but success is not always achieved. Indeed, misplaced enthusiasm to harness the benefits of potential high productivity while ignoring social, economic, and environmental factors has led to many failures.

Fast growth

As the ease and advantages of harvesting natural forests diminishes, there has been a shift away from countries with the greatest forest resources to countries where environmental conditions favour the highest levels of forest productivity (Brown 2000). The most striking advantage of many tropical plantations is rapid growth, which enables relatively quick economic returns. Forest plantations in the moister tropics have some of the highest natural productivities in the world (Lieth 1977). Figure 1.1 (a) and (b), and Figs 2.5 (a)–(d) illustrate examples. In the humid tropics, pulpwood plantations have taken advantage of the very fast growth rates and suitable wood properties of some tropical hardwoods e.g. *Eucalyptus* and *Acacia* species, and softwoods, for example, *Pinus caribaea*. Cossalter *et al.* (2003) have referred to fast-growing industrial plantations as 'fast wood forestry'.

In reporting very fast growth rates one must beware of assuming that a whole plantation will match yields produced in trial plots. Often spectacular performances are on peculiarly favourable sites or where plots have received optimum treatment perhaps as part of research work. For example, in Cameroon one research plot of *E. urophylla* had a mean annual growth rate to 8 years of 83 m^3 ha^{-1} but the estimated average growth rate of a plantation on a suitable site is 30 m^3 ha^{-1} year^{-1} (Eldridge *et al.* 1993). Moreover, though genuinely high mean annual increments can be achieved by close spacing and very short rotations they are meaningless for most production purposes, other than firewood, since trees need to be grown on to a usable size. Nevertheless, in the moister tropics, average productivities of extensive plantations, covering thousands of hectares, are mostly much higher than elsewhere in the world. Indicative growth and yield information of the more important plantation species is given by Brown (2000) and FAO (2001*a*). The data in Table 2.2 are average yields of merchantable wood actually attained.

In general, mean annual increments attained become less as rotations become longer or conditions drier. Teak (*Tectona grandis*) grown for 60–80 years in India mostly has a mean annual increment of about 4–8 m^3 ha^{-1} but under moister conditions and shorter rotations in Central America and the Caribbean mean annual increment is 8–18 m^3 ha^{-1}. Yields of *E. camaldulensis* plantations in the drier tropics are often about 5–10 m^3 ha^{-1} year^{-1} on 10–20-year rotations, whereas in moister regions up to 30 m^3 ha^{-1} year^{-1} may be achieved. *Azadirachta indica* typically achieves 5 m^3 ha^{-1} year^{-1} in the semi-arid Sahel of Africa.

The significance of all the above figures is seen in Table 2.3 where comparison is made with other forest types.

In many parts of the tropics plantations yield several times the quantity of wood of most natural tropical forest and most forest in temperate regions. Some species can produce very high yields of small-sized wood (biomass) on extremely short rotations. If the species coppices well then plantation development can be an efficient means of firewood production. Impressive yields are frequently reported for such species as *Calliandra calothyrsus*, *Sesbania sesban*, *Gliricidia sepium*, *Leucaena leucocephala*, and others. Kerkhof (1990) cites *Calliandra* and *Sesbania* as yielding more than 60 and 80 t ha^{-1} (fresh weight),

respectively, in 1 year in woodlots in Kakamega, Kenya, and Bhatti *et al.* (1989) report annual production of 64.8 t ha^{-1} of forage for *L. leucocephala* at close spacing in a trial in Pakistan. Care is needed with all such figures, but annual biomass production of 20 t ha^{-1} (dry weight) is easily attainable in the moist tropics but only about one quarter of this in the dry tropics, unless irrigated. Maguire *et al.* (1990) report an annual yield of aboveground biomass of 40 t ha^{-1} for irrigated *Acacia nilotica* in Pakistan.

Economic benefits and rural development

Industrial plantation establishment, as well as tree-planting for social and environmental objectives, remains central to forestry development strategies of most tropical countries. Plantation forestry in the tropics can significantly aid economic development, especially through earning foreign exchange from exports of forest products or import substitution. The highly successful use of government incentives in Chile and Brazil to encourage plantations has made these countries become leading exporters of wood pulp. While governments may support plantations for financial benefits alone, more often it is for broader economic reasons and to generate social and environmental benefits. Employment opportunities through developing new forest industries, watershed protection, enhanced landscape amenity values, recreational opportunities, and land rehabilitation are some of the justifications used for government involvement in plantation development, either through state forestry companies or the provision of a range of incentives (see Chapter 6). However, recent policies of some governments and many bilateral aid agencies have directed support to conservation and small-scale community tree-planting rather than large industrial plantations and forest industries.

Major socio-economic benefits of plantation forestry are listed below.

1. Creation of a resource to meet demand for wood products and provide environmental services. Well-conceived projects can yield attractive economic rates of return (10–15%) and returns from social forestry and watershed projects can be even higher (Spears 1987).
2. Development of a flexible resource able to yield many kinds and sizes of product for internal consumption or for export or both. Plantations for poles, pulpwood, sawn timber, veneer logs, etc.
3. Use land often of little or no agricultural value.
4. Provide employment in rural areas.
5. Enable use of skills already common in agriculture and most additional skills can be acquired from 'on the job' training.

6. Bring development of an infrastructure of roads, communications, services, houses, shops, schools, etc., often to remote areas.

Collectively 1–6 above indicate how tree-planting can directly impact poverty eradication, especially in forest-poor regions of the tropics.

Rural development forestry

A major contribution of forestry is to rural development through the benefits brought to villagers and farmers. In many areas exploitation of natural forests has been accelerated to unsustainable levels and so farmers and other landholders are increasingly growing trees on-farm for their own and local use (Fig. 2.6). Farmers are most interested in planting trees to provide construction wood, poles, posts, fruit, etc., especially when these can be sold for cash. Fuelwood may be valued as a secondary product, and shade, shelter and boundary demarcation are also important for some farmers. Those with insufficient land for woodlots often grow a few trees that can be harvested and sold when they experience food shortages. The value of plantations and social forestry to economies and rural development is discussed more fully in Chapter 6. Many of the benefits relate to agriculture, which emphasizes the importance of integrating forestry and agricultural development. Integration of tree-planting with farming (agroforestry) is considered in Chapter 20.

In recent years there has been an enormous expansion in rural development (social) forestry programmes. Types and size of these programmes vary from country to country but their funding now equals or exceeds that given to traditional industrial forestry development by most multi- and bilateral-agencies. Many countries, for example, Indonesia and India, now have policies and incentives to assist small-scale and community-based forestry (Herbohn *et al.* 2001). In addition, numerous NGOs promote tree-planting principally for this purpose, and networks, such the Overseas Development Institute's (ODI) Rural Development Forestry Network, provide valuable information.

Environmental forestry

No discussion of why planted forests are established is complete without mention of their role in protection. Stabilizing soil, preventing erosion, controlling water runoff in catchment areas, providing shelter from wind and heat and against sand and dust storms are all roles for which trees are widely planted (and much needed) in the tropics. In many countries loss of forest cover followed by bad land management has led to increased flooding, accelerated soil erosion, faster siltation of reservoirs, and more rapid desertification. Extensive re-establishment of tree cover, best illustrated by the enormous efforts in China, is an effective means of reversing these degrading processes. And, one of the principal benefits of such protective afforestation

Figure 2.6 Trees and woodlots of eucalypts and *Grevillea robusta* in the highlands of Ethiopia planted and maintained for village and local needs.

is to raise agricultural yields by lessening environmental hazards.

Afforestation for protective purposes need not exclude production of wood. The *Eucalyptus microtheca* and *E. camaldulensis* plantations around Khartoum (Sudan) supply poles and fuelwood as well as providing shelter from dust storms (Musa 1977). The Ndirande forest reserve in Malawi both protects Blantyre's water catchment and yields much pine sawtimber and firewood. Sand-dunes along the South China Sea have been stabilized with *Casuarina equisetifolia* plantations that provide fuel and poles for local communities (Turnbull 1983). Where natural cover has been destroyed, the role of tree-planting is recognized as an important contribution to watershed management, for example, in 1998 in China the Natural Forest Conservation Program's logging ban in natural forests was accompanied by a plan to establish 21 million ha of timber plantations in the upper reaches of the Yangtze River and the upper and middle reaches of the Yellow River (Yang 2001). Such protective plantations must go hand in hand with other soil conservation practices, such as strict control of livestock grazing and general maintenance of vegetation cover, but their role should not be exaggerated or seen as a panacea.

Rehabilitation and restoration

Deforestation and/or forest degradation has resulted in losses of forest productivity in terms of wood and non-wood products and environmental services. Biodiversity has been greatly reduced, especially in tropical rain forests, and few forests can recover unaided. Different levels of site disturbance include: (i) disruption or removal of the native plant community, without severe soil disturbance; (ii) damage to both vegetation and soil; (iii) vegetation completely removed and the soil converted to a state outside natural conditions (Aber 1987). Three approaches, 'restoration', 'rehabilitation', and 'reclamation', are commonly applied to reverse the degradation process and assist recovery of these different degrees of disturbance. Tree-planting is playing an increasing role in all three approaches.

Foresters need to plan forest plantation development to provide diversity within the landscape (Chapter 4) and modify plantation practices when rehabilitation and/or restoration are their prime objectives. A holistic approach for planning and implementing restoration and rehabilitation of degraded tropical forests, taking into account other landscape components in a particular locality, is recommended (ITTO 2002). The landscape context helps planners and resource managers identify management strategies and options that meet the local communities' needs. This topic is discussed in more detail in Chapter 22.

Carbon storage

The increase in atmospheric greenhouse gas concentrations due to fossil fuel burning and deforestation has caused concern about its impact on the world's climate. Vegetation plays a major part in the global cycling of carbon dioxide (CO_2). The gas is taken in by plants during photosynthesis, some of the carbon is stored in the plant, some is released during respiration and decomposition of plant parts. Inclusion of vegetation in the Kyoto Protocol adopted in 1997 was because it offered important options for flexible, low-cost abatement of greenhouse gases (Keenan and Grant 2000). It has been estimated that plantations accumulate carbon at an average annual rate of $0.4–8$ t ha^{-1} depending of the species, site, and management inputs (Schlamadinger and Karjalainen 2000). One estimate suggests forestry has the potential to offset about 15% of the world's greenhouse gas emissions (Brown 1996) but others have made more modest claims. This has implications for forest plantation development as carbon stored (sequestered) in plantations on previously cleared (carbon-poor) land could count towards meeting a country's commitments to emission control or used for emissions trading. However, many of the rules and guidelines regarding the inclusion of carbon in vegetation for the purposes of legally binding emission reduction targets have not yet been decided.

Voluntary forestry-based carbon offset projects started even before the signing of the UN Framework Convention on Climate Change in Rio de Janeiro in 1992 and have since been evolving towards a market-based trading system. The first forestry project, based on agroforestry and planting woodlots, specifically incorporating carbon storage as an objective commenced in Guatemala in 1989 supported by an American electricity company (Trexler *et al.* 1989). It aimed to plant 51 million trees over a 10-year period on 186 000 ha and an evaluation in 1994 indicated at up to 10 million tonnes of carbon could be sequestered. In the early 1990s the Dutch Electricity Board created the Face Foundation (Forests absorbing carbon dioxide emissions) with a budget of US$180 million to establish a portfolio of forestry projects in various parts of the world that would plant enough forests to absorb carbon dioxide equivalent to the emissions from a medium-sized coal-fired power plant in the Netherlands (Moura-Costa and Stuart 1998).

Between 1992 and 1997 (Kyoto) there were many carbon-offset projects involving millions of hectares throughout the tropics. Investor companies paid the full costs of carbon saving in return for the promise of carbon credits generated from the forestry activities (Moura-Costa and Stuart 1998). As part of the Kyoto Protocol in 1997 a number of more flexible methods were approved to facilitate emission reduction credits. There has been development and implementation of institutional and legal mechanisms to allow efficient trade in carbon credits. In Australia, a trading systems has been set up and state legislation has created legal carbon sequestration rights (Follas 2002). Also, NSW State Forests has designed investment packages combining carbon sequestration and timber production and in 2000 signed an agreement with a Japanese electricity company for the carbon rights for 40 000 ha of new plantations over a 10-year period (Brand 1998; Keenan and Grant 2000).

To what extent tropical plantations will be used for carbon sequestration in the future is very uncertain and there are varying perceptions of the availability of land, and social and economic constraints (Dabas and Bhatia 1996; Winjum et al. 1997; Watson et al. 2000). Potential benefits and risks to local livelihoods from forestry-based carbon-offset opportunities have been examined by Bass et al. (2000) and Smith and Scherr (2002). Most tropical countries will need new policies to provide conditions that attract investors to forest carbon projects that will benefit local people. Even if tropical plantations are used to sequester carbon they will have a relatively minor impact on the global warming problem compared with the introduction of cleaner fuels and improvements in energy efficiency.

Conclusions

Some account has been given for the rapidly increasing interest in planted forests in the tropics. Put simply, wood can be produced quickly, often on land hitherto unused or marginal for agriculture. Firewood and numerous other domestic products can be grown efficiently. By planting trees, shelter and shade is provided and ground cover is re-established, which will halt and eventually reverse the degrade of much of the tropical environment. But it is naïve to assume that all planting schemes develop only for the reasons suggested. Large regular plantations are politically impressive, they are clear evidence of development in a perhaps otherwise remote area, and a large sawmill, plywood factory, or pulpmill complex is prestigious. Some plantation projects have developed more from political motives than for reasons of silviculture, land-use, or economics, and by no means all plantations have been successful.

Forest plantations in the tropics will clearly play a very important role in future world wood supply. They also have great potential to improve the livelihoods of rural communities through providing a range of products and services locally and to meet the international objectives for conservation of biodiversity, protection of the environment and mitigation of climate change. There are many positive answers to the question 'why planted forests?' and they combine in these ways to achieve what Westoby (1989) advocates: 'making trees serve people'.

CHAPTER 3

Planted forest in tropical countries

Development of plantation forestry

Since the 1960s there has been an enormous increase in forest plantations in the tropics. By 2000 almost 40% of the global estate of 187 million ha of plantations was in the tropical and subtropical areas we are considering in this book (FAO 2001a). Though some countries commenced plantation establishment earlier than others, the recent upsurge in planting has occurred nearly everywhere and most countries have begun social forestry programmes or are committed to large afforestation projects.

This expansion cannot be viewed in isolation since projects today necessarily draw on silvicultural information from scattered trial plots and small plantations established in the past. In addition, in the last 50 years, much stimulus has come from the political and economic independence of countries in the developing world. Also, this period has seen a new internationalism in world affairs, particularly in aid and development such as the United Nations Organizations of Food and Agriculture (FAO), the Development Programme (UNDP), and the World Food Programme (UNWFP), the World Bank, regional Development Banks, bilateral aid programmes between rich and poor countries, direct commercial investment by industrialized nations in developing countries and enormous growth in non-government organizations (NGOs), indigenous and international, concerned with the Third World. Whether, within the forestry sector, this great increase in aid and investment has led to the hoped-for benefits is open to dispute (e.g. Carrere and Lohmann 1996) but, nevertheless, one of the consequences has been the increasing development of plantations. As background to the present status of forest plantations in the tropics, we have provided an outline of their historical development.

Origins of planting

People have been planting trees for thousands of years for food or other non-timber products, shelter, ornamental, ceremonial, or religious purposes. The first woody species to be selected and planted as early as 4000 BC may have been the olive tree (*Olea europaea*), and it has been cultivated in Greece at least since the Minoan era (3000 BC). The temple of Queen Hatshepsut constructed in 1500 BC at Thebes, Egypt, has depictions of myrrh trees (*Commiphora myrrha*), introduced from Somalia, being planted as sources of perfume, and Theophrastus reported trees of frankincense (*Boswellia* sp.) and myrrh being planted on private estates in southern Arabia in the fourth century BC. There are also several biblical references to tree-planting dating to 2000 BC or earlier, such as the Old Testament record of Abraham planting a tamarisk tree to commemorate the treaty of Beersheba (Gen. 21: 23).

Tree-planting was practised in ancient times in Asia. The Chinese cultivated fruit trees, and grew pines for ornamental, religious, and ceremonial purposes as long ago as 2000 BC (Valder 1999). During the early part of the Chou Empire (ca. 1100–256 BC) the Emperor established a forest service with the responsibility for preserving natural forest and reforesting denuded lands. The Han and Tang Dynasties (208 BC–AD 256) encouraged people to plant trees important for both food and timber production. In the Sung Dynasty (AD 420–589) direct planting of tree seeds for reforestation was widely practised and public land reforested by farmers became their property. Monographs were also prepared describing methods of planting and protection of the tung tree (*Aleurites* sp.), bamboos and other woody species (Wang 1988). Ji Han's *Account of the plants and trees of the southern regions* is dated AD 304. Chinese fir (*Cunninghamia lanceolata*) has a cultivation history of over 1000 years. In Korea, during the Shilla Kingdom, starting 57 BC, trees were being planted around kings' tombs, in royal gardens, along roadsides, as shelterbelts, along rivers and on the coast for erosion control. In Sri Lanka, during the reign of the Sinhalese King Vijaya (ca. 543 BC) village communities planted home gardens with flowering and fruit bearing trees, and under King Dutugemunu (161–137 BC) forests were planted and rules made for forest protection and use of forest products (Winter 1974). Probably the oldest planted tree still living, of which a reasonable historical records exists, is in the tropics, in Sri Lanka, where the Bodhi tree (*Ficus religiosa*) is recorded as being 'planted' miraculously at Anuradhapura in about 220 BC.

Plantations before 1900

Development of tropical plantations can be traced back to sixteenth and seventeenth centuries with the expansion of European influence by the colonial powers. The colonizers encouraged an exploitive timber export trade, often seriously damaging natural forests. But until the late 1800s there was generally timber available and little need to plant trees in the tropics for industrial wood production. The scientific study of plants and animals led to systematic collections, botanical gardens flourished, and the domestication of several tropical tree species began (Turnbull 2002). Plantation activities before 1900 included the introduction and testing of exotic species, especially teak and eucalypts, and the introduction of taungya and irrigated plantations. The establishment of government agencies, use of trained foresters, and definition of forest policies and legislation provided an institutional framework on which the extensive forest plantations of the twentieth century could be based.

There is a long history of planting teak (*Tectona grandis*) in the tropics. In Asia it was extensively planted for timber in Java under the control of the Sultans in the fifteenth century. With the arrival of the Portuguese in fifteenth century and the Dutch in seventeenth century the demand for the durable teak timber for general construction and ship-building intensified. By 1748 the Dutch East India Company controlled all teak forests and monopolized teak trading. Influenced by three German foresters, all the Javanese teak forests were brought under regular management in the late 1800s. The taungya regeneration method was introduced in 1873 and from 1895 almost all teak forests have been regenerated by this system. Taungya is a system in which farmers plant tree seeds or seedlings to make a forest plantation and tend them in association with their food crops. The term 'taungya' originated in Burma (Myanmar) and its application in government forestry is usually attributed to Dietrich Brandis, a forest officer in-charge of Burmese teak forests from 1856 to 1862. In Myanmar, teak planting began in 1856 and continued into the twentieth century. However, most plantations were damaged during the Second World War. Planting was started again on a modest scale in 1963 but increased to more than 10 000 ha per year after 1984. Teak was successfully introduced into Sri Lanka by the Dutchman, van Rhede as early as 1680 (Perera 1962) and, as teak was in short supply for shipbuilding in India during the early 1800s, the Collector of

Malabar in 1840 suggested teak should be planted and the first plantations were established in the Nilambur hills (Puri 1960). Between 1841 and 1855 some 600 ha were established and since 1840 has been planted both within and outside its area of natural occurrence. The taungya method was generally used to establish plantations in Karnataka, Kerala, Uttar Pradesh, West Bengal, and most parts of Assam and Tamil Nadu, although many forests were naturally regenerated.

As early as 1790 several eucalypts were planted in the Palace garden at Nandi Hills near Mysore and became the seed source for widely planted Mysore gum (*E. tereticornis*). In South America eucalypts were introduced in 1823 to Chile, and specimens of *E. robusta* and *E. terericornis* in Rio de Janeiro, Brazil, date back to 1825 (Jacobs 1981). *Eucalyptus globulus* was one of the first eucalypts to be used for plantations. By 1900 it could be found in Europe (Italy, Portugal, Spain), Africa (Ethiopia, Kenya, South Africa), Asia (China and India), and South America (Bolivia, Chile, Colombia, Peru). At that time it was primarily planted for ornamental purposes or fuelwood. Eucalypts, which grew fast, provided fuel for the wood burning locomotives in Brazil, East Africa, India, and South Africa, but were inferior for sawn timber production. Other species, such as *E. robusta* and *E. tereticornis*, were planted in many countries and formed the base on which large industrial plantations developed in twentieth century, when eucalypts became the most planted broadleaved species in the world.

Plantations were started in South Africa in the latter part of nineteenth century. The first wattle trees (*Acacia mearnsii*) were planted for tan bark in 1864 and the first pine plantations in 1884. Pines were not commonly planted before 1900. *Pinus patula* was introduced to New Zealand in 1877 (Wormald 1975) and probably other countries at about the same time, though it was not introduced into South Africa until 1907. *P. caribaea*, now widely planted in the tropics, was little known botanically let alone planted until 60 or 70 years ago. The lack of records of early introductions of pines may be because many quickly failed owing to no suitable mycorrhizas being available on the new sites.

Two important silvicultural practices saw their introduction during the latter part of the nineteenth century. Use of the taungya system to plant teak has already been mentioned and is further described in Chapter 20. The other was irrigated planting, which is generally associated with arid sites, where the annual rainfall rarely exceeds 200 mm, or semi-arid areas with a very short rainy period. Irrigated plantations are usually close to major rivers. In the Indus basin of Pakistan irrigation of *Dalbergia sissoo* and other species has been practised for more than

Figure 3.1 Fourth rotation of *Dalbergia sissoo*, which has been grown using irrigation at Changa Manga, Pakistan. It has just been cut for firewood and timber.

100 years (Fig. 3.1). The first plantings were in the Punjab in 1866 to supply firewood for a new railway and fuelwood for Lahore and other cities. Notable irrigated plantations have subsequently been established in Iraq, Egypt, and central Sudan. Species such as eucalypts, casuarinas, and poplars are commonly grown in these irrigated systems.

Before 1900 there was no need to plant trees extensively as an industrial resource in the tropics, though several European countries were concerned about their own lack of natural forest. The main contribution of the pre-1900 period to tropical plantation silviculture today was introduction and trial of exotic species, notably teak and some eucalypts, and the inception of taungya and irrigated plantations.

1900–45

This period saw the first extensive plantings of industrial tree crops, mostly in countries with little utilizable natural forest and where there had been an early influx of European settlers. Notable examples are South Africa, which by 1945 had 180 000 ha of plantations of *P. patula, P. elliottii, P. taeda,* and some *Eucalyptus* spp. in Mpumpalanga (Eastern Transvaal), and Queensland (Australia) with *P. elliottii* and *Araucaria cunninghamii* plantations covering about 9800 ha at this time (Ryan and Shea 1977). In India, by the Second World War, 80 000 ha of teak had been planted (Griffith 1942) and many trials of *Eucalyptus* spp. set up (Khan and Chaudhary 1961).

Before 1945 the main plantations in tropical America were on a small scale for protective purposes around cities, and for fuelwood, railway sleepers, and pit props. The most extensive early plantations were in the State of Sao Paulo, Brazil from 1905–15 at the instigation of Edmundo Navarro de Andrade, head of forestry services for the Paulista Railway Company. In 1950 Brazil had close to half a million hectares of planted eucalypts, a larger area than any other country.

Between 1900 and 1945, although most of the world's tropical plantations for wood production were made up primarily of pines, eucalypts, and teak, there were major plantings of trees for non-timber forest products. Paramount was the rubber tree (*Hevea brasiliensis*); first planted in Malaysia in 1898 it boomed in about 1910 when it was enthusiastically taken up by coffee and tea planters whose commodities were experiencing difficulties. Indonesia and Malaysia are currently the world's largest rubber producers. In the period 1920–30, the private sector began extensive plantations of the Australian black wattle (*Acacia mearnsii*), for tan bark to supply tannin to the leather industries. In 1921 there were about 115 000 ha in South Africa and 25 000 ha in Kenya. It was also planted in Zimbabwe, Tanzania, India, and Brazil. In recent years the value of the tannin has declined relative to the value of the wood, which makes good charcoal and excellent paper pulp.

In the Indian subcontinent many pioneering developments in tropical silviculture took place and Broun (1912) described afforestation and plantation practice. However, in most tropical countries plantation forestry was still in its infancy though many silvicultural developments took place, which were later to play an important part in plantation work.

1. The taungya system became widely used in many tropical countries. For example, it was tried in Kenya as early as 1910 (FAO 1967*b*), and in the 1920s was the main method of establishing teak in Trinidad (Lamb 1955).
2. For many countries most species later to be successful were introduced in this period and planted in trial plots. The first differences in provenance

were identified and the first specific seed collections for plantation projects undertaken.

3. Development of thinning and pruning schedules in South Africa in the 1930s (Craib 1934, 1939, 1947) revolutionized much traditional plantation practice, which until then was largely based on European silviculture. Policies of wide spacing, heavy thinning, and pruning for wood quality mostly derive from Craib's work.

4. The first serious attempts were made to augment natural regeneration by enrichment planting in groups and lines; Broun (1912) described one approach used in India and Sri Lanka and Eggeling (1942) recorded all the main methods of enrichment planting used as standard practice in Africa.

5. Declining areas of natural forest began to stimulate special plantings, to augment local wood supplies, called 'compensatory plantations'. Arguably, the best early example was the 4000 ha of *E. globulus* planted on the amphitheatre of hills around Addis Ababa, Ethiopia, between 1900 and 1920, which made up for the loss of indigenous forest in the previous decade. Wood harvested from successive coppice rotations of the current 15 000 ha of eucalypts has provided a sustained supply of fuel for the city to this day (Fig. 3.2).

1945–65

The total plantation area in tropical countries in 1950 was estimated to be 680 000 ha, most of which was teak in Indonesia (Lanly 1982). The main developments influencing plantation forestry in the 1950s and 1960s were silvicultural and those reflecting changes in the world order. Three main trends can be identified.

Internationalism. The proliferation of international agencies after the Second World War and their important influence has already been referred to. But, in addition to their general stimulus through provision of funds, development aid, and technical expertise, substantial direct investment in industrial plantations also began. Both the major plantation projects in Swaziland date back to this period and were funded mainly by private companies (Evans and Wright 1988).

Also in this period, several institutions were founded with regional responsibility or influence to assist forestry development, in particular plantation establishment, for example, Centre Technique Forestier Tropical in many French speaking countries, the Institute of Tropical Forestry in Puerto Rico, and Centro Agronomico Tropical de Investigacion y Enseñanza in Costa Rica. Many university and college courses in forestry in developing countries were introduced at this time.

Awareness of silvicultural potential. As they matured, the earlier trials and older plantations demonstrated whether or not plantation forestry was worthwhile. The general observation of fast growth, compared with temperate plantations, appeared true of many species on many sites, though the best matching of species or provenances with particular sites was not always achieved.

Figure 3.2 Young girl gathering twigs and dry leaves (forest litter) for kindling from *E. globulus* plantations near Addis Ababa, Ethiopia. Note pile of leaves and full sack on left in front of young coppice growth.

Evidence of the increasing awareness of the silvicultural potential of plantations—and also the new internationalism—is seen in the resolutions and subject matter of international conferences and meetings. At the Fourth World Forestry Congress (1954) in India it was recommended that an international commission be set up on the use of exotic species for planting in the tropics. In 1957 the Seventh British Commonwealth Forestry Conference resolved that a book be published about experience with exotic species in the Commonwealth. Its subsequent publication (Streets 1962) led to the founding of the Unit of Tropical Silviculture at Oxford (England) in the mid-1960s to work on fast-growing species suitable for plantations. In 1956 the Teak Sub-Commission was set up by FAO and much of its work has concerned problems of plantation teak. The increasing importance of eucalypts for plantations was indicated by publication in 1954 of the first edition of *Eucalypts for Planting* by FAO and the establishment of a 'Eucalypt Clearing House' service in 1962 by the Forestry and Timber Bureau in Australia to provide information and well-documented seeds for research and commercial plantations. And, according to Lamb (1973), the FAO Seminar on Tropical Pines in Mexico in 1960 more than anything else awakened tropical countries to the value of these species.

Plantation forestry around the world became more important in the early twentieth century and management practices intensified. It was realized that the genetic principles being applied to agricultural crops could be used to improve the productivity of trees in plantations. Leaders in the emerging field of forest genetics in the 1920s and 1930s were Oppermann and Syrach Larsen in Denmark, Johannsen, and Bertil Lindquist in Sweden and Ernest Schreiner in the United States. Foresters in the tropics also recognized the need to improve seed quality of plantation species and some designated special seed collection areas. Forest geneticists from 1910–45 made a major contribution by elucidating patterns of variation in commercially important trees, and achieving some basic understanding of pollination biology and vegetative propagation. They provided the basis from which the forest industries in the early 1950s could launch with some confidence the large programmes of applied genetics to improve wood production in plantations (Turnbull 2002).

Independence and the urge to develop. In many ways, India's independence in 1947 ushered in more changes in world order than the tragedy of the Second World War. Declining colonial influence and the newly emerging independent states altered economic and developmental pressures in many tropical countries. Development became a politically important process: the key to meeting the people's aspirations in the new countries. Plantation forestry can claim no great part in development programmes of this period, but in several countries tree-planting increased substantially as part of the overall development process. Though, as King (1975) pointed out, almost every tropical country has at one time to another considered establishing industrial plantations for pulpwood.

Achievements in 1945–65 period
It is impossible to list comprehensively the plantation programmes that took place but, because the period immediately preceded the great expansion in activity over the next 40 years, we will illustrate some of the developments.

Without doubt the largest programme was the huge afforestation effort in China. Annual planting rates ranged from 100 000 to 450 000 ha during this period (Ministry of Forestry 1985), much of it in the southern, subtropical provinces to create industrial timber plantations of Chinese fir (*Cunninghamia lanceolata*). By 1958 Africa's largest block of man-made forest (41 000 ha) had been planted forming the Usutu forest in Swaziland. In fact, total afforestation in Swaziland between 1945 and 1965 was more than 80 000 ha or 5% of the land surface. Fiji commenced planting mahogany (*Swietenia macrophylla*) on logged-over rain forest in the late 1950s to grow veneer quality timber. In Papua New Guinea (PNG) a regular planting programme began in 1951 to provide a replacement source of peeler logs for plywood after natural stands of *Araucaria* were logged out. Many countries increased the rate of planting during this period for all purposes (industrial, protection, firewood) and many new schemes were being planned, for example, the Viphya Pulpwood project in Malawi and the Turbo project in Kenya.

In 1965 the approximate area of plantations in the tropics was 3.5–4.0 million ha (FAO 1967a) excluding those of southern China (Table 3.1). The total area of plantation in 1965 was probably over five times that of 20 years before, but it was still quite small compared with the absolute increase in plantations in the 40 years that followed.

1966–80

In a very real way this period was ushered in by the FAO World Symposium on Man-Made Forests and their Industrial Importance in early 1967, which testified to the increasing emphasis on plantations and their expanding role. Many of the trends initiated in the 1950s and 1960s accelerated, new projects multiplied, afforestation became an important part of national forest policies, several countries began to provide incentives to plant trees, and the importance of trees and forest in the environment became more widely recognized.

Between 1966 and 1977 the annual rate of planting in Brazil rose from 40 000 ha to nearly half a million (Fig. 6.1). In 1968, the Jari venture commenced with the aim of converting 400 000 ha of poor quality Amazon forest to plantations of *P. caribaea*, *E. deglupta*, and *Gmelina arborea*. By 1980, 100 000 ha were planted and a pulp mill was in operation. A country as small as Fiji planted 28 000 ha of *P. caribaea* between 1971 and 1979 (Fig. 4.2). The area of plantations in the Congo, mainly of eucalypts, increased ten times between 1965 and 1980 to 15 000 ha; and in the Sudan about six times to 180 000 ha. India established some 2.8 million ha of new plantation between 1966 and 1979 and investment in afforestation in the 1966–79 period was six times higher than the 1951–65 period (Johri 1978). In the five southern, tropical provinces of China eucalypt-planting expanded rapidly in the 1960s (Haishui 1988), but the bulk of the 6 million or so hectares of all plantation forest in 1980 still consisted of *Cunninghamia lanceolata* and *P. massoniana*.

Much of this expansion was afforestation for industrial purposes, pulpwood, sawtimber, and to a small extent plywood veneers. But planting also increased to meet direct human needs (firewood, shelter, building poles, fodder for grazing animals) and for environmental protection. Precise figures are rarely available, but nearly every tropical country greatly expanded the supply of tree seedlings for extension purposes—the distribution of trees to farmers and villagers usually free of charge. In 1980, of 11.5 million ha of plantations established in tropical countries, 7.2 million were for industrial and 4.3 million for non-industrial purposes, respectively (FAO 1988*b*).

These wider objectives for tree-planting were seen in the themes of the World Forestry Congresses 'The forest and socioeconomic development' (1972), and in 'Forests for people' (1978). Literature similarly reflected the changing emphasis, for example, *Forestry for rural communities* (FAO 1978), *Trees, food and people—land management in the tropics* (Bene *et al.* 1977), and *Forest energy and economic development* (Earl 1975).

This expansion in planted forests, combined with extensive use of exotic species, led to increased international cooperation, especially in seed collection and distribution, and tree breeding. Examples included the Thai-Danish international teak provenance trials (Fig. 3.3), the Commonwealth Forestry Institute's Programme with lowland tropical pines, seed collection, and

international distribution by Centre Technique Forestier Tropical of West African hardwoods (*Terminalia*, *Khaya*, *Lovoa*, etc.) from Côte d'Ivoire, and the centralized production and distribution of improved *P. caribaea* seed from Queensland. Numerous new working groups were organized within the International Union of Forest Research Organizations (IUFRO) such as breeding tropical eucalypts, the productivity of fast-growing species in the tropics and subtropics, and plantation forestry in the neotropics.

In the 1970s many organizations came to play a part in encouraging tree-planting in the tropics, particularly with non-industrial projects. Many of these concerned development of village woodlots, agroforestry, and tree-planting for environmental protection purposes. And, as already mentioned, policies of development banks were revised specifically to include plantation development for both industrial and environmental purposes. In 1978 the World Bank stated that 60% of future lending would be for rural forestry projects primarily to encourage village development and only 40% to help finance large-scale industrial plantations. But even the World Bank had only begun financing plantation forestry projects in the late 1960s; its first such loan was to Zambia in 1968.

Plantation forestry expanded rapidly in all aspects. In 1980 the total area of plantations in the tropics, as defined in this book (Chapter 1) was approximately 21 million ha. However, while this period saw expansion of plantation forestry, and in several countries a switch to it from attempts at natural forest management, there were failures. Valuable plantations were neglected or abandoned, for example, in Cameroon, Gabon, Liberia, and Zaire, owing to budget cutbacks and an inability to meet the expenditure required to maintain the resource. And, equally regrettably, many early attempts at social forestry were unsuccessful often for sociological rather than technical reasons.

Developments since 1980

The relative importance of plantation forestry has increased greatly in the last two decades. Globally, FAO resource inventory data suggest that the plantation estate has increased from 17.8 million ha in 1980 and 43.6 million ha in 1990 to 187 million ha in 2000 (Carle *et al.* 2002). However, the estimated totals for the different years are not strictly comparable as not all countries were included in the earlier estimates and the 2000 plantation assessment was the first global estimate with a uniform definition of forest plantations. Therefore, the rate of increase in plantation development, although very significant,

Figure 3.3 Part of a provenance trial of teak (*Tectona grandis*) in PNG.

is probably less than suggested by these statistics. The most dramatic change has been the recent extensive planting in Asia. While plantations are having an increasingly important role in substituting for wood and non-wood forest products from natural forests their impact on reducing deforestation may be much less significant.

The many roles plantations can play in rural development and environmental protection were recognized in the strategy of the Tropical Forestry Action Plan (now renamed Programme), in the specific mandate of the International Tropical Timber Organization (ITTO) to promote tree-planting in the tropics, and an objective of the World Bank's forest policy and strategy approved in 1991 was 'to slow deforestation in the tropics and to ensure adequate planting of new trees and management of existing resources'. In addition, this period will be remembered for the emergence on to the world scene of the NGOs, perhaps contributing 20% of all tree-planting in developing countries. There has been a rapid rise of people-orientated forestry (social forestry/community forestry) and this is reflected in both international and national policies and planning (Chapter 7). And over this period governments have encouraged the private sector to play an increasing role in industrial plantation forestry often by providing subsidies and other incentives (Chapter 6). This has been the case in China, where forestry is regarded as a major tool for rural development and poverty alleviation, and the Government is promoting household and corporate plantings. Between 1990 and 1997 the National Afforestation Project was implemented and established 1.3 million ha of well-stocked plantations in 16 provinces at a cost of US$560 million including US$330 million of World Bank loans, financed. This massive project is expected to yield social, environmental, and economic benefits (State Forestry Administration 2001).

Several silvicultural trends have also become apparent, by far the largest of which is expansion of tree growing on farms in one of the many agroforestry systems or small-scale woodlots for industrial wood production. Increasing needs for local sources of wood for on-farm use, diversification of income generating opportunities, and land tenure issues have contributed to this trend, as have the impact of the International Centre for Research in Agroforestry (ICRAF), now the World Agroforestry Centre, and the distribution of relevant information (Chapter 20). Attention has focused on the type of tree suitable for farm and village planting (e.g. Evans 1987; Raintree 1991; Hocking 1993; Doran and Turnbull 1997; Roshetko and Evans 1999). Various networks and agencies further promoted trees suitable for farm planting through workshops and publications. These included: the Nitrogen Fixing Tree Association, British ODI's Rural Development Forestry Network, USAID and Winrock's F/FRED (Forestry/Fuel Research and Development Project), and the agroforestry network (AFRENA) in Africa. Traditional industrial wood producing species, such as eucalypts and pines, are increasingly grown by smallholders in various partnership and outgrower arrangements with pulp and paper companies but higher value species for veneer, especially teak, are also being grown.

Of particular importance to industrial plantation forestry has been the successful introduction of clonal propagation techniques to eucalypts, *Gmelina arborea* and hardwoods such as *Triplochiton* (Leakey and Newton 1994) (Chapter 11). More generally, the potential of tropical phyllodinous acacias, for example, *A. mangium*, *A. auriculiformis*, and *A. crassicarpa*, which combine rapid growth, good stem form, and tolerance of infertile sites in the humid tropical lowlands, has been realized (Turnbull *et al.* 1998*a*). These now form large industrial plantations in

Indonesia, Malaysia, and Vietnam. Advances in wood utilization technologies have also seen species not previously considered as forestry species, for example, rubber, *Hevea brasiliensis* (9.7 million ha) and coconut palm, *Cocos nucifera* (12 million ha), being harvested and used for timber products (Carle *et al*. 2002).

Environmental and social concerns about tropical plantations emerged in the 1980s and 1990s. The most celebrated involved eucalypts in social forestry programmes. In India and Thailand further planting of eucalypts was, at times, seriously questioned. Although the debate is frequently framed in terms of the beneficial and negative ecological impacts of eucalypts, the major problem in fact relates to land availability, tenure, management, and the capture of benefits by the relatively better off segments of society (Raintree and Lantican 1993; Saxena and Vishwa Ballabh 1995). Investigation of the effects of eucalypts on water supplies and soil fertility show that they are not as harmful as environmentalists have often portrayed them but in some situations they can have negative hydrological impacts (Calder 2002).

This debate is symptomatic of a wider concern about environmental issues generally resulting from development, especially commercial, of which extensive plantations are one obvious and very public form. A second example comes from South Africa where afforestation of open high veld with pines, eucalypts, and wattle has been called by environmentalists 'green cancer' and companies such as Anglo American Corporation of South Africa advertise in the British Press (The Independent, 12 November 1990) their commitment both to the future in promoting large planting programmes and to conservation in explaining how and in what ways land is or is not afforested.

While most attention necessarily focuses on deforestation of natural forest in the tropics, plantations are certainly not immune from environmental concern. Anti-plantation sentiments are held by some NGOs including Greenpeace, the Environmental Investigation Agency, Native Forest Network, and World Rainforest Movement. All support many of the criticisms expressed in *Pulping the south: industrial tree plantations and the world paper economy* (Carrere and Lohmann 1996). However, most observers would agree with the conclusion of Cossalter *et al*. (2003): '*Fast wood forestry is neither inherently good nor inherently bad. It is a neutral technology which, when poorly planned and executed, can cause grave problems; and which, when well-planned and executed, can deliver not just large quantities of wood, but a range of environmental and social benefits.*' The question of sustainability of plantation forests is discussed in Chapter 24.

Not all plantation projects in the tropics in the 1980s and 1990s were successful. Globally, only 70% of new plantations established annually (about 3.1 million ha out of 4.5 million ha) are judged successful (FAO 2001*a*). In India government plantings on public land have an average survival of 60–70% but private plantings vary from 25% to 50% (Bahuguna 2001). Failures, frequently due to lack of post-planting maintenance (Chapter 13), protection from browsing, or wildfires, in many countries tended to be associated with larger-scale social or community forestry programmes where provision of the local, domestic product (firewood, poles, fodder, etc.) was not matched by adequate local involvement, input, or understanding of land tenure needs. Active promotion of sustainable management practices with community participation was an important strategy to reduce the incidence and impact of fires. Yields from tropical plantations in general have often been very low, frequently less than 50% of their productive capacity and the yield initially planned (Pandey 1995). Factors contributing to this situation included: Political decisions to expand forest plantation areas without proper feasibility studies and the low priority given to technical aspects, such as matching site with species, controlling the quality of the planting stock, tending and monitoring, protection and research support all contributed to this situation.

One other development of recent years needs highlighting. Certification schemes and related instruments to ensure compliance with best practices for sustainable forest management have been increasingly applied to forest plantations, including some in subtropical and tropical areas (Grace 2000). For example, almost the whole of South Africa's forest plantation estate has been certified. This development adds a new layer of management but also should ensure high standards of silviculture and management.

Present status of plantations in the tropics

The most recent data presented in Table 3.1 are mainly derived from that collected by the FAO in 2000 (FAO 2001*a*). The results of this plantation assessment were the first global estimates with a uniform definition of forest plantations and therefore cannot be compared directly with previous estimates, for example, FAO (1981), Pandey (1995). The 1990 data have been updated from the second edition (Evans 1992) principally using FAO's 1990 assessment data (Pandey 1995). The figures in Table 3.1 should only be considered as approximate as the quality and quantity of plantation data are

Table 3.1 Approximate areas of planted forest (000s ha) including woodlots by countries in the tropical and warmer subtropical regions[a]

Country/Region	1965	1980	1990	2000
Africa				
Angola	(88.2)	160	195	141
Benin	6.5	9.7	11.0	112
Botswana		0.3	0.5	1
Burkina Faso	0.6	5.5	46.0	67
Burundi	(42.8)	70.0	80.0	73
Cameroon	(8.8)	16.0	(19.5)	80
Central African Republic	(0.6)	1.0	(1.3)	4
Chad	0.6	(51.0)	(1.3)	14
Comoros		0.3	1.0	2
Congo	1.3	15.0	17.5	83
Cote d'Ivoire	13.7	55.0	86.0	184
Democratic Republic Of Congo (Zaire)	(10.0)	22.5	55.5	97
Eritrea				22
Ethiopia	16.0	135	270	216
Gabon	14.0	30.0	(35.0)	36
Gambia	(0.7)	1.6	1.8	2
Ghana	(11.1)	22.3	76.0	76
Guinea	4.1	20.0	(26.0)	25
Guinea Bissau	(0.3)	0.5	(0.7)	2
Kenya	89.0	157	190	150
Liberia		0.4	9.0	119
Madagascar	(147)	260	295	350
Malawi	25.1	80.3	156	112
Mali		0.4	3.9	15
Mauritania				25
Mauritius	6.0	10.9	12.1	13
Mozambique	(12.9)	33.0	45.4	50
Namibia				0.3
Niger		0.2	16.4	73
Nigeria	(100)	170	259	693
Reunion	(5.3)	9.0	(11.7)	3
Rwanda	(17.1)	90.0	110	261
Senegal	5.8	44.0	145	263
Seychelles	1.1	16.0	2.0	5
Sierra Leone	4.6	6.0	6.7	6
Somalia	2.0	6.0	7.5	3
South Africa[b]	500	697	795	1050
Sudan	26.6	188	330	641
Swaziland	82.0	101	108	161
Tanzania	18.6	90.0	(100)	135
Togo	(3.5)	6.0	23.0	38
Uganda	16.7	28.4	32.0	43
Zambia	4.7	50.0	65.0	75
Zimbabwe	90.0	115	125	141
Regional total	*1378*	*2724*	*3773*	*4566*
Asia				
Arabian peninsula states		5.0	10.0	11
Bangladesh	28.3	153	380	625
Bhutan	(1.0)	6.0	13.0	21

Continued

Table 3.1 *(Continued)*

Country/Region	1965	1980	1990	2000
Brunei		0.1	0.3	3
Cambodia	6.0	(10.2)	(13.3)	90
China[c]	2400	5800	9000	19400
India	954	3800	14000	32578
Indonesia	(706)	1918	3700	9871
Laos		0.5	1.9	54
Malaysia	1.0	26.0	90.0	1750
Myanmar	56.6	87.3	340	821
Pakistan[b]	(43.3)	73.6	102	(98)
Philippines	100	318)	100	753
Sri Lanka	45.0	99.0	195	316
Taiwan	(44.2)	75.0	(97.5)	422
Thailand	21.1	256	560	4920
Vietnam	11.8	(403)	(616)	1711
Regional total	*4421*	*13047*	*29245*	*73444*
Oceania				
Australia				
Northern Territory	0.5	3.9	3.9	7
Queensland	46.0	144	202	191
Fiji	4.9	53.8	85.0	97
French Polynesia				5
New Caledonia	(3.2)	8.0	23.0	10
Papua New Guinea	5.2	24.5	44.3	90
Solomon Islands	(0.3)	18.0	25.3	50
Samoa			4.8	5
Tonga	–	–	–	1
USA, Hawaii	(9.3)	14.6	20.9	(21)
Vanuatu		0.4	1.8	3
Regional total	*70.4*	*269*	*420*	*480*
Central America and Caribbean				
Belize	(1.1)	2.9	4.0	3
Costa Rica	(2.4)	4.1	40.0	178
Cuba	143	(243)	(316)	482
Dominican Republic	1.5	10.9	12.0	30
El Salvador	(1.7)	2.9	(3.8)	14
Guadaloupe	(2.3)	3.9	4.9	4
Guatemala	1.0	3.0	(25.8)	133
Haiti	0.1	1.0	(3.0)	20
Honduras		2.0	5.0	48
Jamaica	4.8	11.3	16.5	9
Mexico	50.0	159	263	267
Nicaragua	0.1	7.9	37.5	46
Panama	1.0	(3.0)	(10.0)	40
Puerto Rico	1.3	1.7	2.7	4
Trinidad and Tobago	7.9	21.4	22.0	15
USA/Florida	1.0	7.0	18.0	(18)
Regional total	*219.3*	*486*	*786*	*1311*
South America				
Argentina[b]		182	(300)	200
Bolivia	23.7	29.2	36.9	46
Brazil	500	3855	7150	6500

Continued

Table 3.1 *(Continued)*

Country/Region	1965	1980	1990	2000
Chile[b]	9.5	22.9	22.9	(24)
Colombia	16.3	95.0	250	141
Ecuador	(4.0)	8.5	60.0	167
French Guiana	(0.6)	1.1	2.0	1
Guyana	0.4	0.7	0.8	12
Paraguay	(0.6)	1.1	7.4	27
Peru	20.0	127	272	640
Suriname	2.8	13.1	18.0	13
Venezuela	1.5	112	350	863
Regional total	*579*	*4448*	*8470*	*8634*

  [a] Estimated areas between 27°N and 27°S of equator.
  [b] Only includes part of country north of approximately 27°S.
  [c] Only includes part of country south of approximately 27°N.
Figures in parentheses are very approximate estimates. A space indicates no data available and in most cases the total area of any plantations will be very small. The 2000 figures include rubber plantations.

Table 3.2 Areas of planted forests (000s ha) including woodlots in tropical and subtropical regions[a]

Region	1965	1980	1990	2000
Africa	1378	2724	3773	4566
Asia including southern China	421	13 046	29 245	73 444
Northern Australia + Pacific Islands	70	269	420	480
Central America + Caribbean	219	486	786	1311
South America	579	4448	8470	8634
Total	*6667*	*20 973*	*42 694*	*88 435*

  [a] Estimated areas between 27°N and 27°S of equator.

very dependent on the capacity of national forest inventory systems to collect and analyze data. Different sources often cite widely divergent figures for any one country. Some countries in official reports equate annual planting achievement with numbers of tree seedlings produced with no check on what land has actually been successfully planted; this invariably results in overestimates. In other cases the full ground area of poorly stocked plantations is also included which is misleading in terms of the quantity of standing timber.

Table 3.2 summarizes the regional and world trends from data in Table 3.1.

Current programme of planting

Globally new forest plantations are being established at rate of 4.5 million ha annually. About 48% of these plantations are to produce material for the wood processing industries, 26% of the remainder is for non-industrial uses (fuelwood, soil and water protection, etc.) and the balance is for unspecified purposes (FAO 2001*a*). Table 3.3 shows an estimate of the areas of industrial plantations in tropical Africa, America, and Asia in 1995. Tropical Asia has the largest area of industrial plantations and is planting greater areas than the other regions. Within the tropics China and India have the largest plantation areas followed by Indonesia, Brazil, and Thailand (Table 3.1) but in Asia a large percentage of plantations is for non-industrial uses.

Undoubtedly, in the decade 1990–2000 the greatest expansion of tree-planting, for social, protection and industrial purposes, has been in China and India, although many of the plantations are not highly productive due to low site quality, or poor stocking and/or inadequate maintenance. It is likely that the greatest area of plantations will be developed in these countries in the immediate

Table 3.3 Estimated areas of industrial plantations in tropical Africa, America, and Asia in 1995

Region	Total plantation area ('000s ha)	Annual establishment rate ('000s ha)	Industrial (%)	Industrial area ('000s ha)
Tropical Africa	2434	120	52	1270
Tropical America	5973	230	76	4540
Tropical Asia	19 098	1300	45	8590
Total	27 505	1650		14 400

Based on Pandey and Ball (1998). Does not include rubber plantations.

future. India has a total reported plantation area of 32.6 million ha and an annual planting rate of 1.5 million ha. Non-industrial plantations make up 20.6 million ha (63%) of the total, and 19.6 million ha (60%) of all plantations are in public ownership (Carle *et al.* 2002). Evidence of the scale of commitment is that India's National Forestry Action Programme plans to expand forest areas substantially throughout the country and estimates it will cost US$32 billion over 20 years (Bahuguna 2001).

China similarly has a major commitment to increasing its forest cover from 14% to 17%. Much of this will be through establishing at least 21 million ha of forest through plantations and aerial seeding The Tenth Forestry Five Year Plan and the 2015 Draft Development Plan aims to increase the fast-growing plantation area to 9.48 million ha by 2015. Almost 5 million ha will be new fast growing plantations and the balance will be replanting poorly stocked and unproductive areas (Durst and Brown 2000; Xu 2000; Yang 2001). Eucalypts are being favoured in south China due to their very rapid growth rate and potential for short rotations.

Elsewhere in tropical Asia there are significant plantation programmes in progress. Indonesia is vigorously promoting industrial plantations to support established pulp mills with an annual planting area of 270 000 ha, and annual planting is averaging 225 000 ha in Thailand (FAO 2001a). Vietnam has a 5-million ha reforestation programme to be carried out by 2010 including 3-million ha of intensively managed industrial wood plantations (Durst and Brown 2000).

In tropical South America, Brazil has the largest area of plantations and an average annual planting rate of eucalypts and pines of 135 000 ha. Venezuela has been planting 50 000 ha a year, mainly industrial pine afforestation of its eastern grasslands, while most other recent major plantation developments have been in more temperate areas of Argentina, Chile, and Uruguay, etc. The extent to which tropical plantations will compete with those of temperate areas in South America and elsewhere will depend on their economic

performance as subsidies and other incentives are phased out.

Although some African countries have significant areas of plantation forest, for example, Nigeria (690 000 ha), Sudan (640 000 ha), and South Africa (1.5 million ha), annual planting rates have generally been relatively small, rarely above 20 000 ha and more often below 10 000 ha. Brown (2000) suggests that the absence of a strong infrastructure is likely to remain a significant competitive disadvantage for industrial plantations in many countries although non-industrial plantations may accelerate to provide fuelwood where there is high population pressure.

The accelerating increase in planting in recent decades is evident in Table 3.1. Figures show that about 1.6 million ha of industrial plantations (Pandey and Ball 1998), large areas of non-industrial plantations, and trees outside the forests being established annually in the tropics (FAO 2001a). If these plantations are planted on favourable sites and managed effectively their contribution to world wood supplies will become even more important. They will help meet the increased demand from the world's growing population and substitute for the declining wood supplies harvested from natural forests. Sedjo (2001) has suggested that '*By 2050 most industrial wood could come from a small area of plantation forests, much of it in subtropical and tropical countries, while natural forests could remain for environmental and other non-wood services.*' The extent to which tropical plantations will play a role in the sequestering and storage of atmospheric carbon remains unclear and much depends on international agreements and initiatives.

Species used

Four genera, *Acacia, Eucalyptus, Pinus,* and *Tectona,* accounted for the majority of tropical planted forests in the tropics in 2000 (FAO 2001a). The species and genera listed in Table 3.4 remain the most important (though the relative proportions

Table 3.4 Species used in tropical planted forest

Genus/group	Species[a]	%[b]
Eucalyptus	grandis, camaldulensis, globulus[c], tereticornis, urophylla saligna, robusta, citriodora, exserta, deglupta, various hybrids, other	50
Acacia	nilotica, mangium, auriculiformis, crassicarpa	17
Tectona	grandis	10
Pinus	caribaea, patula, elliottii, oocarpa, kesiya, merkusii, massoniana, other	23
Other conifers	cunninghamia lanceolata, Araucaria cunninghamii, A. angustifolia, Cupressus lusitanica, other	
Other hardwoods	Gmelina, Leucaena, Grevillea, Meliaceae spp., Terminalia, Paraserianthes, Prosopis, Casuarina, Cordia, Triplochiton, other	

[a] Species ordered very roughly according to extent of planting.

[b] Percentage figures are very approximate estimates based mainly on 1990 data (Pandey 1995) and are given only for the major genera.

[c] E. globulus is only planted in tropical highlands, mainly in China, Colombia, Ecuador, Ethiopia, India, and Peru.

may now differ a little). In tropical Africa pines and eucalypts occupy about 50% of the plantation area, and in tropical America these two genera account for more than 80% of the total. In tropical Asia eucalypts, acacias, and teak are the favoured genera with pines planted to a lesser extent (Pandey 1995). There are an estimated 5–6 million ha of planted tropical pines (Barnes *et al.* 2001) with the majority of plantations in tropical America.

'Non-forestry' species such as rubber, coconut, and oil palm are of increasing importance for the supply of timber and industrial fibre. The reported area of these species (26.5 million ha in 1995) is about one-third of the reported area of forest plantations in the tropics and subtropics (Pandey and Ball 1998) and rubber is already being grown in Malaysia as much for its timber as for latex (Thai See Kiam 2000).

Much planting in recent decades has involved multipurpose species, commonly nitrogen-fixing species, especially *Leucaena leucocephala* and several *Acacia* and *Casuarina* species. Establishment of high quality hardwood species remains limited though significant increase may occur (Grainger 1988*a*).

Purpose of plantations

It has been emphasized that tree-planting in the tropics is no longer almost wholly for industrial purposes. Purpose and ownership of plantations vary considerably between regions. Non-industrial plantations make up 55% of plantations in tropical Asia, 48% in tropical Africa, and 24% in tropical America (Pandey and Ball 1998).

Planting for firewood, other village needs, in agroforestry development, and for protection (to reduce soil erosion, control water runoff, combat desertification, provide shelter and shade) are all becoming increasingly important. Trees planted outside the forest in cities, on farms, along roads and in many other locations are not designated forest plantations but make a major contribution to the environment and rural livelihoods. Where forest resources are scarce, especially in densely populated areas, trees outside the forest are a major source of food, fodder, and fuelwood. For example, in Kenya, it was estimated that farms produced about 18 million m^3 of wood in 2000 (FAO 2001*a*).

Types of plantation development

Four main kinds of tree-planting projects can be identified, and each has significantly contributed to the development of plantation forest.

1. Planting by private companies. Almost entirely carried out for commercial purposes; it includes all plantation development in Brazil and much of it in countries such as Indonesia, the Philippines, South Africa, Swaziland, and Thailand. In some countries governments initially established a critical mass of plantations but many are now devolving ownership and responsibility for plantation establishment and management to the private sector. In many cases some government incentive is provided to assist with the costs of plantation establishment.

2. Specific developmental projects. These are often for creating an industrial resource but organizations

funding projects are as much concerned with aiding investment and development in developing countries as with the direct commercial profitability of the operation. Examples include British aid to Malawi for the Viphya Pulpwood Project, and the World Bank loans to Madagascar for the Mangoro project in the late 1970s (for plantations to produce sawlogs and pulpwood) and to China for its 1 million ha National Afforestation Project (1990–97). These projects are usually implemented by or through national forest departments or by specific quasigovernmental bodies set up for the purpose. India's parastatal afforestation agencies also fall into this category.

3. National afforestation programmes. Tree-planting often forms part of a nation's planned national expenditure and invariably includes tree-planting for non-industrial purposes. Protective afforestation, shelterbelt planting, rural employment schemes, environmental improvement projects, firewood planting, and promotion of rural development forestry generally are mostly part of governmental programmes at national or regional level. This does not preclude external funding. Most aid agencies are now giving greater support to such projects than to industrial plantations (Winterbottom 1990) and governments are encouraging greater private sector investment in industrial plantations. The largest remaining government forest plantation establishment programmes are in India and China. About 70% of plantation establishment in India is carried out with State funding (Brown 2000).

4. Small-scale tree-planting projects promoted by private companies and NGOs. The proliferation of village or individual farmer plantings represents a new dimension in afforestation effort drawing on support and funds from within and outside countries involved (e.g. Mayers 2000).

Conclusion

This chapter has sketched out the development and present status of plantation forestry in the tropics; it is evident that a new tree and timber resource is being created that will contribute significantly to future wood supply and meet a range of economic, social, and environmental objectives. Undoubtedly, the world's planted forests appear destined to supply an increasing proportion of wood requirement. In 2000 the supply of industrial roundwood wood from plantations was estimated at 35% and is projected to be 44% by 2020 with increased production in all regions (ABARE-Jaakko Pöyry 1999).

The potential of tropical plantations to contribute to this increasing supply will depend on many factors including: global and country policies on trade and sustainable forest management; the development of markets, and improved management, silviculture and processing to make tropical wood competitive with wood from other sources. It is highly likely that trees outside the forest will contribute more to the production of fuelwood and some other non-industrial products in the tropics. While there is potential for large areas of planted forest to be developed as sinks to mitigate high carbon dioxide levels in the atmosphere, this measure will have only a relatively minor impact on the problem and similarly, even if plantation forestry continues to expand it will compensate only in small measure the rapid loss of natural forest to other land-uses that is continuing to take place (Chapter 2).

CHAPTER 4

Plantation organization and structure

In Chapter 1 the words 'plantation' or 'planted forest' were defined and described in outline. But it is necessary to examine more closely the organization and structure of plantations and planted forest as background to their development and management. Although most of this chapter is concerned with industrial forest plantations, much of it also applies to small-scale woodlots, though the relative importance of some matters may differ, for example, species composition, age-class structure, stocking, design, and shape of the plantation.

Plantation organization

Organization

The business of planting trees, though a straightforward operation itself, involves much organization and planning. Many different operations are needed both to produce seedlings for planting and to manage the plantation they form. Even provision of tree seedlings for an extension programme requires that a suitable species be first chosen, that sufficient quantities of seed are collected, perhaps stored, dispatched to the nursery, and sown in prepared beds or containers. Newly emerged seedlings need regular watering, tending, and protection, and they must be distributed once they reach the right size. If all these silvicultural operations alone are needed for raising seedlings clearly some thought must be given to the correct siting of a nursery, the provision of facilities such as water supply, accommodation, and storage sheds, and the needs of the labour employed. The apparently simple task of providing seedlings for extension planting requires careful organization; the planning and implementation of a complex major afforestation scheme necessarily requires even greater organization of time and resources.

Organization is clearly important but is often a difficult idea to convey in cultures where foraging in the forest for all kinds of produce (food, medicines, fuel, etc.) has always been practised and the

forest viewed as inexhaustible. The concept that trees can be grown in an orderly and organized fashion like a planted food crop, and, also like a food crop, are left untouched until harvested, can be one of the biggest obstacles to social acceptance and ultimately success of planted forests in developing countries. It requires participation of local communities and other stakeholders, negotiation, and attention to tenure and access rights and other matters before establishing plantations.

Space

Plantation forestry is an extensive activity. Trees need space for proper development, and many trees together in a plantation require large areas of land. Natural forests have been cleared for farming ever since sedentary agriculture began, and it still continues. Land is a finite resource and is increasingly under pressure from alternative uses not only for agriculture, but also for urban expansion, designation as reserves, road building, mineral exploitation, settlement of transmigrants, and so on. The forest plantation is another competitor for land, and one which requires large areas over long periods. Similarly, in rural development forestry conflict can arise because tree-planting may reduce food production or affect adjacent crops, and growing trees for wood production is not often a high priority of villagers and farmers. The land availability factor is central to plantation development and is the subject of Chapter 5.

Industrial plantations may cover tens of thousands of hectares. In the tropics a pulpwood forest supplying a small mill would occupy 20 000–40 000 ha and large pulp mill may need to draw on wood from 100 000 ha for economic operations. But trees are generally less site demanding than agricultural crops, and can often be grown on degraded lands or land too steep or with rocky, shallow soils to be of value for anything else. Plantations require much land, but much of it may only be suitable for growing trees (Fig. 5.1).

Time

Even in the tropics, and even for the production of posts, poles, and firewood, growing trees still takes a few years. Any plantation development will tie up land and capital and will involve continuing oversight for several years.

Depending on end-use the rotation may vary from less than 1 year (fuelwood and alley cropping) to 60 years or more (veneer); see Table 2.2 and Chapter 18. But even the rotation length does not necessarily delimit the period land is committed to a plantation project.

1. Many crops are grown on short rotations using the coppice system and replacement of stumps (the logical time to change land-use unless de-stumping or poisoning are done) is rarely done at less than 20-year intervals.
2. Most projects have several age-classes which extend the project life over many more years than the rotation length of one stand.
3. Access to land is needed some time before planting and after felling, for ground preparation and site clearance.
4. Economically, the second rotation, established on land with an existing management infrastructure— roads, tracks, nurseries, fire dams, fencing, etc., is often much more profitable than the first crop. Few enterprises plan having only one rotation.

This time-factor is important since it makes the decision to plant trees very different from that for cultivating most food crops: (i) the grower is committed to a course of action without real opportunity for change for a long period; (ii) returns on the investment of land, time, and capital are delayed several years; (ii) land may be unavailable for other uses for a generation or more, though there are many possibilities of land-use integration.

The inflexibility the above factors introduce cause many landowners to resist forest plantation development, in particular the time-scale often seems too long and the benefits too remote. Plantations are mostly developed by governments, their agencies, development organizations, or large commercial enterprises. Tree-planting to create an industrial resource has rarely been undertaken by a private individual, Jari in Brazil being an exception, private sector planting of black wattle (*Acacia mearnsii*) and eucalypts in South Africa, and *Gmelina arborea* and *Paraserianthes falcataria* (Fig. 5.7) in Mindanao (Philippines) was on a small scale. Recently companies in the private sector have shown increasing interest to form partnerships with communities and individuals to grow industrial wood. Many of the largest companies producing pulpwood now operate an outgrower scheme or provided extension services to smallholders to grow trees for them (Roberts and Dubois 1996). The costs and benefits of these arrangements are discussed in more detail in Chapter 5. Alternatively, in social forestry planting by private individuals is important because benefits of fodder, shade, fuel, or fencing are quickly realized where the individual has assured rights to the trees (Chambers 1987).

Plantation characteristics

Orderliness, regularity, and relative ecological simplicity show a forest plantation to be man-made and clearly distinguish it from natural forest. The economic advantages of these qualities are greater efficiency in many operations and a more uniform product. Nevertheless, though nearly all plantations exhibit these characteristics, they bring some disadvantages compared with natural forest. This is the subject of regular versus irregular silviculture which has greatly exercised foresters' minds for a century or more; see discussions and analyses in the main silvicultural textbooks. (Champion and Seth 1968; Daniel *et al.* 1979; Smith D.M. 1986; Zobel *et al.* 1987; and Matthews 1989.) Below, only the main characteristics of plantations are considered.

Species composition

Nearly all tropical plantations are grown in monoculture, one species being planted over a large area. Exotic species predominate in both industrial plantations and rural development planting and the reasons for this are discussed in Chapter 8. The overriding reason for monocultures is the economic benefit of simpler silviculture and harvesting, and product uniformity. Experience is gained in handling the one or two chosen species, operations become regularized, instructions and training are more easily given, and great familiarity with a species speeds detection and identification of problems such as disease or nutrient deficiency. For a fuller treatment of the question of mixed and pure forest plantations in the tropics refer to FAO (1992).

Though individual stands and usually whole compartments are one species of one provenance or clone, it is less common for an entire forest to be of only one species. Nevertheless, several pulp companies have planted extensive areas, often over 100 000 ha, using only *Acacia mangium* on the relatively uniform, lowland sites in Sumatra, Indonesia. At Jari Celulose S.A. in Brazil it was formerly the practice to plant *Gmelina arborea* on all the more fertile sites underlain by clay residual soils, *Eucalyptus urophylla* on infertile clays and loams, and

Pinus caribaea on infertile, more sandy sites (Welker 1986) but since 1992 this company moved to planting selected clones of *Eucalyptus* almost exclusively (McNabb and Wadouski 1999). Often two or more species are planted to suit broad site types and for fire protection purposes. In the Usutu forest, Swaziland, about 65% is *Pinus patula*, 25% *P. elliottii* for hotter and drier sites at low elevations, and the remaining 10% comprises *P. taeda, E. grandis, E. saligna,* and *Acacia mearnsii* with the latter three mostly planted in firebreaks (Fig. 4.1). Chinese foresters plant the more tolerant pines such as *P. massoniana* on upper slopes and ridges and use *Cunninghamia lanceolata* on the more fertile, deep lower slopes and valley bottoms. Similarly, *Araucaria cunninghamii* is favoured on the most fertile sites in Queensland while *P. caribaea* var. *hondurensis* and *P. elliottii* are used on poor, sandy sites. In East Malaysia, Sabah Softwoods plant *Gmelina arborea* on deeper, more fertile soils, *A. mangium* on poorer, infertile sites and grassland, and use *Paraserianthes falcataria* for shading cocoa. Rarely are species mixed in one stand in commercial plantings.

Two possible risks are associated with monocultures, susceptibility to diseases and pests, and maintenance of long-term productivity; these are discussed in Chapters 19 and 24. Also, there can be a problem which might be termed 'over familiarity' where a species is planted extensively simply because it is well known even though it is inferior to another potentially available species or because seed is readily available. The most widespread recent example are selected, highly productive strains of *Leucaena leucocephala* which have been introduced everywhere and often on sites far from suitable for the species simply because it became well-known as a 'wonder tree'. Another, example has occurred in Indonesia where early plantings of *Acacia mangium* relied heavily on a local seed source at Subanjeriji (Sumatra). Subsequent introductions

showed seed sources from Papua New Guinea were two to three times more productive (Turvey 1995). In Brazil, widespread success of *E. grandis*, ideal for wood pulp, has displaced other very productive eucalypts in charcoal production even though species such as *E. cloeziana, E. paniculata, E. citriodora,* and *E. maculata* have a higher wood density and produce a better yield and quality of charcoal. They are less widely planted because they are not well known.

Other aspects of species' choice are considered in Chapter 8.

Age-class distribution

Because of the logistics of planting, all trees in a stand are usually of one age, or one planting year. A few months' variation may be introduced when replacing failures after planting, but this does not significantly change the even-aged character of the stand. Even-aged stands are simple to manage since operations, such as planting, tending, thinning, and clear-felling, can be undertaken at one time over the whole crop rather than irregularly in different parts of it. The difficulty of creating an uneven-aged stand artificially is another reason why even-aged stands are established in plantation forestry. However, an uneven-age condition may be encouraged where a plantation regenerates naturally and is primarily for protection purposes to provide continuous ground cover.

Although within a compartment or stand all trees are the same age, over a whole forest this will not be the case. Neither the practicalities of creating a large industrial plantation nor the needs of the end-user would ever require a whole forest to be all the same age. A major project takes time to develop: land to be acquired, nurseries established, ground prepared, roads built, labour recruited and trained, etc.

Figure 4.1 Compartment A9 on the slopes of Mangqongqo in the Usutu forest, Swaziland, showing pure stands of the main species (*Pinus patula*) and *Eucalyptus grandis* planted in a firebreak.

As this infrastructure grows, a steadily larger area of planting each year can be achieved (see Fig. 4.2) and a range of age-classes arises. Also, at the end of a rotation, differential rates of felling further increase the spread of age classes as will unplanned events such as fires and cyclones. Initially yields from the forest are often below the maximum possible since harvesting methods take time to develop and demand from the end-user may expand slowly while a new factory is commissioned or markets established. Even when full production is possible the out-turn of wood from a forest is rarely constant as demand fluctuates from year to year. The effect of this, and recent fires, is clearly illustrated in Fig. 4.3 which shows the changing pattern of age-classes in the Usutu forest plantations since 1966.

Stocking

Stocking was briefly referred to in Chapter 1 in defining a plantation, but it is an important characteristic and further comment is needed. In a plantation, all trees are planted for some purpose and have value, and all the land allocated for timber production is stocked with the chosen tree species. This contrasts with most natural forest where, because of species characteristics or tree form, only a few trees per hectare may be marketable.

However, though plantations are normally planted so that all possible areas have trees growing on them, the site may not be fully occupied. Clearly this is so at planting when trees are small, but also

very wide initial spacings lead to much less than the fullest use of the site ever being made. A point comes when trees are planted so far apart that the 'stand' appears no longer like a plantation, for example, line enrichment planting and many agroforestry systems, with lines often more than 10 m apart.

There is a wide range of planting density in the tropics reflecting the many purposes for which trees are grown. Farmers often plant trees in their small woodlots at spacings closer than 1×1 m (10 000 stems ha^{-1}) to enable them to harvest sticks for vegetable supports, house building, and fencing. Most industrial plantations are planted at spacing

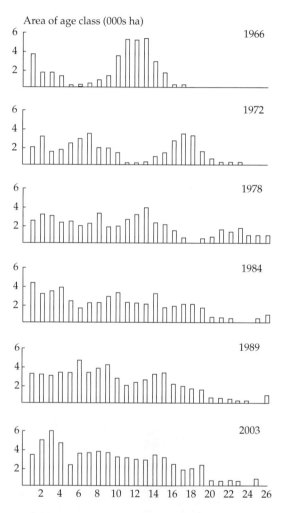

Figure 4.3 The changing pattern of pine age-classes in the Usutu forest, Swaziland. Note the gradual development to almost equal areas of each age-class by 1989. In 2003 above average recent plantings caused by catastrophic fires in late 1990s.

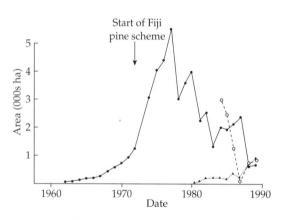

Figure 4.2 Development in annual planting programme of *Pinus caribaea* in Fiji.

of about 2.5–3.0 m spacing (1000–2000 stems ha^{-1}). There is no minimum stocking but, in round figures, stockings at planting of less than 600 ha^{-1} for fast-growing eucalypts and less than 900 ha^{-1} for pines would be inadequate for proper plantation development. ('Proper' does not imply 'correct' since in agroforestry trees may be deliberately spaced far apart to allow room for food crops or maintenance of grass swards or other forage for grazing.) There is a tendency for clonal plantations to have wide spacing as good stem form can usually be guaranteed, for example, ECO sa's clonal plantations of *E. urophylla* × *E. grandis* in the Congo are now planted at 2.7 × 4.7 m (800 stems ha^{-1}).

Where the number of trees at planting is well below the above figures a plantation is only partially stocked, or poorly stocked if below 300 stems ha^{-1}. Where planting augments natural regeneration, as in enrichment systems, it was noted in Chapter 1 that a forest was defined a plantation, if more than 50% of the intended final crop was planted. However, this definition often results in less than full stocking: the extensive mahogany stands in Fiji, line planted at 3 × 9 m^2, qualify as plantations but the sites are not fully occupied by the planted species for most of the rotation (Fig. 1.5).

Stocking does not remain constant during the life of a plantation. Some trees die naturally and others may be thinned out (Chapter 16). Also, no plantation fully occupies a site all the time, in the first year or two at least, crowns and roots will not have spread sufficiently to begin competing with neighbouring trees for light, moisture, and nutrients.

Silvicultural system

A silvicultural system is the method by which a tree crop is grown and regenerated. Matthews (1989) provides an account of the many systems possible.

Nearly all tropical plantations are planted rather than directly sown and, at maturity, clear-felled and replanted rather than naturally regenerated. Where practicable, and where crops are grown on short rotations, regeneration from stump shoots (coppicing) is important. Other regeneration systems tried are mostly experimental, for example, two-storey high forest, or to accommodate special conditions, for example, taungya.

The dominance of clear-felling and replanting, a system whereby at rotation age all trees on a site are felled at one time (Fig. 4.4), is because it is simple and cheap compared with the alternatives. Also, it is a quick, reliable way of achieving uniform regeneration which is important for fast-growing crops. If satisfactory natural regeneration can be obtained, and it is common in pine plantations and can even be invasive with some acacias (Fig. 18.7(b)), replanting is often still desirable. Frequently by the end of one rotation, tree improvement work has made available planting stock that is superior to the old crop in growth, form, or disease resistance. Clear-felling and replanting allows full advantage of this to be taken. Where a species coppices well, that is, the cut stumps produce new shoots which will grow into full-size trees (Fig. 4.5(a)), it is suited to short rotation crops grown for firewood, poles, pulpwood, or mining timber. Several million hectares of eucalypts and teak are managed on the coppice system, and commonly two or three crops are harvested before replacement of the stumps is needed. In various parts of the tropics a modified form of coppicing, called pollarding (Fig. 4.5(b)), is used with trees regularly cut at 4 or 5 m above ground so as to keep new shoot growth above browse height of livestock. Though typically applied to naturally occurring *Acacia*, *Erythrina*, *Combretum*, etc., it is quite feasible to manage planted trees of virtually all broadleaved species in this way to yield fuel-wood, fodder, etc. Pollarding of *Grevillea robusta* planted on farm

Figure 4.4 Clear-felling pulpwood stands in the Usutu Forest, Swaziland.

Figure 4.5 (a) Coppice shoots on a stump of *Eucalyptus grandis*. (b) Pollarded trees in Ethiopia. (c) Shredded stems of oak (*Quercus* spp.) in Northern Iraq.

boundaries in the highlands of Kenya is a very common practice. Related to this is the quite common practice of 'shredding', where all branches are cut from the entire length of the stem every 3–7 years (Fig. 4.5(c)).

Where natural closed forest in the tropics is logged and the remnant is not preserved, direct replacement by plantation is often expensive because of the need to clear remaining trees, luxuriant weed growth, and generally difficult access. Two kinds of silviculture have been introduced to overcome this: the taungya system and enrichment planting. The taungya system is one of many ways of growing food and trees at the same time (agroforestry) and, in its simplest form, is where local inhabitants (often shifting cultivators) are allocated land for their food crops on which they are also required to plant and care for trees (Chapter 20). The other alternative is not to establish pure crops from the time of planting, but to enrich the remnant forest with a desirable species with the intention that it will form the bulk of the final crop; see earlier and Chapter 1.

In protection forestry, good natural regeneration is sought or there is a system of gradual replacement of the crop to avoid a break in ground cover, a practice today called 'continuous cover' forestry. Silviculture in arid areas does not differ greatly from elsewhere, except for the greater use of pollarding and shredding noted earlier, though emphasis is on choice of hardy species, conservation of moisture, and full protection from browsing damage and interference from man, especially the excessive gathering of firewood and fodder.

Plantation design

Plantations are ordered and regular in appearance because trees are planted in lines and boundaries and internal breaks are often straight (Fig. 4.1). In plantation design three aspects need to be considered at the same time: shape, layout and internal subdivision, and lines of communication.

Shape

The outer boundary of any plantation will be determined by its legal position with respect to neighbouring land, but it is not essential to plant right up to that boundary if it is awkwardly shaped or an unplanted strip is left as an external firebreak (Fig. 12.1) or for reasons of good landscaping or conservation. Two cost considerations, providing access and maintaining the boundary, influence desirable shape and both make the compact rectangular or square plantation usually the most efficient. A long ribbon-like shape, makes good access costly. To extract timber, either a long road is needed which only serves small areas or long extraction tracks for dragging the logs have to be accepted. Moreover, if protection along the boundary is needed, for example, against a firebreak or for a fence against livestock, maintenance is very costly because the boundary is long but only a small area of plantation is enclosed. This elementary point is stressed as it is not always appreciated (Fig. 4.6).

The above points also apply to the internal subdivision of a plantation forest. A plantation with numerous small compartments will have a much greater length of internal breaks than one of the same area but with only a few compartments.

The importance of efficient shape will often be far less in social or community forestry plantings. Indeed, shape will be largely determined by land available or being set aside. In the Bilate project in southern Ethiopia more than 600 ha of woodland have been planted in a 200–300 m-wide strip 22 km long, beside the Bilate river. This was the land set

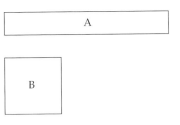

Figure 4.6 The importance of shape on boundary length. A and B are of the same area, but the boundary (perimeter) of A is over twice as long as that of B.

aside by the people and, by being a long ribbon, the project benefits several kabeles (village communes) rather than just one (Evans 1989). Protective windbreaks on farms may also be long and narrow. The *Casuarina equisetifolia* plantations stabilizing sand dunes along the coast of southern China vary in width from 1 to 5 km depending on land availability.

Plantation layout and subdivision

In a large plantation project, primarily for convenience of management and organization, the forest does not steadily expand from one centre, but usually develops around several centres each forming a self-contained unit. For example, before the Viphya plantations in Malawi were placed on a care and maintenance basis in the 1980s, the 32 000 ha established by 1978 were divided into seven administrative units of forest blocks each with its own work programme, labour, and compound, and run by a qualified forester.

Within a forest block there is a subdivision into compartments. A compartment is the 'address' to identify part of a forest and is the operational unit for most kinds of work (inventory, silviculture, harvesting) with the whole compartment treated at the same time and usually in the same way. Normally compartments are bounded by a road, track, or firebreak. Their size ranges from 20 to 200 ha, but 40–50 ha is probably about average.

Roading and access

The regular layout of industrial plantations in conveniently shaped blocks is not easy in practice because forest land is often rough ground in rugged and broken terrain. Consequently, gaining access to and inside a plantation (can roads and tracks be readily constructed?) should primarily determine where planting for commercial purposes is done. Access limited to four-wheel drive vehicles may enable planting to be done but will be wholly inadequate for extraction of timber at the end of the rotation. Land with poor access should not be planted if trees are grown for industrial purposes. It is just as important to assess an area in terms of the difficulties in harvesting the crop as its suitability for growing trees.

Forest roads are an important aspect of tropical plantation management because of their high cost and the potential negative environmental impacts due to erosion and sedimentation. A useful practical manual on the whole subject of earth roads and access for forestry projects in tropical countries is provided by Morris (1989). Roads should provide convenient low-cost access to the plantations. Road density will depend on the type of plantation, topography, the cost of construction and maintenance, and the cost of other harvesting operations. Evans and Hibberd (1993) recommend the following best practices for forest roads:

1. Low impact roads which avoid very steep terrain and wetlands are not only relatively inexpensive but can provide environmental benefits.
2. Where possible roads should be constructed of local materials. Where this is not practical, the impact of road materials on nearby vegetation and watercourses should be investigated in the planning stage.
3. Wide clearance will provide opportunities to manage roadsides and associated forest edges to provide shelter, sunlight, and warmth, thus adding diversity to plantation habitats for wildlife.
4. Bridges and culverts must be constructed to dimensions capable of taking maximum predicted water flow.

Biodiversity conservation

The impact of plantation development on biodiversity conservation depends on what is replaced. If tropical rain forest is cleared to make way for plantations—a practice that is not encouraged—there will be a great loss of biodiversity, yet on a highly degraded site planting trees may increase biodiversity. As the plantation area becomes more extensive it is necessary to design and manage plantations to contribute to biodiversity in the landscape. Small changes in design will often contribute significantly to biodiversity conservation, provide habitat for some species and connectivity of patches of tree cover. With careful planning, management and integration of natural features and access routes will not only provide a structure for plantation development, but also incorporate corridors and refuges for wildlife conservation. At the landscape level, retention of natural forest among plantation areas can have a beneficial effect on plantation stability as Aracruz Celulose S.A. found when they retained 25% of natural forest on their plantation land (Campinhos 1999). Landscape diversity may also be enhanced by creating a mosaic of plantation monocultures of different species matched to local site conditions or mixed-species plantations (Lamb 1998). While such designs may present management challenges, they have the potential to increase diversity at either landscape or local levels.

Although plantations share many attributes of intensive agricultural systems, they are managed on longer time scales and have more diverse structure, often including an understorey. The catalytic effect of forest plantations on natural regeneration and the possibility of restoring degraded landscapes has been documented, for example, Parrotta and Turnbull (1997), and Bernard-Reversat (2001) has reported on the positive effects of eucalypt and other exotic tree plantations on plant diversity and soil fauna in the savanna area of the Congo.

Ways in which industrial plantation design can be modified and plantations managed to achieve some level of biodiversity conservation are discussed more fully in Chapter 22.

Plantation life-history

A plantation is a growing resource and central to its management is that the forester or owner does not allow stands to grow from year to year untended. By appropriate and timely intervention, the pattern of crop development can be manipulated to modify both the quality and the quantity of the end-products.

Though there are many different end-uses which require somewhat different silvicultural prescriptions, all plantations follow a similar life-history. Also, as has been noted, the length of life of a plantation, the rotation, varies greatly, but this itself is one of the tools used to influence the end-product. Plantation life-history, the silvicultural choices available, management options, and the decisions to take are shown schematically in Fig. 4.7.

Management alternatives which directly affect quality of end-product are: (i) species choice, including seed source; (ii) initial spacing; (iii) pruning; (iv) thinning; (v) rotation length. All other silvicultural factors, though influencing wood quality in a small way, largely affect rate of growth and crop health. In general, there is no choice between fast or slow growth, the best possible performance is sought from the species planted, but (i) to (v) above represent real choices at different stages in the life of a plantation which the owner, villager, or manager can make according to the purpose of growing the trees considering especially end-use and economics.

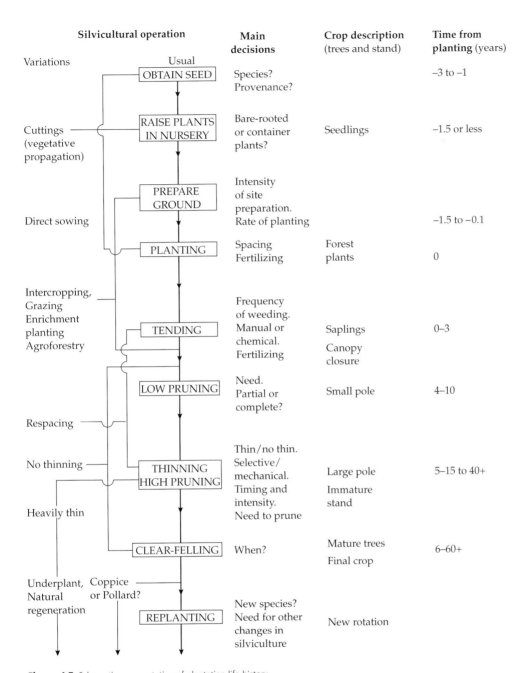

Figure 4.7 Schematic representation of plantation life-history.

Plantation monitoring

Continuous forest inventory to obtain information regularly about resources, growth and yield, pest and disease problems, and other matters in forest plantations, small, large, or nationwide, provides the essential data for proper management. While monitoring the biological aspects of the plantation is essential, on-going evaluation of environmental, economic, and social impacts of plantations is also very important. Assessment of change in a plantation and its surroundings implies that that there is some model or preconception of how the plantation is structured and will function so that key variables for measurement can be selected (Adlard 1993).

Such monitoring of stands and forests involves a variety of methods, especially field surveys, which are increasingly being complemented by remote sensing techniques. Monitoring may be used, for example, (i) to map plantings—date, location, species, area; (ii) to assess how far management lags behind prescription; (iii) to plan production forecasts from a recently developed plantation estate, for example, the pine and eucalypt afforestation in Brazil (Hosoka and Schneider 1984); and (iv) to establish sample plots to obtain yield model data for species to help in yield estimation, financial appraisals, and planning of future afforestation as described by Brister (1990) for *Pinus caribaea* in Sri Lanka, and so on. Gathering this information is an essential characteristic of plantation forestry since such development necessarily represents a major investment by a nation, company, or individual.

Moreover, afforestation in the tropics is often characterized by periods of activity and then neglect, dictated by the project-by-project nature of so much of development, or abandonment and then resumption of programmes because of wars or changes in administration. Obtaining statistics on success of past programmes, evaluating the existing condition of plantations, monitoring achievement of targets and so on, should be a precursor before new investment is made. Inventory provides this information.

Monitoring is now occurring in many plantations on another level. Where plantations are certified for sustainable forest management, evaluation and monitoring of all operations—technical, human resources, economic, etc., play a key role in the process of achieving the agreed standard. This is important, since increasingly the world's forest plantations are participating in certification and labelling schemes to demonstrate compliance with best practice and to safeguard markets.

PART II

Land, social, and economic factors, and planning in plantation development

CHAPTER 5

Land and plantation development

Most tree-planting enterprises, especially industrial plantations, require large areas of land. Therefore, land availability and quality is of first importance in plantation forestry. This chapter considers: (i) land capability; (ii) land-use; and (iii) the ways land is brought under plantation development and tree-planting encouraged.

Aspects of this topic were introduced in Part I but in considering land and land-use further, it is emphasized that developing the potential of a nation's land, removing the constraints of inaccessibility, aridity, and other physical or ecological factors, is one of the cornerstones of development. Although land is a resource which often can be improved and used again and again, there is always the danger that serious misuse may not only waste its present value but almost irreparably damage its future usefulness as eroding slopes and deserts testify. Large areas of former rain forest land in Amazonia abandoned after cattle-ranching failed, demonstrate that developing the potential of land in the tropics requires more than imposing essentially temperate farming systems.

Another point is that the existing occupier of land and the local community have often been overlooked in land-use planning and development schemes. Though much land in the tropics may appear under-used, 70% of people living in developing countries work on the land. In the past, due to inadequate communication and differences in cultural background, land planners may not have fully appreciated local ideas of land tenure, or realized how well local inhabitants know the productive capability and limitations of land in their area. Conversely, the local inhabitants may have a variety of perceptions and attitudes to developments such as planting exotic trees for industrial purposes. The conflict arising from differing perceptions of trees has become a social syndrome of plantation forestry and communities may burn plantations they perceive as harmful (Morrison and Bass 1992). There has been a greater appreciation in recent years that developments are more readily accepted if there is a participatory approach at the planning stage. The purpose, value, and proposed activities of all forestry plantation projects need to be discussed with those affected (stakeholders).

The community must be able to accommodate the development in the present and future land-use pattern in the district. If a plantation deprives individuals or the community of ownership and access to their land without their agreement and compensation, the resentment and instability caused can seriously threaten the plantation's sustainability (Howlett 1993). Where the government granted large concession areas classified as 'forest land' to companies to develop plantations of *Acacia mangium* for pulpwood in Sumatra (Indonesia), some local communities with land tenure concerns have disrupted operations and in some instances substantial areas of plantation have been burnt.

Land-use planning or land resource management is a key issue in development of rural areas. This was recognized by the United Nations Conference on Environment and Development (UNCED), which in 1992 proposed an integrated approach to planning and management of land resources involving environmental, social, and economic issues and the active participation of local communities (Chapter 10 of Agenda 21). FAO (1995) has promoted the concept of 'interactive land-use planning' based on three guiding principles:

1. Conservation of the production potential of land. This aims to allocate land for uses for which it is suitable and prescribes appropriate management practices to minimize adverse environmental impacts.
2. Promotion of equity of land access and use. This is a participatory process recognizing the rights of all stakeholders.
3. Negotiated decision making. This devolves decision making to the lowest possible level that is consistent with the ability for implementation, usually by village institutions, within the broader framework of government land-use policy.

Resource management domains

Part of modern land-use planning is the idea that there is an optimum use for a parcel of land. Traditionally, planning for natural resource management by governments has been based primarily, if not exclusively, on characteristics such as climate, soil, and topography which have been integrated into land capability classifications. Social and economic factors were not part of the classifications and policy makers and planners considered issues such as land tenure separately. However, in recent years society has become increasingly involved in resource management issues and sustainable land management planning now requires the integration of biophysical and socio-economic information.

A variety of landscape and stand level forest management guidelines have been developed to enhance integrated forest management. These guidelines cover factors such as terrain stability, visual quality, stream bank areas, wildlife habitat, road construction, size of clear cutting areas, and biodiversity. In Australia, some forest management plans integrate different land-uses such as timber production, conservation of biodiversity and recreation, for example, the 'preferred management priority' classification in New South Wales allocates all lands in State Forests to broad classes of long-term management intent and land-use as a basis for management actions and priorities (Curtin *et al.* 1991) and in Queensland adoption of management priority area zoning provides a framework for integrating forest values and potential uses into a State Forest Land-Use Plan (Francis and Shea 1991). Integrated resource management in some cases has moved away from 'multiple use' of a piece of land to a system of zoning with (i) designation of specific areas for specific uses; (ii) clear management rules appropriate to societal values for each zone; and (iii) effective institutional arrangements to provide incentives for those managing each zone (e.g. Binkley 1999).

Socio-economic data have rarely been used in defining forestry resource areas because forestry usually takes place in areas that are relatively sparsely populated but there is an increasing need to gather such data to support forestry feasibility evaluations. For example, an area otherwise well-suited for establishing a large area of pulpwood plantations may be rejected if the existing population is high, tenure arrangements are complex, and participation by local communities cannot be assured. The application of resource management domains in forestry plantation development should be a function of biophysical, agro-climatic, and socio-economic factors that describe the actual situation of a particular area, its capability and potential for meeting the plantation objectives.

Development of most modern forestry enterprises involves a systems approach in which problem solving is assisted by a procedure which investigates the entire system step by step (Pancel and Wiebecke 1984; Pritchard 1989). In the development phase there is usually a pre-feasibility study, a feasibility study followed by detailed planning. The availability of a land-use plan is essential at an early stage of the planning process. Chapter 7 provides a more detailed discussion of planning a plantation enterprise.

The delineation of 'domains' or 'zones' provides a practical basis for planning land-use generally and within forest boundaries. It involves identification of the predominant values for specific areas which can then be used and managed according to those values. The process of identifying zones mainly involves field surveys and the development of maps.

Previous systems used in resource management planning relied on static maps but major advances in information technologies such as Geographic Information Systems (GIS), remote sensing from satellites and aircraft, relational database management systems, and various modelling approaches with advanced computer techniques have provided the opportunity for integration and the production of more dynamic and flexible outputs. The term 'Geographic Informations System' (GIS) includes computerized storage, processing, and retrieval of geographically referenced spatial data, such as various types of maps, and corresponding statistical and other attribute information. Combining maps, such as those of soil and rainfall, known as 'overlaying' is one of its most important functions that can assist in land-use planning. There are numerous accounts describing the use of GIS as a tool for spatial analysis to assist natural resource management (e.g. Aronoff 1989; Rajan 1991; Jones 1998).

These technologies allow development of resource management units at any scale, international, national, regional, forest or forest compartment. The term 'resource management domain' (RMD) has been used to refer to this concept of integration for land-use planning (Antoine *et al.* 1998; Dumanski and Craswell 1998). An RMD is defined as a spatial unit (landscape) that offers opportunities for identification of problems, analysis of alternative solutions, and application of resource management options to address specific issues in order to increase the sustainable use of one or more resources. It is derived from geo-referenced biophysical and socio-economic information. The RMD framework may be a better approach than traditional approaches such as classifying natural resources by agro-ecological zones because (i) it provides a multidisciplinary focus to natural resource problems; (ii) multi-dimensional resource problems in both time and space can be handled on a regional basis; and (iii) data sets developed for a particular RMD can respond easily to tailor solutions to specific issues and problems (Babu 1997).

Land capability

The first stage in evaluating land and preparing a land-use plan is to gather data to classify land according to what it may be able to grow. This is the study of land capability. Gelens (1984) summarizes such approaches to land evaluation for forestry and it is detailed in FAO (1984*b*). Land capability, also called 'land suitability', is primarily concerned with the potential biological productivity of land and this is determined by four main components of the environment. In descending order of importance they are: climate; local topography (ruggedness, steepness, exposure) which cause local variation in climate and disposition of soil type; soil; and existing vegetation including the effects of destroying it and replacing it with another crop.

Climate

Climate is the dominant influence on land-use in the tropics. But, it is not easy to determine what are the most useful descriptors of climate for classifying the productive potential of land. Several systems of climatic classification have been developed, such as those of Köppen (1923), Holdridge (1947) and Thornthwaite (1948). They mostly apply at the regional level but tend to break down if applied to specific localities. On the subregional level, grouping into similar climatic zones is a practicable compromise. Climate-based silvicultural zones indicating suitability for afforestation have been delineated in South Africa (Poynton 1971) and Brazil (Golfari *et al.* 1978). These approaches are referred to as bioclimatic zones, and when also incorporating soils and other site data to relate to crop production potential, as agro-ecological zones. Where exotic species are introduced to a new location, such classifications are some guide to suitability but more detailed local study of potential evapotranspiration,

seasonality of rainfall, temperature maxima and minima, occurrence of frost, etc., is invariably necessary for species to be matched closely with sites (Chapter 8).

The overriding influence of climate on the potential of land for plant growth is readily demonstrated. The desert sands of central Australia yield very little, while, at the same latitude, the coastal sands of Queensland support productive *Pinus caribaea* plantations, the difference is simply due to adequate rainfall in the coastal area. It is sometimes possible to overcome the limitations of climate by techniques such as irrigation, for example, Fig. 3.1, and shading.

Topography

The influence of topography on land capability is best seen in the increasing restrictions of possible alternative uses as terrain becomes more broken and rugged. Steep slopes, ridge-top exposure, waterlogging in gullies, ground strewn with rocks and boulders, all severely limit or prevent farming. Much land unfit for cultivation because of the constraints of local topography is used for forest plantations (Fig. 5.1).

Soil

Soil fulfils three essential requirements for tree growth: supply of moisture, nutrients, and provision of mechanical support. Inherent soil fertility, physical condition, and rootable depth are the primary considerations. However, different species grow best under different conditions of nutrient supply, moisture regime, acidity, etc., so there is no one optimum soil. Much work has concentrated on species-soil matching for trees and crops.

Figure 5.1 Compartment B16 in the Usutu Forest, Swaziland, typical of much ground used for afforestation in Africa. This land has yielded three times an average of 380 m^3 ha^{-1} of pulpwood from pine crops grown on 18-year rotations.

Soil survey for land evaluation is complex (FAO 1984*b*) but is essential for forestry. In Queensland, comprehensive surveys are carried out by the State forest service to provide site suitability and site capability assessments for tropical pine plantation establishment. The information provides a basis for site preparation design, species selection and fertilizer prescriptions (Foster and Costantini 1991*a*). The objectives of any survey are essentially to assess both the three functions of soil noted above and more general features affecting its management.

1. *Rootability and rootable volume*. These are assessed by measuring:
  (a) depth of soil to impeding layer, that is, bedrock, stoneline, water table;
  (b) quantity of rocks and boulders in the soil matrix;
  (c) soil texture and structure, especially the proportions of sand, silt, clay, and organic matter; and
  (d) compaction and bulk density.
2. *Nutrient supply* (fertility). In total terms this is affected by rootable volume, but several direct assessments are made which attempt to measure nutrient supply:
  (a) concentrations of plant nutrients in extracted soil solution;
  (b) cation exchange capacity (CEC)—how well nutrient ions are held and stored in the soil;
  (c) soil acidity—pH;
  (d) organic matter and carbon to nitrogen ratio (C/N); and
  (e) geology of underlying rock or parent material.
3. *Moisture supply*. Many of the above factors, especially soil texture, and slope and position of site, influence moisture supply, but specific measurements may occasionally be made of:
  (a) soil moisture content;
  (b) fluctuations in depth of water table.
4. *Soil erosion*. It is also important to gather data about the risk of soil erosion and soil loss. Tropical soils vary greatly in their dispersive characteristics and ability to absorb moisture (infiltration capacity), which together largely determine erodibility. Subsequent site preparation, and land management as a whole, for replanting and plantation forestry must seek to conserve soil to minimize erosion from both wind and especially surface water runoff.

Soils of similar appearance, particularly development of horizons down a profile, are grouped together because they have many of the above properties in common. This is the basis of soil classification into families, series and sets, and these are used to describe and map types of soil in land capability assessment.

Two points must be made concerning use of soil survey information in the tropics.

1. As indicated in Chapter 1, the luxuriance and richness of much tropical forest does not primarily depend on fertile soil but on efficient recycling on nutrients in the ecosystem. Thus, on some sites, soil is of less influence on site productivity than in temperate regions, as its role as a nutrient reservoir is of secondary importance. This explains why forest clearance for long-term agricultural development is often not successful, the nutrients in the ecosystem are mostly in the forest biomass not the soil. For example, in Guyana even many years after clearance of 'wallaba' forest on very infertile white sands only scattered bush is able to grow. Some deforested parts of northern Brazil are similarly unproductive.
2. The actual capability of a soil to grow particular trees or crops will change according to the availability of suitable technologies to work the soil and the extent to which it is practicable to use fertilizers etc.

Existing vegetation

Vegetation may provide a useful indicator of climatic conditions where meteorological stations are sparse. Surveys of existing vegetation can also be integrated with information on climate, topography, and soils in land-use mapping as occurred in land-use mapping in tropical Australia (e.g. Christian and Stewart 1953).

The presence of vegetation can be a constraint on land capability, though both growth and richness of species may help to indicate potential site productivity. The constraint on land capability is the need to clear vegetation before introducing another land-use. Although, in theory, clearance is always possible it may not be done, first, because of the risk of physical damage to the soil and erosion; second, because it may be very expensive (Chapter 12); and, third, because of wildlife conservation. No undisturbed natural ecosystem in the tropics, but especially existing forests, need be cleared to provide land for plantation forestry since there are already many degraded areas with potential for afforestation.

After clearing vegetation or excluding fire from grassland, the weed problem, which is largely determined by the previous vegetation, may be so serious that tending costs make some land-uses uneconomic. The generalization that lowland tropical pines are more suited for grassland afforestation and eucalypts for cut-over forest sites, reflects the response of these genera to competition from different kinds of weeds.

Site assessment and classification

In land capability studies the above factors influencing site productivity are rarely considered in isolation. Most or all aspects are included in a survey and their relative importance analysed. An ecological framework for resource management may be developed at several levels, for example, a broad regional level based on climate and physiographic features, and zonal and site levels based on more detailed ecological parameters. These different scales of classification can satisfy a variety of uses and objectives. The regional level has been used to provide a guide for the management of some wildlife species; the site level has been used to guide silvicultural activities such as choice of species for planting, methods of site preparation, and stand tending (MacKinnon *et al.* 1992). In South Australia a classification based on natural vegetation and soils was used for evaluating the site suitability for *Pinus radiata*, to predict site productivity and to decide whether or not to fertilize plantations (Lewis *et al.* 1976). Another example of qualitative assessment of land capability was a study carried out to decide where trial plantings on land near Lake Malawi would be most representative of the land to be used for extending the Viphya plantations. Data were collated on geology, soils, climate, and vegetation. Analysis first defined major land forms from which six main site types were identified (Massey 1976). Subsequently, large-scale trial plantations were established on each of these sites with the reasonably certain knowledge that they were representative of conditions likely to be encountered.

A more precise way of assessing land capability is to relate crop yields with measurable site characteristics. Necessarily, there must be a crop to assess, but where there is, site: growth relationships can be identified which can help predict growth on other areas. The example below comes from a site assessment survey in *P. patula* plantations in the Usutu forest, Swaziland (Evans 1974). Sixty-one plots were established in 11–13-year-old stands. On each plot top height was measured as the index of growth and numerous soil and topographic parameters were assessed to describe the site, for example, soil set, pH, bulk density, plot elevation, aspect, etc. Regression analysis generated the following significant relationship.

$$Y = -18.75 + 0.0544x_3 - 0.000022x_3{}^2 + 0.0185x_4 + 0.0449x_5 + 0.5346x_{11}$$

(R = 0.81)

Y = top height at 12 years (m);

x_3 = elevation of site above sea level (m);

x_4 = position on slope between ridge top and valley bottom (%);

x_5 = slope steepness (%);

x_{11} = soil set numerically coded in order of fertility.

The above equation shows that growth of *P. patula* is related to certain features of topography and soil; similar results have been obtained by Grey (1979) for the same species in Transkei, South Africa. Though significant relationships may not be causal, defining them assists prediction of growth on similar land nearby and so aids rational decision-making on land suitability for afforestation. A classification and regression tree analysis (CART) enabled an effective classification system to be developed for yield prediction in *P. radiata* in southeastern Australia based on geology, landform, soil profile measurements (texture and depth to the B horizon), depth to rock, acidity, and soil morphology from about 50 sites (Hollingsworth *et al.* 1996). This quantitative approach has been used for important plantation species, for example, *Gmelina arborea* in Nigeria (Chijioke 1988), black wattle (*Acacia mearnsii*) in South Africa (Schönau and Aldworth 1990), *Cordia alliodora* in Colombia (Giraldo *et al.* 1981), and for pine afforestation in Venezuela (Vincent 1986).

Appropriate site survey procedures and predictive modelling to select the best land for plantations can have a major influence on plantation profitability. The cost of site survey has little effect on plantation profitability as other factors such as tree-growth rate, land cost, harvesting costs, and the value of wood sales have a far greater influence (Battaglia *et al.* 2001). It follows that whatever level of survey is required to give accurate predictions of potential plantation productivity should be implemented.

Land-use

The main land-uses in the tropics were described in Chapter 1, and the two general characteristics of under-use and inefficient use noted. Ideally, development should lead to each area of land having a land-use plan to help make the best use of it in the widest sense, including setting aside areas for environmental or wildlife conservation reasons. This would increase the value of land as a resource and this role for land-use planning is specifically encouraged in many countries.

Although land capability assessment determines what can grow, in theory at least, usually there will be several alternative possibilities. But it will be evident that different land-uses will differ in benefit to the local community, effect on the soil, or in productivity or profitability. Deciding the best land-use depends on what can be implemented, which includes not only what will grow (land capability), but on environmental, social, and economic impacts including land and tree tenure, the impact of other land-uses, national land-use policies, and the demand for what the land might produce, for example, food, timber, water, wildlife conservation, etc. This broader analysis to identify the most suitable land-uses is called 'land evaluation' (FAO 1984*b*).

In the past decade there has been a change in emphasis in the forestry sector of most countries in the tropics away from large industrial plantations to smaller scale tree-planting, often directly involving farmers, and for less direct involvement of the government in plantation development. Privatization is now often viewed as an essential part of economic development. However, private companies are finding that it has become increasingly difficult to find large areas of land, not subject to tenurial problems and in strategic locations for marketing, in heavily populated countries, especially in Asia. In contrast farmers control very large areas of land in the tropics and industries are placing greater reliance on wood produced by rural populations. With the greater involvement of farmers in plantation forestry greater attention must be paid to socio-economic issues in land-use planning, a fact that has been stressed by Potter and Lee (1998) in Indonesia.

Social and economic factors are considered in more detail in Chapter 6 but here we introduce in outline broader issues important for plantation forestry as a land-use, especially in the development of sustainably managed plantations.

Land and tree tenure

Lands with potential for plantation establishment are often already used in some way by the local population and their rights to the land are of the greatest importance in planning plantation development. The term 'tenure' is used to refer to the set of rights to land or trees held by a person or a public or private entity. Land tenure is enormously complex and can present difficult obstacles to development in the tropics. The question of tree tenure is equally so. Rights to trees, whether planted or

not, are not always clear, and confusion or uncertainty over who can cut what and when can be a major disincentive in rural development forestry programmes (Chambers and Leach 1990). Tenurial rights can be complicated by the conflict of customary rights and rights supported by the law and it can be that the formal ownership structure does not reflect who actually makes the decisions over land-use (Morrison and Bass 1992).

This short discussion is confined to noting some types of ownership in the tropics to show the kinds of problems involved in using land for forest plantations. From the time teak was first planted in Burma in the mid 1800s there have been disputes and difficulties over rights (MacGillivray 1990) and conflicts over the use of land that continue to reduce the economic returns from plantations, for example, in the Philippines and Indonesia (Fraser 2000). It is recommended that developers identify in advance the tenurial status of potential plantation land and determine how governments intend to obtain land for plantation development (Bass 1993). Sorting out land-rights for local people is the key to sound environmental management and successful social forestry (Shepherd 1990) but there may be less need to increase security of tenure and access rights to encourage private tree growing than is often assumed (Arnold 1997*b*). The roles of land and tree tenure in forestry and agroforestry are addressed and examples given that demonstrate their complexity in, *Land, trees and tenure* (Raintree 1987), *Whose trees? Proprietary dimensions of forestry* (Fortmann and Bruce 1988), *Community forestry: rapid appraisal of tree and land tenure* (Bruce 1989), *Contemporary uses of tree tenure* (Bruce and Fortmann 1999) and *Property rights and participatory forest management: an overview* (Moeliono 2002).

Customary ownership

Customary tenure may pre-date government-initiated land tenure codes and customary rights may be in conflict with legal rights. It is not uncommon for land allocated by governments for industrial plantation development to have customary tenure claims and for subsequent conflicts to arise. Traditionally many communities did not concern themselves with landownership rights as rights of access and use of the resources was considered more important. This view is changing as indigenous people realize rights of access do not provide sufficient protection of their resources (Moeliono 2002).

In eastern Africa (Kenya, Tanzania, Zimbabwe, etc.), most land is held under customary law, with the ultimate ownership of the land vested in the State. Generally a farmer is given land by a village headman or council to cultivate and use its natural resources. If the land is not used it becomes available for redistribution (Warner 1997) but those who plant trees are assured rights to the produce even when they no longer control the land on which the trees grow. However, local control of public forest resources in Africa has been reduced as large areas of land have been taken under government control. In some places customary tenure landholdings are becoming recognized as individual ownership and distinctions between tree tenure and land tenure will probably decline (Warner 1997). In Latin America some indigenous people demanded titles to land for collective ownership which is inalienable and cannot be sold, leased or mortgaged (Colchester 2001 in Moeliono 2002).

In many Pacific island nations, including Papua New Guinea (PNG) and Fiji, nearly all land is under customary ownership and usually recognized by the country's constitution and common law. Land, however remote or apparently uninhabited, belongs to someone or some clan or village. Though no land-use may be apparent and no boundaries evident, every piece of land will be claimed if significant land-use change is suggested. Ownership is claimed from long use and association with the land. It is usually shared ownership, thus to acquire customarily owned land, such as for a plantation project, involves negotiation with numerous interested parties all of whom must agree to the proposal.

Two examples of land tenure problems associated with customary ownership illustrate some of the difficulties when seeking to change land-use. In PNG unwillingness by customary owners to lease land for reforestation in the Gogol valley was the single greatest obstacle to replanting even though the owners were willing for their rain forest to be harvested for wood chips. The reluctance is not so much losing rights to the land for a long period (25 years), they receive rent for the land and royalties from the crop grown, but because the specific act of tree-planting has long been associated with staking a claim to landownership. Bass (1993) refers to the general lack of understanding of the concept of a 'lease' which restricts some of the owner's rights. This can lead to continual negotiation on the use of the land and is a reason for very little investment in plantation forestry in PNG. A similar situation has been reported in Fiji (Vise 1990 in Bass 1993) although leasing land from the customary owner for afforestation by Fiji Pine Limited (formerly Fiji Pine Commission) has mostly been straightforward. However, one problem has occurred frequently which reflects a difference in attitude to land tenure. When preplanting work on leased land begins, notably road-building, many owners immediately want back parcels of better land because of the improved access to it!

Where customary owned land is acquired for planting, two principles help to overcome difficulties. First, every effort is made to meet the wishes of the local people regarding land-use proposals. Second, involvement of the landowners in the project is actively encouraged. This was expressly stated as a primary objective of the Fiji pine plantation scheme and a Division concerned with Landowner's Affairs was established to actively promote business development and training and landowner companies, not only for logging and haulage but also for silvicultural work (Fiji Pine Commission 1988, 1989). The company now also provides an annual grant to the Landowner Business Development Trust Fund which makes grants to landowning villages and surrounding rural communities for community development projects (Fiji Pine Limited 2001).

Private ownership
A very different situation obtains where land is a commodity which can be bought and sold, but tenure problems can still arise. A developing plantation project usually needs to acquire many areas of land to form sizeable and contiguous blocks, and this depends on the property market. Steady acquisition of land is often not possible, and land has to be purchased when it becomes available quite often regardless of the planned rate of planting.

A second type of tenure problem is 'squatting'—the use of land by someone without right to it. The Usutu Forest was established on land purchased directly from European settlers who had acquired large areas of Swaziland high veld for sheep farming. Expansion of the forest in the mid-1970s, under the Swazi Nation Afforestation Project (SNAP), encountered a squatting problem. The land to be planted, purchased from absentee farmers living in South Africa who mostly were not using it, was occupied by Swazi farm-dwellers. The only solution the government felt justified to take was to resettle these 'squatters', and over a quarter of the new land (4131 ha) had to be set aside for this purpose. Since 1989 ownership of all Usutu forest land reverted to the Government of Swaziland, who then granted a 100-year lease of the land to the Usutu Pulp Company later to become South African Pulp and Paper Industries (SAPPI) (Usutu).

In many countries traditional forest dwellers and long-term settlers in forest areas controlled by the state are classed as 'squatters' and are subject to resettlement if the government allocates the land to plantation development. Squatting is widespread but usually reflects a social disorder (locally high population densities, unemployment, displaced people) rather than a malicious act. There have been major conflicts over resettlement, for example, in areas selected for the development of eucalypt plantations in Thailand (Lohmann 1990). Attempts to legalize squatting, such as Thailand's 'right to cultivate' certificates (*So Tor Koh*) which gives people moderately secure tenure rights on national forest reserve land, usually exacerbates forest clearance and in Thailand's case is in conflict with the government's reforestation goal to increase forest cover from 29% to 40%. Land for plantation development is especially prone to the problem of squatting because it may be unplanted for many years and, if it is remote, will only be visited occasionally.

In some countries there is a perception of secure land tenure and what is planted on the land. In the eastern Terai of Nepal, farmers tend to plant trees only on land where there is no confusion over rights and a tenant farmer who plants trees has no legal tenurial rights to them (Subedi *et al.* 1993). Security of land tenure itself arising from private ownership may not alone encourage planting and care of trees. Attempts to change tenure can be counterproductive, as the prospect of change introduces uncertainty and may inhibit investments in long-term activities such as tree growing (Arnold 1991). In China, forest tenure has experienced frequent changes over the last 50 years. Private forests were confiscated in the early 1950s and redistributed to all rural households on equal basis. These private forests were then collectivized in 1956, were further transferred to communes in 1958, and in 1961 ownership was devolved from commune to production brigade. In 1982, collective forests and forestlands were distributed to farmers in an attempt to encourage farmers to plant trees on wastelands and improve management of the existing forests. The frequent changes in forest tenure had at least two negative impacts. It resulted in a crisis of villager trust in the government's policy for tenure. Farmers, concerned that forests and trees might be taken back by government if the policy was reversed, often harvested the trees prematurely and were also discouraged from investing in tree-planting (Liu 2000). A similar crisis of trust in land tenure documents granted by the government prevailed in Vietnam before the Doi Moi reforms in 1989 (Byron 2001).

State control

In many countries the rights of the individual or community over the land where they live are subordinated to the needs of the state. Often under such government land, like air, is deemed to have no monetary value. Moving villages and settlements to make way, directly or indirectly, for changes in land-use is not uncommon, for example, the 'villagization' programmes of Tanzania and Ethiopia in the 1980s. The converse also occurs where people are resettled, such as the transmigrants within Indonesia and along the trans-Amazonian highway in Brazil, and are used as instruments of land development in sparsely populated areas. Under these conditions, what fragile land-tenure system may exist has little influence on land-use plans and policies. In Indonesia, forestry law holds that all forest that has no clear ownership title is owned by the state, an interpretation that ignores traditional forest land claims as well as the presence of people living within the state forest (Kartodihardjo and Supriono 2000). Lands made available by the state for plantation forestry in Indonesia commonly have traditional tenure claims, for example, 60 000 ha out of the 280 000 ha allocated to the Riau Andalan Pulp and Paper company in Sumatra are regarded as 'problem areas' with community land rights conflict (Jenkinson 2000). To displace people living on the concession area a company has to pay compensation but negotiations are often protracted and the residents dissatisfied with the amount paid. In the resulting conflict the displaced residents may retaliate by deliberately burning new plantations (Saharjo 1997).

Tree tenure

Many tenure systems confer property rights in standing trees quite separately from the land on which they grow. A tree tenure system may distinguish between planted and wild trees and even between species (Bruce 1989). Other factors can influence how trees are managed and encourage or discourage maintenance or planting of them. For many rural poor, trees have or can have tremendous value as savings or security quite apart from their many benign influences. Trees are planted to meet future needs, for example, casuarina in South India for daughters' dowries, eucalypts and pines in western Kenya to pay school fees and, with a cooperative plantation in Benin, to provide support in old age (Chambers and Leach 1990).

Where rights to harvesting and the products are adequate land rights may be unimportant. The separation of rights in trees to rights in land has been central to major community forestry programmes developed in Asia, Africa, and Latin America in the 1990s (Bruce and Fortmann 1999). For example, in India's Joint Forest Management programme community participation in afforestation of degraded

Figure 5.2 Fuelwood plantation established near Malawi's capital, Lilongwe.

State forest land involves local groups being given a level of control over planting and harvesting of trees but landownership is retained by the State (Saxena 2001).

Planting is inhibited where uncertainty over ownership and harvesting rights of trees exists. Laws may be unclear or bureaucracy may restrict what owners can do with their trees. Indecision over policy may prevail, making a villager's or community's rights uncertain, for example, trees planted on kabele ground in Ethiopia. In China, although farmers may own the trees on their land the harvesting and marketing is controlled by the government and this can be a disincentive for tree-planting. In some instances the state may control tree harvesting and impose conditions to protect the public interest, for example, in critical watersheds to limit soil erosion (Bruce and Fortmann 1999).

Recognition of the importance of rights over trees (tree tenure) is relatively recent. It is clearly important, especially in social forestry programmes. Chambers and Leach (1990) conclude 'Trees owned and grown by the poor are not a panacea, but the evidence assembled indicates that they have more potential for reducing deprivation than has been recognized . . . they are like savings bank accounts . . . Where ownership and rights to harvest and sell are secure, poor people plant more and harvest less than expected.' Reforms to secure rights over trees for poor farmers 'would at a stroke provide incentives to plant and protect trees, to the benefit simultaneously of the poor, the national economy and the environment'.

Impacts of other land-uses

Future land-use is much affected by present land-use practices. For example, the outward expansion of towns, which affects an increasingly large hinterland, is seen in the steady clearance of forest to make way for food gardens and plots, and to provide fuelwood. Urban development is a major force leading to establishment of fuelwood plantations (Fig. 5.2) and in the case of Addis Ababa (Ethiopia) it was the establishment of eucalypts around the city that was responsible for its continuing existence in its present location. Similarly, industrial development where wood is the energy source, such as for iron-ore smelting in Brazil, or tobacco curing in Africa, has led to natural forest clearing and extensive planting of trees for industrial charcoal and fuel. As noted in Chapter 2, past and continuing loss of natural forest, whether from agricultural development, direct exploitation, or some other cause is an important reason for planting trees and forests today. Frequently a forest industry develops using the natural resource and converts to plantation produce after some years, for example, plywood production in Bulolo, PNG and pulp production by Sabah Forest Industries at Sipitang. Pressure of one land-use; namely, exploitation of forest, itself creates a need for another.

Land-use of one area cannot be considered in isolation; it not only affects land-use in the future but will also influence that of a neighbouring area in the present. A simplistic example is seen in the remote Jari project in Brazil where the primary objective was to convert a large area of forest and scrub to plantations, but a major cattle-ranching scheme had to be set up within the plantations and adjacent savanna to help feed the employees. Also, as with all large plantation developments, some land is used for schools, housing, amenities, and industrial sites. The main change from the one previous land-use of natural forest to land under plantations resulted in several new land-uses developing.

Land-use is not isolated but part of a pattern of integrated relationships. The creation of forest plantations reduces, though need not exclude altogether, the land's food-growing potential and so may indirectly stimulate greater agricultural effort elsewhere. Conversely, forest clearance will frequently generate a regional need, on the same or some other land, to create new plantations. These kinds of relationships underpinned the Tropical Forestry Action Plan (now renamed the National Forestry Action Plan) concept of buffer zones of intensified agriculture and agroforestry around natural forest; increased productivity of one (land-use) may diminish pressure on the other. Present land-use of an area influences both the use of land nearby and the uses to which land may be put in the future.

Population density

The world's population is increasing and most of the increase is in developing countries. This leads directly to increasing demand on land to grow food, other products for clothing, shelter, and keeping warm, and is a potent force for land-use change. This is the case when traditional systems of shifting cultivation, which worked well with low population densities, break down and deteriorate because of shortening cycles, and the semi-nomadic habits of the shifting cultivator can no longer be sustained. Food, fuel, and shelter are still needed, and more intensive farming practices based usually on sedentary agriculture and more integrated land-use (Chapter 20) are required.

The change in land-use this causes is not only one of quantity, such as forest or savanna being converted to large farms, although this kind of alienation of forest land is occurring throughout the tropics, but also one of quality. Low intensity land-use practices to obtain food and wood from natural and semi-natural forest give way to more intensive land-uses: the managed forest and plantation, sedentary agriculture, agroforestry, controlled grazing, etc.

Increasing population is both a cause of land-use change and a reason for needing rational and planned use of land, but the case must not be overstated. It is facile to blame all problems in developing countries on rapid population increase. Even with a global population of 6000 million there is more than 0.2 ha of potentially productive land per person and India, with over a billion people, is largely self-sufficient in food.

Geographical considerations

Geographical location can influence land-use. For example, although almost any piece of fertile ground can be farmed with benefit, at least at subsistence level, development of a pulpwood plantation requires a large amount of land not too widely scattered. If a pulpmill is also included a perennial water supply is needed along with adequate access, services, and land suitable for small town development.

Not only do geographical considerations affect potential land-use directly, but they influence marketing of produce. Perishable food must usually be grown on land close to the consumer since, in much of the tropics, refrigeration and preservation are largely impracticable and food is easily desiccated in transport. In forestry, proximity to market is not essential because wood can be transported long distances without degrading. However, transport is costly and if the market for forest products is not nearby, and especially if it is necessary to export through other countries, product cost is increased and continuity of supply is less assured. This marketing problem has been a serious obstacle to developing pulping projects in land-locked African countries.

National land-use policies and institutions

The biggest obstacle to developing a coherent national land-use policy is that many sectors and disciplines are involved: agriculture, forestry, sociology, surveying, water resource management, etc., a point acknowledged clearly in India's *National Land-Use Policy* (1986) 'afforesting 5 million ha is related to the work of several Ministries and Departments' of which eight are described as being involved in a major way. Similarly in Jamaica's *Forest Land-Use Policy* (1996) the roles and responsibilities of nine agencies involved in forest land-use are defined. The need to bring together many different disciplines for any successful land-use management plan was stressed by Rao (1986).

Both land shortage and achieving the best land-use necessitates integration of all land-uses. Ideally, specific land-use proposals, including forestry policies and projects, should not be formulated in a vacuum, but conceived as an integral part of national development plans. This is because a national development plan is usually the only all-embracing policy document which seeks to evaluate together the various sectors of the economy and allocate resources for the future, hence the examples cited in previous paragraphs are from such plans. Thus, in theory, all factors affecting land are considered so that a foundation is laid for sound, integrated land-use planning, taking into account all aspects of national life and not just individual projects. This should provide the opportunity for implementing land-use policies for the greatest benefit of the people directly affected and ultimately the nation as a whole. In practice other 'higher order' national policies exist and the forest policy needs to be aligned with them. For example, the goals and priorities of *Forest Land-Use Policy* (1996) in Jamaica have been aligned with the *National Land Policy* (1996) and the *National Industrial Policy* (1996) (Headley 2001).

The nature and structure of existing institutional frameworks are having a major influence on the extent, magnitude, and sustainability of forestry projects, and are important in shaping the forestry sector in a country (Gregersen *et al.* 1993). 'Institutions' are defined as sets of rules by which societies live and government, non-government, and private organizations at local, regional, and national levels formalize and implement them through legislation, contracts, etc. In most tropical countries demands on forestry institutions for a range of goods and services are increasing while resources are declining. Public pressure for biodiversity conservation, protection of watersheds, agroforestry, management of non-timber forest products as well as wood production are placing increasing demands on forestry organizations which are frequently under funded and short of trained staff. Improving the institutional framework for forestry projects is essential if sustainable forestry is to be implemented within national policies for land-use and natural resource management. Institutional and organizational constraints mean that land-use plans are not always implemented nor always successful, and implementation of a poorly formulated national land-use plan is the cause of numerous problems in many countries.

Governments rarely legislate directly to dictate how land is used. Exceptions are Rwanda's policy to cap every hilltop with trees (see Fig. 23.5) and, in contrast, Swaziland's 'Control of Tree Planting Act, 1972' which severely limits new plantations of more than two hectares, except on marginal land unsuitable for agriculture and with an average slope of more than 15%. Planting on all other land requires a permit and is only granted where there is

no conflict with any agricultural land-use. Conflict between forestry and water interests is extreme in South Africa, where an average annual rainfall over the whole country is only 400 mm (Calder 2002) and has led to very strict controls of where forests can be planted. Few nations enact such explicit land-use laws which affect tree-planting. However, land-use policies are often implied in laws designed to encourage certain industries, particularly in the agricultural sector, for example, tobacco growing and ranching. In addition, forest policies invariably incorporate planting targets and often an optimum forest cover percentage, for example, 33% for India (India 1989), or explicit aims affecting land-use, such as Burundi's policy objective of 300 trees per homestead for fuel and pole supply. Moreover, government policy may even include provision for land acquisition, as have successive Fiji Development Plans for further planting by Fiji Pine Ltd. Also, for conservation and erosion control, clear policy statements are usually embodied preventing land-use change or limiting the land-uses permitted in certain areas, for example, the protected hill-slope and protected strip next to watercourses defined in the Malawi Forest Act.

Integrated land-use

It will be clear that land-use in one place at one time is itself not isolated and also it affects land-use practices elsewhere and in the future. The planning approach at national level which takes into account all the factors involved in deciding optimum land-use is called 'integrated land-use planning'. It is important that developments in plantation forestry are included in such planning because often forestry departments and planting programmes have existed in virtual isolation from other land-uses.

There is also a need for planning at the landscape level so that tree-planting and plantation developments can be integrated with other land-uses such as agriculture, water catchments, and biodiversity conservation. There has been a major shift in the focus of forestry projects in recent years in tropical countries away from government controlled industrial plantations to projects targeting food security and welfare benefits for the rural poor, and environmental protection. Greater emphasis on rural development forestry, agroforestry, and natural forest conservation and restoration have resulted in the need for greater interaction of forestry with other sectors to achieve more sustainable development and better land-use. Even private companies, for example, Aracruz Florestal in Brazil and some Indonesian companies, such as PT Musi Persada Hutan in Sumatra, have diversified their land-use beyond intensive industrial forestry plantations to include agroforestry and/or conservation of natural forest areas within their boundaries for wildlife conservation and other environmental values. With this change of emphasis, the opportunities for integrated land-use planning including forestry have expanded greatly.

Land evaluation surveys

Land evaluation, to provide a basis for rational integrated land-use planning, involves a complex and wide-ranging study. Any decisions need to be based on comprehensive, relevant environmental, social, and economic information. The principles, concepts, and procedures are outlined in *A framework for land evaluation* (ILRI 1977), *Land evaluation for forestry* (FAO 1984b), and *Assessing forestry project impacts: issues and strategies* (Gregersen *et al*. 1993). Where the analysis of spatial resources is rendered in digital form, *Principles of geographical information systems for land resource assessment* (Burrough 1986), and *Geographic information systems—a management perspective* (Aronoff 1989) provide useful background.

Data collection in a survey follows a sequence from lesser to greater detail. Usually a reconnaissance survey is made first followed by more intensive land evaluation in those areas meriting closer scrutiny.

At the reconnaissance level, as well as gathering environmental, social, and economic data, land is classified and mapped into broad types. In many parts of the tropics existing maps are inadequate and the areas under survey are often so large that remote sensing must be used, that is, aerial photographs, satellite imagery, airborne and satellite radar imagery, and airborne video imagery. The

value of photography for reconnaissance surveys is well established, but photographs are not always available, may be out-of-date, and in some countries acquiring aerial photos can be a lengthy process requiring security clearance. One of the main advantages of the use of remote sensing in forest survey is the relatively short time in which most of the information needed can be obtained and satellite imagery has greatly improved the means available for initial land survey in the tropics. Hussin and Bijker (2000) have reviewed the use of remote sensing systems for measuring, describing, and mapping in forestry (Table 5.1).

An increasing amount of environmental and resource information is being obtained from sensors operating in the microwave part of the electromagnetic spectrum. This type of sensing is referred to as 'radar'. These microwaves can penetrate atmospheric conditions such as haze, light rain, clouds, and smoke, which cannot be achieved by optical and infrared parts of the spectrum. This makes it particularly useful in equatorial countries where persistent cloud cover is widespread. Airborne radar has been available for more than 20 years, for example, with Side-Looking Airborne Radar (SLAR) a signal generated on an aircraft is bounced off the ground back to the plane. The strength of the returned signal is determined by the kind of ground surface (water, soil, vegetation, rock, etc.), the moisture content of the reflecting surface, the slope of the terrain, and the roughness of the surface. The variation in signal strength is reproduced in grey tones on a film to create a 'photograph' of the ground. SLAR was used in terrain classification in the Colombian Amazon (van Zuidam 1977) and by Nigeria's forestry department to identify 45 different vegetation and land-use types in a survey of the country's vegetation resources and land-use patterns. Since the

Table 5.1 Application possibilities of remote sensing systems

Topic	Aerial photos	Airborne scanning	Satellite scanning	Airborne radar	Satellite radar	Airborne laser	Airborne video
Land and forest cover	1	1	1	1	2	2	1
Forest degradation	1	1	2	1	2	2	1
Forest types	1	1	1	1	2	2	1
Biodiversity	1	1	3	2	3	3	2
Biomass data for carbon sequestration	1	1	3	1	3	2	2
Fire damage	1	1	1	1	2	3	1
Fire detection	1	1	1	x	x	x	1
Forest health	1	1	3	x	x	x	1
Forest products	2	2	x	x	x	x	2

Source: 1 = wide range of applications; 2 = moderate range of applications; 3 = limited application; x = no application. Table based on work of Hussin and Bijker (2000).

early 1990s satellite radar has become available on a routine basis, for example, from the European Space Agency (ERS-1/-2), Japanese Space Agency (JERS-1), the Canadian Radarsat and the Indian Remote Sensing Satellite (IRS).

Multi-spectral satellite imagery began in the early 1970s with the Earth Resources Technological Satellite (ERTS), and is a remote sensing tool provided by a platform in space. The Landsat series of satellites with a multi-spectral scanner (MSS) were launched by the United States from 1972 and by 1982 a thematic mapper (TM) was added. This technology provided continuous coverage of the earth every 16 days at a spatial resolution of 30 m × 30 m in seven band (TM) or 18 days at a spatial resolution of 80 m × 80 m in four bands (MSS). In 1999 several earth sensors were launched including Landsat 7, Earth Observing System (EOS) and shuttle radar topographic mapping (SRTM) which will offer better products and enhanced distribution of information (Bergen *et al.* 2000). In 1986 the French launched the SPOT satellite which provides vertical coverage every 26 days in either multi-spectral or panchromatic mode with a resolution of 10 m × 10 m or 20 m × 20 m. Lower resolution imagery (1 km pixel size versus 30 m for TM) suitable for large areas, but not for local scale, is provided on a daily basis by the NOAA advanced very high resolution radiometer (AVHRR) satellite launched in 1979. Using satellite imagery major landforms and vegetation types can be identified and different forest types and plantations are normally easy to distinguish. The images built up can be reproduced as a photograph (Fig. 5.3 (a) and (b)). However most examples of the use of satellite-based remote sensing are at broader scales than for the individual plantation unit and aerial photography is still the only remote sensing data regularly used for operational site-specific mapping.

Because of regular coverage since 1972, satellite imagery has the great attraction that images taken in different years can be compared and information obtained on the change in land-use with time. Landsat imagery in Thailand in the mid-1970s revealed that since the previous assessment 15 years before the proportion of forest land had diminished from 55% to 38%; subsequent image analysis (1986) revealed only 29% forest cover. The state and change in global forest cover is assessed by FAO using both Landsat TM and NOAA AVHRR (Päivinen *et al.* 2000). Satellite imagery is relatively inexpensive and is now usually the first step in reconnaissance of land, including that for a major plantation project. In India, satellite-based remote sensing technology is used for land-use and land cover survey (e.g. Balaji *et al.* 1994) and forms a major part of forest resource survey including monitoring afforestation (e.g. Anon. 1989).

At the forest level, these modern technologies can provide information for planning plantation development (Dykstra 1997) and are used, for example, in Indonesia (Hippi and Rissanen 1996) and in the Philippines (Williamson 1993). In South America they are used for forest production planning (e.g. Kazmierczak and Shimabukuru 1994). Until all the advantages move to these new technologies there will remain a role for recent, good quality, black and white or colour, aerial photographs of a suitable scale as practical tools for mapping geographic features and vegetation types (Caylor 2000). They have many other uses in planning roads, forest protection, and silvicultural operations. Commonly used scales are 1 : 20 000 or 1 : 25 000. Satellite imagery can be more useful than aerial photography for mapping if coverage is limited or difficult to obtain because of cloud cover. Recent imagery and computer mapping systems provide good coverage and high quality maps.

The World Wide Web is increasingly useful for accessing information on imagery resources. The satellite-based global positioning system (GPS) is a navigation system that can provide users with accurate position data anywhere in the world. Principles and applications of this technology are described by Kaplan (1996) and examples of its use to assist forest surveying and mapping operations are provided by NZFRI (1992) and Rodríguez-Pedraza *et al.* (1994).

Ground survey is usually limited to areas needing more intensive study and draws on information from soil surveyors, ecologists, sociologists, hydrologists, etc., the results of research trials, local farming and land-use experience, and so on. Because land evaluation is both complex and interdisciplinary many surveys are carried out by international agencies specifically set up for the purpose. This is especially so where there is a diversity of resource data that can only be handled effectively in digital form using GIS. Overlaying maps, such as those of soil and rainfall, is one of the most important functions of GIS that can assist in land-use planning. Both thematic data (geology, soils) and derived maps such as areas with conservation values, erosion risk, and plantation suitability can be presented. Although GIS is a powerful tool for forestry and land-use surveys it has limitations. Howlett (1993) recognized that these include:

• no matter how good the software and hardware it is only as good as the quality of data input;
• it must be based on a topographic base map at a scale suitable for plantation management; and
• maintenance of hardware and support for software must be available and this may be difficult in remote locations.

(a)

(b)

Figure 5.3 Early Landsat images of the Usutu Forest, Swaziland. (a) 7 December 1972 (part of 8113707l93500 Band 5), (b) 15 June 1978 (part of 83010207072X0 Band 5).

Note: (i) the clear evidence of man-made forest—straight lines, sharp boundaries, (ii) changes in forest due to clear-felling, restocking, etc., in some compartments, (iii) greater impression of relief in Fig. 5.5(b) taken in winter-time with sun at a lower angle.

Source: (a) and (b) Eros Data Center. Imagery: US National Aeronautics and Space Administration (NASA).

The first point also applies the land-use decision-making process. GIS and other information technologies are tools that enable stakeholder participation in decision-making and developing consensus on land allocation and management, but they have the potential to make rudimentary data seem sophisticated and this may unduly influence decision making (Cassells 2001).

Land used for forest plantations

A survey of 62 countries in 1978 (not including China and India) showed there was a predominance of grasslands and savanna (<50%) in the broad categories of land used for plantations, followed by conversion of natural forest and woodland (30%) and other land types less than 20% (Evans 1982). Today, clearing natural forest and woodland to provide land for tree-planting has become less acceptable. The IUCN, UNEP, WWF, and World Conservation Strategy emphasizes that plantations should be in addition to natural and modified forest; they should not replace them (Munro and Holdgate 1991) and increasingly, degraded land unsuitable for intensive agriculture is seen as more appropriate for afforestation. Such afforestation is a central part of the programme of India's National

Wasteland Development Board (Chowdhry 1987) and *Imperata* grasslands and degraded forest are targeted in the allocation of land for tree-planting for wood production in Indonesia (Effendy and Hardono 2000).

Land used for tree planting in rural development

Land for tree-planting for essentially domestic needs or local environmental improvement is influenced by very different criteria from the broad-scale considerations discussed above; decisions are by the individual or the community (village) and concerns allocation within the land they control. In Ethiopia in the community-based Bilate project villagers allocated the most eroding land, on steepest slopes and least suitable for grazing or cropping, for afforestation (Evans 1989). In Kenya, Shepherd (1988) reported remarkable consistency by farmers for saving trees on a new farm site and the locations of planting new species. *Commiphora* and *Euphorbia* spp. were planted on the compound boundary, papaya and mango within the compound, *Grevillea robusta* along the field boundary, and mango and *Senna siamea* in the field.

Obtaining land for plantation development, encouraging tree-planting and extension

Where a national forest policy is defined, and afforestation programmes are planned for whatever purpose (production, protection, social forestry, amenity, etc.) there are different ways to implement them and encourage tree-planting.

Land reservation

In many tropical countries acts dealing with forestry allow for direct reservation of land, often referred to as 'gazetting', for forestry purposes, including plantation development, with or without the consent of the traditional landowner. Usually there will be consultation with existing landowners or occupiers and compensation paid. Compensation may be for the land or, where land is considered to have no intrinsic value, only for any improvements the previous land-occupiers may have undertaken such as drainage, buildings, crops, etc.

An example of direct reservation is provided by the Forest Act of Malawi:

The Minister may, by order and in accordance with the Land Act, declare any public or customary land to be a forest reserve, provided that in the case of customary land the following provisions shall apply:

(a) any chief within whose jurisdiction such land . . . shall be consulted prior to such proclamation;
(b) any person who satisfies the Minister that any right or privilege lawfully enjoyed by him will be adversely affected . . . will be compensated.

The application of direct reservation of land for plantation forestry will depend on a country's constitution, the nature of political power, the vesting of all unalienated land in the nation, and, most importantly, the existence of a forestry authority able to undertake the plantation programme. It is becoming less common for state forestry departments to be directly involved in plantation establishment; in countries such as Brazil virtually all programmes are executed by private organizations. However, in many countries the State still lays down policy, reserves land, and does most of the planting, for example, Australia (Queensland), India, Nigeria, and Zimbabwe. In Ghana, lands constituted as forest reserves are vested in the state in trust for 'stools' (community, tribe, or family) and prospective investors in plantation development have to deal with both the Forestry Department and the landowners concerned (Odoom 1999).

Acquiring land by freehold purchase

This alternative, providing equally secure tenure as direct reservation, depends on the land tenure situation in a country. Most afforestation in Swaziland has been on land acquired by freehold purchase, the actual purchases being made by non-Swazi organizations, notably Peak Timbers Ltd and the Commonwealth Development Corporation. Similarly, the extensive private plantations in Brazil are an example of using freehold land bought from either a private or customary owner, or the government. However, increasingly in many countries, political awareness and national pride is creating an unwillingness to sell land, in the large areas needed for industrial plantations, either to the nation's government or large national or international organizations capable of purchasing it.

Leasing land for planting

Leasing allows someone the right to grow trees on a piece of land for a fixed period while the ownership of the land remains unchanged. This method of obtaining the use of land for plantations is becoming more widespread. When the Usutu project was bought by South African Pulp and Paper Industries (SAPPI) in the late 1980s the Usutu forest land was no longer held freehold but leased from the Swaziland government for 100 years. South Africa's latest forest policy (1996) aims to move the responsibility for managing commercial forests from the public to the private sector and leasing has emerged as a central element in the process of privatization of plantations (Foy 2001). In the Philippines, a number of initiatives have been taken by the government to encourage tree planta-tions, for example, in 1997 the Industrial Forest Management Agreement was launched (Aggangan 2000). This is a contract between the government and a private individual or company to develop a plantation of 500–40 000 ha state forestland and to harvest, that is, utilize all planted trees for 25 years with a renewable period of 25 years.

In the Pacific region, reforestation following chipwood logging in the Gogol valley (PNG), is almost wholly on land leased from customary own-ers. The lease is for a minimum of 20 years to allow for two rotations of *Eucalyptus deglupta*, but 30 years is now usual. However, even this short period in forestry terms is too long for many owners who have been mostly reluctant to lease their land despite the prospect of cash rental and a share in the revenue from timber sales. In Fiji, leasing is more successful and land to be afforested is obtained in this way through the Native Land Trust Board. The lease duration is 50, 60, or 99 years and landowners receive an initial payment when leas-ing is agreed and an annual rent for each hectare planted. They also receive 3% of stumpage revenue derived from sales of logs from the leased area. By 2000, Fiji Pine Ltd had acquired 68 100 ha (86% of its total landholding) through such leasing arrange-ments. There is a mechanism to transfer ownership of the plantations to landowners when the com-pany makes sufficient profits to enable redemption of shares held by the Fiji government (Fiji Pine Limited 2001).

Extension and encouragement of tree-planting by existing landowners and rural communities

Almost every country, with even the most rudi-mentary forest policy, encourages local landown-ers or land dwellers to conserve and to plant trees. The success of such policies depends not only on laws and their enforcement but very much on enthusiasm for such rural development or social forestry by the State and non-government organi-zations (NGOs). The means of transmitting this enthusiasm to potential growers or encouraging existing aspirations of potential tree growers is an evolving and complex task.

'Extension' is used to describe government pro-motion of desirable rural developments including forestry. Landowners and villagers are encouraged to plant trees by being given advice, training, and sometimes assistance with planting. In some coun-tries the trees also are provided free or the cost of seedlings is subsidized. Commonly, extension oper-ates through a network of small rural nurseries where the forester in charge also acts as the local forestry extension officer. Usually the extension officer provided a one-way conduit for information from researchers and professional foresters to the private grower.

However, the contribution of extension methods in achieving forestry goals varies enormously. It can sometimes be spectacularly successful in pro-moting tree-planting and awareness, for example, Kenya's Green Belt Movement under the auspices of the National Council for Women (Bradley 1991) led to tens of millions of trees being planted, and in India thousands of NGOs promote tree-planting as part of the country's social forestry programme of afforesting nearly 1 million ha annually. The more successful outgrower schemes provide farmers with high quality seedlings and support them with sound advice on appropriate growing techniques (Mayers 2000). Much of China's notable record of afforestation and tree-planting (Fig. 5.4) has been due to a well-organized extension effort by the agriculture and forestry bureaux at various levels of government. In parts of China, township forest extension ser-vices may train a few villagers who apply a tech-nology on their land which then serves as a model for other villagers.

Extension has sometimes failed entirely. Skutsch (1985) in analysing 'Why people don't plant trees' in Tanzania reported that between 1973 and 1978 only 6437 ha of woodlots were established in a pro-gramme designed to cover 8000 villages and with an annual target in excess of 100 000 ha. Similarly, in many Sahelian countries, simplistic solutions to fuelwood and fodder shortages in the 1970s resulted in little more than providing eucalypt seedlings. Tree survival was often less than 10%, and disenchantment inevitable. Most forestry extension in the past adopted a 'top down' approach assuming tree growers' efforts were con-strained by a lack of knowledge about new tech-nologies related to forest management. Even when trees have been grown successfully, the absence of a market can render the enterprise a failure from the grower's viewpoint. Without a market, the cash reward that is the promise of much extension work is absent. The prospect of a good economic return was essential for securing participation in social forestry in West Bengal (Shah 1988) and was

Figure 5.4 Four-around planting near a village in China. Planting along roads, beside canals, around houses, and around villages. (Photo D.R. Johnston.)

at the heart of enthusiasm for pine planting in Fiji (Lind and Martel 1986). In Vietnam, farmers were found to spontaneously increase tree-planting in agroforestry systems when these provided a greater benefit to the household than alternative uses for the land. Major driving forces were market forces, a supportive institutional and policy environment, and a degraded natural forest resource (Woods 2003).

Often the extension effort itself is at fault, with inadequate involvement and participation of local people right from the start, no provision for follow-up visits or insufficient time for proper training and advice. Rural people in South Africa showed little interest in woodlot development despite government efforts over many years. This was due to insufficient local participation, an assumption that fuelwood was one of their major needs, and a failure to appreciate complex social, political and economic issues of rural people (Ham and Theron 1999). Appropriate participation by all stakeholders is critical and is one of several principles for successful plantation development suggested by Evans (2001b). Bass (2001) discusses how to identify stakeholders and the social science tools used to ensure successful stakeholder interactions. In southern China, Cai *et al.* (2003) describes how involving farmers, local government officials, extension workers, and researchers was beneficial in identifying options and implementing activities to improve farmers' livelihoods through tree-planting on degraded lands and management of degraded forests.

Developing communication at an early stage through training of teachers and students has become a priority for the Ministry of Education and Culture in Ecuador (Desmond 1989). At village level, SOS Sahel has achieved great success with puppet shows to communicate key points about windbreaks and shelterbelts and the important role of trees in the village environment in northern

Sudan. In Amazonia, village theatre and travelling shows were used to complement illustrated booklets and posters to communicate extension messages (May and Pastuk 1996). Even more effective can be to begin by encouraging villagers to teach outsiders how to plant trees and care for them—'everyone enjoys teaching and showing others their skills and then the outsiders can offer suggestions for improvement if appropriate: development can be fun' (Chambers 1990). In Australia, Schirmer *et al.* (2000) discuss factors affecting adoption of plantation forestry on farmers such as biophysical requirements, opportunity costs incurred, infrastructure, and access to information, and they quote Forge and Black (1998) as identifying deficiencies in providing information about marketing and potential economic benefits of farm forestry.

Analyses of both failed and successful small-scale planting programmes have shown that the one-way, narrowly focused, technical approach is ineffective. Simply training people to use a particular species or a specific technology devised by researchers elsewhere is not enough. That is because most of the constraints to successful small-scale forestry are not technical. In the Philippines it was found that capacity building through training was important but that local people tended to maintain their efforts in community-based forestry projects when they had secure land tenure, ready access to capital for the enterprise and good access to markets for their products (Acosta 2000). Constraints in Australia relate to harvest rights, availability of markets, the long wait for returns, satisfaction with current land-use, tax disincentives, and the traditional dominance of the forestry sector by state forestry agencies (Herbohn 2001). Hence, there are a number of preconditions that must be met if tree-planting is to be an economic and social success and having an appropriate technology is only one of these preconditions. Byron (2001) has identified a number of critical factors, 'keys', that must be in place for successful smallholder forestry in the tropics. They are:

- secure access to land on which to grow trees;
- a viable production technology with all necessary inputs, such as seeds, fertilizer, technical knowledge, and credit;
- control over risks such as fire, pests and diseases, theft and expropriation;
- access to markets, reflected by attractive prices; and physical access to market by road, river etc.

Attending to these local institutional and legislative constraints as part of encouraging tree-planting is essential and will greatly simplify the task of extension. When this was achieved prior to a project to supply domestic fuelwood in Niger it enabled relatively simple technical measures to improve woodland management to be implemented (Peltier *et al.* 1994).

At the farmer level, the extension worker needs to be more a 'facilitator' than a 'messenger' or 'teacher' with the role of providing background information and technical options to stimulate discussions and encourage farmers to experiment with new ideas and options (Enters and Hagmann 1996). This requires a participatory approach involving men and women exchanging ideas, knowledge, and techniques which then leads to attitudes and practices that improve forest and tree management (Anderson and Farrington 1996). Without doubt, encouragement of tree-planting enterprises by landowners and rural communities requires the extension agent to not only be a good communicator and motivator, but also to be a good listener and able to foster the participatory approach to problem solving at the on-farm level.

General tree-planting laws and national tree-planting programmes

National extension activities to encourage tree-planting by the whole population are frequently encouraged. Major examples are the 'Green China Fund', India's 'A tree for every child programme' built around a schools' forestry programme, and Indonesia's 'greening' movement. Most countries have national tree-planting days to raise general awareness. On a much smaller scale in the PNG highlands since the 1950s the effects of enforced tree-planting, as a penalty for grass-burning, and tree-planting programmes like '*Plantim diwai Yar*' (Plant casuarina trees) in 1966, can be seen today since denuded hills are now well-wooded (Fig. 6.3).

More direct encouragement of tree-planting has been attempted by enacting laws. In the Philippines, Presidential decree No. 1157 (1977) required every citizen over 10 years of age to plant one tree every month for the next 5 years. In China the People's Congress in 1981 made it obligatory for every able-bodied person to plant three to five trees every year (FAO 1982). Such planting laws generally contribute little to development of an industrial resource and are largely directed at public education, environmental improvement, and meeting widespread needs for firewood, shelter, and shade, and food and fodder trees. Planting laws have an important role in increasing public awareness of forestry and its benefits, but grandiose, nationwide programmes have rarely resulted in effective afforestation.

Difficulties with tree-planting laws, apart from their enforcement, are provision of enough seedlings suitable for planting on the sites available, advice available about planting and caring for the trees properly, and arrangements for proper maintenance, such as weeding in the years after planting. Skutsch (1985) found inability to deliver seedlings from the nursery to the village at the critical time a key constraint to effective extension. In China many of the plantings achieved by 'mobilizing the masses' had survival rates of less than 20% due to poor species-site matching and lack of post-planting weeding.

For all projects aimed at encouraging tree-planting by rural communities, both local extension efforts and national programmes, there are fundamental social, economic, and institutional requirements which vary in importance depending on the country and local conditions. Many of these have been discussed in detail with examples in publications such as: Arnold (1990), Sargent and Bass (1992), Arnold and Dewees (1997), Harrison and Herbohn (2000), Evans (2001*b*), and Niskanen and Väyrynen (2001). They include: local participation, project objectives acceptable to community and villagers, tenure security, market information, appropriate funding arrangements, and availability of technical support. A more comprehensive list and detailed considerations are provided in Chapter 7.

State forest corporations, cooperatives, and contract planting

In recent years national governments have been less directly involved in plantation development and have encouraged greater private investment in both plantations and downstream industries. They have also encouraged more small-scale forestry planting by community or regionally based organizations which are neither part of national forest services nor commercial companies.

Some governments set up forest development corporations, such as those in each state of India, primarily to initiate and encourage commercial forestry, including plantation projects. In the Pacific, Fiji Pine Ltd, which undertakes plantation development, is almost wholly government owned. The autonomy of these organizations has allowed a freedom of action not possible with traditional government forest services, and the development emphasis, for example, rural employment, may not be restricted by the requirement to maximize profit as in a wholly private enterprise. These two features can lead to more effective and efficient progress in development—orientated forestry projects.

Cooperatives are communal systems where the resources of several families or tribal groups are pooled so that a worthwhile project can be tackled or a marketable quantity of a commodity produced. But, as Edwardson (1978) stresses, the essential characteristic of a cooperative is the encouragement of self-help to improve one's lot in life. In plantation development, the cooperative approach allows many land-occupiers, with rights over only small parcels of land, to combine to form a resource large enough to supply an industry. In India and Sabah (Malaysia), where land is state-owned, cooperatives have been successfully used to develop land, mainly by clearing forest and establishing horticultural and forestry plantations.

Contract schemes in which state land is allocated to landless peasants or poor farmers specifically to establish trees have been established in some countries such as India (West Bengal, Gujarat, Rajasthan, etc.). The rural poor in West Bengal's Social Forestry project harvested and sold their trees after 5 years and with the proceeds were able to buy a small piece of land suitable for agriculture (Chowdhry 1987). However, many contract schemes with laudable objectives have failed due to inadequate financial, social, and tenurial arrangements or because the common land was already used by the very poor and landless for fuelwood collection and/or fodder for their animals. It is also important that the intermittent cash flows from contract planting are compatible with the livelihood needs of those involved (Arnold 2001). Analyses of successful and unsuccessful contract reforestation programmes in the Philippines highlight many of the pitfalls of these schemes (Lechoncito 1987; Pasicolan 1996).

Private sector/community partnerships

Companies in the private sector have shown increasing interest in forming partnerships with communities and individuals to grow industrial wood. In 1996 a worldwide survey of 18 of the largest corporations producing pulpwood showed that 60% either operated an outgrower scheme or provided extension services to smallholders to grow trees for them (Roberts and Dubois 1996). These forest farms illustrate how many small units can be combined to create a sizeable resource.

The main arrangements for outgrower and joint-venture schemes are contracts which specify combinations of land, capital, management, and market opportunities directed towards production of a tree crop. There is a sharing of costs, benefits, and risks between the company and the grower. Often the company will provide seedlings and, fertilizer and other inputs, and will purchase the wood produced. Growers benefit from a guaranteed market and technical assistance. Mayers (2000) notes the wide variation in arrangements with companies paying the market price for the wood on delivery and exercising little control over production to those schemes in which wood prices are fixed and the company strictly controls all aspects of production. Guizol and Cossalter (2000) have questioned whether many of these agreements are sustainable, especially when large international companies are dealing with local communities whose residents may have little appreciation of what an agreement entails.

Perhaps the best known and most quoted outgrower scheme was established by the Paper Industries Corporation of the Philippines (PICOP) (now PICOP Resources Inc.) in 1970 and by 1981, 4500 farms, each between 5 and 10 ha within the PICOP concessional area, were supplying about one-third of PICOP's wood from about 23 000 ha of *Paraserianthes falcataria* on their land. Four-fifths of a farmer's land was planted with trees and the rest used to grow food crops (Fig. 5.5). PICOP guaranteed to buy all wood produced and this enabled farmers to obtain loans from the Philippine Development Bank. Seedlings were acquired at cost from PICOP which, along with repayment of loans, were only paid for when farmers' had income from thinning and fellings. PICOP has diversified the scheme with *E. deglupta* being grown for poles, sawn timber, and pulpwood and *P. falcataria* for sawn timber as well as pulpwood (Aggangan 2000). The scheme was initially very successful but problems arose later when other industries offered higher prices to growers for their wood. Some of the problems and benefits of the PICOP scheme are described by Morrison and Bass (1992) and Arnold (1997*a*). The PICOP model was tried elsewhere, for example, agroforestry and woodlot development for cash tree crops near Sipitang, Sabah, for Sabah Forest Industries. Another large outgrowers scheme is in South Africa where the large pulp company, SAPPI, obtains wood from 88 000 ha of land of which 17 000 ha is owned by 8000 black

Figure 5.5 Tree farm in Mindanao (Philippines) with *Paraserianthes falcataria* in foreground maize in background—part of a scheme involving many smallholders in plantation development.

smallholders and the remainder by 260 white farmers with larger holdings. The growers receive free seedlings, silvicultural training and advice, advance payment for their work, and a guaranteed market for their wood at current market prices (Mayers 2000). In addition to this, and other well-established individual outgrowers schemes in South Africa, there is interest in developing partnerships with community groups on a similar basis or possibly through joint-venture contracts involving equity sharing in which the community provides its land as equity and shares profits with the company in proportion to the equity stake (Foy and Pitcher 1999).

In Brazil, China, India, and elsewhere in South Africa there are outgrower schemes for pulpwood (de Freitas 1996; Arnold 1997a; Cellier 1999; Mayers 2000; Lal 2001), in Ghana indigenous hardwoods, such as *Triplochiton scleroxylon*, are grown for timber (Mayers 2000), and in Brazil 20 000 smallholders grow *Acacia mearnsii* for industries using tannin (Higa and Resende 1994). In Kenya, where farm woodlots have been established since the late 1920s to produce tan bark, a reduced demand for this product has resulted in the woodlots being used for fuelwood, charcoal, and poles (Dewees and Saxena 1995).

Cellier (1999) has stressed that critical elements of an outgrower woodlot scheme are: (i) participatory arrangements with strong, representative grower's committees; (ii) a flexible process to deal with a grower's specific circumstances; (iii) empowerment of growers through training and on-farm research; and (iv) long-term commitment by both the company and grower. Commercial tree growing in woodlots by individual farmers is likely to be an attractive option for farmers where there is an assured market, access to technical advice, and land available in excess of that needed for food crops and other basic needs (Arnold 2001). Security of land and tree tenure, access to financial support while the trees grow and arrangements for participation with the company

and ability to appeal to third parties in disputes are also important for farmers to benefit from outgrower schemes (Roberts and Dubois 1996). Such conditions clearly do not apply to very poor or landless farmers. For such people, poverty alleviation through contract tree-planting on public lands may be more feasible. Bass (2001) has provided some preliminary guidance on objectives and activities for partnership arrangements that might be included in a forest organization's management plan. He suggests an illustrative goal might be 'To ensure commercially viable forestry operations which integrate social values and thereby contribute to local livelihoods and development.'

The interventions, promotion, and factors influencing the outcomes of small-scale forest plantations (woodlots) are also discussed in Chapters 6 and 7.

Reforestation after logging

Although clearing natural forest simply for conversion to tree plantations is deprecated, one way of trying to ensure adequate reforestation after logging is to require concessionaires to replant logged-over land as a condition of their logging agreement. However, this has often proved ineffective since many concessionaires are slow to invest money which will not be realized for 10 or 20 years until the plantation is felled. Moreover, many economically weak developing countries are unwilling or afraid to enforce replanting conditions against large international companies, since there is the threat that the company would pull out and the country become unpopular with foreign investors.

In Malaysia, the National Forestry Act 1984 requires forest licensees to implement a reforestation plan. If they do not comply the licence is revoked and the State authority paid compensation which is made over to the Forest Development

Fund. The State authority can use this fund to engage any person to carry out the reforestation plan. In Indonesia a reforestation levy is charged on each cubic metre of timber harvested, and in Liberia the Forest Management Plan required concession-aires to reforest 0.4 ha or pay a levy for every 70 m^3 harvested. There are four main problems in ensuring reforestation after logging.

1. Generating the necessary income by taxing logs or altering royalty rates, and ensuring it is channelled into planting work.
2. Providing a competent reforestation authority able to undertake the work efficiently and at a rate which keeps pace with forest clearance.
3. Gaining the agreement of local landowners for replanting to take place, especially since many older concessions only provided for timber harvesting.

4. Enforcing the laws effectively when many of the officers who have to do it are inexperienced, poorly paid, and have to deal with large and influential companies.

Generally satisfactory reforestation after logging has met with indifferent success. According to Repetto (1988) typically only between 10% and 40% of logging rents are paid or collected and in the Philippines, for example, probably only half of all companies complied with industrial reforestation regulations. The problems of overly prescriptive regulations in forest operations and the need to review policies in Indonesia are highlighted by Bennett (1998). There is need to adjust policies and achieve sustainable use of forests subject to logging and replanting. The general adoption and require-ments of certification of forestry operations should assist compliance with good practice.

CHAPTER 6

Social and economic factors in tree-planting and plantation development

Introduction

Development cannot bypass the lives of the people concerned whether they are directly involved as future employees on a plantation project or are those living nearby who see their familiar environment changed and new goods and services becoming available as a result of social forestry programmes. All development costs money, most developing countries are short of funds, and money raised from harvesting natural forests is frequently used to develop other sectors of higher national priority. This shortage often results in heavy reliance on official development assistance (ODA). Grayson (2000) notes that globally the flow of funds from rich to poor countries has been more than US$180 billion annually, of which ODA has been one-third and the balance has come from private sources. Of the US$60 billion of ODA only US$20 billion has come from multilateral agencies, the rest being from bilateral aid and national governments. In forestry the proportion from public sources (bilateral and multilateral) has been greater.

Most investment in forest plantation development has been in the form of low interest loans from international or regional development banks. An objective of the World Bank's forest policy and strategy approved in 1991 was 'to slow deforestation in the tropics and to ensure adequate planting of new trees and management of existing resources'. Between 1992 and 1999 the World Bank had a portfolio of forestry projects of US$3.5 billion (mainly for degraded areas in India and China) and US$578 million in its International Financing Corporation (IFC) for pulp and paper industries (Knudsen 2000). Since 1990, the Asian Development Bank has provided loan finance to Indonesia, Philippines, and Laos for industrial plantation development (Fraser 2000) and similar loans have been made in other regions by the African Development Bank and the Inter-American Development Bank. United Nations agencies provide relatively little funding for plantation development but FAO's Investment Centre assists countries in mobilizing investment for forest development and between 1991 and 1995 prepared projects valued at US$850 million (Ball 2000).

As many of the funds for development come from outside the country, they commonly have constraints on how they are used and so a government may spend money on development, which affects many people, but over which it has far from complete control. External aid is most effective where there is good policy and institutional environment within the country (World Bank 1998) and a triangle of factors (economics, social impacts, and development) all interrelate and largely determine the success or failure of plantation forestry and tree-planting in general in the tropics, because they are undertaken as externally funded development projects. The importance of these factors, and the problems they may present, are highly correlated with the scale of development.

This chapter does not attempt to provide a full economic or sociological analysis of industrial plantations or social forestry and their contribution to development in the tropics but examines three aspects which influence their development: (i) scale and rate of development; (ii) which projects may have impacts on the local community, particularly social and cultural life; and (iii) contributions that plantation forestry may make to the economies and rural development of developing countries.

Factors affecting the scale and rate of planting and plantation development

Mention has already been made of the large areas of land required for industrial plantations, but that is one end of the spectrum and this factor of scale needs to be examined more closely because size itself largely determines the magnitude of social, cultural, and economic impacts. Also, the rate of development is closely related to size. A coffee crop or small farm woodlot, planted in one year on 1 ha of ground, while improving a farmer's income a few years later, hardly compares with the impact of an industrial plantation project which takes many years to establish, requires much land, and needs a sizeable labour force. Moreover, the large size of most industrial plantations brings with it responsibilities to the local community, problems of funding, and considerations of marketing which may, in the long-term and for the developing country, even be seen as diseconomies of scale. Having many small- to medium-sized projects or plantings is a more attractive tool of development, for example in outgrower schemes where there is greater smallholder participation and risk sharing (Chapter 5). However, as Roche (1986) points out, both the large-scale industrial development and the use of forestry as an instrument of rural development (numerous small-scale activities) can and should be complementary in the tropics.

The size of a country's plantation programme

The reasons for expansion of planted forests were analysed in Chapter 2. Essentially, the size of a country's plantation programme reflects forest policy. A forest policy with emphasis on tree-planting programmes, reflects needs such as replacing highly degraded natural forest, controlling erosion, satisfying rising internal wood consumption, reducing dependency on imports, improving balance of payments by increasing exports, etc. The overall interaction and influence of these factors were considered by Husch (1987) and formed a central part of the analysis for the Tropical Forestry Action Plan (FAO 1985a). Detailed analysis of forest sector planning and the criteria assessed when evaluating forestry programmes, including plantation projects, has a long history and can be found in Watt (1973), Fraser (1973, 1976), Pansel and Wiebecke (1984), and Gregersen et al. (1993). Appraisal and planning of individual projects are further considered in Chapter 7.

It is sufficient to note that plantation programmes result from factors of local demand for wood and tree-related benefits, yield from natural forest, export/import possibilities, land and labour availability, species and yields, environmental protection, and finance. Sarawak (Malaysia) with most of its land under forest, no shortage of wood, and a sizeable timber export industry, has only recently initiated industrial plantation development. In contrast, Swaziland in the 1950s, with three times Sarawak's population density, had only 5% forest cover and imported almost all of its small timber requirement.

Since then over 100 000 ha of private afforestation has entirely changed the situation; roundwood, wood pulp, and sawn timber are now Swaziland's third largest export and the industry is a major sector of its economy (Evans and Wright 1988). Between these extremes is Nigeria with some natural forest and sizeable plantations, but continuing exploitation of natural forest and rising domestic demand for wood is necessitating further expansion of plantations (Adegbehin 1988). In South America most countries still have significant areas of natural forest but Argentina, Brazil, Chile, Uruguay, and Venezuela are all implementing large-scale industrial forestry programmes through private tree plantations which benefited from government financial incentives, at least initially (IDB 1995).

Demand for wood is high in India where there is only 0.08 ha of forest per capita. Most wood produced comes from plantations supplemented by annual imports of about 1.5 million m^3. There was a plantation boom from 1979 after the setting up of Forest Development Corporations in the States and the initiation of externally funded Social Forestry projects. The annual planting rate in 1980–85 was about 1 million ha and when the National Wasteland Development Board was established in 1985 this increased to 1.78 million ha during 1985–90. The National Forest Policy (1988) of India determined that forest-based industries should acquire their wood from farm forestry and agroforestry sources. The current National Forestry Action Programme (1999) aims to expand the area under plantations but, due to a decrease of funding from the central government and termination of many social forestry projects, the annual planting rate has decreased to 1.5 million ha (Pandey 2000; Bahuguna 2001).

Factors which determine the scale of plantation projects

Scale may be measured in several ways; the area of land for planting; amount of money invested in a project; the number of people employed, and so on. The relative size of these components inevitably varies from project to project but broad relationships exist which relate primarily to the purpose of growing trees and, to a lesser extent, to political and social influences.

Purpose for growing trees
The object of creating a forest plantation, whether for commercial, rural development, or environmental purposes, is the overriding factor determining project size. At one extreme, planting

trees for firewood, shelter, or fodder will cost very little, can be done by one person, and provides the desired commodity without the need for processing. At the other extreme, is the pulping complex which, owing to the high initial cost of the factory, usually requires much wood to produce large quantities of pulp to keep unit costs at a reasonable level. A pulp mill requires a large volume of wood, a major investment, and a substantial number of employees. Within the tropics, the approximate order of size of these components is shown in Table 6.1.

Economic factors mainly determine the minimum size of an industrial plantation project and a broad correlation exists between scale, processing complexity, and capital required for setting up a mill. Individuals can plant trees for firewood, building poles, or shelter but only governments, large companies, or multinational funding agencies can usually afford the finance and take the risk with an integrated pulping project. The purpose for growing trees determines the investment needed for the appropriate processing industry. The higher the cost of setting up that industry the greater must be the annual through-put of timber if unit costs are to be constrained. This is the simple relationship called 'economy of scale'. But projects requiring high through-puts of timber to achieve economy of scale also need much land for plantations, much equipment for silvicultural and harvesting operations, a large labour force, and more infrastructure (houses, schools, etc.). Thus the objective of the plantation project largely determines its scale and impact.

Apart from plantations for protection purposes, from which no readily identifiable income is expected, most cost-benefit evaluations of plantations have, until recently, ignored or understated off-site social and environmental costs and benefits (Laarman and Contreras 1991; Gregersen *et al.* 1993). What are the impacts, who benefits, and who pays? These questions must be addressed.

While sustainable forest management has been a basic objective of public forest management for 100 years or more, values other than wood production were usually not explicit, and concealed in terms such as 'multiple-use'. The sustained yield of wood was often used by foresters as a surrogate measure of sustainability. While sustainable management is still viewed as an ultimate goal, it is generally recognized as an evolving concept which depends on the environmental, social, and cultural objectives society wishes to achieve through the use of forest resources. Impacts of development of plantations, especially large plantations, are now subject to greater scrutiny. Certification schemes for sustainable management of natural forests and plantations are a reality and demand attention to the impacts of forests on water, biodiversity, people's livelihoods, etc. (Bass 1993; Howlett 1993; Ferguson 1996). Some of these factors influence scale and the way in which plantations may be distributed in the landscape. The size of a company's plantation may be limited by the amount of land it can purchase or lease and if it requires additional resources, it may have to make agreements or partnerships with other landholders to provide the wood. Biodiversity considerations may require areas of natural forests

Table 6.1 Approximate minimum viable size of resources for a new plantation project

Project/operation	Volume of timber (m³ year⁻¹)	Net plantation areaª (ha)	Investment (millions US$)	Manpower
Fuelwood/social forestry	No definable minimum			
Integrated sawmilling	15 000	1000	2	30
Integrated sawmill and plywood	100 000	7000	15	200
Integrated pulpingᵇ	500 000	25 000	500	2000
Woodchip export	100 000	5000	10	60
Environmental projects	Minimum area to meet protection need and maintain satisfactory crop			

ª Assumed mean annual increment for fuelwood, pulpwood, and woodchips is 20 m³ ha⁻¹ year⁻¹ for sawtimber and plywood 15 m³ ha⁻¹ year⁻¹. Also, the *total* area needed will be 10–20% greater to allow for roads, firebreaks, land for mill site, villages, etc.

ᵇ Smaller scale operations may be justified if pulp is not intended for world markets. Also, thermomechanical processes offer scope for lower rates of production and capital requirements.

among plantations to be left untouched and so reduce the area available for planting.

Table 6.1 indicates the order of size involved under present-day conditions. The optimum project size is determined by the pre-investment feasibility study (Chapter 7).

It is questionable whether this kind of relationship must always exist between end-use and scale, especially when many developing countries can ill-afford the outlay needed for a large project despite the benefits it would appear to confer. One of the assumptions has been that both tree growing and wood processing occur in the same place (country), for example, the Usutu Pulp project in Swaziland, and Aracruz Florestal and Jari projects in Brazil, but this need not be so. Though the economic benefits of maximizing the 'value-added' processes before export are real (exporting high value pulp or plywood) many countries now accept cheaper alternatives to reduce initial capital outlay. In many cases export of intermediate products, woodchips, and plywood sheets, or even primary products such as peeler logs, sawlogs, and pulpwood are being preferred, at least from plantations, (though not in the case of high-value logs from natural forest). Pulpwood logs from ECO sa's (Eucalyptus du Congo Société Anonyme) eucalypt plantations in the Congo are sent to France and chips from *Eucalyptus globulus* plantations in Western Australia are shipped to Japan. Although this results in less industry in the wood-producing country, and consequently less economic development and a smaller multiplier effect in the economy from the plantation project, and for log exports possibly greater opportunity for malpractices, there are some advantages.

1. Much less finance is needed.
2. Pollution from large factories, for example, a pulp mill, is avoided.
3. The operation is suited to plantation programmes built around many small growers, who perhaps sell through a cooperative, where regularity and timing of timber supply need not be rigid.

The diseconomies of scale, benefits of small size, and low levels of processing, are attractive to countries in the tropics anxious to capitalize on the fast growth of plantations to undertake rural development and provide employment, but which are short of capital, or unwilling to allow foreign investment, or have environmental constraints, such as water availability, for a major industrial complex.

Recognition that small-scale industries may have advantages in development is of relatively recent origin, and arises mainly from the seminal work of Schumacher (1973) and the failure of many early industrial projects to come up to expectation. In several cases pulp mills were built but the planned wood supply from plantations failed to materialize. A parallel re-evaluation of development theory is the emphasis on appropriate technology (see later), where simple improvements in tools and methods are preferred to sophisticated equipment and machinery to raise output, and sustainable development (Letourneau 1987; Carr 1988). In forestry the benefits of small-scale operations and modern technologies, for example, new methods of pulping, small peeling machines for plywood, portable sawmills, are: (i) small and effective capital outlay; (ii) wider dispersion of activity; (iii) more labour intensive; (iv) less prone to under-use and less affected by temporary shortfalls in supply; (v) greater opportunity for development of basic technical and managerial skills; (vi) closer utilization of production resulting in less waste; and (vii) less severe ecological impact.

Political influences
A large, well-maintained plantation is impressive, conspicuous, and is necessarily located in a rural area. Politically, plantation development can have considerable prestige value. Unfortunately, this sometimes results in a scale of development bearing little relation to local or even regional requirements for its produce. Particularly important, in countries where regionalism is strong or programmes are ambitious, is the political pressure always to report satisfactory rates of new planting. This statistic, which is not a good measure of forestry achievement, can take on such importance that it creates a need to continue with planting regardless of sensible practice. To be seen to be doing something and to report increasingly large areas planted are poor reasons for pursuing plantation forestry, yet both occur and influence some plantation programmes. Strong political pressure during the massive afforestation programmes in China in the period 1950–70 was responsible for local officials misreporting plantation area statistics.

Also, for some countries, a project may take on a prestige role where the government or Head of State is anxious to see a certain type of development. The scale and complexity of a woodpulp industry is impressive and plantation projects have been initiated in the tropics with this intended end-use, for example, Madagascar, Malawi, and Zambia, and even mills constructed, for example, Cameroon, where other factors of location, communications, and marketing make the venture of questionable profitability.

Perhaps the most extreme example of political influence has occurred in Indonesia where since the late 1980s the government has promoted the development of industrial pulpwood plantations (*hutan tanaman industri*, or HTIs) on a grand scale to support pulp and paper industries. In 2000, pulp and paper products generated almost US$3 billion in export earning and made up over 50% of Indonesia's forest-related exports. Thirteen 'priority' HTIs, accounting for 2.9 million ha, are eligible for subsidized funding from the government's Reforestation Fund. With considerable government support from this fund and state banks, the private sector has expanded the capacity of pulp mills by 700% to more than 5 million t year^{-1}. So far (2000) only 8% of wood has come from plantations and until the plantations mature the government has permitted harvesting of large tracts of natural forests. Much wood has been obtained from illegal operations and over 800 000 ha of natural rain forest has been cleared since 1988. The strong political support of the forest industry has resulted in the development of a major export industry but at high cost to the environment, and only time will tell if the industry will be sustainable (Anon. 2000a; Barr 2001).

Political influence can even affect species choice. Owing to concern expressed in India about the competitive effects of eucalypts on adjacent farm crops, in the late 1980s the King of Thailand proposed banning them from new plantings in his country. In Kenya there were controls imposed on the numbers of non-native tree species (mainly eucalypts) that could be raised in nurseries. For further discussion of the eucalypt controversy see Chapter 24.

Social responsibilities

Social factors need to be treated just as seriously as technical and economic factors as they can have a major impact on scale of a plantation development and on its long-term sustainability. Harcharik (1997) has stated that to many foresters 'the socio-cultural dimension of modern forest management provides the newest and greatest challenge'. Sustainable forest management will only be achieved if the needs and aspirations of rural communities are considered and conflicting situations created by competition for land and forest resources are acknowledged and resolved (Desloges and Gauthier 1997).

Developers must ensure they understand local cultural and social conditions and place a high priority on the participation of local people in the planning and implementation of plantation development. There are several cost-effective appraisal methods of securing information about the socio-economic and cultural situation of communities in advance of plantation development that will help development plans fit rural people's realities and aspirations. They include, Rapid Rural Appraisal, Agroecosystems Analysis, Diagnosis and Design and, more recently,

Participatory Rural Appraisal (Nair 1993a; Chambers 1997; Rocheleau 1999). Many impact assessment issues are discussed by Gregersen *et al.* (1993). Conflict will undoubtedly arise during a major plantation development and conflict management processes will need to be used. These attempt to reconcile differences by accommodating the respective needs of the parties involved (the stakeholders) to arrive ultimately at a mutually acceptable outcome. Armitage (1998) lists the main features of conflict management in relation to forests as:

- development of trust, understanding, and communication;
- ownership of both the problem and the solution;
- exploration of indigenous knowledge systems;
- promoting equitable dialogue under unequal conditions;
- appreciation of the role of ethnic minorities and women;
- agreeing on practical outcomes capable of implementation.

Where there are serious problems in obtaining sufficient land for planting the development of outgrower schemes to complement wood production on company land may be a solution. This has occurred in many such situations (Mayers 2000). Local people's land rights need to be respected and it is very much in the developer's interest that there is agreement and demarcation of boundaries even if this requires giving the local people the opportunity to veto planting on some areas.

Development in the tropics frequently carries with it responsibilities far in excess of similar investments in developed countries where there is usually basic provision for people's welfare. There must be some continuity of work for employees, particularly where housing and amenities are provided and where workers are recruited from another region and have left their homes. However local people should be given priority in employment opportunities and the provision of goods and services to the project. It may also be necessary to positively discriminate to provide employment opportunities for poor and landless people in the local communities. Both the need for long-term security and a sense of belonging are just as important in the tropics as elsewhere. Not only must a project provide some continuity of employment and on-the-job training but often every service and need its employees require for an adequate standard of living. Health and safety standards must also be an appropriate standard. The responsibility to employees is discussed more fully later in this chapter.

In many places the developer will need to construct roads, bridges, and other infrastructure. Some of these will impact on the local community and they need to be involved in the planning phase. It is essential that some formal arrangements be made for community consultation as this should also ensure on-going feedback that will assist the developer to meet social responsibilities.

Factors affecting the rate of plantation development

The rate of plantation development is usually closely related to the scale of a project. Some large pulpwood plantation projects in Indonesia have had annual planting rates of 30 000 ha or more aiming to plant their concession areas in 5–6 years. Each of Shell's pulpwood plantation projects proceeded at an average annual rate of 3000–4000 ha (Richardson 1989). It is useful to distinguish this time element since frequently there is a gap between the ideal project size and what is achieved in the period allocated. Clearly the rate of tree-planting itself can determine the scale. If resources only allow a 300 ha annual planting programme then a 20 000 ha short-rotation, pulpwood scheme is not feasible. This is what happened, though not by design, in the earlier stages of the plan to reforest the Gogol valley in Papua New Guinea after logging the rain forest for chipwood. The factors responsible for slow progress at Gogol are common in many countries they included problems with finance, limited infrastructure, slow acquisition of the right to plant trees (buying or leasing land), bureaucracy, and difficulties with labour and equipment. Widespread corruption in some countries has also inhibited timely plantation development.

Finance
Substantial capital investment is needed for expansion of the world wood industry and forestry development in general. Funds come from four main sources: (i) international financial institutions; (ii) local and foreign equity capital; (iii) private commercial banks; and (iv) non-government organizations (NGOs) with non-commercial interests. In developing countries, obtaining finance for industrial plantations mostly comes from the first two sources because of the scale involved, the long time before returns are realized, and the rather greater risks involved, notably from political instability. Heavy reliance on overseas aid, bilateral support programmes, and low-interest-rate loans from development banks is inevitable. The alternatives and problems have been discussed by McGauhey (1986) and McGauhey and Gregersen (1988).

Lack of funds is frequently cited by governments as the main reason for the poor performance of the forestry sector including plantation development, for example, in Africa (Blanchez and Dubé 1997). Maydell and Gregersen (1976) stated a long time ago the reasons why private enterprise, and ultimately private commercial banks, are cautious about investing in forestry in developing countries include: (i) high investment costs; (ii) capital is tied up for a long period and the return may not be high; (iii) there tends to be more uncertainties over policies, and supplies of raw material, spares, labour, energy, etc.; (iv) risk of future discrimination against the investment; (v) taxation uncertainties; and (vi) uncertainties about freedom of decision making, about employment, repatriation of salaries and profits, etc. Sedjo (1999) suggests that in recent years the two main impediments to investment in plantation development have been concerns over political stability and objections to plantations by some environmental groups.

In many countries plantation development began as a government activity and in the 1950s and 1960s most plantations in the tropics were made by public institutions on public lands. Governments may support plantations for financial benefits alone but it is more often for broader economic reasons and to generate social and environmental benefits. Employment opportunities through developing new forest industries, watershed protection, enhanced landscape amenity values, recreational opportunities, and land rehabilitation are some of the justifications used for direct government involvement in plantations, often through state forestry companies (Brown 2000). More recently, plantations have also been supported as a means of catalysing regeneration of native forests on degraded forest lands and for carbon sequestration. However, recent policies of some governments and many bilateral aid agencies have directed support to conservation and small-scale community tree-planting rather than large industrial plantations and forest industries.

In general the primary motivation for the private sector to invest in plantations is to generate financial benefits from wood sales. They may also seek to ensure a reliable wood supply for a processing plant, secure land tenure through planting, or for other indirect reasons (Brown 2000). Given some of the constraints to investment listed above, governments in a number of tropical countries have provided direct and/or indirect incentives to assist profitability and encourage private investment in plantations. In the 1980s attractive rates of return from industrial and some other kinds of plantations were reported from tropical and subtropical regions (e.g. Sedjo 1983; Spears 1984; McGauhey and Gregersen 1988). This situation continued through the 1990s (Sedjo 1999) and there has been more private-sector financing of industrial plantations. He points out that while political instability may deter foreign investors local investors may still be prepared to support plantation expansion, as occurred in Brazil's Jari project (McNabb *et al.* 1994).

Incentives are 'public subsidies given in various forms to the private sector to encourage socially desirable actions by private entities' (Gregersen 1984). Commonly governments have justified subsidies on the basis of import substitution, developing exports and/or reduction of rural poverty through employment generation but there has been considerable debate about the merits or otherwise of government incentives for establishing plantations. Over the past 30 years most Latin American countries have provided incentives to encourage private investment in forest plantations. Keipi (1997) gives the following reasons for this approach:

- accelerating initial development of plantations for industrial or social forestry purposes;
- establishing a critical mass of plantations to stimulate forest industries;
- reducing cash flow problems, risks and uncertainty that arise from the long period of investment;
- increasing the financial attractiveness of investments that have low profitability but offer societal benefits; and
- modifying social bias against forestry investment by farmers who have traditionally considered forests as an impediment to agricultural development.

Examples of direct incentives are in the form of government grants, favourable taxation regimes, subsidized low interest loans, loan guarantees, provision of land as concession areas, protection agreements, etc.; indirect incentives include extension services, research, and training. Chile and Brazil provide examples of the highly successful use of government incentives to encourage plantations which have resulted in these countries being the leading exporters of pulp from South America (Sedjo 1999). As Brazil provided the catalyst for the introduction of incentives in other Latin American countries it is described below.

In many countries a tax concession allows money spent on plantation development to be offset against subsequent revenue from sales until the whole development expenditure has been recovered. However, in Brazil tax concessions were widened to allow some of the tax from profits on any business to be devoted to plantation establishment and it proved a powerful stimulus.

Since 1966 the area of plantation in Brazil has increased from some 500 000 ha to about 5 million ha in 2000 (FAO 2001*a*), a rapid expansion almost wholly due to two fiscal incentives introduced in 1966 and 1970. The first law, No. 5.106 introduced in September 1966 and phased out in 1977, allowed a company to use up to 50% of any tax owed for investment into projects such as tourism and afforestation (approved by Instituto Brasilero de Desenvolvimento Forestal). The actual percentage was later varied according to region to encourage investment in poorer states, especially in northeast Brazil. In 1970 a second law, No. 1.134, was passed allowing tax owed by anyone to be reduced by up to 20% provided that amount of money was invested in tree-planting. In applying these incentives to afforestation, the allowable maximum expenditure per hectare, to cover all expenses for the first 4 years (ground clearance and preparation, nursery work, planting, maintenance) varied over the years between US$300 and US$800.

In the first 10 years some 9000 applications were considered and, in the whole period 1967–86, 5.5 million ha of forest plantations were established, mainly eucalypts (52%) and pines (30%), along with 0.78 million ha of fruit trees and palms. Figure 6.1 shows the progress in afforestation. By 1980 Brazil had become a major pulp exporter. In 1986 incentives were withdrawn, except for the poor north-east region, and then afforestation averaged much less than 100 000 ha year^{-1} (Shimizu, personal communication). Outside the incentives programme only few hundred thousand hectares were planted, most notably the 150 000 ha at Jari in the Amazon.

Early effects of these incentives on rural life, national and project economics, the environment, and on institutions were analysed by Victor (1977). One important consequence has been inflation of land prices. Near Bom Despacho, Minas Gerais, the cost of land for plantation development rose from US$20–30 ha^{-1} in 1970 to US$200–400 ha^{-1} in 1977. One unexpected benefit of this inflation was that people living near plantation projects became openly favourable towards afforestation, not only because it brought employment and better services, but because they could sell their poor quality agricultural land for a very good price! Overall, though, as Westoby (1989) concludes, the fiscal arrangements simply enriched the already rich because so little of the resource released went to the impoverished northeast. Indeed, only 2.6% of the fiscally assisted plantations were in this region.

Whatever the criticisms, the Brazilian tax concessions brought into use much hitherto unproductive land, mainly second-quality cut-over forest, cerrado (savanna), and poor grassland. Nevertheless, Brazil is a huge country and the large programme so far has only afforested less than 1% of the land area. Now, emphasis on new planting is changing to replacement of older plantation forests with more productive trees, often hybrid clones, and

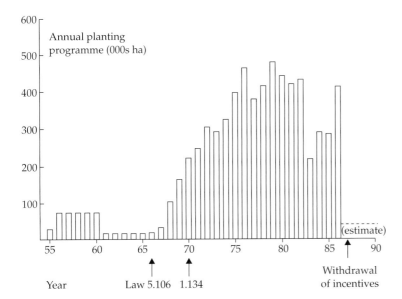

Figure 6.1 Annual planting programme in Brazil showing the dramatic increase following introduction of fiscal incentives and rapid decline after they were withdrawn. Planting between 1966 and 1986 includes a total of 790 000 ha of fruit trees and palms, but exclude tree-planting unsupported by incentives, for example, Jari. New planting since 1990 continues much reduced—Pandey and Ball (1998) quote a total of 230 000 ha year^{-1} for all countries in tropical America.

improved silvicultural practices. More wood is being grown on the same land area.

Experience with plantation incentives in other Latin American countries have been described, for example, in Costa Rica and Nicaragua (Morell 1997), Ecuador (Southgate 1995), Colombia (Gaviria 1997), and Chile (Williams 2000). In a review of financial incentives for plantation development in Chile, India, and Indonesia Williams (2000) concluded that they had proved effective but needed supporting policies and conditions such as secure tenure and technical advice, to be successful.

Despite the qualified success of incentives, in recent years many have argued that direct subsidies for industrial tree plantations should be discontinued in line with wider market and structural adjustment policies. Based on case studies in Latin America, Stewart and Gibson (1995) recommended (i) removing export bans on forest products not in danger of extinction; (ii) elimination of all tariff and non-tariff barriers on forest products; (iii) removal of export subsidies and forest product consumption taxes, other than general sales tax; and (iv) adequate financing for forest research and extension organizations.

In the Asia-Pacific region, direct incentives are most likely to be important when a national forest plantation estate is being developed so as to increase the scale and rate of plantation establishment, and to provide adequate supplies for a developing wood processing sector (Enters *et al.* 2003). These authors suggest that in time, as the wood growing and processing sectors become mature and self-sustaining, direct financial support should

be replaced by variable incentives and services, such as research and extension. Many Asian countries, for example, Indonesia (Potter and Lee 1998) and India (Williams 2000), are redirecting their assistance subsidies to small-scale and community-based forestry.

Generally, incentives in the form of direct subsidies are now regarded as less desirable and where applied they should be short term, targeted, and flexible. Indirect or enabling incentives such as, a well-defined land and tree tenure policy, appropriate marketing arrangements, good governance, and national security are often just as important for assisting plantation and forest industry development.

Financial backing from international sources for forestry projects including industrial plantations and rural development planting comes from bilateral and multilateral organizations, NGOs and private foundations. In Africa, bilateral sources provide 70% of forestry funding, either directly or through FAO trust funds and NGOs, and the World Bank and African Development Bank are the major development bank lenders providing long-term low interest loans (Blanchez and Dubé 1997). There are other sources of international funding, such as the World Food Programme which has supported tree-planting in 'food for work' projects in food-deficit countries, for example, in Vietnam where from 1992–96 it supported the largest reforestation project in the country (Ahlbäck 1995). Similar arrangements exist elsewhere with forest development supported by the World Bank and Inter-American Bank in Latin America, and The World Bank and Asian Development Bank in Asia. In the

1990s the Asian Development Bank committed about US$70 million in loan money to major industrial forest plantation projects in Indonesia, Laos, and the Philippines (Fraser 2000), and China received World Bank funding to establish over 1 million ha of plantations in 16 provinces.

Loans to developing countries are of two main types, those that attract commercial rates of interest and those where repayment is delayed (period of grace) and the interest charged is nominal (soft loans). Both the type and profitability of the project itself and the economic conditions in the country as a whole influence the form of loan granted but those for plantation establishment usually fall into the soft loan category. Tree plantations with a long payback period and relatively high risk are unlikely to be attractive to commercial banks, although pulpwood production on a short rotation supplying an assured market may receive more favourable consideration. Development bank soft loans are usually for 15–20 years with a grace period of 3–5 years and a variable but low rate of interest. Both the rate of interest charged and the period of grace ease the debt burden, at least until the project has a positive cash flow.

Loans rarely provide all capital needed and afforestation projects must compete with other demands for internal revenue. But owing to the factors noted earlier, it is difficult to attract funds from a country's own developmental investments. As described above, often direct assistance is given through reforestation funds from logging royalties and tax concessions of many kinds may be used as incentives to encourage plantation development.

Little has been said so far about continuity of finance over time. However, this is important because aid or loans are rarely guaranteed right up to the time of harvesting when cash begins to be generated internally. Usually, as schemes progress, new injections of capital are made depending both on the needs and prospects of the project. For example, the Commonwealth Development Corporation (CDC) loan of £5 million to the Fiji Pine Commission, negotiated in 1975 and increased in 1977 to be repaid in 18 installments after 1983 when felling began, was intended to finance planting for a further 4–5 years. A further loan of $5 million was made by CDC in 1985. The 1975 and 1977 loans were at 5% interest while the 1985 loan was interest-free and repayable in 20 installments which illustrates the different degrees of soft loan.

Consideration of finance above has dwelt on the needs of larger projects predominantly for industrial purposes. However, more international and national funding is for social forestry and in addition, the proliferation of NGOs may account for a significant percentage of all funds channelled into essentially social or environmental tree-planting in the tropics. NGOs rarely operate loans, apart from revolving credit schemes, but mainly work with local communities by supplementing their resources by giving cash, materials, food for work, technical expertise, appropriate training, etc. In tree-planting schemes the importance of continuity of support is well recognized.

The rate at which a plantation programme proceeds, inevitably and primarily depends on the initial availability and continuing supply of funds. Heavy reliance on external funding in many cases can make orderly development and steady expansion rather precarious.

Land availability
The land factor in plantation development was considered in Chapter 5. The need to acquire or lease land for planting at least 2–4 years in advance, to allow legal title and surveys to be done (Chapter 12), is important for regular planting to proceed unhindered. Where land is already under control of central government or the afforesting agency availability is not a problem, unless there is squatting. In contrast, obtaining the use of customary owned land can be the most intractable problem in maintaining a steady planting programme. In community forestry projects often only waste land is available or set aside by the community though much can be achieved by identifying the kinds of places where planting can be done, such as China's four-around scheme—planting beside roads, rivers and canals, around houses, and around villages (Fig. 5.4), and in compounds and along field boundaries across the tropics.

Difficulties in obtaining large areas of land is a major reason why companies in the private sector have shown increasing interest in forming partnerships with communities and individuals to grow industrial wood on their land. The majority of large corporations producing pulpwood in the tropics now either operate an outgrower scheme or provide extension services to smallholders to grow trees for them (Roberts and Dubois 1996). These partnership arrangements have in most cases provided socially acceptable solutions to the vexed question of acquiring land for plantation development.

Labour
Plantation projects in the tropics are frequently in remote places and in countries where there may be insufficient local labour and/or qualified national staff able to oversee and implement them. As a result, the need to import labour and staff into a region, can be an important constraint on project development. Labour requirements are site-specific

and will vary through the life of a plantation. The amount of labour required for the initial establishment and subsequent silvicultural operations will depend on the nature of the terrain, the previous vegetation and the degree of mechanization employed.

There can be a further problem that labour employed on a new project is totally unfamiliar with silvicultural practices and sometimes even the discipline of routine daily work. On-the-job training of workers is often a time consuming but necessary part of a plantation programme.

In the development of the Jari project in northern Brazil most of the industrial labour force was recruited from the poor northeastern states of Piaui, Ceara, and Maranhao where unemployment was high and subsistence farming difficult. Nevertheless, the remoteness of Jari made it unattractive and many would leave a few weeks after starting for this reason, or because of an inability to adapt to job discipline. A similar problem of retention occurred among managerial staff who disliked the remoteness.

In rural development forestry, communities and farmers do their own planting and, while there are many factors which affect how quickly a project progresses, it is clear that, especially in Africa, the fact of women's central role and interest is the main 'labour' element influencing success or failure. Women often do most of the manual work and may be beneficiaries if, as occurred in the West Bengal Social Forestry Project, firewood collecting becomes less arduous.

Responsibilities of plantation developers to their employees are discussed later in this chapter.

Infrastructure

Large, and especially remote, plantation projects need to provide housing, schools, medical facilities, recreational areas, and perhaps shops and transport for their labour. An extreme example is the Jari project in the Brazilian Amazon where the company had to provide almost all the facilities in the remote location (McNabb *et al.* 1994). The development of infrastructure, usually as forest villages (Fig. 6.2(a–c)) and provision of administrative and

(a)

(b)

(c)

Figure 6.2 Forest villages to house labour employed on plantation projects. (a) Usutu Forest, Swaziland. (b) Silva Villa, Planalto, Jari, Brazil. (c) Chongoni, Dedza, Malawi.

operational facilities (offices, stores, workshops, roads, nurseries, etc.) necessarily takes time and money and can be another constraint on plantation progress. However, where reforestation follows logging, the infrastructure already established is conveniently carried over to the reforestation project.

In conclusion, though a desirable project size may be agreed and operational plans in place, numerous factors can hinder orderly development and slow the rate of afforestation. Frequently, progress is slowed owing to bureaucratic delays. Difficulties over release of funds, permission to import certain goods and equipment, unavailability of foreign exchange facilities, delays in approval of land title or lease, etc., all occur. Abnormal unfavourable weather conditions and even such simple things as difficulties in procuring plastic containers for raising seedlings can impact negatively on progress. Smaller, non-governmental projects may be particularly affected by such events.

Marketing influences on plantation development

It has already been stated that the purpose for growing trees: sawn timber, pulpwood, poles, firewood, etc., is the main factor influencing project size. Although the projected end-use is a very important factor in planning plantation development, it may have to be re-considered at a later date in response to changes in local or international markets. For example, when the Shiselweni Forestry Company in Swaziland commenced planting in the mid-1960s the intention was to produce eucalypt mining timber but 10 years later there was a glut of this material and it was sold mainly for particle board. In Kenya, plantations of black wattle (*A. mearnsii*) were planted to produce tanbark for the tannin industry but selling their wood to make charcoal for urban dwellers in Nairobi has become more profitable. In China, eucalypts were planted mainly to produce round timber for the mining industries until the 1980s when government policy changed and priority was given to producing pulpwood. This has required a change to plant species more suitable for pulping.

In India, part of the concern, even antagonism, expressed by some over planting eucalypts is because trees planted on communal land and thought to be destined for local domestic use have been harvested for industrial purposes providing little benefit to local villagers whose food crops have sometimes been depressed. In some parts of India, even when the plantations are under local

villager's control, eucalypts have little value for timber, the pole market is soon saturated, and selling small quantities of pulpwood from scattered woodlots is impractical Saxena (1991). Many trees are simply sold for firewood at poor prices. The product is there but not the markets in practical terms. The same point about lack of organized market is seen as a major obstacle to increased tree-planting in Uganda, despite official backing promoting afforestation (William 1990).

Lack of reliable marketing information is one of the constraints frequently cited by farmers involved in farm forestry. However, market conditions may change during the life of a plantation, and forward planning has risks, so some flexibility in growing the crop must be maintained. Provided the final decision is not delayed too long, no silviculture is likely to be so restrictive, apart from an unusual choice of species, to preclude subsequent conversion to an alternative end-use, though some loss in profitability may occur.

Another consideration is where development is piecemeal. Often successful small-scale species trials lead to more extensive plantings followed perhaps by establishment of 200 or 300 ha blocks all without a clear end-use, apart from meeting local firewood and roundwood needs. A resource is created, too small to support an industry but too large to ignore, particularly if growers have been encouraged to plant trees as part of an active extension programme and are looking for a cash return; see Chapter 5. If promises are unfulfilled and there is no market for the wood grown, enthusiasm for planting soon diminishes and what has been achieved is largely wasted. The basic problem usually stems from lack of a clear long-term objective: there is a general encouragement to plant but no definite market or specific end-use (Fig. 6.3). Where a ready market exists, private planting can be very successful, for example pine woodlots in Fiji (Lind and Martel 1986). In Vietnam, the introduction of a free-market economic system in the mid-1980s resulted in farmers having greater access to the market for bamboo poles with a result that prices increased and farmers planted more bamboo (Woods 2003).

It is unsatisfactory to establish plantations in a piecemeal fashion without clear market opportunities but sometimes it is an inevitable stage in development when many uncertainties surround initial plantings. The uncertainties so often arise at the stage when there are insufficient plantations to support an industry and the absence of a ready market discourages further planting. However, with government planning and encouragement a sizeable plantation resource may be created which is viable and will eventually attract industry.

Figure 6.3 Small plantations and woodlots in Papua New Guinea highlands. Such development was encouraged, but initially there was no market for many of the trees grown.

Where there is less of a commercial imperative, as is the case in some social forestry activities, this kind of conflict should rarely arise. People's participation from the start helps ensure that planting only occurs where and when it is wanted and it is now widely recognized that this only comes once there is a need felt. It is no good planting for firewood if poles are really what are needed!

Social and cultural effects of plantation development

So far plantation development has been viewed from the standpoint of the investor, manager, or planner with some mention of direct social responsibilities. However, any development, especially in the rural tropics, which is not related to local needs and not receptive to the attitudes of the community, is failing in one of its most important roles. Moreover, the question has to be faced by the forester, as much as by anyone else, is development really wanted? The easy assumption is that the development brings benefits (better health, nutrition, education, employment, and greater access to goods and services) and is therefore what people want. But, people are rarely fully aware of the consequences of development and there are often both winners and losers. Nevertheless, planners should make themselves fully aware of the social and cultural context of any proposed development through pre-project appraisals and then involve stakeholders to the fullest extent in the planning and implementation. Some of the pre-planning activities are discussed in Chapter 7.

This book cannot adequately deal with this question of development strategy and ethics. Whether development is closer to economic imperialism or helping a deprived community or nation, will depend much on a person's own standpoint and all should read the analysis by Westoby (1989).

Industrial-scale plantations, like many development projects, radically change the environment over large areas of land. The impact on and response by the local community in terms of employment prospects, cash incomes, new ambitions, and also on the cultural and social life cannot be ignored. What is abundantly clear is that a failure to acknowledge, to seek to understand, and to respond sympathetically to the ideals and wishes of the community whose land and life are encroached by development, is not only bad economics but irresponsible and unethical.

These considerations, which in the past were sometimes dealt with summarily, are today receiving more attention. This is evident in the emphases in recent national forest policies, certification schemes, allocation of funds to social forestry and environmental needs, the priority now given to village and community based initiatives, and the overall greater participation of stakeholders in forest planning processes. See, for example, the emphasis placed by Higman *et al.* (1999) in their handbook on sustainable forestry, including plantations.

Social and cultural changes

The social impact of development, such as industrial plantation establishment, on a community depends on many factors. The developer rarely appreciates more than superficially the culture of the local inhabitants, while the inhabitants may not readily comprehend the changes that will take place. In remote areas, the inhabitants may have had little contact with the outside world, and, through fear or force, may be displaced from areas they have used traditionally for hunting and gathering non-timber forest products. Alienation of land claimed by local communities for plantation development does occur. This is often due to lack of land registration, poor definition of boundaries, and complex ownership patterns. Often those affected are politically weak or unable to effectively articulate objections. Furthermore, respect for traditions, customs, and beliefs about land such as sacred groves or spirit homes, are easily overlooked when large tracts of land are acquired. The problem is often one of the perceptions and motivation of local people who can have an entirely different value system from the developer.

Morrison and Bass (1992) identify the impacts of a plantation on local people as depending on:

- its size;
- its boundary in relation to adjacent land-uses;
- rate of establishment;
- purpose of the plantation, especially the species used and their familiarity to the community;
- relative economic, political, and legal power of local people and the developer; and
- degree of lifestyle change the enterprise will bring through employment opportunities and changes in social benefits.

An extreme example of these kinds of problems occurred in the State of Bihar in India. The Forest Corporation came up against great tribal opposition when trying to plant teak on logged-over land previously dominated by *Shorea* spp. Not only was grievance caused when the logging denuded hills and displaced wildlife, but it was not realized that local tribes-people considered *Shorea* so valuable for food and fodder that the trees were worshipped, while teak was considered valueless. Morrison and Bass (1992) quote an example in Thailand where local people have expressed fears that eucalypts give off 'poison gas' that burns the ground and that the leaves cause damage to nearby crops and fish in waterways. Conversely, in parts of Chad villagers are happy to plant exotics, such as *Azadirachta indica*, *Parkinsonia aculeata*, *Prosopis* spp. and eucalypts, to prevent wind erosion and provide firewood, but not the potentially valuable *Faidherbia albida* which they believe is an 'evil' tree.

Another area of social change is in village and family life. Often the traditional division of labour will change and where formerly the wife may have been the major cash earner, by selling the food she has grown or non-timber forest products she has gathered, the husband may now earn wages from the plantation enterprise. Moreover, there can be difficulties in adjusting to the regularity and discipline of full-time employment. For example, one of the main labour problems for the Usutu Pulp Company is the Swazi's preference for leisure rather than high wages. This preference, which is widespread in southern Africa, creates great difficulties in scheduling work. The company has partly overcome the problem by paying an attendance bonus if an employee completes his work task on five successive days and an additional bonus if such performance is maintained for a whole month. More recently a wholesale shift to contracting out almost all forest operations has proved more effective.

A great many other changes come with development, but the above examples highlight some of the problems that have affected plantation projects.

Reducing the impact

Development inevitably causes change and the developer has a responsibility to minimize the impact. Ways of doing this include: (i) consulting with and involving the local people; (ii) employing those who are displaced from the land acquired; (iii) adapting practices to suit local skills and abilities; (iv) phasing development through many small-scale developments; and (v) using technologies not needing highly specialized skills.

Consultation with and participation of all stakeholders must be sought if the impact and disruptive effect of a plantation development project is to be reduced. This principle needs more than lip service and local involvement must be more than just discussion about the project. Chambers *et al.* (1989) noted that government projects in India were on the basis of 'I manage, you participate'. P.T. Finnantara Intiga in Indonesia adopted a more sensitive social and environmental approach as they sought to develop a large plantation resource in West Kalimantan to supply a planned pulpmill (Potter and Lee 1998). The company offered to plant truly degraded lands willingly relinquished by villagers and provide a package of benefits to villagers willing to participate. Those willing to provide land for the project retain control over 25% of the land that is developed at the company's expense. The villager's holding is planted with improved rubber trees, with multipurpose trees to meet subsistence needs and improve biodiversity, and *Acacia* spp. to provide pulpwood for the villagers' benefit. In addition, the company would develop infrastructure and assist in setting up credit and loans organizations to enable improvement in agriculture. This approach appeared to be satisfactory to the local people although at high cost to the company. However, it may be that the requirements of a commercial pulp company are irreconcilable with such environmental and social initiatives (Mayer 1996).

There are a number of approaches, such as outgrower schemes and various agroforestry systems, that can directly involve local people (Chapter 20) and be beneficial to both the grower and the commercial enterprise. A form of involvement, which reduces the impact of a project, is to use methods not greatly different from practices already familiar to local people. Plantation establishment by taungya, shamba, or tumpangsari is a good example since the traditional practices of the shifting cultivator that led to their development do not need drastic change. Shifting cultivation is a widespread land-use in the tropics and therefore is a life-style of many people most of whom have little contact with their government. The very impermanence of the taungya system, that is, cultivating food and growing trees on several pieces of land at once, and clearing a new area and abandoning an old one each year, makes it similar to shifting cultivation in its demands on the person working the system. Not only is the life-style of the shifting cultivator little altered, but there is always the possibility of full-time employment at a later stage. Taungya and outgrower schemes are examples of how a large resource can be built up from numerous small-scale developments. This is not the piecemeal development discussed earlier but orderly phasing of project expansion which will usually have more positive social and environmental impacts than a single large plantation development.

While the introduction of new technologies to improve productivity is desirable there are both advantages and disadvantages of mechanized methods compared to labour-intensive methods. The relative merits will vary according to locality but Table 6.2 lists some advantages of the two approaches.

A means of increasing local participation is to use appropriate technology. In the rural tropics this is development of productive work methods using simple tools and equipment needing skills that can be readily learnt (Fig. 6.4). Appropriate wood-harvesting techniques in plantation forests are particularly important (FAO 1987). Increased output

Figure 6.4 Appropriate technology in forestry: oxen extracting *Araucaria* thinnings in Papua New Guinea.

and reduced physical stress in tree-felling in Pakistan's Changa Manga plantations was achieved by changes from axes to bow-saws and peg tooth to raker tooth saws (Ayaz 1987). Portable sawmills and chainsaw milling which enable trees to be processed on site are used in several countries, for example, Brazil, Peru, Philippines, Papua New Guinea, and these have an important role to play in small-scale forestry (Smorfitt *et al.* 2001).

Responsibilities towards employees

The role of plantation forestry in providing employment, particularly in rural areas where there is often serious unemployment and poverty, is generally an important consideration in assessing developmental value (Letourneau 1987). In many large industrial plantations employment is the commonest and greatest benefit to local people both directly and through employment in local wood processing industries. The extent to which operations are mechanized can influence the number of people employed. Sometimes, the plantation work is organized to maximize the employment value such as using animals to extract timber, for example, mules in Zimbabwe and Swaziland (Heinrich 1987); see Fig. 17.4.

Local people should be given priority in employment opportunities and the provision of goods and services to the project. It may also be necessary to positively discriminate to provide employment opportunities for poor and landless people in the local communities. Silvicultural activities are often seasonal and of short duration, but continuity of work for employees is important, particularly if it is necessary to recruit labour from another region.

Table 6.2 Advantages of mechanized and labour-intensive methods in developing countries[a]

Method	Advantage
Mechanized	Demands less human energy
	Productivity per unit time is high
	Fewer accidents
	Able to handle heavy loads in difficult terrain
Labour-intensive	Employs large numbers of people
	Uses locally available resources, for example, labour
	Do not require foreign currency
	Minimizes soil and tree damage
	Do not need high investment for equipment and training

[a] based on Abeli (2000).

The application of computer-based optimization models to schedule silvicultural operations can assist managers in stabilizing labour force numbers (e.g. Pinto *et al.* 2001).

Employers should consider ergonomic principles such as physical working capacity and workplace conditions and then ensure forest workers are provided with appropriate tools, protective clothing, shelter, and trained in safe work practices. Manual forestry work in the tropics is often strenuous, energy-demanding and may have a significant level of accident risk. Sundberg (1978) reported a 65% fall in productivity from heat stress for men doing heavy forest work in hot and very humid conditions. One of the benefits claimed for the PICOP tree-farm scheme in the Philippines is that the farmer can look after his food crops in the morning and evening and tend his tree plantation in the shade in the middle of the day, see Chapter 5. While some large plantation developments provide health facilities for their staff and dependants, there may be health problems associated with the work. Inadequate training and precautions for workers handling hazardous herbicides and insecticides are not uncommon in the tropics. On-the-job training is essential, especially to ensure the health and safety of workers. These issues must be taken into account during planning and design of plantation operations: they are more fully discussed by Abeli (2000).

It is an employer's responsibility to facilitate the provision of services needed for their employees to have an adequate standard of living. Housing is usually the single greatest development needed. A common system is to establish forest villages (Fig. 6.2), with each village housing workers employed in the forest in the immediate vicinity or who all work at a particular operation, for example, timber extraction. Where the development involves a large factory workforce it is usually necessary to create a complete company township. Not all plantation developments in the tropics need to include housing. Small projects or those in well-populated areas often employ labour living in surrounding villages or towns who walk or are transported to work each day. Provision of fresh water and medical facilities are especially necessary and in many places the developer will need to construct roads, bridges, and other infrastructure. Some of these will impact on the local community and they need to be involved in the planning phase. It is desirable that some formal arrangements be made for community consultation as this should also ensure on-going feedback that will assist developers to meet all their social responsibilities.

In addition to providing physical amenities of housing, hospitals, and schools it may be necessary even to supplement the diet by providing rations (Hurtig 1987). As well as ensuring a balanced diet and better health, this can be an important attraction of the project to local people. Finally, an employer needs to be sensitive to local customs and allow, wherever possible, time for traditional ceremonies. To an outsider these may seem purely superstition, but failure to consider spiritual needs or the forceful breaking of local taboos can do much harm to labour relations. Frequently, local traditions disappear with the coming of a project and some filling of the spiritual void is essential.

The number of full-time equivalent jobs generated by industrial forestry in developing countries is about 2.7 million (Poschen 1997). In Brazil, the Aracruz company has about 4800 employees, of whom 1800 were employed directly and the rest outsourced, for 144 000 ha of plantations and its pulp mill. The company also claims to generate 50 000 indirect jobs (Aracruz 2002). In South Africa, Mondi has 600 direct employees, and an average of 15 000 contracted workers depending on operations in progress, to manage 540 000 of land including 327 000 ha of plantations in 2002 (Mondi 2002). The Jari plantation project (Monte Dourado Forestry Company) in northern Brazil directly or indirectly supports 60 000 people on company property of which 12 000 live in Monte Dourado town and 6000 are direct employees. It is claimed that these people have better living conditions and more professional, economic, and educational opportunities than those in most areas of the Amazon Basin (McNabb *et al.* 1994). In Indonesia, the Riau Andalan Pulp and Paper company has supported community development projects in local villages including building a mosque, providing drinking water, building bridges, and training villagers to profitably cultivate unused land with a budget of US $2 million allocated to these activities in 2000 (Jenkinson 2000).

The value of large-scale plantation enterprises for rural employment has been overstated according to some NGOs, for example, The Nature Conservancy, although the full impact, both directly and indirectly, of these industries on employment and community development is not easy to determine. It is clear that not all plantations have positive social effects through employment. A problem in Asia has been the displacement of rural labourers in more densely settled agricultural areas, for example, in India where absentee landlords have planted eucalypts they have often reduced agricultural employment opportunities for the landless poor. Other social disruptions may result from diverting labour from other local activities, upsetting the prevailing wage structure and withdrawing employment after the establishment phase is completed (Bass 1993). When people are moved within a country to provide labour for plantation development, as has occurred in Indonesia's transmigration programme, there may be serious ethnic conflict with long-term residents of the area. Gregersen *et al*. (1993) also point out that setting up a large, modern industrial plant in a rural area can have negative impacts on the local community, for example, if the employment created is semi-skilled and workers are brought in from urban areas the influx of more highly paid outsiders may place pressures on local services and prices.

It can be concluded that development of plantations and associated industries must be carried out responsibly with sensitivity to their impacts on the welfare of local communities and employees who form part of these communities. Issues will change from country to country and place to place but they will need to be addressed.

The contribution of planted forests to economies and rural development in the tropics

Potential benefits of plantation forestry to the economies of developing countries were listed in Chapter 2. Obtaining these benefits has been elusive (Westoby 1978, 1985), the main difficulty being how to measure the whole impact of plantation development or a social forestry programme. Of course, there may also be social costs, such as above-average accident rate in bush work and water use by forests reducing water tables and river flows.

Direct use values include usable products such as timber, pulpwood, poles, fruit, and tannins and services such as recreation sites. Indirect uses and values are associated with watershed protection, soil protection, mitigation of strong winds, and carbon storage. These environmental benefits need to be viewed beyond their immediate physical and biological impacts to their impacts on human welfare. Direct values of products are relatively easy to assess in monetary terms based on market prices but indirect values are less easy to measure.

The essential role of industrial plantation forests in satisfying material needs is well-recognized and is of increasing importance as the world's wood supplies becomes less dependent on shrinking natural forests. Brazil is the largest producer of eucalypt pulp and its plantations yield annually over 2 million t of bleached pulp out of a total world production of about 5 million t (World Resources Institute 1998) and makes a substantial contribution to the country's export earnings. Another example of direct benefits to a country from plantations is in Swaziland, where afforestation of 6% of the land

surface since the 1940s has created an industry which is currently the third most important component of the country's gross national product, provides the government with much tax revenue, and is a major exporter yielding annually US$80 million from dried woodpulp and sawn timber. Large industrial plantation projects usually provide further benefits through construction of roads, houses, schools and hospitals, and employment opportunities. Other goods and services have also been developed because of the afforestation effort. At Usutu alone it is estimated that over 20 000 men, women, and children (4% of the population) depend on the company's prosperity (Evans and Wright 1988). These 'externalities' are immeasurably important. Small-scale plantation forestry such as outgrowers schemes, can provide more opportunities for farmers to derive income from growing trees for industrial purposes (Chapter 5).

It is now recognized that a major contribution of forestry to development is benefit brought to villagers and farmers as an essential ingredient of rural life and hence the enormous expansion in social forestry programmes. Population growth has led to a greater demand for fuelwood, poles, and other wood products and to meet these needs exploitation of natural forests has been accelerated to unsustainable levels. The scale of wood shortages in many countries is such that farmers and other land holders are increasingly becoming involved in providing wood from new on-farm plantings (FAO 1985*b*). Farmers plant trees for a variety of reasons including (i) shade and shelter; (ii) wood products; (iii) non-wood products such as fruit, fodder, honey, chemicals; and (iv) savings and security. Although fuelwood may be valued as a secondary product, farmers are more interested in trees as a source of construction wood, poles, posts, fruit, etc., especially when these can be sold for cash. Tree-planting for shade, shelter, and the marking of farm boundaries is also important for some farmers. The use of trees as a living bank account, to be harvested when there is a need for cash, is widespread. The rural landscapes in Ethiopia, China, India, Peru, and elsewhere are often dominated by eucalypts established to a very large extent by farmers. Those who have insufficient land to have woodlots nevertheless often grow a few trees which can be harvested and sold when they experience food shortages. Figure 6.5 shows these benefits of rural development forestry. Many of them relate to agriculture, which emphasizes the importance of integrating forestry and agricultural development (Fig. 6.6). Integration of tree-planting with farming (agroforestry) and other land-uses is discussed further in Chapter 20.

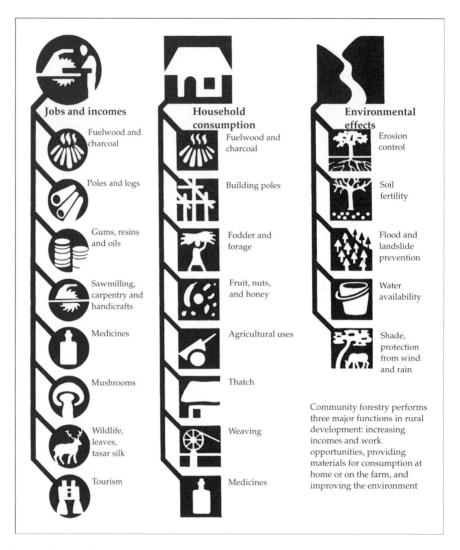

Figure 6.5 Illustration of the benefits of community forestry.

Source: United Nations Food and Agriculture Organisation FAO.

Many forestry activities provide social and environmental services though how to quantify their economic contribution effectively and comprehensively remains a key issue (World Bank 1986). Underestimation of the value of forest goods and services has been one of the key constraints for sustainable forest management (IPF 1997). One problem is that values can be dynamic, for example, economic aspects are usually emphasized in developing countries trying to modernize but social, cultural, environmental, and other non-economic aspects receive more attention as the population becomes more affluent (Simula 1998). Most efforts at valuation seek to express values in economic terms in relation to their impact on human welfare. Kengan (1997) notes that despite the importance of valuation in the decision-making processes it does not ensure that sustainable forest management will be preferred to alternative land-uses. The valuation of forests is discussed in detail by Gregersen *et al.* 1995 and Kengan (1997).

The potential contribution of forestry, including plantation projects, to development is readily apparent, but realizing this potential to improve

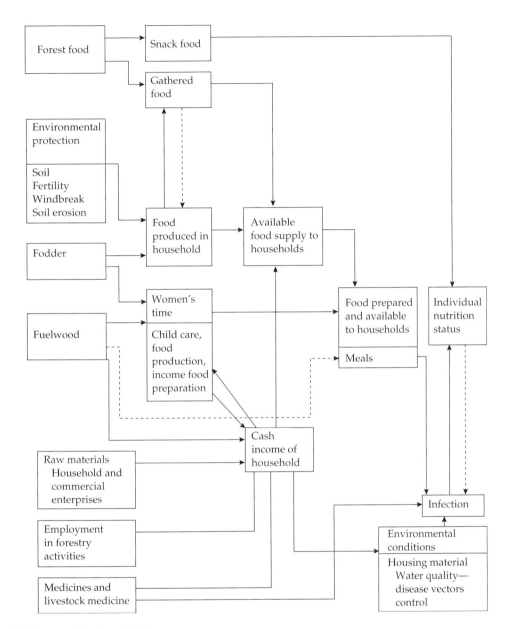

Figure 6.6 Forestry and food security: linkages.
Source: Falconer and Arnold (1988).

people's welfare, even in well-planned and integrated rural developments, depends on actually implementing them as planned. This, and the previous chapter, have indicated the need for and the ways of mobilizing interest and participation at the local level. It is, however, a regrettable fact that the words penned more than 25 years ago by Westoby (1978): 'in many countries true rural development is stifled by the power of landlords, of moneylenders, and of the agents of foreign capital' remain true. Many studies have confirmed that most problems in community forestry programmes are not those of technical production, but issues of distribution and equity: how really to help the poorest of all?

CHAPTER 7

Planning the plantation enterprise

Introduction

Most of the previous two chapters, indeed most of this book, impinges on the subject of planning. Forest planning identifies activities and their timing to achieve the goals of forest management. It is at the foundation of any project, industrial, environmental, or social (rural development) forestry because it is all those processes and steps, which involve data collection, analysis, projections of supply and demand, estimates of equipment, manpower and training needs, and so on. Planning is primarily the responsibility of managers. It introduces order into any development to avoid haphazard and ill-conceived (unplanned) activities. Put simply, planning is the thinking behind the doing of a project to help it succeed.

The influence of planning in policy making, silvicultural practice, inventory, marketing, and other aspects of forestry was demonstrated by Johnston *et al.* (1967) in their classic book *Forest planning*. Moreover, planning has a role at all levels of decision making including international, national, regional, and on-site. Hence, planning cannot be considered fully in one chapter and readers are referred to more specific texts, for example, Johnston *et al.* (1967), Watt (1973), Fraser (1973), Ferguson (1996), and Armitage (1998), for more adequate treatment of planning in forestry, and particularly plantation development.

This chapter outlines major areas of planning at different levels needed in plantation development. It also aims to show that all forestry projects, however large or small, should be planned holistically to avoid narrow, compartmentalized thinking in which economics, social impacts, inventory, harvesting, and especially silviculture are considered in isolation.

Planning can be viewed in two dimensions: (i) vertically as different levels of scale—international, national, regional, on-site, and operational; and (ii) horizontally in relation to time—the progress and implementation of a project, or weekly work programme, or national policy as they develop. This chapter refers briefly to planning at international and national levels but is mostly concerned with planning on-site at project level on a timescale basis. Because scale largely determines the complexity of planning, planning in large-scale projects, such as commercial pulpwood ventures or government-backed afforestation schemes, and in small-scale plantations such as woodlots, are considered separately. Of course, such a division is not clear-cut and planting projects cover the whole spectrum of size. It is also recognized that sometimes many small-scale projects may coalesce to form a single large one—the approach to development advocated in Chapter 6 to reduce the impact of change.

In the past much planning at national and regional levels has not been entirely effective as it has failed to properly take account of the influence of the political system and powerful vested interests. Local planning has also sometimes been ineffective due to failure to fully canvass and address the needs of all stakeholders. New projects provide an opportunity for adequate planning from start to finish with the goal of greater efficiency, orderliness, and sensitivity to people's needs.

Policy planning at international and national levels

International planning

Strong economic development after about 1950 was followed in the 1970s by increasing concern for the environment. The United Nations Conference on Environment and Development (UNCED) met in 1987 to debate sustainability issues and this was followed in 1992 by UNCED's Earth Summit in Rio de Janeiro. Although a convention on forests was not produced at this conference, a non-legally binding authoritative statement of 'Forest Principles' was agreed. Some of the 27 principles set out in the Rio Declaration to guide states and peoples in global environment and development matters relate to forestry, for example, for environmental protection to be an integral part of development. Agenda 21 provided an action plan, and Chapter 11, 'Combating deforestation' has specific references protection and sustainable management of forests, including greening of degraded areas, afforestation and reforestation, and an appeal for countries to formulate and implement national forestry programmes or plans using international frameworks such as the National Forestry Action Plan. The Rio conference also agreed on the terms for three Conventions on biodiversity, climate change, and combating desertification all of which have implications for forest management. To monitor progress and identify problems in the implementation of Agenda 21 the UN Commission on Sustainable Development (CSD) and in 1995 this body recognized the need for an intergovernmental mechanism to focus on forestry issues. The Intergovernmental Panel on Forests (IPF) was established for this purpose. A large number of meetings were convened to feed into the IPF and CSD covering issues such as rehabilitation of degraded forest ecosystems, sources of finance for sustainable forestry and criteria and indicators for sustainable forest management. In 1997 IFP's final report was endorsed by CSD and considered by the United Nations General Assembly. The policy dialogue on forests was continued in an *ad hoc* Intergovernmental Forum on Forests and after 2000 by the United Nations Forum on Forests. This marks a new stage in implementing action proposals for sustainable forest management but political will, appropriate financial mechanisms, fair access to global forest products markets, and the transfer of environmentally sound technologies are needed (Wang 2001). Meanwhile the impetus for a legally binding forest convention has slowed.

Greater awareness of environmental issues has resulted in a number of initiatives to produce standards for the sustainable management of forests. One of the objectives of the International Tropical Timber Organization (ITTO) is to encourage industrial tropical forest reforestation and management and it has produced guidelines for the establishment and sustainable management of planted forests (1993). Globally, major efforts to identify measurable criteria and indicators to assist sustainable forest management and conservation of biological diversity have been made through international initiatives such as the Helsinki and Montreal Processes and research by the Center for International Forestry Research in Indonesia. There has been a movement towards product labelling from forests certified as sustainably managed. Third party assessment use schemes such as that developed by the Forest Stewardship Council or self-regulation using standards compatible with the International Organization for Standardization's ISO 14001 for Environmental Management Systems. An Environmental Management System is a continual cycle of planning, implementing, reviewing, and improving the actions an organization takes to formulate and meet its environmental policy and

objectives. Most of these international policy and planning initiatives are described by Grayson and Maynard (1997), Ruis (2001), Söderlund and Pottinger (2001), and Wang (2001).

Of special relevance are two UN intersessional meetings in Chile (1999) and New Zealand (2003) concerning the role of planted forests globally and where they fit into the overall forestry scene. One clear conclusion is the recognition that plantations will become the dominant wood fibre source in the foreseeable future.

The supply of forest products is heavily dependent on international trade. The European Economic Community and Japan import large quantities of wood products, much of it from the tropics. Identifying major producers and forecasting supplies from them forms an important part of international planning to try to match supply and demand as far as practicable. Planning at this level cannot be directly implemented and one nation cannot insist on the pattern of forest developments in another, nevertheless, international organizations such as FAO, UNCTAD, ITTO, the World Bank, and regional development banks play an important role in collating data, preparing forecasts, advising governments, and often providing financial assistance to encourage projects, which are consistent with the planning goals conceived by these organizations as appropriate in the forestry sector. Thus international pressure on the course of development is brought to bear by the lending priorities of these institutions.

In 1985 FAO, UNDP, World Bank, and the World Resources Institute recognized the need for a more holistic approach to forest development when they co-sponsored the Tropical Forestry Action Plan, subsequently expanded to non-tropical countries and re-designated as the National Forestry Action Programme (NFAP). International organizations, donor agencies, (non-governmental organizations NGOs) participated with national agencies to develop national plans. This mechanism provided a critical link between international and national planning. Unfortunately, with some exceptions, this ambitious forest planning exercise involving more than 100 countries generally failed to match expectations (Contreras-Hermosilla 1997). There were gaps between estimated funding requirements and availability of funds (Chandrasekharan 1997) but other problems also limit the effectiveness of national planning as indicated in the next section.

National and sectoral planning

Planning at national level in most tropical countries is primarily through preparation of development plans, commonly of 5 years' duration. These plans analyze each sector of the economy and national society and project planned developments according to government policies. The importance of any one sector, such as forestry, depends on its contribution to national welfare, and its incorporation into a national plan may vary from almost nothing to being a major instrument of development. National planning must also be made in the context of international agreement to which the country is a signatory. The United Nations' International Tropical Timber Agreement (1984) and the Convention on Biological Diversity (1992) are examples of such agreements that may influence national forest planning.

The importance of national forest policy and the development of effective land-use planning are essential parts of overall national forestry sector planning (see Chapter 5).What is actually included in a development plan usually arises from within-sector policy and planning studies. They include national targets and strategies for reaching them. It is important that the forest sector plans are linked to the national development plans and contribute to the country's overall goals of economic and social development (Ellefson 1991).

As policies change in response to political pressure so do the plans. For example, in 1976 the National Commission on Agriculture in India made a full analysis of forestry, including strategies for development and numerous recommendations on all aspects. One recommendation was that 48 million ha of land should be under forest, to provide for timber needs projected for the year 2000, of which some 15 million ha should be plantations. This major policy document provided the basis for planning forestry developments in the Sixth Plan (1980–85). However, since then a full forest policy revision has taken place and a new National Forest Policy was enacted in 1988. This re-stated the importance of substantially increased forest cover and emphasized the central role of trees and forests in environmental and domestic (social) needs. From this emerged plans for industrial requirements for raw material to be supplied from farm forestry and agroforestry though agreements with industry and for the Joint Forest Management Programme to involve local communities in the management of forests (Bahuguna 2001). In West Africa, Senegal's Forest Action Plan (1993) reflected a change in policy to increase participation by local people in management of natural resources and an integrated management strategy for village lands. It resulted in the establishment of more village and community plantations and relatively few state-owned plantations (Diouf *et al.* 2001). In Australia, a new National Forest Policy Statement (1992) had a goal of ecologically sustainable

management and use of Australia's forests, an adequate and representative reserve system, and long-term stability of forest-based industries. Regional forest agreements achieved after comprehensive regional assessments and community/stakeholder consultations, were seen as the main way of implementing the policy. A further development was a policy document *Plantations for Australia: the 2020 vision* (Plantation 2020 Vision Implementation Committee 1997) envisages a partnership between State and Federal Governments and the private plantation and wood processing industries to triple the size of Australia's plantation estate in tropical and temperate areas by 2020. These examples from India, Senegal, and Australia illustrate major forest policy changes that are being reflected in regional and local planning, and ultimately in plantation development.

An outline of forest planning at national level, with special reference to plantation development, provided by Oseni (1976) remains relevant.

1. *Need for planning*: planning seeks to assess the needs of a country and to set objectives for supplying them.
2. *The national development plan*: most countries have a national development plan whose objectives are the outcome of political processes. The task of planners is to formulate the methods to achieve these objectives.
3. *Planning in the forestry sector*: the planning unit must be organized to be capable of: (a) collecting and analysing data; (b) formulating alternatives for attaining the objectives; (c) evaluating choices available; and (d) frequently laying down guidelines, for example, Harwood (1985) for fuelwood plantations in the Pacific Islands.
4. *Plantation planning within the forestry sector*: possible projects are identified and each analysed in detail. The criteria for selecting the best projects vary widely but may be profitability (variously defined), employment, volume production, or social benefit, etc.
5. *Implications of a national plantation programme*: these need to be evaluated in the national planning process since effects are often felt outside the realm of forestry. Plantation development can be used by government to raise rural standards of living by providing employment, and by encouraging agroforestry to improve self-sufficiency of the rural population. It is necessary for national planners to take account of the interactions and interdependences of forestry.

National forest planning cannot be static and some recent important advances have been listed by Contreras-Hermosilla (1997):

1. Social, economic, and environmental dimensions of planning are now more effectively incorporated into formal models for forest management and

attention is given to impacts beyond the life of a particular project or sector plan.
2. Incorporation of environmental aspects, including the use of resource accounting methods.
3. More intense consultation with stakeholder groups, with greater analysis of political power and attention to indigenous knowledge and values.
4. Greater attention to macro and structural adjustment policies, trade and developments in related sectors, such as agriculture.
5. Development of new institutional structures to create mechanisms for joint implementation, independent certification, etc.

Despite these advances, the outcomes of planning have not always been satisfactory due to inadequate attention to the constraints of decision makers and through failure to predict the implementation capacity of the country. Inadequate recognition of corruption can also be an important factor in explaining unsatisfactory outcomes from planning processes (Contreras-Hermosilla 1997). Planning also often proceeds on the basis of incomplete knowledge or poor predictions of market factors. A series of six-country studies (Costa Rica, Ghana, India, Pakistan, Papua New Guinea (PNG), and Zimbabwe) and an overview paper describe the processes that make and manage good policies, and contribute to a greater understanding of the forces at work, the winners and losers, and factors that affect national policy outcomes (e.g. Ahmed and Mahmood 1998; Mayers and Bass 2000; Watson *et al*. 2000).

At national level, forest plans must be linked to plans of other sectors and there should be a clear linkage between local and regional plans and the national forest and national development plans that address broader issues of economic and social development.

Regional planning

In large countries, planning by regions is important because of considerable political, social, economic, and environmental differences that exist. In Brazil, the northeast region is much poorer than the south, it is densely populated but too arid to support a high level of subsistence farming, and consequently receives special fiscal support to encourage development including plantation forestry. Australia's Regional Forestry Agreements reflect this diversity within the country. Even in smaller countries, different regions merit different development strategies owing to differences in climate, geography, past land-use, local culture, tribal structure, etc. Planning at regional level seeks to implement national objectives but methods and practices

will often differ between regions, for example, species' choice and silviculture. The incorporation of forest uses other than timber production, such as conservation, recreation, and water management, may be better quantified at the regional level and incorporated into regional planning models (Ferguson 1996). It is at this level that public participation in the planning process may be most appropriate and effective.

Planning in large projects

Different approaches to afforestation were discussed in Chapters 5 and 6. The choice of who will implement a project usually reflects a mixture of national economic policy, local needs, and the nature of the project. The main alternatives are private companies, national afforestation authorities and commissions, regional agencies or corporations, and non-government organizations usually involving local people. The following discussion is focused on developing large industrial plantations by the private sector or government agencies.

It is usual for project identification, pre-feasibility, and feasibility studies to precede and form a basis for plantation design and preparation of a detailed management plan.

Project identification

A large plantation development should be consistent with national policies on land-use, forests, people and the environment. Its prime purpose should be decided from the outset and be consistent with the organization's strategic plan. The primary function of strategic planning is to analyse broader issues, concerns, opportunities, allocate lands, adopt standards and guidelines, establish production levels, and identify environmental effects (Barber *et al*. 1996). Typically strategic plans deal with planning horizons of two or more rotation lengths. A clear objective aids clear thinking and permits more orderly medium- and short-term tactical planning to implement the strategic plan. Of course, several subsidiary objectives may be set for a project some of which may in part be incompatible such as (i) to maximize yield of woodchips; (ii) to protect biodiversity and other conservation values. This common situation of partial conflict is resolved by setting priorities, which are largely a matter of policy. This may be achieved by the designation of specific areas with different management priorities or reaching a compromise on practices for a specified area.

Examples of integrated forest management to meet several objectives were described in Chapter 5 and the question of whether the hierarchies of strategic and tactical planning can be collapsed into a single planning process has been discussed by Sessions and Bettinger (2001).

When the objective has been decided the next step is to select locations where the plantation will potentially be successful and sustainable. These must be sites where the trees will grow well and where political, environmental, financial, socio-economic, and marketing factors are favourable. A project proposal can then be developed based on this preliminary appraisal.

Project appraisal

When a project proposal has been developed, a detailed analysis of all relevant information is undertaken before investment is made. The project needs to be carefully formulated (thoroughly planned) and it is important that issues are addressed logically. The FAO training guide for agricultural and rural investment projects identifies five main steps: preparation for project formulation, reconnaissance and preliminary design, project design, analysis of expected results, and writing the project document (FAO 1986b). Very useful guidelines have been developed by Shell International/World Wide Fund for Nature on planning plantation projects in a series of publications (e.g. Bass 1993; Evans and Hibberd 1993; Howlett 1993) and we have drawn heavily on these in considering two key aspects of the appraisal, that is, is the project feasible, and will it be sustainable in terms of its social, environmental and economic impacts?

Feasibility studies
A feasibility study investigates and, where possible, gathers data about all aspects of a project. This should be an iterative process often with a pre-feasibility study making assessments of one or more potential sites and recommending whether planning should proceed to the next stage. For example, the pre-feasibility study in a proposed pulpwood project may examine the potential species capable of producing the desired wood quality on the sites available and produce a short list. This list will then be reviewed during the feasibility study on the basis of costs and benefits and many other factors. During the pre-feasibility phase a multidisciplinary team will examine maps, reports, remote sensing data, conduct preliminary surveys, and consult with local authorities and communities covering a range of issues including the following (based on Howlett 1993):

1. *Land tenure*: Identify different types of land tenure and any constraints on plantation development.
2. *Land-use*: Identify major land uses and users and any degraded, unused or under-utilized lands in the area.
3. *Land cover*: Determine extent of primary or secondary forest, which may be protected for environmental purposes.
4. *Physical features*: Identify areas where climate, soils, and topography favour plantation development and examine potential effects of plantings on water resources in the area.
5. *Ecology*: Identify protected areas and sensitive ecosystems and endangered flora or fauna that will constrain or require careful development.
6. *Socio-economics*: Identify cultural and historic sites, determine community attitudes to plantation development, and assess labour availability.
7. *Legislation and policies*: Identify all local land-use plans and development policies and assess if there are any conflicts.
8. *Infrastructure*: Identify existing infrastructure and determine whether plantation infrastructure development will have negative environmental or other impacts.

If the recommendation of the pre-feasibility study is to proceed to the next stage then more detailed feasibility studies and baseline surveys can be made. These detailed studies will form the basis for final selection of the plantation area, species to be used, project design, plantation management plans, impact statements, and design of monitoring systems and audits.

The feasibility studies cover similar issues to the pre-feasibility study but in greater detail, for example, there will be soil, topographic vegetation and hydrological surveys and mapping of the actual plantation area instead of using pre-existing data. This can be used to make final selection of the species for planting and for planning other management operations. Accurate socio-economic surveys will be made with participation of local communities. Infrastructure surveys of roads, bridges, etc. will be made and preliminary design and layout of new infrastructure undertaken. Environmental and social impact assessments of the new development will be made. A project justification is also prepared, this is usually an economic analysis but if profitability is not the only criterion factors such as employment value, environmental protection, import saving, and social benefits may be included.

Moving to this level of detailed assessment and planning without the preliminary stages risks wasting time and incurring unnecessary costs should major constraints to plantation development be identified.

The feasibility study gauges the likely success of the project itself; an impact study examines all the other consequences and effects it may have. Positive financial predictions of plantation development are critical and computer models are increasingly used to examine the profitability of forestry plantation investments. An example is the Forest Plantation Profitability Model developed in 1999 by FAO and freely available (Whiteman 2002). The project may directly or indirectly affect the environment, community life, health and welfare, etc., and they all should be evaluated. The impact study will fulfill two purposes: first, to indicate all the effects of a project, good and bad, and, secondly, to provide base data against which the actual effects of the project can be monitored once it begins. So important is this aspect of project development that several countries have enacted forest laws which make environmental impact assessment (EIA) an explicit part, for example, Indonesia, Malaysia, and PNG, and some insist on an integrated, environmentally sound approach to forest soils and water. The importance of evaluating social consequences was noted in Chapter 6.

An impact study attempts to forecast all changes a project will cause and how these will affect the community, the environment, and the region as a whole. Inevitably this will include some negative aspects, for example, pollution from a mill, loss of soil from erosion, destruction of ecosystems, disruption of village traditions, and an attempt is made to evaluate this 'damage'. Issues and strategies in assessing forestry project impacts, especially on people, have been discussed by Gregersen *et al.* (1993) and Bass (1993); and essential features of baseline surveys and environmental audits are given by Howlett (1993).

Together, the feasibility studies and the impact studies, form the appraisal which all projects should undergo during formulation. Sometimes this appraisal also includes recommendations on how to proceed. Often feasibility and impact studies and recommendations are reported together in one document. They provide the basis for an informed decision on whether to implement a project and are the basis for subsequently evaluating performance.

It will be evident from the above that most of the work in a feasibility study is gathering appropriate data. Particularly important in plantation projects are reliable data about growth rates of the likely main species, communications (roads and transport), land tenure, availability of labour, and local participation and support, and an indication of political stability if outside commercial interests are involved. However, all the work involved, the problem of inadequate data and the pressures to deliver certain results can make feasibility studies little more than exercises in wishful thinking (Kilander 1985). Adopting a properly planned and structured approach to project formulation will help guard against this.

Implementing the project

If, after appraisal, it is decided to proceed with a project at a particular location, the next stage is to plan how best to carry it out. This involves (i) design of the project; (ii) preparation of a management plan; and (iii) preparation of impact statements and design of environmental monitoring systems and audits.

Most of the decisions to be taken will be influenced by the findings and recommendations during the formulation stage and in the appraisal. If, for example, a feasibility study concludes that *Pinus caribaea* is the most suitable species for afforestation to produce long-fibre pulp, then planting will at least commence with this species. The appraisal will have proposed the main strategies for the project; in implementing the programme short- and medium-term tactical plans are made and decisions have to be taken.

Accurate and up-to-date information is essential for strategic planning and for shorter term tactical planning. While traditional information sources such as maps, surveys, and inventories remain essential for planning they can be complemented newer information technologies. Aerial photography has been a tool for may years but development orthophotographs, which correct displacements due to camera tilt and topographic relief, can be assembled into very accurate photographic maps. A Geographic Information System (GIS) is a powerful tool for general planning and a Global Positioning System (GPS) has many applications. Operational management electronic data recorders, harvesting machines with on-board computers, and vehicle-mounted GPS can provide timely information for very short-term planning. The application of these technologies in forest management planning has been reviewed by (Dykstra 1997). All these technologies require capital investment, maintenance, and operator training so managers must review their value in providing accurate information against their costs.

With the development and widespread use of microcomputers, it has become easier to use computer software as a decision support tool to examine alternative silvicultural scenarios in terms of financial costs and benefits in advance of implementation.

These software tools can assist in determination of optimum rotation length, thinning schedules, and other management practices. Examples are a Forest Plantation Evaluation model to assist managers in Malaysia to appraise operations and evaluate management options (Leng 1994); decision support systems used in Queensland to assist tropical pine plantation management (Catchpole and Nester 2002) and an integrated plantation management and yield regulation system (PLYRS) used in pine plantations in southern Australia to assist management planning and operational control (Strandgard *et al.* 2002). The use of computers to facilitate a systems approach in the planning and control of forest operations and further examples of the use of modelling (simulation, optimization etc., are described in many publications, for example, Pritchard (1989), Vanclay (1994), and Finn (2002). It must be emphasized that computer models are only aids to decision-making, they are not decision makers and they have limitations.

In the design of plantations and preparation of the forest management plan due attention must be paid to existing guidelines for sustainable management, such as the *ITTO Guidelines for the Establishment of Planted Tropical Forests* (ITTO 1993) and national Codes of Forest Practice (Turnbull and Vanclay 1999). These are sets of regulations or guidelines developed by governments or other organizations to assist forest managers select appropriate practices to follow when carrying out forest management and harvesting operations. The codes assume technically and economically feasible practices that can be identified to conduct operations in a manner consistent with the needs of sustainability, for example, for harvesting and road design (Dykstra and Heinrich 1996b; Applegate 2002).

The forest management plan
All the important factors concerned with implementing a project are usually brought together in the first forest management plan. This document is similar to a forest working plan, and usually consists of three parts: policy and objectives, basic information, and present state and future management. A forest management plan is a well-coordinated operational programme for a forest and for regulating forestry activities for a set period of time through the application of prescriptions that specify targets, action, and control arrangements. It regulates planting, protection, inventory, yield determination, harvesting, monitoring, and other silvicultural operations (Armitage 1998). It sets out management objectives and specifies actions and communicates these to those responsible for its implementation. Chapman and Allan (1978), FAO

(1974, 1986b), and Fraser (1973, 1976) provide outlines of plantation management planning that are still relevant.

The goal and specific management objective(s) for each forest management unit are essential features of a management plan. Where, for example, wildlife conservation has the highest priority in a management unit, decisions on choice of species, use of herbicides and pesticides, and other silvicultural operations need to be compatible with this objective. Armitage (1998) defines objectives as 'measurable activities, or outputs, which state specific results to be achieved during a specified period of time'. He suggests each objective should be clear about:

• What activities will be undertaken?
• Where they will occur?
• Who is responsible for the action?
• When the action should be taken?
• When the action should be completed?
• Specific quantitative statements concerning outputs.
• Why an activity will contribute to achievement of an objective?

The forest management plan is required to provide continuity in management, to formalize administrative arrangements, and to provide the basis for monitoring operations. It is an essential tool in sustainable forest management. Through the plan, a dynamic approach to plantation management can be achieved integrating the many facets of operations, social needs, protection, and other requirements. It is important that there is feed back from the field and that the plan is flexible, responsive to change, and regularly reviewed (Evans and Hibberd 1993).

Some important factors in implementing large plantation projects are discussed below.

Supply of resources
To implement a plantation project effectively and in an orderly way full provision, with appropriate agreements and firm commitments, must be reached in several key areas: (i) funding for several years; (ii) land acquisition, site survey, and site preparation well in advance; (iii) recruitment and training of the work-force at all levels, professional, artisan, and unskilled labour; (iv) supply of equipment and machinery along with a full stock of spares and adequate servicing arrangements; (v) purchase of seed and development of nurseries 1–2 years before planting begins; (iv) construction of accommodation, schools, health centres etc., before labour is recruited; (vii) provision of adequate communications,

notably roads, telephones/radios, purchase of vehicles for supervisors, etc. Full stakeholder participation in all this planning is essential.

The important planning activity of preparing a budget will necessarily bring out these factors. Also, the above list highlights the roles of long- and short-term planning when, for example, housing developments are planned to expand around a nucleus over many years, but seed purchase only sufficient for 2 or 3 years planting. The supply of resources for a project must be coordinated, with all needs being satisfied, since shortfall in any one can jeopardize progress of the whole. This again is aided by preparing a forecast of expenditure on all items (the budget) over 1 year, or for longer periods for major capital investments.

Silviculture to achieve project objectives

Silvicultural decisions of what to plant, whether to thin, what spacing to use, and so on are determined by: (i) the purpose of growing trees; (ii) the degree of mechanization planned and the importance of labour-intensive working; (iii) site fertility and the need to cultivate and fertilize before planting; (iv) the tree species itself, seed availability, and growth characteristics.

The silviculture applied to plantations to produce high quality sawtimber, such as for *Cupressus lusitanica* and *Pinus patula* in Kenya, will be different from that for *Eucalyptus grandis* plantations in Brazil intended for pulp or charcoal production. The former require regular thinning, high pruning, long rotations, and regeneration by planting or seeding, the latter will be neither thinned nor pruned, will be grown on a short rotation, and usually regenerated by coppicing. These different silvicultural treatments have different planning needs both in the type of development and the time-scale. The use of decision support computer software to assist managers in examining alternative silvicultural operations has already been mentioned.

The appraisal stage usually indicates species, probable growth rates, and possible rotations, but in implementing a project more information is needed on optimum spacing, provenance-site interactions, fertilizer responses, mensurational data on yields and stand assortments, effects of pruning, stump mortality in coppicing, and so on. At no point can there ever be full information on every point, nevertheless neglect of adequate planning is a major factor in plantation failure and decisions must be made in the light of what is known. It is an important part of a project to have continuing assessment and feed-back of information from current operations and to lay down trials and carry out research to examine aspects of proposed silviculture where there are uncertainties. For example, comprehensive trials of species, spacing, cultivation, and fertilization have been established in the Maryborough-Gladstone region of Queensland to provide the basis for a major hardwood plantation development to produce pulpwood (Dickinson *et al.* 2001).

Three examples clearly illustrate the benefit of such research. In much of southern and central Brazil introduction of *E. grandis*, following trials of new provenances, and more recently clonal propagation, led to growth improvements of 100–300% compared with the previously widely planted species (Chapter 8). Second, evaluation of *Acacia mangium* provenances has shown PNG to be the best source of seed resulting in productivity increases of 70–80% in Indonesia (Chapter 8). Third, fertilizer trials in many countries have shown very significant productivity increases although new eucalypt plantations in Brazil often will not respond to N fertilization mainly because sufficient N is available from mineralization of organic material in the top soil (Chapter 14). However, research takes time and in the early stages of a project the manager must depend on other sources of information: (i) research plots already established in the area, often by the national forest service; (ii) experience with the species elsewhere on similar sites; (iii) published silvicultural information, national, and international.

Harvest planning

It is in the early stages of plantation planning that decisions about harvesting must be made. Too often harvesting planning is given too little attention before plantations are established. Harvesting systems must be silviculturally acceptable, economically feasible, and physically possible, requirements that may be difficult to satisfy in wet or mountainous terrain, or if the product value is low (Aulerich 1991). He also suggests that the planning harvesting at an early stage will contribute greatly to efficient harvesting and to the overall social and economic performance of the plantation by:

1. Identifying areas that should not be planted (from a harvesting perspective);
2. Identifying system and equipment alternatives; determining operation compromises between establishment and harvesting operations;
3. Projecting realistic demands on manpower and machines;
4. Identifying the required levels of expertise and manpower training needs;
5. Assuring adequate raw material supply at the processing plant.

Rate of planting and labour planning

Although the rate at which planting proceeds is controlled by such factors as nursery production capability, land acquisition, finance, etc., the rate of planting itself dictates the timing and intensity of almost every subsequent operation. Thus, rate of planting, that is, the size and development of the annual planting programme, is a powerful tool in labour planning.

The amount of planting each year can be manipulated to provide continuity of employment of labour, or to help balance the total work load in a forest over a year. If plantations could be established quickly and easily, like a farmer sowing a field of maize, it would seem sensible to plant all available land at once, since there is no silvicultural reason for deliberately spreading planting over several years just to create a wide age range, and land lying unplanted is not growing trees. However, afforestation requires much investment in land, labour, and machinery, and an enormous amount of money and a vast production of seedlings would be needed to do all the planting in 1 year. This would be difficult to finance and in succeeding years much of the labour force and machinery would not be needed. This approach, even if feasible, would be irresponsible. The usual development of a project is for planting to increase steadily each year, as resources and production capability expand, until an optimum rate is reached and then maintained (Fig. 4.2). This not only applies to industrial projects but all large projects executed over a number of years. The Bilate community forestry project, in southern Ethiopia (Fig. 7.1), aimed at soil erosion control and pole, firewood, and fodder supply, has reforested more than 800 ha over 10 years, beginning with 50 ha in the first year and peaking at 140 ha.

Sometimes within a project, small areas of land are acquired during the development phase or there are outlying parcels of ground. Normally it is desirable to plant these immediately. Delay is only justified if additional areas of land are expected in the vicinity. Delay in planting small parcels of unused land results in: (i) loss of profit; (ii) increased risk of squatters and trespass; (iii) ground appearing unmanaged which is a poor advertisement for the project.

It was stated above that rate of planting is an important tool in labour planning. Only where labour can be recruited and dismissed freely may this become less important, but this would greatly add to problems of training and supervision. Normally a company or national forest service employs a semipermanent labour force, or operates a carefully controlled system of contractors, where rapid changes in size are both impracticable and undesirable for good labour relations and all the ancillary factors of housing, supervision, training, etc. For a new project it is

Figure 7.1 Part of Bilate community forestry project in Ethiopia showing good tree growth and revegetation to the south of the river after just 4 years. Its success was helped by phasing the rate of planting.

desirable to expand the labour force steadily and then maintain it relatively constant. It may be supplemented by casual labour at particular times of the year. Below is a worked example for development of a 50 000 ha pulpwood afforestation project.

Assumptions

1. *Management*
   (a) Normal rotation of 15 years.
   (b) Average annual yield (mean annual increment) 20 m^3 ha^{-1} year^{-1}.
   (c) Two thinnings: (i) 7 years a 1-in-3 line thinning yielding 25 m^3 ha^{-1}; (ii) 11 years a 50% mechanical thinning yielding 100 m^3 ha^{-1}.
   (d) Clear-felling yield is 175 m^3 ha^{-1} at 15 years.
   (e) Tending required in years 1 and 2.
   (f) Protection against fires and grazing trespass required for first 5 years.
   (g) Low pruning at 5 years, then access road development for first thinning.

2. *Marketing*: Pulpmill size is initially capable of producing 150 000 t of dry pulp year^{-1}, but with capacity to expand. Annual wood requirement is a minimum of 850 000 m^3.

3. *Special factors*: Mill operation is scheduled to start 12 years after commencement of the project. Some of the earliest plantings receive no second thinning but are felled prematurely at 12 years to increase output for start-up of mill: yield of these 220 m^3 ha^{-1}.

4. *Labour productivity*: For simplicity all operations have been shown as the number of hectares one person could theoretically cover in 1 year assuming 240 working days.

  (a) Establishment (nursery, site
      preparation, planting).   10
  (b) Tending (weeding, fertilizing, etc.).   40
  (c) Protection.   400
  (d) Brashing and roading (labour
      component of construction).   10
  (e) Thinning.   15
  (f) Clear-felling, extraction, and haulage.   5

Labour productivity is assumed to rise for all operations at 3% per year for the first 5 years, 2% for the next 5 years, and 1% per year thereafter.

The areas worked and the build-up in the labour force are shown in Table 7.1. This simple model is for illustration only but it highlights two common problems. (i) What should be done with the produce from early thinnings before the mill is in operation? (ii) How can labour requirements be scheduled to reduce fluctuations in labour force numbers?

On-going management

Management has two main aims as a plantation project becomes established . First, to ensure that planned programmes and operations proceed smoothly. Second, the complementary need to maintain sufficient flexibility with a programme or operation so that changes can be introduced in response to unexpected events. For example, an annual planting programme of 1000 ha may, despite all usual precautions, be suddenly held up by lack of plants owing to torrential rain washing out a road or low-lying nursery. This will confront the manager with at least three problems.

1. What to do with the men originally scheduled to do the planting?

Table 7.1 Model of the build-up in labour force in an afforestation project

Operation	Year														
	1	2	3	4	5	6	7	8	9	10	11	12	13	14	15
Planting etc.															
Area (ha)	2000	4000	6000	8000	8000	6000	4000	3000	2000	3000	4000		2000	2000	3000
Labour	200	400	600	800	800	600	400	300	200	300	400		200	200	300
Tending															
Area (ha)	2000	6000	10 000	14 000	16 000	14 000	10 000	8000	5000	5000	7000	4000	2000	4000	5000
Labour	50	150	250	350	400	350	250	200	125	125	175	100	50	100	125
Protection															
Area (ha)	2000	6000	12 000	20 000	28 000	32 000	29 000	23 000	18 000	16 000	12 000	11 000	11 000	11 000	
Labour	5	15	30	50	70	80	72	58	45	40	30	28	28	28	
Pruning															
Area (ha)				2000	4000	6000	8000	8000	6000	4000	3000	2000	3000	4000	
Labour				200	400	600	800	800	600	400	300	200	300	400	
Thinning															
Area (ha)							2000	4000	6000	8000	8000	7000	7000	8000	8000
Labour							133	267	400	533	533	467	467	533	533
Clear-felling etc.															
Area (ha)												2000	3000	3000	3000
Labour												400	600	600	600
Total employees	255	565	880	1200	1470	1430	1463	1639	1583	1603	1548	1297	1545	1761	1968
Productivity factor		0.97	0.94	0.91	0.88	0.86	0.84	0.82	0. 80	0.78	0.77	0.76	0.75	0.74	0.73
Employees required	255	548	827	1092	1294	1229	1228	1344	1266	1250	1192	986	1158	1303	1450

Table 7.2 Lines of responsibility within Section I, Usutu forest, Swaziland

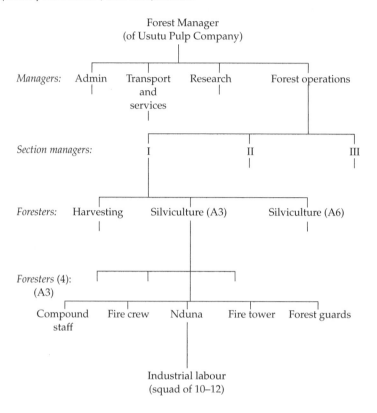

2. Whether to try and obtain plants from elsewhere to complete the programme?
3. Whether to repair the damaged nursery or abandon it and establish a new one?

Surplus labour from 1 could be used for 3 during the time it takes to make enquiries, place orders, and receive fresh consignments of seedlings in 2. Inevitably the year's programme will need modifying; deciding the most suitable changes is one of the manager's tasks and an important part of on-going management.

Staff relations
Management is first and foremost a matter of working with people. Technical competence and administrative ability are important, but good staff and industrial relations are the basis of successful ongoing management. It can be helpful to view staff as key stakeholders.

Management of a project involves a hierarchy of responsibilities and levels of supervision. Table 7.2 shows the lines of responsibility for silvicultural work: planting, nurseries, tending brashing, etc. in a forest section in the Usutu forest, Swaziland that continued until the late 1990s when the company switched to mainly contract work.

Smooth running of an organization depends on good relations between these different levels of supervision. This is achieved primarily by ensuring proper communication so that staff know what is going on and what is required of them. It is facilitated by: (i) clear job definitions stating what is expected; (ii) regular meetings and discussions between different staff levels; (iii) staff knowing why a job needs doing or is to be done in a certain way, and providing the necessary training; (iv) having a sensitive mechanism whereby grievances can be aired without recrimination or penalty.

This subject is elaborated because often plantation projects in the tropics are remote and this remoteness brings pressures which frequently are manifested as dissatisfaction with the job. The staff are stakeholders and their full involvement brings better understanding of the situation, greater job satisfaction, and does much to alleviate problems.

Work programming

An important tool of management is to set targets to be achieved on a daily, weekly, monthly, and annual basis. Programming work is a management responsibility and the longer the time-span involved the more senior the manager concerned. A nursery foreman sets daily work tasks, for example, for tube filling, while a general manager lays down an annual planting programme in line with project development over a 5-year period. Work programming invariably forms part of budgeting and usually the approved budget becomes the standard against which to measure progress.

However, for programmes to be set, data must be available about areas, volumes, yields, terrain, labour supply, equipment capability, machinery output, etc. Gathering these data and keeping them up-to-date is another component of management. Forest inventory (Chapter 4) is the obvious example of a system of data generation, which allows predictions to be made and programmes to be set. But, any system of weekly, monthly, and annual reporting of progress provides the information for future planning as well as a means of direct management control of operations.

Where data are missing or are inadequate, or new possibilities emerge, formal research may be required. Usually the research programmes of a project aim to improve some specific aspect, such as timber yield or harvesting method, rather than finding out essential information needed for project implementation.

A complementary management activity to data gathering and research, is to satisfy the ongoing needs of an operation as it proceeds. Anticipating the need to recruit labour, to purchase equipment and machinery, to implement programmes are all an essential part of management. Even maintaining the optimum level of stocking of spare parts for a project's plant and machinery may, in financial terms, be as important as determining the optimum initial tree spacing at planting. The point is that forest plantation management involves much more than growing trees.

Monitoring

A forest plantation manager needs a regular flow of information on the performance and effects of operations being undertaken to effectively control the implementation of a plantation project. Monitoring forms the basis of transparent accountability of operational activities. It also allows control of current management activities and evaluation of the plantation's performance (Armitage 1998). Only by monitoring the impact of harvesting and other silvicultural operations can actions be taken to improve practices and satisfy the management objectives. Comparisons between actual achievements and planned targets, and between expenditures and budgets need to be made at specified, regular intervals. Continuous monitoring of some key activities may be necessary, for example, the rate of planting so nursery and transport operations can be optimized and as required for certification purposes.

The extent of to which monitoring is needed will vary from project to project depending on its location, size, and objectives. Generally it will include attention to whether operations are proceeding according to schedule and within the budget allocated to them, and information on plantation growth trends, potential pests and disease problems, undue site disturbance etc. Information relating to the social acceptability of plantation activities and their broader environmental effects can be important to the manager's decision-making as well as information from biological and economic monitoring. Of course the extent and detail of monitoring must be kept within budget limits and concentrated only on relevant activities but it must be accurate and reliable. The precision of information required on which to base detailed decisions largely on the questions posed and amount of time and money the manager is prepared to spend to get it. Adlard (1993) and Gregersen *et al.* (1993) provide useful guidelines and recommendations for monitoring the effects of tree plantations on their physical and social environments.

Planning is concerned with the orderly and efficient use of resources in appraising, implementing, and managing a project. Thinking through problems and developments before they occur is at the heart of planning. A plan, incorporating the thinking behind the doing of a project, provides the framework into which all else, including plantation silviculture, fits. But planning is also more than this, for planned work, with the estimates and forecast made, provides a basis against which to gauge the success of what is actually achieved. Moreover, results achieved are then incorporated in the next phase of planning, thus improving the accuracy of the next forecast. It is a process of continuous assessment and planning revision.

Planning for small-scale tree-planting

Small-scale forestry differs from country to country but typically comprises a single or a number of planting blocks, non-professional management, often a lack of silvicultural skill and little planning for future marketing (Herbohn 2001). It is not a scaled-down version of large-scale industrial plantation forestry but a distinct enterprise with a range of objectives, species, plantation layout, management, harvesting, and marketing (Harrison and Herbohn 2001). It is often tree growing on farms in the form of small woodlots, windbreaks, shelterbelts, and a range of agroforestry systems. There are many kinds of such developments: (i) village woodlots for firewood and poles; (ii) trial plantings to evaluate species on different sites; (iii) plantations arising from extension work and rural development forestry programmes; (iv) farm plantings for shelter, fodder, windbreaks, local wood production, or amenity; (v) scattered plantings arising from national or regional campaigns; (vi) use of multi-purpose trees in rural development; (vii) much tree-planting done for protection is only small-scale, to contain localized or incipient problems. In recent years there has been more support by governments and aid agencies for small-scale forestry than for large industrial plantation projects because of its potential to improve the welfare of poor farmers.

These small-scale plantation enterprises, and every tropical country has some of them, may have several features in common which influence their planning and management.

1. Lack off economies of scale can mean higher establishment and maintenance costs but the importance of costing and economics is usually much less because most of the work of site preparation, planting, and tending is done in the grower's own time or as a community project where specific payment for work is not made or is used only as a general incentive (see Gregersen 1984). This means that such developments often require little cash and there is limited financial risk involved.
2. Estimation of financial returns is more difficult because more than one species may be planted, silvicultural methods vary, and predicting accurate yields is not easy.
3. Although government may support small-scale forestry it is often through provision of planting stock, extension services etc. with less assistance in the harvesting and marketing phase. Some regulations on harvesting and marketing, and local taxation on sales may reduce return on the grower's investment.

These factors show that for a single small plantation the planning phases of project identification and formulation, appraisal, implementation, and on-going management are quite different from those required for large-scale industrial plantations. The planning process will generally be much less formal. Of course, the individual villager or farmer will have considered each of these aspects in relation to the availability of credit, allocation of the family's labour resources, other sources of income, access to markets, assessment of risks etc. Essential

conditions for small-scale forestry by farmers to be successful were discussed in Chapter 5.

Where small-scale plantations form part of a larger, perhaps community-based, project or where an external funding agency, including NGOs, is involved, consideration (planning) must be given to layout of tree planting, location, species' choice, scheduling of work, etc., to the point of preparing an informal plan. Indeed NGOs, just like more commercially orientated funding agencies, carefully appraise projects to ensure money is not wasted and sensible development takes place. Even though there is often no profit motive of their own, there is responsibility towards those who give to them to disburse money carefully as well as not disrupting the local economy by ill-considered or misguided funding. Thus all agencies seeking to encourage tree-planting schemes of this type, require planning no less than in a large plantation project.

In Chapter 5 extension methods and national planting programmes were discussed and it was noted that their success or failure depends very largely on the planning undertaken. There are a number of fundamental social, economic, and institutional requirements for all projects aimed at encouraging tree-planting by rural communities, which vary in relative importance according to the country and local conditions. They include:

1. Participation and involvement of local people including women, local leaders, and other stakeholders.
2. Adequate sociological and economic information about the rural communities to be involved, gained by participatory rural appraisal and other techniques, so that the programme or project will be both acceptable and realistic and thus contribute positively to living standards.
3. Project objectives that meet the expressed wishes and needs of the community and reasonable expectations of villagers.
4. A short period of time between the proposal being put forward and its implementation.
5. Integrated land-use planning so that all needs are met—food, fuel, fodder, posts and poles, timber, shade, erosion control, etc.
6. Security of land and tree tenure and clear rights to harvesting and marketing the trees.
7. Reliable market information for the products of the project.
8. Commitment and financial provision by the funding agency, or availability of credit, over a long period as immediate results cannot be expected.
9. Provision of technical expertise.
10. Institutions capable of carrying out programme requirements, including availability or training of suitably qualified personnel.

11. Adequate facilities to meet requirements, for example, nurseries.

Four important aspects of planning for small-scale plantations are: (i) involvement of all concerned; (ii) promoting awareness of trees and forests in the environment; (iii) availability of seedlings; and (iv) provision of advice and help.

Involvement of all concerned—the stakeholders

Essential to any community, rural development or social forestry programme is full involvement of all concerned. This includes the people on whose land the trees will be established, those who may do the work, for example, women (especially in Africa), village leaders, local and regional administrators, interested and participating NGOs such as church groups, international charities, and so on. The importance of this preparatory step cannot be over-stressed and Fig. 7.2 illustrates this for the Bilate project referred to earlier and featured in Fig. 7.1. Many writers emphasize the need for NGOs to make these links to avoid any criticism of being viewed as anti-government (Guggenberger *et al.* 1989). Higman et al (1999) in their *Sustainable Forestry Handbook* deal at length with the important issue of stakeholder involvement.

Promoting awareness of trees and forests in the environment

A successful programme of small-scale plantations, social or agroforestry is usually preceded by some form of rural appraisal to identify needs, followed by a campaign to inform and educate the people whom it is hoped will plant the trees, especially if the initial interest has not come from them. Such campaigns involve planning when to publicize, how best to communicate, for example, the successful puppetry programme in Sudan, use of drama in Zambia, what materials to make available (say) to schools and community leaders, how many demonstration visits to make, etc. This requires coordination of activities and commitment of resources. Some of the pitfalls and successful approaches for extension and encouragement of tree-planting by landowners and communities are discussed in Chapter 5.

Availability of seedlings

This is the most obvious requirement since few plantations can rely on either direct seeding or wildings gathered from existing forest. If a programme is to be developed a steady and continuing supply of seedlings of suitable species and of the right size must be available. Unless there is a good

Figure 7.2 A meeting in 1982 at the inception of the Bilate community forestry project near Durami in Ethiopia showing many of the key participants, that is, the stakeholders. From left to right: local farmers (far left) on whose land the planting will take place, a missionary, Kale Hywet church staff (center)—a local NGO providing support, the first author (on behalf of Tearfund, an international Christian relief and development agency), a local government administrator (right foreground), and Ato Joseph Wodebo (far right) forester with the local church who has overseen the whole project. Missing from the picture are women who fulfilled many key roles including work in the forest nursery and tending operations.

national distribution network, the seedlings must be made available near to where they are to be planted; a conclusion seen as the most important of all in an analysis of farmer needs and project sustainability in Tanzania (Guggenberger *et al.* 1989). Typically, this is the main role of extension nurseries or small private nurseries (Chapter 10). A range of species should be available to meet rural and domestic needs of food, fuel, fodder, poles, and shelter as well as products for sale.

In Bangladesh not only was lack of seedlings reported as the prime constraint but the range must include fruit trees, such as mango and jack fruit and other multipurpose species (Khaleque 1988); thus planning must identify the tree species needed and address seedling supply. In Ethiopia, a shortage of plastic bags has presented a different constraint and has led to successful bare-root cultivation of the ever-popular blue gum, *E. globulus*. It suggests that planning should consider flexibility in seedling production methods. In Haiti, an agroforestry out-reach project, supported by the Pan-American Development Foundation and CARE, assisted, 110 000 farmers to plant 25 million tree seedlings through a network of 39 nurseries. This carefully planned programme was assisted by sociological studies of past failures, implementation by NGOs, extension agents who knew the project participants personally, and who both gathered information and provided advice, use of many small nurseries to produce small, easily transportable seedlings, and promotion of agroforestry leading to supply of more than 40 species (Conway 1988).

All extension nurseries must plan for the simple task of raising seedlings, and there are at least four main activities:

1. *Species identification.* What species are needed by the local communities?
2. *Seed procurement.* Can seed be collected locally or must they be purchased from elsewhere? Fruits must be picked at the right time, the seeds carefully extracted and properly stored.
3. *Plant production.* This involves all the nursery operations of site preparation, fencing, irrigation,

obtaining and filling containers, fertilizing, weed control, etc.
4. *Distribution and monitoring.* How will plants be sold or distributed? If project vehicles are used, the purchase and storage of fuel and spares for these and other motor driven equipment must be provided for, and so on.

Provision of advice and help

Advice about what to plant, when to plant, and even how to plant, is almost invariably needed by people wanting to plant trees and providing it must be planned for.

The local nursery forester often fulfils this role of extension officer and environmental educator, for example, in Haiti (Josiah 1990). The nursery forester stimulates interest in tree-planting by visiting schools, talking to village leaders, and generally advertising the availability of seedlings. This requires a participatory approach involving men and women exchanging ideas, knowledge and techniques, which then leads to attitudes and practices that improve forest and tree management. Today's forester must help improve people's welfare as well as possessing skills in tree management. As already noted in Chapter 5, at the farmer level, the extension worker needs to be more a 'facilitator' than a 'messenger' or 'teacher' with the role of providing background information, technical options and market information. He or she has to be a good communicator, motivator, listener to assist planning and problem solving in tree-planting enterprises at the farm level. The successful end result of all such interactions is that local farmers and villagers want the project to succeed not just those promoting it.

It has been well said that two of the best questions to ask are: (1) How would you hope your village and surroundings will look in 5 years time? (2) Can you show me how you plant trees? Both questions convey a real interest in the people's well-being and enquiring about another person's way of doing things is a great compliment.

PART III

Plantation silviculture

CHAPTER 8

What to plant?

Introduction

When a plantation project is initiated, which species to plant is the most important decision. The species selected not only influences silvicultural practice, but also greatly affects the management and utilization of the crop. Moreover, once the choice is made it must remain until the end of the rotation; the cost of cutting prematurely one crop to replace it with another of superior species is formidable.

Choice of species depends on three basic questions.

1. What is the purpose of the intended plantation?
2. Which species are potentially available for planting?
3. What will grow on the sites available and how well will they grow?

The order of the above questions is the most logical, but it is not always followed. Occasionally, plantation programmes may start without a clearly defined end-use, or change in intended end-use may occur part-way through the life of the crop.

This chapter considers separately each of these questions concerning purpose, potential choice, and the planting site, though in reality they are closely related and usually considered together. Also, the choice of species necessarily occurs at the beginning of a project, when information is often limited, little experience is available and end-use plans may be uncertain. Thus the decision on species cannot be absolutely final, while the great number of alternatives possibly rules out a categorical assertion that the selected species or provenance is actually the very best choice. Figure 8.1 illustrates diagrammatically the important interactions and influences affecting species' choice.

Intensive industrial plantation forestry in the tropics now requires greater refinement in matching species, provenances and even individual clones with sites. There is also increasing emphasis on agroforestry, non-timber forest products and environmental services from tree-planting. In choosing species it is necessary to consider more than ever before the economic, environmental and social consequences and the interactions between them. Priorities and perspectives are changing as the value placed on trees and tree cover increases. The conservation of biodiversity, restoration of degraded forests and forest lands, and the potential use of trees for carbon sequestration are issues receiving greater attention. It remains essential that species and sites are properly matched and that the resulting forest must be economically viable, socially acceptable, and ecologically sustainable.

Purpose of the plantation

Plantations are established for one of four main purposes.

1. Industrial wood production—fuelwood, charcoal, mining timbers, pulpwood, sawn timber, and panel products such as plywood and other boards.
2. Domestic wood production—notably firewood, but also rough roundwood for poles, stakes, etc.
3. Environmental protection—check soil erosion and water runoff, stabilize soil surface, windbreaks, rehabilitation of wasteland/industrial sites.
4. Rural development—agroforestry and social forestry with tree-planting as an integral part of providing amenity, shade, shelter, fruits and nuts, animal fodder and browse, and to enrich the soil through leaf fall and nitrogen fixation etc.

What is the most suitable species to plant will differ between each of these purposes and between different end-uses. Where plantations are established to satisfy more than one objective species' choice will depend on priorities or be varied to suit different purposes. For example, afforestation of upper mountain slopes in East Java (Indonesia) is with *Pinus merkusii*, *Acacia decurrens*, and *Calliandra calothyrsus* and lower slopes with *Swietenia macrophylla* and *Maesopsis eminii* with the main aim of soil protection. However, where possible timber is extracted, though harvesting operations are always secondary to the forest's protective role. Also the *C. calothyrsus* stands are coppiced for firewood to relieve the pressure from the local population on what remains of natural woodland on the highest slopes where protection is even more important.

One further point needs to be made. Though, in general, the purpose for growing trees dictates

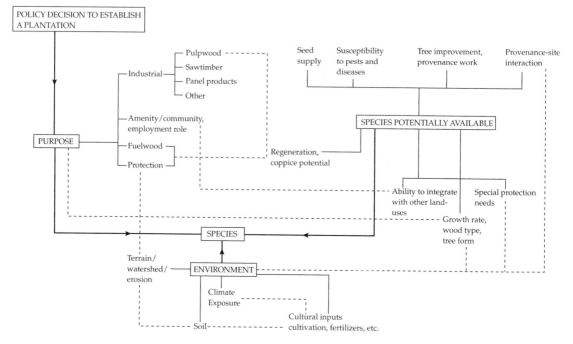

Figure 8.1 Factors influencing choice of species.

species' choice, this is not always so. Trees may occasionally be planted to promote rural employment, with the assumption that forest produce will always be of value, and often the most readily available species is used. Another situation is where some forms of agroforestry are used, for example, taungya. While the ultimate aim is to grow a tree crop, the most productive species for the site may not be selected if it is known to interfere with the food plants. The choice of species is therefore modified, because the most important factor in the establishment of the plantation, perhaps for social, land settlement, or food supply reasons, is the use of agroforestry methods and not primarily the purpose of the tree plantation.

Industrial forest plantations

The main industrial uses of plantations were noted previously. Other produce such as posts, poles, resins, and foliage may also be produced from an industrial plantation but will rarely form the main purpose of the crop. Important exceptions are gum arabic from *A. nilotica* and *A. senegal* plantations in the Sudan, tannins from wattle bark (*A. mearnsii*) widely planted in eastern and southern Africa, growing mulberry (*Morus alba*) for silkworms in China and other parts of Asia, and distillation of essential oils from eucalypt foliage (mainly *Eucalyptus citriodora* and *E. globulus*) in Brazil, China, India, and elsewhere.

The desirable species' characteristics for different end-uses are listed in Table 8.1. However, one general point needs stressing. For a commercial plantation, though the species chosen must yield a marketable commodity at a competitive price, almost all plantation development is a long-term investment, which increases the burden of interest charges. Even if commercial rates of interest are not charged, as may happen with loans from aid and development agencies, the long period of time that elapses before any sizeable income is forthcoming means that fast growth, and hence short rotations, is a most important criterion to satisfy when selecting a species, always providing the wood produced is of acceptable quality. Balsa (*Ochroma lagopus*) and *Cecropia peltata* grow exceedingly fast but their very low density timber is useless for most industrial purposes.

Domestic fuelwood, poles, and other wood products

The urgent need to grow trees for firewood, much the most important source of domestic energy in the

Table 8.1 Important factors in species' choice for industrial end-uses

	Fuelwood/charcoal	Wood pulp	Sawn timber	Plywood/veneer
Growth and silviculture	Fast growth with early culmination of maximum growth rate. Large tree size unimportant or disadvantageous Tree shape unimportant. Should be easy and cheap to grow; coppicing ability desirable	As for fuelwood, but straight stems are important to aid rapid debarking	Moderate to fast growth with ability to grow to large size. Good form important, ease of pruning and freedom from butt rots highly desirable	As for sawn timber, but growth to large size important. Good natural pruning with rapid wound occlusion desirable
Wood properties	Quick drying. Low ash content. Burn steadily without smell, sparks, etc. Moderate to high density	Fibre length, colour, extractives content, density. Papermaking quality	Strength, dimensional stability, wood uniformity. Good seasoning, preservation, working and finishing properties	Peeling or slicing quality. Figured grain. Knot-free. Good adhesive bonding for plywood. Good burning characteristics for matches
Examples of species chosen for large-scale industrial plantations	*Casuarina equisetifolia* (India), *Eucalyptus grandis* (Brazil), *Gmelina arborea* (Malawi, W. Africa), *Leucaena leucocephala* (Philippines)	*Pinus patula* (Africa). *P. caribaea* (Australia, S. America) *G. arborea* (W. Africa, Brazil) *Paraserianthes falcataria* (Philippines) *Eucalyptus* spp. (Brazil, Congo) *Acacia mangium* (Indonesia)	*P. patula, Cupressus lusitanica* (Kenya) *Tectona grandis* (Indonesia, India) *Triplochiton scleroxylon* (W. Africa), *P. caribaea* (Australia), *Cordia alliodora* (Central America)	*Swietenia macrophylla, Alstonia scholaris* (matches) (Fiji) *Araucaria cunninghamii* (Papua New Guinea)

tropics even in urban areas, was described in Chapter 2. While many villagers gather firewood from natural forest, there is a widespread practice of establishing village woodlots and small plantations, often with exotic species, for fuelwood and poles as part of social and community forestry programmes. This is now essential in many regions, often to satisfy urban demands as well as local needs. The main requirements of tree species for domestic firewood plantations are: (i) available seed and ease of cultivation using simple techniques because all the work is done by the villager; (ii) rapid early growth; (iii) good survival under adverse conditions, especially aridity; (iv) resistance to browsing and grazing damage, although some species may be chosen if their foliage can also be used for fodder; (v) ability to coppice; (vi) multiple-use potential (Burley 1980; National Academy of Sciences 1980; Boland 1997).

Tree form and ability to grow to large size are unimportant for firewood. In fact, the ideal size of firewood is sticks or small poles that can be easily cut or snapped to length and carried. Quick drying, freedom from smell and spitting when burnt, moderate wood density for steady burning, and low ash content are all desirable wood qualities, but most important is to use a species, which grows easily on sites available.

Species used for firewood mostly reflect what will grow in the region concerned, and what traditionally has been used. Examples are *C. calothyrsus* in Indonesia, *Leucaena leucocephala* in the Philippines and throughout much of the tropics, *Senna siamea* in Burkina Faso, Mali, Niger, and other Sahelian countries, *Eucalyptus camaldulensis* and *Azadirachta indica* in West Africa, Sudan, and Ethiopia, *Dalbergia sissoo* in India and Pakistan, *Casuarina equisetifolia* in China and India, *E. globulus* in the Andes of Peru, Ecuador, and Colombia and *Gliricidia sepium* in Central America.

In rural areas supply of poles for building is frequently the most pressing need. For many purposes pole diameter of 5–10 cm is required along with a smooth surface, freedom from snags and resins and good natural durability. Eucalypts are very widely used, for example, Fig. 12.11, but any reasonably straight-stemmed species is suitable, a characteristic that can be encouraged by close spacing at planting and in the many shoots from coppice stools.

Environmental protection

Trees are planted for many other purposes than to provide wood products. As perennial, hardy plants with deep, extensive root systems they are of great value in controlling erosion, providing shade and shelter from strong winds, and rehabilitating land degraded by salinity, acidity, and other factors. The implications and interactions of trees in land-use for protective purposes are considered in Chapter 21. Frequently the sites encountered are difficult to revegetate owing to extreme infertility, harsh climate, moving soil, or pressure from grazing, and firewood collection. But of overriding importance for protective afforestation is to ensure continuity of ground cover. This objective more than any other determines the choice of species. In many situations the ability to regenerate naturally to maintain cover is essential. Other important characteristics include the ability to colonize the site rapidly, persistent side branches, resistance to grazing and fire, tolerance of infertile sites, high levels of soil acidity and salinity, and hardiness to extremes of climate.

In selecting species for protective afforestation it is often best to choose one already growing in the area, since it will be adapted to the environment and able to regenerate naturally. Although growth may be slower than for an exotic species, this is of less importance since wood yield is not the primary purpose. Also, it should not be overlooked that once an area of newly planted trees is protected, natural regrowth may occur and can itself be used to advantage. However, many species that have the characteristics, which make them effective in providing protective cover also have the potential to become weeds in adjacent areas. Great care must be exercised to ensure this does not occur.

Tree-planting in agroforestry and for rural development

The scientific approach to modern agroforestry has involved great attention to the use of selected trees and shrubs for use in particular farming systems (Sanchez 1995). The choice of species is more complex in contrast to industrial forest plantations. The focus is not primarily on wood production, the species chosen must be compatible with farm crops and produce valuable products and environmental services, such as soil improvement. Socioeconomic issues generally have a greater impact on species selection than in industrial forestry (Raintree 1991; Franzel *et al.* 1996). The small scale of agroforestry, often involving a single farmer on a few hectares of land, brings in wider considerations of stability, security and risk reduction, micro-site matching, product quality and timing of production in relation to seasons, market participation, self-sufficiency and autonomy (Hughes 1994). Who selects the trees can

also be important and men and women may have different priorities (Shepherd 1997). In Bangladesh when women selected the trees for planting around homesteads the survival rate improved dramatically (Hocking *et al.* 1997a). There is a great diversity of species in use and many more with potential for incorporation into agroforestry systems. The ecology of these species is often poorly known and effects of the interactions between the trees and the crops are still difficult to predict. In the 1970s and 1980s a number of exotic multipurpose tree species were strongly promoted for use in agroforestry but more recently there has been greater emphasis on using a wider range of indigenous species. These local trees may have advantages in being less invasive, well-adapted to local conditions, culturally accepted by local people and supported by indigenous knowledge.

While it is highly desirable to promote local species and to include them in trials, the long history of successful exotic introductions for both forestry and agriculture indicates that there is still a role for both well-tried and new exotic species (Boland 1997). Many valuable multipurpose species have been identified. The African Sahelian zone species, *Faidherbia albida*, provides shade in the dry season when most other trees are bare, it yields edible pods when little alternative fodder is available, it enhances soil fertility through nitrogen fixation and is a source of fuelwood. Many tropical legumes in the genera *Acacia, Albizia, Calliandra, Cassia, Gleditsia, Gliricidia, Leucaena, Prosopis, Sesbania* etc. have found a place in agroforestry. *Grevillea robusta* and the nitrogen-fixing *Casuarina* species, which give shade shelter and wood products, and fruit trees, such as jack fruit (*Artocarpus heterophyllus*) and mango (*Mangifera indica*), are examples of well-known and widely used agroforestry species.

Chapter 20 discusses silvicultural aspects of agroforestry, Chapters 21–23 include discussions of species choice for arid regions, mining sites, protective afforestation, restoration, etc.

Species potentially available

When the purpose of a plantation is known the choice of species is narrowed; for example, for an industrial pulpwood plantation the choice would no longer include a branchy, slow-growing legume with edible seed pods. Nevertheless, taking the whole world, there may be several suitable species. Indeed, in the moist lowland tropics all the following broadleaved species are grown in plantation to produce opaque, white, short-fibre pulp: *Paraserianthes falcataria* in Indonesia and the Philippines, *Anthocephalus chinensis* in Southeast Asia, *Eucalyptus deglupta* in Papua New Guinea (PNG) and the

Philippines, *Gmelina arborea* in West Africa and Brazil, *Acacia mangium* in Indonesia and *Maesopsis eminii* in Africa. This choice, and there are other possibilities, is further increased if provenances and varieties within species are also considered. The approach to find the right species lies in the axiom of matching species with site—to select the species that grows best on the sites in question. The factor of site is considered later, considered here are the possibilities and limitations in species selection under the headings indigenous species, exotics and species introduction, and genetic variation and tree improvement.

Indigenous species

An indigenous species is one that grows naturally in the country concerned, though not necessarily in all parts and certainly not suited to all sites. With such species there are no political or quarantine problems to obstruct its use in a plantation programme, and there are some important biological advantages.

1. Growth of natural stands provide some indication of possible performance in plantation.
2. The species is adapted to the environment and already filling an ecological niche. This may render it less susceptible to serious damage from diseases and pests since controlling agents (predators, viruses, climatic factors) are already present.
3. Indigenous species, even in monoculture, have generally greater biodiversity conservation value than exotics. They can provide habitat for native wildlife, especially birds and insects, and help maintain the local gene pool. They form an integral part of the natural landscape and environment and are frequently preferred for cultural reasons.
4. The timber is likely to be known to local wood using industries and artisans, and may be the

preferred firewood source. In Africa, local species of *Acacia*, *Balanites*, *Brachystegia*, *Combretum*, *Pterocarpus*, etc. are used for fuelwood and charcoal and better management of existing natural *Acacia* or miombo woodlands may prove a viable alternative to plantations of eucalypts, *A. indica*, or other exotics.

For these reasons if an indigenous species grows well in plantation on the sites for afforestation there is no compelling reason to widen the choice. Examples of major plantation programmes using native species are *Pinus massoniana* and *Cunninghamia lanceolata* in China, *P. merkusii* and *A. mangium* in Indonesia, *E. deglupta* in the Philippines and PNG, *Araucaria cunninghamii* in PNG and Australia (Fig. 1.4), *Tectona grandis* and *Dalbergia latifolia* in India, *Terminalia ivorensis*, *Nauclea diderrichii*, and *Triplochiton scleroxylon* in West Africa, and *Cordia alliodora*, *Swietenia macrophylla*, and *Cupressus lusitanica* in Central America. Many recent trials have been laid down to extend further the use of indigenous species. In Costa Rica, trials with species from the humid tropics have demonstrated the potential of *Hyeronima alchorneoiodes*, *Virola koschnyi*, *Vochysia ferruginea*, and *V. guatemalensis* for planting on degraded sites (Butterfield and Fisher 1994; Gonzales and Fisher 1994) and in cooler moister areas *Terminalia amazonica* is preferred to exotics by farmers (Nichols 1994) (Fig. 8.2). On the Caribbean island of Trinidad, Homer (1997) suggests commercially important local species such as *Brysonima spicata*, *Mora excelsa*, *Sterculia caribaea*, and *T. amazonica* should be considered as indigenous alternatives to introduced *P. caribaea*. In Indonesia valuable local timber species *Anisoptera marginata*, *Eusideroxylon zwageri*, and *Shorea balangeran* grew well when planted under an *A. mangium* plantation (Otsamo and Kurniati 1999) and on grasslands *Peronema canescens*, a local substitute for teak, has also performed satisfactorily (Vuokko and Otsamo 1996). On the Pacific the island of Western Samoa, where exotic trees have been

Figure 8.2 Part of a trial of native species compared with common exotics near La Selva, Costa Rica—see Butterfield (1990) for details.

susceptible to cyclone damage, native species *T. richii* and *Flueggia flexuosa* show promise (Pouli *et al.* 1995). Native arid zone trees have been less well studied for planting but compilations of information on their characteristics and uses (e.g. Maydell 1986; Hocking 1993) have assisted in species selection. Indigenous species have been identified for use in arid and semi-arid areas of East Africa, for example, *Melia volkensii* in agroforestry in Kenya (Stewart and Blomley 1994) and *Acacia* spp., *Albizia anthelmintica*, and *Zizyphus hamur* in Somalia (Helin 1989).

However, restricting species' choice to those naturally occurring within a country does have limitations. First, there is a problem with the definition of 'indigenous'. Geopolitical boundaries of a country in no way relate to plant distribution. For example, in Brazil *Araucaria angustifolia* occurs naturally in the south but if planted in the Amazon basin it would still theoretically be classed as indigenous, even though its natural range is south of 20°S and biologically it would be as much an introduction to the Amazon as *A. cunninghamii* from Queensland, Australia. The important corollary is that if a small country only uses native species it suffers greater limitation of choice simply because it is small, a factor that bears no relation to what species will in fact grow well. Second, there is the question of whether there is any virtue in restricting the choice to indigenous species. In the past many foresters argued against introducing new species because: (i) it was claimed that an exotic was unlikely to be superior to a native one already adapted to the environment especially if it was the natural climax species; (ii) many early introductions failed or grew badly with poor form.

Today these points are discounted. The first is not true, as the success of *P. caribaea, P. patula, Gmelina arborea, T. grandis,* and numerous eucalypts as exotics testify. The second point, although true, arose because many early introductions were unplanned, or badly sited, and with little appreciation of a species' natural provenance variation. Nevertheless, even failed plots have management value in allowing conclusive elimination of a species (Fig. 8.3). Other causes of poor growth and disappointment with exotics arose because of a failure to detect a critical nutrient deficiency, for example, boron with eucalypts in central Africa, or an absence of mycorrhizas, especially on *Pinus* spp.

One should also be aware that site conditions can change substantially after natural forest is cleared and species that previously grew there may no longer be as well-adapted as other exotic species. Whether to use indigenous or introduced species is important but it is no longer an either/or alternative. As a rule, where a native species meets the need there is no reason to choose an alternative.

Figure 8.3 Very poor growth of *Pinus taeda* (foreground) compared with *P. caribaea* (background) on a lowland equatorial site in the Amazon.

Indeed, for reasons of biodiversity conservation, if the choice lies between two species of comparable growth and quality, one of which is native and one exotic, such as could be true of *P. merkusii* and *P. caribaea*, respectively, in Indonesia, the native species is to be preferred. Of course, where no suitable native species exists trials of introduced species must be laid down.

Exotics and tree introduction

The majority of successful plantations in the tropics comprise species of *Acacia, Eucalyptus, Pinus* and *Tectona* and most are planted as exotics. There are several factors which can lead to great success if a carefully chosen exotic is used and these are elaborated in detail by Zobel *et al.* (1987).

1. The much wider choice of species, in theory from anywhere in the world, increases the chance of finding a species well suited to the planting site. The very need for a planting programme is sometimes evidence that there is no really satisfactory native species.

2. An exotic species, away from its natural habitat, is often free from diseases and pests, for the first rotation at least, since insects and fungal pathogens in the recipient region may not find the 'alien' species a suitable host. For eucalypts outside Australia,

Pryor (1978) points out that 'One of the reasons for fast growth when they are planted as exotics is the absence generally of widely prevalent leaf-eating insects, which abound in their native habitat'. However, this freedom from pests and diseases also means that natural predators and controlling agents are likely to be absent as well. This can render the exotic, especially in monoculture, more susceptible to massive uncontrolled attacks once a serious pathogen colonizes the species. The devastating psyllid defoliation of *L. leucocephala* was assisted by extensive areas of genetically very uniform plantations. Attacks of the stem borer *Phoracantha semipunctata* on eucalypts in the Mediterranean region are facilitated by many of the stands being drought-stressed. Plantations of *E. camaldulensis* in humid areas of Thailand and Vietnam have been severely defoliated by fungus *Cylindrocladium* sp. Ecologically an exotic may be both initially more free from pathogens and pests and also more vulnerable to serious, uncontrolled damage once they occur. However, the risks are greatly reduced if there is good species/provenance-site matching so that the trees are not stressed, high genetic diversity within the plantation providing variable resistance, and effective management and hygiene. From a study of nine commonly planted species in the tropics, it was concluded that monocultures resulted in an increase in pest problems but the risk of an outbreak was not solely dependent on the exotic or indigenous status of a species (Nair, 2001). The greater economic investment and returns in densely planted, even-aged, and genetically homogeneous clonal tree plantations designed to maximize wood production allows for more active monitoring and intensive control of pests and diseases, including selection and breeding for resistance (Turnbull 2000).

3. When exotics are used worldwide, for example, *P. caribaea*, *E. grandis*, and *T. grandis*, already-existing research and experience allows them to be used with some degree of certainty. From the outset seed is available and dependable silvicultural methods can be used, and mistakes and problems avoided, which so easily occur with a little-known species.

Though most tropical industrial plantations use exotics it is important to emphasize that not all tree introductions are successful. Quite the opposite is true; most countries evaluate a large number of exotic species of which only a few are usually found to grow well. Over 300 exotic *Eucalyptus* species have been introduced into China but only about 10 species have been widely cultivated in plantations (Wang *et al.* 1994). Not surprisingly most of these (e.g. *E. camaldulensis*, *E. citriodora*, *E. globulus*, *E. grandis*, *E. urophylla*) are widely planted elsewhere in the world. An example from

PNG illustrates a second important point: use of exotics does not automatically lead to success. At Kunjingini, on degraded grassland sites of low fertility and poor drainage, many species have been tested. For example, Table 8.2 shows the performance of 12 species after 4 years' growth. The two introduced pines grew moderately well with heavy fertilization but they were inferior to several native species. And, after 8 years, both pines were in poor condition and, of all the species, only *A. auriculiformis* and *E. tereticornis* were growing satisfactorily (Lamb 1975) and then only the plots, which received fertilizer.

Factors influencing tree introduction and domestication

Wright (1976) and Zobel *et al.* (1987) discuss this subject in detail, but it is important to outline the main considerations to indicate the care required when planning species introduction, and to illustrate further the main principles behind species selection.

Genetic variation—the importance of provenance, variety, and land race
So far, what to plant has been confined to considering alternative species, but it has become abundantly clear that more than just choosing the species is needed to take full advantage of available genetic potential. Most species grow naturally over a range of sites and locations, and some, such as

Table 8.2 Height growth of trees at Kunjingini, PNG

Species	Height without fertilizer (m)		Height with fertilizer[a] (m)	
	Year 1	Year 4	Year 1	Year 4
Pinus caribaea[b]	0.46	0.76	1.46	4.20
P. kesiya[b]	0.30	0.43	1.11	3.96
Eucalyptus deglupta	0.79	1.49	3.67	7.65
E. tereticornis	0.94	2.04	2.77	4.78
Terminalia brassii	0.52	1.13	2.50	4.69
T. complanata	0.40	dead	0.76	1.34
Tectona grandis[b]	0.21	dead	2.60	3.04
Acacia auriculiformis	1.77	3.99	2.68	9.66
Pterocarpus indicus	0.67	0.73	1.31	1.61
Intsia bijuga	0.58	0.36	0.49	0.97
Prosopis juliflora[b]	poor	dead	poor	dead
Fraxinus spp.[b]	poor	dead	poor	dead

[a] Fertilized with 500 kg ha^{-1} NPK (17:5:22) in Year 1.
[b] Exotic species.

Cordia alliodora, which grows throughout Central and South America from 25°N to 25°S, are very widely distributed. This wide separation and often isolation of stands may have led to genetically different populations within one species. Where these exhibit distinctive phenotypes the term 'variety' is used. However, even within a variety, owing to differences in adaptation to local conditions, there will be variation between stands; this variation is recognized as differences of 'provenance'. And, it has been found that different varieties and provenances often perform very differently when tested together on one site (Fig. 3.3). The concept of provenance has been explored by Turnbull and Griffin (1986) and its importance is illustrated by three examples.

1. *Pinus caribaea* in Queensland, Australia. *Pinus caribaea* has three varieties—*P. caribaea* var. *hondurensis*, *P. caribaea* var. *caribaea*, and P. *caribaea* var. *bahamensis*, which occur naturally in mainland central America, Cuba, and the Bahamas respectively.

When grown together their form and rates of growth usually differ markedly. Table 8.3 shows these varietal differences, as part of a comparison between four species of tropical pines, from a trial at Kuranda, Queensland (17°S).

This trial shows, first, the general superiority of *P. caribaea* over other tropical pines, and, second, the growth superiority of *P. caribaea* var. *hondurensis* over other varieties though form and windthrow susceptibility are poorer, a matter amplified in Nikles *et al.* (1983). This growth superiority of var. *hondurensis* has been observed in many varietal trials in the tropics (Greaves 1981; Gibson *et al.* 1988) as has the greater windthrow susceptibility of *P. oocarpa* compared with *P. caribaea* (Bell *et al.* 1983; Liegel 1983; Birks and Barnes 1990).

In Queensland the study of *P. caribaea* has been taken much further (Nikles 1996). *P. caribaea* var. *hondurensis* displays considerable variation, var. *bahamensis* little variation and var. *caribaea* hardly any. Differences in volume production, resistance to windthrow and other characteristics among *P. caribaea* var. *hondurensis* provenances are shown in Table 8.4. This again shows the growth superiority of variety *hondurensis* but also differences between provenances. In addition, the relative growth of different provenances of *P. caribaea* var. *hondurensis* is closely related to site; superiority of one provenance in one trial does not mean it will be best on all sites. Nevertheless, the improved *P. caribaea* var. *hondurensis* developed in Queensland has proved very adaptable to sites in many countries. International collaborative provenance trials show clearly that seed of *P. caribaea* var. *hondurensis* from Queensland seed orchards would be an excellent choice for pine planting in many tropical and subtropical countries, and would be

Table 8.3 Comparison between mean growth rate, form, and stability of *Pinus* spp. at 4.5 years

Species/variety	Height (m)	Bends (%)[a]	Lean (%)[b]
Pinus oocarpa	2.87	7.1	16.0
P. kesiya	2.30	6.4	3.6
P. merkusii	1.05	(too small to assess)	
P. caribaea var. *hondurensis*	4.82	12.0	9.4
P. caribaea var. *bahamensis*	3.59	6.3	3.1
P. caribaea var. *caribaea*	3.14	4.0	0.8

[a] Inherent form character.
[b] Indication of wind susceptibility after a cyclone.

Table 8.4 Performance of different varieties and provenances of *P. caribaea* at 9.1 years of age at Byfield, Queensland

Variety	Provenance	Mean dbh (cm)	Predom. height (m)	Volume (m³ ha⁻¹)	Windthrow (%)	Stem straightness[a]	Foxtailing[b] (%)
bahamensis	Grand Bahamas	15.8	14.3	90	0	2.56	0
caribaea	Cuba	16.0	13.0	81	1	2.60	0
hondurensis	Mountain Pine Ridge, Belize	18.5	16.4	111	3	2.10	7
hondurensis	Dona Maria, Guatemala	18.7	16.4	120	7	1.99	1
hondurensis	Cristina Qu, Guatemala	17.5	15.1	90	13	2.01	7
hondurensis	Poptun, Guatemala	17.7	15.5	96	0	2.11	4

[a] Individual tree assessment, 1 = poorest form, 6 = best form.
[b] The habit of producing continuous leader growth without side branches.

Table 8.5 Performance at 4 years of age of *Acacia auriculiformis*, *A. mangium*, and *A. crassicarpa* at two sites in Sabah, East Malaysia (Sim and Gan 1991)

Species	Provenance	Height (m)	
		Site 1	Site 2
A. auriculiformis	Sepilok (local)	9.9	11.5
	Balamuk, PNG	12.1	14.0
	Iokwa, PNG	14.9	13.1
	Bula, PNG	11.6	12.3
A. mangium	Tawau (local)	11.5	9.6
	PNG	15.6	13.4
A. crassicarpa	Wemenever, PNG	17.5	15.8
	Oriomo, PNG	15.4	17.4
	Woroi/Wipim, PNG	15.0	17.7

beneficial for incorporation into local breeding programmes (Cockford *et al*. 1990; Nikles 1996).

As well as selection of the best provenances it is possible, of course, to create artifically new genetic combinations through tree-breeding (see later). In Queensland the F₁ hybrid between *P. elliottii* var. *elliottii* and *P. caribaea* var. *hondurensis* was first planted in 1958 and since then it has been developed for use across a range of sites. In 1985 the use of *P. elliottii* var. *elliottii* was discontinued for routine plantings on poorly drained sites in favour of the F₁ or F₂ hybrids and in 1991 the hybrid replaced *P. caribaea* var. *hondurensis* on well-drained sites in the major pine growing area in subtropical southeast Queensland. The growth rate was equal to or better than the pure species and the stem straightness and windfirmness were better on all sites (Powell and Nikles 1996). A clonal planting programme with the hybrid will be fully implemented with an estimated annual planting of 3000 ha (Walker *et al*. 1996).

2. *Acacia auriculiformis, A. mangium* and *A. crassicarpa* in humid tropical Asia. These phyllodinous acacia species occur naturally in northern Australia, parts of Indonesia, and PNG and trials as exotics in recent years have stimulated much interest in them. In 1966 *A. mangium* was introduced into Sabah (Malaysia) from Queensland as a firebreak for pine plantations but it grew so well, that from 1973 it began to be widely planted for wood production. *A. auriculiformis* had long been used as an ornamental but crooked stem form and heavy branching has typified plantings in Asia and prevented its wider use. *Acacia crassicapa* has grown better than *A. mangium* and *A. auriculiformis* on peaty sites in Indonesia, on poorly drained tin mine tailings in Malaysia and on weedy and sandy soils in Sabah (Table 8.5). Since 1989 it has been used in industrial plantations and

one pulp and paper company in Sumatra, Indonesia had planted over 40 000 ha by 2000.

As with many acacias, these species are able to grow on infertile soils and degraded *Imperata* grasslands and offer much promise both as industrial crop species and in agroforestry to assist soil enrichment. Research results from China, India, Malaysia, Thailand, and Vietnam provide clear evidence that knowledge of provenance is essential to utilize the great potential of these species. Results from two species/provenance trials on differing soils in Sabah are shown in Table 8.5, which compares local sources of *A. mangium* and *A. auriculiformis* with new provenances from PNG and with three provenances of *A. crassicarpa*. The new introductions are significantly taller and have greater wood volume production than the local seed sources and that has been the same in similar trials in several countries. The poor growth of local sources has probably been the combined result of inbreeding depression and use of suboptimal provenances. The stem form of most provenances of *A. auriculiformis* from PNG and northern Queensland has generally been superior to local sources. Similar poor growth of plantings of *A. mangium* in Sabah and Indonesia has been attributed to the poor quality seeds from local seedling seed orchards established in the early 1980s using very restricted base populations. Loss of productivity in plantations using Subanjeriji (Indonesia) seed has been estimated as high as 70–80% (Turnbull *et al*. 1998*b*). International provenance trials of *A. mangium* show PNG origins are the best on a range of sites.

Differences in provenance performance of *A. auriculiformis* are less clear cut. Generally the Queensland provenances have straighter stems and light crowns, PNG provenances have heavier crowns and higher biomass production and Northern Territory provenances appear to be more drought tolerant. Local sources usually perform poorly compared with the best new introductions (Tables 8.4 and 8.5). In comprehensive provenance trials in Vietnam superior provenances in terms of productivity were from throughout the natural distribution in Queensland (Coen and Morehead Rivers), Northern Territory (Manton River and Elizabeth River), and PNG (Old Tonda Village and Mibini) (Nghia and Kha 1998). In Thailand, a second generation progeny trial had considerable improvement in stem form with families maternally descended from Queensland provenances the most productive and those descended from the Thailand races least productive (Luangviriyasaeng and Pinyopusarerk 2002).

In common with several pines and eucalypts, there has been major progress in recent years in developing hybrid combinations of acacias. Natural hybrids between planted *A. auriculiformis* and *A. mangium* have been recorded in several countries.

Some hybrid individuals exhibit heterosis and produce more wood than the pure species, they may have finer branching and other desirable characteristics. Development of technologies for vegetative propagation and controlled seed production to take advantage of the benefits of acacia hybrids began in Malaysia in the 1980s. In Vietnam the development of hybrid clones of *A. auriculiformis* and *A. mangium* has added to gains already achieved with species/provenance selection (Kha 1996, 2001) and similarly in India gains of 25–30% using selected provenances are being enhanced by the careful matching of hybrid clones and specific planting sites (Varghese *et al.* 2000) (Table 8.6).

The above emphasizes the importance of species/provenance research, the risks of building tree improvement programmes on a limited genetic base, and the need to test on a range of sites. The use of *Acacia crassicarpa* and other potentially productive tropical *Acacia* species and hybrids on specific sites suggests that the choice of species and site is still a critical decision and that clonal hybrids can be used to capture gains achieved from provenance selection and tree breeding.

3. *Eucalyptus grandis* in the coastal lowlands of Brazil. This species is probably the most widely planted eucalypt for industrial wood production, with very large areas in South America and southern Africa. It generally grows exceptionally fast on good sites in subtropical and warm temperate areas but is less suited to the lowland humid tropics. The early eucalypt plantings by the Aracruz Florestal in the coastal lowlands of Espiritu Santo, Brazil (19°S), where the climate is humid and tropical, grew relatively slowly and were severely infected by the canker disease *Cryphonectria cubensis*. Further south in Brazil the large plantation companies were using the Coffs Harbour provenance from the main part of the *E. grandis* natural distribution (26–31°S) in Australia and this was tried by Aracruz. Although growing faster, canker damaged 40% of the trees. The company sent its forest research manager, Edgard Campinhos Jr., to Australia to inspect the natural stands of eucalypts and determine which species and provenances were suitable for their pulpwood plantations. He returned with seed from the isolated natural stands of *E. grandis* from the Atherton Tableland in north Queensland (17°S) and provenances of *E. urophylla* from Timor and Flores in Indonesia. Provenance trials were established and the Atherton provenance of *E. grandis* achieved an annual growth rate of 35–40 m^3 ha^{-1} and was significantly less susceptible to canker. This provenance, the Zimbabwe land race and selected provenances of *E. urophylla* from Indonesia were the main species planted in the late 1970s. Hundreds of progenies of individual trees, mainly from the Atherton area, were the basis of a recurrent selection to provide select parents for interspecific crosses, principally with *E. urophylla*. Aracruz's research indicated that further gains could be achieved using these hybrid clones and from 1979 the company gradually replaced its plantations derived from seeds with clonal plantations. The superior clones, combined with intensive cultural practices, have resulted in very substantial increases in productivity, wood quality, and lower harvesting costs (Campinhos 1999). The amount of product (bleached wood pulp) grown annually on each hectare of forest has almost doubled, from 5.9 t ha^{-1} to 10.9 t ha^{-1}.

The Aracruz experience illustrates the benefits of correct matching species and provenances and the value of visiting the natural stands of a species to better understand its ecology and potential adaptability. The combination of optimum genetic material with intensive management practices can result in highly productive industrial plantations.

Tree improvement

The introduction of new species, careful selection of variety and provenance, formation of hybrids, and clonal propagation are all kinds of 'tree improvement'. However, in the tropics fast growth and early flowering of many plantation species allow direct

Table 8.6 Growth of four provenances of *A. auriculiformis* at Agilubagilu, Karnataka, India, at 5 years of age (Bulgannawar and Math 1991)

Seedlot[a]	Provenance	Height (m)	Diameter (cm)	Volume (m^3 ha^{-1})	MAI (m^3 ha^{-1})
13684	Balamuk, PNG	12.4	10.0	91	18.1
13861	Springvale, Qld	12.2	10.4	96	19.6
13854	Oenpelli, NT	10.3	6.7	34	6.7
—	Local source	10.3	7.2	39	7.8

[a] Seedlot number of CSIRO Forestry and Forest Products, Australia.

improvement of trees through selection and breeding of superior genotypes in a relatively short time, 10–20 years. Genetic improvement from breeding may conservatively yield gains of 10% in volume growth in the first generation, though Van Wyk (1983) and Ladrach (1983) have reported much higher figures with *E. grandis* and *C. lusitanica*, respectively, and perhaps more in second-generation seed orchard progeny. Often high heritabilities for stem straightness, fine branching, and wood quality occur, for example, *Gmelina arborea* (Wong and Jones 1986)—see Fig. 8.4(a) and (b), *E. camaldulensis* (Quaile and Mullin 1984), *P. patula* (Wright *et al.* 1996), and *T. grandis* (Harahap and Soerianegara 1977). Such improvements, along with greater pest-resistance and superior wood quality, are benefits that may be realized in moderate periods of time in the tropics, increasing wood recovery by up to 50% in the first generation (Kanowski *et al.* 1991).

Tree improvement work has four main objectives: (i) to improve resistance to diseases and pests; (ii) to improve growth, form, and other desirable tree characteristics; (iii) to identify new provenances, varieties, and land races better suited to sites; (iv) to breed a tree suited to a specific site type such as improving drought resistance. Much work of this kind has been undertaken on all the important plantation species; indeed, tree improvement, including in recent years clonal propagation and the use of hybrids (Chapter 11), has probably received more research effort, both with national and international programmes, than any other aspect of plantation forestry in the tropics. For literature on the subject readers are referred to the following important references: Anon. (2000*b*),

Barnes (1995), Boyle *et al.* (1997), Cotterill and Dean (1990), Dieters *et al.* (1996), Eldridge *et al.* (1993), Gibson *et al.* (1989), Griffin (1990), Libby (1973), Matheson (1990), Namkoong *et al.* (1988), Simons *et al.* (1994), and Zobel and Talbert (1984).

Every large plantation project in the tropics should undertake improvement programmes in its plantations and/or be actively associated with other improvement programmes involving its main species. The benefits are likely to be considerable.

1. Significant heritable gains in growth rate, form, wood quality, disease and pest resistance, and suitability to special sites have been demonstrated for several species.

2. Despite the greater cost of genetically superior seed, if it yields even a small improvement it is usually worthwhile trying to obtain it. For example, assume genetically improved seed of *P. patula*, when grown on a 12-year rotation for pulpwood, gives a small but significant increase in yield of 5.9% such that ordinary seed would result in a standing volume of $340 \, m^3 \, ha^{-1}$ and superior seed $360 \, m^3 \, ha^{-1}$ when felled. If the standing value of pulpwood is $US\$15.00 \, m^{-3}$ then the extra revenue from using superior seed is $20 \, m^3 \, ha^{-1} \times 15 = US\$300 \, ha^{-1}$.

One kilogram of *P. patula* seed produces about 100 000 viable seedlings, which is sufficient to establish about 30 ha of plantation at normal pulpwood spacing, including wastage from nursery losses and blanking. Consequently, the extra revenue earned per kilogram of superior seed is $300 \times 30 = US\$9000$. Discounted to the time of seed purchase (13 years previously) this gain in revenue equals $9000 \times 0.29 = US\$2610 \, kg^{-1}$, using a 10%

Figure 8.4 *Gmelina arborea*, aged 3 years, in Sabah, Malaysia: (a) from unimproved seed source; (b) first-generation improvement using cuttings from selected phenotypes showing good vigour, stem form, and fine branching.

rate of interest. Thus one can afford to pay up to this amount extra for improved seed in this example. In 2003 SETROPA Ltd., a Dutch seed merchant, sold seed orchard quality *P. patula* seed at less than twice the cost of ordinary seed, about US$320 and US$175 kg^{-1}, respectively.

The small volume gain in this example is clearly profitable. Usually, other benefits would also accrue from using superior seed, such as better tree form, which would lower costs of debarking and slightly reduce the amount of compression wood thus improving pulp quality. Moreover, it needs emphasizing that the cost of seed per hectare represents a tiny proportion of total establishment cost, frequently less than 1%. Improved seed should always be used whenever it is available.

The above example illustrates the benefit of a conventional (sexual) breeding strategy, but it is clear that rapid genetic uplift can be obtained by vegetatively propagating superior phenotypes, and this is becoming increasingly used, for example, eucalypts in Brazil, Colombia, Congo, South Africa, etc., *Gmelina arborea* (Sabah), pines in Queensland, *T. sceleroxylon* and *T. superba* in West Africa (Chapter 11). Its value is shown by the eucalypt breeding, selection and clonal programme of the Mondi company in South Africa where plantation productivity has been more than doubled from an initial mean annual increment (MAI) of 10 t ha^{-1} over a 10 year rotation to 21 t ha^{-1} over a 7 year rotation (Harvett 2000). There is no doubt that this strategy is now a major tool in the tropics to harness the gains of good individual performers within an existing tree-improvement programme of provenance selection, land race identification, breeding, etc.

The selection of species for improvement in industrial plantation forestry is usually relatively straight forward. The species have usually been planted on a significant scale and are economically important. Most tree improvement programmes have been conducted for species grown in plantations for a few specific end uses by public sector organizations or large private companies. Developing breeding programmes for tree species used in agroforestry plantings can be more problematical. Farmers grow a diverse range of tree species for a range of purposes including marketable products (e.g. poles and fruits), domestic products (e.g. firewood and fodder), and environmental services (e.g. wind protection and shade). Prioritizing the species for improvement is critical and guidelines to do this involving defining a target user group, identifying products and determining the priority species have been drawn up by the International Centre for Research in Agroforestry (ICRAF) and International Service for National Agricultural Research (ISNAR) and national research organizations (Franzel *et al.* 1996). Strategies and

practices for species testing and breeding multipurpose trees for agroforestry have been discussed by Briscoe (1990), Raintree (1991), Simons *et al.* (1994), Pottinger *et al.* (1996), and Boland (1997).

Limitations in availability of species
Ideally a forester would wish to have unrestricted access to the whole world's tree potential—indigenous species, exotic species, varieties and provenances, improved seed, etc. Unfortunately, this is rarely the case, at least in the short term. In the long term there is clearly a responsibility upon foresters to evaluate the world's species and an excellent example, greatly extending knowledge and evaluating the potential of Australia's lesser-known species for fuelwood, land rehabilitation and farm forestry, mainly in the genera *Acacia, Casuarina*, and *Melaleuca*, has been the ACIAR-CSIRO systematic seed-collection and field-testing of over 100 species (Boland 1989; Doran and Turnbull 1997). The programme has resulted in great economic and social benefits through, for example, the extensive use of *A. crassicarpa* and *A. holosericea*, and better provenances of *A. auriculiformis* and *A. mangium*. A parallel programme by CTFT has shown the potential of several Australian acacias for planting in tropical dry zones (Cossalter 1986; Thomson *et al.* 1994). However, several factors prevent the widespread use of a species or particular provenance even though plantation trials may have shown it to be a most suitable tree.

1. Problems of seed supply. Problems may be biological, political, or due to difficulties of access and communications. These are discussed in Chapter 9.
2. Nursery difficulties. Occasionally difficulties in raising seedlings have led to abandonment, at least temporarily, of otherwise promising species, for example, *Pterocarpus indicus* in Southeast Asia.
3. Some species appear ill-suited to plantation conditions. Many rain-forest species, for example, dipterocarps, are extremely difficult to raise in plantations and often fail completely, though not always (Fig. 8.5). The main reason is that most rain forest trees are dominated by late secondary or climax species in ecological succession, whereas species with good plantation potential are generally invaders of gaps, pioneers, or early secondary species. Also a few species, though fast-growing as individuals, do not thrive in a closed plantation. *Octomeles sumatrana* is an example; it has an open, spreading crown, which appears to suffer from any form of branch contact with neighbouring trees. Another example is *Grevillea robusta*, which appears to suppress growth of its neighbours and grows best when planted singly or in lines.
4. Quarantine restrictions may complicate or prevent seed imports particularly from countries where economically important diseases occur which are not

present in the recipient country such as guava rust (*Puccinia psidii*) and vigorous strains of *Phytophthora cinnamomi*.

5. In some countries foreign exchange is so limited that there is no money to import seed or it is only made available to government agencies.

6. Susceptibility to diseases and pests. Sometimes a species may show good growth and have valuable timber but be of little value for planting owing to great susceptibility to damage. By far the most important example in the tropics is the almost universal damage to Meliaceae species by the shoot borers *Hypsipyla robusta* (Fig. 8.6) in Asia and

H. grandella and *H. ferrealis* in the neo-tropics when trees are planted in pure stands (Floyd and Hauxwell 2001). These pests have greatly restricted the planting of cedars and mahoganies—*Carapa, Cedrela, Toona, Khaya*, and *Swietenia* spp. in many tropical countries. There is some evidence that damage is reduced by planting under shade. Other important examples are Iroko gall bug (*Phytolyma* spp.) damage of *Chlorophora (Milicia) excelsa* when planted pure in West Africa, widespread psyllid (*Heteropsylla cubana*) defoliation of *L. leucocephala* outside the neo-tropics, and *Dothistroma* needle blight of *P. radiata* when planted in the highland tropics (Fig. 8.7).

Figure 8.5 Plantation of dipterocarps (*Shorea negrensis*) in the southern Philippines.

(a)

(b)

Figure 8.6 Shoot borer damage to Meliaceae species grown in plantation: (a) borer damage by *Mussidae nigridenella* (Pyralidae) to *Khaya nyassica*, in South Africa. (b) stem deformation of young *Carapa guyanensis* caused by *Hypsipyla* spp. (Pyralidae) in Ecuador.

Figure 8.7 *Pinus radiata* plantation in Ecuador suffering from Dothistroma needle blight.

Matching species with site

The development of tree-planting and plantation forestry in the tropics has been described in Chapter 3. Trees have been moved from their native habitats to many parts of the world to meet the diverse needs of people for wood and non-wood forest products, for amenity, and soil protection and other environmental uses. Each year foresters and farmers plant more and more trees. Predicting where and how native and exotic tree use will grow is fundamental to forestry and agroforestry, and global in extent. Making a correct choice can double or triple yields for virtually no additional establishment cost. So two important questions are: 'which tree species, provenances or clones, will grow at this particular site, and how well will they grow?' (Booth 1996).

Much of silviculture is concerned with achieving the best match between species and the planting site (including site manipulation such as adding fertilizer, cultivation, and weeding). The problem is not simple; there are numerous species and an infinite variety of sites and it is now clear that individual trees or provenances may respond differently relative to each other in different environments. This is known as genotype × environment interaction, and breeding strategies must recognize this feature (Matheson and Raymond 1986). At provenance and variety level, interactions are principally with climatic parameters; at family or clonal level they are with edaphic factors (Barnes 1984). This knowledge underlines the need to define site accurately, as well as knowing as exactly as possible the genetic origin of what is to be planted.

A relatively few descriptors of the environment (rainfall, temperature, soil type, etc.) broadly define a site's potential. Once a reliable description of a tree's environmental requirements has been developed it can be used to assess the potential of possible planting sites. These estimates can then be used as the basis of socio-economic evaluations. The environmental conditions in the species' natural habitat can initially provide the best guide to the sort of conditions the planting site should have. Hence the importance attached to describing the environment in which a seed source grows, as well is the stand and trees themselves. However, there will always be a need for properly conducted field trials for precise matching of species and provenances to particular sites, even if only to validate predictions and refine prediction techniques (Eldridge *et al*. 1993). This problem of matching species with site is of little importance with indigenous species occurring naturally in or near the planting area, but it is of primary importance when an exotic is to be planted.

Climate of a site

Eleven climatic factors for assessing the climatic conditions of a site were recommended by Burley and Wood (1976). They included rainfall, humidity, temperature, and wind factors but two components of the climate, rainfall and temperature, are of overriding importance when selecting a species to plant. Webb *et al*. (1980, 1984) used six factors to characterize the climate of a site:

- mean annual rainfall (mm);
- rainfall regime (uniform/bimodal, winter, summer);
- dry season length (consecutive months less than 40 mm rainfall);
- maximum temperature of the hottest month (°C);
- mean minimum temperature of the coldest month (°C);
- mean annual temperature (°C).

These six factors have generally proved to be useful discriminators of regions climatically suitable for growing particular trees. However, it is sometimes necessary to add the absolute minimum temperature (Booth 1996).

Studying natural distributions can provide a first impression of a species' climatic requirements, but many species can grow well in the conditions, which are somewhat different from those within their natural habitat. Sometimes the natural habitat of a species is restricted for reasons other than climate, for example, occurrence of fires, and the species may have potential to grow in a much wider range of environments than any habitat would suggest. Examples include *P. caribaea*, extensively planted in tropical lowlands, and *E. globulus* in tropical highlands. Additional information from plantings outside their natural range can help provide a more accurate estimate of their climatic requirements.

Amount and distribution of rainfall

Annual rainfall in the tropics varies from almost nothing in deserts to many thousands of millimetres in equatorial regions and is an important influence on species' choice since water requirements of trees differ considerably. For example, *E. deglupta, A. mangium* and other species from moist lowlands, which grow naturally where there is rainfall in almost every month and over 2000 mm annually, will not survive in semi-arid regions with similar average temperatures and only 500 mm of rain. Conversely, drought hardy species found in regions of lower rainfall (e.g. *A. nilotica, A. indica, Prosopis juliflora*) will not grow in areas with high rainfall and humid conditions. However, a few species, particularly those with an extensive natural range, may show adaptations to a wide range of rainfall. For example, *E. camaldulensis* has an extensive distribution in Australia and occurs mainly in areas with a mean annual rainfall of 250–600 mm, while a few areas receive up to 1250 mm and some as little as 150 mm. However, *E. camaldulensis* is a riverine species and may derive part of its moisture requirements from seasonal flooding or a high watertable. Nevertheless, by selecting specific provenances it can be grown in both semi-arid Mediterranean and dry tropical climates (see Table 8.7).

Annual rainfall is only a first approximation of the supply of water for plants as rainfall effectiveness is modified by several factors including seasonal distribution, intensity, evaporation, surface run-off, and seepage or subsurface flow. These directly influence soil moisture availability, which in turn depends on the water holding capacity of the particular soil type. A large fraction of high intensity rains may run-off when they exceed the infiltration capacity of the soil surface (Brown *et al.* 1997). Even gently sloping soil surfaces with a poor structure may take up little moisture from rainfall of moderate intensity, and soils of poor structure are common in many parts of the tropics. In an established plantation not all rainfall reaches the ground. Rainfall interception by *Eucalyptus* species is about 12%, teak (*T. grandis*) and mahogany (*Swietenia macrophylla*) is about 20%, *P. caribaea* and *P. merkusii* above 20% in lowland tropical areas and vigorously growing *A. mangium* plantations up to 30% (Bruijnzeel 1997). Hence records of mean annual rainfall may give a poor indication of the requirements for successful establishment.

A first simple division of species can be made according to annual rainfall characteristics (Chapter 1). A broad grouping recognizes four classes with examples of species that have grown well in these zones (Table 8.7).

There is, of course, some correlation between total rainfall and rainfall distribution; a region with four humid months will generally have less rainfall

Table 8.7 A classification of climate according to the amount and distribution of annual rainfall with examples of suitable species for tropical and/or subtropical regions

Annual rainfall (mm)	Climatic type	Examples of species suitable for planting
More than 1800	Humid	*Eucalyptus deglupta, E. pellita, Acacia mangium, Swietenia macrophylla, Paraserianthes falcataria, Pinus merkusii*
1000–1800	Monsoonal wet/dry	*E. camaldulensis, E. tereticornis, Senna siamea, Albizia lebbeck*
700–1800	Subtropical humid and subhumid	In drier areas *E. camaldulensis, E. tereticornis*. In wetter areas *E. grandis, E. urophylla, Gmelina arborea, P. caribaea, A. auriculiformis, A. mearnsii, Tectona grandis*
>200–600	Arid and semi-arid	Few species will grow without irrigation in the most arid areas. Elsewhere *Prosopis* spp., *Acacia* spp., *Azadirachta indica, Parkinsonia aculeata*

in total than somewhere with ten humid months. Nevertheless, as well as total amount of rainfall, tree species often show a marked preference for a specific pattern: teak will survive even in a uniformly moist tropical climate but it grows best where there is a dry season of 3–4 months. Few eucalypts will tolerate a tropical humid climate where there is no distinct dry season as they become highly susceptible to leaf diseases.

Away from equatorial regions, and where rain is not uniformly distributed, whether rainfall falls in the warmest or coolest times of the year (summer or winter) becomes important. *P. patula* only grows well where annual rainfall is above 1000 mm and when this falls in summer as in the highlands of southeastern Africa. In other areas, such as the PNG highlands where there is high rainfall but little seasonality, this species grows less well. In contrast, *P. elliottii* and *P. taeda* grow well in both summer rainfall areas and where rainfall is uniformly distributed. Eucalypts from winter rainfall areas rarely thrive where summer rainfall predominates and *vice versa*. Provenances of *E. camaldulensis*, which occurs naturally in both summer and winter rainfall areas, are adapted to the rainfall pattern and those from winter rainfall regions grow poorly when planted in summer rainfall areas. Few tropical species show preference for a winter rainfall pattern, which is a Mediterranean characteristic, and occurrence of this kind of climate is mostly outside the tropics.

Water balance

The input of moisture (rain) is only one part of the equation; moisture also evaporates. The rate of evaporation from any surface (lakes, the soil, tree leaves) depends on temperature, relative humidity, and wind speed. The rate will be high on hot, dry days with a breeze blowing. If the amount of evaporation that could take place, called potential evapotranspiration, combined with water losses due to percolation and runoff, exceeds the quantity of moisture available from rainfall or groundwater a moisture deficit develops. All but the wettest climates have a mean annual deficiency in moisture. The work of Thornthwaite (1948), elaborated by Thornthwaite and Mather (1955), shows that the estimation of seasonal changes in this water balance is very useful for classifying climate ecologically. By plotting mean monthly rainfall and evapotranspiration potential, deficits become clearly evident. Such graphs show at a glance the total amount of rainfall, the pattern of distribution, and the deficit periods—the main determinants of the water balance of a particular climate. Climate characterized in this way can be used to compare

a species' known habitat with that of the proposed planting site, and this is an important step in matching species with site (Fig. 8.8). Using such data, and evidence from trial plantings, Golfari and Caser (1977) have been able to define broad ecological zones for northeast Brazil and make recommendations about which species to plant.

The use of water balance and climatic classifications were used effectively to match climates and develop maps where particular species could be planted in places such as Brazil and South Africa (Golfari *et al.* 1978; Poynton 1979). However the computing technology now available can generate a climatic description for each species or provenance and almost instantly generate a detailed map showing where matching environments occur for that species (Booth 1990, 1996). The new technologies are considered later under 'computer-assisted climatic analysis'.

Other forms of precipitation

Occurrence of mist or low cloud, while not condensing into rain, can augment moisture supply and reduce evapotranspiration. It may be locally important and some species, such as *P. patula*, benefit from misty conditions. Needle morphology studies suggest that some pine species may be able to absorb moisture condensing on the needle. Coastal mists appear to assist growth of eucalypts planted in arid coastal areas of Mexico and South America.

Hail occasionally affects species choice, owing to the susceptibility of some species, for example, *P. patula*, to infection from *Sphaeropsis sapinea* (*Diplodia pinea*), which can be lethal. In the Usutu forest, Swaziland, hail-prone areas are planted with *P. elliottii* or *P. taeda* rather than *P. patula*.

Temperature

There are few places in the tropics where temperature extremes limit vegetative growth though at high altitudes frost may preclude the use of many tropical species. The main influence of temperature is on evapotranspiration, high temperature accelerating evaporation and therefore stress. However, quite apart from this effect, species do differ in their requirements for warmth. *P. patula* and *E. grandis* grow best in the cooler tropics, *P. caribaea* and *E. deglupta* in hotter areas. The poor growth of *P. taeda* in Fig. 8.3 is probably due to high night temperatures typical of the equatorial tropics; day : night contrast in temperature appears a requirement of many, more subtropical, species.

For species choice the mean temperatures of the hottest and coldest months should be compared between the natural habitat and the planting

Region 1   Humid tropical

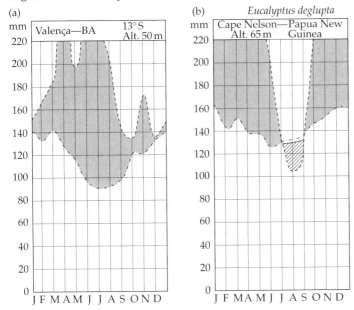

Region 4   Semi-arid tropical/subtropical

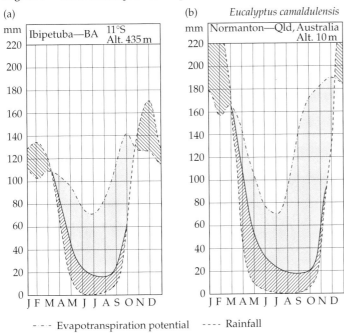

- - - Evapotranspiration potential

☐ Water surplus

▨ Soil moisture use

---- Rainfall

▧ Soil moisture recharge

☐ Water deficiency

Figure 8.8 Graphs of rainfall and evapotranspiration (the water balance) for two ecological zones in northeast Brazil (a) next to those of the native habitat (b) of eucalypt species considered well-suited for planting in the zone (from Golfari and Caser 1977).

area (Fig. 8.9). Diurnal temperature variation and occurrence of frost are also important when matching species with site. Frequency, intensity, and duration of frosts in relation to the degree of hardening and inherent frost tolerance of the species are the determinants of the extent of frost damage. This complexity has made species-site matching difficult in some frost-prone areas.

Some regions that have similar mean minimum temperatures of the coldest month experience different extreme frost events. Subtropical areas of southern China and southeastern United States occasionally have severe drops in temperature, which can be catastrophic for tree growth of more sensitive species. Generally it is the severity of frosts that limits eucalypt in many countries of the northern hemisphere (Turnbull and Eldridge 1983).

Computer-assisted climatic analysis
The use of climatic mapping programs to select tree species and provenances for introduction is based on the concept of homoclime analysis. Homoclimes have been used for many years to identify areas where there is a high probability that an introduced species will grow satisfactorily. One problem was that forest sites, especially in the tropics, are frequently remote from meteorological stations and so accurate climatic data were not available. Computers and associated database software have now made it feasible to store, retrieve, and manipulate large climatic data sets. The INSPIRE database, developed by the Oxford Forestry Institute, was an early attempt to match species requirements described by Webb *et al.* (1980) with specified climatic conditions at a site.

A further development was the interpolation of climatic data from meteorological stations to sites where no climatic data were available. This made it possible to estimate reliably climatic conditions for most locations around the world and to analyse climatic conditions experienced by a species throughout its natural distribution and at any sites where it has been planted outside its natural range. PC-based climatic mapping programs can use the descriptions of the climatic requirements of a species and generate maps showing climatically suitable and unsuitable areas for planting it. Climatic mapping

Figure 8.9 Matching of northern Australian provenances of *Eucalyptus camaldulensis* with climatically suitable sites[1] elsewhere in the tropics using climate analysis (reproduced by kind permission of T. H. Booth).

Note: [1]Climatic limits: mean annual rainfall: 250–2500 mm rainfall regime: summer; dry season: 2–8 months (consecutive months <40 mm precipitation); mean maximum of hottest month: 28–40°C; mean minimum of coldest month 6–22°C; mean annual temperature: 18–20°C.

programs have been developed for individual countries including Australia, China, Indonesia, Laos, Philippines, Thailand, Vietnam, and Zimbabwe. Programs for major regions such as Latin America and Africa are also available. Examples of the application of these techniques to assist species and provenance selection are given in Boland (1997), Booth (1996, 1998), and Booth *et al.* (2002).

At this stage of their development the climatic mapping programs are very useful tools, which can remove some of the guesswork that is often applied to tree species selection (Hackett 1996). Nevertheless they have some limitations and need to be applied correctly. For example, climatic information from the natural distribution may be misleading if the species has access to water other than incident rainfall. Factors such as fire, pests and diseases, and competition from other tree species may limit the extent of the natural distribution of a species. For these reasons the climatic requirement data of a species based on its natural distribution must be iteratively improved by data from planting sites. While some programs attempt to take into account the effects of low temperatures, the dynamics of frost resistance and the small-scale variability in the incidence of frost have been difficult to incorporate. This problem can be addressed integrating GIS and various climatic and biological models to assess and map frost risk for planting particular species. This has been done for *E. grandis* and *A. mearnsii* in southern China (Yan 2001). Finally, where there is significant provenance variation in a species the climatic requirements need to be defined for a particular provenance or group of provenances (Davidson 1996).

Soil

Soils in the tropics are highly diverse and variable. Some soils (e.g. Oxisols, Ultisols, and Alfisols) are low in inherent fertility and these make up about 50% of soils in the tropics. Alfisols and Inceptisols have severe physical constraints such as compaction and susceptibility to erosion; and Aridisols and Vertisols are drought-prone and erode easily (Lal 1997). The forester is primarily concerned with the properties and local variation of soils within a project area rather than with the historical development of soil profiles or systems of classification. The main soil properties affecting species' choice also determine what pre-planting ground preparation is required. They are depth, physical structure, fertility, soil reaction, and salinity. For selecting species for planting, soil differences may lead to alteration in choice but only from among the species suited to the climate, except of course where irrigation wholly changes the soil moisture regime. A few soils are so poor that attempting tree-planting is questionable.

Soil depth

Rooting depth is of first importance. Shallow soils may oscillate between being swampy in the wet season and arid in the dry, and the poor conditions for root growth reduce tree stability, resistance to drought, while the small volume of soil may lead to nutrient shortages. Ascertaining the depth to which roots will be able to grow in the soils to be planted, and whether any depth limitations such as a hard iron pan can be overcome, for example, by ripping, forms an important part of site evaluation and the preplanting survey (Chapter 12).

Shallow soils should be planted with drought resistant species. For wet or swampy sites a species must tolerate anaerobic soil conditions, for example, *Terminalia brassii*, *A. auriculiformis* and *Melaleuca cajuputi*.

Physical structure

Soil structure affects retention and movement of water in the profile, aeration, fertility (cation exchange capacity—CEC), and penetrability for roots. Soils vary from heavy clays to coarse sands, the intermediate loams are usually most favourable for tree growth. Sandy soils are often infertile and freely draining and an undemanding species should be planted: in the lowland tropics, *P. caribaea* var. *hondurensis*, *Casuarina equisetifolia*, and *A. auriculiformis* will often be suitable. In contrast, very clayey soils, which are mostly more fertile, frequently have poor drainage, which leads to waterlogging or, if exposed, hardening and cracking under conditions of alternate wetting and drying. Few species thrive in such conditions, though some acacias, for example, *A. stenophylla* and *A. victoriae*, and a few eucalypts, for example, *E. microtheca* grow well on heavy clays. However, some of the best soils for tree growth, found in several central American countries, are in fact clays, but they are well aggregated and have good internal drainage.

Fertility

High soil fertility is not as important in forestry as in agriculture but soil fertility levels may influence species choice. The nutritional demands of most tree species are only moderate and many species have symbiotic relationships with nitrogen-fixing bacteria and mycorrhizal fungi to enhance their ability to grow on infertile soils. There is considerable variation among species in adaptability to soils of low nutrient status. In Queensland, hoop pine (*Araucaria cunninghamii*) is planted on soils with a higher fertility level than *P. caribaea* var. *hondurensis*. Similarly in China, *Cunninghamiana lanceolata* is favoured on lower slopes where nutrient levels are usually higher than on the ridges where pines (e.g. *P. massoniana*) are more satisfactory. In Sabah,

Gmelina arborea has been planted on the most fertile soils, *E. deglupta* and *Paraserianthes falcataria* on the soils of moderate fertility and *A. mangium* on the poorer soils, reflecting their differing nutritional requirements. The application of fertilizers can widen species choice, for example, on the depauperate soils in southern China the very tolerant, but slow growing, *E. exserta* was extensively planted without fertilizer but since fertilization has become a common practice more productive species such as *E. urophylla* and *E. grandis* are replacing it.

Tree nutrition is considered in Chapter 14, but it is noteworthy that unsuspected nutrient deficiencies or toxicities have in the past modified species' choice. Common nutrient deficiencies in tropical and subtropical plantations are phosphorus, nitrogen, zinc, copper, and boron (Gonçlaves *et al.* 1997; Dell *et al.* 2001). Once detected, most deficiencies are easily corrected and should not today influence species' choice.

Soil reaction
Soil acidity is important because it influences availability of nutrients. The higher the soil acidity the less base cations are available. Soil acidity also influences the amount of organic matter stored in the soil by retarding decomposition processes. Soils can be classed in terms of pH as moderately acidic (5.5–6.5), highly acidic (4.5–5.5), and very highly acidic (<4.5). Soils developed from calcareous substrates are often neutral to alkaline (6.5–8.0) and highly saline or sodic soils (8.0–10.0). Different tree species are adapted to different levels of soil acidity and alkalinity. Most pines and eucalypts are adapted to acidic soils and grow poorly when planted in alkaline conditions but there is considerable variation in tolerance between and even within species. *P. caribaea* var. *hondurensis* grows poorly on alkaline soils, but var. *bahamensis* is more tolerant and is used on calcareous soils in Jamaica. There is great variation in tolerance among *Acacia* spp., for example, *A. ampliceps* and *A. salicina* require alkaline conditions whereas *A. mangium* and *A. mearnsii* only grow well in acidic conditions. In the drier tropics *A. indica*, *Casuarina junghuhniana*, and *Santalum album* are well-adapted neutral to alkaline soils, and *Cordia alliodora* and *Swietenia macrophylla* to calcareous sites in the humid tropics.

Soil salinity
Significant regions of the tropics are composed of salt-affected soils, either naturally saline or arising from many years of irrigation. Salt-affected soils are usually found in countries, that are seasonally dry in tropical climates (e.g. Thailand, Sri Lanka), semi-arid and arid climates (e.g. Pakistan, northern India, Australia), and semi-arid coastal climates (e.g. Thailand, India). Also, many countries have coastal mangrove systems. Salt-affected soils have usually been adversely altered for the growth of most trees by the action of soluble salts (saline soils), the exchangeable sodium percentage (sodic or alkali soils) or both (saline-sodic soils). Halophytes, that is, plants that grow and complete their life cycle on saline or sodic soils, have evolved specialized physiological adaptations for coping with high salt levels. Halophytes have growth stimulation at low levels of salt but growth reduction at higher concentrations. Most trees and shrubs are non-halophytes. They show reductions in growth with increasing salt concentrations. For example, *A. stenophylla* has its growth increased at a salinity level (electrical conductivity) of 5 dS m^{-1} and reduced by about 50% at 20 dS m^{-1} whereas *E. camaldulensis* may have its growth reduced by 80% when the salinity level reaches 15 dS m^{-1}. Provenances within a species can grow very differently on saline soils such variation has been found, for example, in *E. camaldulensis*, *C. glauca*, and *A. saligna* (Marcar *et al.* 1995). Waterlogging is often associated with saline soils and the degree of plant damage and growth reduction varies with the duration of the waterlogging event and the adaptation of the species. Only a limited number of species can be considered for planting on salt-affected sites. Severe salinity (EC > 8) is tolerated by species such as *A. ampliceps*, *A. stenophylla*, *Casuarina glauca*, *Melaleuca leucadendra*, and *M. quinquenervia*. The tolerance of species to soil salinity, alkalinity, and acidity is given in many species descriptions, for example, Webb *et al.* (1984), Hocking (1993), Marcar *et al.* (1995, 1999), and Marcar and Khanna (1997)-see also Chapter 23.

Some species have specific soil requirements while others tolerate a wide range of conditions. Many species are intolerant of flooded soils and high levels of salinity or alkalinity. Inferences on preferred soil characteristics can often be drawn from a study of site conditions in the species' natural habitat but species trials on different soils are the only certain way of establishing preferences. A survey to identify soil suitability is commonly made at the project appraisal stage and should always be done before planting to enable species to be matched with soils or arrangements made to modify the site by appropriate cultivation and fertilization measures.

Other factors

Although climatic and soil conditions are the primary factors to consider in species-site matching, other factors may also need serious consideration. Some examples are given as illustration.

Flooding

Few species will tolerate prolonged flooding (Gill 1970). But there are exceptions. In the Mekong delta in Vietnam *M. leucadendra* grows satisfactorily in areas periodically flooded to a depth of 1 m (Chuong *et al.* 1996). Most eucalypts require relatively freely drained soils but species such as *E. camaldulensis* and *E. robusta* will survive inundation especially if the water is moving. In a flooded trial of 18 species in Thailand, *A. auriculiformis* survived over 6 months of permanent water, *A. mangium* and *E. camaldulensis* about 4 months, *Albizia procera*, *C. equisetifolia*, and *Samanea saman* about 2.5 months. Other species in the trial were severely damaged within 6 weeks of inundation (Yantasath 1987). Species that grow naturally close to rivers are often flood tolerant, for example, some *A. nilotica* stands in Sudan are regularly flooded for several months by the Blue Nile. Mangrove communities occurring in the intertidal zone on the coast or in river estuaries are highly adapted to their waterlogged environment and species, such as *Avicenna marina* and *Bruguiera* spp., have root adaptations, pneumatophores, to facilitate gaseous exchange (Saenger 2002).

Fire

Species with light crowns, which produce little litter and are resistant to fire damage can be selected for use in firebreaks or in plantations where there is a high risk of fire. Many trees growing in areas where recurrent fires are a feature of the environment have developed adaptations to enable them to survive. Some acacias, eucalypts, and melaleucas have aerial buds protected by bark in the upper parts of the tree that ensure rapid regeneration of the crown following fire (Turnbull 1997). *A. mangium* was originally introduced to Sabah for firebreaks to protect pine plantations in *Imperata* grasslands before it became an industrial plantation species (Sulaiman 1987). *A. auriculiformis* and eucalypts were also planted for this purpose in wide belts between blocks of *P. caribaea* in Leron plains, PNG. In grassland afforestation the risk of fire may require plantings of species that recover well from fire even when young. Some pines, e.g. *P. merkusii*, and many eucalypts regenerate well from lignotubers or buds in the stem after fire damage.

Weed competition

Sensitivity to weed competition affects species choice. Both *P. caribaea* and *Cassia siamea* can grow through dense grass untended although they are easily suppressed by vines climbers and herbaceous weeds on cutover rain forest sites. Eucalypts are particularly sensitive to weed competition, and other fast growing hardwoods such as species of *Anthocephalus*, *Maesopsis*, *Cedrela*, *Chlorophora*, *Cordia*, and *Terminalia* grow poorly in grassland. Usually initial weed competition can be removed by cultivation or herbicides but the rate of growth and density of the crown cover may be critical in preventing re-invasion of weeds (Vuokko and Otsamo 1996). Fast growing, dense-crowned species, such as *A. mangium*, *E. pellita*, *Cedrela* spp., *Gmelina arborea*, and some *Terminalia* spp., close canopy quickly and are good weed suppressants whereas the open crowns of other species, for example, *E. camaldulensis*, *E. tereticornis*, and *Paraserianthes falcataria* are readily re-invaded by weeds in moist tropical conditions. Species with stiff leading shoots, for example, *Khaya* spp. may be favoured where vines are abundant.

Wind

Parts of the tropics, notably the western Pacific Rim and the Caribbean are prone to severe cyclones and hurricanes. In these high risk areas it is essential to consider sensitivity to wind breakage and wind throw when selecting species. Even if the trees survive they may take a long time to recover. Observations after Cyclone Hollanda on Reunion Island in the Indian Ocean noted endemic species were well-adapted to cyclonic conditions and that there was a range of resistance amongst exotic species in plantations, for example, *Albizia* spp. were very resistant whereas *G. arborea* was easily damaged (Tassin and Hermet 1994). Both choice of species and provenance can reduce damage levels. Species with dense crowns are particularly at risk. *E. urophylla* and *A. mangium* are avoided on Hainan Island and exposed coastal areas of south China where *E. tereticornis* and *C. equisetifolia* are more resistant to cyclone damage. For pine planting *Pinus oocarpa* is not considered wind firm and *P. caribaea* var. *bahamensis* is more resistant than var. *hondurensis*.

Animals and insects

There are many circumstances where species will be selected because of their resistance to grazing animals or damaging insects. Thorny acacias in dry areas resist browsing by goats and camels and unpalatable eucalypts may be preferred to more palatable species such as *L. leucocephala* where cattle graze freely. However, eucalypts may be avoided in favour of *Senna siamea*, *Grevillea robusta*, *Pinus* spp. or some acacias in areas where termites are abundant.

Weediness and invasive potential

As concern has grown over the past decade for the protection of indigenous flora there has been strong pressure to choose species for planting with little or no weed potential (Hughes 1994). Serious problems with aggressive woody weeds, often called invasives, such as *Acacia nilotica* in Australia, *A. mearnsii, A. melanoxylon,* and *A. saligna* in South Africa and *Melaleuca quinquenervia* in Florida illustrate the potential dangers of some exotic trees. Leguminous trees and shrubs with hard seeds that persist in the soil have a relatively high risk. In many places the extent of the weediness problem is only just being realized. A cautious approach is recommended with a risk assessment before any new introduction is made and careful monitoring during the testing phase. For species that have already been introduced, the lists of recommended species must be closely scrutinized to avoid weedy species (potential invasives) or care taken when planting to minimize the chance of escape.

Growth prediction

Identification of where particular species will grow on the basis of climate and soils and other environmental variables is useful but further discrimination is needed to provide an indication on how well they will grow at a particular site. The PLANTGRO model (Hackett 1991) is a software package on PC which has been applied to predict tree growth. Its strength lies in the core programme, which can be applied directly in any location. The soil, climate, and plant files are the variable components which can be taken from existing databases or compiled by users specifically for a particular task. Plantgro provides semi-quantitative predictions of growth as well as detailed evaluations of limitations due to moisture, temperature, light, and 12 soil factors (including pH, phosphorus and nitrogen availability, and salinity). It indicates which conditions are most suitable for growth and which are less suitable. It uses Leibig's Law of the Minimum in which the most limiting factor determines the level of tree performance. For some species with extensive distributions and very variable environmental conditions it is critical that the part of the species' range for which the file is being made is specified. Separate files are desirable for low and high altitude sources of *Eucalyptus urophylla*, or for tropical and temperate provenances of *E. camaldulensis*. Davidson (1996) and Pawitan (1996) describe how Plantgro was used in the development of suitability ratings and expected yields and linked to a GIS for the National Master Plan for Forest Plantations in Indonesia. A test of Plantgro predictions for the growth of *E. camaldulensis* in Central America found prediction values were reasonably correlated ($r^2 = 0.63$) with actual field growth (Fryer 1996). Effective use of Plantgro requires detailed information on soils, which may not be readily available in some tropical countries.

Making the choice

How is a species selected in practice? First the purpose of the planting must be clearly defined as this influences subsequent actions. Often the purpose is determined by government policy, which dictates whether the land is used to produce timber, fuelwood, or other commodities or to provide protection of watersheds, soil, or other non-industrial purposes. Whatever the purpose, the type of product or service will determine the choice of species to plant. Species with the potential to produce the desired products and/or services then must be matched with the site conditions (see also Chapters 5, 7, and 12). In the unlikely situation that little is known about which species are suitable for planting the land available, it will be necessary to gather information about potential species and then to set up field trials. Ultimately species selection is a combination of subjective judgement about future markets and other longer-term variables and a critical assessment of available information. The following summary suggests sources of information and types of trials that can be established.

Information sources

There is a large and growing knowledge about tree species and their potential uses. The literature covers their ecology, silviculture and site requirements, and experience with them in many countries. Reviews, monographs, conference proceedings, journal papers and other literature are available in libraries but information is increasingly available in electronic form such as CD Rom or on the Internet. Examples of the different sources are:

General compilations. These are often produced for conferences, workshops, or as guidance manuals. There are annotated bibliographies, some of which cover older literature, for example, *Lowland tropical hardwoods. An annotated bibliography of selected species with plantation potential* (Fenton *et al.* 1977). *A guide to species selection in tropical and sub-tropical plantations* (Webb *et al.* 1984) is a comprehensive listing of species which also suggests species for particular situations such as alkaline soils, high altitudes, etc. The *Multipurpose tree and shrub seed directory* (Von Carlowitz 1986) provides information on sources of seeds, site adaptability, uses and management of a many species. In different regions there are compilations of the ecology, silviculture, and site requirements information, for example, in Africa, *Trees and shrubs of the Sahel. Their characteristics and uses* (Maydell 1986); in India, *Trees for dry lands* (Hocking 1993) in Australia, *Australian trees and shrubs: species for land rehabilitation and farm planting in the tropics* (Doran and Turnbull 1997); and in the Caribbean, *Bwa Yo: important trees of Haiti* (Timyan 1996). There are the two volumes of *Firewood crops* (National Academy of Sciences 1980, 1983) and *Forest production in tropical America* (Wadsworth 1997).

Monographs, bibliographies and reviews of a species, genus or family. Examples include: *Eucalypts for wood production* (Hillis and Brown 1984), bibliographies on teak (Krishna-Murthy 1976), *Eucalyptus pellita* (Harwood 1998), *Grevillea robusta* (Harwood 1989), *Casuarina* spp. (Pinyopusarerk and House 1993) and tropical acacias (Pinyopusarerk 1990; Thomson 1994); reviews such as *Acacia mangium* (Awang and Taylor 1993), *Cordia alliodora* (Greaves and McCarter 1990), *Gliricidia sepium* (Stewart *et al.* 1996), *Erythrina* spp. (Westley and Powell 1993), *Pinus caribaea* and *P. oocarpa* (Greaves 1980), teak (Tewari 1992; Wood 1993) and dipterocarps (Appanah and Turnbull 1998). There are also FAO's Forest Genetic Resources Information series and monographs on fast-growing tropical species for plantations produced by Oxford Forestry Institute, Oxford.

Species' choice for well-defined regions. There are numerous examples—*Tree planting practices in African savannas* (Laurie 1974), *A general silviculture of India* (Champion and Seth 1968), *Handbook of eucalypt growing for southern Africa* (WRI 1972), *Choice of species in China's plantation forestry* (Guofang 1990), *Teak* (Tectona grandis Linn. F.) *provenances of the Caribbean, Central America, Venezuela and Colombia and a practical field guide for care and management of teak plantations in this region* (Keogh 1980a, 1987). *Zoneamento ecologico da regaio nordeste para experimentacas florestal for north-east Brazil* (Golfari and Caser 1977). *Trees for rural Australia* (Cremer 1990), *The use of trees and shrubs in the dry country of Australia* (Hall *et al.* 1972).

Local country statements. These include Skelton's (1981) analysis of all plantation work and species trials in PNG—*Reforestation in Papua New Guinea; The cultivation and management of commercial pine plantations in South Africa* (Marsh 1978); *Gmelina arborea* in Nigeria (Adegbehin *et al.* 1988); Tree improvement activities in Tanzania (Madoffe and Chamshama 1989); Performance of exotic tree species in Puerto Rico (Francis 1986); Chaplin's (1990) *Silvicultural Manual for Solomon Islands*, and *Studies on clinal and ecological variations of some tree species used in plantations* (in Vietnam) (Chuong and Dien 1987).

Choice of species by end-use. Many reports describe the properties and characteristics of tropical plantation species for various end-uses. For fuelwood: Burley (1980). Pulp and paper-making qualities: FAO (1975). Pulping characteristics of many tropical species have been published by the Overseas Development Natural Resources Institute, UK, and Commonwealth Scientific and Industrial Research Organization in Australia. The species monographs in (2) above usually have a section devoted to utilization, for example, *Pinus caribaea* (Plumptre 1984). An excellent series of books with reviews of information on Southeast Asian species for particular end-uses have been published by PROSEA, for example, timber trees (Soerianegara and Lemmens 1993; Sosef *et al.* 1995), bamboos (Dransfield and Widjaja 1995), and essential-oil plants (Oyen and Dung 1999).

Choice of species for specialized purposes. There are now numerous species lists for agroforestry applications (Von Carlowitz 1986) and other publications of International Centre for Research in Agroforestry (ICRAF), country experiences with multipurpose trees, for example, in Withington *et al.* (1988), and arid zone species: Kaul (1970), Burley *et al.* (1986), and so on.

To find out what has been published on a species or the silviculture of a region reference can be made to information storage and retrieval systems such as the Commonwealth Agricultural Bureau abstracting service and the publications *Forestry Abstracts, Agroforestry Abstracts*, and *Forest Products Abstracts*. Where the facility is available a simple way of finding out about recent publications is to use databases where numerous abstracts are stored on computer file and made available on CD-Rom or on the Internet. Examples are the CABI's Forestry Compendiums (CAB International 1997, 2000), the TROPIS database at the Center for International Forestry Research (CIFOR), Indonesia and the MIRA database in the Centro Agronomico Tropical de Investigacion y Enseñanza (CATIE), and the Multipurpose Tree and Shrub database. Version 1.0. International Centre for Research in Agroforestry, Nairobi, Kenya (Von Carlowitz *et al.* 1991).

Visits from specialist foresters, bilateral cooperation, association with FAO, reports of consultants. These will all provide some guidance about species choice. But in recent years a particularly useful source of information has arisen through establishment of networks linking people with common interests in a topic, for example, Winrock International's Forest, Farm and Community Tree Network (FACT Net) and its predecessor the Nitrogen Fixing Tree Association, Rural Development Forestry Network (Overseas Development Institute, London), FAO's TEAKNET and the IUFRO Working Group on Nitrogen-Fixing Trees.

Laying down field trials

Species and provenance selection cannot be made exclusively from study of literature and site data, a trial of more promising ones must be carried out. It is a fact that most plantation programmes require a choice of species/provenance and accurate prediction of performance relatively quickly. These two goals may conflict but there are several approaches that can be taken. The choice of method depends on factors such as the species, what traits are important, the range of sites under consideration and how much is already known about the potential species and their performance in the proposed plantation area. As previously indicated, selecting species for planting is still a very subjective process largely reliant on personal knowledge, judgement and experience, augmented by literature reviews but methodologies have been evolving and in recent years computer-assisted support systems have been developed.

There have been an enormous number of field trials in the tropics over many years. Conventional field trials are relatively straightforward and can be carried out without specialized equipment. However, they are relatively long-term experiments and subject to numerous hazards such as fire, cyclones, and pests and diseases. They require significant investment of money but provide accurate results slowly. Using early results can be misleading and economically risky if a large plantation programme is envisaged. The large plantations of *Acacia mangium* in Asia were started with minimal and inadequate information on the performance of different provenances on a range of sites over the proposed rotation period of 7–15 years. Even when several hundred thousand hectares had been planted, very few trial plots or plantations had reached rotation age. Production in most of the initial plantations was suboptimal and less profitable due to incorrect choice of seed sources and inappropriate silvicultural practices.

Conventional field trials will continue to be a critical part of any species/provenance selection programme. Guidelines for field testing species in the tropics have been described in manuals by Burley and Wood (1976) and Briscoe (1990). There normally are several stages for the selection of species for industrial plantations.

1. Species elimination: evaluate many species, eliminate failures, and identify promising ones.
2. Species refinement: examine genetic variation within promising species, in particular compare provenances.
3. Industrial scale trials: large-scale trials to provide stand growth data, to test methods of cultivation, and to evaluate the likely species on the range of sites encountered in the project.
4. Tree improvement: identification of land races, breeding, clonal propagation, etc., to create better forest stands for later plantings and subsequent rotations.

Following this sequence takes time, and usually a species will be chosen based on 1, or from the evidence of neighbouring plantations where these exist. Stages 2 and 3 are mostly carried out while a project is being established, and stage 4 is an ongoing commitment once significant investment in a project has begun. The time and area of land required for field trials for the species elimination and refinement phases may be reduced by the careful use of existing data. Some of the computer modelling techniques that have been developed to process data, which can assist the setting up of trials and complement their results, are discussed in the following text.

The design and management of field trials to select species for small-scale woodlot planting or agroforestry by farmers and private entrepreneurs may differ markedly from those set up to choose species for industrial plantations. The domestication of new (wild) species requires the same degree of scientific input but a more locally based strategy. The choice of species is often more complex because interactions with other crops, risk factors, and socio-economic issues must be considered carefully. The growers are very heterogeneous and may use many species to produce a variety of products, only some of which will be traded. The process of species/provenance selection involves identification of farmers' priorities, a similar approach for exploration and seed collection for both industrial and agroforestry species but a rather different process of evaluation and involving additional input from other disciplines and from the farmers (Pottinger *et al.* 1996; Boland 1997). The testing of candidate species in agroforestry systems will involve the farmers directly and any additional breeding, and seed production and diffusion is likely to differ significantly from industrial plantation species (Simons *et al.* 1994; Stewart *et al.* 1996). The major steps in the evaluation and domestication of agroforestry species are:

- exploration of the natural resource and collection of seeds;
- establishment of on-station trials to identify superior species and their provenances;
- on-farm evaluation of the most promising provenances;
- establishment of seed production areas and distribution of seed to users.

CHAPTER 9

Seed collection, supply, and storage

Introduction

Growing seedlings in a forest nursery is the main way of raising planting stock in the tropics, though increasingly some plantation species are propagated vegetatively. Large quantities of seeds are needed each year for industrial plantations and smaller-scale planting by farmers, villagers, smallholders, and other land managers. There is abundant evidence from around the world (e.g. Jones 1994) to show that the availability of good quality seed of the best adapted species is one of the most critical factors in ensuring healthy, productive, and profitable plantations.

An understanding of the biology and ecology of seeds in general provides a good basis for applying tree seed technologies in seed collection and nursery practice. Baskin and Baskin (2001) have synthesized information on seed germination ecology. Detailed accounts of all aspects of tree seed technology in the tropics are found in Bonner *et al.* (1994) and Schmidt (2000) and in a series of lecture notes and technical notes published by the Danida Forest Seed Centre (list available on www.dfsc.dk). There are also compendia for specific groups of species, for example, eucalypts (Boland *et al.* 1980), dry-zone acacias (Doran *et al.* 1983) and dipterocarps (Krishnapillay and Tompsett 1998). Aspects of tree seed supply and management for agroforestry and farmers are covered by Gonslaves (1990) and Mulawarman *et al.* (2003).

Seed quality

Seed quality is a broad term embracing seedlot characteristics such as genetic constitution, germination percentage, vigour, and purity. It is common to refer to the *genetical, physical*, and *physiological* states of the seed, although the precise definition depends on the use to which the seed will be applied. Genetic quality refers to the inherent capacity of a seed to produce a tree adapted to the environmental conditions at the site where it is planted and having the potential to produce the desired products and services economically. Choice of provenance, families, and individuals can have important influences on genetic quality, as do contamination by foreign pollen and inbreeding. Seeds with good physiological quality for industrial plantations have high germination percentage and vigour. Physical quality refers to seed size, infestation by pathogens, etc. Many factors can influence the physiological and physical quality of seed and all affect the economics of producing seedlings in the nursery. In general, the more mechanized the nursery system the higher the quality of seed needed. The advantages of good physiological and physical seed quality are:

- improved storage life;
- minimal seed wastage; and
- uniform seedlings in the nursery.

Seed quality is usually at its maximum at seed maturity, thereafter deterioration will occur due to normal ageing, the effect of environmental conditions, and any damage sustained during collection, processing, and storage (Lauridsen 1996). The rate of deterioration depends on the type of seed and the conditions to which it is exposed. To minimize deterioration total quality control must be applied to all phases of seed collection, handling, processing, and storage to produce high quality seeds (Poulsen 1993, 1996; Turnbull 1996).

Quantity of seed needed

The rapid expansion of planted forests in recent years has led to large increases in seed requirements. Companies and government agencies with large planting programmes may require tonnes of seed each year. In the 1980s Brazil was using 16 tonnes of tiny eucalypt seeds annually and in recent years it has not been unusual for some Indonesian companies to use over 1 tonne of seeds to undertake an annual planting programme of 30 000 ha of *Acacia mangium*. Plantation managers must consider seed supply at least 1–2 years in advance of the nursery phase. This is especially the case if seed has to be imported from another country or if the species concerned exhibits periodicity in its seed crops with heavy crops several years apart. Failure to allow adequate time for seed procurement can place the success of plantation development at considerable risk. It may also tempt the manager to use seeds of suboptimal quality in order to meet seedling production requirements.

Seed often costs hundreds of dollars per kilogram and so it is sensible to calculate carefully the quantity needed. Nevertheless the cost of seed rarely exceeds 2% of direct establishment costs so it is false economy to buy cheap inferior seed when higher quality seed is available (Turnbull 1984; Midgley 1990). It costs just as much to establish trees from poor seed as it does from seed of the highest quality but there can be an enormous difference in productivity and profitability. This is equally true whether in large-scale industrial plantations or in small-scale farm woodlots or even single trees. Normally, at least double the amount of seed needed for an annual planting programme is stored so that temporary difficulties in obtaining seed will not hold up a project.

Seed requirement is calculated from the following: (i) size of annual planting programme in hectares for each species = Ha; (ii) intended number of seedlings to be planted per hectare = N; (iii) an estimate of survival after planting, for example, 90%, and the need to fill gaps = S as a decimal; and (iv) estimated number of plantable seedlings per kilogram of seed = PS. To obtain the weight of seed needed for 1 year's planting of a species the equation is:

$$\text{Weight of seed needed in kilograms} = \frac{\text{Ha} \times \text{N}}{\text{S} \times \text{PS}}$$

Information about the planting programme (i), spacing (ii), and likely survival (iii), is usually readily available, but an estimate of plantable seedlings per kilogram of seed (S) is often more difficult to obtain. It is determined using the relationship:

$$\text{PS} = \text{N} \times \text{G} \times \text{P} \times (1 - \text{M})$$

N = number of seeds per kilogram (Table 9.1);
G = germination percentage, as a decimal;
P = purity percentage, as a decimal, indicates proportion of seed which is other matter (impurities), for example, chaff, husks, etc.;
M = estimate of seedling mortality during nursery life, as a decimal.

Factors N, G, and P are mostly assessed at a national seed testing centre, such as a forest research station, before distribution to nurseries. M, caused by disease, pests, and other damage, must be estimated from past experience, but is best overestimated, that is a high decimal, as an insurance against unexpected losses. The level of seedling recovery varies between species and nurseries but

is commonly about 70% for conifers (M = 0.3). Commercial eucalypt 'seed' is often a mixture of seed and chaff which may be very difficult to separate and seed testing figures refer to either the germination or viability percentage per unit weight of seed and chaff.

The variation in seed weight shown in Table 9.1 may be due to: (i) provenance differences, for example, *Pinus caribaea* var. *hondurensis*—Honduras provenances 72 000 kg^{-1}, Mountain Pine Ridge (Belize) provenances 52 000 kg^{-1} (Lamb 1973); (ii) fruit size—for example, large capsules of *Swietenia macrophylla* have heavier seeds with higher viability than small ones (Busby 1968) but this is only true sometimes for teak (Kumar 1979); (iii) crown class of parent tree; (iv) seed maturity; (vi) site fertility, etc.

Table 9.1 Seed weight of some common plantation species

Species	Seeds kg^{-1}
Acacia mearnsii	60 000–80 000
A. mangium	60 000–70 000
Anthocephalus chinensis	Several million
Araucaria angustifolia	100–180
A. cunninghamii	2400–4000
Azadirachta indica	4000–6500
Casuarina equisetifolia	200 000–800 000
Cedrela odorata	45 000–60 000
Cordia alliodora	20 000–30 000
Cupressus lusitanica	170 000–320 000
Eucalyptus camaldulensis	650 000–750 000
E. deglupta	1–2 million
E. globulus	70 000–90 000
E. grandis	550 000–700 000
E. maidenii	155 000–175 000
E. tereticornis	600 000–700 000
E. urophylla	400 000–500 000
Faidherbia albida	20 000–40 000
Gmelina arborea	700–1400
Leucaena leucocephala	27 000–34 000
Nauclea diderichii	300 000–900 000
Paraserianthes falcataria	38 000–44 000
Pinus caribaea var. *hondurensis*	52 000–72 000
P. elliottii	30 000–37 000
P. kesiya	55 000–62 000
P. merkusii	58 000–62 000
P. oocarpa	41 000–55 000
P. patula	100 000–160 000
Senna siamea	30 000–45 000
Swietenia macrophylla	1300–3700
Tectona grandis	1000–1900
Terminalia ivorensis	5500–6600
Toona sureni	Many million
Triplochiton scleroxylon	2800–3100

Seed certification, records, and dispatch procedures

Full records for all lots of forest seed should be obtained and kept. It is most important to correctly identify and label every seedlot.

Seed certification

It is the responsibility of those collecting seed to identify correctly the species and provide information on the collection location. Most important is for every seedlot to be correctly identified and labelled. Full records about all lots of forest seed should be made and kept. Figure 9.1 is an example of a form used for recording seed collection information. Seed purchasers should insist that they receive adequate documentation on the seed source. This is usually provided using a certificate of seed origin examples, which can be seen in Schmidt (2000). This certificate shows the recipient of the seed how the species was identified, where and when the seed was collected, and other site and stand information about the seed source. Thus the recipient knows what he is getting and can trace back a seedlot if problems develop later. In the past use of uncertified seed, sometimes even inadequately labelled, has often made it impossible to trace the origin of seed which has produced (say) a promising stand meriting further trial or one of bad form and to be avoided.

There has been international pressure for seed certification to be more than just source identification such that the genetic identity of all reproductive material used in forestry is known (e.g. Barber 1969). The Organization of Economic Cooperation and Development (OECD) operates a scheme for forest reproductive material (OECD 1976) which several countries use (about 20), but no worldwide system yet exists and it is rarely used for tropical tree seeds. The OECD system recognizes three basic categories of genetic value of reproductive material:

1. *Source-identified*: from good natural stands and plantations registered as seed sources by a Forest Department (Designated Authority) and which controls collection, processing, and storage of seed obtained.

CSIRO Forestry and Forest Products
PO Box E4008, Kingston,
Canberra, ACT 2604

AUSTRALIAN TREE SEED CENTRE
SEED COLLECTION DATA SHEET

CSIRO

Species:	Lat: ° '	Long: ° '	Seedlot:
Location:			
		State:	Alt (m):

Habitat:		Provenance name for Database:			Koeppen Climate Class
Veg'n structure:	Soil texture:	Association includes:	Freq:	Ht (m)	Comments:
Sp freq:	pH:				
Aspect:	Soil colour:				
Slope:	Geology:				
Seed crop:	Predation status:				
Bud:	Root sucker:				
Flowers:	Coppice:	Map name:			

Colln No	Bot sp	Film No	Ht (m)	Age	Bole dbh (cm)	Bole Form	Den	Crown Brn	Crown Wdt	Ht (%)	Description/notes:	Seed weight (g)	Viab/ 10g
Team:							Date:				Collected as Bulk: Individuals: Total:		

Figure 9.1 Seed collection record.

2. *Selected*: stands meeting criteria of uniformity, size, isolation, phenotypic quality, etc.

3. *Certified*: from clonal trees in seed orchards or elite trees whose genetic superiority is proven from progeny or provenance testing.

Statement of analysis and seed testing

Source identification is one important component of certification, but national and international certification systems exist to control the phytosanitary condition, physiological and physical quality, and genetic composition of plant propagules, including seed, though few developing countries at present participate in such schemes (Burley 1987; Zobel *et al.* 1987). Seed-testing provides many of these additional data needed about a seedlot: viability, moisture content, purity, etc., and the analysis is normally shown on the seed consignment form sometimes called the seed quality test certificate.

Seed testing is a science itself with many different ways of assessing seed physical and physiological characteristics. Standardized rules for seed testing have been prepared by the International Seed Testing Association (ISTA) but the main emphasis is on seeds of agricultural, and horticultural crops and temperate trees. The principles used to achieve accuracy and reproducibility provide an excellent basis for testing tropical trees and shrubs (ISTA 1993; Gosling 1996). Good general accounts of testing tree seeds are provided by Gordon *et al.* (1991) and Schmidt (2000). Standardized procedures for testing Australian eucalypts, acacias and other genera have been developed (Gunn 2001). The information can be accessed on the Australian Tree Seed Centre's website (www.ffp.csiro.au/tigr/atscmain/index.htm).

The minimum testing required is direct germination tests on representative samples of seed taken from the batch to be sown. Ideally, this is done under fully controlled conditions of light, temperature, and humidity, but a covered Petri dish with moist filter paper can be used if nothing else is available (Fig. 9.2). The germination percentage is ideally the maximum germination obtained under optimal conditions but in practice time may be limiting or the optimum conditions not known. Thus, instead, the proportion of seed that have germinated satisfactorily after a fixed number of days is often used, for example, 28 days is usual for *Pinus caribaea*. For large seeds a sample of the seedlot can be cut open to determine what proportion is empty, since this is the main reason for failure to germinate. Gupta and Kumar (1976) describe the technique for teak. However, this cutting test tends to overestimate germination percentage. Nevertheless, this simple

Figure 9.2 Germination testing.

test greatly helps the nursery manager decide how much to sow.

It must be emphasized that laboratory germination and viability tests predict the maximum germination percentage attainable under laboratory conditions and not nursery emergence. Germination capacity in the laboratory is usually an overestimate of seedling emergence in the nursery and viability tests are usually an even greater overestimate. This is because tree seeds are sensitive to environmental conditions in the nursery, which are suboptimal compared to the laboratory, and because the pretreatment of large seedlots is usually less effective than pretreatment of smaller lots for a seed test. As indicated earlier, the factor by which the laboratory germination estimates must be reduced has to be estimated by past experience with the species and the nursery. Generally the laboratory germination estimate is closer to the nursery germination rate for larger seeds than for smaller seeds.

Seed procurement

Depending on the choice of species, seed may either be obtained locally or imported. For exotics, initially imports must be made from either countries where the natural stands occur or from countries that have a surplus of seed from plantations, for example, *P. caribaea* (Fiji) or *E. grandis* (South Africa, Zimbabwe). Importing seed is costly, the quantities available are often inadequate, and the quality unpredictable. Any quarantine and administrative delays can substantially reduce the viability of some species.

Procurement of seed from local sources may require less lead-time but considerable efforts are required to survey seed crops, undertake collection and extraction, and provide adequate storage conditions.

Importing seed

The source of supply depends to a large extent on the quantity of seed required, the use to which it will be put and the funds available. Large quantities of seed of single species for use in routine industrial plantations are usually obtained through commercial seed companies whereas small quantities for purposes such as research are best obtained from more specialized agencies. Almost all countries import at least some seed to help meet their requirements. Most forestry services have seed sections or officers responsible for importing (and exporting) arrangements including issue or checking of phytosanitary certificates, to show that seed has been treated to prevent inadvertent transmission of pests and diseases. Forestry services, private companies, and research establishments are all potential suppliers, and information about seed availability can be obtained through FAO, reference to compendia such as Von Carlowitz (1986), Von Carlowitz *et al.* (1991), Gonslaves (1990), and national seed centres.

Local collection of seed

Few countries rely wholly on imported seed. Therefore, it is important that local collections both make the best of the genetic potential available and are carried out efficiently to minimize costs. Schmidt (2000) gives a thorough account of all stages of the process. Making seed collections requires a systematic approach and involves a number of stages as shown in Fig. 9.3.

Planning seed collections
When planning the seed collection it is essential to consider the following factors:

● species selection;
● the quantity of seeds required;
● where to collect including seed source and a seed tree selection;
● when to collect; and
● the collection method to be used.

    The last 3 bullet points relate to the immediate objective of seed collection, that is to provide the appropriate quantity of seed with high physiological and genetic quality at the lowest cost. Some species are easy to collect from and bear abundant and regular seed crops which remain on the tree for a long

time but other species are not so easy. Several species produce small crops during a prolonged season but little can be harvested at any one-time, for example, *P. merkusii*. Others have abundant seed crops but only at long intervals, and the seeds of other species are dispersed or destroyed by predators.

    For collections to be successful the seed crop to be harvested must be studied so that in one operation the maximum quantity of fully ripe seed is obtained. Before collections can be made the forester must know the species' reproductive habit and how it may be affected by environmental factors. In general the following information is the most important.

Seed-bearing age
There is great variation in the age at which trees begin to reproduce; and this is influenced by both genetics and the environment. The age of the tree at first flowering is important in tree breeding as it determines the speed of generation turnover and hence the rate of genetic improvement. It also determines the period that must elapse before local collections can be made in plantations of exotic species. Generally pioneer species have short life cycles and reproduce at an early age, whereas climax species have long life cycles and reproduce at a late age. Fast growing tropical pioneer species such as *Leucaena leucocephala*, may produce flowers and fruits as early as 1–2 years old and *Acacia mangium* starts flowering and seeding in 2–3 years under plantation conditions. Many eucalypts also flower from an early age, for example, *E. deglupta* often starts flowering by 2 years of age and *E. grandis* within 2–3 years but a few others, such as *E. dunnii*, produce little seed before 10 years of age. Neem, *Azadirachta indica*, typically starts bearing seeds at about the age of 5 years. Mahogany, *Swietenia macrophylla*, begins regular flowering and fruiting at age 10–15 years, while many dipterocarps do not start flowering until after 20 years of age (Ng 1981). Many bamboos do not flower and fruit until immediately before they die. This one-time fruiting does not occur in conifers or dicotyledonous trees. In *Araucaria cunninghamii* the female flowering starts about 12 years while the male flowers do not appear until 22–27 years (Haines and Nikles 1987). The production of substantial quantities of seed may be some years after flowering starts. Male and female flowers have been observed on *P. patula* after 2–3 years but a worthwhile seed crop does not usually occur until the trees are about 8 years old (Barrett and Mullin 1966) and in *Grevillea robusta* the first fertile seeds may be produced at 6–8 years of age but the heaviest crops come from trees 20–40 years old (Harwood 1992). Generally, seed production increases with the age of the tree

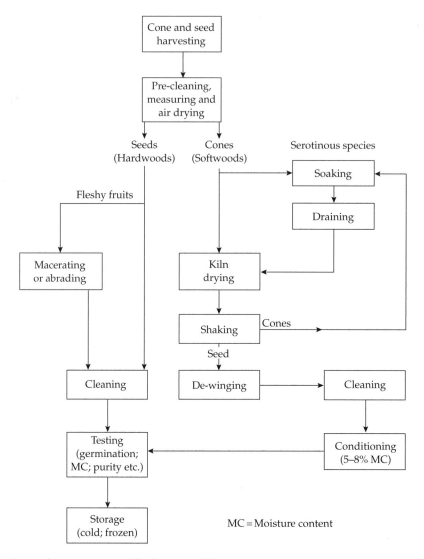

Figure 9.3 Flow diagram of seed processing (modified from Wang 1977).

and with the size of the crown until it peaks and remains at a high level until the over-mature tree starts to decline.

Flowering and seeding habits
The flowering and seeding patterns of tropical trees are very variable and exhibit great complexity (Bawa and Hadley 1990). Some of these patterns can have a major effect on seed production and on the timing and execution of seed collection. They include:

1. *Whether male and female reproductive organs are borne separately on the same tree (monoecious), together in the same flower (hermaphrodite), or on separate trees (dioecious).* Most gymnosperm genera (e.g. *Pinus, Taxodium*), and many angiosperms (e.g. *Quercus, Artocarpus*) are monoecious; only angiosperms show hermaphroditism (e.g. *Acacia, Eucalyptus, Dipterocarpus, Tectona*); and dioeciousness is characteristic of some gymnosperms (e.g. *Ginkgo, Podocarpus*) and angiosperms (e.g. *Calamus, some Casuarina, Populus*).

2. *Extent of seasonality and periodicity in flowering and fruiting.* Gregarious flowering and mass fruiting at long intervals in response to climatic or other factors is common in rain forest trees (Wright and Cornejo 1990). In Asian forests, dipterocarps flower every year but at varying intensities, then at intervals of several years whole regions may experience mass flowering and abundant seed production (Ashton *et al.* 1988; Nguyen-The and Sist 1998). In West Africa the heavy flowering of *Triplochiton scleroxylon* occurs at intervals of 4–7 years after extended dry periods (Jones 1974; Ladipo *et al.* 1994) while in Malaysia *Acacia mangium* and *A. auriculiformis* flower throughout the year although there is usually a seasonal peak (Zakaria 1993). *Pinus kesiya* generally bears abundant seed crops every year but species such as *P. caribaea* and *P. oocarpa* are more variable.

3. *Mechanisms and timing of pollination, fertilization, and seed development.* Gymnosperms often have a very long period between flowering and seed maturity. Araucarias have a reproductive cycle of almost 3 years, several tropical pines such as *P. caribaea* and *P. kesiya* take about 2 years and a few species like *P. merkusii* complete seed development in about 12 months (Schmidt 2000). In some species, for example, *P. merkusii*, cones in different stages of development can be found on the same branch and this can complicate seed collection and extraction. In contrast to the gymnosperms the reproductive cycle of angiosperms is usually fast, especially in tropical pioneer species. In *Gmelina arborea* the flower buds appear only 10 days before the flower expands and the fruits are mature after 2 months (Lauridsen 1990). *Tectona grandis*, requires about 120 days from pollination to fruit maturity with maturation in the final 70 days when the fruits may have reached full size but are still undergoing physiological and biochemical changes (Hedegart 1976).

4. *Its new environment can affect flowering habits of an exotic species.* Poor seed yields of *P. caribaea* is a common feature in lowland, humid tropical areas (Gallegos 1981) whereas *E. citriodora*, which is a relatively shy seeder in native stands in Australia, flowers and seeds prolifically each year in Brazil (Brune 1990). *Eucalyptus camaldulensis* on Leyte, Philippines and *Acacia crassicarpa* on Melville Island, Australia have produced little seed in these environments. Such variation can influence the siting of seed production areas.

Seed dispersal

A good understanding of the development and maturation of seed-bearing organs and seed will enable those planning seed collections to choose the best time for collecting (see later) to ensure that seed at the best level of maturity is obtained. This is the time when most seeds are mature but only a few have been dispersed or lost to predation. The processes of fruit and seed maturation and dehiscence and abscission have been summarized by Schmidt (2000) and the biological constraints in the reproductive cycle that influence seed production are described by Owens (1994). In environments that lack seasonality some species have more or less continuous flowering, and fruit and seed at any stage of maturity may be on the tree at anytime of the year. This means seed collection is not restricted to particular times but also means that only a small quantity can be picked on each occasion. There may be some peaks of flowering and seeding (e.g. in *A. auriculiformis*) when seed collection is best carried out.

There is a group of species with a definite, sometimes short, seed maturation and early dispersal. This facilitates the collection of large quantities of seeds but makes timing very critical. The seed of *Pinus strobus* var. *chiapensis*, a species with potential for tropical highlands, is hard to obtain simply because rapid seed dispersal by wind follows within 24–48 h of full cone ripeness. Premature collection of cones yields low viability seed, delayed collection yields little seed. Eucalypts complete the reproductive cycle relatively quickly but there is considerable variation between species in the time between seed maturation and seed shed. Species with thin-walled fruits, for example, *E. deglupta*, may begin seed shed within days of the seeds maturing (Eldridge *et al.* 1993).

A third group has fruits that persist on the tree for a long time after maturity. These usually present little difficulty for seed collection although seed predation may be significant and seed extraction may be more difficult. The fruits of *Terminalia ivorensis* stay viable on the tree for a long time although they may suffer damage from weevils. Some eucalypts, such as *E. cloeziana* (Boland *et al.* 1980) and species of *Allocasuarina* and *Melaleuca* may retain mature seeds in woody fruits on the tree for 2 years or more.

The above has only touched on some of the important factors; fuller accounts will be found in Matthews (1963), Pijl (1982), Bawa and Hadley (1990), and Schmidt (2000). Species' descriptions and monographs usually include information on reproductive systems and phenology. With a little-known species the habits of specimen trees must be observed, but often people living near where it grows can provide much of the information needed. Keeping records when collections are made is essential.

Seed sources

The term 'seed source' applies to the stand of trees where seed is collected. Potential seed sources are selected during the planning phase of seed collections but the actual trees may only be chosen during the seed collection. One aim of seed collection is to obtain seed of the highest genetic quality economically. The seed sources can be natural stands, thinned plantations, especially managed seed production areas or seed orchards. However the higher the selection intensity applied in the seed sources the greater is the genetic gain but also greater is the expense. It is particularly important to avoid collecting seeds from related individuals or inbred populations and from phenotypically inferior trees.

Seed from natural stands

A poor phenotype may be the result of environmental stress and the progeny may perform well under plantation conditions. *Eucalyptus camaldulensis* growing on the banks of Emu Creek, Petford in north Queensland have poor stem form due to flooding but when planted in the tropics this seed source has trees with straight stems and rapid growth. Such a response can only be determined by planting trials. In the absence of such information, detrimental genetic effects can be avoided by:

1. selecting vigorous trees with desirable characteristics such as straight stems, fine branching, absence of damage by pests and diseases;
2. avoiding stands where the seed crop is sparse or confined to a few isolated trees;
3. collecting from the largest number of widely dispersed trees (preferably more than 10 spaced at least 100 m apart).

Seed from thinned plantations

Collections from final crop trees of good phenotype in managed plantations are an inexpensive way of obtaining seed of slightly improved genetic quality. Seeds can be collected easily and cheaply when the final crop trees are felled. After thinning has taken place to retain the best phenotypes for the final crop it may be assumed that there has been some positive genetic selection. However it is important to know something of the genetic history of the stand. Many exotic plantations have been raised from imported seed originating from just a few mother trees. With this narrow genetic base the progeny are likely to have inbreeding depression and perform more poorly than their parents. Such seed sources should be avoided.

Seed from seed production areas

Seed production areas are stands of suitable provenance raised from seed of a reasonably large number of unrelated trees (at least 10) and managed specifically for seed production. Such seed production areas may be planted with seed production as the main objective but the quickest and most common way is to convert an existing stand of good growth and genetic quality. Early and heavy selective thinning will favour the best phenotypes and allow them to develop full crowns and assist seed crop development (Eldridge *et al.* 1993). An example of a successful seed production area is that of Champion Papel e Celulose S.A. in Brazil. The company planted more than 250 ha of plantations with a selected Australian provenance of *Eucalyptus grandis* at 1600 trees ha^{-1}. After 4 years 100 ha was thinned to 500 stems ha^{-1} and after 7.5 years the stocking was reduced to 160 stems ha^{-1}. The seeds were harvested by climbing the trees (Freitas *et al.* 1983; Zani and Kageyama 1984).

Seed orchards

A seed orchard is a special plantation of highly selected trees, isolated to minimize contamination with pollen from outside sources, and managed for abundant seed production. Seed orchards may be of seedling or clonal origin. A seedling seed orchard is established from seeds of selected families (plus trees) whereas in a clonal seed orchard the selected trees are vegetatively propagated as grafts or cuttings. A seedling seed orchard begins as a progeny test and is gradually converted into a seed orchard by removing the poorest trees within each family or whole families until only the seed producing trees remain (Fig. 9.4). Zobel and Talbert (1984) discuss the relative merits for the two types. Seedling seed orchards are highly flexible, easy to establish and manage, and economical (Varghese *et al.* 2000). They are very commonly used for eucalypts, acacias, and other fast-growing and early fruiting species. Clonal orchards are relatively expensive to establish but are used to speed up seed production in species that are late flowering. They are often used for *Tectona grandis* and *Araucaria cunninghamii*, and *Pinus* species. Problems associated with clonal orchards include graft incompatability, non-synchronized flowering, greater incidence of wind damage, and less genetic diversity.

Figure 9.4 A seedling seed orchard of *Acacia mangium* at Subanjerjii, Sumatra, Indonesia.

The rate of genetic improvement increases in sequence from seed production areas, through unpedigreed seedling and clonal seed orchards to tested advance-generation orchards. It is desirable ultimately for all seed for industrial plantations to be of seed orchard quality but such genetic improvement is not likely to be economically feasible for the diversity of species used in agroforestry and non-industrial plantations. Cossalter (1994) suggests that when considering whether a seed orchard and associated breeding programme is justified economically the following questions need to be addressed.

1. Is there a large annual planting programme?
2. Will the planting programme continue for a number of years?
3. What is the fruiting habit of the species including minimum seed bearing age and periodicity of seed crops?
4. What is the expected seed yield per hectare of the seed orchard?

Whether a seed orchard programme is justified or not, the importance of regular, reliable seed supplies must be emphasized; control of seed production areas and/or seed orchards will enable the use to influence the quality, quantity, and cost of the seed.

Seed crop forecasting

Prediction of quantity and quality of a seed crop and prediction of the optimum harvest time is essential for species with variable seed crops from year to year and with a short harvest period.

Forecasting quantity, quality, and timing of a seed crop is subject to the problem that the earlier the assessment, the better it can be incorporated into the work plan but the more unreliable the prediction.

Seed collecting is usually only worthwhile if there is a heavy crop; an exception is high value orchard material all of which should be gathered. Because timing of collection is critical, it is important to forecast how good the crop is likely to be. A common way to get an approximate estimate is to count flowers or ripening cones on a sample of trees by eye or using binoculars. For example, in pines the cone count seen from the ground multiplied by three gives a reasonable whole tree estimate, for teak each flower head (panicle) usually contains about 60 seeds, and for mahogany, *Swietenia macrophylla*, fruits can be counted about 4 months before ripening and 40 seeds per capsule assumed.

Timing of seed collections

Correct assessment of seed maturity is critical in timing collections and has a major effect on the physiological quality of tree seeds. Knowledge of the reproductive biology and ontogeny is essential and the lack of definitive information on the phenology of flowering and maturation of fruits and seeds is one of the major seed problems for tropical seeds (Bonner 1992). The seed collector needs to be able to judge the state of maturity of the potential seed crop. The period between when a seed becomes germinable and its dispersal may be very short for some species and the effects of climate in a given year may displace the dates of seeding by several weeks from the average. It is therefore necessary each year to check the correct timing of the collection by examination of the maturity of the crop (Stubsgaard and Baadgaard 1989).

The visual signs of approaching maturity vary between species and may need careful observation. Changes in colour, moisture content, and abscission zone development of the fruit or cone are common indicators. Colour changes are often accompanied by hardening of cone scales or of the pericarp of dehiscent woody fruits. For example, the ripe fruit of *Anthocephalus chinensis* is deep yellow or orange, *E. deglupta* seed capsules should be collected as they turn from greenish-yellow to brown, *P. caribaea* cones should be gathered when they are grey-green with brown-coloured scales. The relationship of fruit colour to seed maturity for *Gmelina arborea* is described by Woessner and McNabb (1979) and colour grading of fruits during transport and handling is desirable to maintain seed quality (Lauridsen 1986). Examination of the appearance of larger fruits can be supplemented by cutting tests to examine the development of the embryo and the firmness of the seed coat and endosperm.

The appearance of distinct dehiscence lines of the valves of the eucalypt capsules is a good example of a maturity index for this genus. This can be confirmed by a cutting test as mature seeds have a firm white embryo within a dark seed coat while non-viable immature seeds have a pale seed coat and an embryo that is milky and rather soft when squashed (Boland *et al.* 1980). As the moisture content of fruits decreases with maturation, specific gravity and density (ratio of unit weight to unit volume) will also decrease. Specific gravity indicators have proved useful for some conifers, for example, *P. caribaea*, the guideline is that if more than three-quarters of a freshly collected cone sample floats in water, then collection can start (Stubsgaard and Baadsgaard 1989).

It is often necessary to collect immature seeds if no estimate has been made of maturity and the seed collection team arrives too early, or in species where the time between maturation and dispersal is very short leaving too little time for collections. Seeds may also be harvested early to minimize losses from predation, avoid development of dormancy or reduce pre-processing damage. It is possible to collect fruits prior to ripening and to store them in relatively cool well-ventilated conditions which permit after-ripening of seeds within the fruits or cones (Stubsgaard 1993).

Methods of collection

The choice of seed collection method is usually dictated by the location of the seed source, the size of the trees, and the type of fruits or seeds and their maturity. Seed collection methods and their applicability have been described comprehensively by Schmidt (2000). When undertaking seed collections it is important that those concerned act responsibly and with due attention to safety. Codes of practice for tree seed collections which seek to provide a minimum standard of behaviour during collection operations have been developed, for example, by the Australian Tree Seed Centre (Midgley 1996).

Collection from the ground
Species which produce large heavy seeds or fruits, which normally fall to the ground, can be collected by spreading a tarpaulin or polythene sheet under the tree. This is widely done for teak, *Gmelina arborea*, *Triplochiton scleroxylon*, and many dipterocarps notably *Shorea* spp. In seed stands and orchards, manual or mechanical shaking of trees may be used to increase seed or fruit fall (Fig. 9.5(a)). Seeds of many species begin to germinate as soon as they fall, for example, *T. scleroxylon*, *G. arborea*, *Azadirachta indica*, and need to be sown within days of collection. Collection from the tree just before seedfall followed by depulping or drying can delay this immediate germination and permit longer storage. The main advantage of this method is its low cost and simplicity, especially for large trees where climbing is difficult. Predation by insects and ground-foraging animals may be a problem especially with fleshy fruits.

Seed can also be collected from the ground in association with normal felling operations which may be timed to coincide with seed maturation. It is desirable that individual trees with good phenotypes are selected and marked prior to the logging operation. In addition to collecting seed during normal felling operations, special fellings for seed may be made. In the past this method was used for eucalypts and species growing in rainforests in Africa and elsewhere. Although an individual tree may provide large amounts of seed, this is not a good practice as the trees are killed in the process and unless a large number of trees are felled the seed will have a narrow genetic base. Such fellings will also be unacceptable in National Parks, forests of religious significance and close to habitation.

Collection from standing trees
Collections from the ground can be made with long-handled tools or with a flexible saw attached

(a)

(b)

Figure 9.5 Collecting tree seed. (a) Manual gathering, shaking, and catching ground fall of seed in an *Acacia mangium* seed orchard, Chachoengsao, Thailand. (b) Climbing with a tree bicycle to collect cones in a *Pinus caribaea* seed stand, KwaMbonambi, KwaZulu-Natal, South Africa.

to ropes when the trees are small or low branches of tall trees are accessible. Climbing has to be used when the trees are tall and the crowns inaccessible from the ground. This is a common method for bulk collections in plantations of species of *Pinus, Eucalyptus, Araucaria, Swietenia*, etc. Climbing is hard work and time consuming. It is also potentially dangerous even with trained and well-equipped crews (Fig. 9.6). The climber removes the cone or fruit using a long-handled tool, such as a bamboo pole or a cone hook. Supervision is needed to ensure safe working and that seed is collected from the best phenotypes rather than from trees that are easy to climb. Forest dwellers are often skilled climbers and use little equipment but most seed collection crews use spurs, a tree bicycle or ladders to reach the crown. A technique for climbing tall trees and trees with stems unsuitable for climbing involves securing a rope over a high branch and using this 'advanced line' to pull up a rope ladder pull the climber into the crown using a block and tackle.

Climbing nets, shooting down seed with a rifle or shotgun, extension platforms, and other costly methods are mostly restricted to tree-improvement work and research.

Pre-extraction handling

Handling practice in the period between collection of the seed and its extraction from the fruit or cone can be critical in maintaining the quality of tree seed. It is important to maintain the identity of seedlots to protect genetic quality by careful attention to labelling. Physiological damage is a major hazard in this period. If high temperatures are experienced or there is infestation by insects or fungi significant loss of viability and vigour may occur. Cones and fruits that have just been collected have a high moisture content and are highly susceptible to such damage if not handled appropriately. The sooner the seeds can be processed the better. However if there are

Figure 9.6 A well-trained and properly equipped tree climber prepared for collecting seed from tall eucalypts at Iringa, Tanzania.

spp., *Tectona grandis*, and dipterocarps) the seeds are stored and sown with the entire fruit.

Seed processing always carries with it the risk of seed damage by machines, heat, chemicals, or prolonged soaking in water. The potential severity of the damage depends on the extraction procedures and seed type and must be minimized while ensuring the end product is seed of high physical and physiological quality. It is emphasized that damage sustained during processing will inevitably reduce the seeds' storage life and vigour.

In forestry, seed rather than cones or fruits is normally stored though there are a few exceptions, for example, dry teak drupes. Once the fruit is collected the seed is separated from other parts of the reproductive organ, this is called seed extraction. The importance of this varies with species. For example, seed left in mahogany capsules (*Swietenia macrophylla*) begins to rot in 2 or 3 days and *Chlorophora* fruits start to ferment if stored for any length of time, whereas the 1–4 seeds in a teak drupe are not normally extracted but either germinate *in situ* before planting or sown with the whole drupe. Occasionally a short period of fruit storage, in dry conditions with good air circulation, enables seed collected prematurely to mature but as a regular practice this procedure needs careful evaluation for each species.

unavoidable delays some intermediate processing may be necessary. Removal of twigs and leaf material from the bulk collection, depulping fleshy fruits, reducing the moisture content of orthodox seeds, and storing the collected material in well-ventilated cool conditions sheltered from rain and out of direct sun are examples of the measures that should be taken depending on the species. The container used for temporary storage and transport should allow good ventilation but be fine enough to retain any seed that is released. Loose weave hessian bags are frequently used and Lauridsen *et al.* (1992) discuss other packaging materials.

Seed extraction and cleaning

Seed processing includes a number of handling procedures prior to storage. The processes include some or all of the following depending on the species: pre-cleaning, pre-curing, extraction, dewinging, cleaning, grading, and adjustment of the moisture content (Schmidt 2000). Only seed extraction and cleaning are considered here. While the seed of most species are extracted before storage, the seeds of some species (e.g. *Pterocarpus* spp., African *Terminalia*

Indehiscent (pulpy or fleshy fruits)
Examples of fleshy fruits include: *Anthocephalus chinensis, Azadirachta indica, Aleurites* spp., *Chlorophora excelsa, Cinnamomum camphora, Diospyros* spp., *Gmelina arborea, Santalum* spp., and *Ziziphus mucronata*. Most fleshy fruits can be soaked overnight in water, then macerated on a wire screen while submerged to remove the pulp. Alternatively they can be gently pounded in a pestle and mortar. Generally the pulp separates readily from the seeds but species with a relatively dry pulp, for example, *Santalum* spp. may need to be soaked for several days. Several methods of seed extraction are available and the most appropriate depends on the fruit type, quantity of fruits to be processed and equipment available. Schmidt (2000) describes the following methods:

• individual manual extraction: for small seedlots, very fragile seeds (e.g. *Syzygium cumini*) or very large fruits (*Artocarpus* spp., *Durio* spp.);
• washing in deep bowls or drums: where pulp separates easily from the pulp (e.g. *Melia volkinsii, Prunus africana*);
• washing on wire mesh screens: widely applicable to most fleshy fruited species, for very small seeds

for example, *Anthocephalus chinensis*, very fine screens must be used;
• concrete mixer with abrading material: fruits are mixed with gravel or other abrading material and water and rotated for various periods of time. Not suitable for fragile seeds;
• mechanical depulping: used for large quantities of seeds with relatively hard seed coats (e.g. *Gmelina arborea*) or when the flesh remains firm. The 'Dybvig' depulping machine is a very effective macerator (Stubsgaard and Moestrup 1991; Schmidt 2000).

Extracted seeds should be dried to a safe moisture content for storage.

Indehiscent (dry fruits)
Some species have fruits that will not open on drying and must be mechanically broken to extract the seeds, for example, *Acacia* spp. and *Prosopis* spp. The pods are normally dried to make them brittle before being threshed manually or mechanically. Mechanical threshers are described by Stubsgaard and Moestrup (1991). Where the seeds are embedded in tissue that is impregnated with gums, for example, *Prosopis cinerea*, more pods may need additional treatment and several threshing events.

Dehiscent (dry fruits)
Examples of species with dehiscent fruits that open readily when dry include *Casuarina* spp., *Eucalyptus* spp., many leguminous species, and conifers. There are some species with serotinous fruits that are morphologically dehiscent but which require exposure to high temperature to open, for example, *Allocasuarina* spp. Most dehiscent fruits open when their moisture content falls below 25% and extraction usually involves drying and removal of impurities. In the tropics sundrying is normally adequate—capsules or cones are laid out in the open on sheets or in containers sheltered from wind. In more subtropical locations and when collections are made during the wet season artificial heating or special solar cone kilns may be needed for some species, for example, *Pinus patula*. Robbins (1985) describes a low-cost kiln developed in Honduras for opening cones of *P. caribaea* and *P. oocarpa*. Care must be exercised in both sun drying and kiln drying not to overheat the seeds while they have a high moisture content. The release of seeds may be assisted by shaking or tumbling and if this proves ineffective by threshing.

With eucalypts, opening the capsules by drying releases most of the seed so care must be taken to trap the fine seed before it is blown away. If capsules are either immature or old, vigorous shaking is needed to release all the seed. All pine cones need tumbling to get the seed out, and various mechanical devices are used. Seed, is usually collected on plastic sheets under the drum. Granhof (1984) reported that for *P. merkusii* just 2 min of tumbling reduced extraction time from 6 to 3 days, with only 3% loss of viable seed. *Araucaria* cones disintegrate readily and do not need special drying or shaking; they are best left on a shelf under cover for a few days. After extraction and before any further processing, orthodox seeds should be dried to an adequate moisture content and recalcitrant seeds surface-dried. Dry seed facilitates cleaning and upgrading (Lauridsen 1996). The theory and practice of tree seed drying have been reviewed by Stubsgaard and Poulsen (1995) and extraction trays, drying kilns, and tumblers illustrated in Schmidt (2000).

Seed cleaning
After extraction the seed is still mixed with loose wings, fruit parts, fragments of pulp, dust and other debris. Cleaning separates this extraneous material and often upgrades the seedlot by removing empty and damaged seeds. The aim of cleaning is to eliminate foreign matter to reduce bulk, facilitate storage, and make the seeds easier to handle. There are many types of equipment available for cleaning, but typically air blowing and sieving are used (Fig. 9.7). Winnowing is a common manual method and flotation in water or other liquids of different specific gravity is applicable to some species. Schmidt (2000) describes all these techniques.

Some seeds, such as those of *Pinus* spp. and *Swietenia* spp., may also be dewinged either mechanically or by hand. In Honduras, dewinging of pine seed (*P. caribaea* and *P. oocarpa*) was successfully carried out with a little water in a cement mixer rotating at 20 rpm for about 1 h; the seeds subsequently dried. Teak drupes may be dehusked, not because it benefits germination, but to reduce their storage volume. Seed grading may be a result of the cleaning process as small and light seeds are often removed with other impurities. It can assist nursery operations by giving a more uniform germination rate and seedling growth, and facilitate precision sowing.

Safety issues
Seed processing may pose hazards to the operators and they should be aware of the risks and the safety precautions. Respiratory, eye, and skin irritations are often a hazard. Some species, for example, *Acacia mangium*, release large amounts of irritating dust during threshing or even in manual winnowing and

Figure 9.7 Seed cleaning with a blower at Morogoro, Tanzania.

the use of dust masks and extractor fans can help protect staff. Well-maintained machines, safe operating procedures, and training can reduce risks with mechanical equipment.

Seed storage

The quantity of seed of a species that must be stored depends on the annual demand for seed, the interval between times when good quality seed can be collected and the length of time the species will survive under the storage conditions available. The ability of seed to remain viable under natural conditions varies greatly. Hard, impermeable seed-coats of many legumes such as *Acacia* spp., confer long viability on the seeds under almost any conditions provided they are kept dry and free from insect or rodent damage. In contrast, many large-seeded species are reported to have short periods of natural viability: *Shorea*

robusta and *Azadirachta indica* lose viability in a matter of weeks (Kandya 1987). With some conifers loss of viability in the open can also be rapid: all seed of *Araucaria hunsteinii* is dead within 8 weeks if left in the open and allowed to dry out.

Many factors influence longevity of seeds in storage and it must be stressed that long-term storage is only feasible if the seed is mature and free of damage when placed in storage.

Recalcitrant and orthodox seed

Roberts (1973) distinguished between seeds that can be dried to low (2–5%) moisture content and can be stored at low temperature, 'orthodox', and those which cannot, 'recalcitrant'. Recalcitrant seeds usually have a high (> 30–50%) moisture content at maturity and are sensitive to desiccation below 12–30%. In practice the storage physiology of seeds covers a continuum ranging from extremely recalcitrant in which viability can only be maintained for a few days to extremely orthodox where the seeds can be stored for centuries. A group of species that can be dried to a moisture content low enough to be considered orthodox but sensitive to low temperatures has been termed 'intermediate' (Ellis et al. 1990).

In the tropics, most species used in industrial plantations including pines, eucalypts, and acacias are orthodox. Recalcitrance is primarily associated with climax species in moist tropical rain forests, Dipterocarps consist almost entirely of recalcitrant or intermediate species and most mangroves (Rhizophoraceae) are recalcitrant (Ellis and Hong 1996). About one-third of species in Araucariaceae are recalcitrant including *Araucaria hunsteinii*, *A. angustifolia*, and *A. bidwillii*, while *A. cunninghamii*, *A. heterophylla*, and *Agathis australis* are orthodox. There is a similar proportion in Meliaceae with species of *Trichilia* and *Guarea* recalcitrant and species of *Cedrela*, *Entandophragma*, *Khaya*, and *Swietenia* orthodox (Tompsett 1994). Recalcitrant species are not found in the tropical dry areas although a few, for example, *Azadirachta indica* (Scande and Hoekstra 1999) display intermediate behaviour. Many species that have not stored well have been labelled as recalcitrant just because they lost viability quickly, for example, *Triplochiton scleroxylon* and *Khaya senegalensis* but many are orthodox and their storage life can be greatly extended by improved harvesting and processing techniques.

Recalcitrant species have presented major problems to those who would grow them in industrial plantations. Although plantations of

Araucaria hunsteinii and some dipterocarps have been established in their areas of natural distribution, there is no appreciable trade in tropical recalcitrant forest tree seeds. This is primarily due to difficulties in seed storage, but also to problems of seed collection and control of seed-borne pests and diseases (Bonner 1996).

Storage conditions

General prescriptions for seed storage (based on Schmidt 2000) are:

- store seeds at lowest possible temperature and moisture content that will not damage the seeds;
- eliminate as many pathogens as possible before storage and provide protection from them during storage;
- store in the dark;
- store orthodox and intermediate seeds with low moisture content in airtight containers; and
- store recalcitrant seeds in material permeable to gases but with retention of moisture

Storage of orthodox seeds

Seed moisture content
Orthodox seeds can generally be stored at ambient temperatures for several months or even years without significant loss of viability if they have been properly dried and are protected from insects and other predators. Seeds with a hard seed coat, such as most *Acacia* spp., when dried to a moisture content of 6–8% have a viability period of several years. In general, orthodox seeds respond to reduced moisture content with extended viability with an approximate doubling of storage life for each 1% reduction in moisture content from 14% down to 4–5%. For long-term storage, control of temperature and moisture content are important, the optimum being –18°C and 4–6% moisture content. In general, orthodox seeds should be dried down to at least 5–10% moisture content, at this level there is little or no fungal activity, before storing in sealed, airtight containers or plastic bags. Cloth bags are not suitable for longer-term storage in the tropics, especially if the relative humidity is high. Stubsgaard (1992) gives examples of the types of seed storage containers in use.

Temperature
Most eucalypts, pines, legumes, and casuarinas have been stored successfully at ambient temperatures for several months or even years (e.g. Boland

et al. 1980; Doran *et al.* 1983). Nevertheless for seeds of the same moisture content deterioration will be faster in the high temperatures of the lowland tropics than in the cooler conditions of the tropical highlands. As a rule of thumb, the mean viability period doubles each time the temperature is lowered 5°C between about 50°C and 0°C (Stubsgaard 1992). Cold storage of seeds is expensive and only seeds likely to lose their viability significantly during the storage period are normally stored at low temperatures. The benefits of low temperature storage for species such as *Eucalyptus deglupta, E. microtheca, Flindersia brayleyana*, and *Khaya senegalensis* has been demonstrated by Doran *et al.* (1987). Most orthodox seeds will maintain their viability for many years at temperatures of –10 to –15°C.

Atmosphere
Orthodox seeds do not need oxygen when storage temperature and seed moisture content are low and replacement of the storage atmosphere by carbon dioxide or nitrogen is feasible. As carbon dioxide is relatively cheap, it provides a safe way of controlling the development of seed insects and micro-organisms (Sary *et al.* 1993).

Storage of recalcitrant and intermediate seeds

Seed moisture content
Recalcitrant seeds are very sensitive to desiccation. The lowest safe moisture content is not known for many species but for some very recalcitrant species it is as high as 60–70% while some intermediate species will tolerate drying to 12–17%. For dipterocarps, Tompsett (1992) found *Dipterocarpus intricatus* could be dried to 10% whereas species of *Shorea* and *Parashorea* had a lowest safe drying limit of 30–50%.

Temperature
Tropical recalcitrant seeds are generally damaged by low temperatures and many must be stored above 20°C. Often the seeds most sensitive to desiccation are also the most sensitive to low temperatures. If the species will tolerate low temperatures it may be possible to extend the storage life for 1–3 years. Finding the temperature that inhibits germination which at the same time does not cause chilling injury is the key to extending the storage period. For storing *Shorea robusta*, Tompsett (1992) suggests a temperature of 16°C and a moisture content of 41%. The effect on viability of storing *Araucaria hunsteinii* seeds under different temperature and moisture conditions is shown in Table 9.2.

Table 9.2 Viability of *Araucaria hunsteinii* seed under different storage conditions

Storage method	Length of time before viability dropped to 50% or less
Open container at room temperature	2 weeks
Sealed container at room temperature	2 weeks
Open container at 4°C	6 weeks
Sealed container, dried seed, at 4°C	9 months
Sealed container, moist seed, at 4°C	18 months

The viability of many seeds is maintained longer if they are stored at constant rather than fluctuating temperatures.

Atmosphere

Recalcitrant seeds with a high moisture content continue respiration in storage and consequently need oxygen. It is therefore important that they are well-ventilated while maintaining adequate moisture content. Cotton or hessian bags may be suitable for intermediate seeds but are less suitable for recalcitrant seeds. Polythene bags with a wall thickness of 0.1–0.25 (Bonner 1996) or perforated polythene bags (Stubsgaard 1992) have been found suitable for preventing water loss while permitting gaseous exchange. Moist sawdust, moist coconut dust, and perlite have proved suitable for storing some species (Schmidt 2000).

Storage facilities

Because of the importance and value of seed of pines, eucalypts and other orthodox species in tropical plantation forestry, most tropical countries have erected special seed stores which allow full control of temperature and humidity. However, seed can be stored temporarily anywhere by keeping it in airtight containers in a refrigerator at 0–4°C. Schmidt (2000) discusses the management of seed stores and provides guidelines for their location and construction.

Where seed is readily available and shortage in any 1 year highly unlikely, and it has good natural longevity, there is no need to invest in a sophisticated seed store. For example, teak does not need expensive storage facilities. Flowering and seeding is usually precocious and dry drupes can be satisfactorily stored for at least 2 years in hessian or jute sacks, with no control of temperature or humidity apart from keeping the sacks dry, in the shade, well-ventilated, and

Figure 9.8 Sacks of teak seed in a well-ventilated shed at Brown River, Papua New Guinea.

off the ground to prevent damage from rodents (Fig. 9.8).

Packaging and transport of seed

The transfer of seed either domestically or internationally involves measures to ensure that it retains its quality in transit. Poor packaging and handling of seed in transit is an important cause of deterioration. Numerous factors may affect viability including the maturity of the seed, packaging, the temperatures experienced during shipment, mode of travel, transit time etc. The use of suitable packaging materials to provide protection from both environmental and mechanical damage is essential (Lauridsen *et al.* 1992) and the shortening of transit time is especially crucial for recalcitrant and other sensitive seeds. Schmidt (2000) has reviewed various measures needed to successfully trade and transfer seed.

Seed supply and trade problems

Greater demands for seed for both industrial and other plantations in the tropics have highlighted several problems, notably the inadequate supply of high quality seeds, maintenance of viability in storage, and the transfer of pests and diseases.

Supply of seed

To implement a tree-planting project in an orderly way requires agreement and commitment in several key areas, one of which is the procurement of high quality seeds well in advance of the proposed planting date. Many factors determine species choice but if seed of the best seed-source is unavailable or not used then the project result will be suboptimal or in the extreme case it may fail completely. However, when great attention has been paid to using top quality seed it has led to reduced rotation lengths and increased productivity, for example, in the *Pinus caribaea* and *Araucaria cunninghamii* plantations in Queensland (Keys *et al.* 1996). The supply problem is usually most severe soon after the best seed sources have been identified for a species that is becoming popular for industrial planting. The lack of attention to seed sources and supply of tree seed in many tropical forestry projects is remarkable and disappointing especially when one considers the widespread use of improved varieties in agriculture and the many tree improvement programmes in progress. There is a tendency to use readily available and cheap seed irrespective of quality if the supply of the best seed source is restricted and the price is high. The problem is particularly acute in some public-sector plantations where poor quality seed, mediocre nursery practices and inadequate post-planting maintenance have contributed to low productivity (Jones 1994).

Seed is frequently difficult to obtain from countries suffering internal upheavals, war, or of opposite political persuasion to the potential recipient. Elsewhere, seed supply may be poor simply because of an inability to collect, or inefficient bureaucratic systems which hold up seed in transit, local animosities, etc. This particularly affects small users such as villagers and farmers and many non-governmental organizations (NGOs) involved in social and community forestry programmes. Clearly, the lesson is always to order seed as soon as possible and well in advance of time of sowing.

Seed importation may be restricted owing to lack of foreign exchange and reliance on local collections, particularly in largely uncontrolled ways in widely dispersed social forestry programmes, can lead to very poor genetic material, collection of immature seeds, and so on. Too often seed is simply gathered from the nearest convenient tree with scant regard for, or ignorance of, the consequences.

The following examples illustrate some seed supply problems:

1. *Acacia mangium*. In the mid 1970s *A. mangium* was recognized in Sabah, Malaysia as a species having good plantation potential. In the following 20 years very large plantation programmes were initiated in Malaysia, Indonesia, and some other Asian countries with varying degrees of attention to the seed supply. In Sabah an attempt was made to use the trees from the initial introduction as a seed source and to select trees for a seed orchard. Unfortunately the narrow genetic base of the original introduction resulted in significant inbreeding effects, deterioration in genetic quality, and reduced productivity (Sim 1987). In Indonesia a similar situation arose when the Subanjerijii (Sumatra) plantations, derived from several provenances of *A. mangium* from the southern part of its range in Queensland, were used for seed production. By this time provenance trials had demonstrated clearly that the best seed sources were in the far north of Queensland and Papua New Guinea (PNG) (e.g. Harwood and Williams 1992; Otsamo *et al.* 1996). Unfortunately these seed sources were in remote locations and the cost of seed collection was high. So seed was expensive and demand far exceeded supply. As a result, many plantation managers used genetically inferior seeds that were available and cheaper. The result (Table 9.3) was that plantations from derived populations in Sabah and Sumatra (Subanjerijii) were about 45% less productive that those from PNG sources. The economic consequences of this has been recognized by some companies such as PT Musi Hutan Persada, one of

Table 9.3 Comparison of growth of *Acacia mangium* plantations derived from a natural stand and plantations

Seed source	Mean annual volume increment (m³ ha⁻¹)	Decrease in volume growth (%)
Wipim, PNG	57	0
Sabah seed orchard	30	47
Subanjerijii, Sumatra	31	45

Source: Based on Turvey (1996).

the largest plantation managers in Indonesia, which has imported better seed sources and is developing its own seed production programme (Hardiyanto 1998).

2. *Pinus caribaea*. In the 1980s adequate supplies of this important lowland tropical pine were sometimes difficult to obtain, particularly of varieties *bahamensis* and *hondurensis*, apart from seed for research and trial purposes (Gibson and Barnes 1985). Most natural stands had been cut over, seed crops were poor, and collection was poorly organized. Worldwide demand had increased prodigiously. Few plantation projects were old enough to have their own seed stands and some early seed stands and orchards established in equatorial latitudes in humid, equable climates produced only poor seed crops. However, the supply situation has steadily improved as plantations become older and start yielding fertile seed and programmes for increasing seed supply, often of orchard quality, are pursued. The progress in moving from unselected seed from natural stands to seed self-sufficiency and ultimately to seedling and clonal plantations based on hybrids is well-documented in Queensland, Australia (Nikles 1996; Keys *et al.* 1996). Initially bulked seed from natural stands of *P. caribaea* var. *hondurensis* in Central America was imported without the benefit of the results of provenance trials. Mass selection in the plantations took place in the 1960s and clone banks were established. Queensland became self-sufficient in *P. caribaea* var. *hondurensis* in the late 1970s. A second wave of introductions was made between the early 1970s and mid-1980s as new provenances became available from internationally organized seed collections and information from provenance trials appeared. Extensive seedling seed orchards were established in the late 1970s and early 1980s to back up the clonal seed orchards. In the 1990s monoclonal seed orchards were developed to produce hybrid seed of *Pinus elliottii* var. *elliottii* and *P. caribaea* var. *hondurensis* for commercial plantations.

3. *Triplochiton scleroxylon*. There has been restricted use of this species in West Africa in the past due to a shortage of seed. Flowering and fruiting were erratic, seed yields from fruits were often poor. The seed would begin germinating as soon as it fell and it had a reputation for poor storage and low viability. Clonal seed orchards were established in Ghana and Nigeria but flowering and fruiting have been poor. Because the species has great potential for plantations, vegetative propagation techniques to root cuttings were developed to circumvent the seed supply problem (Ladipo 1985). Since then, clonal selections have been made and the scale of vegetative propagation using both misting systems and low technology non-misting propagators has been expanded to enable the mass propagation of improved rooted cuttings for plantations (Ladipo *et al.* 1994).

Seed storage problems

The maintenance of seed viability in storage should not be a problem for orthodox seeds providing there is careful handling of the seed through the processing phase and due attention is paid to the application of well-proven storage technologies outlined by Schmidt (2000). Bonner (1992) stated that the most challenging problem for seed science in the tropics is how to handle and store recalcitrant seed. Since then a significant effort has been devoted to this problem and the results have been reported in several international works, for example, Ouedraogo *et al.* (1996), Marzalina *et al.* (1999). The International Plant Genetic Resources Institute (IPGRI) and the Danida Forest Seed Centre are coordinating a major international effort to screen the desiccation tolerance and storage behaviour of a large number of tropical forest trees. In recent years progress has been made in detecting recalcitrance and determining the lowest safe minimum temperatures for storage, but overall the problem of determining improved methods of short- and medium-term storage and for handling recalcitrant species remains. To some extent this problem is being circumvented by the development of clonal technologies (Chapter 11).

Seed health and transfer of seed

An important aspect of seed trade is that seed may have to be imported and with it the risk of transferring dangerous seed-borne pests and diseases into areas where they are not already found. If introduced, these organisms may attack closely related native species or exotic species that were previously free of these pests and diseases. The risk does not apply only to the seed itself but to leaf fragments, twigs, and other contaminants. All seeds carry micro-organisms, mainly on their seed coats but these are not all pathogenic and some may even be beneficial. Most seed storage fungi occur widely throughout the world and are not considered a quarantine risk in the movement of seed. But some seed-transmitted fungi have a restricted geographic distribution and are of great economic importance. Imported seeds have been responsible for the spread of some tree pathogens, for example, *Sphaeropsis sapinea* on *Pinus elliottii* seeds into Uruguay (Romero 1997). Nevertheless the transfer of seeds is still a low phytosanitary risk compared with the risks associated with movement of scions, cuttings, and rooted plants (Chin and Diekmann 1999).

The first tactic of defence is to exclude the threat by banning or controlling movement of the seed. Most international transfers of seeds are subject to phytosanitary legislation and seeds may have to be fumigated or dusted with pesticides. In Australia, *Pinus* spp., *Araucaria* spp., and most other coniferous seeds are fumigated with methyl bromide if live insects are detected and surface sterilized with 1% sodium hypochlorite solution if the seed has not been heat extracted. Seeds of species of *Eucalyptus, Melaleuca*, and other genera in the Myrtaceae family are dusted with a seed fungicide due to the concern that the serious fungal pathogen, *Puccinia psidii* (guava rust), which is widespread in the tropics, will be introduced. The IPGRI has developed protocols for the safe movement of *Eucalyptus* spp. germplasm in international trade (Ciesla *et al.* 1996).

Quarantine procedures are often applied irrespective of the type of seed, and seed can be damaged or destroyed. Those importing tree seeds should comply with quarantine requirements and contact quarantine authorities to make the necessary arrangements for the safe importation of their seeds well in advance of seed dispatch.

CHAPTER 10

Forest nurseries

Introduction

Each year many millions of tree seedlings are raised in nurseries for use in plantations throughout the world. In the tropics there is a great variety of forest nurseries. Size alone ranges from a villager's own plot of perhaps only 50 or 100 seedlings in an assortment of containers, to large, highly organized, nurseries covering several hectares and producing several million plants annually. The use of nursery-raised plants is generally the most efficient and effective way of establishing a plantation, although direct seeding is practised successfully by aircraft in inaccessible areas and where a protective tree cover is more important than wood production.

Seedling quality is the integrated effect of genotype, phenology, and nursery cultural practices (Ritchie 1984) and these factors have a major impact on survival and subsequent growth in the field. Many seedlings produced in nurseries are of poor quality and fail to survive when they are subjected to drought, high temperatures, and other environmental stresses in the field. This problem has been evident for many years (e.g. Stone 1955). Planting healthy, vigorous seedlings or cuttings ensures that the plants have the opportunity to produce new roots quickly to access soil nutrients and water and thus cope with normal environmental stresses. Many factors that produce good quality nursery stock are under the control of the nursery manager and investment in producing high quality plants is justified.

Low quality seedlings grow more slowly after transplanting and add to weeding and maintenance costs. They are more susceptible to diseases and insect pests and have reduced wood production. Economic aspects of seedling quality in industrial plantations have been stressed by Dunsworth (1997) who states: 'Seedling quality has paid for itself by ensuring that we spend our planting dollars effectively by planting trees that will survive.' Three nurseries in Vietnam producing *Eucalyptus camaldulensis* seedlings of different quality have been compared; the seedlings from the nursery using best practices cost five to six times the poorest seedlings but their growth in the field was 9 m^3 ha^{-1} year^{-1} compared to 2 m^3 ha^{-1} year^{-1}

(Poynton 1996). The cost of seedlings is an insignificant part of total plantation costs and if better quality seedlings raise the average yield of plantations by only 1 m^3 ha^{-1} year^{-1} it is well worth paying double the price for them (Valli 1996).

Determination and adoption of the best nursery practices, within reasonable limits of time and investment constraints, to suit local conditions can be critical to the success of the plantations they support. Serious attention to producing high quality planting stock has seen an evolution of nursery practices over the past 30 years. In Brazil, for example, a typical eucalypt nursery in the 1970s was using unselected seed, direct sowing into subsoil substrate in plastic bags, with nursery beds on the ground and with little protection and attention to nursery quality control. By 2000, highly selected seeds or cuttings were being raised in rigid plastic tubes filled with vermiculite and organic compost, controlled fertilization and irrigation, with the plants grown on elevated benches with great attention to protection, conditioning and nursery quality control (Stape *et al.* 2001).

Most forest nurseries produce plants for a specified programme, thus they must be: (i) of the required species; (ii) ready at the right time, for example, beginning of the wet season; (iii) of the best quality in terms of size, sturdiness, root development and vigour; (iv) produced in sufficient numbers for the planting programme.

This chapter considers how these requirements are achieved, within the overall objective of any nursery to raise good quality, healthy plants at the lowest cost. Only seedling propagation systems are considered as vegetative propagation is covered in Chapter 11. An understanding of the biology and ecology of seeds in general provides a good basis for applying tree seed technologies in nursery practice and Baskin and Baskin (2001) have synthesized information on seed germination ecology. There an extensive literature on forest nursery practices in the tropics, for example, Liegel and Venator (1987) and Gonçalves and Benedetti (2000) for the Caribbean and Latin America; Wightman (1999) for community nurseries with emphasis on Latin America, Donald (1979) and Quayle and Gunn (1998) for southern Africa; Chakraborty (1994) and

Rai (1999) for India; Mantayla (1993) for Indonesia; Anon. (2001) for the Philippines, Oliver (1992) for the South Pacific, Carter (1987) and Doran (1997) for Australian trees; and Shanks and Carter (1994) for small-scale nurseries.

Planning forest nurseries

The first decision is whether to establish a formal nursery or to obtain plants from elsewhere. Factors of remoteness, cost, and availability of a site are all important, but usually an annual requirement of at least 10 000 plants is the minimum needed to justify starting a small nursery except where encouragement of villagers' own nurseries is part of a social or rural development forestry programme.

Type of nursery

There are various types of forest nursery. A permanent forest nursery which supplies plants for many years and usually in large numbers. A temporary or 'flying' nursery which is established for a short period, mostly less than 5 years, to meet a specific but temporary local need. There has been much discussion whether to have several small scattered, temporary nurseries, or one large permanent one: Greenwood (1976), Fatimson (1989), and Jagawat and Verma (1989) have summarized the arguments. However, the two types of nursery should not be seen as exclusive alternatives but ones suited to different needs. Moreover, in a major plantation project both may have a place. The third type, an extension nursery, is used in social and community forestry programmes to provide plants of many species for local needs such as fruit trees, shade, amenity, firewood, fodder, posts and poles, etc.; see Chapters 5, 7, and 19, and Fig. 10.1. There is a fourth type where individual villagers or farmers develop their own nurseries to provide trees for their own farm, for sale, for a government department, or other institution on a contract basis.

Permanent forest nursery
Major plantation projects have at least one permanent nursery often also supplying temporary nurseries at outstations. Permanent, centralized operations enable easier planning, record maintenance, and stock control which allows better forecasting of production and costs. However, there are many questions to be considered before setting up a permanent nursery and these include:

1. Does the scale of present and anticipated future plant production justify the capital investment and will it result in overall cost savings as well as improving the quality of the plants produced?

2. Is a site available with secure tenure in a location with favourable weather conditions, adequate natural drainage, a supply of good quality water, access to the components used in the growing media, access to all-weather roads and relatively close to proposed planting sites?
3. Are staff and accommodation for them available near the site?
4. Is electrical power available at the site and if not, can it be cheaply and easily connected?
5. Is the site secure against vandalism and theft?

Temporary, satellite, or flying nursery
Both because of remoteness and the relatively small requirements for plants, much of early plantation development in the tropics was based on temporary nurseries.

The temporary nursery is located near or within the planting area on flat ground where soil is workable and a regular water supply (perennial stream) available. This confers several advantages.

1. Nearness to the planting site improves survival because transit time between nursery and plantation is short and plants suffer less from overheating, windburn, and loss of soil and bruising due to vibration on long journeys.
2. Transport costs of bulky container plants is much reduced. For example, the Bilate community forestry project in Ethiopia (Fig. 7.1) had become so extended by 1988, a ribbon of planting of over 22 km, that two satellite nurseries were constructed in addition to the main one to supply the continuing planting at the north and south ends; the transport available to bring all the plants from the original nursery during the few critical weeks for planting was simply inadequate. Also, as distances increase there is more chance of part of the route becoming impassable, especially in the wet season, thus jeopardizing plant supply.
3. The capital investment needed is low.
4. By having many small nurseries isolation of disease and other damage is much easier. The main disadvantages, and the reasons why the trend is towards large permanent nurseries for industrial projects, include: (i) higher costs per plant arising from small scale of operation and low-level supervision; (ii) the lack of permanent installations limits species grown or propagation techniques used to ones not requiring special attention; (iii) greater risk of damage and theft owing to less on-site supervision.

Extension nurseries
There can be several types of extension nurseries. Most produce a wide range of species including

(a)

(b)

Figure 10.1 (a) Small extension nursery at Kalacha, Kenya on the north side of the Chalbi desert. Species of *Prosopis, Parkinsonia, Acacia*, and *Leucaena* are grown for use in erosion control on the Hurri hills and for distribution to the semi-nomadic Gabbra peoples. Note the very necessary fence (foreground) to keep out livestock, and the seedlings kept in the shade of a spreading *Acacia tortilis*. (b) Communal nursery, Sagula, Gamo Gofa, Ethiopia—filling plastic tubes for seedlings is a time for talk as well as work(!), and note soil sieving in background.

fruit trees and multipurpose species for individual and community planting. Some are relatively large and run by government agencies or international non-governmental organization (NGOs) but others are small and produce only a few thousand trees each year. A study in the Kenyan highlands identified (i) small private nurseries, usually run by men and generating cash income, (ii) nurseries run by women's groups providing seedlings for the local community and not often making a profit and (iii) school nurseries used to educate students in tree planting and to develop school woodlots to generate income (Nieuwenhuis and O'Connor 2000). In this study, average annual nursery production was about 10 000 seedlings and women's nurseries tended to have a greater variety of multipurpose tree species than the private nurseries. Common problems were lack of a reliable water supply, seed shortages, lack of fertilizer, and poor expertise in nursery management techniques. These problems are the same as experienced in small extension nurseries in many developing countries. Benefits of small extension nurseries include timely availability of seedlings, employment of local people, and motivation of farmers in adjacent areas (Fatimson 1989). Practical guidelines for community nurseries are available, for example, Wightman (1999), but such information frequently does reach those with the smallest tree nurseries.

Much of this chapter concerns permanent nurseries but much of what is written also applies equally to temporary and extension nurseries.

Nursery management

Successful nursery operation depends on many factors: selection and development of a suitable site, efficient supervision and administration, adequate planning, forecasting, and control procedures, orderly timing of operations; use of appropriate cultural methods, and protection from pests, diseases, and other damage. Timing is particularly important in tropical nursery management and operations must be scheduled to coincide with the dates proposed for planting. Advanced seed procurement and the scheduling of sowing to provide plants of the optimum quality at planting time are essential. The timing of nursery operations is crucial to maintaining seedling quality. The smaller the seedling container the more precise must be the scheduling of nursery operations. Nursery activities are perhaps more complex than any other forestry operation, with planning, organization, and control needed to produce good quality planting stock and avoid large surpluses or shortfalls in production.

The main operations in a nursery include: (i) planning, controlling, and recording all stages from receipt of seed to consignment of plants to the forest; (ii) seed storage and pretreatment or preparation of cuttings; (iii) soil preparation in the seedbed, container, or medium for inserting cuttings; (iv) basal fertilizer production and topdressing to control nutrition; (v) sowing seed and/or rooting cuttings; (vi) operations of pricking out, standing out, undercutting, lifting, transplanting, stumping or preparing striplings, etc.; (vii) inoculation with mycorrhizas and nitrogen-fixing bacteria as required; (viii) control of weeds; (ix) protection against climatic damage irrigation, shading, frost protection—and against fungi, insects, and animals; (x) packaging and despatch of plants.

A qualified forester normally has charge of a large permanent nursery. In most cases a core of specialized labour is permanently employed familiar with all routine nursery operations.

Nursery size

While an annual requirement of at least 10 000 plants is the minimum to justify starting a nursery, there are also practicable upper size limits for annual production and the amount of land needed to raise a given number of seedlings.

Upper size limit
In the tropics there are some very large plantation programmes and a project with an annual planting target of 10 000 ha could, in theory at least, have one huge nursery supplying the 10–20 million plants needed. But even in the unlikely situation that all land around the nursery was plantable, after 5 years afforestation, the minimum transport distance would be 10–12 km and after 10 years about 18 km. In practice, some land may be unsuitable for planting or may not be available, thus in a few years a nursery producing 10 million seedlings would be sending planting stock 30 or 40 km. Though such a distance is common in temperate countries, poorer roads and harsher conditions in the tropics often makes long-distance transport undesirable. This, combined with considerations of forest administration and territorial division, makes an annual production of 5–8 million plants about the upper desirable size for a nursery.

A common nursery size is an annual production of 1–3 million plants designed to meet the steady planting (or replanting) needs in a 10 000 to 20 000 ha forest block administered as one unit. In Indonesia the government has established a central nursery in each province to produce good quality seedlings economically and to a planned schedule. The twelve nurseries completed have an annual production capacity of 92 million seedlings (Valli 1996).

Individual nursery area
Examples of nursery size are shown in Table 10.1; it is mainly determined by three factors.

1. Level of annual plant production.
2. Method of raising plants—containers occupy more bed space than bare-rooted plants, but the latter are often worked on crop rotation and fallow areas have to be allowed for.
3. Nursery life of plants in the tropics, there is a wide variation in the time different species take to reach a suitable size for planting. In the Philippines *Anthocephalus chinensis* is raised in four weeks and several crops are grown in one year, whereas *Araucaria* spp. in Papua New Guinea (PNG) have to be grown for 18 months to two years in the nursery.

As a guide, the total area needed for an annual production of 1 million seedlings, including space for sheds, access tracks, storage areas, etc., is: 4.0 ha where the nursery life of plants is more than 1 year, 1.5–2.0 ha where it is between 4 months and a year, and about 1.0 ha where it is less than 4 months and several crops are grown in 1 year.

Table 10.1 Examples of nursery size and annual production for industrial plantation projects

Location	Species	Growing method	Area (ha)	Seedling nursery life (months)	Annual production (million)	Notes
Usutu forest, Swaziland	*Pinus patula*	Planter flats	2.3	6–8	6.0	Seedlings raised on tables
Beerburrum, Queensland	*P. caribaea* and *P. caribaea* x *P. elliottii* hybrids	Root trainers (Queensland native tube)	6.6	5–6	3.0	Nursery raises mainly cuttings (on tables)
Pulpwood companies, Brazil	*Eucalyptus grandis*	Rigid plastic tubes	3–5	3.0–3.5		Seedlings raised on steel mesh tables
Bukidnon forest Industries, Philippines	*Eucalyptus* spp., *P. caribaea*, *Gmelina arborea*, *Acacia mangium*	Root trainers and poly bags	—	3.5 6 3.25 3.75	1.2	Seedlings raised on steel mesh tables.

Nursery location and design

The nursery site

Careful siting of a nursery is important; not only is it where substantial investment and development takes place but it has strategic value having a large concentration of labour available for rapid deployment in case of forest fires or other emergencies. Choice of location is determined by factors of management and silviculture.

Of first importance for management is good access to and within a nursery at all times. Second, a continuous water supply for irrigation, and other services such as electricity and telephone are usually essential. Where water is not supplied by mains services a nursery must be sited near a perennial stream. Third, there should be ready access to local labour and the nursery manager should be housed within or near the nursery. Commonly, nurseries are sited near townships.

Several important silvicultural factors must be considered. Extremes of climate should be avoided such as frost prone areas in the highlands and, in the lowland tropics, where local topography renders the site excessively hot. Exposed sites known to suffer from hail, persistent or squally wind, lightning, or at all at risk from flooding, landslips blocking the access road, are all unsuitable.

Where it is planned to raise bare-root planting stock, it is critical to choose a nursery site where the soil provides the optimum environment for root growth and function. The soil for nursery beds or filling containers must be light enough to provide free drainage and easy lifting of plants without damaging the roots but heavy enough to hold water and nutrients. The best soils are sandy loams or loamy sands whose organic matter content can stabilize soil structure and maintain pore size. There should be a high resistance to soil compaction. Loams with more than about 20% clay and silt will be unsuitable for most potting mixes and very sandy soils do not retain moisture. High natural fertility is not necessary but acidity should be about pH 5.5–7.0 for broadleaved species, and pH 4.5–6.0 for conifers. High organic matter content facilitates working but soil should be relatively inert and free of weed seeds and pathogens. The majority of modern nurseries raise plants in containers and use 'soil-less' mixes as the growing medium. The soil has been replaced by materials such as washed river sand, composted organic matter, pine bark, and peat. This is mainly due to the poor physical properties and pathogens associated with soil-based mixes. Before establishing a permanent nursery the soil should be physically and chemically analysed.

Nurseries should be substantially weed-free, Fig. 10.2. A new site should be located away from waste ground where weeds proliferate and on land where previous use has not led to an abundance of dormant weed seeds near the soil surface.

Local topography is important and ideally the site should slope gently (1–3°) to allow rainwater to run off but not cause erosion. Though sites heavily shaded for a large part of the day are best avoided, partial shading is desirable in the tropics to prevent excessive day temperatures. Avoiding strongly desiccating conditions is an important part of nursery silviculture. In general, hilltop and valley bottoms are unsuitable, locations on middle to lower slopes are preferable.

Figure 10.2 A clean, weed-free nursery in East Kalimantan, Indonesia.

Figure 10.3 Nursery bed straddled by a tractor undercutting *Pinus caribaea* seedlings, Beerburrum, Queensland, Australia.

Nursery design and layout

A nursery should be compact and approximately square to minimize length of boundary, especially if protective fencing is required. Orderly internal layout is important where machines are used. There should be no waste ground, where weeds will grow, though occasionally unused land has to be included if expansion is planned.

The nursery bed, for growing bare-rooted plants or for standing containers, is the basic unit. Normal bed width is 1.0–1.2 m to facilitate hand tending or use of tractors (Fig. 10.3). Length is unimportant but if machines are used there must

be space for turning at the end of a bed. For calculation of bed area a density of 100–200 bare-rooted seedlings m^{-2} is normal, but for container-grown stock it depends on container size (see below). Paths between beds should be wide enough for walking and access by equipment. Time and labour will be saved if the layout enables operations to flow logically through the nursery. The extent of any movement of potting mixes and plants should also be minimized and from a hygiene view point the natural drainage of the site should direct water away from media mixing and germination areas so that any diseases that develop in areas with older plants are carried away from newly germinated seedlings (see Fig. 10.4).

Many buildings and special installations are needed in a permanent nursery; most are shown in the nursery plan in Fig. 10.4. With increasing use of vegetative propagation in mass production of plants, sheds and rooms for preparation and insertion of cuttings and facilities for misting or fogging during the critical rooting phase are essential.

Growing tree seedlings

Seedling quality and raising methods

Seedling quality has two main aspects, the genetic quality of the seed and the physical condition of the planting material immediately prior to planting in the field. The genetic and physical aspects of the seed have been discussed in Chapter 9. Nursery practices and the effect of stresses on nursery

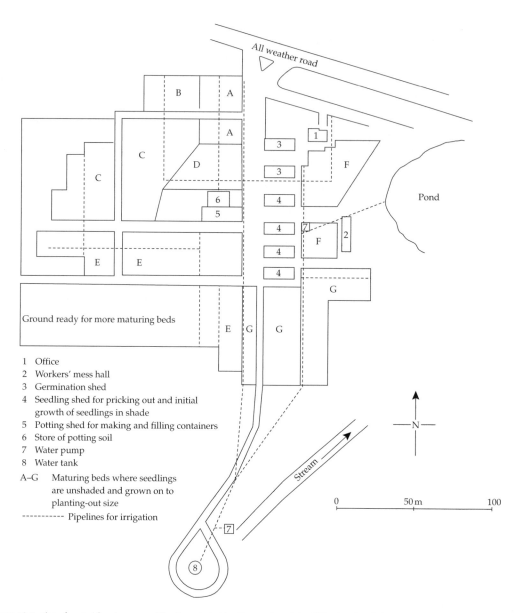

Figure 10.4 Plan of central forest nursery of the Paper Industries Corporation of the Philippines in Mindanao.

stock during transport to the planting sites have a significant effect on the survival and subsequent growth of the trees. A rigorous definition of seedling quality has proved to be difficult (Puttonen 1995) but in general high quality planting stock is disease-free, has a sturdy stem and a fibrous root system free of deformities with an optimal balance between the root and shoot mass, has been conditioned to withstand direct sunlight and a short period without

water, has good carbohydrate reserves and mineral nutrient content, and has been inoculated with symbiotic micro-organisms if required (Keys *et al.* 1996, Wightman 1999, Stape *et al.* 2001). It is emphasized that performance and survival of good quality nursery stock can be severely impaired by exposure to high temperatures, desiccation, root loss, and rough handling during the phase between leaving the nursery and planting.

The two general methods of raising seedlings are bare-rooted and containerized. Bare-rooted stock is raised in beds where the seedlings are lifted and planted with their roots free of soil. Containerized stock is grown in individual containers or multi-cavity trays which are taken to the planting site and the seedlings planted with a ball or plug of soil around the roots. Comparative studies on the growth and survival show little difference between the performance of container-grown and bare-root stock although container stock are generally favoured on dry and less fertile sites (McKay 1997). A comparison between container and bare-rooted methods of raising seedlings is given in Table 10.2. The choice of which to use depends on the species and the environment in which it is to be grown. A combination of bare-root and containerized pine stock is used in Queensland to extend the planting season with container stock being planted in the less favourable conditions (Baxter 2000).

There are many variations in both methods arising from their historical development in the search for simplicity and convenience. Initially, tropical practice followed that of temperate countries raising seedlings in open beds and planting them out bare-rooted. Survival was often poor owing to high temperatures, arid conditions, and the unpredictable and variable start to a wet season. However, survival greatly improved when seedlings were taken from the nursery bed to planting site with soil still attached to the roots. Early methods used locally available materials as containers, bamboo pots, earthenware pots, and baskets (Champion and Seth 1968). Alternatively nursery beds were cut between

Table 10.2 Comparison between container and bare-rooted methods of raising seedlings

	Container systems	Bare-rooted system
Materials	Need as many containers as seedlings. Import or manufacture plastics if used. Supply of good soil for potting mix.	Nursery site with easily worked soil suitable for bed cultivation.
Equipment	Container filling device, soil sieving screen. Tubing shed. Tables for multi-cavity trays and hard plastic tubes.	Tractor and several implements for ploughing, rotovating, bed formation, seed sowing, undercutting, lifting, etc.
Labour	i) Individual containers. Labour intensive, not easily mechanized. Much labour needed for container filling, seed sowing, weeding and container removal at planting. Typically 10–20 persons per million seedlings produced. (ii) Tray and batch systems. Partial mechanization possible. On large scale, 2–5 persons per million seedlings.	Well suited to mechanization. Most labour intensive component is lifting and packaging, but even these may become mechanized in the future.
Transport	Bulky and heavy to transport, costly over long distances. Not easily damaged.	Plants easy to transport over long distances if carefully packed.
Silviculture	Direct sowing occasionally difficult. Fixed spacing between plants in nursery. Excellent survival at planting. Overgrown plants become pot-bound and suffer serious root deformation and later instability.	Spacing between plants in bed dependent on precision of sowing. Good survival depends on careful timing of lifting and planting, to coincide with wet weather, and adequate conditioning of plants. Gives poorer results where climate is unreliable.
Supervision	Easier to grow satisfactorily, timing of operations not too critical, but may suffer more from casual neglect of watering, shading, overgrowing etc.	Requires a high degree of supervision to ensure proper timing and regularity of operations.
Protection	Fresh soil in every container reduces chance of build-up of pathogens or soil pests. Diseased seedlings easily isolated and discarded. Weed control tedious.	Reuse of same soil may lead to build-up of pathogens or soil pests. Pest and diseases more likely to affect all seedlings in a bed.
Cost	High labour intensity with individual containers or high container and associated equipment cost with tray and batch systems tends to produce more costly seedlings.	Capital intensive but at high levels of output unit costs are low.
Suitability	Individual container systems: all smaller nurseries and especially: (i) for good survival in arid conditions; (ii) when many different species are raised; (iii) where plants are distributed to the public and post-planting care is likely to be poor, for example, extension nurseries. Tray and batch systems: as above for large-scale production where more sophisticated equipment available.	Large production nurseries raising only a few species for planting and where climate is dependable; (ii) raising 'stump' plants, 'striplings', and as a cheap method for hardy species.

each seedling so that a block of soil remained attached to the roots, for example, for raising eucalypts in southern China (Turnbull 1981) or pines in Swaziland. The advent of cheap polythene for making tubes or bags led to almost total use of container-grown seedlings in the tropics. This has remained the case in many small nurseries in the tropics but most larger nurseries have adopted preformed multi-cavity trays such as styro-blocks and planter flats and sets of rigid plastic tubes or 'books' such as 'root trainers' which are units of cavities in rows formed from thin sheet plastic. In South Africa, virtually all of the 240 million plants produced annually for commercial tree plantations are produced in containers and since the 1980s multiple cavity containers have been favoured (Zwolinski and Bayley 2001).

Today many types of containers are used (Figs 10.6(a)–(f)). Container methods conserve valuable seed, provide versatility to the planting operation, and most importantly improve the chances of the plant surviving in stressful environmental conditions. However, they also have drawbacks; they require high quality seed, have to be filled with soil or other potting mixes and are bulky to transport, and must be purchased. The use of some types of plastic bags and pots with an unsuitable growing medium have resulted in poor aeration and waterlogging. The seedlings produced have poor root development and often the roots spiral with consequent poor field survival. Root trainer systems with internal grooves to direct the roots downwards and suitable potting mixes can overcome some of these problems (e.g. Josiah and Jones 1992) but for some species well-conditioned, bare-rooted stock is still a viable option. The sequence of operations in each method is shown diagrammatically in Fig. 10.5.

Container systems receive fullest treatment as they are most widely used in the tropics. However, before considering the two methods, the preparation of the seed prior to sowing is discussed.

Preparation of seed

Each consignment of seed should have been tested for germination percentage and cleaned before it arrives in the nursery (Chapter 9). This aids the decision on sowing rate—number of seeds to be sown per container, or along a seedbed drill, or what yield to expect per gram of seed in a germination tray. However, the growing environment in a container or seedbed is never as favourable as under test conditions, therefore it is important to maximize germination of all seed sown, in particular to separate out empty seeds. For tiny seeds, such

as *Nauclea* and *Anthocephalus* species, or some *Eucalyptus* species, for example, *E. cloeziana*, where the seed and chaff are identical size and shape, all the seed mixture is sown. With larger seed, for example, *Pinus* spp., it is feasible to separate full from empty, immature, damaged, or dead seeds.

In many species seed size is positively correlated with rate of germination and vigour although the effect may disappear in time. Grading seeds into several size classes, by means of sieves, and sowing each size class separately can be useful to obtain uniform germination rate and seedling growth in each size class. This makes management of the seed beds easier and may result in fewer culled seedlings. Having seed of uniform size also facilitates the operation of sowing machines, especially where precision sowing is used.

Pre-sowing treatment

The seeds of some species germinate slowly or incompletely even after thorough cleaning and the removal of empty and damaged seeds. This is because they are completely or partially dormant. Seed dormancy is the state in which viable seeds fail to germinate when provided with conditions normally favourable to germination. Treatments have been developed to overcome the various types of seed dormancy before the seeds are sown. Examples are used to illustrate five main pretreatment methods but it is important to understand the basis of the dormancy and to apply the best treatment for each individual species. The basis of dormancy and detailed descriptions of pretreatment methods are provided by Bewley and Black (1994) and Schmidt (2000) but species' manuals and compendia of species descriptions include specific recommendations, for example, CAB International (1997, 2000), Von Carlowitz (1986), Doran (1997), National Academy of Sciences (1980, 1983), Rai (1999), and Webb *et al.* (1984).

1. Mechanical scarification is one of several methods devised to overcome physical seed coat dormancy. All methods are based on the same principle of piercing the seed coat to allow imbibition to occur. Manual methods treat individual seeds by nicking, chipping, filing, or burning the seedcoat but are very labour intensive. Mechanical methods can handle large quantities of seeds and usually involve tumbling and abrading the seed (Poulsen and Stubsgaard 1995). While much faster than the manual treatment the level of serious damage to seed can be higher. The seeds are best treated immediately before sowing. It is effective on many hard seeded legumes such as *Acacia, Albizia, Prosopis*, and *Pterocarpus*.

Bare-rooted cultivation

1. Cultivate and raise beds in nursery

2. Sow seed, usually in drills

3. Care and tend seedlings over whole bed

7. Plant

6. Seal in bags and keep under shade

5. Dip to protect roots from drying cut

4. Undercut and sidecut roots to condition seedlings; lift when ready for planting

Container growing (using individual bags or sleeves)*

1. Obtain soil

2. Fill containers

3. Place in beds

4a. Sow small seed in germination tray

4b. Sow large seeds direct

Prick out

8. Plant

7. Remove container just prior to planting

6. Pack in cartons/ crates/boxes for transport

5. Care and tend

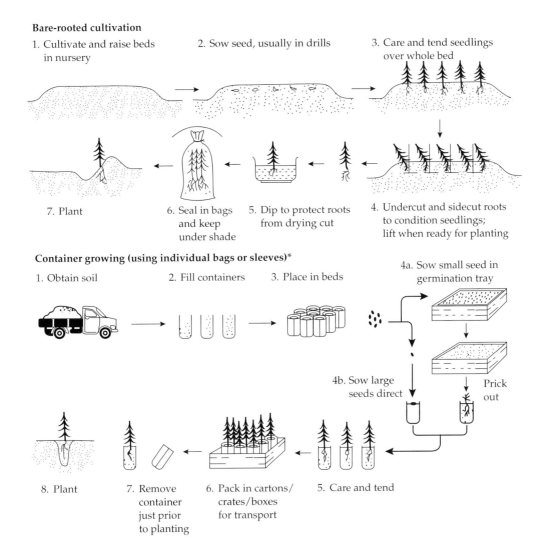

Figure 10.5 Diagrammatic representation of seedling cultivation methods.
*Preformed trays follow a similar sequence and can be considered as a block of containers.

2. Hot water is also commonly used to break physical dormancy in *Acacia, Paraserianthes*, and other species in the Leguminosae family but is not needed for some legume species with thin seed coats, for example, *Cassia fistula*. Sim (1987) describes breaking seed coat dormancy of *Acacia mangium* by immersing the seeds in boiling water (100°C) for 30 s. The boiling water is then poured off and replaced with cool tap water in which the seed is soaked for 24 h. The ratio of seed to boiling water is important with 1 part seed to 10 parts boiling water being optimum. The seeds can then be dried, stored in air-tight containers in cool conditions (4–10°C) for

up to 2 years and then sown without further treatment. The details of application of the method differ slightly between species and sometimes between different seedlots of the same species. Seeds are often immersed in boiling water for 1–2 min and occasionally 5 min. The seeds are sometimes allowed to remain for 24–48 h in the water as it cools. A simple test to check germination of small subsamples in slightly different pretreatments will quickly indicate the optimal conditions for a particular seedlot. Dry heat or mechanical scarification can have a similar effect as hot water treatment in breaking seed coat dormancy (Schmidt 2000).

3. Teak (*Tectona grandis*) is a species which has a combination of physical dormancy and chemical inhibitors in the fruit and obtaining consistently good germination has been difficult. Alternate soaking and drying of the fruits is an effective method although proponents differ in the recommended duration and number of cycles. Keiding (1993) and Willan (1985) list the following: (a) Four soaking and three dry periods of about 30 min for scarified seed (b) Five to ten cycles of soaking for 1 day and 3–5 days drying and sun-baking (c) alternate 24 h soaking and drying for 2 weeks. In India, Rai (1999) recommends alternate wetting and drying at weekly intervals over six weeks while Yadav (1992) found prolonged soaking to be as effective as the alternating treatments. Fruits stored for 1 year germinate better than freshly collected fruits but in some countries, for example, Trinidad and Tanzania, locally collected fruits appear to germinate well without pretreatment. Chemical inhibitors can prevent the germination of *G. arborea* seeds and extraction of the seed from the fleshy pulp followed by thorough washing to leach out inhibitors enhances germination (Schmidt 2000). Some species from fire-prone environments respond to treatment with smoke, normally by soaking in smoke-infused water, by faster germination and enhanced germination rates. A technique for smoking seeds directly is described by Quayle and Gunn (1998).

4. Cold, moist stratification. Low temperature thermodormancy is found in many temperate species and in some pines and eucalypts from highland tropical environments. Germination is improved when imbibed seeds of *Pinus patula* (Daniels and van den Sijde 1975), and some seed-lots of *E. nitens* (Turnbull and Doran 1987) are stored at about 1–4°C for 2–3 weeks or longer. The seeds are soaked for 24–48 h, and usually mixed with a medium that retains moisture, for example, sand, peat, vermiculite or a combination of these, and stored in a refrigerator or cool room. In Swaziland it was found that simply storing *P. patula* seed in cold conditions for 4 weeks improved germination percentage. The seeds of other tropical pines do not need cold stratification but benefit from soaking in water for several hours before sowing to speed up imbibition compared to that in a nursery bed or container.

5. Concentrated sulfuric acid (95%, 36N) is commonly used to break physical seed dormancy in African acacias and other legumes but can also be applied to non-legumes. However, the acid is highly dangerous and the method is not recommended for general nursery use. If other methods have failed it may be used with great care and following very strict safety rules. Gloves, protective clothing, and eye protection are essential; plenty of water must be close by to rinse accidental spills on clothing or skin, and operators must be warned never to add water to undiluted acid. Schmidt (2000) lists other safety precautions. It is one of the most effective treatments for species with very hard seed coats such as *Acacia nilotica* and *Faidherbia albida*. It can be carried out quickly, requires no special equipment and the seeds can be stored after treatment. Examples of the length of soaking are 15 min for *A. nilotica* (Rungu 1996), 15–60 min for *Prosopis juliflora*, and 30–180 min for *A. tortilis* (Tekatay 1996).

Other pretreatments to promote germination include chemical or hormonal applications such as potassium nitrate, thiourea, and gibberellic acid. Polyethylene glycol (PEG) is used as a priming fluid to allow seeds to develop to a stage where they are ready to germinate as soon as conditions are favourable for germination. Priming is not a dormancy breaking treatment but promotes rapid and uniform germination of seeds (Schmidt 2000).

Seed coating and pelleting

A major reason for pelleting seed by coating with an inert substance is to increase size and uniformity. This can assist some operations such as machine sowing. Other materials, such as fertilizers, fungicides, and insecticides, and rhizobia and other microsymbionts may be incorporated into the coating material. Pelleting is more commonly practised when seeds are to be directly sown in the field rather than in a nursery. Generally a sticker, for example, gum arabic, methyl cellulose, is applied directly to the seed coat or mixed with the coating material to assist the coating process (Schmidt 2000; Roberts 2001). The role of microsymbionts, mycorrhizal fungi, and nitrogen-fixing bacteria (e.g. *Frankia* and *Rhizobium*), and the inoculation of seedlings with them to improve establishment and growth are discussed later in this chapter. Seedlings are usually inoculated by incorporating the symbionts into the nursery bed or potting medium.

Time of sowing

Sowing should be scheduled with the aim of having seedlings of the optimum size at the time of the planting season. The rate of germination and seedling development depends very much on the species and the growing conditions. Table 10.1 gives examples of seedling nursery life. Most tropical pioneer species, for example, eucalypts and acacias, germinate and grow very fast whereas others, such as araucarias, are slower. There may be local preferences for the size of plant and this can alter the duration seedlings remain in the nursery. For example, mahogany (*Swietenia macrophylla*) seedlings usually take about 4 months to reach 30 cm and 6 months to reach 60–90 cm while some forest managers like small plants (30–45 cm) for easier handling and transport, others prefer tall plants (60–100 cm) to reduce the cost of vine cutting and weeding after planting (Mayhew and Newton 1998). Each nursery must determine its own schedule of sowing to satisfy local requirements.

The container system

Comprehensive coverage of methods, problems, and developments of both traditional (individual) containers and multi-cavity trays and sets will be found in Liegel and Venator (1987). They describe the great variety available, some of which are illustrated in Figs 10.6(a)–(f). Only a general account of container systems is given here.

Choice of container

Individual containers are of two kinds, impervious or pervious, but multi-cavity trays or books provide an impervious container environment for raising sets of plants. Each is outlined briefly.

1. *Impervious containers removed or slit at planting.* These are usually plastic (polythene) or metal and either bags with holes or tubes (sleeves). Usually, thin (0.04 mm) plastic is used and is black to prevent algal development; the container is thrown away at planting. If thicker gauge is used, the container may be reused several times. Metal tubes were once widely used in Australia and PNG but were expensive and had to be reused 15 or 20 times to make them economical. Other impervious containers include whalehide pots (tarred paper), bamboo pots and tubes, veneer sleeves, and reinforced paper.

Black polythene bags and tubes are most commonly used in the tropics but, as with all impervious containers, they have two disadvantages. First, time must be taken to remove or slit the container at planting, otherwise root growth is distorted which leads to instability of the tree (Fig. 10.7). The only exception is where the sleeve is sometimes left in place deliberately to afford some protection from termites. Second, seedlings left too long in impervious containers become 'pot-bound'. The roots, unable to grow outwards, coil round inside the container, a habit which continues even after planting. Normal lateral root development is impaired and the same instability problem occurs. This tendency to coil or

spiral begins at an early stage and modern stiff plastic containers used in book form, for example, root-trainers, and in sets incorporate ribs and grooves to encourage downward root growth (Fig. 10.8(b)).

Choice of container depends on availability, cost, and convenience. Where a suitable local material is plentiful and cheap this is often preferable to importing plastic. From time to time shortage of plastic in Ethiopia cuts off supplies of polythene bags without warning, making nursery planning difficult. In the Philippines, waste veneers from local peeling operations and bamboo tubes cut and supplied by local shifting cultivators have been used. Folded banana leaves are still used in extension nurseries in Rwanda. In Brazil, from about 1970 to the mid 1980s wood veneer tubes and plastic bags were used extensively to raise pine and eucalypt seedlings but then were largely replaced by conical tubes of rigid plastic (Valeri and Corrandini 2000).

2. *Semipervious or pervious containers not removed at planting.* Because impervious containers restrict roots there have been many attempts to produce containers, mostly paper based, which, while still easy to handle, are not so impenetrable as to deform root growth. Ideally, the container should last just a little longer than seedling nursery life so that at planting there is no barrier to lateral root growth. The only semipervious or pervious containers sometimes used in the tropics are Japanese Paper Pots, which are honeycombs of paper cells joined together with a water-soluble glue. They were developed for temperate conifers but have been used for raising pine seedlings in Central American countries. Though the idea of not having to remove the container at planting is a great advantage, the easier root penetration of the container sides brings two problems. If seedlings are grown for too long or in the wrong size container roots may grow through the sides and become matted together and have to be torn apart at planting. In addition, containers often become weak by the time of planting which makes them difficult to handle.

3. *Cavity trays, rigid tubes, and other container sets.* Batches of seedlings are raised together, taken to the planting site as a unit, removed and planted with a plug of soil attached, and the tray or set of containers returned for reuse. Their great advantage is ease of mechanization for filling and sowing, regular spacing between plants and ease of handling to facilitate root pruning. The principal disadvantage, as with all container systems, is that they are bulky to transport.

Root trainers have been used to grow good quality seedlings in the tropics for many years. They are available in a variety of sizes and shapes but all have vertical ribs and a large hole at the bottom

Figure 10.6 Examples of containers used for growing seedlings. (a) Black plastic (polythene) bags (PNG); (b) folded banana leaves (extension nursery, Rwanda); (c) tubes made of a veneer offcut (PICOP, Philippines); (d) individual rigid plastic tubes—note ribbing to prevent root spiralling (Department Forestry, Queensland); (e) block of 24 rigid tubes (Chachoengsao, Thailand)—note use of empty block to guide fine eucalypt seed to correct position on filled block below; (f) planter flat (styroblock)—polystyrene trays of 128 cavities (South Africa).

(Josiah and Jones 1992). They are supported on a frame above the ground to permit air-pruning of the roots. Rigid plastic tubes (dibble tubes) either as a single moulded set, for example, *E. camaldulensis* in Thailand [Fig. 10.6(e)], or as individuals inserted into frame or other support, called 'dibbling' or 'Queensland native tube' in Queensland and used

for raising pines. In Indonesia, eight nurseries producing 6–10 million seedlings each have been designed to use square-shaped pots arranged in trays. Each tray contains 28, 45, or 77 cavities and each has a 1.5 cm diameter hole in the bottom and four right-angled corners to prevent root spiralling. The trays are easy to handle and provide savings in

Figure 10.7 Ten-year-old *Pinus patula* blown over owing to severe root distortion because the polythene sleeve was left in place around the seedling at planting (Swaziland).

labour and transportation (Adjers and Srivastava 1993). Styroblocks and planter flats which are polystyrene trays with cavities of various shapes [Fig. 10.8(a)] to hold the growing medium, are used for *P. patula* in Swaziland and *E. grandis* in South Africa.

Container design and size

Local preference largely dictates the type and size of container used, and to some extent this reflects the desired size of the seedling (see Table 10.3). As a rule of thumb, the height of the plant at planting time should be one to two times the height of the container. Wide pots are more stable but take up more space. Narrow, deep containers may assist establishment in drier areas as the roots will go deeper into the soil but the design should allow root training and pruning. Most containers currently used in advanced nurseries have vertical ridges to direct the roots downwards. In less advanced nurseries polyethylene bags are preferred to rigid containers as they are cheaper but they use more potting mix and have a greater tendency for root spiralling.

(a)

(b)

Figure 10.8 Seedlings removed from containers. (a) *Pinus patula* taken from styro-block (planter flat) of ideal size for planting with first mature needles just developed. (b) Effect of ribbing on rigid plastic tube (rigi-pot) preventing root spiralling-note downward growing root along depression, and also note lignotubers just above root collar on this young *Eucalyptus camaldulensis* seedling.

Table 10.3 Examples of dimensions of containers (plastic bags, tubes or sleeves, and root trainers) used for *Pinus* species

Location	Species	Dimensions			Size of trees when planted (cm)
		Diameter (cm)	Length (cm)	Volume (cm³)	
Fiji	*Pinus caribaea*	3.8	15	170	15–25
PNG	*P. caribaea*	4.8	20	362	25–40
Queensland, Australia	*P. caribaea*	5.0	12.5	220	
Swaziland	*P. patula*	6.4	10	322	20–35
Malawi	*P. patula*	6.5	15	498	
Indonesia	*P. merkusii*	7.5	20	884	60–70

Containers should have good drainage and permit root pruning in the air or manually. The volume of the container should match the planned seedling size. Cavity trays, rigid tubes, and container sets typically range in capacity from 40–250 cm³. In Brazil the dominant container used in eucalypt nurseries is a black, rigid conical tube with internal ridges and with a capacity of 50–70 cm³. Tropical hardwoods, for example, some *Acacia* spp. (Ryan *et al*. 1987), tend to require larger container sizes than pines and eucalypts. Although large containers (e.g. 1 L) may appear beneficial, they are more costly per plant to buy, fill, store, and transport. With good nursery practices and appropriate potting mixes, smaller containers can be used and give satisfactory seedling growth and survival. Small seedlings raised in containers frequently survive better than larger seedlings after planting in South Africa, possibly due to root malformation that occurs when seedlings are left too long in small containers (Zwolinski and Bayley 2001). A relatively recent development is the 'jiffy pellet', comprising compressed peat surrounded by a biodegradable net. The peat expands when wet and allows good root penetration. Support trays are required but the nursery does not have to purchase containers or substrate.

Substrates

A potting medium with good physical and chemical properties is essential for the success of seedling production. It should be light enough to allow ready penetration of water and for unimpeded root growth but heavy enough to hold water and nutrients. Soil is used in many nurseries in the tropics but soil-less mixes, including materials such as compost, pine bark, sawdust, coconut fibre, sugar cane bagasse, carbonized rice hulls and peat, are being used increasingly. In large commercial nurseries in Brazil a mixture of vermiculite and organic compost is often used, especially in clonal nurseries

where it assists rooting (Stape *et al*. 2001). Where soil is used it is usually mixed with varying amounts of sand and compost to improve its texture and water holding capacity. Different plants have different optimum pH ranges but most grow well within the range pH 5.5–6.5.

Growing media often contain micro-organisms and weed seeds detrimental to the growth of seedlings. Treating the mix by sterilization with a chemical, such as methyl bromide, or pasteurization with steam or using the heat of the sun will reduce or eliminate this problem. After such treatment beneficial organisms such as mycorrhizas and nitrogen-fixing bacteria may be added. Potting mixes high in organic matter will probably contain sufficient nutrients for the seedling, if not, artificial fertilizer can be added (see Table 10.4). Tube container systems for eucalypts and pines in Brazil usually have capacity of 50 cm³ and use organic compost substrates mixed with secondary components to increase porosity and drainage. A benefit of using a substrate with a low fertility is that seedling growth can be controlled by manipulating the timing and concentration of nutrient supply (Valeri and Corrandini 2000). Nutrients can be rapidly leached from the potting medium in areas where there is heavy rainfall and incorporation of slow release fertilizer can be beneficial (see discussion on seedling nutrition later in this chapter). Practical details of the preparation of potting mixes are given in most nursery manuals, for example, Leigel and Venator (1987), Carter (1987), Quayle and Gunn (1998), Wightman (1999), and Quayle *et al*. (2001).

Container filling

Individual bags are usually filled manually with the aid of a scoop. Tubes are either filled individually by hand, often with the aid of a funnel, and with the first soil put in firmly consolidated to form a plug, or as a long 'sausage' which is subsequently cut up to the right length. Rigid containers and

Table 10.4 Examples of suitable soil mixtures used to fill containers

Location	Species	Potting mix	Fertilizer	Comment
Australia	Australian species	Composted pine bark: coarse washed sand : peat in ratio 3 : 1 : 1	Slow release NPK fertilizer, Micromax, coated iron	Mix sterilized before fertilizer addition (Doran 1997)
Australia (Queensland)	Tropical pine hybrids	Perlite and pine bark peat (manufactured from pine bark) in ratio 1 : 1.	Slow release fertilizer, Osmocote	Mix is sterilized (Baxter 2000)
Brazil (many companies)	*Eucalyptus* spp.	Vermiculite plus burned rice hulls or composted eucalypt or pine bark	Slow release fertilizer, fertigation or traditional inorganic fertilizers	Data based on survey of 20 large nurseries in 1999 (Stape *et al.* 2001)
South Africa (Mondi)	*Eucalyptus* spp., *Acacia mearnsii*, *Pinus* spp.	Composted pine bark	Superphosphate and urea each at a rate of 4 kg m^{-3} of compost	
Fiji	Broadleaved species	Forest soil : sand in ratio 4 : 1	NPK granules added when mixed	Soil not sterilized
Haiti	Various Agroforestry species	Composted sugar cane: unmilled rice husks: silty clay soil in ratio 14 : 3 : 3	Some NPK added	Josiah (1990)
Indonesia	*A. mangium*	Local peat and rice husks in ratio 7 : 3	A basic dose of triple superphosphate at 10 g m^{-2} in a 1% solution 2–3 days before pricking out followed by a total 225 g m^{-2} of NPK (15 : 15 : 15) applied in a 1% solution twice weekly for 19 weeks	Peat pH may be adjusted by adding lime, Adjers and Srivastava (1993)
Sarawak, Malaysia (Borneo Pulp and Paper)	*A. mangium*	Oil palm fire, coconut husk, coarse river sand, rice husk, Peat Gro, peat moss in ratio 7 : 3 : 3 : 3 : 2 : 2	Nurserycote 4 month 13 : 12 : 12 + trace elements at a rate of 2 kg per 200 L of potting mix and gafsa rock phosphate at a rate of 3 kg per 200 L of potting mix	This is one of several mixes under trial, Aken (personal communication)

trays are stood together on a bed and filled, or placed on conveyors and filled as they pass the operator. Additional compacting of the medium may be needed.

Though rate of filling depends on container size, the efficiency of this operation varies greatly. In different countries daily quotas range from a few hundred to several thousand. High filling rates have been achieved by introducing partial mechanization even for individual polythene bags (Fig. 10.9) while for multi-cavity trays and batches of rigid tubes tens of thousands can be filled and sown in a day by one person. For large nurseries this high efficiency makes tray and batch systems cheaper and more economical than traditional individual containers (Wilson 1986).

The nursery bed

Surfaces on which containers may be placed include: (1) low tables (Fig. 10.10) or a raised surface

Figure 10.9 A hopper formerly used for filling polythene bags at Aracruz in Brazil—one operator could fill up to 8000 bags in 1 day. (Bags have now largely been replaced in Brazil by trays of black rigid conic tubes filled with a mix of vermiculite and one or more organic materials.)

(a)

(b)

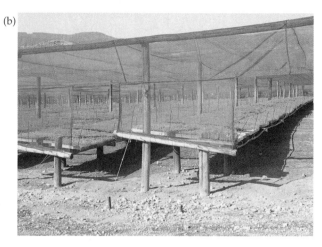

Figure 10.10 Support of containers. (a) Individual polythene tubes placed on a bed and supported by wire netting (PNG); (b) planter flats (styroblock trays) on low tables to assist air pruning of roots—overhead netting prevents hail damage (Swaziland).

of wire mesh or wooden slats about 10 cm above the ground which permit good drainage, air pruning of roots, and ease of handling, (2) a layer of clean, sieved crushed coral, crushed bricks or gravel, and (3) concrete or black plastic sloped to give good drainage. Placing the containers directly on soil is not recommended as drainage can be a problem and seedling roots can penetrate the soil increasing the possibility of invasion by pests and diseases, and requiring root pruning (Carter 1987).

If filled impervious containers are rigid or are packed tightly they need no special support (Fig. 10.6(f)) though there is evidence that growing seedlings too close weakens growth. Heavy gauge wire netting to support filled containers provides this small separation.

Sowing seeds

Containerized seedlings are raised either by sowing seeds directly into the container or cavity, or are first sown in special trays or germinating beds and later 'pricked out' into containers. Both methods are widely used but direct sowing is usually simpler and cheaper and is preferred unless, seed viability is low, maximum germination is needed, the seed is very fine or susceptible to insect and fungal damage, or where lack of control of nursery watering has the potential to wash seeds from containers.

Even very fine seed may be directly sown when diluted with very fine sand or sawdust and applied from a seed shaker made from a jar with holes punched in the lid. The major problem with pricking out is that there is a high probability of causing root deformities unless the operation is performed carefully by skilled staff. Species that rapidly develop a tap root, for example, many arid zone acacias, may suffer transplanting shock in the pricking out operation. Guidelines for pricking out are given by Doran (1997) and Wightman (1999). An advantage of sowing in beds or trays is that the quality of the germination substrate, usually sterilized soil, vermiculite or washed river sand, and the germination conditions can be more easily and better controlled. Many eucalypts have very fine seeds, for example, *E. deglupta* and *E. camaldulensis*, and are pricked into containers when they have one or two pairs of leaves above the cotyledons (Fig. 10.11).

A problem of sowing seeds directly into containers is the risk that either no seedlings emerge or that several do if more than one seed is sown, hence the importance of using high quality seed. Knowledge of the laboratory germination percentage guides how many seeds to sow per container. Quayle and Gunn (1998) recommend three seeds per container if the germination percentage is 40–60, two seeds if 60–90, and one seed for higher percentages. If the germination percentage is less than 40 it is better to sow the seeds in germination trays or beds. If the

Figure 10.11 *Eucalyptus camaldulensis* seedlings being pricked out into containers at Ba Vi, Vietnam.

germination is 60%, by sowing two seeds on average only 16% of containers will be empty, 36% will have two seedlings, and the remaining 48% one seedling. Where two or more seedlings exist the surplus may be pricked into empty containers. A variation of the two-stage method is to pregerminate the seeds between sheets of paper towel or other suitable materials in a controlled environment and transfer them into containers when the radicle has just emerged. This technique has been used successfully with acacias, eucalypts, pines, and some other genera (Doran 1997).

Fine seeds sown in a germination tray are normally germinated on the surface or only very lightly covered. Larger seeds are sown, either by dibbling with a stick or finger, to a depth equal to its shortest dimension or laid on the surface and covered to this depth.

Seed covering material should be inert, free of weed seeds, not prone to caking or capping, and near neutral in reaction. Well rotted sawdust, washed river sand, vermiculite, rice husks, and even dead pine needles have all been found suitable though loose organic coverings need to be applied to greater depth than indicated to conserve moisture (Fig. 10.12).

Raising bare-rooted seedlings

In this system seedlings are not grown in separate containers but together in a bed with thousands of others (Fig. 10.3). It is best suited for species with medium- to large-size seed, for example, *Pinus* spp. Raising bare-root plants is more common in temperate areas and the well-developed practices (e.g. Dureya and Landis 1984) may assist those choosing to raise this type of planting stock in the tropics.

Soil preparation

Because the ground in the nursery itself is cultivated, when planning to raise bare-rooted seedlings, selection of the site becomes extremely important. Of overriding importance is for the soil to be easily worked and of good physical condition.

Bed preparation must commence several months before sowing to bring it to a suitable condition. The standard practice in major forestry nurseries raising tropical and subtropical pines at Beerburrum and Toolara in southeast Queensland before they were converted to container—only nurseries in 1999 illustrates the steps that may be needed. Details of the plant production techniques in these nurseries are given in Baxter (2000):

1. Crop rotation was practised and, at any one time, two-thirds of the nursery was fallow and planted to a ley crop grass (*Panicum maximum*) to reduce *Phytophthora* build-up and aid good soil structure. Thus any one part of the nursery was used for growing pine seedlings only once every 3 years.
2. Eight months before sowing pine seed, the fallow was heavily slashed and turned in using a rotary hoe and ploughed. During the subsequent decomposition period glyphosate and rotary hoeing was used to control weed growth. It was ploughed again and deep ripped to break up any plough pan 6–8 weeks before sowing.
3. Soil was sampled periodically to determine nutrient status and fertilizer added to replace losses from leaching and harvesting crops of seedlings. Half the nitrogen and phosphorus requirements were applied to the bare bed and the fertilizer incorporated using the rotary hoe 4 weeks before sowing.
4. Two days before sowing a final cultivation was carried out with a rotary hoe and roller to produce a fine tilth.

(a)

(b)

Figure 10.12 (a) Shaded bed of *Pinus merkusii* showing needle mat to conserve moisture and reduce surface temperature around newly emerging seedlings (Indonesia). (b) Heavy shading of newly sown containers in small extension nursery in Mali.

5. Finally, after the seedling crop was lifted the soil was returned to fallow, NPK fertilizer added and the ley crop sown. While fallow, the grass was rotary slashed as required and top dressed with nitrogenous fertilizer to maintain good growth and so maximize organic matter production for subsequent incorporation.

Although the above is not a complete list of nursery operations, it demonstrates the main aspects of good soil management, that is, maintenance of good physical condition, regular checks on nutritional status, planned control of weeds, and prevention of build-up of soil pathogens.

Bed preparation

In most cases beds are formed about 10 cm above ground level by a moulding machine attached to a tractor. This produces a smooth, lightly compacted seed bed with side ridges.

Seed sowing

Seeds are sown broadcast or in drills (lines) (Fig. 10.13). A precision sower is used for drill sowing in large nurseries where operations are mechanized. The machine controls drill alignment and spacing, sowing depth, and seed covering after sowing. A combined implement to form the bed and drill the seed is often used.

The number of drills across a bed, usually between 5 and 12, depends on bed width and drill separation. Seedlings are spaced 4–5 cm apart in the rows. About 150 seedlings m^{-2} is the optimum for most *Pinus* species. About 250 m^{-2} results in weak, poor quality seedlings while below 100 m^{-2} is wasteful unless large plants are needed.

Figure 10.13 Drill-sown *Pinus patula* (South Africa).

Lifting

This operation is the removal of seedlings from the ground in readiness for taking to the planting site. The object of lifting is to separate the seedling's roots from the nursery bed soil quickly and without damage. It is usually done by hand though preparatory loosening of the soil may be carried out mechanically by drawing a bar under the bed just below root-pruning depth (wrenching; see Fig. 10.16). Once lifted, the roots are at great risk from drying out, and such bare-rooted plants should be placed in polythene bags immediately and then always kept in the shade. Ideally, lifting and planting should be done on the same day. This susceptibility to desiccation, at lifting, during transportation, and just after planting, is the greatest drawback of the bare-rooted method. Exposure of bare-root plants to even 10–30 min of drying conditions can damage their fine roots. However, advances in root dipping using clay slurries (Bacon and Jermyn 1977; Bell and Evo 1982) or even plain water (Venator *et al.* 1977) have greatly improved survival levels. Other methods used to enhance survival of bare-rooted mahogany (*Swietenia macrophylla*) seedlings include: wrapping the roots in bundles of dry grass and soaking, putting bundles of seedlings in cloth bags with wet sphagnum moss or in wet sacks, and keeping the plants cool and in the shade (Mayhew and Newton 1998).

Small nurseries

The bare-rooted method, although mostly used in large mechanized nurseries, can be used in small-scale operations if the species is hardy, for example, *Casuarina equisetifolia* in Vietnam, *Swietenia macrophylla* in Fiji, *Cedrela odorata* in the humid tropics of Latin America, and much of *Pinus merkusii* planting stock in Java. Moreover, it has been the traditional method of raising planting stock for the Sahel-Sudan zone in West Africa using *G. arborea*, *Azadirachta indica*, *Senna siamea*, and *Dalbergia* spp. (Delwaulle 1977*a*) though in dry years survival was often poor. Also, in the cool tropical highlands of Ethiopia most *E. globulus* is raised as bare-rooted stock and planted out very successfully.

Species planted out using 'stumps' are stripped of most foliage before planting—striplings, are always sown and raised in open beds; these are described later in the chapter.

Care, protection, and conditioning of seedlings

Watering

Provision of adequate moisture at all times is essential for a nursery in the tropics. Even in humid equatorial regions, with high, well-distributed rainfall, 1 or 2 days' drought can kill young seedlings. But, though supply of water is of primary importance, quality also matters. Water should be of pH less than 7, of low salt content, and not cloudy with matter in suspension. The newly sown container, seed tray, or nursery bed must never dry out, but be kept continually moist though not soaking wet, that is damp to touch but not dripping water when the medium is squeezed. With small fine seeds sown on the surface water is applied by mist spray or from capillary rise by standing the tray in water.

Soil should be maintained moist for several days after germination or pricking out, by lightly watering six or eight times each day. This frequency is reduced in damp, overcast weather and as seedlings get larger watering is restricted, usually to no more than watering in the early morning and early afternoon. A study of seed sowing and watering schedules for eucalypts in moist tropical

conditions in India found once per day was sufficient once germination was complete (Bahungda
et al. 1987). A more exact and less wasteful method
of estimating water requirements of nurseries is to
check weight loss from a container, with the aim of
maintaining soil moisture at more than one-third
above temporary wilting point (Donald 1971*a*). In
general, check soil frequently for moisture needs,
especially the rooting zone 2–12 cm below surface,
and even lift a few seedlings to check root growth
by looking for fine white roots. Over-watering
should be avoided, as this will increase the risk of
damage by fungal pathogens. Gradually reducing
the amount of water in the 3 weeks prior to planting in the field will help condition the plants (see
section on 'Conditioning or hardening' later in this
chapter).

Most smaller nurseries water plants manually
with a hose or watering can but this often results in
uneven watering. Using an adjustable spray nozzle
on the hose or a watering lance to direct the water
to the base of the plants enables better distribution
and control of the water. For large nurseries, producing more than 500 000 seedlings year^{-1}, there is
usually some form of automatic irrigation such as
sprinklers. Several methods of irrigation are practicable (Fig. 10.14(a) and (b)); see Greenwood (1976),
Liegel and Venator (1987) and Landis *et al.* (1989)
for details.

Shading

Shading is related to watering since lower plant
and soil temperatures reduce evapotranspiration
stress. Though this is the main benefit of shading,
for young seedlings not casting much shade soil
temperatures alone can rise to near lethal levels and
impair growth; Table 10.5.

The need for shading differs according to species,
stage of seedling development, and nursery location. Tiny seedlings of small-seeded species are usually very tender and full shade is needed over the
seed tray, containers, or nursery bed (Fig. 10.12(a)
and (b)) but, depending on species, it is quickly or
gradually reduced. For most pioneer species, for
example, eucalypts, pines, and acacias initial moderate shading (50–70%) after emergence or pricking
out can be followed by gradually increased exposure and removal of all shade within a few weeks.
Araucaria hunsteinii is an example of a sensitive,
though large-seeded species, which in PNG requires
full overhead shade after emergence, 75% shade for
the first 4 months, and 50% shade for the next
3 months. Pines are shaded while seedlings emerge
or immediately after pricking out, and some shade
is usually needed for a further few weeks, but full
light must be provided thereafter otherwise plants
become spindly and weak. Mahogany seedlings in
Fiji are grown without shade but in other countries

Figure 10.14 Nursery irrigation. (a) Germination shed at the small Kalacha nursery (Fig. 10.1) showing 5 cm deep troughs to allow 'flood'
irrigation around containers where they stand. This conserves moisture, so important in the very dry locality, but irrigation is still done by watering
can (on right). (b) Piped irrigation used in the old Block A nursery, Usutu forest Swaziland—today all plants are raised in one central nursery with
permanently installed sprayers for irrigation—seen in Fig. 10.10(b).

Table 10.5 Soil temperatures in a nursery bed in PNG at 13.00 h on a hot, sunny day with an intermittent 3–5 knot wind

Depth (cm)	Temperature (°C)			
	Unshaded bed	Shaded with sarlon	*Pinus caribaea*[a]	*Eucalyptus deglupta*[b]
1	49.0	35.5	27.0	28.5
4	40.5	33.5		
10	32.5	29.0		
20	26.0	24.0		
30	25.0	23.0		

Note: The nursery bed soil was a dark, gritty loam friable to 35 cm. Sarlon is plastic woven netting.

[a] 40 cm tall *P. caribaea* fully shading the bed surface.

[b] 80 cm tall *E. deglupta* providing dense mottled shade of the soil surface.

shade is provided for young seedlings (Mayhew and Newton 1998). Teak does not need any shading and grows satisfactorily from the time of germination in an open bed.

Any system of shading must be flexible to accommodate species differences, changes in the weather, time of day, etc. A nursery bed normally has its own supports and almost any covering material is suitable including banana leaves, grass, wood or bamboo strips, veneer offcuts, netting, etc. Plastic shade cloths are available that cut out 30–95% of sunlight. Natural shade from trees in and around the nursery is often favoured in small nurseries because it is cheap and easy to manage. Species like *Gliricidia sepium, Erythrina poeppigiana*, and *Grevillea robusta* can be heavily pruned or pollarded to control the extent of shading. A disadvantage of using shade trees is their potential to compete with seedlings for water and nutrients in nursery beds.

Shelter

Seedlings are tender and, as well as shading, may need shelter from four climatic phenomena.

1. Wind physically damages seedlings, accelerates desiccation, erodes dry soil around the root collar, and upsets irrigation spray. Choosing a sheltered nursery site is usually sufficient precaution but where strong persistent winds regularly occur, or where sandstorms flare up some direct shelter is essential. Screens of matting, coarse cloth, or hedges are suitable.

2. Large hailstones break and crush seedlings. Where hail is a problem, overhead shelter with fine wire netting (see Figs 10.10 and 10.14) or simply using the material for shading is usually adequate.

3. Frost is only important at high elevations. In certain locations past experience is usually the

only guide. In Malawi at Chongoni forest nursery (Fig. 6.2(c)) severe frost damage to *Pinus kesiya* in 1962 has led to continuous use of protective screens ever since.

4. Heavy rain causes physical damage, erodes soil around seedlings, and may lead to local waterlogging. Polythene sheeting is widely used when a storm threatens. However, Liegel and Venator (1987) report that heavy rain rarely seriously damages seedlings and excessive precautions are unnecessary.

Seedling nutrition

Production of healthy seedlings depends on an adequate supply of plant nutrients. Poor plant nutrition will result in poor seedling development, reduced growth in the field, and greater susceptibility to pests and diseases. The need to fertilize depends on the nutrient content of the potting medium or soil in the nursery bed, the size of plant, and the duration of their stay in the nursery. Replacement of nutrient losses from nursery beds due to leaching and uptake by seedlings is essential if soil fertility is not to decline. Fertilizer is applied either before sowing, when the bed is prepared or potting mix made up, or sometimes after germination when the seedling is past its most tender stage. Often fertilizing is done at both times. Fertilizer applied before sowing is usually in granular or powder form; but once seedlings are growing it is applied by foliar spray or by watering with a very dilute solution. Foliar fertilizers can be applied through automated irrigation systems (fertigation) but this should complement nutrients in the substrate. Phosphorus fertilizer is usually required, as many soils are deficient in this element, and nitrogen encourages leaf and shoot growth, however excessive use of these fertilizers will discourage the development of mycorrhizas and nitrogen-fixing bacteria. Micronutrients (trace elements) may be added through slow release fertilizers, such as Osmocote, incorporated in the potting mix or applied in a foliar spray.

Slow release fertilizers provide nutrients to the plant over a long period and reduce leaching losses. The fertilizer is encapsulated in resin and generally releases no more than 75% of nutrients in 28 days (Trenkel 1997) although the rate may vary with different brands. Average temperature and frequency of irrigation chiefly determine the rate of nutrient release. Adding high levels of slow-release fertilizer to the nursery media, beyond that which is required just to produce adequate seedlings, can be of benefit to seedlings for a considerable time after planting in the humid tropics (Reddell *et al.* 1999) but this 'nutrient loading' should be considered carefully in relation to the field conditions. Australian experience with pines and eucalypts suggests nutrient loading may lead to better performance where soil fertility is low, or competing vegetation is a significant problem, but should be avoided where there is the possibility of frost damage or high levels of insolation (Hunt 2001). See also the note on conditioning nursery stock later in this chapter.

The questions of what fertilizer and how much to apply are decided by local experimentation for the species and soil concerned. Quayle and Gunn (1998) suggest a balanced fertilizer is usually appropriate, for example, N : P : K in proportions 5 : 1 : 5 and Table 10.4 (p 165) gives further examples. The value of determining an optimum substrate and fertilizer mix has been known for many years (Table 10.6). The assumption that fertilizer is always beneficial is not always valid, for example. Chamshama and Hall (1987) in Morogoro, Tanzania, found with their nursery soil mix neither nitrogen nor potassium fertilizer was necessary for raising *E. camaldulensis* seedlings.

Regimes of post-germination foliar sprays and watering on fertilizer vary widely. In PNG solutions of fertilizers containing all essential plant nutrients are watered on once a fortnight. In Indonesia, *A. mangium* is watered with NPK (15 : 15 : 15) in a 1% solution twice weekly for 19 weeks after pricking out (Table 10.4). It is better to apply liquid fertilizer to the soil not to the leaves which are easily burnt if fertilizer stays on them. After foliar applications normal irrigation is used to wash the leaves and avoid fertilizer scorch. However, foliar fertilizers that are absorbed through the leaves and sometimes include an agent to assist them stay on the leaves. These sprays are often expensive but may be used to get nutrients to the plants quickly to correct deficiencies. By contrast, no fertilizer is needed at Tai Levu nursery, Fiji, for growing *Swietenia macrophylla* or at Chongoni nursery, Malawi, for *Pinus kesiya*.

It is important that fertilizer is applied uniformly. Uneven application produces patchy growth which upsets nursery production schedules and may prolong nursery life of some seedlings. Also, there is a tendency to over-fertilize to ensure seedlings grow well. This is wasteful and only the minimum required for good health and growth should be applied. Finally, in most tropical countries, fertilizers are imported, and obtaining them may be both costly and administratively difficult. In these circumstances nurseries may decide to use fertile soil, compost, or local manures (blood and bone meal, cow dung, organic wastes).

Nutrient disorders

Nutrient problems should not arise if there have been trials of potting mixes and fertilizer regimes and the recommendations implemented. Disorders that do occur are easily confused with fungal and insect damage, over-watering, frost, or other

Table 10.6 Effects of potting mixtures and fertilizers on growth of *Eucalyptus camaldulensis* seedlings at the Savanna Forest Research Station, Nigeria[a]

Potting mixture	Fertilizer added (g seedling^{-1})		Mean height after 96 days (cm)
	N	P	
Sand	Nil	Nil	5.1
Sand	Nil	Single super 3.0	3.2
Sand	Urea 1.3	Single super 3.0	15.0
Sand : soil (3:2)	Nil	Single super 3.0	10.1 (90 days)
Sand : soil (3:2)	Urea 1.3	Single super 3.0	22.4 (90 days)
Sand : cowdung (4:3)	Nil	Single super 3.0	50.3
Sand : cowdung (4:3)	Urea 1.3	Single super 3.0	80.3

[a] From Jackson (1977*a*).

Figure 10.15 Black polythene sheeting laid beneath containers to prevent weeds (*Anthocephalus chinensis* seedlings in Fiji).

environmental factors. Any of the following symptoms may indicate poor nutrition: chlorosis (pale colour of foliage, emaciated appearance), poor needle/leaf retention, stunted or deformed growth, purplish tints to foliage and other discolorations, needle tip burn, and patchy growth in a nursery bed (Fig. 14.3). It takes practice to recognize nutrient deficiencies in seedlings but a familiarity with general common deficiency symptoms (see Wightman 1999) or more specific symptoms, for example, for eucalypts (Dell 1996; Dell *et al.* 2001) will be helpful (Chapter 14).

Weed control

Competition from weeds for moisture, nutrients, and light, depresses seedling growth, and if not controlled can lead to seedling death. Weed control is time consuming though, as discussed previously, selecting a weed-free nursery site and keeping it clean greatly reduces the need for direct weed control. However, no nursery is entirely free from weeds and so weed control is always necessary except where seedling growth is so fast that weeds do not have time to reach a harmful size, for example, where nursery seedling life is less than 60 days (Table 10.1).

Weeds growing in containers are best uprooted by hand while small so as to cause least disturbance

to the seedling. In the humid tropics hand-weeding of containers may need doing once a week or fortnight, but less frequently in drier areas. Where weeds grow up from beneath and between individual containers laying a strip of impervious material such as thick black polythene will overcome the problem and also prevent the roots of seedlings penetrating the soil beneath the containers (Fig. 10.15). This problem does not occur with tray and batch systems. Hand-weeding is also used in small nurseries raising bare-rooted seedlings where beds are not large.

The main alternative to hand-weeding is to kill weeds chemically with a herbicide (weedicide). However, the use of herbicide once the seedlings have germinated is largely restricted to nurseries raising conifer seedlings as the chemicals used to kill weeds (grasses, herbs) will also kill broadleaved species. But in any nursery herbicides can be used to control weeds in open areas, along paths between beds, and in beds or container before sowing.

Chemical weed control is usually cheaper than hand-weeding because much less labour is required. Nevertheless, equipment and chemicals have to be bought and greater care and supervision is required. In applying herbicide checks have to be made of: (i) likelihood of damage to seedlings from the formulation; (ii) concentrations and dosage rates; (iii) timing; (iv) prevailing weather conditions; (v) safe methods of application, because herbicides are poisons. Four classes of herbicide are recognized in nursery use.

1. A pre-ploughing herbicide used to kill rank weed growth before land is cultivated, for example, glyphosate (Roundup) and paraquat (highly toxic to humans). Formulation or strength of herbicide are not absolutely critical since no seeds or seedlings are present.

2. A pre-sowing herbicide applied before sowing to kill all existing weeds. The chemical must volatilize or become inactive in the soil before seeds germinate and have no residual effects on the trees. Residual herbicides containing chlorthal, oxyfluorfen, linuron, or diphenamid as active constituents and applied at low rates have controlled weeds with little damage to seedlings in nurseries raising open-rooted eucalypts (Fagg and Cremer 1990).

3. A pre-emergent herbicide is applied in the period between sowing and emergence of the seedling. The herbicide must not interfere with germination and a check has to be kept on seedling development since spraying should not be done with less than 3 days to emergence. Baxter (2000) reports that chlorthal and propazine applied at seeding and 6–8 weeks later controlled weed growth in tropical pine nurseries in Queensland for about 15 weeks. Other possible herbicides include diphenamid, terbumeton, and nitrofen though Lewty and Frodsham (1983) found the latter two to decline in herbicidal activity after about 7 weeks. In pine nurseries in Venezuela linuron and dacthal have been used for this purpose (Fernandes and Arrieche 1982).

4. A post-emergent herbicide is a very mild chemical, e.g. white spirit with 22% aromatic content or diphenamid, used to kill small weeds among conifer seedlings. Application is not done during the first few weeks after emergence. Good pre-emergent and earlier weed control treatments will often eliminate completely the need for this phase or only a requirement for localized hand weeding or spot herbicide application. In Queensland pine nurseries, grasses developing late in the seedling production phase have been controlled by spraying a selective herbicide (fluazifop-p-butyl) at intervals as required until lifting starts.

Hand-weeding is likely to remain the main method of direct control in most tropical nurseries because it is safe, simple to carry out, and suitable for all species.

Protection against fungal diseases

Fungal diseases are important because: (i) a whole bed of seedlings can be affected; (ii) only a small change in the environment can cause the problem, for example, continuing to water seedlings at a high rate during cool, damp weather. Management of nursery diseases should be viewed as one aspect of the nursery operations involving watering, nutrition, shading, and insect and weed control. Poor cultural practices are frequently the cause of, or contribute to, the incidence of soil-borne and water-borne fungal diseases. It is far better to avoid disease than to apply controls when the disease has broken out.

Numerous diseases can damage tree seedlings. Gibson (1975) deals with those affecting many species commonly planted in the tropics in the families Leguminosae (*Albizia, Acacia, Leucaena*), Myrtaceae (*Eucalyptus*), Meliaceae (*Cedrela, Khaya, Swietenia, Toona*), and Verbenaceae (*Tectona, Gmelina*). Ivory (1975*a, b*) deals with the pathology of *Pinus, Araucaria*, and *Agathis* in west Malaysian nurseries. Brown (2000) has reviewed management of disease during eucalypt propagation, and experience in many tropical countries is reviewed in Sutherland and Glover (1991), Perrin and Sutherland (1993), and Day *et al.* (1994). Serious damage to mahogany seedlings is rare (Mayhew and Newton 1998) but production in teak nurseries in India may be disrupted by attacks of leaf spot diseases *Phomopsis* spp.

and *Colletotrichum gloeosporioides* (Balasundaran *et al.* 1995). This section does not cover individual diseases but discusses briefly three widespread non-specific fungal problems, that is, damping-off, moulds and powdery mildews, and outlines main methods for controlling fungal diseases.

Damping-off diseases

A number of non-specific fungal diseases collectively referred to as 'damping-off' diseases are the most serious threat to seedlings in the tropics. Both newly germinated and older seedlings of any species can be affected. Seedlings die from rotting of tissue near the root collar which causes the typical symptom of seedlings toppling over. Many fungi may cause damping-off including species of *Pythium, Fusarium, Cylindrocladium, Rhizoctonia, Penicillium*, and *Phytophthora*. A related problem, sometimes called seedling blight, may affect stem and foliage tissue and be caused by several of these fungi and by *Colletotrichum* or *Glomerella*.

Serious outbreaks can cause high mortality if unchecked, for example, in northern Vietnam it is responsible for the loss of 10–15% of pine (*P. caribaea* and *P. merkusii*) seedlings and sometimes up to 80% (Pham Van Mach 1995). Similar problems are reported from Indonesia (Hadi 1978) and India (Sharma *et al.* 1985). The disease usually first occurs in patches and then spreads. As the name suggests, very wet conditions, often induced by over-watering, excessive shading or poor ventilation, are conducive to the disease.

Moulds and powdery mildews

Moulds affect the foliage of seedlings of any size and are favoured by low temperatures (below 20°C), high humidity (above 85%), and where plants are over-crowded. Moulds are caused by *Botrytis* and *Penicillium* species. Susceptibility varies between species but considerable damage can occur to both pines and eucalypts. *Botrytis cinerea* is one of the most frequently reported nursery diseases of eucalypts throughout the world (Brown and Ferreira 2000).

Powdery mildews produce white powdery growths on the surface of leaves or young shoots. They can cause significant losses on a wide range of hosts in nurseries when not recognized and treated. A species of *Oidium* is found in most nurseries where tropical acacias are grown and damage can be severe with up to 75% mortality reported in Thailand (Old *et al.* 2000). The disease rarely kills eucalypt seedlings but can cause severe leaf distortion, shoot discolouration, and reduction in growth of nursery stock (Brown and Ferreira 2000).

Control of fungal disease

Several cultural practices can reduce the risk of damage.

1. *Nursery cleanliness and quarantine*. No part of a nursery should be allowed to lay waste for long periods where weeds, pests, and fungal pathogens can multiply. Cleanliness should extend to nursery procedures in handling containers, thoroughly cleaning tools, not allowing water to pond, use of bench tops and sloping hard surfaces rather than (say) placing containers on the bare ground (Hendreck 1985). In Brazil changes in nursery practices have had a major effect in controlling damping-off in eucalypts. Until 1970 the seed bed system resulted in severe damping-off, from 1970–84 direct seeding of containers on the ground reduced the incidence of the disease and since 1984 direct seeding into containers suspended above ground covered with coarse gravel or crushed rock has virtually eliminated the problem (Ferreira and Muchovej 1991). These authors also recommend a mixture of chlorine, Thiram and Captan, to sterilize racks and plant containers. Where seed beds are used repeatedly for bare-rooted seedlings experience in pine nurseries in southern Queensland has shown that the application of strict hygiene principles can keep a nursery free of soilborne pathogens without the need for costly sterilization treatments (Brown and Baxter 1991).

2. *Sterilization and pasteurization*. Treating the soil and potting mixes before sowing by sterilization with a chemical, such as methyl bromide, or pasteurization with steam or using the heat of the sun will reduce or eliminate nematodes, insects, fungal pathogens, and weed seeds detrimental to the growth of seedlings. It is essential to sterilize soil used for germinating small-seeded species in seed trays or germinating beds since the environment of full shade and very high humidity, favourable for germination, is also conducive to damping-off. Small quantities of potting media can be pasteurized by direct heating over a fire or boiling. Solar heating of moist nursery soil under a clear plastic sheet can control pathogens and weeds (Doran 1997). A whole nursery bed can be sterilized using fumigants such as methyl bromide gas, or steam, or drenching with formalin. Methyl bromide is a very poisonous chemical. Its application requires trained staff, it is costly and potentially hazardous to humans, animals, and to the environment, and its use is likely to be phased out (Hartmann *et al*. 1997). Recommendations for its use by Liegel and Venator (1987) and Hartmann *et al*. (1997) should be consulted. After such treatments beneficial micro-organisms such as mycorrhizas and nitrogen-fixing bacteria may be added to the soil or potting mix.

3. *Careful control of moisture levels*. Avoidance of over-watering will reduce the incidence of disease. Shading, ventilation, irrigation, and soil physical conditions, notably texture and drainage, all play a part.

4. *Chemical control*. It is far better to institute practices to prevent fungal diseases than attempt to control them with fungicides. Numerous products exist which are mostly copper based, such as Bordeaux mixture (copper sulfate), cuprox (copper oxychloride), or benzimidazoles such as benlate (benomyl) and are useful for controlling foliage diseases. Brown (2000) provides a comprehensive list of fungicides used in eucalypt nurseries.

Protection against insect pests

As with diseases, there are many potential pests but polyphagous caterpillars and white grubs are the general cause of damage to seedlings in the nursery. It is usually the larval (grub) stage of an insect's life-cycle that is harmful since this is when most feeding on leaves, shoots, roots, and boring in stem tissue takes place. However, an important exception to this rule is the weevil group in which adult feeding mostly causes damage. In general, a particular insect problem is localized to nurseries and species. Exceptions are the leaf-cutting ants (*Atta* spp.) of South America, termites in central Africa, and cutworms, white grubs (*Leucopholis* and *Melolontha* spp.), and some borers, for example, *Diapreps abbreviatus*. The latter two are soil inhabiting insects which cause much damage in nursery beds if uncontrolled. In 1974, in some nurseries in Maharashtra State (India), white grubs annihilated teak seedlings and *Diapreps* has killed entire beds of mahogany (*S. macrophylla*) seedlings in Puerto Rico. In the Philippines white grubs can cause mortality rates of 50–80% in seedlings of pine, eucalypt, acacia, and *Paraserianthes falcataria* (Braza 1987). By contrast leaf skeletonizers and defoliators, for example, *Hyblaea puera* on teak and *Ozola minor* on *G. arborea*, though impairing growth rarely kill seedlings (Lapis and Genil 1979). Nursery seedlings of *A. mangium* are attacked by a number of insects including plant bugs, grasshoppers, and bagworms, causing variable damage (Wylie *et al*. 1998).

Control of insect pests in nurseries is through soil sterilization and directly using insecticides. It is important to identify the cause of the problem and choose an appropriate insecticide. This choice has become complicated owing to the undesirable ecological effects of organochlorines notably Aldrin, Dieldrin, and DDT, even though they break down more quickly in warm tropical conditions, while pyrethrum-based insecticides are not always suitable. The final choice depends on local availability and what is permitted in the country. The organochlorine gamma HCH, the organophosphorus Malathion and the pyrethroid, Permethrin, are widely used but the first two should be avoided. An example of recommended insecticides in a nursery in southern Africa (Quayle and Gunn 1998) is: caterpillars, grasshoppers, and weevils (Dimethoate or Maldison), scale insects and mealy bugs (Maldison or Methidathion), leaf rolling thrips and leaf and twig webbers (Maldison). Insecticide is applied either before sowing as a drench to control soil-inhabiting insects, or as a spray against leaf eaters while seedlings are growing. Also, in the nursery, mixing of insecticide with the potting mix for container plants can reduce termite damage after planting. Such treatment of the root ball can confer protection for 3–4 years (Wardell 1987).

Many pesticides for insect and disease control are very toxic to humans and animals and strict safety precautions should be enforced. The operator must wear protective clothing, follow the product instructions carefully, take care when mixing and applying chemicals, dispose of excess chemicals properly, clean spraying equipment, and wash thoroughly when the spraying and clean up is finished.

Damage from animals and birds

Domestic grazing animals must be excluded from a nursery either by erecting a strong fence around the boundary or having several forest guards always on site.

Rats and mice may cause damage by eating seeds and digging around seedlings. The habitat of windbreaks and areas around the nursery should not favour rodents. Control is by trapping, laying poison baits, or some local method—hawks are used in Kenya, and sometimes pythons in PNG. Monkeys and baboons are often a nuisance but rarely cause serious damage.

Snails can cause serious defoliation of broad-leaved seedlings. In PNG metaldehyde poison baits are used to control them.

Birds eat seed and peck shoots, but the presence of people in and around nurseries is a deterrent and extensive damage is rare though germination trays should be in huts or covered. However, some

seeds are unpalatable (mahogany), too hard-coated (teak), or too tiny (*Anthocephalus chinensis*, *Nauclea diderichii*, and most eucalypts) to suffer much bird predation.

Preparation for planting out

Root pruning

For good survival after planting, seedlings need a balanced root:shoot ratio and an adequate root collar diameter. In a nursery a large shoot can be supported by a small root system because water is plentiful, but at the planting site watering is rarely possible. Bare-rooted plants, before they have become established, are particularly susceptible to desiccation and it is important not to plant seedlings with too much shoot in relation to roots. This is prevented by cutting roots of seedlings while in the nursery (root pruning) when shoot growth is slowed and a compact fibrous root system encouraged. Root pruning is an important tool in managing root development and assists in physiological conditioning of the plant and controlling its height. It encourages development of a strong, compact, and fibrous root system and improves lateral root growth. Root pruning has been a standard nursery practice for open-rooted *Pinus caribaea* and *P. elliottii* seedling production in Queensland (Baxter 2000), it is commonly used for *S. macrophylla* (Mayhew and Newton 1998) and was beneficial for field survival *E. camaldulensis* in Tanzania (Chamshama and Hall 1987).

Root pruning is also important for container stock. A root:shoot imbalance is a problem if the plants are kept in the nursery too long with the shoot continuing to grow but the roots restricted by the container. With open-ended containers roots will readily grow into the ground beneath. This is prevented by regular undercutting [Fig. 10.16(a) and (b)], pot turning, by standing containers on impervious material, or by keeping them off the ground to air-prune the roots [Fig. 10.10(b)]. One of the big advantages of using suitable containers (e.g. root trainers) away from the ground is to ensure air pruning of the roots with no extra work. For arid-land plantings, Romero *et al.* (1986) found that the best measure of plant quality was the root:shoot ratio, the optimum being 0.45–0.65, and this was achieved by root pruning: chemical 'pruning' with copper carbonate was used. Bayley and Kietzka (1997) found strong correlations between mortality and height to root collar diameter ratio ($r = 0.76$) in *Pinus patula* and it is generally agreed that seedlings with a large stem diameter at planting have a better post-planting performance.

(a)

(b)

Figure 10.16 (a) and (b) Simple undercutting tool used in Malawi—taut wire drawn beneath containers.

There are several methods of root pruning in open nursery beds. The commonest is undercutting where a wire, sharp plate, or spade cuts through the soil, usually about 12–15 cm below the surface (Fig. 10.3). Additional side-cutting may also be achieved by slicing between rows of seedlings to sever lateral roots. Another operation, wrenching, is like undercutting except that the whole seedling is slightly lifted to sever fine roots as well as cut the taproot. Seedlings regularly root pruned develop many short fibrous roots, a greater capacity to produce new ones, and a more balanced root : shoot ratio, and withstand transplant shock better. Satisfactory root pruning depends on it being done regularly starting from about halfway through a seedling's nursery life.

Rate of growth determines frequency of pruning and for pines undercutting every 1–3 weeks and side-cutting every 4–6 weeks is typical, whereas eucalypts in the humid tropics may need weekly undercutting. Immediately before and after root pruning the bed should be thoroughly watered to soften the soil and reduce subsequent desiccation of seedlings. Tropical pines in Queensland are subjected to three distinct pruning operations: regular severing of tap roots at a preset depth (progressively increasing from 11 to 14 cm) with a reciprocating blade, lateral pruning at 6-weekly intervals and bar wrenching by a large blade pulled under the nursery bed at a depth of 15–18 cm which causes subsoil aeration and induces a drought hardening response

(Keys *et al.* 1996, Baxter 2000). Bacon and Bachelard (1978) discuss the physiological consequences. Napier (1980, 1983) and Liegel and Venator (1987) describe root pruning development and practice for the neo-tropics.

Leaf stripping and top pruning
Both operations modify the root : shoot ratio and reducing transpiration stress at planting by removing most leaves. These striplings are tall plants, often 2–3 m high, which are used in areas where they are prone to animal damage, where stem borers are a problem and in underplantings where weed competition is severe (Parry 1956). Making striplings is common with species of Meliaceae in West Africa (e.g. *Cedrela, Entandophragma, Lovoa,* and *Khaya*) when used for enrichment plantings. In Fiji and several other countries all but a few leaves of *Swietenia macrophylla* seedlings are removed just before planting (Mayhew and Newton 1998).

Top pruning reduces the shoot to 15–25 cm to prevent tall seedlings suppressing smaller ones and to improve shoot:root ratio. It is common practice with pines raised by bare-root methods using a tractor mounted rotary mower but is used for many species, for example, eucalypts (Jacobs 1979), *Acacia mangium* (Adjers and Srivastava 1993), *Swietenia macrophylla* (Mayhew and Newton 1998). It is a low cost and easy way of reducing transpiration losses and may have application when seedlings are planted out in harsh and unpredictable climatic conditions in the semi-arid tropics (Chamshama *et al.* 1996).

Stump planting
A 'stump' is a term applied to a type of cutting prepared from a seedling. It is usually 15–25 cm long with about 80% root and 20% shoot with the shoot stripped of leaves and the lateral roots trimmed off (Fig. 10.17). Root collar thickness is usually from 1 to 2.5 cm. Stumps are most commonly used for *Tectona grandis* and *G. arborea* plantations. In India it has become common to transfer them to containers and allow them to shoot. Good results have been achieved by planting pre-sprouted stumps, 60–70 cm tall, especially on sites that are somewhat degraded or unburnt (Rai 1999). In Brazil, *G. arborea* stumps have been extensively used (Woessner 1980). The stumps are prepared from seedlings 4–6 months old, they are about 2 cm diameter at the root collar, and have 4 cm of shoot and 10 cm of root.

Many important broadleaved species can be planted as stumps having the ability to produce new roots and shoots when inserted into moist soil, for example, *Acacia saligna, Azadirachta indica, Bauhinia* spp., *Chlorophora excelsa, Cordia alliodora,* *Dalbergia sissoo, Erythrina* spp., *Gliricidia sepium, Gmelina arborea, Leucaena leucocephala, Nauclea diderichii, Paraserianthes falcataria, Pterocarpus* spp., *Senna siamea,* and *Tectona grandis.* Many of these species have such regenerative power that even branches or poles, 3–5 m long, cut from a tree and inserted in the ground to at least 1 m, will frequently become established and grow. Jolin and Torquebiau (1992) describe the use of the large branch cuttings of many species in Latin America. A common variant is to insert short, thick sticks (truncheon planting), and is often used in Africa with *Ceiba pentandra* and *Pterocarpus angolensis.*

The great advantage of using stumps is ease of storage, cultivation, and transport. Stumps are ideal for extension nurseries since they can be cut and, at least in the case of teak, stored for up to 5 months in cool, dry sand (Kaosa-ard 1982), and distributed as demand requires. The main steps in preparing and planting stumps are shown in Fig. 10.18. Woody teak and *Gmelina* stumps can often be split lengthways into two just before planting and so double the number available.

Conditioning or hardening
Root pruning starts the process of ensuring the production good quality planting stock suitable for planting out. A few weeks before seedlings or cuttings are planted it is common practice to further

Figure 10.17 Teak stumps suitable for planting once side roots are trimmed.

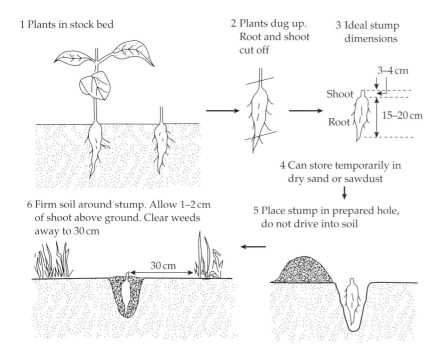

1 Plants in stock bed

2 Plants dug up. Root and shoot cut off

3 Ideal stump dimensions

3–4 cm

Shoot

Root

15–20 cm

4 Can store temporarily in dry sand or sawdust

6 Firm soil around stump. Allow 1–2 cm of shoot above ground. Clear weeds away to 30 cm

5 Place stump in prepared hole, do not drive into soil

30 cm

Figure 10.18 Diagram of stump preparation and use.

condition or 'harden' them so they are better adapted to the harsher conditions at the planting site. Hardening is achieved by progressively reducing nitrogen fertilizer and watering, and exposing the seedlings to full sunlight. Some shade may be retained for species that will be planted under a cover crop. Hardening leaves by successive drought cycles decreases leaf water potential and increases stomatal resistance. It assists survival where the plant will be water-stressed after planting but may not be beneficial under optimal post-planting conditions. For example, 'hard' and 'soft' seedlings of *Pinus patula* did not show significant differences in survival when planted on three contrasting sites in South Africa (Bayley and Kietzka 1997). Reducing nitrogen fertilization can result in poor nutritional status of seedlings and can reduce growth after planting so attention is being given in some places to pre-shipping fertilization and fertilization at planting to minimize the problem (Stape *et al*. 2001).

Mycorrhizal inoculation
Two main types of mycorrhizal fungi are important in plantation management. They are ectomycorrhizas, which increase the surface of the root, and vesicular-arbuscular (VA) mycorrhizas that are primarily located within root cortical cells.

Ectomycorrhizas are mainly found in moist environments where soils are low in nitrogen and VA mycorrhizas where phosphorus (P) is low and in arid and semi-arid areas (Vogt *et al*. 1997) although ectomycorrhizas also enhance tree growth by taking up P in P-deficient soils (Grove and Le Tacon 1993). It is well established that exotic pines grow poorly on many sites if not inoculated with suitable mycorrhizal fungi, but ectomycorrhizal associations occur in many families important for tropical plantations including Fagaceae, Dipterocarpaceae, Pinaceae, Myrtaceae, and many Leguminosae (Newman and Reddell 1987). VA mycorrhizas occur in the majority of families and unlike ectomycorrhizas show little or no host specificity. They can co-occur with ectomycorrhizas (e.g. in eucalypts) and/or nitrogen fixing bacteria (e.g. in acacias and *Alnus*).

Inoculation of seedlings with mycorrhizas in the nursery improves initial seedling establishment through increasing the uptake of nutrients, especially phosphorus, and protecting the roots against soil pathogens. Inoculation provides benefits in the nursery where beds have been fumigated or potting media sterilized. In the field, specific ectomycorrhizal fungi may be absent when an exotic species is planted, for example, eucalypts being established on infertile hill soils in south China formerly occupied by *Pinus* spp. lack the broad range

Figure 10.19 Mycorrhizal mother trees left to aid inoculation of *Pinus caribaea* seedlings (Bulolo, PNG).

of ectomycorrhizal fungi they are associated with in their native Australia (Malajczuk *et al.* 1994). In Liberia, *Pisolithus tinctorius* inoculation of *Pinus caribaea* led to the best mycorrhizal development and enhanced foliar nitrogen levels at the end of the first growing season (Marx *et al.* 1985). In fast-growing plantations there is a high requirement by the tree for nutrients in the period up to canopy closure and mycorrhizas may play a crucial role in meeting this need. The longer term persistence and role of inoculated mycorrhizas is less well known.

While mycorrhizas can benefit tree growth in the nursery and after outplanting, whether or not to inoculate selected fungi in the nursery will depend on the particular circumstances. It should be considered when exotic species are being planted for the first time on degraded and eroded sites or mine spoils. The decision-making steps required to evaluate the potential benefits of mycorrhizal inoculation in a specific situation have been set out by Brundrett *et al.* (1996). When a decision has been made to inoculate nursery stock there is a choice of several forms of ectomycorrhizal inoculum and all have advantages and disadvantages. Using soil collected from under the host tree has given inconsistent results and risks introducing pathogens. Inoculation with spores is widely practised as they are relatively easy to apply by mixing with soil or watering on to nursery beds and containers. The puffballs, *Pisolithus* spp. and *Scleroderma* spp. are commonly used because they fruit abundantly and contain large numbers of spores. Spore pellets have also been manufactured and used (de la Cruz 1996). Cultured mycelium, applied in peat-vermiculite

mixes, encapsulated in alginate beads and in liquid media, has also been very effective means of inoculating roots (e.g. Marx *et al.* 1989). Unlike ectomycorrhizas, VA fungi have not been isolated in the laboratory making mass production of inoculants for nurseries difficult. Trap plants, such as corn or grasses, are used to multiply the fungi and their potting soil together with spores and infected roots are used as inoculants (de la Cruz and Yantasath 1993) or inoculated 'nurse seedlings' are planted in the nursery bed at intervals of 1–2 m before the seeds are sown. Potting mixes and fertilizer regimes may have to be adjusted to favour good inoculation. Techniques for obtaining inoculum, its application and managing mycorrhizal plants in the nursery with special reference to eucalypts are well-documented by Brundrett *et al.* (1996). Inoculation of pines (Fig. 10.19) in bare-root nurseries and containers is described in Marx *et al.* (1982), Duryea and Landis (1984), and Landis *et al.* (1990); and general microsymbiont management by Schmidt (2000).

Inoculation with nitrogen-fixing bacteria
Symbiotic nitrogen-fixing associations are formed with the bacteria *Rhizobium* and *Bradyrhizobium* by many leguminous tropical trees, such as *Acacia*, *Albizia*, *Leucaena*, and *Sesbania* spp., and with *Frankia* by some non-legumes, for example, species of *Casuarina* and *Alnus*. These bacteria can improve seedling establishment and growth, for example, the positive effects of inoculating *Acacia mangium* with selected *Bradyrhizobium* strains in Côte

d'Ivoire persisted several years with trees obtaining up to 90% of their nitrogen from symbiotic fixation of atmospheric nitrogen (Galiana and Prin 1996). These useful organisms are often absent in modern nurseries, on eroded or depauperate soils, and where a species is being planted for the first time, and so must be introduced to the seedlings through inoculation. Often the nodules are collected from established plants, finely chopped and mixed with the potting medium after sterilization or applied to the nursery beds. There are varying degrees of specificity and interactions between the rhizobial and *Frankia* bacteria and the tree (e.g. Dart *et al.* 1991; Reddell *et al.* 1996*b*) and so selection of the most effective bacterial strains may be warranted. Commercial strains of selected rhizobial bacteria can be purchased. They are usually mixed with a carrier such as peat and can be stored under cool, dry conditions. The inoculant can be applied by mixing it in the soil or potting medium, suspending it in a clay slurry and applying it directly to the roots of bare-rooted seedlings or by coating or pelleting the seeds (Schmidt 2000). In contrast to rhizobial inoculation the inoculation of *Frankia* in pure cultures is quite recent. Selected *Frankia* strains can be applied as liquid culture, absorbed on to peat or imbedded in alginate beads (Reddell *et al.* 1996*a*) but are not so readily available as rhizobial inoculants.

Quality control, dispatch, and transport of plants

Quality control

To ensure good survival after planting only plants which are healthy, growing well but not vigorously, the right size, and with a balanced root:shoot ratio should be sent from a nursery. At the time of lifting or packing for transport, all poor plants not satisfying the above requirements are rejected; this is called 'culling'. Visual inspection ensures that any plant which is damaged, of poor colour, or too straggly is discarded. However, quality control over plant size is more difficult, not because of measurement, but because of deciding what is the optimum for a particular species.

Smaller plants may not be rejected if they are otherwise perfectly satisfactory, similarly with tall plants if they can be cut back without harm, for example, *G. arborea*, though the increasingly common conditioning practice of 'topping' greatly reduces the incidence of oversize. Culled plants should be burnt; it is invariably wasteful and expensive to try and improve them by growing on or repotting.

Although it is easy to measure, height and shoot size are poor indicators of seedling survival. Root collar diameter is a better indicator partly because it is related to the size of the root system and partly because a large stem diameter makes the seedling studier and increases the stem area available for water transport (Hawkins 1996). Based on studies of loblolly pine (*P. taeda*), South *et al.* (1995) concluded that the use of seedlings with large root collar diameters (4 mm and over) is one establishment option with the potential to increase volume production and reduce the per unit cost of wood. At Jari in Brazil *P. caribaea* seedlings are rejected if the root collar diameter is less than 4 mm. Several seedling traits work together to produce good survival and growth in the field. The Weyerhauser company recognizes this by aiming to have loblolly pine seedlings that have height (20–25 cm), diameter (4 mm), and root volume (>3 mL) with high root growth potential (Rose *et al.* 1990). Enso in Indonesia aim to produce healthy seedlings of *Acacia mangium* 30–50 cm tall, with a firm, strong stem, a root collar diameter of 3 mm and a strong, compact rootball (Malmivaara 1996). South African experience suggests that grading seedlings leads to significant yield improvements especially when integrated with genotype selection, and improved silvicultural practices (Zwolinski and Bayley 2001).

Packaging, transport, and dispatch of plants

The time and effort involved in raising quality plants in the nursery can be quickly wasted if care is not taken when transporting them to the field. Field survival is very dependent on the plant being able to produce new roots and avoid water stress (Margolis and Brand 1990) so the roots must be handled carefully and not physically damaged or allowed to dry out. Plants in individual containers are normally transported in trays or boxes: tray size is determined by what one man can carry. Batch container systems may be taken to the planting site or plants first lifted from the plug, tubes or book, and 'food wrapped' before dispatch. However, the disadvantages of container plants are seen at this stage: (a) the space required to transport them (Fig. 10.20), and (b) the need to return plastic trays, preformed containers, etc., for reuse. Some economy in space can be achieved with individual containers, especially polythene bags, by stacking them on their side if only a short journey (<5 km) is made. Lifting and packaging of bare-rooted plants were mentioned earlier.

Handling of plants should be kept to a minimum and during transport they must be protected from wind. Bare-rooted plants in polythene bags should be covered and well shaded. Accurate records

Figure 10.20 Much room needed for transporting containerized seedlings. Packing *Acacia mangium* into boxes for transport to the planting site in Mindanao, Philippines.

should be kept of all plants dispatched from a nursery as part of stock control. This enables a stand of trees to be related back to the original nursery stock and in turn to the seed batch. Also, since lifting, culling, boxing, etc., have to be done in advance of planting, knowing the rate plants are dispatched helps with organizing these operations.

Temporary storage of lifted plants

Plants must be kept cool and storage at moderate temperatures for longer than 2 weeks can be damaging (McKay 1997). In warm to hot areas stored plants must be shaded. Storing plants at moderate temperatures may increase the risk of attack by fungal pathogens. In Queensland, staff sort, cull, root prune to 15 cm, root dip, and package 350 to 400 pine seedlings into a plastic crate. The crates are stacked into pallets and stored under refrigeration at 2°C and a relative humidity of 95% until required, usually in only 1–2 days (Baxter 2000). This allows lifting and dispatch of plants to proceed in an orderly way and enables plants to be immediately available when suitable weather occurs.

CHAPTER 11

Clonal plantations

Introduction

Vegetative propagation is being increasingly used to multiply selected planting material in plantation forestry and agroforestry in the tropics. It is a way of raising planting stock not involving seed and includes cuttings, layering, budding, grafting and, recently, micropropagation. Propagating plants vegetatively results in clones, with each clone retaining the genetic characteristics of the tree at the point from which the buds or cuttings were obtained. It must be stressed that vegetative propagation is only a means of multiplying planting material, it is not a breeding method and does not result in genetically improved plants. However, it is commonly used to mass propagate trees that have been improved genetically through selection and breeding.

Extensive plantation crops of rubber and oil palm, developed after selection, breeding and cloning, have been established in the tropics for many years and the total technology has resulted in very substantial increases in latex and oil yields. Cloned multi-purpose trees and shrubs, and fruit trees, have also been widely used in agroforestry and urban forestry. Trees have been cloned for wood production for centuries in Asia, Europe, and the Middle East. Nevertheless, large-scale planting of clones of eucalypts, pines, and other species (mainly for pulpwood production) did not start until the 1970s and 1980s but is now common, especially for eucalypts in the tropics and subtropics.

Clonal forestry implies the large-scale use of a limited number of clones in operational plantings (Burdon 1989) and is an important option for species that can be vegetatively propagated easily. Libby and Ahuja (1993) considered 'clonal forestry to be a complicated discipline that has a thin history, an exciting present and a bright future'. Clonal plantations offer a number of benefits compared to those developed with seedlings. Cloning enables genetic gains from selection and breeding to be captured quickly; is a cost-effective way of using hybrids; permits easier use of desirable characteristics such as high pulp yields and disease resistance; and produces a uniform material for processing. It is emphasised that these benefits will only be fully captured if there is integrated planning and imple-

mentation of plantation strategies including tree breeding, clonal testing, operational propagation and clonal deployment.

Development of clonal plantations

History

Horticulturists and foresters have vegetatively propagated trees over a long period. Fruit trees, such as figs (*Ficus* spp.) and olives (*Olea europea*), have been propagated by grafting for thousands of years. Rooted cuttings of sugi (*Cryptomeria japonica*) have been planted in Japan since about AD 1400 (Ohba 1993). In China, *Paulownia* spp. have a cultivation history of over 2000 years. They are usually raised from cuttings of roots and there are very extensive clonal plantations, mainly of *P. tomentosa*, in central and eastern regions (Zhang 2000). Poplars (*Populus* spp.) and willows (*Salix* spp.) have been grown from cuttings in Asia, Europe, and the Middle East for 300–500 years (Zsuffa *et al.* 1993). In Europe, clonal forestry as a system using grafting was proposed in 1906 by G. Andersson in Sweden but was generally not adopted because of cost factors and lack of knowledge of species and varieties (Libby and Ahuja 1993). Grafting and cuttings were used to develop seed orchards and as a research tool to appraise tree genotypes in 'tree shows' in tree improvement programmes (Larsen 1956). Most of these developments were in temperate areas but Chinese fir (*Cunninghamia lanceolata*) has been cultivated in a traditional system in subtropical areas of China for about 1000 years with seedlings and cuttings (Sheng 1991). Throughout Asia bamboos have been traditionally propagated from offsets, rhizomes, culms, and branch cuttings or layering (Aala-Capuli 1993; Mascarenhas and Muralidharan 1993) due to the infrequent periodicity in their flowering.

The rise of clonal forestry in the tropics and enthusiasm for it since the early 1970s has mirrored the development of short-rotation, fast-growing plantations to provide raw material for the pulp and paper industries. Since the 1970s, millions of hectares of these plantations have been established, initially from seeds and, after 1980, with increasing use of rooted cuttings. The feasibility of mass propagation for industrial plantations came with the development of operational cloning systems, usually with rooted cuttings, and the greater awareness of the benefits of clonal technologies (Foster and Bertolucci 1994).

Major advances in developing rooted cuttings from mature trees for commercial plantations were made in the Congo by researchers from the Centre Technique Forestier Tropical between 1969 and 1973. Substantial increases in growth and wood production were achieved, initially with naturally occurring hybrids in plantations, and subsequently with artificially manipulated crosses (Delwaulle 1983; Martin 1991). First Brazilian studies to develop a rooted cutting technology were made by Poggiani and Suiter Filho (1974). A Brazilian private company, Aracruz Forestal S.A., recognized the potential role of clonal forestry in increasing productivity in its eucalypt pulpwood plantations. The company's research indicated that gains in volume production and wood quality could be achieved using hybrid clones, for example, *E. grandis* × *E. urophylla*, and in 1979 decided gradually to replace its plantations derived from seeds with clonal plantations (Campinhos *et al*. 1993). The mass production of clonally propagated planting stock soon spread to other areas of Brazil.

Although these technologies were initially applied to eucalypts they were later used for tropical pines (e.g. Walker *et al*. 1996), teak (Monteuuis and Goh 1999), and other tropical hardwoods (e.g. Mascarenhas and Muralidharan 1993; Ladipo *et al*. 1994; Mesen *et al*. 1994; Kha 2001). Many major industrial plantation projects in the tropics now use vegetative propagation with cuttings to develop clonal plantations. Important examples are: with eucalypts, at Aracruz in Brazil (Fig. 11.1(a)), ECO sa at Pointe Noire, Congo (Fig. 11.1(b)), ITC Limited in Andhra Pradesh, India, Mondi Forests and Sappi in South Africa; with *Gmelina arborea* at Sabah Softwoods in East Malaysia (Figs 11.2 and 11.3); with pines in Queensland, Australia (Fig 11.4); and with *Acacia mangium* and its hybrid with *A. auriculiformis* in Vietnam, Indonesia, etc. On a lesser scale there are reports of vegetative propagation by stem cuttings of tropical hardwoods such as *Triplochiton scleroxylon* and *Terminalia superba* in Nigeria, Ivory Coast, Cameroon, and other West African countries (Ladipo *et al*. 1994), dry zone species in Africa (Milimo *et al*. 1994), rain forest dipterocarps in Asia (Smits *et al*. 1994; Sakai *et al*. 2001), mangroves (*Sonneratia apetala*) in Bangladesh (Serajuddoula *et al*. 1996), *Casuarina junghuhniana* in India and Thailand (Surendran *et al*. 1996), and so on. Leakey *et al*. (1994) list over 70 tree species from throughout the tropics that have been propagated

vegetatively by the Institute of Tropical Ecology, Edinburgh.

Although the practice of rooting stem-cuttings has been very important in establishing clonal plantation forestry, it has its limitations, such as the loss of rooting ability due to ontogenetic ageing of the hedges or other sources of cuttings. New technologies, involving rooting of micro- or mini-cuttings, have improved rooting potential, rooting speed, and root system quality, have reduced costs, and have shown their potential to substitute for rooted stem-cuttings (de Assis 2001). Development of these highly intensive cloning systems marks a new phase of mass vegetative propagation of *Eucalyptus* and other woody species.

Benefits of clonal plantations

Advantages of clonal plantations compared to using seedlings include:

Overcoming inadequate seed supply. Some tree species either produce very little seed, produce it at long intervals or have recalcitrant seeds that remain viable for only a short time. Dipterocarps are an example of species whose plantation potential has been limited by seed supply and for which vegetative propagation by cuttings is a promising option for plantation development in Malaysia and Indonesia (Ismariah and Aminah 1994; Chai *et al*. 1998). Problems in producing sufficient high quality seeds in seed orchards have stimulated interest in developing techniques for cloning of teak (Monteuuis and Goh 1999).

Capture of genetic gains. Cloning enables genetic gains achieved in tree breeding to be captured maximally and directly used in operational plantings. It enables the capture of a greater part of genetic superiority of the parents than is possible via seed orchards. Cloning permits utilization of non-additive variation and more rapid availability of genetically improved planting stock compared to seedling-based tree improvement programme (Leakey and Newton 1994). MacRae and Cotterill (2000*a*) state that creation and utilization of genetically superior seedling families are ultimately the cornerstone of any breeding strategy aimed at producing new and better clones for clonal plantation forestry.

Clonal planting using hybrids derived from controlled pollination has realized gains in dry mass production of about 100% compared to improved non-hybrid seedlings in Brazil (de Assis 2000). Cloning enabled the Aracruz Celulose S.A. company in Brazil to utilize the results of its selection and breeding programme and was a key factor in nearly doubling wood production in its eucalypt plantations and raising air-dried pulp yields from

(a)

(b)

Figure 11.1 Clonal plantations (a) *Eucalyptus urophylla* × *E. grandis* aged 32 months at Aracruz Celulose S.A., Brazil (photo K.G. Eldridge), and (b) a hybrid eucalypt clone at ECO sa, Pointe Noire, Congo (photo C. Cossalter).

5.9 t ha^{-1} year^{-1} to 10.9 t ha^{-1} year^{-1} (Campinhos 1999). Clonal *E. grandis* plantations of Smurfit Carton de Colombia have meant annual production rates of 35 t ha^{-1} compared to 25 t ha^{-1} for seedling plantations (Wright 1997). Significant operational gains in plantation productivity of pines have been achieved by using a clonal strategy to rapidly multiply and plant superior genetic material

(e.g. Balocchi 1996; Walker *et al*. 1996). In China, clonal selection and propagation of *Cunninghamia lanceolata* is expected to result in increases in yield per unit area of over 50% compared to seed orchard seed stock (Zhou *et al*. 1998).

Hybrids. Cloning has been used for many years to deploy selected poplar hybrids and the difficulties and costs of producing hybrid seed in other tree

(a)

(b)

Figure 11.2 Vegetative propagation of *Gmelina arborea* in Sabah. (a) Ideal cutting ready for insertion in rooting medium. (b) Cuttings in propagation beds where they are kept under controlled conditions of humidity and light while rooting takes place.

Figure 11.3 Part of a *Gmelina arborea* clone bank developed from selected phenotypes in plantations of Sabah Softwoods. The tall stumps, 1.5–2.0 m high, are regularly trimmed of their new shoots as the source of cuttings.

species means that vegetative methods are likely to be the only cost-effective way of mass propagating hybrid plants for commercial planting (Zobel and Talbert 1984). Cloning of eucalypt hybrids is being used increasingly to rapidly capture genetic gains and produce uniform raw material. This can have a positive influence on the competitiveness through increased plantation productivity, better product quality, and reduced production costs, for example, at the Brazilian pulp company Riocell S.A. (de Assis 2000). Similarly in the Congo

hybrids of *E. urophylla* × *E. grandis* propagated clonally have resulted in very large genetic gains compared to that of pure species and natural hybrids (Vigneron *et al.* 2000). The potential for hybridization to increase pine plantation yields in Queensland was fully recognized by the 1970s but commercialization of the hybrids could not be realized because of the need to make crosses by controlled pollination, low yields of viable seeds, and the lack of methods for vegetative propagation (Nikles 1996), but development of techniques to mass propagate *Pinus elliottii* var. *elliottii* × *P. caribaea* var. *hondurensis* by cuttings cleared the way for extensive plantations of this hybrid (Walker *et al.* 1996) (Fig. 11.4). Cuttings were recommended as the best propagation option for deploying hybrids of *Acacia mangium* × *A. auriculiformis* (Haines and Griffin 1992) and cloning to utilize hybrid vigour and superior wood qualities of this hybrid in plantations in Vietnam is described by Kha (2001).

Uniformity. Seedling populations derived from seed collected in natural forests or from plantations where hybridization has occurred with closely related species can be highly variable in growth rate, stem form, and wood properties. This results in reduced plantation productivity potential as slower growing individuals are suppressed and make harvesting and processing less efficient. These factors prompted Aracruz Celulose S.A., (mentioned above) to embark on its tree improvement and clonal programme (Campinhos 1999).

Figure 11.4 A superior F1 hybrid clone of *Pinus elliottii* var. *elliottii* and *Pinus caribaea* var. *hondurensis* aged 7 years at Toolara, Queensland, Australia. The clone shows notable uniformity of stem straightness, very fine branching; and good control of site (photo M. McDonald, D.G. Nikles and R. Peters).

At Mondi in South Africa, clones of *E. grandis* are screened for uniformity in wood properties to ensure high pulp yields (>50%) and a basic wood density of 480–520 kg m^{-3} to maintain pulp quality (Duncan *et al.* 2000).

Adaptation. There is an opportunity to match selected clones with the conditions prevailing at a particular location. When the unit of management is an individual clone, sets of clones can be deployed to specific sites whereas seedlings from a seed orchard, where a range of genotypes are produced, are suitable for planting on a variety of sites (Eldridge *et al.* 1993). Clones respond differently to different factors, such as soil fertility levels (e.g. van Wyk *et al.* 1991) suggesting that selection and deployment for different sites will give the best yields. When different genotypes planted across different sites do not maintain the relative differences in growth between them there is said to be a genotype by environment interaction. Where this occurs the grower can either select a few

'stable' clones that can be planted across all sites, recognizing that on some sites there will be some productivity loss, or attempt to select 'reactive' clones adapted to particular sites recognizing that some site factors are not readily predicted. In South Africa, the SAPPI company recommends using stable, high yielding eucalypt clones until the causes of genotype × environment interaction are better understood (Retief and Clarke 2000). Smurfit Carton de Colombia has had a strategy of matching of clones to specific site conditions and has achieved an operational improvement in wood yield of 40% with clones of *E. grandis* compared with seedlings since its clonal programme was initiated in 1986 (Wright 1995). A similar approach has been suggested for clonal plantations of *E. grandis* × *E. urophylla* in Venezuela and the need for a detailed site classification to assist the clone-site matching is recognized (Rojas *et al.* 1999). The strategy in Queensland with hybrid pine has been to identify both stable and reactive clones for general planting and for particular plantation sites (Walker *et al.* 1996).

Cloned hybrids can provide an opportunity to extend planting to sites unsuitable for the pure species. *Eucalyptus grandis* is poorly adapted to lowland sites in the humid tropics but its hybrid with *E. urophylla* grows well in this environment. In South Africa, the adaptability of *E. grandis* has also been enhanced by hybridization with other species, such as *E. camaldulensis*, enabling planting of sites previously considered too dry or too cold (Arbuthnot 2000).

Disease resistance. Clones of both pure species and hybrids can be selected for resistance to damage by pests and diseases and this will form an important part of the protection strategy for many tropical plantations (Wingfield *et al.* 2001). In response to the stem canker, *Cryphonectria cubensis*, plantation managers in parts of Brazil selected, propagated, and planted resistant clones. Soon after planting *E. grandis* on the Zululand coast in South Africa it was damaged by fungal diseases and the Mondi company successfully deployed resistant hybrid clones of *E. grandis* × *E. camaldulensis* and *E. grandis* × *E. urophylla* to combat them (Harvett 2000).

Reduced wood production costs. Fewer plants are needed for clonal plantations than for open-pollinated seedlings because it is possible to rely on the performance of each ramet. This means savings in site preparation as there are fewer rows to cultivate, and in other costs incurred on a per plant basis such as nursery stock, fertilizers, and herbicides. Better form and uniformity of log sizes may substantially reduce the cost per unit volume of mechanical harvesting.

Problems in developing clonal plantations

Rooting difficulty

Some species are difficult to root, for example, eucalypts such as *E. citriodora* in the subgenus *Corymbia* (Catesby and Walker 1998), and some clones within species that are generally amenable to rooting present rooting problems. Cuttings taken from physiologically mature trees have proved very difficult to root and this problem has made it difficult to vegetatively mass propagate many major plantation species. Older tissues may contain chemical rooting inhibitors, as was demonstrated for *E. grandis* (Paton *et al.* 1981). An appreciation of the key physiological factors that interact to influence rooting is helpful in developing protocols for rooting cuttings and some of these processes are discussed by Leakey *et al.* (1994). Techniques to reverse the maturation state to improve the rooting ability of tissues from older trees is referred to as 'rejuvenation'. Rejuvenation by coppicing, serial production of cuttings over several generations, repeated grafting and micropropagation has been successful for some eucalypts (Eldridge *et al.* 1993).

The level of rooting ability of a species or a clone is its most important propagation characteristic. Failure to achieve a high level of rooting ability can result in many promising genotypes being excluded from the clonal programme on practical and economic grounds. So major research efforts have been directed to developing satisfactory techniques to achieve high rooting levels in commercially important plantation species. It has been difficult even to get cuttings from young seedlings to root for species such as *Acacia mearnsii, E. globulus,* and *E. nitens*. Advances such as rejuvenation of adult material through meristem culture, have improved the prospects for *A. mearnsii* (e.g. Roux *et al.* 2000) and progress has also been made with *E. globulus* by taking advantage of variation in rooting ability (Wilson 1998). In Chile, Portugal, and Spain companies are producing millions of conventional cuttings each year at costs two or three times that of seedlings but justifiable on economic grounds. Mini-cutting technologies are being developed and are likely to be the preferred method in the future. Micropropagation systems remain a problem and are not yet operational even for producing mother plants.

Physiological ageing has been a particular problem in developing clonal options for most *Pinus* spp. It has generally been very difficult to root cuttings from trees over 3 years old. Because most tropical pines do not coppice on felling and are perhaps 10 years old before they can be reliably selected it has been necessary to start cloning programmes with genetically improved seed to raise seedlings as a source of juvenile cuttings for clonal testing and for hedge establishment. By the time clones have been tested adequately they are often too old to propagate economically.

Physiological ageing reduces the length of time clones can be used. There is much research in progress on how to retain juvenility in pines through cool storage of organs/tissues, cryopreservation, and somatic embryogenesis plus storage. Techniques for large-scale production of rooted stem cuttings of *P. taeda* are still being developed (e.g. LeBude *et al.* 1999), but for *P. radiata, P. elliottii,* and hybrids of *P. elliottii* var. *elliottii* × *P. caribaea* var. *hondurensis* protocols for rejuvenation by embryogenesis using apical meristems have been developed (Smith 1999). These new developments will assist the management of selected clonal material and could also make it possible to make selections from near mature trees in the field on the basis of individual wood properties and other important traits.

Plagiotropism

In some species propagules taken from older trees and from parts of the tree such as branches do not grow into trees with normal form. This has been a problem with araucarias although rooted cuttings can be produced readily from young and older trees if cutting material is taken from young orthotropic shoots (Haines and Nikles 1987; Nikles 1996).

Risk of plantation failure due to genetic uniformity

The high risk of serious damage or even failure of single clone plantations due to pests, diseases, or other agents is recognized (e.g. Zobel 1992) and mixed clone stands or mosaics of monoclonal stands are sometimes used to increase the genetic variability of the forest to counter this problem. How many clones should be used is discussed later. However, unless sets of clones are selected for uniformity of wood properties it can create a major problem for planning and marketing. This happened in Congo where some clones are unsaleable and each buyer tends to specify which clones they do or do not want (Griffin, personal communication).

Availability of high quality trees to propagate

The cost of rooted cuttings is usually higher than for seedlings. But production costs vary considerably depending on the scale of the nursery operation, labour costs, species being propagated, etc. Some huge nurseries in Brazil can produce planting material of hybrid eucalypt clones nearly as cheaply as seedlings. Advances in technology development continue to reduce costs of cuttings, for example, in the early 1990s eucalypt cuttings produced in the large Aracruz nurseries in Brazil were 40% higher than for seedlings (Campinhos *et al.* 1993). Mass propagation by cuttings of a species with rooting problems, such as *E. globulus*, may be three times as expensive as for seedling production. Nursery managers usually aim to produce cuttings that are not more than twice as expensive as cuttings.

Vegetative propagation can only be justified if sufficient gains are achieved from clonal plantings and it follows that it is important that cuttings are obtained from genetically superior stock. This may be a problem for enterprises that are too small to justify a selection and breeding programme. Growers may acquire new clones from other tree breeding programmes through purchase or exchange. These could be from other companies or from regional or national tree improvement programmes (White 1993). This can be very important in enabling a new plantation enterprise to quickly develop clonal plantations with a high level of genetic improvement and diversity. It has been a very common practice in Brazil.

Choosing a clonal plantation strategy

The financial wisdom of embarking on a clonal planting programme aimed at rapid increases in productivity requires careful consideration and analysis of key factors. Effective use of clones requires, (i) an understanding of interactions between genotype, site, and age, (ii) an appreciation of the impact of unmanaged variation in wood properties from the pulpmill perspective, and (iii) a commitment to quality management through the subsequent silvicultural operations without which genetic potential will remain unrealized (Griffin 2001).

Returns on the investment of a tree improvement programme can be measured in different ways but the cost of operations, genetic gains from alternative strategies, the time frame in which the investment is made, and the size of the benefits are all critical in an industrial forestry plantation enterprise (Griffin *et al.* 2000). While factors such as the genetic quality of the clones, site-genotype interactions, silvicultural practices, and especially nursery operations determine the potential success of the clonal approach (Adendorff and Schön 1991), other factors may be equally or more important in determining financial outcomes.

This was illustrated by Walker and Haines (1998) with a simple cost/benefit analysis for a clonal plantation of a tropical acacia that examined the sensitivity of key performance measures to variation in factors such as size of the plantation estate, rotation length, stumpage price, productivity gains, total clonal programme cost, and plant propagation cost. The internal rate of return (IRR) was 18.6% for their base scenario of 100 000 ha, a project life of 22 years, rotation length 8 years, a stumpage price of US$10 m^{-3}, an investment of US$500 000 year^{-1} in a clonal improvement programme yielding average productivity gains of 30% for clones planted in years 5–10 and 40% for clones planted after year 10, and cuttings costing US$0.06 to produce.

The outcome was quite sensitive to the size of the plantation programme. It was attractive for a 100 000 ha estate but marginally affordable for 50 000 ha. It was very sensitive to the stumpage price and to the size of the productivity gains but less sensitive to rotation length within the range of 6–9 years. The cost of the clonal improvement programme is an important factor but reducing the amount spent carries a greater risk of failing to achieve the necessary productivity gains. The financial outcome was sensitive to plant propagation costs. While the cost of US$0.06 per plant was considered realistic, if the cost rose to the exceptionally high US$0.50–0.70 per plant, a yield above 12% IRR could only be achieved by productivity gains of 80–100%, which is probably unrealistic in the life of the project. This type of analysis using likely stumpage prices and productivity increases, and realistic production costs can provide valuable insights and guidance to managers contemplating the merits of moving from a seedling to a cutting based operation.

Vegetative propagation methods

Common techniques of cloning are rooting cuttings, air layering (marcotting), budding, and grafting. More recent approaches involve micropropagation through the application of cell and tissue culture technologies. Mass propagation by cuttings is applied extensively in clonal forestry whereas grafts, micropropagules, etc., have all found application but only on a small scale and for specialized tasks. Using cuttings avoids incompatability problems often encountered in grafting. Many species can be propagated easily and relatively cheaply by leafy stem cuttings but several of the other techniques require skilled labour and/or high capital investment, and may have a low rate of multiplication.

Propagation by rooted cuttings has so far proved the most practical way of mass propagating trees for clonal plantations. Understanding the key factors influencing rooting is fundamental to successful mass propagation of cuttings. Four main difficulties have to be overcome.

1. All conifers and eucalypts and most other species can only be propagated effectively from young (juvenile) material.
2. Ease of rooting of cuttings varies between clones of one species as well as between species, and even the season in which cuttings are taken can be critical.
3. For roots to form on cuttings precise environmental conditions of humidity, warmth, auxin amendment, and potting medium are usually needed.
4. For good field survival plants must be given careful conditioning.

Space does not allow full consideration of this complex subject. It is discussed more fully in other texts, for example, Leakey and Newton (1994). Outlined below are typical nursery techniques for mass production of trees. This is followed by comment on micropropagation methods, including the newer mini-cuttings technologies, and the impact of advances in biotechnology on clonal forestry.

Vegetative propagation using macro-cuttings

The first commercial clonal forest plantations used rooted cuttings (macro-cuttings) obtained from field plantings of stool beds of selected clones. These systems are still used although new technologies involving indoor clonal hedges to yield mini-cuttings are being introduced. There are several stages in developing a cuttings programme including securing superior individuals selected in plantations or the products of tree breeding to propagate, setting up and managing clone banks, and rooting and conditioning (weaning). Discussion in this section concentrates on clone banks (hedges) and mass propagation of cuttings from them with special reference to eucalypts and pines. For details of techniques we have drawn heavily on the accounts of Walker *et al.* (1996) and Baxter (2000) for pines and Eldridge *et al.* (1993) for eucalypts. Figure 11.5 shows the pathways for developing clonal plantations by using controlled pollinated seed from selected individuals or rejuvenating vegetative material from these individuals.

Selecting superior genotypes for propagation
Phenotypically superior parent trees are selected in routine plantations or from the results of tree breeding. Selection intensity varies in plantations; for example, at Aracruz 5000 trees were initially selected from 36 000 ha of plantation, of which 150 clones were identified as potentially suitable, and just 31 of the very best actually used in the plantation programme (Ikemori 1987); at Sabah Softwoods 100 superior trees ha^{-1} were identified throughout one 30 ha block of good quality, Philippine origin *Gmelina arborea* (Wong and Jones 1986). The fact that

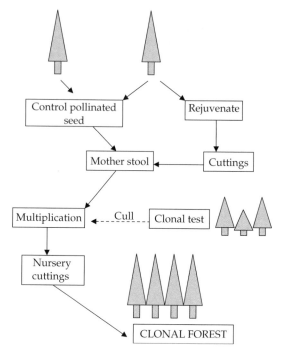

Figure 11.5 Clonal forestry options. Pathways for developing clonal plantations by using controlled pollinated seed from selected individuals or rejuvenating vegetative material from these individuals.

the selection of superior genotypes for propagation in all clonal programmes is dynamic with rapid application of research results is illustrated by the hybrid pine clonal programme in Queensland. In the first clonal test (Series 1) planted on two sites in 1986, there were 250 clones from 14 F_1 families of little-known potential; in 2003 (Series 4 tests) there were 1950 clones from 22 (17 F_1 and 5 F_2) purposefully-produced families (all with predicted high breeding values) planted using a sophisticated test design on each of 9 sites from central Queensland to northern New South Wales) (Nikles personal communication).

The number of clones that should be available for operational plantings to maintain genetic diversity and minimize pest and disease risk is discussed below.

Setting up and managing clone banks
When the selected individual trees (ortets) have been identified buds or shoots are removed and these ramets (individual members of a clone) are taken to the site where rejuvenation and multiplication will take place. This is often sited for convenience near the cuttings nursery. The rejuvenation technique applied to material from older trees depends on the species.

In Queensland's hybrid pine programme controlled pollinated seed is sown and the resulting seedlings are hedged and cuttings taken from these for field testing. Based on the test results a subset of about 70 clones is established to form a hedged clone bank. The retention of juvenility and multiplication of clones occurs in hedged stool beds. The stool plants have a life of 3 years and a third of the plants are replaced each year. About 60 000 stool plants of hybrid pine were established annually and the clone bank provided material for over 3.5 million rooted cuttings each year.

In eucalypts that are non-coppicing or when the original tree cannot be felled, the ramets are taken after air-layering, or after scion material for grafting has developed to become the source for cuttings. A more destructive technique is to cut down the tree leaving a stump less than 20 cm high. The numerous coppice shoots which emerge are normally large enough to make cuttings 6–12 weeks after the tree is felled. Slightly lignified shoots up to about 1 m long are suitable, succulent shoots and shoot tips do not make satisfactory cuttings in most eucalypt species. First generation coppice shoots usually have a lower rooting success rate than later harvests. Alternatively, growth of epicormic shoots can be stimulated by ring barking or wounding the tree. Cuttings with one or two nodes are generally preferred and the conditions for rooting depend very much on the species and the local environment.

These cuttings are planted out to form a clone archive. First generation rooted cuttings from selected trees should preferably be field tested before they are used operationally as not all good phenotypes produce good trees. But rooted cuttings from superior phenotypes are likely to be better planting material on average than some alternative seed sources. The initial clone archive comprises first generation cuttings, air layers or grafts that generally still need rejuvenation by grafting, hedging, or micropropagation to improve rooting. Following rejuvenation one or more clone banks may be established.

As with pines, a eucalypt clone bank for mass producing operational planting material comprises many selected clones replicated many times. These clone banks may contain 100 clones and thousands of ramets and cover many hectares. They need to be managed to remain vigorous. Fertilizing is essential, and irrigation and pest and disease control may also be necessary. Selective harvesting of ramets in the clone bank on a 10–14 day cycle reduces stress and increases production (Denison and Kietzka 1993).

The high cost of establishing and maintaining extensive outdoor clone banks is a major disadvantage of the traditional methods of producing rooted stem cuttings. They require large areas of land, significant quantities of water and nutrients, and a high labour input. Cutting production in clone banks has evolved from widely spaced (3 m × 3 m) field plantings to closely spaced (0.5 m × 0.5 m) field planting to nursery-based hydroponic systems. In the future cuttings may come from intensively managed, indoor hydroponic mini-clonal hedges that have high productivity and low requirements for space, water, and chemicals (de Assis 2001).

Nursery techniques
Most clonal nurseries produce containerized planting stock but open-root stock can also be raised successfully with some species.

1. *Pines*. The techniques used to produce both container cuttings and open-rooted cuttings of *Pinus elliottii* var. *elliottii* × *P. caribaea* var. *hondurensis* hybrids in Queensland (Baxter 2000) are given as an example of mass propagation of pines.

Containerized cuttings can be produced throughout the year with 70–80% of shoot-sets meeting planting specifications. Having a good growing environment for container stock is important and the Queenland operation features the following conditions:
(a) Site protected from wind.
(b) Cutting area shaded by 50% shadecloth and sides protected from direct sunlight.
(c) Good airflow maintained under drip-free shade structures.
(d) Temperature kept below 35°C.
(e) Site free of weeds, algal growth, and moulds.
(f) Only chlorinated irrigation water used.
Following the preparation of mix and filling of trays the following procedures are adopted:
(a) Filled pots are placed on benches at waist height.
(b) Pots are irrigated to full capacity.
(c) Holes are centred and dibbled to a uniform depth of 4 cm.
(d) Cuttings are selected, placed, and firmed with one action into the dibble hole. Rooting hormones are not required.
(e) Pots are drenched within 10 min to reduce stress and assist with compaction.
(f) Newly set cuttings are protected from wind and sun by shading.
Cuttings are irrigated for at least 2–5 min each hour during the day. A root assessment is carried out to determine strike rates at 16–20 weeks after setting. When a strike rate of 80% is achieved the cuttings are conditioned by removing shade cloth and applying slow release fertilizer. When stock height averages 12 cm the plants are moved to a reduced irrigation conditioning area to harden

them and encourage mycorrhiza development and root consolidation. They are conditioned for at least 1 month before planting.

Open-root cuttings are grown in raised beds. The beds are irrigated to maximize soil moisture and then a pre-emergent herbicide application is applied just prior to setting. The cutting setting includes several operations.

(a) Cuttings from the hedge production site are placed into cold storage in crates and drenched with water 4 times daily.

(b) Shoots are placed to a depth of 4 cm into predrilled holes in the bed and gently firmed around their base. They are shaded immediately after setting.

(c) Irrigation for 5 min every hour during the day (equivalent to 6 mm day^{-1} of rainfall) is applied after the setting operation.

(d) A weedicide (Oxyfluorfen) is applied within 1 week of setting.

(e) Shoots generally take approximately 16 weeks to strike roots. When 90% rooting has been obtained shade covers are removed, slow release fertilizer applied, and irrigation reduced. Subsequent operations are similar to those used for open root seedlings (Chapter 10).

Fertilizer additions of NPK and micronutrients maintain the health and vigour of planting stock. Rooted cuttings take up less of the available nutrients than seedlings but respond well to slow-release forms. Root pruning is an important tool in managing root development and assists with the physiological conditioning of planting stock and control of height. All open-root cuttings receive several root prunings as this helps develop a strong, compact and fibrous root system, and improves lateral rooting. Height growth is controlled by topping with a tractor slasher. This induces diameter growth and gives a uniform plant height. Only soft new growth is cut as topping too low into woody tissue results in multiple stems.

About 65% of open-root cuttings will normally meet planting specifications. They are lifted, packaged, and placed in cold storage until transported to the field. It is important to minimize plant exposure in this phase to ensure maximum survival after planting.

2. *Eucalypts.* Most eucalypt cuttings are rooted in containers. Usually several cuttings with one or two pairs of leaves are taken from the base of a coppice shoot. The leaves are trimmed to reduce transpiration and the cuttings treated with a fungicide solution (e.g. Benlate) to prevent decay. The base of the cutting is dipped in an auxin (plant hormone) powder to assist rooting. Auxins such as indolebutyric acid (IBA) and napthylacetic

acid (NAA) at concentrations of 10–30 ppm often stimulate rooting. It is frequently necessary to identify the most suitable auxin and its concentration for a species, for example, IBA (3%) for *E. deglupta* (Davidson 1974).

The cuttings are set in containers similar to those used in seedling production (Chapter 10) as determined by local experience. Cuttings are inserted into a sterile medium, usually of fine washed river sand, peat, vermiculite, or perlite or a mix of these, which should be free-draining but have good moisture retention. Rooting success depends on a proper balance of aeration and moisture within the medium. Fertilizer can be applied either as slow-release pellets incorporated in the substrate or applied as a solution during irrigation. Environmental conditions determine how quickly roots appear but it is usually within 6–12 days.

During the rooting phase the cuttings are usually grown under shade and with continuous or intermittent mist (e.g. 20 s every 5 min for 28 days). As the rooting progresses the time between misting is progressively increased and eventually the shade is removed to assist the conditioning process. Normally it takes from 8 to 12 weeks for the plants to be ready for transfer to the field, at which time they can be 20–30 cm tall. More detailed accounts of raising eucalypts by cuttings are available, for example, Kijkar (1991) and Eldridge *et al.* (1993); see Fig. 11.6.

Vegetative propagation using mini-cuttings

Two highly intensive techniques, micro- and mini-cuttings, have been developed for cloning *Eucalyptus* and some other commercial plantation trees on a commercial scale. Mini- and micro-cutting techniques are very similar in concept and operational procedures. The main difference is in the origin of the initial propagules. Micro-cutting uses the apices obtained from micropropagated plantlets (see next section), while the mini-cutting is based on the rooting of axillary sprouts from rooted stem-cuttings. In both systems the plants are managed intensively to produce mini-cuttings (Fig. 11.7). Traditionally the term 'micro-cutting' has been used for rooting of shoots produced *in vitro*. But some propagators (e.g. de Assis 2001) suggest a system of vegetative propagation *ex vitro* based on mini-propagules can be termed 'mini-cutting', including where initial propagules originate from a micropropagation system.

Field clonal hedges are being replaced by indoor hydroponics mini-hedges, which provide high degree of juvenility of micropropagated plantlets

(a)

(b)

Figure 11.6 Clonal propagation of eucalypts. (a) A eucalypt clone bank at Aracruz Celulose S.A., Brazil (photo. E. Campinhos, Jr). (b) Raised beds of eucalypt micro-cuttings (left) and macro-cuttings (right) at Champion, Amara State, Brazil (photo C.E. Harwood).

or rooted cutting. In the early 1990s the development of micro-cutting technology for *Eucalyptus* contributed significantly to the progress made in systems for large scale production of vegetative propagules *ex vitro*. Originally the system was based on mini-hedges established through rooted mini-cuttings, grown in small containers. The idea of hydroponics, in an operational indoor system based on drip fertigated sand beds, was introduced by Higashi *et al.* (2000) in Brazil. Campinhos *et al.* (2000) used the same concept in a highly efficient intermittent flooding system, where containers of the mini-stumps are immersed in a nutritive solution for fertigation. These systems can produce annually about 25 000 propagules of *E. grandis* $\times$ *E. urophylla* hybrids compared about 120 propagules m^{-2} for

conventional clone banks or hedges (Campinhos *et al.* 2000; de Assis 2001).

Compared to stem-cuttings, the rooting of micro- or mini-cuttings improves rooting potential, rooting speed, root system quality, and reduces costs. In general these mini-cuttings root better than juvenile stem-cuttings of the same clones produced in clonal hedges in the field with their success partly dependent on their optimal nutrient concentration. With minor adaptations this mini-cutting technique can be applied to other, broadleaved and coniferous species such as *Acacia mearnsii and Pinus taeda* (de Assis 2001). These very intensive cloning systems mark a new phase in development of mass vegetative propagation of *Eucalyptus* and other tree species.

Figure 11.7 Mini-cuttings of hybrid clone of *P. elliottii* var. *elliottii* and *Pinus caribaea* var. *hondurensis* at Toolara, Queensland, Australia. (a) Field hedges for mini-cuttings, (b) mother stool for mini-cuttings, (c) mini-cuttings. (Photos M. Baxter.)

Micropropagation and tissue culture

Micropropagation techniques use fragments of differentiated plant tissue, such as buds, shoot tips, hypocotyls, roots, or embryos to develop clonal lines. The fragments (small propagules) are referred to as 'explants'. They are placed in sterile culture and induced to grow shoots. Subsequent rooting is then stimulated in various conditions. Strictly defined, 'tissue culture' refers to the culture of cells or callus tissue in sterile culture but the term is often used more broadly to cover a range of micropropagation techniques and facilities. Commercial tissue culture laboratories produce many agricultural and horticultural plants and have been used for some forestry species.

To micropropagate a plant, a fragment is sterilized and its actively growing tip cut off and placed in a multiplication medium in a sealed jar (*in vitro*) and kept in a well-lit, sterile, environment at carefully controlled temperatures. This induces the fragment to grow shoots which can be separated and used, if necessary, to raise more propagules, a process that, in theory, can be repeated again and again. Shoots for planting stock are transferred to a rooting medium and transferred to a container with normal soil when rooting is satisfactory, and finally hardened off. Sometimes the multiplied shoots are removed and these micro-cuttings rooted on a medium such as river sand or pasteurized bark under less controlled glasshouse conditions, for example, eucalypt hybrids in South Africa (Watt *et al.* 1995), and may or may not require auxin treatment.

Micropropagation *in vitro* provides a means for very rapid multiplication of plant material, and is especially valuable for species that are hard to root by conventional means. Also, *in vitro* plants appear to have root systems that resemble more closely those of a seedling than do plants derived from macro-cuttings, and *in vitro* propagation can contribute significantly to clonal production programmes (Watt *et al.* 2003). It can be a way of rejuvenating older material which can then be used as stock to produce shoots for macropropagation by cuttings.

Research into micropropagation techniques gained momentum in the 1980s and considerable success was achieved with *Eucalyptus camaldulensis, E. grandis, Gmelina arborea, Swietenia* spp., and *Tectona grandis*. Research on tropical pines, *P. caribaea, P. kesiya, P. oocarpa*, and *P. tecunumanii*, was also promising (Purse 1989; Halos and Go 1990). Since 1990 a number of problems have been overcome, protocols have been developed for some species and plantlets have been field-tested.

The high cost of production remains a barrier to widespread use of plantlets from micropropagation. In eucalypt breeding and clonal programmes, *in vitro* culture systems are being used for the rapid multiplication of selected genotypes, rejuvenation of physiologically old shoots, germplasm conservation, and other applications. Yet, with notable exceptions, the potential commercial benefits of using *in vitro* technology for mass propagation of selected genotypes have yet to be realized (Watt *et al.* 2003).

The term 'somatic embyrogenesis' is used for micropropagation techniques in which embryos are extracted from seeds and multiplied in containers (*in vitro*); simply a process of embryo cloning. This is a promising technology for long-term storage and large-scale propagation of trees from selected clonal lines (Yanchuk 2001). Protocols are being developed for particular species, for example, for *E. grandis*, (Tsewana *et al.* 2000), *E. tereticornis, Dalbergia sissoo* and *Azadirachta indica* (Gill and Gossal 1996), and significant progress has been made with protocols for rejuvenation by embryogenesis of *P. radiata* and *P. elliottii* using apical meristems (Smith 1999). While there are still many problems with somatic embyrogenesis, these new technology developments have the potential to assist clonal planting of superior pine clones as it will be possible to make selections from mature trees in the field on the basis of individual wood properties and other important traits.

Micropropagation and *in vitro* embryogenesis are becoming increasingly important tools for tree improvement. As reliable protocols become available it is now feasible to develop both short-term and long-term *ex situ* conservation of valuable germplasm and *in vitro* germplasm conservation may be especially important for trees with recalcitrant seeds (Mudge and Brennan 1999). There have been problems with cell or callus cultures not producing normal plants due to unwanted genetic changes 'somaclonal variations'. Such problems need careful monitoring if these *in vitro* techniques are to be fully exploited.

Recent changes to the traditional tissue culture technology of semi-solid cultures to liquid cultures could make the technique more attractive for propagating superior genotypes (Harvett 2000). For example, Shell Forestry (United Kingdom) has patented 'Bulk-Up', a liquid tissue culture system providing rapid multiplication of germplasm and faster deployment of superior genotypes. Plantlets can be rooted *ex vitro* and a pilot laboratory in Paraguay is able to produce 75 000 plants per year (Watt *et al.* 2003). The 'Bulk-Up' method can improve multiplication rates by ten times those of semi-solid systems and will become more widely used (Griffin, personal communication).

Micropropagation is expensive relative to macropropagation due to the need for skilled labour and more sophisticated facilities. In New Zealand the cost of micropropagation of pines is five to seven times that of a seedling (Smith 1999) but where labour costs are lower it may be an option. It has been used on a limited scale for mass propagation of selected material, for example, in Vietnam hundreds of thousands of plantlets of acacia hybrids have been produced by meristem tissue culture (Kha 2001). In southern China selected clones of the hybrid *E. urophylla* × *E. grandis* and have been micropropagated for routine plantations and Zhang (2000) reports *Paulownia* spp. can be propagated *in vitro* on a large scale. Protocols have been devised in Sabah for the mass propagation of *Tectona grandis* using micro-cuttings rooted in a nursery under a mist system which are cost effective and efficient for the scale needed (Monteuuis and Goh 1999). Micropropagated plantlets of salt tolerant *E. camaldulensis* have grown successfully in Australia, although their wider use in protection planting will depend on their benefits compared to the extra cost of producing them relative to seedlings (Bell *et al.* 1994). Micropropagation *in vitro* will play a critical role in propagating genetically modified (GM) trees (see below).

Clonal deployment

Concerns about potential serious failures due to genetic uniformity of clonal plantations have resulted in much debate about how to deploy clones within the plantation estate. Failures have occurred occasionally in clonal plantations. In Europe, monoclonal stands of hybrid poplar were severely damaged by the leaf rust fungus *Marsonnina brunea* in the 1960s while in the tropics the danger of growing trees with a very narrow genetic base was illustrated by the catastrophic effects of the psyllid infestation on *Leucaena leucocephala* (Nair 2001). Risks of planting a single genotype over a wide area are significant but the more clones that are needed to reduce the risk the fewer are the advantages of adopting a clonal strategy (Zobel 1992).

Key questions are:

1. How many clones should be used in an industrial plantation project?
2. How many should be used at one site?
3. Should the clones be planted at one site in a mixture or monoclonal stands established in mosaics?
4. What is the basis for arriving at a decision on clone deployment?

Number of clones

The situation is complex as the optimum number of clones to use is related to species, length of rotation, environment, and the genetic make up and adaptability of the clones. So far there is no simple scientific model to determine the ideal number of clones to plant in an area. An early attempt was made by Libby in 1982 on theoretical grounds to determine the number of clones in a clonal plantation that would be as safe as a natural seedling forest. He proposed between 7 and 30 unrelated clones would be sufficient (Libby and Ahuja 1993). Risk analysis of different scenarios has shown that risk can decrease, remain almost constant or even increase as the number of clones increases (Bishir and Roberds 1999). Even with an extremely simple model, the cases they simulated resulted in 'required' numbers from 1 to 100 but the authors concluded that situations requiring more than 40 clones are not common.

There has been legislation in some European countries to control the level of risk by specifying the minimum number of clones to be used. These have ranged from 20 to over 100 depending on species and other considerations. From a risk perspective, the range of genetic diversity among clones is more important than the actual number used. Genetic diversity associated with survival and adaptation traits needs to be balanced with uniformity among production traits, such as wood density (Foster and Bertolucci 1994). Generally forest managers decide on the number of clones based on a subjective assessment of the risks involved and the potential benefits to production and profits.

Deployment of just a few superior clones of plantations of *Populus deltoides*, *Eucalyptus* hybrids, *Paulownia* spp., and other species still occurs in some countries and is a high-risk strategy. More responsible enterprises usually have between 50 and 100 genetically superior clones available for their production plantations and where an effective breeding programme is being undertaken concurrently new selections are added to the pool and those performing less well are replaced. Tasman

Forestry Ltd in New Zealand had an initial pool of 100 clones but was testing another 2000 clones (Gleed 1993), Aracruz Celulose S.A. in Brazil had about 80 eucalypt clones in its production population and was using subsets of these to plant at different planting sites. Initially the company planted large areas with a single clone and experienced few problems, possibly because about 25% of the planting area remained as natural forest. Now a mosaic distribution of different clones is used and care taken to use clones well-adapted to local conditions (Campinhos 1999).

The practical reality is that it is difficult to work on a large scale with more than 5–10 clones in one season in the nursery. Nevertheless, it makes sense to have a production population of at least 20 clones that can be rotated in successive planting years. It must be emphasized strongly that a clonal plantation programme must be backed by an active breeding programme so that new material becomes available. The plantation manager needs to review the list of operational clones to be planted every year and be able to substitute new clones for any existing clones that have a problem.

Despite the identification of genetically improved material in clone trials established in the 1990s, most nursery production of some companies is still based on a few clones developed and tested in the 1980s. The reason for this is that many nursery managers prefer using clones that are well tried and tested as far as nursery performance (ease of rooting etc.) is concerned and are reluctant to replace them with new clones, even though these have better field performance (MacRae 2003). Better communication between tree breeders and nursery managers may help resolve this problem and it is essential to integrate breeding and propagation strategies.

Clonal distribution

Most forest managers consider the risk of deploying a single clone over a large area is too high. They are then faced with the choice of planting intimate mixtures of several clones or mosaics of different clones in monoclonal blocks.

The potential advantages of mixtures of clones are increased yields through different genotypes exploiting different parts of the site, greater stability in available environment, and less risk of serious damage by pests and diseases. Experimental results have generally failed to support the expected superiority of intimate clonal mixtures (Foster and Bertolucci 1994). Zobel (1992) suggests that in a 20 clone mixture of *E. grandis* about 20% of the trees will be suppressed and possibly die, a situation that will be unacceptable to managers practising intensive clonal forestry.

Most managers growing short rotation crops with intensive management consider that planting a mosaic of single clone blocks provides sufficient genetic diversity within their forest area to provide an acceptable level of risk. Hybrid pines in the Queensland plantation programme and eucalypts in many countries are planted as single clone blocks. Monoclonal blocks enable observations to be made on the growth and health of clones and replacement of those that are not performing well. Harvesting blocks of single clones also provides good control of the quality of wood being sent to the processing plant.

The size of the monoclonal blocks will depend on species and location. Usually clones are planted in 10–20 ha blocks which permit efficient operations (Zobel 1992). In the Congo about 15–20 clones (mainly *E. urophylla* × *E. grandis*) are planted in monoclonal blocks of 20–50 ha.

Arbez (2001) suggests that the risks associated with the lack of genetic diversity in monoclonal blocks can be reduced by:

• decreasing the rotation length
• limiting the plantation area and the period a given clone is used
• increasing the number of commercially available tested clones.

Initial planting density

Seedling plantations for many years were established at stocking rates of about 1600–2200 trees ha^{-1}

(3.0 × 1.5–2.0 m spacing) which anticipated losses due to mortalities and poor growth of some trees. In eucalypt plantations the availability of highly productive clones, the potential to reduce harvesting costs by growing bigger trees, and lack of water on some sites has resulted in initial stocking being reduced to between 800–1100 trees ha^{-1} (3.0 × 3.0–4.0 m). Spacing to achieve these densities varies according to management practices. Planting at this density requires good establishment practices to reduce mortalities and ensure a stocking of 1000 trees ha^{-1} for pulpwood production and 800 trees ha^{-1} for sawlogs (Stape *et al.* 2001).

Small-scale clonal plantings

Small-scale industrial plantation enterprises

Use of clones in small-scale plantations has the potential to achieve the same dramatic benefits as have occurred in large industrial plantations. The small-scale forestry sector faces the difficulty of having to obtain selected material to propagate. It is unlikely that small-scale operations can justify their own tree breeding programmes so they must make arrangements to use material developed by larger companies or government agencies, or by joining tree improvement cooperatives where the costs of tree breeding and selection can be shared. In India, ITC (Bhadrachalam Paperboards Ltd) have successfully promoted eucalypt plantations on marginal agricultural lands since 1992 (Fig. 11.8) and now supply over 2 million high quality rooted cuttings annually to farmers (Lal 2001) (Fig. 11.9).

Figure 11.8 Propagation greenhouse at Pragati Biotechnologies, Punjab, India, producing 0.75 million rooted cuttings of selected hybrid eucalypts annually for farm planting (photo P. Lal).

Figure 11.9 A selected clone of *Eucalyptus tereticornis* aged 5 years in a farm forestry plantation in Andhra Pradesh, India (photo P. Lal).

Small-scale operations will generally have limited staff resources and low technology facilities so the propagation system is likely to use cuttings rather than plantlets from micropropagation culture. Several propagation systems suitable for producing rooted cuttings of tropical trees have been developed. In the tropics, only minimal improvements in facilities and work standards may be needed to convert a conventional seedling nursery into a cuttings nursery. Technical aspects of the mass propagation of eucalypts on a small scale are discussed by Kijkar (1991) and White (1993). Low technology, non-mist propagators have been developed that are more effective than a misting system for semi-arid species, such as *Acacia tortilis* and *Prosopis juliflora*, but are also suitable for tropical moist forest species, such as *Cordia alliodora* and *Gmelina arborea* (Leakey *et al.* 1990, 1994). Foster and Bertolucci (1994) suggest that small-scale growers may have to use novel strategies to mitigate risks when they use clones.

Those considering using a clonal strategy for producing industrial wood on a small scale are recommended to undertake a simple marginal benefit/cost analysis of the type described by Walker and Haines (1998) before making a final decision.

Clonal propagation in agroforestry

Farmers in the tropics have also used vegetative propagation in traditional agroforestry systems for centuries. One of the simplest traditional vegetative propagation systems is to collect root suckers, which arise spontaneously of from roots that are wounded. Many species, such as *Faidherbia albida*, *Inga feullei*, and *Melia volkensii*, naturally throw root suckers (Mudge and Brennan 1999). Farmers in Latin America, Africa, and Asia have traditionally used large cuttings to establish living fences to produce fuelwood and fodder, and to mark the boundaries of their land (Jolin and Torquebiau 1992).

Stem cuttings of various sizes have been a common method of vegetative propagation of trees and shrubs by farmers as they require only simple tools and minimal skill. Direct planting in the fields of large cuttings is preferred. Planting is usually by insertion of hardwood cuttings 3–7 cm in diameter and about 1.5 m long, but they can be up to 15 cm diameter and 2.5 m in length (Jolin and Torquebiau 1992). The basal 20–30 cm is inserted into prepared ground so as to leave at least 1 m of shoot above ground. If inserted at the start of the wet season rooting will begin almost immediately and a first flush of foliage occurs about 6–8 weeks later. Minimal maintenance is essential but when tall enough, typically about 2 m, regular pollarding or pruning is required, often just before the rains commence, to maintain vigorous regrowth of shoots.

Root cuttings have been widely used in China to mass propagate selected clones of pure species and hybrids of *Paulownia*. After 1- or 2-year-old trees have been lifted from the nursery the remaining roots are collected. Roots 15–18 cm long and 1–4 cm diameter are collected, stored until the following spring and then lined out in the nursery to raise saplings 4–5 m tall. These saplings are planted out at wide spacing and intercropped with tea, wheat, corn, cotton, or vegetables (Chinese Academy of Forestry 1986). Other cloning techniques, such as marcotting or air-layering, and grafting, are used by farmers for multi-purpose trees and fruit trees and examples are given by Mudge and Brennan (1999).

Farmers usually use several species and prefer simple and inexpensive technologies. This is why cuttings that can be easily collected locally and set directly in the field are attractive to them. In most countries there is little coordination of agroforestry germplasm production or release, so while many farmers are familiar with using clonal material the pathways for distributing genetically improved stock are not well-defined (Simons 1996).

Role of molecular genetics in clonal forestry

Recent advances in biotechnology through molecular genetics and gene modification are being applied to tree breeding and Haines (1994) suggests 'clonal forestry is the gateway to major applications of known biotechnologies in tree improvement'. Molecular genetic technologies include molecular markers that can be used for clone identification and registration (DNA fingerprinting), genetic diversity assessment and management, and genome mapping. All these can be applied to improve conventional tree breeding and enable it to produce superior genotypes for clonal plantations, for example, Carson *et al.* (1996); MacRae and Cotterill (2000*b*).

Possibilities for the addition of new genes to selected genotypes through the use of recombinant DNA and gene transfer technologies are providing new avenues for genetic improvement of trees. Cloning is a prerequisite for the production of such genetically modified trees. Genes that are added to an existing genotype to add value and produce trees with characteristics that are economically important but could not be produced by conventional means within a reasonable period of time. The traits so far considered for genetic modification of trees include herbicide tolerance, reduced flowering or sterility, insect resistance, and wood chemistry and fibre quality (Griffin 1996; Yanchuk 2001). No GM trees are yet used in commercial planting but, according to the World Wide Fund for Nature, there were 116 confirmed GM tree trials involving 24 species in 17 countries between 1988 and 1999 (Owusu 1999).

Genetic modification or regulation of a tree's own genes may be capable of producing a satisfactory outcome, for example, to reduce lignin content. But the inserted genes may come from totally unrelated genera (transgenes) and this has made the technology highly controversial, especially in food crops. Considerable research is in progress (see Ritter 1999; MacRae and van Staden 1999) and the release of GM trees into the environment is being promoted, providing predictable risks have been tested in risk assessment studies (e.g. Fladung 1999). Technical risks include environmental issues, such as invasiveness, impacts on biodiversity and transfer of genes to other organisms, and the possible vulnerability of long rotations of trees to climatic and biotic factors due to adverse side effects of single genes on fitness (Burdon 2003).

The question of whether genetically modified trees can be safely deployed into the environment is still being debated and some countries have laws and regulations concerning the testing and release of GM plants. There are also issues of equity of access to GM trees, as most are being developed by private companies (Yanchuk 2001). However, some companies, such as Stora Enso in Portugal and Brazil (MacRae and Cotterill 2000*b*), have decided not to use genetic engineering technology to modify trees for their industrial plantations. If GM technology is to be used in plantations it should be as an adjunct to conventional tree breeding and integrated with other approaches (Burdon 2003).

Genetic engineering of forest trees still requires much research to understand, control, and use transgenes to improve the productivity of plantation forestry. It is a challenging task and ultimately it will be national and international bodies that determine if and when GM trees are deployed in clonal plantations.

CHAPTER 12

Plantation establishment

Trees and forests are planted on many kinds of sites and this chapter is concerned primarily with three main ones: grassland, savanna, and recently cleared woodland or forest. Silvicultural techniques for where agroforestry is practised and on difficult sites, such as in arid and/or eroded lands, mining waste, and are considered elsewhere (Chapters 20 and 23). Nevertheless, a number of principles are common to all and much of this chapter is relevant to any plantation establishment regardless of site conditions.

The major principles set out below are based on Evans and Hibberd (1993) and Evans (2001b):

- land should be free of a significant alternative use or designation;
- local support and acceptance (stakeholder participation) of the plantation activity is necessary;
- the long-term nature of forestry requires substantial political stability and support;
- access to the plantation area needs to be related to the end-use of the crop and provide infrastructure for maintenance, protection, and other essential activities;
- the land must have the potential to support tree growth at a level to satisfy the plantation's objectives;
- plantation development is socially and politically most acceptable on land which has been significantly modified by human activities: conversion of natural forests and woodlands should be avoided.

Establishment techniques for tropical plantations have been more widely described than most aspects of silviculture. In many countries local prescriptions already exist and in recent times there have been moves to incorporate them into Codes of Forest Practice. These practices, when correctly applied, should meet standards for sustainable forest management. A manual covering all the main practices, primarily for industrial plantations, published by FAO (Chapman and Allan 1978) remains relevant, apart from an overemphasis of clearing woodland and forest for potential plantation sites, which is now strongly deprecated.

One silvicultural principle applying to all kinds of tree-planting is emphasized at the outset. Sustained productivity is influenced by all management practices throughout the rotation and between crop cycles but the phase between harvesting one crop and establishment of the next is perhaps the most critical. The largest changes in hydrology, nutrient cycling, and other key site factors occur during the establishment phase. This is usually in the first 3 years after clearing but the site is especially vulnerable in the first year when the soil surface is exposed. Many environmental hazards diminish greatly when the plantation canopy closes. Therefore, overriding objectives of tree establishment are good survival and rapid early growth; these depend on satisfying all the requirements for tree growth. Correct choice of species, adequate ground preparation, control of weeds, addition of fertilizer if necessary, and protection against browsing, insects, etc., must all receive attention. Omission of any one, perhaps because of cost, may greatly prolong the establishment phase and be a false economy.

Pre-establishment essentials

The following discussion is primarily directed at large-scale industrial plantation development. However, there is increasing interest among farmers to plant trees to improve farm productivity and reverse environmental damage. As with industrial forestry, there are several important steps that must be taken before the trees are actually planted and tree-planting plans must be developed and incorporated into the overall farm plan. Aspects of these plans are referred to below but reference is recommended to texts dealing specifically with farm planting, for example, Cremer (1990).

Legal title to land and rights to trees

Though it is not necessary for the afforesting agency always to own the land to be planted (Chapter 5) for any area the following provisions must be agreed before planting begins; (i) purchase price or rent for land and how, when, and to whom money is to be paid; (ii) the exact boundaries;

(iii) what rights are retained by the seller or lessor; (iv) adequate rights of access to the land, particularly if entry is gained over someone else's land; (v) ownership of timber and rights to using trees, stumpage payments, distribution of revenues especially with tribally owned land; (vi) the period of lease or right to grow trees when land is not owned; (vii) agreement over compensation and provision of benefits for any displaced people.

Reaching agreement on these matters is usually long and involved, see Chapters 5 and 6, but it is an important stage in project development and numerous problems will arise later if it is not fully covered at the start. As was noted in Chapter 6, delays can occur in acquisition of land for industrial tree-plantations and it is important to maintain a reserve of plantable ground. Without it, the size of annual planting programmes may vary widely and planning of nursery work, labour, finance, and equipment needs will be extremely difficult. While a project is expanding rapidly, ideally 2 years plantable reserve should be in hand.

It is also stressed that successful social forestry programmes largely depend on rights of ownership, harvesting and marketing of the trees. Uncertainty in these tenure matters creates many problems and is a powerful disincentive. In Ethiopia in the 1980s, both the policy of villagization, which involved resettling people into large villages, and confusion regarding ownership of trees planted on kabele ground (communes) deterred tree-planting in some parts since rights to trees in 5 or 6 years' time were too uncertain. Similar disincentives have occurred in China where the government guaranteed tree ownership rights but maintained controls over timber harvesting and marketing.

Boundaries and demarcation

A fundamental principle of forest management in the tropics is to have clear and permanently defined forest boundaries linked to permanent marking, surveying, and mapping of boundaries (Armitage 1998). They are critical for ensuring protection of the plantation and for other sustainable forest management activities.

The external boundary (Fig. 12.1) must be clearly evident: (i) it defines areas and boundaries with respect to the adjacent landholder, this is especially important where grazing and grass-burning are practised; (ii) it stakes a legal claim to deter unwitting trespass or theft; (iii) in the first few years after planting trees are not always obvious and a well-defined boundary will lessen the chance of mistakes and accidents. Where exactly the planting takes place and especially the boundaries, is a crucial factor in social and community forestry projects. Involvement of local people in these decisions, a key stakeholder group, is excellent 'people's participation' and encourages their commitment. In the Bilate project in southern Ethiopia people from each of the local kabeles involved would agree the boundary, and dig a ditch and low wall along it, in advance of each year's work (Fig. 12.2). Generally, the eroding slope to the river was set aside for the project, the flatter land above was retained for livestock and farming—but the decision was up to the local people. Also, they agreed where access to the river should be kept open to water livestock. All these were important ingredients to the project's success, taking priority over purely technical considerations (Evans 1989).

Plantation boundaries are frequently marked by a stout fence, which is erected for protection purposes. Fencing is costly, it should only be constructed to a standard needed to exclude harmful animals, and the length kept to a minimum with awkward corners fenced out. Where protection is not needed some permanent marking, such as concrete or treated-wood posts or low walls, should be erected at least at corners and changes of direction. In Trinidad, red-leaved *Dracaena* spp. have been used to mark forest boundaries; it is cheap and very obvious. In some countries it is common to see notices describing a project on its boundary.

A map showing boundary details markers, distances and angles, and other information such as owners/occupiers of adjacent land, should be kept readily available and not just with the legal documents.

Preplanting survey

A preplanting survey is a detailed study of the planting area to enable a plantation manager, smallholder, or villager to (i) decide what land should be left unplanted for protection, conservation, landscape, archaeological, cultural, or amenity purposes; (ii) select species for planting by site types; (iii) to determine what ground preparation is required; (iv) consider possible harvesting systems; and (v) plan internal layouts of roads, rides, firebreaks, and location of water points, depots, etc.

The preplanting survey differs from a general site evaluation survey to assess afforestation potential (Chapter 5), in that it is a more detailed study of a particular area to be planted. Of course, it will include relevant data from the site evaluation work, which will normally have preceded it. A detailed account is found in FAO (1984*b*).

Figure 12.1 Southern boundary of Block E of the Usutu forest, Swaziland. Note wide firebreak and roads providing access to it.

Figure 12.2 Ground preparation in the Bilate community forestry project in Ethiopia. The ditch, and low ridge planted with aloes, was dug along the line agreed by the villagers as the project's upper boundary—all land to the left was allocated for the project.

The information in Table 12.1 is collected in an orderly way by systematic sampling of the land, study of aerial photographs or other imagery, discussion with former owners, neighbours, local people, and through contact with local or regional government officers, and possibly research staff for specific advice. A special survey team often undertakes the work and the information is presented in both written and map form. A convenient map scale is 1:10 000, which can accommodate most details normally required for planning plantation establishment. Several maps of each area are usually prepared (soils, exposure, communications) and a system of transparent overlays is helpful in subsequent study and analysis. As detailed in Chapter 5 such information can also be digitized and used in geographical information systems (GIS), which greatly enhances planning, delineation of environmental features, biophysical limitations, and so on.

In Queensland, comprehensive surveys are carried out by the State forest service to provide site suitability and site capability assessments for tropical pine plantation establishment. The information provides a basis for identification of plantable areas, design of watercourse protection zones, site preparation design, species selection, and fertilizer prescriptions (Foster and Costantini 1991a; Last 2001). For large projects, survey companies may be contracted. In the mid-1980s in East Malaysia, Sabah Forest Industries contracted soil surveys over many thousands of hectares to record soil characteristics as the basis for identifying land suitable for *Gmelina arborea* with its more exacting requirements compared with *Acacia mangium*. Soil and site studies should concentrate on parameters that have good correlations with tree growth such as root-available soil depth, slope position, impeded drainage, soil texture, acidity (pH and Al-saturation), organic matter

Table 12.1 Information sought in pre-planting surveys and the use made of it

Category	Information details required	Management value
Terrain	Elevations, aspect, slope steepness, ground configuration, rockiness	Limits of economic harvesting, risk of erosion, limit of mechanization. Species choice, ground preparation, etc. Ground configuration may affect location of firebreaks, nursery, depot, roads, etc.
Drainage	Occurrence and network of rivers, streams, lakes. Flow rates, risk of flooding, periods when beds are dry, etc.	This is strictly part of terrain but is especially important in siting roads and extraction systems, designing bridges and culvert size, and locating nurseries, dams for fire protection and watering points
Soils	Soil types, soil chemical and physical status, erodibility, underlying geology and outcrops	Species choice, ground preparation work, fertilizer requirements, need for erosion control, possibly supply of road building materials, etc.
Vegetation cover	Vegetation types, density, species composition, areas or species of special ecological value	Need for vegetation clearance, identify areas for conservation, use of existing vegetation as index of site fertility
Communications and services	Location of public roads and rights of way, telephone and electricity lines, other easements	Assist with infrastructure development, identify unusable land (below) and where rights must be maintained
Special factors which may preclude planting	Areas of special scientific or historical/archaeological interest	Land with important geological, biological, ecological, historical, or traditional value
	Areas or vegetation of religious or cultural significance	Local custom will dictate this. Land or existing trees may be valued for spiritual associations or as source of food, for example, *Pandanus* groves in Papua New Guinea
	Areas of landscape significance	Exclude land, or modify boundary in sympathy with landform
	Easements—power lines, telephones, etc.	Cause restriction on tree height growth and hence value for tree-planting
	Rights of access, grazing, hunting, etc.	Essential for these to be fully recognized and local people involved at this planning stage to prevent conflict of interest and antipathy. Too little attention has generally been paid to this

content and salinity (Zech and Drechsel 1998). In South Africa, soils and surface features are usually surveyed on a 150 × 150 m grid and the soils grouped into relatively homogeneous units according to structure, depth, drainage, and local topography (Schönau 1984b). Such soil surveys are most appropriate where they are most expensive, for example, on large land holdings with a range of land forms and heterogeneous soils. The costs of site surveys have little effect on plantation profitability but the availability of accurate and detailed site information combined with modelling to predict productivity can prevent costly mistakes (Battaglia et al. 2001). Benefits include better choice of species, close matching of clones and sites, and targeted fertilizer applications leading to more uniform and productive plantations.

Internal layout of a plantation

Planning plantation layout and starting construction of roads and facilities follows the preplanting survey and is the last important task undertaken before tree establishment operations begin.

Many decisions have to be made. Where should the nursery or depot be sited? What are the best alignments for roads and tracks with a view to future projected harvesting operations? How should the area be subdivided with internal breaks for fire protection? How can any rights of access through the plantation be accommodated? What species should be planted considering topography and site fertility variations? Which areas need special treatment, for example, draining, fertilizing?

The preplanting survey for a specific project should collect the information to answer these questions but some general guidelines can be laid down.

Nursery or depot location
This is usually close to workers' accommodation and there must be access to a main road and source of water. Locate near to existing services when possible (Chapter 10).

Roading
Road construction is expensive. The following points are important.

1. Plan alignment for the entire network of an area to suit the anticipated extraction system. This is most easily done before the ground configuration is obscured by trees. Roads not constructed before planting can be pegged, marked by a single pass of a bulldozer, left unplanted, and used as fire-breaks.
2. Construct the least possible length of road before planting to satisfy access and protection needs.
3. Construct roads needed for extraction just before commencing commercial thinning.
4. Whenever possible combine roads and fire-breaks; note road in Fig. 12.3.
5. Avoid alignments involving many cuts and fills and generally take great care to minimize soil erosion as a result of road and track construction.
6. Drain water from roads and ditches at regular (100–200 m) intervals.

Subdivision of a plantation
In addition to roads, further artificial breaks subdivide the plantation into compartments mainly for fire protection purposes. These breaks are best sited along natural features (Fig. 12.3) such as ridges and watercourses. Where they are unavoidably located on fertile ground the firebreak can be planted with a fire-resistant species to form a green break, though these are of limited effectiveness.

Frequently, as a plantation programme develops, it crosses traditional paths or routes used, for example, for taking livestock to water. Although it may perpetuate a risk of fire or other damage, maintaining this access does much to reduce perceived negative effects of plantations and removes the impression of a barrier or total exclusion of land for other uses.

Ground preparation

Purpose

Plantation establishment is the formation of a healthy tree crop by planting or direct sowing. As with any crop, some site preparation is usually necessary to achieve reasonable success. Ground preparation is an integral part of establishment with the aim of securing both high survival and rapid early growth. This is achieved by (i) control of competing vegetation; (ii) removal of physical obstructions to tree growth; (iii) cultivation to improve soil structure, primarily to aid root development and to ensure the trees obtain sufficient oxygen, water, and nutrients; (iv) modification of natural drainage, either to improve drainage on wet sites or retain moisture in dry areas; and (v) construction of contour ditches, walls, or other engineering works to reduce soil erosion.

It is rare for all the above measures to be needed on one site to achieve satisfactory establishment. Ground preparation is often the most costly silvicultural operation. The aim is to do the minimum necessary in the most cost-efficient way consistent with securing both high survival and rapid early growth.

Figure 12.3 Fire break following a rocky ridge in the Usutu forest, Swaziland.

Need for ground preparation

The degree of ground preparation needed depends on the purpose of planting, species to be planted, vegetative cover, and site and soil conditions. However, in some circumstances it is not possible to carry out ground preparation, even though desirable, apart from digging and other manual work. This may be due to steep slopes or rocky ground impassable to machines (Fig. 12.4), or lack of finance to purchase equipment, or simply that an area is too small to justify bringing in machinery. In some instances inability to prepare ground adequately may render planting not worthwhile if only a poor crop will result. Also, sometimes ground will be left untouched because its conservation or other value depends on it not being disturbed.

Purpose of planting

Tree-planting to protect soil from erosion, while often requiring construction of bunds or terraces (Chapter 23), must minimize all other site disturbance, especially clearance of existing vegetation. It is pointless removing what protection is afforded, however inadequate, and exposing soil to even greater risk while trees become established.

Species

The degree of ground preparation needed directly reflects the species' ability to compete for light,

moisture, and nutrients on the site in question. This is closely related to matching carefully species with site. If much ground preparation is needed to establish a certain species it may be worth considering changing to another. The examples below illustrate the variation in requirements.

1. Eucalypts require a completely cultivated and weed-free site, often with the addition of fertilizer, for rapid early growth. Their sensitivity to weed competition and need for intensive weeding is emphasized by Eldridge *et al.* (1993) and clearly demonstrated in minimum tending trials on *Imperata* grasslands in Indonesia (Otsamo *et al.* 1995). In South Africa experiments comparing methods of establishing *Eucalyptus grandis* in grassland show that complete soil cultivation is greatly superior to all other methods, including chemical weed control, digging pits, etc. (Schönau 1985; Boden 1991). Similar intensive site preparation techniques are applied for other eucalypt species in other countries, for example, *E. globulus* in Portugal (Pereira *et al.* 1996) and Chile (Prado and Toro 1996). Araucarias also suffer from grass competition and will only grow very slowly and often turn chlorotic. Many small plots of *Araucaria cunninghamii* and *A. hunsteinii* in the Papua New Guinea (PNG) highlands are in this condition due to lack of ground preparation and weed control.

2. Many pines, including *P. oocarpa*, *P. patula*, and *P. caribaea*, are tolerant of grass competition and are able to grow through grass cover without tending. However, slower growing *Pinus merkusii* is not so tolerant (Whitmore 1972). Though some pines

Figure 12.4 Rocky slopes impassable to machines. (Usutu forest, Swaziland.)

suffer relatively little from grass competition, they are checked and sometimes swamped in dense grassland (e.g. Otsamo *et al.* 1997), and on old woodland sites by rapidly growing shrubs and vines (Fig. 13.3) and their growth will be retarded if cultivation, weed control, and fertilization are not applied (e.g. Cannon 1982, Heywood *et al.* 1997). Other species shown to be tolerant of grass include *Acacia* spp., *Gmelina arborea*, *Paraserianthes falcataria*, *Peronema canescens*, and *Senna siamea* in Indonesia (Otsamo *et al.* 1997); and *Alnus acuminata* in Costa Rica.

3. Species such as *Swietenia macrophylla* and *Cordia alliodora* can be planted to enrich existing forest with little direct ground preparation, only needing control of climbers and vines, because they will grow under shade.

4. Tropical acacias such as *Acacia auriculiformis*, *A. mangium*, and *A. crassicarpa* are relatively tolerant to grass competition. However, trials in Indonesia demonstrated that after 30 months weed control in a plantation of *A. mangium* had a greater effect on wood volume production than fertilization and cultivation (Turvey 1996). On stoney and steep sites weeds are controlled by herbicide without tillage but growth is poorer than where complete disc ploughing and harrowing are used to control *Imperata* grass (Terry *et al.* 1997). Mechanized site preparation is expensive so a combination of herbicdes to control weed competition and fertilizer to boost growth are used in many industrial plantations of *A. mangium* (Srivastava 1993).

Vegetative cover and its clearance

Many different types of vegetation may be encountered, varying from patchy grass to dense forest. Combined with differing needs of species, and several methods of clearance this leads to an almost infinite variety of clearance practices. However, clearance of natural forest or woodland for tree-plantations is today generally considered to be unacceptable on grounds of conservation, risk of erosion and, even in semi-arid areas, careful management of natural rangeland and bush can be as productive as planted trees in the dry conditions. Of course, the situation is different if little or no woodland remains. The serious environmental consequences of land clearing have been demonstrated by Ross and Donovan (1986) who stress the importance of choosing an appropriate land-clearing method for each particular site to minimize harm. The actual effort needed in clearance will primarily depend on the vegetation itself, its type and size (grass, small shrubs, large trees), its density—sparse and scattered woodland or dense

forest and regrowth. These factors determine the cost of clearance and also its ecological consequences (Chapter 24).

Four main ways of clearing vegetation, their application, advantages and disadvantages, are discussed below. A useful comparison of environmental and social impacts of methods used in the humid tropics is found in Ross and Donovan (1986). Examples of specific clearance practices are given later in this chapter in the section on 'ground preparation practices'.

Manual clearance

Saws and axes are used to fell trees, cut off branches, stack and pile debris, and clear a site. The advantages of manual methods are that they can be done at any time, few new skills need to be learnt, capital cost is small, temporary employment can be provided for shifting cultivators or villagers, there is no pollution, and damage to the soil is slight. Soil bulk density, water retention, and porosity were all least affected by manual methods compared with mechanical in a carefully designed study in Nigeria (Hulugalle *et al.* 1984). The main disadvantages of manual methods are that they are slow and expensive for clearing dense woodland. In Malawi it can take up to 150 man-days ha^{-1} to clear *Brachystegia* woodland.

Manual site preparation is widely practised in the tropics, especially where wages are low. It is best suited to sites where clearance is relatively easy, for example, cutting planting lines in grass or for enrichment planting, where sites are small and scattered, where damage to soil structure is likely from heavy machinery, and as a supplement to mechanical clearance.

Mechanical clearance

This is usually done using heavy crawler tractors, for example, D4-D8 bulldozers. The front blade shears or pushes over and uproots trees and shrubs. The pressure per unit area exerted by tracked vehicles is lower than for wheeled tractors and they have less impact on soils susceptible to compaction (Bruijnzeel 1997). Where vegetation is not large a heavy chain slung between two tractors and weighted with a steel ball can be dragged over shrubs and small trees and open up swathes of up to 20 m. In Zambia, savanna has been prepared for planting in this way, and 30 to 40 ha can be cleared in a day (Laurie 1974). A bulldozer with a shear blade is used to cut standing trees and move them into windrows but a V-blade is most suitable for clearing stumps in old plantations (Kretzschmar 1991).

Mechanical clearance has several disadvantages: high capital cost; problems of equipment maintenance and obtaining spares; supply of fuel and oil; the need for training and supervision; employment of few, though skilled men; and the risk of serious damage to the soil from compaction. Mechanized operations during harvesting, stump extraction, and redistribution of woody material during windrowing can accelerate soil erosion and runoff, and limit the extent of root penetration and the volume of soil available for roots to tap for water and nutrients (Folster and Khanna 1997). Lal (1997) reports on the effects of several methods of clearing in western Nigeria. Although manual clearing was slowest and most costly, soil erosion was a low $0.4 \, \text{t ha}^{-1} \, \text{year}^{-1}$ compared with the $4 \, \text{t ha}^{-1} \, \text{year}^{-1}$ after clearing with a crawler tractor and shear blade, and $15 \, \text{t ha}^{-1} \, \text{year}^{-1}$ after a crawler tractor and tree pusher/root rake. The amount of surface runoff had a similar trend. Soil degradation due to mechanized clearing through compaction, removal of litter and top soil, and exposure of subsoil can have a long-term influence on site productivity. The extent of the problem varies with soil type, degree of wetness, and type of machinery used.

Because of its great cost advantage, mechanical clearance has been widely used to clear scrub savanna and stumps and debris in logged plantations for industrial plantation establishment. Such equipment has been hardly used at all in social and community forestry programmes or where soil protection is an even greater priority. When used it is essential that equipment does not stand idle for long periods and that soil damage is minimized.

Burning

A key tool of the shifting cultivator, fire has been much used in forestry to reduce the debris on a site after manual or mechanical clearing of woody vegetation to establish tree plantations. It is still a common practice to burn grassland just before planting, but this can stimulate new grass growth. To be successful the vegetative material must be dry and well compacted, the weather calm, and full precautions taken to keep the fire under control. Where grassland is burnt, fire can move rapidly and it is sometimes wise to limit individual burns to less than 30 ha. Burning is cheap and quick, it greatly improves access (Fig. 12.5), the resulting layer of ash is rich in mineral nutrients, and vermin are often killed (the main reason for grass burning before planting pines in southern Africa).

Although burning is an inexpensive operation it can have hidden costs. Burning large areas produces atmospheric pollution that may carry significant health risks. This was the reason for the Indonesian Government banning burning for plantation development following major wild fires in 1997–98. Another cost is the loss of valuable nutrients. It is probable that the major part of nutrients released by burning slash will be lost by volatilization and subsequently by leaching of the ash (Folster and Khanna 1997). High intensity burns consume more biomass and result in greater nutrient losses. Destruction of the surface soil organic matter compounds the negative effects on tree growth of nutrient loss. *Acacia mangium* planted on unburnt areas in Sabah produced twice the aboveground biomass at age 3.7 years compared with trees on a burnt area (Sim and Nykvist 1991). Elsewhere, burning has depressed subsequent growth of *Eucalyptus deglupta* on logged-over rain forest sites in PNG (Lamb 1976) and pine on burnt savanna land (Welker 1986). Other disadvantages of clearance by burning are the risk of the fire getting out of control or flaring up from smouldering debris hours or even days afterwards, the restriction of its use to certain days, the variability of result depending on the conditions, exposure of the soil surface to erosion especially if the burn is

Figure 12.5 Land cleared and burnt in readiness for planting; site preparation at Jari, Brazil in the 1970s.

very hot, and the risk of fungal infection from *Rhizina* spp.

Careful management of harvesting residues will contribute to maintaining productivity of subsequent rotations of plantations, especially where soil nutrient levels are low. Burning harvesting slash will release nutrients and produce a short-term stimulation of growth in new plantations but is likely to be undesirable in the longer term due to loss of nutrients by volatization, leaching, and erosion (Gonçalves *et al.* 2000).

These disadvantages suggest that the use of fire in land preparation should be minimized, and if unavoidable, low intensity burning should be practised. It should never be used where there is a significant risk of erosion.

Use of chemicals

The use of chemical herbicides involves potential hazards to those who apply them and to the environment. Herbicides are poisonous and must be used and stored with great care. Strict adherence to the manufacturer's safety precautions, the use of protective clothing and the application of the correct doses are essential. Transfer of safe herbicide technology to smallholder farmers is one of the critical challenges to be overcome. Herbicides are useful for preparing sites for planting under some conditions (Fig. 12.6). They are used to (i) kill grass along planting lines before planting; (ii) kill remnant overstorey trees (the chemical may then be called an arboricide) in logged-over forest as part of line-enrichment planting or undesirable stump

(a)

(b)

Figure 12.6 Minimum cultivation. In one operation (a) herbicide is applied from a boom mounted on the front of a tractor while (b) behind the tractor a subsoiler cuts a furrow with a disc, subsoils to a depth of 30–40 cm, and applies fertilizer in the furrow. Companhia Suzano de Papel e Celulose, Itatinga, São Paulo, Brazil. Photos J. L. Gava and J. L. M. Gonçalves.

regrowth (coppice). They kill a plant by desiccating the leaves, upsetting its hormonal balance, or interfering with its metabolism (see Chapter 13).

Chemical control of vegetation can be cheap and effective under the right conditions, but it only kills the plants it does not clear the site of woody vegetation. In some instances it may remove the need for mechanical clearance or ploughing preventing site disturbance and possible erosion. Herbicide control of grass is usually more effective than either burning or manual clearance. Glyphosate has become the main herbicide used to control *Imperata cylindrica* prior to planting and its use, spray volume rates and equipment are reviewed by Terry *et al.* (1997). In the tropics both the chemicals and application equipment usually have to be imported and often effects on vegetation control and residual activity are poorly understood. Great care should be taken not to contaminate water supplies or to permit drift to non-target areas. Herbicides are most effective when applied in dry, windless weather.

Soil, terrain conditions, and cultivation

In plantation establishment the aim is to achieve adequate cultivation of the soil at the least cost to ensure good survival and rapid establishment of the newly planted tree by breaking barriers to root growth and optimizing soil aeration and drainage. The minimum 'cultivation' needed to plant a tree is simply to make a single opening in the soil and to insert the roots.

Evidence for effectiveness of ground preparation and tillage operations has been mixed. In some cases survival and growth have been improved by intensive ground preparations. Smith (1998) reports compelling evidence that ploughing as a method of establishment on grassland sites in South Africa is effective for all species planted. However, benefits cannot always be assumed. Smith *et al.* (2001) also report intensive site preparation (e.g. de-stumping, ripping/subsoiling, and discing) has generally not improved growth of *E. grandis* and *Acacia mearnsii* on a range of sites in South Africa and in some cases has reduced site productivity.

The principles and practices of soil cultivation for plantation establishment in Brazil have been fully described and illustrated by Gonçalves and Stape (2002). The intensive cultivation practices including slash burning and harrowing employed in the 1970s and 1980s have given way to 'minimum cultivation' techniques that involve disturbing the soil only to the extent necessary, retaining organic residues and using herbicides to control invasive weeds. On lighter sandy soils one or two rows can be subsoiled to a depth of 30–40 cm and fertilizer applied in the lines at the same time

(Gava 2002), elsewhere there is subsoiling to 100 cm. By 1999 a minimum cultivation system was used for about 60% of eucalypt plantations in Brazil (Stape *et al.* 2001).

Intensive mechanical cultivation had little long-term effect on the growth of *A. mangium* in Indonesia (Turvey 1996) and for *Pinus taeda* the gains were not large enough to make it an economic treatment (South *et al.* 1995). Simply controlling grasses with herbicide on pastured hills in Colombia, where soil compaction is not a problem, improved growth of *E. grandis, Cupressus lusitanica* and *P. oocarpa* more than physical loosening of the soil (Ladrach 1992). Additional ripping of coastal lowland soils for pine plantations in Queensland, despite the suggestion in Table 11.2 of better growth, generally has not been found to enhance growth or crop stability; the best treatments are cultivation and mounding, yielding growth responses not solely attributable to reduction in weed competition (Francis 1984). The problem is to find the level of cultivation, which results in the best growth but is not overly expensive.

The need and intensity of cultivation varies greatly. Most logged-over rain forest sites support good growth without soil cultivation whereas tree growth is only possible on some infertile, compacted soils after first subsoiling and mounding or ploughing; see Table 12.2 and Fig. 12.7.

Soil conditions likely to benefit from cultivation include; (i) surface or subsoil compaction; (ii) surface capping; (iii) impervious layers from clayey or pan horizons; (iv) poorly draining soils. Cultivation is carried out by hand or mechanically though obviously the former is restricted only to localized improvement of the planting spot.

At most, hand cultivation involves digging a hole (pitting), usually no bigger than 30 cm deep and 30 cm across, or making a small mound. The holes may be made either at planting or some months before (Fig. 12.8(b)). When dug in advance, pits are not filled until planting both to show the work has been done and to indicate readily the planting position. Pitting is slow and expensive and it is often not necessary. In Fiji for many years small seedlings of *P. caribaea* were successfully planted by inserting them in holes dug only with a crowbar (Fig. 12.8(a)).

Mechanical cultivation offers great scope for soil improvement. There are three main kinds of cultivation; (i) opening of narrow channels into the subsoil, to depths of up to 1 m, to improve downward rooting by breaking impermeable barriers and aiding drainage, for example, ripping, subsoiling, tining, etc.; (ii) cultivation of the topsoil to 20–30 cm depth either completely over the site or in strips to form furrows and ridges or mounds, for example, ploughing; (iii) complete surface cultivation of the top 10 cm to keep the soil surface friable and to

Table 12.2 The effects of site preparation and fertilizer application on growth of *Pinus caribaea* var. *hondurensis* on an infertile groundwater podzol at Byfield, Queensland, Australia. (Fig. 12.7)

Treatment		Assessment at 12 years			
Cultivation	Fertilizer	Predominant height (m)	Mean diameter breast height (cm)	Basal area (m^2 ha^{-1})	Total volume (m^3 ha^{-1})
Nil	Nil	2.6	—	—	—
Nil	P,N,Cu	11.0	15.7	18.0	64.3
Nil	P,N,Cu,Zn,Mo	12.1	16.9	21.7	79.2
Mounding	P,N,Cu	15.7	18.7	28.0	121.7
Mounding	P,N,Cu,Zn,Mo	16.8	19.6	30.2	139.3
Mounding and ripping	P,N,Cu	16.7	20.0	31.8	146.0
Mounding and ripping	P,N,Cu,Zn,Mo	17.5	20.5	33.7	161.8

Ripping was with a single tine with a wing attached to a depth of 60 cm along planting lines.
Mounding was done using the tilted edge of a bulldozer blade, and carried out over the ripped line when the cultivation treatments were combined.

Figure 12.7 The dramatic effects on improving growth (background trees) of site cultivation and fertilizer addition. Untreated trees in the foreground. All trees are 12 years old. Byfield, Queensland, Australia.

discourage weeds, for example, discing, harrowing, rotovating.

In one operation mechanical cultivation, especially ploughing, confers three benefits; weed suppression, soil structure improvement, and easier planting both in digging the hole and laying out the planting lines. However, compared with hand cultivation mechanical methods have more limitations as well as the difficulties of first buying, then maintaining, the equipment. Mechanical cultivation is impossible on steep slopes; two-way ploughing is restricted to slopes of less than 5°, one-way downhill ploughing on slopes up to 20° with care and provided there is no risk of soil erosion. Other restrictions are caused by logging debris or other material

on a site, rockiness (Fig. 12.4), or inaccessibility to tractors. In practice most ploughing in the tropics is done parallel to the contour to conserve moisture.

Ground preparation practices

Frequently eroding, arid, and reclamation sites have little or no vegetation; examples of these and the problems they present are considered in Chapters 21, 22, and 23. And, of course, small-scale plantings in social forestry encounter a wide variety of conditions dealt with manually. The following examples of vegetation clearance are listed in order of increasing scale of operation.

1. Restocking with *P. patula* after pulpwood crop in the Usutu forest, Swaziland. Establishment of second and later rotations often requires little or no vegetation clearance the previous crop having suppressed all weeds following canopy closure (Fig. 24.1). Logging slash is opened, while still green soon after felling, to locate planting positions and planting done soon after, even in the dry season. The slash acts as a mulch conserving moisture and helping to suppress weeds and contributes to successful dry season planting. No cultivation or ploughing is done, though any slash piles will be spread out, apart from digging small pits (Fig. 12.8(b)).

2. Grassland dominated by mission grass (*Pennisetum polystachyon*) or reeds (*Miscanthus floridulus*) in Fiji for planting *Pinus caribaea* (Fig. 12.9). (Yalimaitoga, personal communication.)

Mechanical cultivation is not possible and the only ground preparation is to burn off grass and other vegetation at the end of the dry winter in August.

Figure 12.8 (a) Minimal site preparation. Opening a hole with a crowbar for planting *Pinus caribaea* in Fiji. Note distancing (spacing) stick. (b) Pitting in advance of planting *P. patula* on a weed-free third rotation site in the Usutu forest, Swaziland.

Figure 12.9 Grassland for afforestation in northeast Viti Levu, Fiji.

Unwanted *Casuarina* trees are either poisoned before planting or cleared mechanically before the area is burnt. At planting small holes are dug [Fig. 12.8(a)]. Intensive weeding by hand or spot-gun treatment with glyphosate follows planting.

3. Grassland dominated by *Imperata cylindrica* in South Kalimantan, Indonesia for planting with *Acacia mangium*.

Preparation is intensive where the slope is not too steep, generally cultivated twice with a disc plough with a 1–2 month interval between ploughings during the dry season and harrowing once before planting. Herbicides, for example, glyphosate, are used on steeper slopes where there are constraints with machinery and increased erosion risk. Weeding is not usually necessary after planting but if required is carried out manually or harrowing between the rows (Otsamo 1996; Otsamo *et al.* 1997).

4. Clearance of open savanna for eucalypt plantations and replanting harvested eucalypt plantation sites in the Congo (Bouillet, personal communication).

Savanna is usually burnt, regrowth chemically controlled using glyphosate and the site ripped with a single tine along the planting lines. Thereafter the stands are chemically weeded (2–4 times), except initially (3–9 months) if the trees are too small and sensitive to glyphosate, inter-row ploughing and/or manual weeding is used. On replanting sites, if there is a fire risk and if the slope is gentle (< 2–3%), the litter is incorporated into the soil with a disc harrow; otherwise the forest floor is left undisturbed. Weeds are treated with glyphosate 1 month before planting. Subsequently the stands are weeded manually if the trees are very small or chemically (1–3 times).

5. Savanna clearance in the Guinea zone of Nigeria for planting pines and eucalypts (Allan and Akwada 1977).

Nigerian practice follows that in Zambia. Where possible vegetation is knocked down using large crawler tractors with an anchor chain; this is done at the beginning of the wet season. Stumps are removed and debris is piled into windrows 50 m apart and the ground in between ploughed. Work stops at the end of the wet season, the windrows are burnt during the following dry season and, at the start of the next wet season, the site is complete ploughed and harrowed ready for planting.

6. Cerrado woodland and poor quality scrub arising from cut-over forest near Belo Horizonte, Brazil.

Companhia Valle Rio Doce (CVRD) near Itabira carries out the following operations to establish plantations of *Pinus patula*, *P. caribaea*, and *E. grandis*.

(a) Three months before planting all trees and shrubs are felled by hand and cut for firewood or made into charcoal; (b) remaining debris is burnt; (c) large stumps are uprooted by bulldozer; (d) all except steep sites are completely disc ploughed (Fig. 12.10); (e) immediately before planting two surface ploughings are done at right angles to layout planting positions.

7. Second and subsequent rotations of eucalypt plantations on former cerrado woodland, at Itatinga, São Paulo State, Brazil.

The Companhia Suzano de Papel e Celulose operates mainly on land with gentle topography and deep sandy soils. It was the first Brazilian company to develop minimum cultivation techniques and carries out the following operations to re-establish eucalypts after harvesting. Harvesting residues are retained on site and distributed along the stump lines. In one operation, a pre- or post-emergent herbicide is applied from a boom mounted on the front of a 75 hp, rubber-tyred tractor. Behind the tractor a subsoiler cuts a furrow with a disc, subsoils to a depth of 30–40 cm, and applies fertilizer in the furrow (Fig. 12.6(a) and (b)). There is no burning of harvesting residues or surface cultivation. Herbicides are used according to local site conditions. Subsoiling is to a greater depth on more compacted soil types and a more powerful (160 hp) tractor is used. The equipment and techniques are described in Gava (2002).

8. Rain forest logged for pulpwood in Sabah, East Malaysia.

Sabah Forest Industries in their concession near Sipitang log forest for pulpwood and then prepare sites for planting with *Acacia mangium*, *Gmelina arborea*, and *Paraserianthes falcataria*. After clear-felling forest for pulpwood, planting ground is surveyed and remaining trees and undergrowth are cut (underbrushing). Debris is lopped and then burned when dry. Re-stacking and a second burning is commonly needed. The site is then 'clean'. Finally, planting lines are pegged, holes prepared, and trees planted.

9. *Araucaria cunninghamii* (hoop pine) plantations in Queensland. (Bubb, personal communication.)

This native species is planted on relatively fertile former rain forest sites. Since the mid-1990s most

Figure 12.10 Disc plough used immediately prior to planting by CVRD in Minas Gerais, Brazil.

planting has been on second rotation sites and the aim has been for all tree debris to be retained on site. Following manual or mechanical harvesting, processing occurs at stump leaving a relatively homogenous residue layer across the site. Site preparation occurs in winter when the weather is drier and the larger woody material is raked into windrows across the slope at 15-m intervals. These are left to decay. An excavator is with a raking head is preferred equipment for forming the windrows on slopes less than 20° making a single pass across the slope between each windrow. On slopes between 20° and 25° or where traction is difficult (i.e. stony areas) the excavator works up and down the slope. Hoop pine is reasonably tolerant of Atrazine and this is applied 'over the top' as a residual herbicide. The area is sometimes treated with glyphosate and other herbicides specific to woody weeds in order to control some problem weeds. Three to four rows are planted manually between windrows and maintained for about the first 2 years after planting. The sites are not cultivated or fertilized.

Planting

Time of planting

Throughout the tropics the rainfall pattern usually determines when trees should be planted. Where there is only a very short dry season planting is sometimes done throughout the year. More commonly there are only a few months during which there are reliable rains and this is followed by a severe and prolonged dry season. Evapotranspiration stress at planting is the main cause of death and is minimized by three practices.

1. Plant seedlings when soil moisture levels have returned to field capacity; this is often only after about 100 mm of steady rain has fallen and the wet season commenced. The onset of a wet season is unpredictable and the first rains are frequently followed by long dry spells; in arid regions planting under such uncertain conditions usually cannot be avoided.
2. Plant on cloudy days.
3. Use high quality, conditioned plants that have been well-watered just before leaving the nursery.

In arid regions it is safest to use container stock and to plant when there are heavy rains during the period with the highest probability of rain. However, this optimum time for tree-planting frequently coincides with sowing and planting of food crops by villagers and farmers; a conflict, which needs to be recognized and allowed for, particularly in social and community forestry projects.

While the importance of planting trees during the wet season is stressed, as noted earlier in cooler subtropical conditions dry season (winter) planting of pines has proved possible where there is little competition for moisture from weeds and the plants are well-conditioned or container-grown. The practice has been used in southern Africa, for example, Cawse (1979), and it has been recommended for *P. caribaea* in Fiji's dry zone area as a better option than holding seedlings over to the next wet season. In Brazil, eucalypts are sometimes planted towards the end of the dry season and irrigated weekly until the rains begin (Ladrach 1992) and in South and Central America *Bombacopsis quintata* seedlings, which are leafless during the dry season, can be successfully planted without irrigation a month before the rains begin (Kane 1989).

Laying-out planting positions

There are two considerations; alignment of rows and spacing between trees (distancing).
 Row alignment should fit in with the intended extraction system and be at right angles to the main extraction roads and tracks in a compartment. Roads and tracks mostly follow the contour, rows run up and down slopes. Where rows follow contour ploughing it is important that they all end at a road and do not terminate within a compartment.

There are many methods of marking the planting position to ensure regular spacing of trees. It is important that the method is simple, easy to apply, and practicable on the kinds of terrain encountered. Extreme precision is not necessary; clearly visible rows and evenly spaced trees are quite adequate.

On ground cross-ploughed at right angles furrow intersections readily mark the place to plant the trees. With strip ploughing, planting positions are spaced along the furrow using a pole of suitable length (2–4 m). Where ditches, bunds, or terraces have been made to prevent erosion, planting is usually carried along their alignment (Chapter 23).

On unploughed ground all methods start by laying out baselines, usually with a compass, about 50 or 100 m apart. Spacing from baselines can be done in several ways, for example, long ropes or wires with markers at fixed intervals, two poles (Fig. 12.8(a)) to measure separation within and between rows, or using one pole for spacing along the row and distant sighting poles for alignment.

Where holes are not dug immediately at the selected planting position, the spot is marked with a stick or peg about 2 m high. Though many sticks are needed the expense is worthwhile. The positions are clearly located for planting and if the stick is left in it simplifies the early tending operations by identifying the position of the young plant. On second and later rotation sites planting position is readily located in relation to the old stumps.

Planting arrangements in social and agroforestry are mainly determined by villagers' or farmers' needs—along boundaries, in compounds, evenly spaced over a field, on ground unsuitable for food crop, etc.

Planting pattern

Two patterns are mainly used. The commonest is square planting, where distance between trees is the same along and between rows. Rectangular planting patterns, where trees are closer in the row than between rows, may be used to aid machine access, allow food crops to be cultivated between the trees, or where trees are planted in spaced lines in enrichment planting or to avoid complete clearance of vegetative cover for protection or conservation. Much teak planted in Trinidad has followed Indian practice, with three rows planted as strips in a matrix of natural forest with strip separation equal to final crop spacing.

Other planting patterns have been proposed from time to time, such as triangular planting for exotic plantations in Malaysia (Sandrasegaran 1966), but all suffer from greater complexity in initial layout.

Initial spacing

The distance between trees, or the number of trees planted per hectare (stocking), is one of the most important silvicultural decisions in plantation establishment; see also Chapters 4 and 16. Table 12.3 lists examples of some spacings currently used and Wadsworth (1997) provides many other examples illustrating the wide variety of initial spacings.

Spacings wider than 5×5 m are largely confined to agroforestry, when there is often no intention to create a forest, to planting for erosion control where large gaps may exist between terraces, and to enrichment planting where only late in the rotation may the planted trees begin to occupy the site fully. In very dry areas, where there is limited water available, wide spacing is used to reduce competition between trees. Wide spacing may also be used in water catchments to increase water yield from the site.

Table 12.3 indicates the variety of spacings used, but the decision on what is the optimum for a plantation project is not always straightforward. Planting density combined with thinning and pruning schedules is used to manipulate the size and quality of intermediate and final crop trees, the timing of harvests, and growing costs and revenues.

Table 12.3 Examples of spacing used in plantations in the tropics

Spacing (m)	Growing space (m² tree⁻¹)	Number of trees ha⁻¹	
1×1 or less	1.0	10 000	*Eucalyptus globulus* poles in Ethiopia
1.8×1.8	3.2	3086	*Pinus caribaea* in Sabah
2.1×2.7	5.7	1764	*P. caribaea* in Peninsular Malaysia
3×2	6.0	1667	*Gmelina arborea* at Jari, Brazil
2.5×2.5	6.3	1600	*P. caribaea* at Jari, Brazil
3×2.5	7.5	1333	*P. caribaea* on poor sites in Fiji
2.74×2.74	7.5	1333	*P. patula* in Swaziland
4×2	8.0	1250	*Paraserianthes falcataria* in Philippines
3×3	9.0	1111	*E. grandis* in Aracruz, Brazil
5×2.4	12.0	833	*Araucaria cunninghamii* in Queensland
4.5×3.0	13.5	741	*P. caribaea* in Queensland
2×7	14.0	714	*Acacia saligna* for erosion control in Ethiopia
4×4	16.0	625	*E. deglupta* in Philippines
2×10	20.0	500	*Swietenia macrophylla* enrichment planting in Puerto Rico
4.5×4.5	20.3	494	*E. deglupta* grown for saw timber in PNG
6×6	36.0	278	*Terminalia ivorensis* in Nigeria
10×10	100.0	100	*Faidherbia albida* in agroforestry with millet, Mali

Spacing and costs

Overall, more widely spaced plantations are probably cheaper to grow, but see Table 12.4. Allan (1977) cites a comparison of eucalypts planted at 3×3 m and 3×1.5 m, and reported that at the closer spacing costs were 83% higher during the first year due to both greater planting costs and weeding costs because of less opportunity for mechanized weed control. Welker (1986) similarly reports greater costs at close spacing with *Gmelina arborea*, typically 1.75×1.75 m spacing cost 80% more per hectare, both in establishment and harvesting, than at 3.5×3.5 m.

Seedling plantations have had to be established at stocking rates of about 1600–2200 trees ha^{-1} (3.0×1.5–2.0 m spacing) to anticipate losses due to mortalities and poor growth of some trees. With the development of clonal forestry there has been the opportunity to reduce establishment costs by using wider spacing. For example, in eucalypt plantations in Brazil the availability of highly productive clones, the potential to reduce harvesting costs by growing bigger trees, and lack of water on some sites has resulted in initial stocking being reduced to between 800–1100 trees ha^{-1} (3.0×3.0–4.0 m) depending on whether the output is aimed at pulpwood or sawlogs (Stape *et al.* 2001).

Spacing and revenues

The effect of spacing on revenues is complex. Initial spacing directly influences total volume production and tree size at least up until the time of thinning. The mensurational effects of different spacings are discussed in Chapter 15; but in summary: (i) wider spacings reduce total volume production, especially in short rotations, since for a longer period a site is not fully occupied; (ii) wider spacings increase mean tree size, which may or may not be advantageous depending on the market; (iii) wider spacing tends to increase stem taper, which may reduce the percentage conversion when the log is sawn; (iv) in broadleaved stands wider spacing usually results in trees of poorer form, with larger crowns, and less strong apical dominance; (v) with wider spacing there are fewer final crop trees from which to choose. These relationships are generalized and while most conifers with strong apical dominance can be grown successfully over a wide range of spacings, many broadleaves require close spacing to assist upward development and reduce overly spreading crowns such as teak, and iroko (*Milicia* (*Chlorophora*) *excelsa*).

All the above effects can influence revenue earned by the crop. If quantity of wood only is of importance, for example, in production of firewood or pulpwood, then closer spacings are desirable.

Conversely, large trees are desired to improve conversion recovery for sawtimber and veneer. For this reason large logs generally command higher prices per cubic metre than small trees—the price-size curve. In widely spaced stands, proportionally more trees are large and the yield of merchantable timber may be greater than from more closely spaced stands even though the total volume is less. And, it will usually be more profitable to obtain a given quantity of merchantable timber from few rather than many trees. Balancing this trade-off between diminishing costs and diminishing yield with wider spacing will result in different answers for each situation. Welker (1986) reported that Jari eventually found 3×2 m spacing was optimum for gmelina rather than wider spacings of 3×3 m or more.

In conclusion, because very close spacings are extremely expensive and very wide spacings grossly underuse a site, the approximate order of spacings are indicated below for different kinds of crops.

Figure 12.11 Closely grown plot of *Eucalyptus camaldulensis*, spacing less than 1×1 m, typical of numerous small woodlots cultivated by villagers and farmers throughout the Ethiopian highlands for sticks, poles, and sometimes firewood.

1. Firewood and domestic   Spacing 1–2 m
sticks and small poles
for building purposes
(Fig. 12.11)-maximize
yield, short rotations, no

Table 12.4 Effect of wider spacing, planting fewer trees per hectare, on the costs of forestry operations

Operation	Effect on unit cost	Effect on period of operation	Overall effect	Comments
Ground preparation	−		−	Fewer pits, wider spaced plough furrows
Planting			−	Fewer plants needed, fewer to plant
Blanking	+		+	High survival per cent more important, therefore blanking important
Tending	−	+		More opportunity for mechanization but fewer trees to tend though usually for a longer time
Brashing (low pruning)			none	Fewer trees offset by thicker branches taking longer to cut
Pruning	+		+	Thicker branches
Fire protection		+	+	Delay in canopy closure may lengthen period of high fire hazard
Thinning			−	Some delay in first thinning but removal of fewer, larger trees
Harvesting (unthinned stands)	−		−	Fewer, larger trees to extract per cubic metre

+ = higher costs, − = lower costs.

small size limit, but large timber difficult to handle.
2. Pulpwood—maximize yield, short rotations, 5–15 years, typical size limits 10–40 cm diameter logs.　　　Spacing 2–3 m
3. Sawtimber and veneer—large log size 30 cm + in diameter, long rotations and regular thinning, loss in total volume compensated by high value of wood.　　　Spacing 2.5–4.5 m

Transport of plants to the planting spot

Normally, a truck or tractor and trailer brings plants as near as possible to the planting area. Plants are then carried to the planting position, which is easily done if they are bare-rooted since many can be carried at one time. However, container plants must be carried in a tray, box, or basket to the planters; see Figs 10.23 and 12.4. This is often a separate operation and used as a way of checking on the number of plants put in by each planter. At all times, plants should be protected from over-exposure and drying out.

Manual planting

Most tropical trees are planted by hand as many sites are too hilly or stony to be machine planted, machines are mostly imported and expensive, and use of manual labour is often a social benefit. Bare-rooted plants and stumps can be planted in a hole or slit dug with a spade, mattock, or even opened up with a crowbar. Container plants are planted in a small pit. Where soils are well cultivated and plants are raised in containers that do not need to be removed, for example, made of paper or peat, a dibbling tool, which makes a hole by removing a core of soil to match the container can be used.

Planting is an important operation and deserves being done carefully; poor practices hastily carried out can lead to high mortality, even with robust species like teak (Prasad 1987). For all planting the following general rules apply.

1. Insert roots into the soil up to the root collar and hold plants vertically.
2. Avoid damaging roots by breaking, bending, or crushing.
3. Firm soil around the roots using the ball of the foot, *not the heel*.
4. Remove impervious containers before planting. However, in several African countries, on termite-infested sites, plastic sleeves are slit but left in place around eucalypt seedlings as a protective barrier and shallow planted with part of the container above ground (Fig. 12.14), though Wardell (1990) does not advocate the practice.
5. On dry sites the planting position should maximize water retention, e.g. furrow bottom, base of mound or microcatchment (Fig. 23.3).

6. Stump plants should not be forced into the ground. They should be placed in specially prepared holes and the soil firmed around them in the normal manner.

7. In arid regions plants may be put in especially deep to ensure that roots reach moist soil and only a small part of the shoot is above ground and subject to transpiration stress. Tall plants are sometimes planted deeply to prevent them being blown over.

Planting is a critical part of establishment and labourers need to be properly trained to use good planting practices commensurate with the good quality plants from the nursery. Poor planting techniques can result in poor survival or retarded growth.

Fertilizers at planting

For some species on some sites addition of fertilizer is essential for satisfactory growth, and its use has become standard practice in several countries. Table 12.2 and Fig. 12.7 illustrate one example. This subject is considered further in Chapter 14.

Watering

Provided planting is done in a rainy period and plants are well firmed in, watering is not usually necessary and is rarely done. However, watering can improve survival and early growth and may also be used to enable planting to continue into the beginning of the dry season. Nevertheless, watering is usually only practicable: (i) where a farmer or villager tends only a few trees and has the water available and the time to apply it; (ii) where sites are flat and accessible to tractors which can obtain water and apply it from a boom or hoses, for example, on the flat plantation sites of Companhia Agricola e Florestais in Brazil *E. grandis* plants are given 3–4 L on the day of planting and again after 3 days and 9 days if the weather is dry. If watering is done apply a liberal quantity (5–10 L) to each tree to soak the soil and not just wet the surface.

The use of irrigation to grow forest plantations is discussed in Chapter 23.

Protection of newly planted trees

An overview of protection is provided in Chapter 19 but it must be emphasized that trees are at their most susceptible when young and newly planted. The sooner canopy closure occurs the safer is the crop, hence the principle of seeking to maximize early growth. Protection from weed competition is considered in the next chapter.

Fire protection is especially important on dry grassy sites since young trees are easily killed if burning occurs. General protection measures provided for the whole forest are usually sufficient to protect a young plantation provided there is adequate weed control. Serious fungal damage of newly planted trees is uncommon, at least on first rotation sites, provided cultivation, weeding, and nutrition are satisfactory. Root rot is a serious threat to second rotation *Acacia* plantations established in the slash and organic residues remaining from previous harvesting (Old *et al*. 2000). However, animals and insects are the two most serious sources of damage to young seedlings.

Animal damage
Generally animals and young trees do not mix and damage from browsing, trampling, breaking, and rodents gnawing bark can be serious. Few areas are completely free from the threat of wild or domestic animals. Although it is often hard to quantify the long-term impacts of browsing damage, economic losses may be substantial. Protection of trees from grazing domestic animals and vermin is often the biggest single cost associated with establishing trees on farms. Protection needs to be anticipated and planned for, rather than reacted to when it is often too late. Vermin need to be controlled before planting or very soon afterwards and before they inflict damage. The problem with domestic animals is most acute in dry countries with sparse natural vegetation where animals turn to planted trees for food (Fig. 12.12). For example, goats, camels, sheep, and cattle in the Sahel, Sudan, and Guinea zones of Africa are a perpetual obstacle to successful afforestation. Only with total exclusion of these animals can establishment of planted forests be successful.

Protection against animals where browsing pressure is high, must be enforced though it is rarely straightforward and often costly. Fast-growing, long-lived trees or shrubs with spines, stiff branches, and unpalatable foliage are commonly planted in the tropics as a stock-proof 'fence' because of the low establishment cost (Chapter 21). Erection of fences, walls, individual tree guards other barriers are expensive but may be the only solution. The use of tree shelters is expensive, has had variable success, and is discussed in Chapter 13. Fencing and other useful practical measures used in Australia are described by Campbell (1990) and many are applicable to other tropical regions.

In some countries it may be necessary to employ villagers to guard plantations. Owners of livestock usually resist any restriction of grazing, however lawful the exclusion, though in community-based projects people's participation often brings the commitment that alleviates this source of friction. In southern China (e.g. Guizhou province), some ethnic minority villages have customary laws to protect newly planted trees and impose fines and other penalties on owners of animals that damage trees.

Figure 12.12 Goat browsing *Acacia abyssinica* in Ethiopia.

Insect problems

Insects can destroy newly established plantations. In many parts of the tropics, especially Africa, termites are the worst problem; but in South America the leaf-cutting sauva ants (*Atta sexdens*, *A. laevigata*, and *Acromyrmex landoltii*) cause most damage (Fig. 12.13).

Protection against termites is most important for eucalypts. In Zimbabwe, for example, deaths of eucalypts due to termites are commonly 30–50% and approach 100% in some areas unless the pest is controlled by insecticides (Mitchell 1989). *Acacia* spp., *Faidherbia albida*, *Gmelina arborea*, *Pinus* spp., *Senna siamea*, and *Tectona grandis* are much less susceptible. Destruction of termitaria, allowing insecticide to build up in the soil of the container during a seedling's nursery life, and applying insecticide to the soil around the planting hole afford some protection, though the latter is inefficient, wasteful, and environmentally damaging. Wardell (1987) recommends using resistant species, overplanting so that some losses can be accepted (this is a common solution in social forestry projects where insecticide is usually unavailable) and good forestry practices, such as weeding, to promote healthy plants. Plant extracts, such as 'azadirachtin' from neem (*Azadirachta indica*), mulching with foliage of certain species and surrounding newly planted trees with *Euphorbia* have all been found to possess some insecticidal ability (Fig. 12.14(b)). Wardell (1990) lists nine suggestions for reducing termite damage, but does not recommend leaving the plastic sleeve around the tree.

Figure 12.13 Leaf-cutting ants in *Cordia alliodora* plantation in Ecuador.

Leaf-cutting ants are particularly troublesome in *P. caribaea* plantings on the eastern grasslands of Venezuela and in eucalypts in parts of Brazil. Initial disc ploughing will reduce populations but laying poison powder or bait, and applying insecticide around the tree at planting may be necessary (Ladrach 1992). They are impossible to eradicate, but selection of trees showing some resistance in breeding programmes may ease the problem in the future.

In general, healthy, vigorous plants are less liable to be killed than weak, stressed seedlings.

Mechanization in establishment

Ground preparation, digging holes, planting, and adding fertilizer, make establishment costly, time-consuming, and labour-intensive. Mechanized ground preparation is well-established for most industrial plantations, but other operations are often still manual. However, considerable mechanization of planting is possible on flat sites and is increasingly used, for example, for eucalypts in the Congo and Brazil and pines in Australia (Shea 1987) and Venezuela. The planting machine is generally a hydraulically linked or towed frame on which one or two planters are seated. A wedge-shaped blade or tine opens up a slit and the planter inserts trees into the slit regular intervals as the machine is pulled along. The slit is closed by two rubber-tyred wheels mounted side by side to form a 'V'. This appears to improve survival because seedlings are firmed down better in the soil.

Another type of machine used in 1970s is shown in Fig.12.15. It was being operated by Companhia Agricola e Florestais in Brazil to establish *E. grandis*. After the site is cleared, ploughed, and harrowed, the tractor-drawn planting machine does the following operations: (i) ploughs a furrow at 3-m intervals; (ii) applies NPK fertilizer along the base of the furrow; (iii) applies Aldrin insecticide near to the plant at a rate of 5 g per plant for protection against leaf-cutting ants; (iv) deposits the seedlings in the furrow at the correct spacing. Planters are spaced out along the planting line to plant the seedlings where they fall using mattocks. Eighteen labourers work with the machine, 15 planting, two feeding seedlings on to the conveyor belt, and one loading seedling trays. In an 8-h day this team plants between 15 000 and 20 000 seedlings. Output is around 1000 seedlings planted, fertilized, and treated with insecticide per labourer per day. This compares with typical manual operations in the tropics where planting alone varies between 100 and 600 plants each person-day depending on site conditions: steepness of slopes, rockiness, debris, and other access impediments.

The relative merits of manual versus machine planting should be assessed from both economic and social aspects in advance of any large plantation project.

(a)
(b)

Figure 12.14 Protection against termites. (a) Young *Eucalyptus camaldulensis* planted with plastic sleeve left around roots—sleeve was slit before planting (Ethiopia). (b) Newly planted *Cupressus lusitanica* with sprig of *Euphorbia* to give protection against termites and browsing (Rwanda).

Figure 12.15 Mechanized establishment.

Special forms of establishment

Direct sowing

Direct sowing (or direct seeding) is where the seed is sown directly into the ground on the plantation site. As an establishment practice in the tropics it has been widely tried but found reliable only with a few species. However, it is often successful with large-seeded species, for example, *Gmelina* spp., *Swietenia* spp., and *Mora excelsa*, and is sometimes used to supplement natural regeneration. Aerial seeding of *Acacia nilotica* and *Prosopis cineraria* on the Indus flood plain in Pakistan has been successful (Kermani 1974) and for reclamation of ravines in Madhya Pradesh, India, with *A. nilotica* and *Prosopis juliflora* (Prasad 1988). Both ground-based direct sowing and aerial sowing from helicopters are used to regenerate eucalypt forests in Australia (Fagg 2001) and direct sowing of *Pinus caribaea* and *P. oocarpa* takes place in poorly stocked forest in Honduras.

The attractions of direct sowing are that a forest nursery is not needed, and no time is spent on planting. Also, it is argued that trees develop more naturally on the site, since they are undisturbed from the time of seed germination, and this may improve early growth and tree stability later. However, in nearly all afforestation projects the potential benefits of direct sowing are more than outweighed by the disadvantages which are: (i) germination survival percentages are usually very low (1–10%), for example, Dalmacio and Banragan (1976) only achieved 4% with *P. kesiya*, therefore much seed is needed to achieve adequate stocking, this is expensive, especially if seed is imported; (ii) in a nursery a seedling receives

daily attention to obtain the best results and stimulate early growth, this cannot be done on a forest site; (iii) newly germinated seedlings are easily smothered by weeds, and the cost of control at each sowing site is very high; (iv) plantations are usually poorly and irregularly stocked even if seeds are dibbled into specially prepared spots at regular intervals; (v) poor weather conditions can kill an entire crop; (vi) to have any chance of success seeds should be pre-treated ready to germinate, and protected with fungicide, insecticide, and other repellents to resist biological damage; (vii) the sowing site must be very well prepared to provide a good germination medium. Rimando and Dalmacio (1978) obtained 75% survival with direct sown *Leucaena leucocephala* where the soil was scalped and cultivated, without this survival was only 14%. As a result of these disadvantages planting typically achieves much higher survival rates than direct sowing; Noble (1985) reported 70% compared with 6% in a trial with *Pinus kesiya* and Sulaiman (1987) 90% compared with 30% for *Acacia mangium*.

One of the most notable examples of direct sowing for raising plantations is *Acacia nilotica* and *A. senegal* in the Sudan. Seed is sown by mixing it with the soil dug out of the 'planting' pits. Where rainfall is higher the soil/seed mixture is left in mounds over the site (Laurie 1974). In Africa, *Acacia mearnsii* for wattle bark is an important crop usually established by direct sowing or by stimulating the seed bank in the soil in the second and subsequent rotations. Direct seeding is also used to establish plantations of cashew (*Anarcardium occidentale*). In the Jari project in Brazil direct sowing of *Gmelina arborea* was the preferred method wherever weed competition was not severe. Two to four seeds were sown in a group 10 cm apart and thinned out to the best plant when 20–30 cm tall (Welker 1986). However, the species is now little used.

Taungya planting and intercropping

Within the tropics, taungya has been an important means of afforestation. However, though it is considered in detail in Chapter 20 it is important to note that much of this chapter is still directly relevant to taungya cultivation, intercropping, and other agroforestry practices particularly important in rural development forestry. Taungya seeks to satisfy both a social need (land for growing food crops) and the establishment of a plantation, thus its difference is largely social not silvicultural. Any silvicultural differences arise from the small scale of operation and the need to cultivate food crops between rows of trees for a few years.

Line and enrichment planting

To save complete land clearance numerous attempts have been made to create plantations by planting trees along cleared lines in natural forest separated at about final crop spacing. Weaver (1986) states that in the neotropics at least 163 species in 12 countries have been established on a trial basis using various forms of enrichment. Because the intervening matrix of trees and undergrowth is cleared only gradually some savings in establishment costs accrue. Occasionally, group plantings in prepared openings or opportunity planting in natural gaps are carried out to enrich natural forest, but these are outside the scope of this book.

Most line enrichment planting is in tropical high forest but no absolutely successful method has been devised—Poore (1989) cites several examples across the tropics but concludes that almost all failed to proceed beyond the project scale, though many thousands of hectares of *Swietenia, Cedrela, Virola, Flindersia*, and *Cordia* enriched forest have been created. Enrichment planting in dipterocarp forest has been widely practised throughout the Asian tropics. It is used where there is insufficient natural regeneration of seedlings. Several dipterocarp species have been successfully planted into natural forests in both moist deciduous and evergreen forests in India and Sri Lanka (Appanah and

Weinland 1996; Appanah 1998). However, overall the success of enrichment planting has been variable and so its efficacy has been questioned (e.g. OTA 1984) and, with some exceptions (e.g. Adjers *et al.* 1995) its promotion and application have declined. Reasons for this include the difficulty of planting supervision, the high cost and timeliness required to weed the seedlings and release them from regrowth, and the absence of a regular supply of seedlings of most dipterocarp species.

The main principles of enrichment planting were outlined by Dawkins in the 1950s based on work with enrichment of high forest in Uganda (Dawkins 1958; Lamb 1969). In addition to normal requirements for healthy plant establishment: (i) there should be little or no demand for thinnings in the area; (ii) species must be fast-growing, naturally straight, and self-pruning, that is, colonizers; (iii) there must be no upper canopy; (iv) regrowth between lines must be non-flammable; (v) browsing animals must be scarce or preferably absent. The key requirement is to undertake the minimum amount of opening up/line clearance and subsequent maintenance, which allows sufficient light to the planted species. Eamus *et al.* (1990) found this optimum with *Terminalia ivorensis* enrichment in Cameroon at 5 × 5 m spacing, to be 50% clearance of overstorey followed by four cleanings in the first year and one in the second year after planting.

(a)  (b)

Figure 12.16 Line enrichment planting. (a) Newly cut line showing staked out planting positions, Gogol, PNG. (b) Line planted mahogany, *Swietenia macrophylla*, age 19 years in Puerto Rico.

Main operations in line planting

1. Land most often considered for line planting is logged-over forest with substantial regrowth and few sizable trees still present such as unusable *Ficus* species.

2. Clear a baseline at right angles to the direction of plantation rows.

3. Cut control lines about 2-m wide through the vegetation at intervals of 100–200 m.

4. Cut and peg planting trace lines, about 1-m wide, at the desired spacing. In Sierra Leone 14-m separation is used (Chapman 1973), in the Solomon Islands 10 m for mahogany but 4 m or 5 m for other species (Oliver 1992), in Puerto Rico 10–12 m and in Fiji 7–9 m.

5. Before planting, matrix trees are poisoned and shrubs near the planting line heavily cut back. Poisoning is usually done by frill girdling and applying arsenic pentoxide.

6. Plant seedlings along the lines at the desired spacing. The distance between trees is 4.5 m in Sierra Leone, 3 or 4 m in PNG (Fig. 12.16(a)), 3 m in the Solomon Islands, 2 m in Puerto Rico (Fig. 12.16(b)). In the Philippines and PNG there has been a practice to plant a cluster of four seedlings in a 1×1 m square every 3 m in the row. The object is to save in establishment but still obtain a normal plantation stocking. Inevitably, poor form and development of much reaction wood result as the trees lean outwards from the cluster, but these drawbacks may not be too serious where trees are grown for pulpwood or if the group is singled to the best stem.

7. Post-planting operations include all usual ones but full weed control and the essential operation of climber cutting are needed to prevent trees being smothered. Tending must always be maintained to ensure full overhead light and sufficient room sideways for the trees to grow unchecked.

8. As the trees grow and begin to dominate the matrix a plantation appearance becomes more evident through the enrichment lines always remain clearly delineated (Figs 1.5 and 12.16(b)).

CHAPTER 13

Plantation maintenance

Several silvicultural operations are needed to ensure that a plantation is adequately established and protected up to the production stage in industrial forestry and canopy closure or desired tree-size in environmental and social forestry. They are often referred to collectively as tending operations.

Immediate post-planting problems

Globally, only 70% of new plantations established annually are judged successful (FAO 2001a). In India, state plantations have an average of 60–70% survival but private plantings vary from 25% to 50% (Bahuguna 2001) and is one reason why yields from tropical plantations in general have often been very low, frequently less than 50% of their productive capacity (Pandey 1995). Death of seedlings and abnormally slow or checked growth are two main factors contributing to these problems.

Death of plants

Not all planted seedlings survive. A few weeks to a few months after planting, depending on growth rate, an assessment is made of how many plants have died; it is often done during the first weeding. A plantation where one in five newly planted trees has died is said to have 80% survival or 20% mortality.

The usual aim of establishment is to form a plantation fully using the site, not for absolutely every planting position to have a live tree. At close spacings, with many trees per hectare, a higher mortality (failure rate) is more acceptable than at wide spacing with few trees per hectare. As a guide, if more than 1250 trees ha^{-1} are planted up to 20% mortality is acceptable, but with a lower stocking only up to 10%. At extremes of stocking local prescriptions may lay down other figures, for example, only 5% of deaths may be tolerated at 4 × 4 m spacing (625 trees ha^{-1}).

Many factors affect initial survival: (i) planting skill, especially firmness of soil around the roots and planting depth; (ii) immediate post-planting weather; (iii) condition of seedling, for example, bare-rooted, container-grown, shoot : root ratio, whether it suffered moisture stress during lifting and transport to site, etc.; (iv) poor soil conditions, especially waterlogging or eroding surface; (v) insects such as termites, *Hylastes* spp., leaf-cutting ants; or fungi such as *Rhizina* sp.; (vi) weed competition (see later); (vii) animal damage, for example, grazing, browsing, and trampling.

If deaths after planting are unacceptably high, the failures are replaced. This operation has many names: blanking, beating-up, refilling, in-filling, etc. In the tropics blanking must be done within a few weeks of planting for fast-growing broadleaved species such as *Eucalyptus grandis*, *E. deglupta*, *Paraserianthes falcataria*, and *Anthocephalus chinensis*, and within a few months for other species. A delay in blanking permanently puts the replacement trees at a disadvantage, which may never be overcome and the resulting stand will show great variation in tree sizes (Fig. 13.1). Such variation in size is less important where the benefit sought is from individual trees

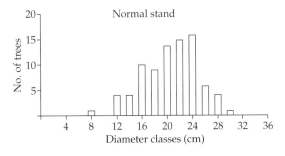

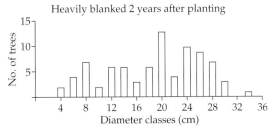

Figure 13.1 Diameter distributions of two *Pinus patula* stands of similar age (12 years) and growth potential, but one originally suffered heavy planting losses and was not blanked until 2 years later.

(agroforestry, shelter, and shade in the village) and failures are replaced as required. Regrettably, this is more of a problem than it should be. Many newly planted trees in social forestry or extension programmes die of neglect in the months after planting: people's participation – full investment of all stakeholders – should stimulate and sustain commitment.

Seedlings used for blanking should be healthy, robust, and a little larger than average with good root development. A fertilizer boost may be given when planted. Blanking should only be done once; if plants continue to die then species or provenance choice, site conditions, establishment practices, protection measures, etc. should all be examined. Jackson (1984) rightly points out that most plantation failures arise from a failure to apply known techniques, not because of any technical impossibility.

Abnormally slow growth

Sometimes a young tree, though not dying, stagnates in growth. This may occur at any time but is most common before canopy closure. Typically the tree just survives, perhaps with small, unhealthy-looking leaves, and grows only a few centimetres in height each year.

Causes
There are numerous causes of very poor growth; below are some of the most common.

1. Poor choice of species or provenance. Examples include species planted in the open which require shade in the early years. On open ground the rainforest species *Intsia bijuga* grows very slowly whereas under shade, with other natural regeneration, it may grow to 2 m or more in 1 year. Other examples of poor species choice were noted in Chapter 8, and see Fig. 8.3.
2. Nutrient deficiency. Lack of a major nutrient such as phosphorus, or a micronutrient such as boron, can significantly reduce tree growth. See Chapter 14. Most nutrient problems are induced by poor soil conditions such as poor drainage, erosion of topsoil, or excessive leaching.
3. Poor soil physical conditions due to compaction or erosion (Fig. 13.2(a)). This is common on logged-over forest or second or later rotation sites where extraction, stacking, and loading of logs has compacted or eroded the soil. Poor root development results poor growth that may last for a whole rotation; see Fig. 18.3.
4. Missing or poor symbiotic associations with mycorrhizas or nitrogen-fixing bacteria can reduce growth of many plantation species especially on infertile soils. See Chapters 10 and 14.

5. Some species have delayed shoot growth such as *Pinus merkusii* (continental provenances) and *P. pseudostrobus*, which initially develop a 'grass tree' stage (Fig. 13.2(b)).
6. Poor maintenance, neglect, termite and other pest damage (Fig. 13.2(c)).
7. Simply use of slow-growing native species may lead to disappointed expectations compared with often vigorous exotics (Fig. 13.2(d)).
8. Other avoidable causes include inadequate weeding (very common), excessive browsing, prolonged overhead shade, etc.

Sometimes there are several causes of poor growth, for example, both infertility and poor physical soil condition can be responsible for very poor growth.

Overcoming slow growth
Identification of the cause often indicates the solutions or at least how to avoid the problem. In general, the first step should be to check that control of competing weeds is adequate. Application of fertilizers to stimulate growth is rarely sufficient or worthwhile unless a specific deficiency is being corrected. However, fertilizing may hasten recovery from check, and it is best done along with other amelioration practices such as weed suppression. Improvement of soil physical conditions after planting, for example, digging drains, is usually difficult to do and costly. If this is the cause of poor growth it is usually best to abandon the first crop and start again with improved ground preparation or set land aside for conservation, for example, to develop a piece of wetland. Pre-planting surveys should aim to identify very poor sites where planting should be avoided unless economic amelioration measures are available.

Weed control

It was stressed in Chapter 12 that control of competing weeds is an important part of plantation establishment. However, the need to control weeds is not restricted to the time of planting alone; nearly all plantations require some weeding during the first few years until the trees are growing well, are approaching canopy closure and are an adequate size to suppress competing weeds.

Two stages in weed control are usually recognized:

1. Ground weed control. Control of grasses, herbs, and shrubs, which directly compete with newly planted trees and which must be controlled from the time of planting.
2. Cleaning and release operations. These are concerned with removal or killing of perennial

Figure 13.2 Four causes of poor or slow initial growth. (a) Ten years old *Eucalyptus camaldulensis* on very rocky, impoverished reclamation site at high altitude in Mexico showing both poor growth and poor survival. (b) Grass tree stage of *Pinus pseudostrobus* in Mexican highlands. (c) Poorly maintained woodlot of *E. camaldulensis* in Mali that also suffered termite damage. (d) Afforestation with native acacias and Euphorbia near Deki Zeru, Eritrea, led to high survival and stocking (>98%) but very slow growth of 5–20 cm per year in this dry environment.

plants, unwanted trees, vines, and creepers likely to smother trees in a young plantation. This work follows ground weed control and may be needed for several years.

The distinction between these two operations is not used here though both are covered. Weed control terminology is somewhat confused. Both 'tending' and 'weeding' are used to describe post-planting weed control and the word 'cleaning', which is usually confined to control of woody weeds especially vines and climbers, is used by Donald (1971b) to cover all post-planting weeding operations.

Need for weed control

Weeds can damage young trees in many ways.

1. Weeds compete directly for light, soil moisture, and nutrients. Table 13.1 shows the strong effect of grass on survival in tropical pines, particularly for some varieties and seed sources. Numerous examples of close correlations exist between weed control and tree growth: in semi-arid regions, for example, Felker et al. (1986), in agroforestry, for example, Chingaipe (1985), and in industrial plantations, for example, Haywood et al. (1997). In general, the better the weeds are controlled in a plantation the better the trees grow.
2. Weeds can smother and eventually kill trees by their cumulative weight, shading, and growth habit (e.g. strangulation by vines). In Southeast Asia and

the Pacific, the 'mile-a-minute' vine, Mikania cordata (syn. M. scandens), commonly smothers newly planted trees and is a major reason for expensive and frequent weeding operations (Oliver 1992).
3. Dense ground vegetation increases the fire hazard and can shelter harmful animals such as rats from hawks and other natural predators.

Intensity of weed control

This varies according to species, site, and climate.

Species
It was noted in Chapter 12 that complete removal of weeds (clean weeding) is essential for good early growth of eucalypts (Fig. 13.3). Other species are more tolerant though all benefit from suppression of competing vegetation except where partial shade or shelter is required by some climax trees species in the early stages of establishment.

Sites
Soil fertility, presence of weed seeds, live stumps and culms, and ground preparation treatment all affect the species and vigour of weed growth. In particular, old forest sites have much woody regrowth after clearing notably of species of Cecropia in America, Musanga in Africa, and Macaranga in Africa and Southeast Asia and lianes and scrambling weeds such as species of Lantana, Scleria, Mikania, and Merremia (Fig. 13.4).

Climate
Rainfall is of most importance. More weeding is needed in wet years than dry years, and as many as eight weeding operation may be needed in humid equatorial regions. However, it is in dry regions where weed suppression around trees is most important to reduce competition for moisture.

Table 13.1 The influence of grass competition for moisture on seedling survival during a drought at the Umi River Plantation, Papua New Guinea

Species	Variety provenance, or seed source	Age (months)	Survival (%)	
			Dense grass	No. grass
Pinus kesiya	Philippines (1)	8	10–20	80–90
	Philippines (2)	8	20	90–95
	Vietnam	8	10	100
P. caribaea	var. hondurensis	8	20–70	90–100
	var. bahamensis	8	10	100
	var. caribaea	8	30–100[a]	100
P. caribaea	var. hondurensis	18	40–100[b]	
	var. caribaea	18	55–85[b]	
P. kesiya	Philippines (1)	20	30–60[b]	
	Vietnam	20	40–50[b]	

[a] Variation in grass cover: dense grass 30–60% survival, moderate 90–100%.
[b] Great variation in amount of grass but no areas completely free.

Weeding practice

Frequency, period, timing, and extent
In addition to the factors above, the frequency of weeding depends on the method of suppression whether it is, manual, mechanical, or chemical, and whether the weed is killed or just cut back. Weed control by applying herbicide often gives better and longer-lasting suppression than other methods because plants are killed. It is also one reason why herbicide weed-control tends to improve tree growth more than hand-weeding, which may simply cut the weed plant.

(a)

(b)

Figure 13.3 Clean weeding between rows of young eucalypts. (a) Clonal *E. urophylla* at Aracruz, Brazil. (b) *E. camaldulensis*, Chacchoengsao, Thailand.

Figure 13.4 Six-year-old *Pinus caribaea* almost overgrown by climbers on a rain forest site in the southern Philippines.

Where weeds grow profusely and there is year-round rainfall, weed control may be needed six or more times in the first year. However, tree growth on such sites is usually fast and weeding,

though very intense, is often only necessary in this first year. For example, in Sabah (Malaysia), Abod (1982) cites hardwoods as receiving three spot-weedings and three interrow slashings and two creeper cuttings in the first year, and pines five spot-weedings and five interrow slashings, but by the second year only creeper cutting for hardwoods and two spot-weedings for pines, are required. Many fewer weedings are needed on sites where the growing season is short, but the associated slower tree growth often requires weeding to be continued for several years.

How long weeding is continued depends on early growth rate, onset of canopy closure, and crown density, such as under many lightly crowned eucalypts or, for example, stands of *Schizolobium parahibum* in Ecuador. Once the canopy of a stand begins to close, weeds become suppressed by shading. However, full weed suppression may not always occur, for example, under some species of lightly crowned eucalypts such as *E. camaldulensis*. At Aracruz in Brazil weed control in stands of *E. urophylla* and *E. grandis* continues throughout the rotation though after canopy closure it is only carried out once a year. Such intense (complete) weed control is quite common in industrial plantations of eucalypts to maximize productivity. However, reasonable growth is obtained by many eucalypts, *Paraserianthes falcataria* and teak on better sites by weed control for only for 6–9 months after planting. For slower growing araucarias in Papua New Guinea (PNG) weeding is sometimes necessary for 6 to 8 years before a plantation is considered safely established. For tropical pines this period is typically 1–3 years.

Weeding should be timed to prevent trees suffering any serious slowing in growth. For example, with pines control of grass becomes essential only when it grows more than 50% taller than the tree though earlier control may further aid growth. For *E. grandis*, Marchi *et al.* (1995) (cited in Gonçalves *et al.* 1997) demonstrated that for every month that elapsed from planting to the application of weed control in the first year there was a decrease in wood volume production and an increase in tree mortality. The intensity of weed control needs to be established by understanding the competition processes (Nambiar and Sands 1993). Weed control is also important for fire prevention in plantations and retention of grass or other vegetation including after spot or strip weeding increases the risk of fire damage.

The extent of weeding, that is, the area around a tree in which weeds are controlled (see Fig. 13.5), has an important effect. In general, the greater the area weeded, the less the competition to the tree, and the better it grows. Typically, spot weeding is less effective than strip weeding with clean weeding the best of all. Lowery *et al.* (1993) concluded from a review of weed control in tropical forest plantations that complete weeding in most cases results in the best growth and survival, but partial weeding in strips along the tree rows may be a good compromise between

making soil resources available to the tree and nutrient conservation. Application of the general principle *'the more weed control the better'* must be tempered by considerations of cost, the risk of exposing soil to erosion, and reduced biodiversity.

Method

There are three main alternatives: manual, mechanical, and chemical control, and three levels of coverage on a site: complete control (clean weeding), strip or line weeding (Fig. 13.6), and spot-weeding just around a tree.

Manual weeding

This is the commonest method for small-scale plantings. Tools such as bush knives, brush hooks, and similar implements are used to cut away competing vegetation. However, for best results cutting alone is not always adequate, and some soil cultivation by hoeing is needed to remove weeds (Allan 1977). Though slow, costly, and labour intensive, sometimes quite small improvements in manual weeding (implement used or method) can yield sizeable increases in productivity (ILO 1979). Control of *Imperata* grass by spot weeding and pressing the intervening grass with a log or plank when it is about 1 m high at the end of the dry season has proved effective in community projects in Indonesia and the Philippines (Terry *et al.* 1997). Manual weeding is straightforward, it requires little skill except for the important task of tool maintenance, little capital outlay, only limited supervision, and can be done on all sites in almost all weather conditions and with all species. Manual methods are restricted to spot or line weed control. The main disadvantage is that cutting only gives temporary control and the vigorous regrowth that follows can be as competitive to the planted tree as the original uncut weeds.

Mechanical weeding

Mechanical weeding is where a machine operates between rows of trees and cultivates the ground, by harrowing, rotovating, or shallow ploughing, or cuts (swipes) the weed growth. The machine is usually pulled by a tractor so rows must be at least 2.5 m apart; weeds in the row tend to get missed. Mechanical cultivation for weed control is extensively used where eucalypts are grown (Fig. 13.3(a)), especially in the first few months, since eucalypts are very sensitive to herbicide. As well as inter-row cultivation, mechanized clearing saws to cut woody weeds have been tried. With the generally low wage rates in the tropics their widespread use is not yet economic.

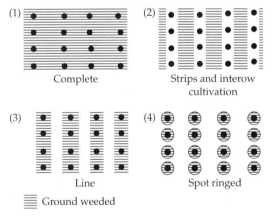

Figure 13.5 The main weeding patterns.

Figure 13.6 Line weeding of young *Pinus caribaea* in Fiji.

Chemical weed control (see also Chapter 12)

Use of herbicides to control weeds is a valuable method in conifer plantations since chemicals, which kill grasses and herbs are unlikely to damage coniferous trees, at least at normal dosage rates. However, their use in broadleaved plantations is more difficult since a chemical that kills an angiosperm weed is equally likely to kill an angiosperm tree. While herbicides are available, which kill grasses but not cereal species, in arable crops none has been developed for forestry use to discriminate between (say) different families of angiosperms. Control of grasses amongst broadleaved trees is possible.

Successful use of herbicides depends on: (i) choosing the best herbicide to control the weed species encountered; (ii) establishing the minimum dosage rate for effective control; (iii) training operators to minimize any health hazard to apply that dosage rate safely; (iv) applying it under favourable weather conditions, so that there is little possibility of the herbicide drifting, volatilizing, or being washed off in the rain; (v) timing the application for optimum kill of weeds.

Detailed information about herbicide action and effectiveness is found in herbicide handbooks by Hance and Holly (1990) and Weed Science Society of America (1994). However, information on products and safety issues is regularly being up-dated and can be accessed at numerous sites on the worldwide web including the Society of American Foresters, Pesticide Action Network UK, manufacturing companies etc. Users are advised to seek the latest information available. Table 13.2 lists a few important herbicides and their use in forestry. The

list is indicative not exhaustive and some products may not be available in some countries.

Herbicides in Table 13.2 are marketed under many different trade names and formulations. In using them the recommendations that come with the chemical should be rigorously followed and full health and safety precautions adopted. A different practice from that recommended on the label should only be adopted if the change is supported by proper evaluation and testing in local research trials. Trials of herbicides are an on-going feature of most industrial plantation programmes; some examples of results for key species can be found in Bell (1980) and Constantini (1986) for *Pinus caribaea*, Felker *et al.* (1986) for *Leucaena leucocephala* and *Prosopis alba*, for eucalypts, acacias, and casuarinas (Hall 1985), eucalypts (Florence 1996), and Amakiri (1983) for teak.

Herbicides are applied by tractor-mounted and hand-held weed wipers, stem injectors, knapsack sprayers, spot sprayers, controlled droplet applicators, or aerial sprayers. A brief description of the main techniques and calibration equipment is given by Fagg and Cremer (1990). The coverage rate, risk of drift, and total operation cost increase with each of the above methods but over large areas the unit cost per hectare diminish.

The advantages of chemical weed control are: (i) one application provides long lasting control; (ii) weeds are usually killed thus completely removing their competitive effect; (iii) dead weeds are left *in situ* acting as a mulch and reducing the risk of soil erosion; (iv) higher productivities are achieved than in manual weeding and it will be cheaper in countries where labour rates are high. The main disadvantages of chemical weed-control are the hazard to

Table 13.2 Common herbicides and their use in forestry

Herbicide	Use
Amitrole	Site preparation for certain herbaceous and woody perennials. Conifers injured. Chlorophyll synthesis inhibited
Asulam	Control of bracken/ferns
Atrazine	Pre-emergence spray for grasses and some broadleaved weeds. Most conifers not injured (triazines may become unavailable)
Glyphosate	Site preparation and post-planting control of grasses, herbaceous, and woody weeds. Conifers not usually affected. Slow acting
Hexazinone	Residual translocated herbicide primarily taken up through the roots for control of grasses and herbaceous broadleaved weeds
Metsulfuron methyl	Effective for controlling bracken and some other broadleaved weeds such as *Rumex* spp. before planting
Oxyfluorfen	Effective control of a range of weeds post planting. Has been relatively safe with eucalypts
Paraquat	Controls both pasture and broadleaved species. Often used where site is to be cultivated by mounding
Simazine	Pre-emergence spray for annual grasses and some broadleaved weeds. Conifers not injured
Terbuthylazine	Often used with atrazine as a foliar and soil-acting herbicide for control of grasses and herbaceous weeds. Some residual action
Triclopyr	Plant growth regulating herbicide absorbed through foliage stems and roots for control of woody weeds
2,4-D	Herbaceous and woody perennials. Grasses not injured. Conifer injury seasonally dependent

health, possibility of damage to crop trees or drift on to neighbouring crops, possible adverse ecological effects including contamination of waterways, the cost of equipment and chemicals, and the need to train and supervise operators.

Chemical weed-control is becoming increasingly important but it cannot be stressed too strongly that in-country trials must be carried out first to test efficacy and, most importantly, gain first-hand experience to avoid all indiscriminate use. Ensuring the safe use of herbicides is one of the greatest challenges to be overcome. For a comparison of weed control methods see Table 13.3.

Other weed-control methods

Biological control, mulching, cover crops, and livestock grazing are four other methods that all have a place in weed control. Biological control involves using another organism to control weeds. Several weed species are being examined (Bennett 1980) but the main success so far is in control of the noxious forest weed *Lantana camara* by using leaf miners, which cause premature defoliation (Diatloff 1977), though in India this miner was also found to attack teak and therefore of potential danger (Misra 1985). Parasitic *Cusenta chinensis* plants have been tried with some success in Sri Lanka and India to

control the climbing strangler *Mikania micrantha* (Nair 1988).

Mulching is when a material is placed around the tree to suppress weeds by physical weight and exclusion of light and has the added benefit of reducing water loss from the soil surface (Fig. 12.7). Mulches of *Senna siamea* and *Azadirachta indica* (neem) foliage inhibit termite attack of newly planted trees. Mulching is usually too costly in large-scale plantings but an ideal weed-control method for village or farm plantings where any reasonably inert matter such as foliage, bark, stones, and old plastic bags can be placed around the tree. The coarser and more woody the organic materials the longer they take to decompose. Finer material decomposes faster and may compete with the tree for nitrogen. Organic material should not be more than 10 cm thick and should be kept at least 10 cm from the stem to prevent rot. Mulches are not always successful with very vigorous weeds. Tampubolon and Hamzah (1988) found black plastic mulch to be most effective in conserving moisture and aiding growth of teak in East Java, Indonesia. Such synthetic films are useful but because they are impermeable they should be laid over a depression so that water runs towards a slit in the film made for the stem. Porous or woven sheets of plastic or fibres, for example, weed mats, control weeds but let water through.

Table 13.3 Comparison between weed control methods

	Manual	**Mechanical**	**Chemical**
Skill and supervision	Low	Moderate	High
Capital cost	Almost nil	High	Moderate
Control achieved for longer	Up to 100% for a short period	Up to 80% for a short period	Up to 100% for long period if effective chemicals are used
Benefits			
Tree survival	Good	Moderate	Good
Growth	Poor to moderate if weeds cut, good if weeds removed by hoeing	Poor if cutting only, good if control by cultivation	Good where herbicide kills weeds
Dangers			
To trees	Small, some risk of cutting	Very low if done carefully else can cause much damage	Broadleaved species Mostly at risk Danger from wrong dosage, and drift
To operators	Very low	Very low provided operators are skilled and adequately trained	Small, but a definite risk of misuse, which is dangerous with toxic chemicals
Productivity	Low	High	moderate-high

Figure 13.7 Mulch of sawdust in polythene bags to control weeds around young *Pinus caribaea* (experimental).

Recently, the introduction of cover crops, that is, artificially using selected plant species to smother unwanted weeds, has begun to look promising not only as part of agroforestry or providing forage for livestock, but as a weed-control system in its own right. *Mucuna pruriens* var. *utilis*, a native of southern Asia and Malaysia, has been effective in controlling *Imperata cylindrica* for 3–4 months in West Africa, South America, and Indonesia, and *Puereria* and *Centrosema* may be suitable for cover crops for longer periods (MacDicken *et al.* 1997). Trials in Nigeria with plants such as *Pueraria, Calapogonium,* and *Centrosema* showed the cover crop weed-control method cheaper in *Pinus caribaea* plantations than hand-weeding and may preferentially benefit growth of trees (Nwonwu and Obiagu 1988).

Use of livestock to control weeds in established plantations is an important indirect benefit in such agroforestry systems (Fig. 20.9). Fire hazard is reduced and even 1-year-old pines are relatively little damaged if livestock are managed properly.

Climber cutting and line opening

Both in direct plantations and line plantings full release of the tree must be ensured. On old forest sites growth of vines, lianes, and climbers is often profuse and fast, and they must be removed or controlled in other ways (previous section) else trees are smothered. Complete unwinding and leaving them on the ground is preferable but costly. When cutting vines, a section of stem is removed; simple severance is not sufficient since regrowth at the cut is fast. On fertile sites it is desirable for early weeding operations to include uprooting and grubbing out vines and creepers. In South America, climbing vines such as *Ipomoea* spp. and *Thevetia* spp. must be cut several times during the first years (Ladrach 1992).

The operation of line opening in enrichment planting ensures that planted trees receive not only full overhead light, but have adequate lateral space as well; see Fig. 12.14(a). In the Solomon Islands widening of lines begins when the trees are 1.7 m high. At this time adjacent vegetation taller than half tree height is cut back to at least 1 m from the tips of the branches, vines are cut and weed trees cut or ring barked. However, these practices have proved very time consuming and only partly control the ever-present scourge of climbers. More recently, closer planting of lines to accelerate canopy closure and suppress climbers is adopted in the Solomon Islands (Chaplin 1990). Traditional line planting was not as successful as once thought.

The cost of these maintenance operations is high if done properly and probably not cheaper in many cases than plantation establishment according to a review of experience in the neotropics (Weaver 1986). The operation is an essential component of enrichment planting silviculture (Chapter 12).

Control of stump regrowth

This is a difficult problem on old forest sites. Many tropical species produce vigorous coppice shoots from a cut stump and these can compete with the planted trees. Uprooting and grubbing out stumps is expensive. Herbicides such as triclopyr in diesel oil can be used on cut stumps to prevent regrowth while hexazinone, glyphosate, or a mixture of triclopyr and picloram can be injected into the stems of advanced growth of eucalypts, acacias, etc. However, such chemicals are expensive and can be a hazard to both health and the environment. Quite apart from the generally widespread availability of non-forested land and the presumption against clearing natural forest simply to grow plantations, this establishment difficulty is another reason for not using former forest land for plantation development.

Sometimes eradication of stump growth is necessary when the previous plantation crop is no longer wanted, has proved unsatisfactory or needs to be replaced with genetically improved trees (see Chapter 18).

Examples of weeding treatments

1. Tropical and subtropical pines (*Pinus caribaea* var. *hondurensis* and *P. elliottii*) in Queensland, Australia (Bubb personal communication). Simazine is used as a method of controlling weeds in pine plantations. This herbicide is applied three times (dependent on weed development) at a rate of 2.5 kg ha^{-1} (active ingredient) either by tractor or manually. It is not applied within 10–30 m of drainage lines.

2. *Eucalyptus* spp. on ex-pasture sites in Australia. Sites are usually ploughed. Before planting grasses and herbaceous weeds may be controlled by glyphosate or amitrole and a pre-emergent herbicide, for example, atrazine. Application rates vary with sites, weed species present, and sensitivity of the planted species. Post-planting oxyfluoren has effectively controlled a range of weeds and is relatively safe with eucalypts (Florence 1996).

3. Grassland planting in Kenyan Highlands when Shamba system is not used (mainly for planting *Pinus patula* or *Cupressus lusitanica*). The following is from the Kenyan Forest Department's Technical Order 1996.

Weeding should start soon after planting and continue for 2–3 years depending on species and site. A minimum thorough weeding is required once a year and should preferably be done soon after planting and at start of the dry season. The following weeding methods are recommended for use either singly or in combinations:

(i) manual hoeing of 1 m around each tree and line slashing of vegetation;
(ii) mechanical harrowing between rows combined with spot hoeing of individual trees give good results;
(iii) chemical weeding can also be combined with manual and mechanical weeding. (Suitable herbicides are recommended by Head of Silviculture at Kenya's Forestry Research Institute.)

4. *Pinus patula* in South Africa (from Goodall and Klein (2000), Zwolinski and Hinze (2000), and Little and Gous (2000) in the South African Forestry Handbook 2000 (Owen, 2000)). Manual or chemical ring weeding is recommended to improve tree performance. The aim of this operation is to create a weed-free zone of 250–500 mm radius around each planted tree. However, sometimes row weeding is more practical than spot weeding. Chemical weed control is the most widely used form of vegetation management and glyphosate the most common herbicide as a foliar spray and Imazapyr or Triclopyr for cut stump or basal stem treatment of woody weeds.

5. *Acacia mangium* on *Imperata* grassland, Kalimantan, Indonesia (Otsamo 1996). The dominant grass (*I. cylindrica*) is controlled by intensive site preparation by ploughing twice with a 1–2 month gap in the dry season and then harrowing once prior to planting. On slopes where mechanical cultivation is impractical the grass is controlled by spraying with

glyphosate at a rate of $5 \, l \, ha^{-1}$ initially and $2 \, l \, ha^{-1}$ if a follow up weeding is required. During the first year a maximum of three weedings conducted manually, mechanically by tractor, or herbicide spraying should start approximately 3 months after planting

6. *Acacia mangium* on *Imperata* grassland, Mindanao, Philippines (Anon 2001). Manual site preparation is very common as most planting in the Philippines is on very steep land. A hole is dug, backfilled with topsoil and then cultivated for a metre radius around each hole. More weeding is required than if mechanical cultivation of the site is feasible. Pre-plant spraying is carried out in a broad band using $4 \, l \, ha^{-1}$ glyphosate to which ammonium sulphate is added to make it faster acting and allow its use in the rainy season. One or two post planting weed control operations using $2 \, l \, ha^{-1}$ of herbicide are carried out. One release spraying with glyphosate is considered as effective as three manual weedings.

7. *Acacia mangium* on *Imperata* grassland, PNG (Yelu 1998). Sites are ploughed before planting. In the first year four manual weedings occur, after 2 months a strip (0.5–1 m wide) is cleared, followed by three complete tendings at 3-monthly intervals. In the second year a further three complete tendings are carried out at 3-monthly intervals. If crown closure has not occurred further weeding continues into the third year and sometimes into the fourth year.

8. *Terminalia brassii*, *Cedrela odorata*, *Gmelina arborea*, *Swietenia macrophylla*, and *Tectona grandis* line planting in logged over forest in Solomon Islands (Oliver 1992). Planting lines 1 m wide are cut through the forest and widened to 3 m at the planting locations. The lines are spaced 4 m or 5 m apart except for *S. macrophylla* (10 m). Weeding is manual. The lines are clean weeded each month for 3 months after planting and then spot weeded around each tree monthly for the next 9–15 months, or until the trees are 1 m tall. Weeding continues in years 3–5 being progressively reduced from 9 to 3 times per year.

Staking

Planted trees are not normally provided with support. However, in exceptional circumstances staking or propping up may be done to repair damage to young trees from strong winds or cattle provided the trees are only leaning and not broken or killed. Staking is slow and very expensive and is usually only applied to better trees evenly spaced throughout the damaged crop. The benefits and problems of supporting trees are well illustrated by the experience in Fiji. Following a severe cyclone in 1972 (Fig. 13.8), an attempt was made to repair the damage to young stands (Table 13.4). Flattened

Figure 13.8 Curved stems of *Pinus caribaea* in Fiji resulting from propping up 4-year-old trees blown over by a cyclone.

trees aged 2 and 3 years grew satisfactorily after propping upright with a clod of earth or forked sticks but 4-year-old trees propped with large forked sticks developed deformed stems. Some 634 000 *Pinus caribaea* trees 3 years old or younger were also straightened and staked after another cyclone (Bell *et al.* 1983).

Tree shelters

The use of tree shelters made out of ultra violet stabilized plastic tubes to improve the establishment and maintenance of young trees was pioneered in the United Kingdom but the technique also has potential to improve tree establishment in tropical environments (e.g. Applegate and Bragg 1989). Tubes greater than 90 cm in length may result in the trees lacking stability when they are removed (Beetson *et al.* 1991). The tubes protect the trees from damage by small mammals, provide shelter from wind, and allow easy and safe application of herbicide to control weeds. The tube can even be used to apply nutrients through fertilizer placed in the cuff spilling out on to the soil after each rainfall event (Crane 1990). However, individual tree shelters are an expensive option and do not always work. Trials of tree shelters in South Africa on newly planted *Pinus patula*

Table 13.4 The result of repairing cyclone damage to a *Pinus caribaea* stand in Fiji

Tree age (years)	Damage	Treatment	Result
0–1	None	None	Unaffected
2	Flattened	Alternate trees propped upright with clod of soil	Full recovery, growth satisfactory, straight stems
3	Flattened	Trees propped upright with forked sticks	Growth satisfactory, more or less straight stems
4	Most trees flattened, some snap	Trees propped up at an angle with large forked stick	Growth continued, but large 'S' sweep developed in stem (Fig. 13.8)
5 +	Much snap	Salvage operation only	

provided neither permanent protection nor enduring growth advantage (Taylor and Perrin 1999), and in Australia they failed to consistently protect newly planted eucalypt seedlings from browsing (Di Stepano and Mazzer 2002).

Singling

Singling is an operation done early in the life of a tree, often when 2–4 m tall, where forked or multiple stems are reduced to a single stem to improve tree-form. It is particularly important in broadleaved crops grown for timber and is usually done during cleaning. Singling is expensive and is usually only worthwhile if there are many poor stems and the crop is for a high quality end-use such as veneer or sawtimber. Where the proportion of misshapen trees is low, say less than 20%, such trees can be removed in thinning. Singling is usually not necessary in crops grown for pulpwood or firewood but it is standard practice in many *Acacia mangium* pulpwood plantations in Asia. It is carried out at 4–6 months after planting when the acacias are 1.5–2.0 m tall and sometimes this is repeated at 8–12 months (Mok *et al.* 2000). As it is expensive and may result in reduced volume growth and increase the risk of heart rot, the need for singling should be assessed on a case by case basis (Tuomela *et al.* 1996). Much forking and multiple leaders in a stand may indicate a nutrient disorder, for example, boron deficiency, extensive insect damage from shoot borers, or use of seed of poor genetic quality.

With teak newly inserted stumps may produce more than one shoot. In India use of stumps 1–2 cm thick at the top was found to minimize this occurring. However, if more than one dominant shoot does develop all but one are removed when they are about 1 m tall. During tending any renewed basal shoot growth is cut back.

While singling is a form of 'crown' pruning to ensure a straight stem, occasionally the leading shoot may be removed by pruning to encourage a more spreading habit in trees planted for shade, shelter, fodder, or other social forestry purpose.

Respacing

This is a 'weeding operation' which involves removal of competing trees of the *same* or similar species as the ones planted. Re-spacing may be undertaken in four situations.

1. Too close initial spacing.
2. Where profuse natural regeneration competes with replanted second rotation trees (Fig. 13.9).
3. As a very early thinning (pre-commercial thinning: see Chapter 16) to encourage rapid growth of the remaining trees or to maintain light levels in the stand as part of agroforestry for good growth of forage plants for grazing.
4. To thin out plants coming up after direct seeding operations.

Plantations are re-spaced either within the first 2 years of planting or at the same time as low pruning. In the Usutu forest, Swaziland, natural regeneration is pulled up from around planted trees 1–2 years after planting.

Low pruning

Low pruning is the removal of branches to a height of about 2 m up the stem (Fig. 13.10) at or just after the time of canopy closure in a plantation; it is sometimes called 'ground pruning'. It is done to: (i) provide access into a stand for inspection, and marking for thinning; (ii) reduce the fire hazard by lessening the chance of ground fires burning up into the crown; (iii) facilitate felling and extraction of thinnings so making these operations safer; (iv) produce knot-free timber at the base of a tree low pruning for this purpose would be the first of several prunings—see Chapter 17.

In rural development forestry often the first operation to provide any firewood, fodder, or other produce from the crop is a low pruning of side branches. In agroforestry, moderate pruning may be carried out to reduce light competition between trees and crops and heavier pruning to reduce the water requirement of the tree (Jackson

Figure 13.9 Dense natural regeneration of *Pinus patula* in the Usutu forest, Swaziland, 18 months after the previous crop was clearfelled. Re-spacing is needed to release the planted trees which are of preferred genetic quality, e.g. *Pinus patula* ssp. *tecunumanii*.

et al. 2000). When low pruning simply involves breaking off small dead branches, as in many eucalypt stands, it is called 'brashing'.

Methods

Low pruning is a hand operation and is very labour intensive. The tools used include an axe or billhook, a pruning saw, or a light chain-saw (4–6 kg).

Pattern of low pruning in a stand

Low pruning can be carried out on all trees in a stand, on only some, or on none at all. The terms complete or partial low pruning are used, or sometimes the proportion of trees pruned is expressed as a percentage. The amount of low pruning needed depends on the natural pruning characteristics of the species, the need for fire protection, and the intended thinning method. If selective thinning (Chapter 16) is planned about 80% of the trees need low pruning to provide good access and allow the trees to be clearly seen when selecting the trees to be thinned. In mechanical thinning, where the trees to be thinned are pre-determined, 25–50% low pruning is usually sufficient. If no thinning is intended low pruning may not be necessary.

Partial pruning is often resorted to because low pruning is costly. As a result, many different patterns of low pruning a stand have arisen.

Figure 13.10 Low pruning to 2 m in 5-year-old *Pinus patula*.

Complete: all trees low pruned except the dead and dying; easy to administer.

Eighty per cent: all trees except smaller sub-dominants, dead and dying; this is the maximum ever needed for selective thinning.

Lane pruning: 25–50% of trees pruned either half of each tree between rows or all trees in one row, for example, every second or fourth row.

Haphazard pruning: usually 40–50% of trees pruned at random by the operator. Difficult to implement but provides access through the whole crop without too much hindrance.

Access pruning: several years before thinning one row in 10 or 20 may be pruned as an inspection rack to assess vigour and uniformity of the stand.

Strip pruning: this is a compromise sometimes applied to unthinned crops where pruning is only needed for fire protection. 'Islands' of trees are left unpruned to reduce the cost per unit area of the operation.

Cost

The cost per hectare is influenced by several factors.

1. Species: the numbers, persistence, and death of low branches.

2. Spacing: number of trees per hectare; though wide spacing results in fewer trees per hectare their branches will be thicker.

3. Proportion of trees to be pruned.

4. Method: a pruning saw is less damaging, slower but safer than an axe.

5. Ground conditions: this affects ease of movement through the stand. A steep rocky slope will greatly reduce the productivity of a pruning gang.

6. Low pruning height: to 2 or 2.5 m, etc.

Timing and intensity

When cutting branches for firewood or fodder there is often a temptation to cut too many and too soon from young trees with their resulting growth impaired. A similar problem can occur in industrial plantations with overzealous, premature low pruning delaying canopy closure and weed supression and poorer crop growth. Some pine crops in both Madagascar and Swaziland have suffered in this way. Low pruning the live crown of *Acacia mangium*

in association with singling is practised in some plantations in Indonesia but may reduce productivity of young plantations by 30% (Tuomela *et al.* 1996). Loss of diameter increment is especially significant if more than 40% of the crown is removed (Nik and Paudyal 1992).

Low pruning in industrial plantations

In general, low pruning is more widely practised in the tropics than in temperate countries because wage rates are low. The practices used vary considerably from country to country.

In PNG the following prescriptions have in the past been used: (i) *Araucaria*: at canopy closure when 75% of trees are over 6 m tall the stocking is reduced to 1000 trees ha^{-1} and all trees pruned to 2.4 m, (ii) *Pinus*: all trees are pruned to 2.4 m when 75% of trees reach 6 m in height; only saws may be used, (iii) *Eucalyptus*: no pruning is needed in *E. deglupta* stands, and (iv) *Tectona*: no pruning except the removal of basal shoots during weeding and cleaning operations.

In Swaziland, *P. patula* is low pruned at 4–6 years of age to a height of 1.8 m. Low pruning takes place about 1 year after full canopy closure. *E. grandis* is stick brashed once lower branches have died and can be knocked off.

In Queensland, Australia, *Araucaria cunninghamii* stands are completely low pruned when average predominant height is 9.5 m, but *P. caribaea* only selectively treated on the trees which will be subsequently high pruned.

Efficient self-pruning is a characteristic of most eucalypts that are grown on plantations but there is variation between species and in some conditions branch shedding and occlusion are delayed. In countries such as South Africa where eucalypts are thinned heavily or planted at wide spacing to produce sawlogs, pruning often takes place in two or more lifts to 6–7 m (see Chapter 17). The chance of decay organisms entering the stem through pruning wounds is much greater in eucalypts and acacias than in pines. If pruning is carried out on eucalypts it should be when the branches are small and still alive, and in dry, cool weather when the risk of infection is minimized (Gadgil *et al.* 2000).

CHAPTER 14

Nutrition of tree crops

Trees, like all plants, require supplies of certain chemical elements for growth. Elements needed in large quantities, called 'macronutrients', are nitrogen (N), phosphorus (P), potassium (K), calcium (Ca), magnesium (Mg), and sulfur (S). Elements needed in minute or trace amounts, 'micronutrients', are iron (Fe), copper (Cu), chlorine (Cl), manganese (Mn), boron (B), zinc (Zn), and molybdenum (Mo). Of course, in addition to these elements, which are mostly supplied from the soil, plants require carbon (C), hydrogen (H), and oxygen (O), the primary constituents of all organic matter, which are obtained from water and air. Atmospheric sources of nutrients in precipitation in general represent a tiny fraction of the requirements for growth. This chapter is concerned with the supply of macronutrients and micronutrients. If any of them is in limited supply, or in excessive quantities, tree growth may be impaired. The survival and growth of trees on nutrient-poor soils depends on mechanisms that enhance nutrient uptake and enable them to be used efficiently and retained within the tree.

Historically, fertilizer use, that is, artificial addition of nutrients, has been much less important in forestry than in agriculture for three reasons.

1. Slow growth and long rotations in temperate forestry practice resulted in little or no net nutrient loss from the ecosystem due to forest operations. Growth was maintained by nutrient additions in rainfall and by nutrient cycling and, in many instances, it was claimed that tree cover improved soil nutrient levels through the accumulation of litter and organic matter at the surface.
2. The failure or poor growth of species planted on some sites, often those long devoid of tree cover, was usually overcome by changing the species rather than improving the site.
3. Fertilizing was considered uneconomic because application was usually needed early in the life of a crop, responses were uncertain, and rotations long.

Now, with greater emphasis on more intensively managed planted forests this situation is changing. Although fertilization is not a practice in all tropical plantations and is nowhere near agricultural levels, it is increasing and other methods of improving tree nutrition are often used, notably recycling harvesting residues. Reasons for this greater attention to tree nutrition in the tropics are:

1. Most tropical plantations are established on highly weathered soils in which the levels of essential plant nutrients, P, N, K, Ca, Mg, Zn, S, Cu, and B, are low or very low.
2. Intensively managed, fast-growing, short-rotation plantations place large demands on soil nutrient reserves and on poor soils nutrient depletion is a real possibility, especially in the moist tropics.
3. Plantation forests are a commercial activity supported by intensive management and short rotations, commonly 7–15 years, which make fertilizing more economic.
4. Forestry enterprises now often use only one or two species for simpler management and more uniform end products. These species may not be suited to all sites available and nutrient input is often necessary to aid their establishment and improve uniformity of growth.
5. The addition of small amounts of nutrients on some impoverished sites has produced spectacular improvements in growth, for example, B for eucalypts growing in many African grasslands (Stone 1990), Zn for pines in Madagascar (Rampanana *et al.* 1988) and N and P for eucalypt plantations in Australia (Kriedemann and Cromer 1996).

Nutrient deficiencies

Causes

Suboptimal levels of nutrients primarily reduce tree growth and do not induce deficiency symptoms. Deficiency symptoms occur when there is a severe reduction in the relative rate of supply of a balanced set of essential nutrients or an imbalance among those available for growth. Conditions leading to nutrient deficiencies are:

1. Impoverished soil of any kind, for example, sand, eroding soils, mine spoil. Most soils in the tropics available for tree-planting, for example, Oxisols, Alfisols, and Ultisols, are naturally low in inherent fertility and deficient in nutrients such as Ca, Mg, K, and P although they contain excessive amounts of Al, Fe, and Mn (Lal 1997).

2. Inadequate rate of nutrient cycling on infertile soils. Excessive litter accumulation can temporarily immobilize a significant proportion of the nutrients available. Nutrient cycling in pines and eucalypts is generally less efficient than most broadleaves, for example, a *Pinus caribaea* plantation in Puerto Rico accumulated twice as much litter as broadleaved secondary forest (Cuevas *et al.* 1991) and a similar situation prevailed in Australia when *Eucalyptus grandis* was planted on a subtropical rain forest site (Turner *et al.* 1989). However, the accumulated nutrients may be critical for growth of trees in the next rotation (Lugo *et al.* 1990).

3. Induced deficiencies caused by other factors.

(a) Very high rainfall leaching out nutrients.

(b) Very low rainfall. Deficiency symptoms commonly appear during a drought.

(c) Lime induced chlorosis from poor Fe and Mn uptake on high pH soils.

(d) Effects of different pH levels. Phosphorus is precipitated in strongly alkaline soils, and also can become bound to Al^{3+} and Fe^{3+} ions in very acid soil and be unavailable.

(e) Interaction with other nutrients. The levels of one nutrient can affect the requirement for others, notably with N and P, and P and K. This kind of relationship is very common.

4. Inadequate mycorrhizal associations and poor nodulation in nitrogen-fixing trees.

5. Excessive weed competition.

6. Loss of nutrients through removal in harvested tree components, burning of logging debris and surface litter layer and deliberate gathering of litter material (litter raking) as the plantation grows (Fig. 14.1)

7. Use of unsuitable species, for examples, *Leucaena leucocephala* on acid soils with pH less than 4.5.

Occurrence

Life of crop

The rate of nutrient uptake of a stand of trees changes over its life and distinct stages can be recognized (Miller 1995) but it is in the first stage in plantations, usually within a few months after planting in the tropics, that deficiencies mostly develop. The main reasons for this are:

1. In the first weeks after planting deficiencies are rare since the seedling continues to benefit from the nutrients applied while in the nursery and the requirement for nutrients is small.

2. Once the newly planted tree is established, which takes a few weeks to a few months, growth accelerates rapidly notably in height, numbers of branches, leaf area, and the size of the root system. Figure 14.2 shows that in 5 years a moderately fast-growing *Pinus patula* produced over 100 main branches. During this period none of the lower branches was suppressed, thus each year's production of new foliage further increased the live crown. By 5 years the crown was vastly greater than the 30-cm tall seedling originally planted. Moreover, foliage has the highest concentrations of nutrients

Figure 14.1 Litter raking beneath *Pinus caribaea* in southern China. Gathered litter is used for bedding for livestock, firing brick kilns, and kindling. The practice interrupts nutrient cycling.

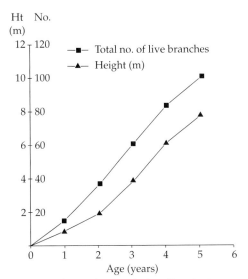

Figure 14.2 Height growth and production of branches in one moderately fast-growing *Pinus patula* tree.

of any part of a tree and may contain 20–40% of the total amount of nutrient in a stand (Table 14.1). Clearly, in the early years of growth nutrient uptake must be considerable to support this rapid expansion of the crown. Similarly in fast-growing eucalypts in Brazil most of the nutrients accumulated are in the first 2–5 years depending on water availability (Gonçalves *et al.* 1997). Nutrient deficiencies occur if sites cannot meet this demand.

3. Very little recycling of nutrients occurs in the first few years because the crown is developing, few leaves are shaded and die, and no branches will fall. Litterfall, except with the few tropical deciduous species, is small and little nutrient is returned to the soil. As the tree matures and the stand closes canopy, crown expansion slows and, on an area basis, nutrient uptake and return through litterfall reach equilibrium. For example, Lundgren (1978) found that 18–20-year-old *P. patula* and *Cupressus lusitanica* stands in East Africa produced 6.2 and 5.2 t ha^{-1} year^{-1} of litter respectively, whereas the total amount of litter produced in the whole of the first 5 years was only about 4.0 t ha^{-1}.

4. Some species, such as eucalypts, can 'reuse' significant quantities of absorbed nutrients by withdrawing N, P, and K, etc. from wood when the heartwood is being formed and from leaves just before they senesce. Kadir and van Cleemput (1995) found *Acacia mangium* translocated 78% of P, 26% of N, and 13% of K from its foliage before abscission. The relative importance of these mechanisms changes with age so that large trees are more highly buffered than seedlings or small trees against changes in nutrient availability (Snowdon 2000).

5. Root development may fail to make full use of site potential because of competition from weeds, poor mycorrhizal development, or physical barriers to growth in the soil.

This high nutrient uptake in early years is illustrated in Table 14.1. At just 7 years the total amounts of aboveground nutrients are already 55–60% of that found in much older (22 years) stands of similar site quality. Also it can be seen that much crown development has already taken place during the first 7 years since between 7 and 22 years foliage only doubles in weight whereas branch-wood and stem-wood increase three or four times. Similarly, data correlating stand development with nutrient demand can be shown for all species, for example, teak (Nwoboshi 1984), eucalypts (Barros and Novais 1996), acacias (Pande *et al.* 1987).

Soils, sites, and species interaction
Many factors influence nutrient availability in soil and it is difficult to predict if deficiency will arise. Usually only local experience will show if certain species fail on some sites because of poor nutrition.

Most soils in the tropics are strongly or moderately P-deficient and low P availability frequently limits tree growth, including eucalypts, pines, and other commonly planted species, and growth response to N will not occur unless P-deficiency is corrected (Gonçalves *et al.* 1997). Examples of nutritional problems include *Pinus caribaea* on infertile white sands in Guyana, Suriname, and Malaysia, while in Madagascar extensive plantations of *P. caribaea* and *P. kesiya* were found seriously deficient in K. Fast-growing broadleaved species such as *Paraserianthes falcataria*, *Cordia alliodora*, *Gmelina arborea*, and *Anthocephalus chinensis* have high fertility requirements and do not grow well on leached soils even if soil physical conditions are good and there is no impediment to rooting. Large differences between species in uptake and accumulation of N, P, K, Ca, and Mg for several tropical plantation species has been found in Costa Rica (Lugo 1992). Miller (1995) has suggested that observed differences between species has little to do with actual nutrient demands but reflect differences in their abilities to obtain nutrients from intractable soil sources. Boron deficiency is relatively common in recently established eucalypt plantations throughout the tropics (Table 14.2) and the intensity of the symptoms is aggravated by seasonal drought conditions (Stone 1990; Dell and Malajczuk 1994). In PNG, on rain forest sites, pines often do not grow well because of weed growth and initially from excessively high levels of mobilized N (Lamb 1973).

Table 14.1 Nutrients (in kg ha^{-1}) in aboveground biomass and stump/root–crown in 7- and 22-year-old stands of *Pinus patula* and *Cupressus lusitanica* in Tanzania

	Pinus patula						*Cupressus lusitanica*					
	Dry weight (tonnes ha^{-1})	N	P	K	Ca	Mg	Dry weight (tonnes ha^{-1})	N	P	K	Ca	Mg
Seven years old												
Stem-wood	105	192	21	118	103	44	69	133	16	111	122	14
Stem-bark	6	101	9	46	91	23	7	71	7	62	197	11
Branch + cones	37	158	15	88	135	32	35	144	19	148	354	29
Foliage	15	329	26	144	104	33	9	192	16	99	252	27
Stump/root–crown	23	69	5	29	7	11	18	47	5	37	82	5
Total	186	849	86	425	470	143	133	587	63	457	1007	86
Twenty-two years old												
Stem-wood	319	415	45	255	224	96	240	288	34	240	264	31
Stem-bark	13	101	9	46	91	23	17	122	12	106	339	31
Branch + cones	99	320	31	176	256	63	86	302	41	320	569	55
Foliage	31	514	40	225	163	52	17	226	19	117	297	32
Stump/root–crown	49	127	10	54	68	19	38	69	8	54	119	8
Total	506	1477	135	756	802	253	398	1007	114	837	1588	145

Source: Lundgren (1978).

Table 14.2 Occurrence of nutrient deficiencies in tropical eucalypt plantations

	B	Ca	Cu	Fe	K	Mg	Mn	N	P	S	Zn
Australia	x		x	x	x		x	x	x	x	x
Brazil	x	x			x	x		x	x	x	x
China	x			x	x			x	x	x	x
India					x			x	x		
Indonesia	x		x		x	x		x	x		x
Philippines	x			x	x	x		x	x		x
South Africa		x	x		x	x		x	x		x
Thailand	x			x				x	x		
Zambia	x										

Source: Dell *et al.* (2001).

Symptoms of nutrient deficiency

Visual symptoms

Sometimes visual symptoms are so distinctive that a useful conclusion can be drawn regarding nutrient status, however, similar visible symptoms may be produced by completely different factors and hence symptoms alone are often inadequate for diagnosis. Where the growth of a stand, of a species well-suited to the climate, is below expectation and trees appear stunted with poor vigour, short leaders and side shoots, and pale or thin crowns a nutrient problem may be indicated. In older stands it may show up as an irregular diameter distribution, and it is frequently reported that stand uniformity improves following fertilizer application, for example, Schutz (1976) for *Eucalyptus grandis* in South Africa.

Foliage colour can be a useful guide for detection and diagnosis of a problem. For example, seedlings of *Araucaria* spp. in PNG suffer acute chlorosis, with upper leaves turning uniformly yellow or even white, under conditions of Fe deficiency induced by alkaline soils of high pH (Basden 1960). With some species it has been possible to induce specific deficiencies and to identify characteristic colour symptoms associated with a shortage of each macro- and micronutrient. Dell *et al.* (2001) have produced coloured illustrations of symptoms of nutrient disorders in plantation eucalypts and a description of symptoms is reproduced in Table 14.3. Field experience has shown that for macronutrients the order of reliability of visual symptoms of deficiency in eucalypts and pines to make decisions on fertilizer applications is: P > N ≥ K ≥ Ca >S > Mg, and for micronutrients: B ≥ Zn > Cu (Gonçalves *et al.* 1997).

Anatomical changes

Malformations of shoots, leaves, etc., occur with nutrient disorders. Probably the best recorded example is with B deficiency. In pines this deficiency causes abortion and dieback of the tip bud, which leads to multiple leaders and stag-headed trees, stem swelling below the leader, and twisted needles (Smith 1976). In eucalypts it causes deformation of developing leaves, dieback of young shoots, pendulous branches, and prostration (Dell *et al.* 2001). Other signs which may indicate a nutritional disorder include sparse foliage and poor leaf retention, which has been observed in Nigeria with *Pinus caribaea* and *P. oocarpa* on P-deficient soils (Kadeba 1978) and for Zn deficiency of *P. caribaea* in northern Australia (Rance *et al.* 1982). Intercostal chlorosis and necrosis of leaves occurs with *Gmelina arborea* and *Azadirachta indica* when K-deficient (Zech 1984a). Zn deficiency in *P. caribaea* causes and resin exudation from stems and shoot-tip fasciation. Nutrient disorders can also change wood properties, for example, the specific gravity of the wood and size of fibre cells changed according to which kind of N fertilizer was applied to *G. arborea* (Zech and Drechsel 1998).

Great care must be exercised when using visual symptoms to identify a nutritional problem. It is important to know what a healthy tree looks like during the various stages of development and also the tree's normal reaction to changes in light, temperature, and amount of rainfall. Knowledge of recent climatic changes and an appreciation of local soil physical and chemical properties can also assist in determining the cause of abnormal growth, fungal and bacterial pathogens, insects, and climatic events may also resemble nutrient deficiencies. Both fungal disease and frost can change leaf colour, though usually the whole leaf is affected and not just some parts as occurs with certain nutrient deficiencies. Shoot-boring insects and hail damage can cause multiple leaders in pines very similar to B deficiency. And, lastly, there is not always agreement on the symptoms of a particular deficiency. This confusion probably arises from observing the deficiency at

Table 14.3 Key to nutrient deficiency symptoms in eucalypts

	Deficiency
A1 Symptoms appear first or are more severe on older leaves	
B1 Leaf colouration is even over all leaf	
C1 Leaves pale green to yellow, small reddish spots may develop secondarily	Nitrogen
C2 Leaves green with reddish blotches or leaves uniformly purple or red	Phosphorus
B2 Leaf colouration forms a pattern	
C1 Leaves with marked interveinal chlorosis	Magnesium
C2 Leaves with scorched margins or interveinal necrosis, sometimes preceded by marginal chlorosis	Potassium
A2 Symptoms appear first or are more severe on expanding leaves	
B1 Dieback present at shoot apex	
C1 Nodes enlarged, proliferation and death of lateral shoots	
D1 Leaves with corky abaxial veins, or apical chlorosis, or malformed with incomplete margins	Boron
D2 Leaves with regular or undulate margins, some interveinal chlorosis	Copper
C2 Nodes normal, leaves buckled due to impaired marginal growth	Calcium
B2 No dieback at shoot apex	
C1 Leaves normal size	
D1 Leaves pale green to yellow	Sulfur
D2 Leaves yellow with green veins	Iron
D3 Leaves with marginal or mottled chlorosis, small, necrotic bleached, or brown spots may then appear	Manganese
C2 Leaves small and crowded	Zinc

Reproduced by permission from Dell (1996). Symptoms can be related to coloured photographs in Dell *et al.* (2001).

different stages of development and on plants of different ages, but it emphasizes the difficulty of basing diagnosis simply on visual features. If a nutrition problem is diagnosed, the symptoms can be compared with coloured photograph showing typical symptoms for a particular species, for example, eucalypts (Dell *et al.* 2001), *Swietenia macrophylla*, and *Cedrela odorata* (Webb *et al.* 2001).

Diagnosis of nutrient deficiency

Visual symptoms may indicate severe deficiency of a particular nutrient but, unless there is much experience with the species and the sites, they should be considered tentative until confirmed by other techniques. While recognizing that field fertilizer experiments are the ultimate test in identifying nutritional constraints, they can be preceded by a series of integrated experiments and surveys involving plant tissue analysis, soil survey, and bioassays designed to answer specific questions. Such experiments and surveys are short-term and can be used to select the most appropriate treatments to test in more operationally relevant field fertilizer trials. This integrated

approach was used by Webb *et al.* (1995) to identify deficiencies in macro- and micronutrients for plantation establishment of tropical broadleaved species in the Solomon Islands.

Plant tissue analysis
The prediction of nutrient status of trees by plant analysis has usually been based on foliar tissues but in some cases tissues other than leaves, for example, twigs, bark, roots, and xylem sap, have been useful in the diagnosis of nutrient deficiencies (Dell *et al.* 2001). Foliar analysis is the chemical analysis of needles or leaves to determine the concentration of mineral nutrients present. Trees suffering deficiency have lower than normal foliar concentrations of the nutrient in short supply. Since sample collection and analysis can be done in a few weeks, foliar analysis allows rapid diagnosis (Weetman and Wells 1990). Foliar nutrient content is usually well correlated with soil variables, for example, Herbert (1990*a*). The foliar samples reflect the demand of the plant, its ability to access nutrients and the potential of the soil to supply nutrients in contrast to soil analysis, which reflects the potential of soil to supply part of its nutrient resource. Foliar analysis can assist in identifying

nutrients that are limiting and in understanding the response of trees to fertilizer applications.

Foliar analysis has two main steps.

1. For each nutrient, critical concentrations are determined from the relationship between the nutrient concentration in the plant tissue and the growth and yield of the plant. Threshold levels associated with deficiency, adequacy, and perhaps toxicity are identified. This is done by sampling many trees of varying vigour and health on a range of sites and measuring the associated foliar nutrient levels. The underlying principle is Liebig's law of the minimum that growth is controlled by the factor (nutrient) in shortest supply and procedures for determining critical concentrations have been reviewed by Smith (1986). Herbert (1991) reported that 78–96% of variation in foliar analysis values could be explained by regressions involving site and soil parameters. A study by Lamb (1977) sampled 116 trees of *Eucalyptus deglupta* and variation in height growth was largely explained by differences in foliar concentrations of N and P. An equation relating these variables was statistically highly significant. Using this relationship, and growth responses in fertilizer trials, he concluded that 2.1% was the critical level for N in *E. deglupta* leaves. Nitrogen concentrations below 2.1% indicated some degree of deficiency. Figure 14.3 shows another example from a study of foliar nutrient levels in an avenue of *Araucaria hunsteinii* (Evans 1980), which included trees with good growth and healthy dark green foliage and others with slow growth and yellow chlorotic leaves. Nitrogen was the only element to show marked variation, and this was closely correlated with both leaf colour and growth. This limited

study suggested a critical foliar N concentration of about 1.2%, which compares closely with the 1.35% for *A. cunninghamii* used in Queensland, Australia.

Simple examination of relative concentrations of nutrients can be misleading and ratios between N, P, K, Ca, and Mg may be a better diagnostic tool to indicate nutrient 'balance' within the plant. For many plants ratios of about 10 : 1 : 5 for foliar nutrient concentrations of N, P, and K in relation to P appear about the right balance, and significant departure from this by one nutrient, for example, N : P ratios wider than 15 : 1 in eucalypts, is a strong indication of a disorder (Schönau 1984*a*). Schönau (1982) describes the importance of such nutrient relationships, also Ca:Mg, and the more complex P : K : Mn balance with *Eucalyptus grandis*. Zech (1984*b*) stresses the importance of P : Al for teak, P deficiency often being associated with high Al (aluminium), and P : B (boron) with *Gmelina arborea*, with B deficiency leading to low P. An extension of the ratio approach is the Diagnosis and Recommendation Integrated System (DRIS) (Beaufils 1973). This is based on the calculation of ratios of several nutrients and their use in combination to indicate nutrient deficiencies. It has been used for eucalypts and pines and more recently for Chinese fir (*Cunninghamia lanceolata*) (Zhong and Hsiung 1993) and teak (Drechsel and Zech 1994). Dell *et al.* (1995) suggest that the greater analytical requirements and the lack of a sound physiological basis for interpreting nutrient ratios is a disadvantage and DRIS has little advantage over concentrations for diagnosing nutrient disorders in eucalypts.

2. The second step in foliar analysis is to sample the stand under investigation, analyze the leaves to determine foliar nutrient concentrations, and compare the results with the threshold or critical values. If one nutrient is much below the critical value it is likely to be deficient.

Foliar analysis appears straightforward. However, in practice, many factors other than the nutrient supply to the tree may cause variation in foliar nutrient levels, and they must be taken into account. Climate, season, soil type, time of day, aspect of the tree, crown class of tree, tree age, position of leaves in the crown, age of foliage, internal nutrient balance, natural between-tree variation, effects of diseases, etc., may all affect nutrient concentrations in leaves. The importance of ascertaining this information cannot be over-stressed. For example, Schönau (1981) suggested that *Eucalyptus grandis* leaves should be sampled in mid-growing season to enhance differences in nutrient concentrations while Payn *et al.* (1989) recommended winter for *Pinus patula* as being the most stable period. Spatial distribution of foliar N and P within the crowns of young *E. grandis* trees in the field had a range of N and P concentrations from 5–30 mg N g^{-1} dry matter and 0–3.5 mg P g^{-1}

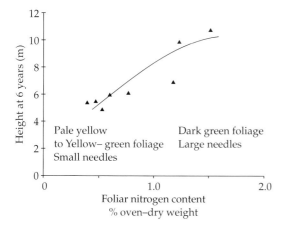

Figure 14.3 The relationship between height growth and the level of nitrogen in the leaves of 6-year-old *Araucaria hunsteinii*.

Table 14.4 Foliar nutrient variation in the crown of a single *Gmelina arborea*

Sample position arranged in order of increasing shade	Nutrient content (% oven-dry weight)			
	N	P	K	Ca
Topmost leaves (fully exposed)	2.10	0.11	1.5	1.0
Outer leaves on upper crown	1.86	0.11	1.6	1.0
Outer leaves on mid-crown	2.02	0.11	1.5	1.1
Outer leaves on lower crown	1.79	0.11	1.5	1.5
Outer leaves at base of crown	1.72	0.10	1.5	1.6
Leaves from inside the upper crown	1.69	0.10	1.1	1.8
Leaves from inside mid-crown	1.62	0.10	1.4	1.8
Leaves from inside lower crown (heavily shaded)	1.39	0.08	1.6	1.7

dry matter (Leuning *et al.* 1991). Table 14.4 shows the kind of variation associated with sampling leaves from different parts of the crown of one tree of *Gmelina arborea* (from Evans 1979). All sampling of the *G. arborea* was on 1 day, with several samples from each position to provide a mean value, and the leaves from the same relative part of a shoot.

In addition to the above, errors may be introduced into foliar analysis by poor handling of samples, delay in drying, extraneous dirt causing contamination, differences in analytical procedures, and so on. All these difficulties are mostly overcome by rigidly applying carefully standardized sampling procedures and analysis. Provided all sources of variation in nutrient concentrations are allowed for, foliar analysis is a useful diagnostic technique.

Most published foliar nutrient data refers to pines and eucalypts, for example, Lambert (1984), Judd *et al.* (1996). There are fewer data on other tropical hardwoods although Drechsel and Zech (1991) reviewed foliar nutrient levels in tropical broadleaved species to find concentration ranges indicating deficient, marginal, adequate, high and toxic levels of foliar mineral nutrients. Such data can only be taken as approximate as results of foliar analyses are so dependent on age and position of the foliage, sampling time, number of trees, etc. However, few critical values have been set and until they have been defined for all essential nutrients and for all commercial species standard concentration ranges provide an acceptable alternative (Dell 1996). Further research on a range of species will enhance the reliability of this powerful technique, which is less expensive than fertilizer trials.

Soil analysis
Since a tree derives almost all of its nutrients from the soil, chemical analysis of the soil might be expected to yield useful data for studying nutritional disorders. Unlike in agriculture, soil analysis

has been much less useful in forestry, at least for determining the availability of specific plant nutrients. This is because tree roots are more extensive and often penetrate deep into the soil so that the amount of available nutrients in the surface soil layer may not correlate well with tree growth. Whereas plant analysis measures the actual uptake of nutrients from the soil, soil analysis attempts to measure the potential quantity of nutrients in the soil for plant growth. Soil analysis provides information about pH, cation exchange capacity (CEC), C : N ratio, soil texture, organic matter content, etc., but as there are few verified guidelines for interpreting 'available' soil nutrient levels for tropical trees only general conclusions are usually possible. Assay for micronutrients can be helpful—in 1988 a soil survey of Fiji Pine Commission's land revealed deficiencies of B and Zn in some areas. The soil sampling strategy should be chosen with due consideration to variation in topography and depth of sampling.

Relationships between growth and soil nutrient status have been obtained, for example, nutrients in the 0–20 cm soil stratum were related to growth of eucalypts in Brazil (Gonçalves *et al.* 1990) and base saturation of B horizons with growth of *Gmelina arborea* in Nigeria (Chijioke 1988), but relating this information to fertilizer requirement or occurrence of deficiency has proved difficult. Interpretation is difficult because: (i) what is measured in an extracted soil solution may bear no relation to what a tree can obtain from the soil under natural conditions; (ii) soil analysis implies one knows how much nutrient a tree requires, and while this is better understood than in the past, a tree is not static or fixed in its demands, both changing with age (Miller 1981) and in the relative importance of internal and external nutrient cycling (Miller 1989); (iii) tree roots may occur in several horizons and numerous samples are needed to analyse fully the volume of soil available for root development.

Though soil analysis has these drawbacks it is still an important tool in plantation forestry and, as mentioned in Chapter 12, forms part of pre-planting survey. In addition to the kinds of information referred to earlier it remains one of the few ways of gaining some idea of the nutrient status of unplanted land (foliar analysis is only possible where a tree crop already exists).

Bioassays

Pot trials involve testing soil from the planting site by growing seedlings in it under controlled conditions in a glasshouse. Pot trials can be used to identify symptoms of single-nutrient deficiencies, define critical values for deficiency and toxicity, and provide initial fertilizer recommendations for tree establishment (Dell 1990). Such trials are short term and relatively inexpensive but are rarely sufficient on their own. The extrapolation of results from pot trials of large, deep-rooting trees to field conditions has often been debated but such trials can yield valuable results that have frequently been confirmed by field experiments, for example, the identification of S deficiency limiting the growth of eucalypts on savanna soils in Brazil (Barros and Novais 1996). In the Solomon Islands, pot trials were used to identify P-deficiency in a deeply weathered kraznozem soil and then rate trials were used to indicate the amount of P fertilizer needed to achieve maximum tree growth (Webb *et al.* 1995). While this methodology has limitations, it can be used for preliminary evaluation of substrate suitability for trees when planning afforestation of new planting areas including industrial spoil heaps, tailings, and other artificial wastes, etc. Pots studies have also been used successfully for screening species for salt and waterlogging tolerance (Marcar and Khanna 1997).

Field fertilizer trials

These are experiments to test the effects of different fertilizers, singly or in combination, on growth and foliar nutrient concentrations (see below) of a species on a particular site. For the basic requirements of any fertilizer trial, see Gentle and Humphreys (1968), and in nursery research, Liegel and Venator (1987). Fertilizer field trials are often combined with other cultural treatments as interactions can occur. For example, Haywood *et al.* (1997) describe a trial combining fertilization, weed control, and litter application on *P. taeda* and an example in Table 12.2 illustrates the effects of a fertilizer trial combined with cultivation treatments for *P. caribaea*.

A fertilizer trial is slow and expensive but usually conclusive in diagnosis of a nutritional problem provided it included the limiting nutrient(s). Because of this last point fertilizer trials are rarely used as the primary diagnostic tool. Normally they are laid down to confirm a deficiency diagnosis based on other evidence, to identify the most efficient treatment of the disorder, and to indicate the size of any response to decide whether it will be worthwhile fertilizing whole stands. In South Africa, growth responses of eucalypts to fertilizer containing N, P, and to a lesser extent K, based on fertilizer trials were reported by Schönau and Herbert (1989) and more recent trials show how a combination of P and K fertilization can significantly increase the growth of *P. patula* (Carlson 1998).

Use of fertilizers

Most tropical soils are highly weathered and levels of P, Ca, Mg, Zn, S, N, K, Cu, and B are low or very low (Gonçalves *et al.* 1997). Often one or more nutrients are in critically short supply to sustain tree growth. Fertilizer application is the simplest way of improving soil fertility and fertilizers are used in plantation forestry in three situations: to correct a specific deficiency; to establish a crop on nutritionally poor land, or to stimulate growth. The distinction between them is not precise and fertilizing may be done for one or more of these reasons. Although increase in wood yield is the main improvement sought, healthy plantations usually are less susceptible to pests and diseases, and more resistant to damage by frost, drought, etc. (Baule 1973).

Correction of known deficiencies

Nutrient deficiencies have been identified in many tropical plantations and very often phosphate is the chief limiting nutrient, particularly for pines, eucalypts, and gmelina, and satisfactory growth depends on applying this nutrient at planting. Lack of micronutrients, such as B deficiency in many parts of Africa, Asia, and South America, and Cu and Zn in South Africa, can also limit growth severely. To achieve satisfactory tree growth these deficient nutrients must be supplied. In Brazil the application of P fertilizer is obligatory in all eucalypt plantations, and on some sites the micronutrients B and Zn are applied mixed with NPK formulations (Barros and Novais 1996).

Establishment on poor land

Very infertile sites may require addition of several nutrients for trees to grow well. The land in Fig. 12.7 was classified as unplantable until the prescription of cultivation and fertilizing was introduced.

As Table 12.2 shows fertilizing alone, without cultivation, greatly increases growth. In Nigeria, young pines stagnate on phosphate-deficient savanna soils and fertilizing with P and sometimes N is used to improve substantially both tree growth and survival (Kadeba 1978). On poor soils dominated by *Imperata* grass in the Philippines and Indonesia fertilization at planting is essential. Boron and Zn, together with NPK, are also routinely applied to tropical acacias, eucalypts, and pines in the Philippines (Anon. 2001) and a similar requirement has been demonstrated for *Acacia mangium* in Indonesia (Otsamo 1996). On sodic soils application of gypsum can supply Ca for exchange with Na, reduce pH and enable other nutrients to become more available to the trees (Singh *et al.* 1989).

Stimulating growth

On many sites tree growth rate is only moderate and it is economical to add fertilizer to boost it. In South Africa, where fertilizer trials have been undertaken for 50 years, there are detailed recommendations for the application of fertilizers containing some or all of the elements N, P, K, Ca, and S, depending on soil type and method of site preparation (Herbert 1996). The wood production of *E. grandis* on a low fertility Oxisol in Brazil increased from 55 to 125 t ha^{-1} at 8 years old with the application of 100 kg K ha^{-1} at planting (Gonçalves *et al.* 1997). In Australia, fertilization of *E. grandis* with N and P increased stemwood mass from 6.9 to 41.4 t ha^{-1} at the age of 3 years (Cromer *et al.* 1993). In Colombia, *P. caribaea* has been found to respond to several nutrients, notably K, P, B, and Mg (Bolstad *et al.* 1988), and in Cuba the application of P fertilizer increased yield of *P. caribaea* by 56–69 m^3 ha^{-1} 13 years after planting (Herrero *et al.* 1988). In *P. taeda* in the United States, application of as little as 56 kg P ha^{-1}, or 56 kg P ha^{-1}, and 40 kg N ha^{-1} may result in optimum growth of newly established trees, and increases of 5–89% for at least 3–5 years can occur when 56–168 kg N ha^{-1} are applied to pole-sized stands on clay soils (Schultz 1999). New eucalypt plantations in the tropics often will not respond to N fertilization mainly because sufficient N is available from mineralization of organic material in the top soil, although intensively managed forests may need N application in subsequent rotations (Gonçalves *et al.* 2000).

Timing of fertilizer application

Fertilizers can be added at any one of four stages in the life of a stand: (i) at establishment, usually within 3 months of planting; (ii) during the post-establishment phase up to canopy closure when deficiencies begin to show; (iii) pole-stage fertilizing to boost thinning response and generally stimulate growth; and (iv) as a pre-felling application, 3–10 years before felling, to add increment before the end of the rotation. For intensively managed, short rotation plantations fertilizing is restricted to before and after canopy closure. Later stage fertilizing is not practised widely but in plantations producing sawlogs thinning creates an opportunity for crown expansion, which benefits from additional fertilizer (Miller 1984; Gonçalves *et al.* 1997). In Queensland, where the typical rotation age for tropical pine plantations is 30–35 years, late stage (about 15 years) re-fertilization with P has yielded a 20% volume improvement in *P. elliottii* (Shea 1987; Simpson and Grant 1991).

The supply of nutrients ideally is timed to coincide with expected growth rate and nutrient demand. Fertilizers may be applied in several doses between planting time and canopy closure to reduce loss from leaching and fixation. The major objective of fertilizer application at planting is to establish a vigorous root system and nutrient demand is relatively low. Where there is adequate water available and growth conditions are favourable the trees soon enter a stage of rapid growth and associated high nutrient demand so additional fertilizer is needed for above ground growth up to canopy closure. In Brazil most of the nutrient accumulated in *E. grandis* plantations is between 2 and 5 years depending on water availability (Reis *et al.* 1987). After canopy closure the leaf mass remains constant and litter breakdown and translocation of nutrients within the tree provide the nutrients necessary for new growth (Miller 1995).

Applying fertilizer

Fertilizer may be applied by hand, from a tractor with spreader, or from an aircraft. Hand application is slow but ensures proper placement of fertilizer near the tree and, theoretically, any area and kind of ground can be covered. Tractor spreading of fertilizer requires moderately flat and even ground and is only suitable for strip or broadcast application. Aerial application is only worthwhile for large areas and where full service facilities are available.

Fertilizer placement is important since it must neither be too close to cause root 'burn' to the young tree nor too far away to be unavailable during the period before it is leached, washed away, or is permanently immobilized in the soil. At time of planting placement of fertilizer dose in a slit or small hole 10–30 cm from a tree is being increasingly used instead of surface application. A combination of high solubility single superphosphate 20 cm deep in the planting hole before planting and low solubility rock phosphate in furrows 60 cm either side of the rows produced the best growth in *E. camaldulensis* at a site in Brazil (Fernandez *et al.* 2000). Placing fertilizer in the planting hole may not be appropriate if N is a major component as root 'burning' may occur, possibly resulting in the death of the seedling. It is best to avoid applying fertilizer during periods of heavy rainfall or very dry conditions.

Most fertilizers are marketed in granular form though some compound fertilizers containing micronutrients are applied as a liquid to be watered on or applied as a foliar spray. Liquid fertilizers may cause seedling mortalities due to root damage (for example, Herbert 1990*b*). The quantities of fertilizer needed depends on the soil, species, and silvicultural objectives and are usually based on results from fertilizer trials. Normally the amount of fertilizer is prescribed in kilograms for the particular element or its oxide. If P is required at 75 kg ha^{-1} (about 68 g tree^{-1} if planted at 3×3 m) then about 577 kg ha^{-1} of rock phosphate is needed, since the content of elemental P in rock phosphate is usually only 13% or 357 kg ha^{-1} of triple superphosphate with 21% elemental P. Typical amounts used in commercial, short-rotation plantations are: N 30–100, P 30–80, K 30–60, B 0.5–1.0, and Zn 1.0–1.5 kg ha^{-1} (Gonçalves *et al.* 1997).

Choice of fertilizer source is important. In South Africa, Herbert (1996) found urea formaldehyde chips were inferior to limestone ammonium nitrate granules when applied to *E. grandis* on sandy soils poor in organic matter due to poor conditions for decomposition but was better on organically rich soils. Also, growth was best with a rock phosphate: superphosphate mixture, and a powder rather than granular rock phosphate. It appears that this eucalypt prefers readily available nutrients. Soluble phosphate is best for high pH soils, mineral phosphate for more acid soils. Some primary fertilizer sources, for example, single superphosphate, may contain nutrients such as Ca, S, Mg, and micronutrients, which will benefit the plantation.

Finally, fertilizing should not be an isolated operation but considered as part of tending operations to aid rapid establishment. There is ample evidence that the presence of weeds usually reduces the size of response to fertilizer application, for example, a combination of herbicide application and fertilization with N and P resulted in higher stem volume than fertilizer alone in a 5-year-old *P. taeda* planting (Haywood *et al.* 1997). Also, in general, application of herbicide and fertilizer gives a superior response compared with application of fertilizer alone even if weeds were not obviously competing with the trees. But there are two other reasons for combining weed control with fertilizing. First, the fertilizer will stimulate weed growth as well as the tree. If this is not checked the benefit of fertilizing can be lost if the tree is swamped by vigorously growing weeds and suffers through competition for site resources, that is, water, nutrients, and light. Second, the stimulus to growth from combining fertilizing and weed suppression hastens canopy closure and usually reduces the number of times weeding is needed, thus the cost of initial weed control and fertilizer may be recouped by reduced weeding costs. This is particularly the case in eucalypt plantations and when acacias are planted on grasslands.

Economics and the importance of growth response

Fertilizing costs money, but it is expected that improvement in the site will be repaid by greater yields of wood. Where a substantial investment has been made in genetic improvement programmes, the introduction of intensive silvicultural techniques, including fertilization, is necessary to realize tree growth potential and improve economic returns. This has been the experience of profitable pulp and paper companies in Brazil (Campinhos 1980). A review of the profitability of fertilizing eucalypts found the real internal rate of return on fertilizing costs was in the range 15–41% per year (Schönau and Herbert 1989) and a 25% return on fertilizing *E. grandis* in South Africa was reported by Boden and Herbert (1986).

To decide whether fertilizing is economically worthwhile, the size of the growth response, in terms of higher yield or shorter rotation, is the most important factor, in addition to the cost, interest rates, etc. This information is obtained from fertilizer trials established on representative sites. Growth response has two components: the size and how long it lasts. In a fertilizer experiment with *E. globulus* in Australia, trees that did not receive NP fertilizer 18 months after planting grew poorly compared with those that did. At 9.5 years the most heavily fertilized plots contained three times more stemwood mass than the unfertilized plots. The fertilized plots closed canopy at age 6 years while this had still not occurred at 9.5 years in the unfertilized plots (Cromer and Williams 1982). Many soils available for plantation establishment in Australia are unable to supply the high rates of nutrient required during canopy expansion so fertilization early in the life of a plantation is recommended as a routine management practice.

It is not uncommon for a sizeable fertilizer response to occur in the years immediately after application but which diminishes with time and finally results in little or no improvement in total yield. For example, Schönau (1977) analysed results from six fertilizer experiments with *E. grandis* in southern Africa and found that initial improvement in height and diameter growth was usually highly significant but in later years growth was comparable with unfertilized stands. Nevertheless, the absolute improvement due to fertilizing, ranging from 17 to 96 t ha^{-1}, was maintained to the end of the rotation and more than covered the cost of fertilizing (Schönau and Pennefeather 1975). This example is a warning against over-optimism when spectacular growth improvements are seen immediately after fertilizing. Nevertheless, many responses to fertilizer are sustained or even continue to widen the difference between fertilized and unfertilized stands (Fig. 12.7), especially where a nutrient deficiency has been corrected during the rotation.

Prediction of the size of fertilizer responses requires more research so no firm rules can be laid down about the long-term improvement in growth of a stand due to fertilizer application. These must be determined from preliminary field trials.

Trends in use of fertilizer

In some countries fertilizing at planting is now standard practice. In Queensland, Australia, the success of more than 130 000 ha of tropical pine plantations is due to a large extent on the use of fertilizers (Simpson 1998). For eucalypts in South Africa fertilizer application is an integral part of silviculture (Schönau 1985); similarly for Brazil (de Freitas 1996). In many parts of Asia P-fertilizer application at planting is standard practice for *Gmelina arborea* and *Acacia mangium* plantations. Overall in tropical countries fertilizer use is rising, though relatively little is used in Africa despite the need (Peter 1989) and fertilizer is not widely applied to tropical pine plantations in Southeast Asia (Simpson 1998).

Mineral fertilizer is expensive; in 2001 one of the cheapest fertilizers, ground rock phosphate, cost about US$130–150 t^{-1}. Also, for most tropical countries, such fertilizers have to be imported. This cost in foreign exchange, combined with uncertainty in response, explains in part the low consumption in some countries. These constraints may continue to limit widespread fertilizer use in plantation forestry in many developing countries where it is likely to be applied only at planting to correct known deficiencies. Even in a developed country such as the United States, there may be inadequate

application of fertilization and other silvicultural practices in the subtropical pine forests due to (i) lack of awareness of opportunities to increase productivity, (ii) inadequate capital for silvicultural investments due to uncertainty of the long-term supply and value of wood, and (iii) infrastructure barriers that slow acceptance and implementation of new silvicultural techniques (Allen *et al*. 1990).

Maintaining soil fertility without using mineral fertilizer

Because fertilizers are costly, alternative methods of maintaining soil fertility are important. This problem is not peculiar to forestry since proper plant nutrition is essential in agriculture, but relying on fertilizers to achieve this is simply not feasible in many poorer developing countries. There are several alternatives available, which can very largely substitute for mineral fertilizer input, but two generalizations need to be made.

1. Non-fertilizer methods of maintaining soil fertility are usually cheaper in capital cost but more time-consuming and labour intensive, and involve careful soil husbandry. Soil fertility is usually only maintained or gradually improved and will rarely lead to large, dramatic growth responses.

2. In the tropics, soil fertility can often be enhanced by using the mutual benefits of combining trees and farm crops in a multistructured ecosystem (agroforestry) and appropriate technologies are being developed (Nair 1989; Young 1997).

Alternatives to fertilizer, and nutrient conservation practices

Change of tree species
If a species will only grow when fertilizer is added one option is to plant a less demanding species. However, on many extremely infertile sites there are often few worthwhile alternatives and if fertilizers are unavailable tree-planting must be abandoned. The need for fertilizers and the performance of species with and without them should be part of the feasibility study carried out before starting a plantation project.

Site preparation practices
On sites where forest is cleared before planting much can be done to minimize the damage to the soil and loss of stored nutrients. Soil compaction and erosion should be avoided during clearing operations. Ground cover and accumulated litter and other organic matter should be maintained as

long as possible to protect the soil surface from erosion and leaching of minerals. Clear-felling and burning the debris worsen these risks. There is increasing interest in and application of minimum soil cultivation techniques involving retention of logging slash and other organic matter on the site, no burning or intensive soil preparation, and hand planting. Good results have been achieved in *E. grandis* plantations in Brazil with this approach (Gonçalves *et al.* 1997) and trials are in progress in several countries (Nambiar *et al.* 2000).

Nutrient cycling

Nutrients in tree crowns, especially in young stands, account for a much higher proportion of total nutrients above ground in the stand than do those of other components (Table 14.1). However, there is variation between species in the proportion of biomass and nutrients accumulated by their crowns. Clear-felling is a major disruption to nutrient cycling. Practices that ensure the tree crowns are left behind after logging (and conservation of organic matter generally) aid the return of the stored nutrients and several aspects of management of plantation logging residues for nutrient conservation have been reviewed by Fölster and Khanna (1997). Debarking in the field and removing only stemwood minimizes nutrient depletion as, for example, short rotation eucalypts have 25% of their aboveground K and 30% of Ca in their bark (Gonçalves *et al.* 1997). Whole-tree logging (the removal of all parts of a tree including bark, branches, and foliage) removes two or three times more plant nutrients than ordinary logging while only marginally increasing wood yields. In agroforestry, appropriate systems can maintain and sometimes restore N fertility through biological nitrogen fixation, deep nitrate capture, and other cycling mechanisms (Sanchez and Palm 1996).

Adding organic waste and ash

There are several sources of organic waste and ash being applied to plantations. This approach has the dual benefit of disposing of waste material and increasing plantation productivity. Urban and industrial sources of organic waste, including pulp mill waste, and ash produced by plants burning biomass for their energy requirements are being used. The dramatic response often observed in Brazil to compost and ash application has stimulated interest in conservation and recycling approaches in plantations (Gonçalves *et al.* 1997). The addition of 20 t ha^{-1} of biomass ash doubled the volume growth of *E. grandis* at 79 months compared with the untreated control and improved the uptake of nutrients, especially K and Ca (Moro and Gonçalves 1995).

Animal dung

In many countries animal dung is burnt to provide heat for cooking and warmth (Fig. 2.3). But, it has been estimated that for every tonne of dung not returned to the soil to improve fertility there is a loss of some 50 kg of food grain. As pointed out in Chapter 2, if trees are planted for fodder and firewood, then a twofold attack is made on this nutrient drain: animal food is provided and dung can be used to enrich the soil. More nutrients are retained within the ecosystem and the rural economy improves.

Plants as soil improvers

Several biological methods of enhancing nutrient uptake may reduce the amount of inorganic fertilizer required for optimum growth.

Nitrogen fixation

Many trees and shrubs have a symbiotic relationship with *Rhizobium* bacteria or the actinomycete, *Frankia*, enabling them to fix atmospheric nitrogen. Most nitrogen-fixing trees and shrubs are in the family Leguminosae and the principal genera containing trees often planted in the tropics are: *Acacia, Albizia, Alnus, Cajanus, Casuarina, Dalbergia, Erythrina, Faidherbia, Gliricidia, Inga, Leucaena, Mimosa, Paraserianthes, Parkia, Pithecellobium, Prosopis, Pterocarpus*, and *Sesbania* (Dommergues *et al.* 1999). The fixed-nitrogen typically supplies 40–80% of the nitrogen used by these tropical trees, with input rates of 50–150 kg ha^{-1} year^{-1}. Environmental factors that favour tree growth also tend to facilitate nitrogen-fixation including adequate moisture, high temperatures, availability of soil nutrients, such as P and Ca, and favourable pH. Plantations of nitrogen-fixing trees generally have higher rates of nutrient cycling for all elements than non-nitrogen-fixing stands but little is known of the long-term influence of nitrogen-fixing trees on soil properties (Binkley and Giardina 1997). If nitrogen-fixing trees are harvested their effect of soil nitrogen ranges from little change to moderate increases. Cases of decreased soil pH, available P and other soil nutrients have been recorded. Benefits of using nitrogen-fixing trees to improve soil fertility and tree productivity will depend on the balance between soil improvement from increased nitrogen, possible depletion of other soil nutrients and their demands, and competitiveness for water, light, and other site resources.

Growing nitrogen-fixing trees and shrubs is practised as one way of improving the growth of other species mixed with them. There is much anecdotal evidence about improved growth of mixtures due to a nitrogen-fixing component but little clear scientific evidence. Some of the best evidence comes from a series of research trials in Hawaii using *E. saligna* and the nitrogen-fixing albizia (*Paraserianthes falcataria*) as pure plots and in mixtures. Eucalypt growth increased as the proportion of albizia in the stand increased from 11 to 66 %. Stem volume per hectare at age 10 years was at least equal to that of fertilized, pure eucalypt. There was an increase in both total and available soil nitrogen under albizia but a significant decrease in soil P despite increased cycling of nitrogen and P in the litterfall of the albizia (DeBell *et al.* 1997). In Puerto Rico, an intensive study of productivity, nitrogen-fixation, and soil nutrient effects on *Casuarina equisetifolia*, *Leucaena leucocephala*, and *E. robusta* as pure stands and in mixtures has been made (Parrotta *et al.* 1994). The eucalypt grew better in mixed stands with the nitrogen-fixing casuarina and leucaena. The results for soil nitrogen availability were more complex and had to be explained in terms of release from the organic sources and immobilization in the microbial biomass (Binkley and Giardina 1997).

There is clearly a role for nitrogen-fixing trees in tropical plantation forestry both as monocultures and possibly in mixed plantings. Tree legumes and casuarinas also feature in many agroforestry systems. MacDicken (1994) suggests nitrogen-fixing species for soil improvement should:

- have the potential to fix more than 20 kg ha^{-1} year^{-1},
- minimize competition with companion crops,
- be easily established,
- have economic value for other uses, and
- foliage that decomposes readily.

Mycorrhizas
Inoculating seedlings in the nursery with selected ectomycorrhizas or VA mycorrhizas can improve initial seedling establishment and growth in the field probably through improved nutrition in this critical stage. Later mycorrhizas may buffer the tree against environmental stress by increasing uptake of nutrients when available and releasing them from storage when scarce (Vogt *et al.* 1997). In addition, mycorrhizal hyphae or rhizomorphs significantly increase the surface area of tree root accessible to immobile elements such as P, Cu, and Zn and so may be particularly important improving the nutrient uptake in acidic tropical soils. Mycorrhizas may enable trees to recycle more P and other elements through the soil than would otherwise have been possible.

Foliage value
Many nitrogen-fixing species have nutrient-rich leaves which readily break down and augment soil

Table 14.5 Nutrient elements in dried *Leucaena* foliage including fine stems

Element	Percentage of dry weight
Nitrogen	2.2–4.3
Phosphorus	0.2–0.4
Potassium	1.3–4.0
Calcium	0.8–2.0
Magnesium	0.4–1.0

fertility (Table 14.5). This role of trees is potentially so valuable that it has been called 'the poor man's fertilizer'. According to NAS (1977) six bags of dried *Leucaena* foliage contain as much N as a bag of ammonium sulphate fertilizer. The use of organic mulches around trees to suppress weeds, improve soil moisture conditions, and augment organic matter as an aid to soil fertility it is used widely in agroforestry. For example, in West Africa *Faidherbia albida* branches are often laid around cassava and other food crops and the foliage of *L. leucocephala*, grown between rows of food crops as a shrub, is a common form of alley cropping yielding regular supplies of nutrient-rich leaves as a mulch (Fig. 20.9). Kang *et al.* (1998) found the prunings from hedgerows of *L. leucocephala* and *Gliricidia sepium* increased soil organic carbon and exchangeable cations, and increased yields of intercropped maize. Hussain and Ibrahim (1987) showed steadily increasing wheat yields and N content of plants with increasing additions of *Sesbania* leaves (2.8% N) to the soil.

Trees as 'nutrient pumps' and 'safety nets'
Deep rooting species, such as *Grevillea robusta*, have long had a reputation for bringing up nutrients from the lower layers of the soil and distributing them in leaf litter of the soil surface. Tree roots can go to great depth, for example, *Prosopis juliflora* and several *Eucalyptus* spp. have been recorded at over 30 m (Stone and Kalisz 1991), and may have great lateral spread, for example, *Acacia tortilis* roots commonly grow out 40 or 50 m from the tree. These extensive root systems are primarily to aid moisture uptake but they also potentially enable an enormous soil volume to be exploited for nutrients. A number of conditions for trees to act as nutrient pumps have been specified by van Noordwijk *et al.* (1996):

1. The tree should have a considerable quantity of fine roots and / or mycorrhizas in deep soil layers,
2. The deep soil layers should contain a store of nutrients in directly available form or as weatherable minerals;

3. Soil water content at depth should be sufficient to allow diffusive transport to the roots.

Moreover, root channels in the soil accelerate weathering processes releasing nutrients from soil minerals. Thus when foliage of trees is fed to animals or prunings and leaf litter decompose on the soil surface, nutrients that would otherwise be inaccessible nutrients become available for plant growth. Hence, nutrients found in low concentrations throughout the soil profile become concentrated near the surface and it is in this sense that trees may be called 'nutrient pumps'.

Where trees or shrubs develop an extensive root system below the root zone of the inter-planted crop in agroforestry systems, the roots may act as a 'safety net' intercepting nutrients leaching from the crop roots zone. They can then be returned to the soil surface through litter fall or prunings. In contrast to the 'nutrient pump' hypothesis the 'safety net' depends on the root distribution patterns of the tree and crop plant and is not restricted to specific soil types (van Noordwijk *et al.* 1996).

Nutrient toxicity

Toxicity is the opposite of deficiency; it is where a mineral(s) is at such high concentrations in the soil that tree growth is impaired. Almost all elements will induce deleterious consequences when present in plants beyond critical levels. Some plants are protected by ectomycorrhizal associations on their roots, which either exclude metals or accumulate them within their mantle sheaths (Vogt *et al.* 1997). The high acidity of many tropical soils make them prone to Al and Mn toxicity as well as low P availability, and mycorrhizas may be particularly important in this situation. Toxicity problems are uncommon in plantation forestry, but do occur in the following circumstances:

1. Naturally high levels of elements. Some soils develop with high levels of Mg, Al, and heavy metals such as Zn, Ni (nickel) and Cu, which produce toxicity symptoms. At Bukidnon Forest Industries, Mindanao, Philippines Ni and Cr (chromium) from ultramafic rocks cause toxicity in eucalypts and the more-tolerant pines are planted on these sites (Anon. 2001).

Figure 14.4 Patchy growth of *Pinus caribaea* in a nursery bed owing to zinc toxicity.

2. Mine waste and other sources of heavy metals. Toxicity occurs on man-made sites where industrial waste contains high levels of heavy metals, for example, copper tailings. In a species trial on copper tailings in the Philippines, *Acacia auriculiformis* grew satisfactorily but most other species were severely affected by toxicity and/or the highly acidic conditions. In Queensland, Australia some of the raised beds proved useless because toxic levels of zinc had built up owing to galvanized tubes having stood on the ground for many years (Fig. 14.4).

3. Addition of incorrect fertilizer. Application of KCl 100 g tree^{-1} of *Pinus caribaea* depressed growth and increased mortality on Nigerian savanna sites (Kadeba 1978).

4. Application of too high a dose. Applying excessive quantities of fertilizer is an uneconomic practice as well as producing toxicity symptoms. Young trees are severely damaged if more than 200 g of borate per tree is applied to correct B deficiency. Excessive doses of N and P fertilizers may alter mycorrhizal relationships with the outcome being decreased growth due to metal toxicities or increased susceptibility to root pathogens (Vogt *et al.* 1997). Simple rate trials can quickly establish a safe dose level of fertilizer.

Dynamics of stand growth

In the tropics, tree growth and stand development are usually more rapid than under temperate conditions. Therefore, it is important that the influences on stand development available to the forester to modify the quality and quantity of timber produced are understood. Similarly, understanding individual tree-growth patterns can help management of farm and village trees to maximize yield of branchwood or fodder or minimize their competitive affect on other crops and so on. After the establishment period, and provided growth is not impaired by deficiency, disease or other damage, the amount of growing space a tree has dominates its growth and development and the yield of a stand as a whole. Initial spacing between trees and the timing and intensity of thinning determine growing space and, after species choice, are probably the most important influences on tree and stand development under a manager's control.

Periodicity and pattern in growth of individual trees

Tree growth is not uniform. It is periodic over short time-spans and follows a definite pattern in the long term. This is well seen in pines which, throughout the tropics, have a multinodal habit producing several branch whorls each year. Chudnoff and Geary (1973) found that in one-year-old *Pinus caribaea* a new growth flush occurred about every 9 weeks and that within the 9 week period height growth was not constant. This kind of periodicity in growth is largely under genetic control though it may be modified by environmental factors such as soil moisture, for example, as in *Cordia alliodora* (Blake *et al.* 1976). Srivastava and Elias (1982) found monthly temperature was related to *P. caribaea* shoot growth, which varied between 7 and 14 cm per month under Malaysian conditions.

Where there is a definite growing season there is periodicity in growth during a year. If growth slows markedly the wood will show annual rings and the stem terminal nodes. However, not all species experiencing the same seasons respond in the same way. *P. caribaea* var. *hondurensis* in

southern Queensland, Australia grows continuously throughout the year though not at a constant rate, while *P. elliottii* grows faster in early summer but becomes dormant in the winter. Throughout the tropics as a whole most active growth of trees occurs during the wettest periods and in monsoonal climates diameter increment is usually closely synchronized with the wet and dry periods, for example, in *Khaya grandifolia* (Ola-Adams and Charter 1980). However, as noted elsewhere, there are exceptions of which *Faidherbia albida* is the best known because it flushes new foliage at the onset of the dry season.

As well as periodicity in one year, trees exhibit a strong pattern of growth during their life. The pattern in relation to age is typically sigmoid (Fig. 15.1), although early growth is frequently very rapid with only a few months elapsing before height and diameter increments are large: Alder and Montenegro (1999) show this well for *Cordia alliodora* in Ecuador, which achieved 16–24 m height in 5 years and then slowed substantially.

Stand growth is dynamic

The dynamic nature of stand growth is a little less obvious than for a single tree since superficially a stand seems a collection of individual trees the

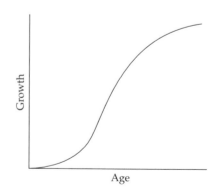

Figure 15.1 Typical sigmoid pattern of tree growth with age, whether measuring growth as height, stem diameter, or volume.

relations between which appear constant. If one considers a small tree in a mature plantation it is clear that this is not so. At planting, this tree will have been about the same size as its neighbours, but must subsequently have competed less successfully for nutrients, light, or moisture to become inferior to its neighbours. If a stand is left untouched from planting its initial uniformity of similar size seedlings with equal growing space progressively disappears (Fig. 15.2). Thus the stand is continually changing, it is dynamic. A stand is not only a collection of individual trees, each with its own genetic potential for exploiting a site, but also a collection of trees that interact and compete with one another. Even in a clonal plantations there is size variation due to site heterogeneity. The operation of thinning influences this interaction and competition and uses this dynamic nature of a stand to achieve certain objectives of timber outturn or volume yield.

Immediate post-planting growth

At planting, trees stop growing temporarily while they become established in the new site; this is called 'planting shock'. Then, growth for the first few years accelerates (Fig. 15.1), but to begin with the tree is an isolated individual competing only against weeds. However, even during this phase, growth of different trees is not identical (Fig. 15.3).

Onset of between-tree competition

This depends on species, growth rate, and initial spacing. Competition occurs when the presence of neighbouring trees begins to slow a tree's own development. Slowing in development may occur first from competition between root systems or only once branches touch and shade one another. Craib (1934, 1947), in his pioneering work on spacing and thinning, initially with *Acacia mearnsii* and later with *Pinus patula*, *P. elliottii*, and *P. taeda*, in general considered root competition to set in well before

crown competition. In arid areas crown contact is rare and it is root competition that causes the relatively uniform but scattered occurrence of plants, each depending on an extensive root system to obtain what little moisture is available. However, on sites where moisture and nutrients are abundant root competition is much less important. The main factor is onset of lateral shading and competition for light as the crowns of neighbouring trees begin to touch.

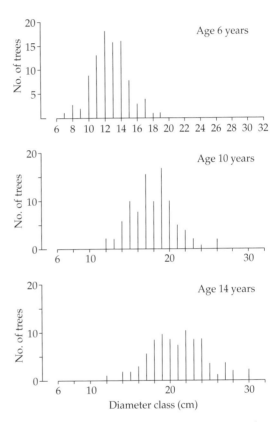

Figure 15.2 The diameter distributions of the same *Pinus patula* stand at 6, 10, and 14 years of age. Source: the author (JE).

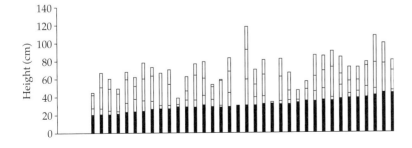

Figure 15.3 Height increments of 36 *Pinus caribaea* trees in one 6 × 6 plot measured 2, 6, and 9 months after planting. Trees arranged in ascending order of size at planting (shown in solid black).

Using slowing of growth as the indicator, Evans (1976) found between-tree competition occurred in *P. patula* stands, planted at 2.74 × 2.74 m spacing, at about 5–6 years of age when the trees were about 7 m tall (Fig. 15.4). For *P. patula*, competition appears to have an effect from about the time of canopy closure. This is generally the case and was found in spacing trials of *Eucalyptus deglupta* in Papua New Guinea (PNG) and *E. saligna* in Hawaii where, in the latter case, diameter increment began to decline at 3 years of age (trees 10 m tall) at 2.4 m spacing but not until 5 years of age (trees 20–22 m tall) at 4.3 m spacing (Walters 1980). Similarly, *Terminalia superba* in Nigeria, averaging 12 m height, showed significant diameter growth reductions at close compared with wide spacings (1.83–6.70 m), which must have begun many years before the time of assessment at 11 years (Okojie *et al.* 1988). Diameter increments of individual trees began to show the effects of between-tree competition at a time, which corresponds closely with the onset of canopy closure, the effects appearing last in the most widely spaced plots.

Differentiation into crown classes

As competition begins the growth of some trees is slowed more than others. These tend to be the smaller trees at the time of canopy closure and competition between trees then reinforces their inferior status. In fact, few trees recover a dominant position after being overtaken by a neighbour. Thus, as a stand develops, a range of tree sizes emerges and, traditionally, these are classified into different crown classes according to a tree's relative position in the canopy. This classification derives mostly

from work in temperate coniferous forest but it has been successfully applied and refined in some tropical countries, notably India.

Four main crown classes are recognized along with a category for dead and dying trees (Fig. 15.5). Table 15.1 formally describes them based on definitions in both the British Commonwealth Forestry Terminology and the Terminology of Forest Science, Technology, Practice and Products published by the Society of American Foresters (Ford-Robertson 1971; Winters 1977).

Not all types of tree in Table 15.1 will be found in all tropical plantations. For example, *E. deglupta* trees can pass from co-dominant status to suppressed in a few months. So sensitive is the species to crown competition that even slight overtopping can cause a tree to become fully suppressed in a short time and then die. Crowns rarely overlap (Fig. 15.8) and very few trees can ever properly be classified as sub-dominant. *E. deglupta*, like almost all eucalypts, is a strong light demander and rapid natural thinning results in stands mostly of dominants or tall co-dominants.

An example of the growth of trees through the three stages of stand development (early growth, onset of between-tree competition, and differentiation into crown classes) is shown in Fig. 15.6. At 3 years of age the trees were the same height but as competition set in their relative performance began to differ. The volume curves show how the dominant tree is continuing to grow well, the co-dominant is now beginning to fall behind, and the sub-dominant, though staying alive, never really grew successfully at any stage.

Physiological basis of stand development

Initial growth

The ability to colonize the planting site and compete against other vegetation determines a tree's

Figure 15.4 Cross-sectional discs cut from the stems of sub-dominant (left) and dominant (right) *Pinus patula* trees growing in an unthinned stand. Note marked reduction in ring width at 5–6 rings from the pith indicating onset of between-tree competition.

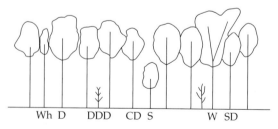

Figure 15.5 Diagrammatic representation of crown classes in a plantation. D, dominant; CD, co-dominant, SD, sub-dominant; S, suppressed; DDD, dead, dying, or diseased; Wh, whip; W, wolf tree.

Table 15.1 Classes of trees in a stand

	Class	Description
I	Dominant	The highest trees of the four canopy layers. Trees have most of their crowns free. Dominants are the tallest, and most vigorous trees but not necessarily the straightest
II	Co-dominant	Upper canopy trees, but below the crown level of the dominants
III	Sub-dominant (intermediate)	Not in the upper canopy, but their leaders still have free access to light
IV	Suppressed (overtopped)	Trees whose branches do not reach up into the branches of the upper canopy, or which have leading shoots under some part of the crown of another tree. They are destined for early death
V	Dead, dying, and diseased	Can occur in any canopy layer but are usually trees in the lower canopy classes
Other kinds of trees include:		
	Pre-dominant	Much taller than the rest of the stand and whose crown has grown above the general level of the upper canopy (also called 'emergent')
	Wolf trees	Often the largest trees in a stand, but ungainly, of bad form, and often with misshapen crown, which damages neighbouring tree
	Whip trees	Tall trees with narrow crowns and thin stems, which sway in the wind and damage neighbouring trees

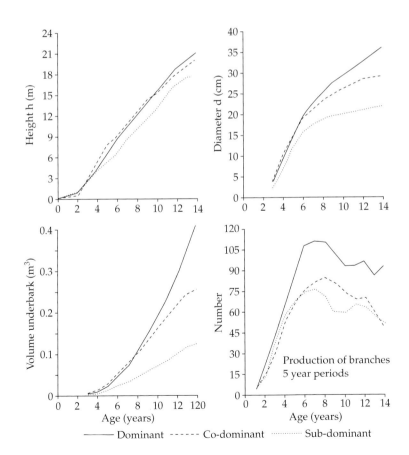

Figure 15.6 Development of three *Pinus patula* trees, growing near each other in a stand, which came to occupy dominant, co-dominant, and sub-dominant crown classes.

success in early growth. In particular, initiation and growth of new roots and resistance to desiccation are important for rapid establishment. Freezaillah and Sandrasegaran (1969) found a very close relationship between increment in root volume and increment in tree volume for *P. caribaea* in Malaysia. With the same species in Nigeria, Egunjobi (1975), found a weak correlation between diameter breast height and biomass of small roots. Moreover, in the early years after planting root biomass is at its highest proportion of the total. For example, with *Shorea robusta*, root biomass constitutes 34% of total plant biomass when trees are young, 16% at the age of 15 years, and 14% at 26 years (Raman 1976). For *P. caribaea* in Nigeria the proportion root biomass formed of the total (about 20%) was constant above 4 years of age (Kadeba and Aduayi 1982).

It is primarily genetically fixed characteristics that determine a tree's early growth relative to others rather than the interplay of chance factors such as minor differences in fertility, weed competition, soil structure, etc. that may occur (Assman 1970). However, there is less certainty about the physiological mechanisms, which enable one tree to grow more vigorously. One or more of the following may be critical: (i) root cells more resistant to desiccation? (ii) a more efficient mycorrizhal association? (iii) greater nodulation or of a superior strain in nitrogen-fixing species? (iv) a tendency to branchiness in rooting? (v) more root hairs per unit length of root? (vi) minute differences in cell osmotic relationships, and ion transport, which favour nutrient uptake? (vii) slightly better water-use or nutrient-use efficiency?

Development of the crown

Chlorophyll in green leaves is the seat of photosynthesis. Though trees of the same species may differ a little in the rate of photosynthesis per unit area of leaf or needle, the main source of variation in assimilation processes is differences between trees in the total photosynthetic area available to intercept light, that is, a tree's crown. Consequently, a relationship is generally found between crown size (depth and spread) and other measures of tree size, such as stem diameter, volume, etc. Dawkins (1963) reported many examples for tropical forest trees. Mayhew and Newton (1998) list several equations relating crown diameter with stem diameter for mahogany (*Swietenia macrophylla*) and Wadsworth (1997) shows that simply using the crown diameter : stem diameter ratio is a helpful way of describing a species' light requirements. He reports ranges from 16 for some eucalypts to 39 for an extreme light demander, the pioneer species *Maesopsis eminii*.

In terms of the growth of individual trees, and later competition between trees, it is crown expansion in the early years, and subsequently its restriction, which is critical. Matthews and Wareing (1971) point out that this rate of increase in total leaf area will depend on: (i) the rate of production of leaf primordia by apical meristems; (ii) the area attained by the individual leaves; and (iii) the number of apical meristems actively engaged in leaf production, as determined by the branching habit of the tree. Figure 15.7 shows the relationship between annual volume increment of a co-dominant *P. patula* in an unthinned stand and the number of branches produced in the preceding 5 years—used as a crude index of effective crown since canopy closure occurs at 4–5 years of age after which lower branches begin to be shaded and die. The close relationship between volume growth and crown development is evident.

A detailed investigation of several *P. patula* and *Cupressus lusitanica* trees in plantations in Tanzania, found high correlations between total foliage weight ($r = 0.95$), live branch weight ($r = 0.96$), and stem diameter (Lundgren 1978). Similarly, Schönau and Boden (1982) report high correlations ($r = 0.96$

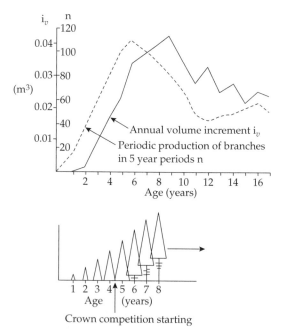

Figure 15.7 The relationship between crown development (simply defined as number of branches produced in preceding 5 years, *n*) and growth (annual volume increment, i_v) in a co-dominant *Pinus patula* tree. The lower diagram illustrates the assumed development of effective crown, bearing up to 5 years of live branches, since crown overlap and competition began at about 4–5 years of age.

to 0.99) between leaf biomass and stem diameter for several eucalypt species.

Competition

Mention has already been made of root competition for moisture. And, in the last chapter, it was noted that great variation in stand diameter distribution may indicate nutrient deficiency. Some trees appear to have roots that are slightly better in obtaining what nutrient is available at the expense of their close neighbours. Reasons why some trees have more efficient roots than others has already been noted, but discovering which is the significant competing factor in any one situation is very difficult.

More readily apparent is above-ground competition for light. A tree's lower branches are shaded by its own upper crown and by the crowns of neighbouring trees. The ability of foliage to withstand shade varies so much between species that two broad classes are recognized: light demanders and shade bearers. Most fast-growing plantation species in the tropics are strong light demanders. This is not surprising since nearly all are naturally pioneer species, which invade gaps in the forest or grow on open ground. However, this means that in plantation, competition for light is strong. Some species, for example, *Octomeles sumatrana* and *Anthocephalus chinensis*, cannot stand any shading and many eucalypts are similarly intolerant, hence their lack of crown class differentiation. This sensitivity to shade (low light levels) and abrasion of naked buds, which leads to very little branch contact for many eucalypts (Fig. 15.8), a characteristic known as 'crown shy' (Florence 1996), has two effects.

Figure 15.8 Very little crown contact (crown shy) in the canopy of a densely planted stand of *Eucalyptus deglupta.*

1. High light levels are found beneath most eucalypt stands and weeds are rarely fully suppressed (Fig. 1.1(a)).
2. Lower branches of most eucalypts quickly die and the genus in general shows good natural pruning, for example, Figs 1.1(b) and 18.8.

By contrast, death of branches and loss of crown due to shading is slower in less strongly light demanding species such as *Swietenia macrophylla*, teak, and *Araucaria* spp. Nevertheless, in time it does occur, and few species remain furnished with live branches down to ground level for many years after canopy closure. However, by altering the light regime a tree experiences, lower branches can be encouraged either to stay alive or to die earlier and thus crown size is influenced; this is the basis of thinning (Fig. 15.9(a) and (b)).

This division of tree species into two categories is, of course, an over-simplification. Some species are more tolerant of shade when young than when mature. And, some species when given full light develop a large open crown, for example, *Milicia (Chlorophora) excelsa* and require dense planting, and attendant side shade, to encourage upward growth and straight stems; thus the degree of shade is used as a silvicultural tool. Species with a strong apical dominance, for example, *Terminalia superba*, and indeed most pioneers, do not require such stand manipulation to achieve good stem-form.

Whether or not competition leads to death of lower branches and so reduces effective crown area, any shading of lower foliage means that less light energy is available for photosynthesis. Assman (1970) quotes as typical that at the point of crown contact in a coniferous stand some 60% of available energy has already been intercepted. Maximum energy consumption occurs in the upper canopy, therefore smaller trees, with smaller crowns and little of them even in the upper canopy, are deprived of energy and so grow more slowly. The poor productivity of the sub-dominant, and the virtual impossibility of it regaining dominance, is easily seen. Similarly, the good growth of dominants, with large crowns and enjoying full sunlight, reinforces their dominance. Interestingly, the total amount of light intercepted by the closed canopy of a tree plantation varies by about a factor of two to three between light-demanders (leaf area index LAI 2–4) and shade-bearers (LAI 4–8+)—it is darker under shade bearers! However, as Beadle (1997) points out in a review of leaf and canopy development in tropical plantations, maximum LAI varies with season and tree age, not only with a species' degree of shade tolerance.

If it is correct that most of the competition between trees in a stand is for light then, assuming

(a)  (b)

Figure 15.9 Two *Pinus patula* stands with very different past treatments. (a) 20-year-old, unthinned, trees with small tight crowns on the top 20% of the stem. (b) 24-year-old, thinned three times, trees still with large crowns and furnished with live branches on the top 40–50% of the stem.

sites of similar fertility, the very tallest trees in an unthinned stand ought not to grow very differently from the tallest ones in heavily thinned stands since in each case the crowns are well exposed. This is exactly what was found in a thinning trial of *Araucaria cunninghamii* in PNG (Fig. 16.3). When the most heavily thinned plots (73 stems ha^{-1} at 25.5 years) were compared with unthinned plots (1006 stems ha^{-1}) the mean diameter (dg) differed by almost 20 cm, 46.4 cm compared with 29.9 cm (Fig. 16.7), but the mean diameter of the 10 largest trees differed by only 8 cm, and the biggest trees in the unthinned plots were almost as big as those in the very heavily thinned plots. Keenan *et al.* (1999) found the same effect in thinning studies of Queensland maple (*Flindersia brayleyana*). The most heavily thinned stands led to a mean DBH almost 10 cm greater than that of unthinned stands by age 22 years, but the average DBH of the 100 largest stems ha^{-1} differed by only 3.4 cm.

Silvicultural implications

Establishment practices of cultivation, weeding, and fertilizing uniformly improve the quantity of growth of all trees. Manipulation of initial spacing and the mostly selective practices of thinning and pruning are used to develop a certain kind of stand and a certain quality of tree growth.

In the development of a stand there are two main alternatives. Either the highest production is sought from the site by maximizing light interception of the whole stand canopy for as much of the rotation as possible, or crown development of selected trees is encouraged to optimize their individual growth. Similarly, the farmer manipulates the balance between tree growth and food crop growth in agroforestry systems not only by the spacing and location of trees, but by the timing and intensity of pruning and trimming of the crowns to control light levels. Indeed, the balance of light distribution between upper and lower storey components will greatly influence productivity potential, and how this is affected by the trees is important information (Cannell 1983; Jackson 1983).

Initial spacing

It was noted in Chapter 12 that wider spacings led to some loss in total volume production per hectare but that individual trees grew larger. It is now apparent why this is so. A stand of trees planted far apart will have a lower photosynthetic surface area per hectare to intercept light in the early years and consequently lower yield—at least to begin with. But a wide spacing enables individual trees to develop and maintain large crowns and for their root systems to occupy a large volume of soil before

(a)

(b)

Figure 15.10 Effect of initial spacing on growth of *Azadirachta indica* after 4 years in adjacent trial plots, Ratchaburi, Thailand. Trees planted (a) 0.5 × 0.5 m apart, and (b) 4 × 4 m apart.

competition starts, both of which enhance growth (Fig. 15.10). Data from two spacing trials illustrate these affects.

- *Eucalyptus deglupta* in PNG (Table 15.2).
- *Pinus caribaea* var. *hondurensis* in Queensland, Australia (Table 15.3).

Another example of how important is the spacing effect is that this parameter alone explained 65% of the variation in volume per hectare at 11 years of age in a *Terminalia superba* spacing trial, which ranged from 1.83 × 1.83 m to 6.70 × 6.70 m (Okojie *et al.* 1988).

The question whether spacing influences height growth is more open. The assumption that it does not is the basis of using 'top height'—the height of the 100 largest stems per hectare—as the best measure to assess site quality or yield class, but there is evidence that it is not true of all species. Keenan *et al.* (1999) indicate that high initial

stocking of *Flindersia brayleyana* results in more rapid height growth than when planted at lower densities and Habiyambere and Musabimana (1992) report the same effect for *Grevillea robusta* in Rwanda.

Timing and method of thinning

Thinning provides the remaining trees with more growing space. However, if it is delayed two serious effects result. First, between-tree competition will have reduced the effective crown (Fig.15.9(a)), and second, the tree will be at a later stage in its natural growth pattern (Fig. 15.1). Both effects mean that where thinning has been delayed trees will take longer to respond to extra growing space when it is finally done. A 20-year-old tree with only a tuft of live branches at the top will inevitably respond more slowly than a 10-year-old tree, still in the

Table 15.2 Effect of initial spacing progressively up to 5 years of age in stands of *E. deglupta*

Initial spacing (m)	Age in years							
	2		3		4		5	
	dg	G	*dg*	G	*dg*	G	*dg*	G
1.83	5.58	3.56	8.63	8.16	12.00	14.35	14.55	17.27
2.44	4.35	1.86	7.67	5.66	11.32	11.24	14.99	16.33
3.05	5.40	1.45	8.92	3.93	13.56	8.92	16.83	12.64
3.66	5.56	1.30	9.41	3.57	14.77	8.67	18.45	12.27

dg = diameter of tree of mean basal area (cm); G = basal area per hectare (m^2 ha^{-1}).

Table 15.3 Effect of initial spacing on yield of *Pinus caribaea* at 9.3 years of age

Initial spacing (m)	Survival (%)	Height *h dom* (m)	Diameter *dg* (cm)	Volume		
				V_{10} (m^3 ha^{-1})	V	V_{10}/V (%)
2.13 × 2.13	88	18.7	16.6	146.14	205.39	71
2.44 × 2.44	86	18.6	17.8	137.12	181.51	76
2.74 × 2.74	89	18.6	18.6	126.85	159.84	79
3.05 × 3.05	90	18.7	20.1	130.33	155.28	84
3.66 × 3.66	92	18.1	20.9	102.60	119.25	86

h dom = top height, mean height of the 100 largest diameter trees per hectare; *dg* = diameter of tree of mean basal area; V = total volume per hectare; V_{10} = merchantable volume (Queensland) per hectare—volume to 10 cm top diameter.

rapid growth stage, and with a large live crown. Delayed thinning, because of no market for the produce, is a common problem in the tropics but the dynamics of stand growth do not permit indefinite delay if flexibility in stand management is to be retained.

The distinction between low and crown thinning methods (Chapter 16) is essentially manipulation of different components of a stand's photosynthetic surface area: the canopy as a whole or of individual trees. But, there can be other consequences. Mayhew and Newton (1998) report much increased damage in mahogany stands where heavy crown thinning opens up a stand.

Selective pruning

When live branches are pruned on selected trees to improve wood quality these trees must be favoured in thinning. Such pruning causes the loss of photosynthetic surface area and will temporarily reduce the growth of the tree compared with its neighbours, hence the need to favour it in thinning.

CHAPTER 16

Thinning

Definition and objectives

Thinning is the operation that artificially reduces the number of trees growing in a stand. It is normally carried out several times and starts a few years after canopy closure. For example, in Kenya *Pinus patula* plantations grown for sawlogs are thinned three times during a 30-year rotation and the number of trees per hectare reduced from 1110 to 250 (Anon. 1996). In Nilambur, Kerala state, India, teak plantations are thinned six times during a 50–55-year rotation, which reduces the number of trees from 2500 ha^{-1} to 100–150 (Chundamannii 1998). A third example from Costa Rica (Camino and Alfaro 1998) shows teak plantations established at 1111 ha^{-1}, which by rotation age of just 25 years have been reduced to 220 ha^{-1}.

The definition of thinning given in Terminology of Forest Science, Technology, Practice, and Products (Winters 1977) is 'A felling made in a stand at any time between establishment and the initiation of a regeneration cutting or clear-felling in which the trees removed are the same species as the trees favoured'. Thinning is done for many reasons of which the chief ones are: (i) to reduce the number of trees in a stand so that the remaining ones have more space for crown and root development to encourage stem diameter increment and so reach a usable size sooner; (ii) for stand hygiene both to remove dead, dying, diseased, and any other trees, which may be a source of infection for, or cause damage to, the remaining healthy ones and to reduce between-tree competition to avoid stress levels, which may encourage pest and disease attack; (iii) to remove trees of poor form crooked, forked, basal sweep, roughly branched, etc. so that all future increment is concentrated only on the best trees; (iv) to favour the most vigorous trees with good form which are likely to make up the final crop; (v) to provide an intermediate financial return from sale of hinnings.

Thinning may also be carried out for a number of minor or only locally important reasons including maintaining light levels beneath a stand to provide vegetation (grass sward) for grazing, encouraging native ground flora, providing poles and posts, promoting flowering and fruiting in seed stands,

reducing the fire hazard next to places of public access, and improving the recreational and amenity value of the plantation.

Thinning is a purposeful intervention in the life of a stand to alter its development, and that of the trees that constitute it, to achieve certain objectives.

Experience with thinning in the tropics

Compared with information about afforestation techniques, less is known about the effects of thinning and what are the most suitable regimes. This is largely because extensive plantation development in the tropics is more recent and there has been insufficient time to gain adequate experience and to carry out long-term experiments over a whole rotation. There are some detailed studies of thinning and the effects of varying initial spacing, for example, *Pinus caribaea* in Surinam (de Vries *et al.* 1978), Puerto Rico (Liegel 1983), and Brazil (Weaver 1986); *P. patula* in East Africa, especially southern Tanzania (Borota and Proctor 1967; Adlard 1980); mahogany (Mayhew and Newton 1998); teak in Nigeria (Abayomi *et al.* 1985) and the neotropics (Keogh 1980*b*) as well as in India, and some of eucalypts (Florence, 1996) and specifically *E. grandis* (Schönau 1984*b*), *E. camaldulensis* in Pakistan (Hussain and Cheema 1987), and *E. deglupta* in Costa Rica (Ugalde 1980). Generally, a shortage of information remains for many species now widely planted in the lowland tropics: *Eucalyptus urophylla*, *Gmelina arborea*, *Pinus kesiya*, *P. merkusii*, and *P. oocarpa*, and the important moist tropical acacias *A. auriculiformis* and *A. mangium*. This lack is even more acute for high quality tropical hardwoods, apart from teak, as Weinland (1998) illustrates: 'With the exception of *Shorea robusta* no thinning regimes have been developed for dipterocarp plantations'.

However, there are two important exceptions where information is not lacking. Some 70 years ago in South Africa, Craib (1934, 1939, 1947) worked on thinning regimes for the subtropical pines, *Pinus elliottii*, *P. patula*, and *P. taeda*. Indeed, Craib can be described as the father of most modern thinning practice. Current recommendations build on Craib's work (Owen 2000). At about the same time considerable research took place in many teak plantations on the Indian subcontinent. Since the 1950s many African countries, notably Kenya, Madagascar, Nigeria, and Tanzania, have used or adapted South African and Indian thinning practices to develop local regimes.

The second case is spacing and thinning research in Queensland, Australia, with *Araucaria cunninghamii*, *Pinus elliottii*, and *P. caribaea*, which has focused on thinning response and control using stand basal area parameters, for example, Anderson *et al.* (1981).

Now, in most countries, standard thinning schedules are in operation for important species.

Effects of thinning

Thinning is really no more than an extension of the kinds of changes produced by differences in initial spacing. In fact, for a particular species and site, the current growth of a thinned stand is identical to that of an unthinned one if both have the same stocking and standing volume per hectare, though the unthinned stand will usually be younger (Marsh and Burgers 1973). How a stand reaches a particular stocking and volume, because of initial spacing or due to thinning, does not significantly influence its present rate of growth. Both distance between trees at planting and thinning affect the same environmental factor, that is, growing space per tree.

Physiological effects

Thinning a stand reduces the number of trees competing for light, soil moisture, and nutrients (Chapter 15). For example, after thinning, more light reaches beneath the canopy and there is usually a resurgence of weed growth, increased breakdown of litter, and sometimes epicormic shoots on tree stems. Also the watertable may rise and the ground become wetter, since there is temporarily both less demand for soil moisture and less interception of rain by the canopy.

Lessening of competition between trees has three main effects.

1. Lower natural mortality. After a thinning, trees near full suppression and death, if still left, are able to continue growing mainly because of greater access to light. Natural mortality of trees in thinned stands is uncommon; trees survive longer and ones becoming suppressed are usually thinned out anyway. In the hoop pine (*Araucaria cunninghamii*) thinning trial, examined in detail later, by 25.5 years of age and top height of 36.0 m, 28% of the original trees in the unthinned plots had died.

2. Deeper crowns on remaining trees. The shaded lower branches of a tree receive more light and remain alive longer (see Fig. 15.9(b)), therefore trees in thinned stands have deeper crowns. In a study in the Usutu forest, Swaziland, unthinned 19-year-old *Pinus patula*, planted at 2.74 m spacing, had a live crown only on the top 29% of the stem whereas in an adjacent stand, where every other row had been thinned out at 9 years of age, trees retained a live crown on the top 40% of the stem. This proportion of the length of the stem furnished with live branches is referred to as the live crown ratio.

3. Crown expansion. The increased growing space surrounding a tree after thinning induces active growth of shoots, foliage, and roots.

The effects of 2 and 3 above result in a greater photosynthetic area on each remaining tree, thus increasing their growth. If, however, thinning is heavy and large gaps occur in the canopy, which only slowly become occupied, the interception of light energy by the stand is less and some loss in total production could be expected; see below.

Mensurational effects

Individual tree growth
The main effect of thinning is greater diameter growth on the remaining trees. Much research has established that height growth of the trees themselves is usually little affected, for example, Karani (1978), though there is some evidence that very heavy thinning may lead to a small diminution in height growth compared with a densely stocked, unthinned stand. However, both Bredenkamp (1984) and Walters (1980) report best height growth of eucalypts at wide spacings and with heavy thinnings; the severe competition of dense stocking suppressing height as well as diameter growth. In contrast, Keenan *et al.* (1999) reported greater height growth when Queensland maple (*Flindersia brayleana*) is grown at high stockings.

This effect of thinning on diameter growth is shown in Fig. 16.1.

The kind of response to thinning in Fig. 16.1 is well-established for all species (Figs 16.2 and 16.3 (a) and (b)): thinning, like wider spacing, produces larger individual trees since they have larger crowns producing more 'wood'. Indeed, there is generally a close relationship between crown diameter and stem diameter as already noted.

A good, if quite old, illustration of the effect of thinning in a broadleaved species (teak) was reported by Iyppu and Chandrasekharan (1961) who compared normal thinning with very heavy thinning based on Craib's recommendations. At age 26 years the mean diameter breast height of trees in normally thinned plots averaged 29.5 cm and in heavily thinned plots it was 39.9 cm.

Since the response to thinning is mainly in diameter growth, with height usually little affected, thinning changes tree shape; overall, the trunk tapers more rapidly. Van Laar (1976) showed that at 14 years for *Pinus* species in Mpumalanga (South Africa) high stand densities (3000 trees ha^{-1}) produced trees with form factors of about 0.565 and at low stand densities (125 trees ha^{-1}) form factors were about 0.495. Other responses to thinning include increased bark thickness and delayed natural pruning, since the extra light allows lower branches to survive for longer. Thus branches grow thicker and knots in the wood are larger.

All these effects of thinning only continue while the tree is expanding into the newly available growing space and before between-tree competition again becomes intense. Therefore, if rapid growth of individual trees is sought, thinning is repeated at intervals during the life of a stand.

Stand growth

Removal of trees in thinning immediately reduces some of the photosynthetic surface area in a stand. An immediate drop in production per unit area occurs followed by resurgence in growth as the remaining trees respond to the extra space by increased production of new foliage and roots. It is difficult to demonstrate this recovery of growth because many factors, notably climate, cause year to year variation in stand performance. Nevertheless, Fig. 16.4 is reproduced to show the annual volume increment of a *Pinus patula* stand for 6 years following a heavy thinning carried out at 11.9 years of age. The measured annual increments follow fairly closely the theoretical response curve.

Figure 16.4 shows an acceleration in growth, occurring in about the third and fourth year. This improvement is not sustained (except in a stand which was previously grossly overstocked and many trees were dying from acute competition) because thinning itself does not raise site growth potential; see Fig. 16.6.

So far the effect of one thinning has been examined, but, as noted, usually several thinnings are carried out. Figure 16.5 shows the actual development of stand volume, in plots of *Araucaria cunninghamii*

Figure 16.1 Cross-sectional discs from dominant (right) and sub-dominant (left) *Pinus patula* trees growing in a stand that was heavily thinned at 9 years of age. Note increase in ring width 9–10 rings from the pith, especially in right-hand disc.

Figure 16.2 Seven-year-old *Gmelina arborea* stand thinned twice. Forty-five per cent of original trees removed at age 3 years, 60% of remainder thinned at 5 years (Jari, Brazil).

(a)

(b)

Figure 16.3 *Araucaria cunninghamii*, age 25.5 years, thinning trial in PNG. (a) Unthinned (stocking 1006 trees ha^{-1}, basal area 57.2 m^2 ha^{-1}). (b) Frequently and heavily thinned to maintain basal area at 11.5 m^2 ha^{-1}.

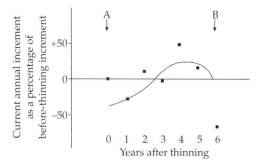

Figure 16.4 Increment recovery after thinning. The points are the volume increments recorded after thinning a sample plot of *Pinus patula* at Peak Timbers, Swaziland. They are shown as a percentage of the plot current annual increment immediately before thinning. At 'A' the plot was 11.9 years old, standing volume was 191.7 m^3 ha^{-1}, and 74.0 m^3 ha^{-1} was removed in thinning. At 'B' the plot was 17.8 years old and standing volume had recovered to 212.9 m^3 ha^{-1}. The solid line shows the theoretical pattern of increment recovery.

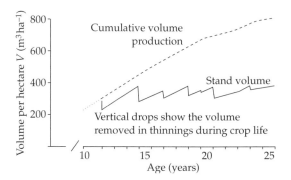

Figure 16.5 Cumulative and standing volume per hectare (*V*) in a stand of *Araucaria cunninghamii* that was frequently thinned to maintain stand basal area at about 29 m^2 ha^{-1}.

thinned regularly to maintain stand basal area at approximately 29 m^2 ha^{-1}. Standing volume rises only gradually though, of course, the volume is distributed on increasingly fewer trees. Total volume production (volume thinned and volume standing) rises steadily.

The typical 'saw-tooth' pattern of stand volume (Fig. 16.5) shows what thinning does. Immediately, there is a drop in stand volume as trees are removed, followed by a steady rebuilding of volume until the next thinning again reduces the volume and the cycle is repeated. This pattern is true of all thinnings, conifers and broadleaves, and Mayhew and Newton (1998) reproduce three examples of the impact on basal area per hectare from mahogany stands growing in Indonesia,

Sri Lanka, and Martinique. For most thinning regimes the total volume of timber removed in thinnings during a rotation amounts to about 40–60% of total production. In Fig. 16.5, by 25.5 years, 52% of the total volume had been cut as thinnings.

To summarize: the effects of thinning are to: (i) stimulate diameter growth of remaining trees; (ii) interrupt stand development by volume reduction, followed by recovery; (iii) redistribute future growth on fewer trees resulting in a stand with a few large trees as opposed to many small ones (Fig. 16.3(a) and (b)); (iv) reduce natural mortality.

Effects of different thinning intensities

So far only the effects of thinning in general have been considered. However, at any one time thinning can remove many trees or only a few, it can be done lightly or heavily or not at all; it can vary in intensity.

Intensity in thinning has three related aspects: (i) how early thinning begins—*the time of first thinning* (considered later); (ii) what proportion of the trees are removed in one thinning—*thinning weight*; (iii) how frequently thinning is done—*thinning cycle*. Thinning weight (ii) and thinning cycle (iii) are not independent in their relation to intensity. For example, if a regime stipulates that 60 m^3 ha^{-1} (weight) should be thinned every 3 years (cycle) then the annual rate of removal is 20 m^3 ha^{-1}. However, this rate of removal (intensity) could also be achieved by felling a greater weight, for example, 100 m^3 ha^{-1} on a longer cycle of 5 years. In each case the thinning intensity is the same. Provided the thinning cycle is not overly long and the weight of thinning not too heavy, thinning intensity can simply be described in terms of volume removed per hectare per year.

However, removal of a certain volume per hectare per year will not be the same thinning intensity for a slow-growing stand as for a fast one. Thinning 300 m^3 during a 30-year rotation from a stand, which only produced 500 m^3 ha^{-1} in total, represents a more intense thinning than if 300 m^3 had been thinned from one which yielded 650 m^3 ha^{-1}. This lack of comparability owing to different growth rates can be overcome by defining intensity as a percentage of the maximum mean annual volume increment a stand will achieve. This approach is used in Britain, but it is not an essential part of thinning practice. For a particular species and site the intensity of thinnings carried out can be considered as the volume of timber removed per hectare per year. In practice basal area per hectare is often used because it is easier to measure.

The effects of differing thinning intensities are illustrated below using data from a trial laid down in an 11-year-old *Araucaria cunninghamii* plantation in Papua New Guinea (PNG) (Figs. 16.3(a) and (b) were photographed in this trial). When the trial started average stand data were: stocking 1399 trees ha^{-1}; top height 21.3 m; basal area 38.3 m^2 ha^{-1}; and average tree volume 0.22 m^3. Six thinning treatments, replicated four times, were applied. Each treatment involved thinning to a prescribed basal area per hectare and then maintaining this level by further thinnings every 2 or 3 years. Five thinning intensities were applied which, over the period of the experiment, held stand basal area (G) at 11.5, 15.6, 21.8, 28.7, and 35.6 m^2 ha^{-1}, respectively; the sixth treatment was an unthinned control. A very heavy thinning was needed to achieve the lowest basal area level of 11.5 m^2 ha^{-1} and 78% of the trees were removed at the first thinning. Other treatments were progressively less severe.

Growth data were analysed 14 years after the trial began when the trees were 25.5 years old. (The progressive development of cumulative and standing volume per hectare shown in Fig. 16.5 is for the thinning treatment where basal area averaged 28.7 m^2 ha^{-1}). Figure 16.6 shows the relationship between thinning intensity and productivity defined as mean annual increment (I_v), that is, standing volume + thinning volume ÷ age, and Fig. 16.7 shows the relationship with mean tree diameter (dg), and volume (V).

Two conclusions may be drawn from Figs 16.6 and 16.7.

1. Over a moderate range of thinning intensities stand productivity is little affected, but under extreme competition or heavy thinning some loss in production occurs. This observation that stand productivity (increment) is little affected over the range of most 'normal' thinning intensities is commonly reported, for example, Schönau (1984*c*) for *Eucalyptus grandis* and Abayomi *et al.* (1985) for teak, though Hussain and Cheema (1987) did not find this with irrigated plantations of *E. camaldulensis* where thinning of up to 45% basal area removal increased stand increment compared with unthinned plots nor Keenan *et al.* (1999) for Queensland maple (*Flindersia brayleyana*).

2. Increasing intensity of thinning results in increasingly large individual trees, up to the point of free growth and no between-tree competition.

These two general conclusions provide the flexibility thinning offers. The extremes are either to maximize total volume production in a stand, for example, for pulpwood, or that of individual trees, for example, for sawtimber. Often both are aimed at and thinning is made as heavily as possible but

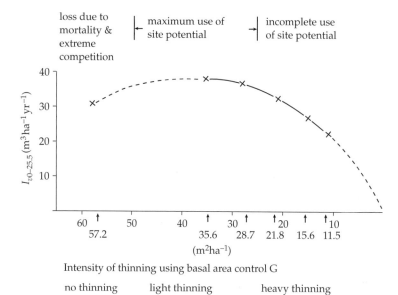

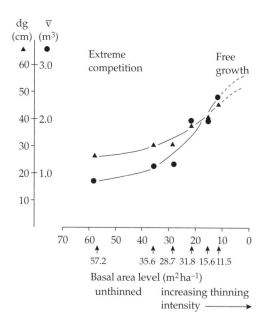

Figure 16.7 The relationship between average tree size (mean diameter, *dg*; mean volume, *V*) and intensity of thinning in *Araucaria cunninghamii* (see text).

without causing significant loss in total yield—this would have been about 30 m² ha⁻¹ in Fig. 16.6. Such a thinning is described as at marginal intensity and is applied in Queensland, Australia, using the concept of limiting basal area (Bevege 1967) though there is now less reliance on commercial thinning to achieve silvicultural objectives (Anderson *et al.* 1981; Shea 1987). Limiting basal area is the minimum stand basal area that will still yield maximum basal area increment, that is, there is no loss in total yield. Thinning at marginal intensity aims to maintain this basal area. Figure 16.8 shows the relationship computed for *Pinus caribaea*.

Effect of constant thinning intensity in practice
If a constant intensity is applied over a series of thinnings, the actual proportion of volume removed to volume remaining will gradually decline. This mensurational fact accords with the growth response of a stand since the rate of basal area growth declines as trees become larger. Therefore, often in later thinnings, the cycle is lengthened so that thinning weight is increased to make the thinning more worthwhile while still maintaining the same intensity.

Methods of thinning

Not only can thinning be made at different intensities, but also in different ways. Two thinnings which remove very different kinds of trees can be of the same intensity. Provided volume removed is the same, intensity remains the same, whether many small trees are cut, or just a few, large trees. In fact,

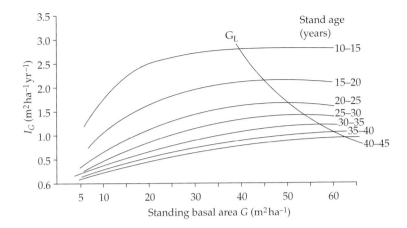

Figure 16.8 Growth model showing the relationship between standing basal area (G) and basal area increment (I_G) for *Pinus caribaea* in Queensland, Australia. The intersecting line crosses the curve of a particular age-class at the point of the minimum stand basal area per hectare needed to maintain maximum increment, that is, limiting basal area, G_L.
Source: Queensland Department of Forests.

Figure 16.9 Line thinning in 12-year-old *Pinus patula* in Malawi.

there are several ways of making a thinning, but two main categories are recognized.

1. Systematic, mechanical, or line thinning. Trees are thinned following an objective and systematic procedure in which individual tree quality is not considered. Removal of every third row of trees is an example, Fig. 16.9. In systematic thinning only intensity can be varied.
2. Selective thinning. Trees are thinned or left, depending on the subjective judgement of the person marking the thinning. There are two main methods, and they determine which kinds of trees are removed, low thinning and crown thinning. In selective thinning both intensity and the kind of tree favoured can be varied. Selective thinning is particularly important in species with generally poor form, so that the best stems can be favoured.

Systematic thinning

In systematic thinning no regard is paid to the canopy class or kind of trees removed. The common-

est form is line or row thinning. Thinning weight is altered by varying the proportion of rows removed.

Systematic thinning is not as common in the tropics as in temperate plantations since low labour costs allow the greater time needed in the more subjective and intensive selection systems. Sometimes early thinnings are mechanical or systematic and later ones selective—this is the practice for thinning teak plantations in Nilambur, India; two mechanical thinnings are followed by four selective thinnings over a rotation of 50–55 years (Chundamannii 1998).

Selective thinning

Low thinning (thinning from below)
This is the commonest form of selective thinning. Low thinning removes mostly trees in the lower canopy classes, that is, the smaller, less vigorous trees and largely speeds up natural processes. Traditionally, four or five intensities have been recognized, though the exact descriptions differ

Table 16.1 Grades of low thinning recognized in India

Grade	Description	Trees removed
A	Light thinning	Dead, moribund, diseased, suppressed trees
B	Moderate thinning	Trees in grade A and defective subdominants, whips, and branchy advance growth which cannot be pruned or lopped
C	Heavy thinning	Trees in grades A and B, all subdominants, any defective co-dominants and dominants that can be removed without making a permanent gap in the canopy
D	Very heavy thinning	Trees in all above grades and many dominants and co-dominants so that the remaining stand consists only of trees with good boles and crowns, well spaced and evenly distributed over the site

Source: Based on (Champion and Seth 1968).

Figure 16.10 Fourth and final selection thinning in a well-grown stand of *Cupressus lusitanica* in Kenya.

from country to country. Table 16.1 gives the classification used in India.

In all low thinning, because smaller, less vigorous trees are removed, there is an element of selection for vigour. The final crop consists of trees with good form which have always grown faster than average (Fig. 16.10).

Crown thinning (thinning from above)
This is more complex than low thinning since crown thinning always involves removal of some dominants and co-dominants to release other trees. Often a number of potential final crop trees are selected and thinning is used to favour their crown development. Crown thinning tends to have two categories of trees, those favoured by the thinning and the rest.

If crown thinning is persisted with throughout a rotation the continual removal of many of the more vigorous trees can lead to some increment loss overall (Johnston *et al.* 1967). Assman (1970) found that the volume increment contributed by subdominants and suppressed trees in a stand was only 3% and 1%, respectively, of the total, and that the increment of these classes per unit area of crown surface is only about half that of dominants. Hence crown thinning, which allows many small trees to be left, even if the total crown surface area is the same as a low thinning, will have smaller growth potential.

Two grades of crown thinning are recognized.

1. Light crown thinning. All dead, dying, and diseased trees are removed along with whips and wolf

trees. Also some upper canopy class trees are thinned to break up groups of dominants. Most co-dominants are left.

2. Heavy crown thinning. Final crop trees are usually selected to achieve a certain stocking, for example, 200 trees ha^{-1}, scattered evenly over the site. Heavy thinning is carried out around selected trees to provide almost full crown release (all competing dominants and co-dominants are removed). In early thinnings many sub-dominants and some co-dominants are retained as a matrix between these selected final crop trees.

Other methods of thinning

Various modifications to the main methods of thinning have arisen of which two are described.

Queensland selection system
This is essentially a systematic method, which includes an element of selection. It has been used in East Africa, PNG, and Queensland, Australia but is now less common. It was applied to teak in PNG in the following way.

1. Before first thinning about 400 good quality trees per hectare are selected.
2. In choosing trees to cut in the first thinning the marker works along each row with the aim of removing three trees out of every group of seven. A form of crown thinning is applied and trees are marked for removal using the following order of priority:
   (a) diseased or badly deformed trees;
   (b) poor dominants or co-dominants competing with selected trees;
   (c) trees with double or multiple leaders;
   (d) any other trees competing with selected trees.
3. The second thinning removes all remaining trees not selected.
4. The third and fourth thinnings remove poorer selected trees according to low thinning principles.

Selection thinning
This differs from selective thinning and is an extreme crown thinning where dominants are thinned (felled) to favour and release trees of lower canopy class. It is not widely practised in commercial plantations and is only suitable for shade-bearing species, but is a way of 'creaming' the largest trees from a stand over a number of years. It is not uncommon in small scale pine and Chinese fir plantations in southern China where farmers take advantage of higher prices for larger diameter trees

but eventually are left with degraded stands (Liu 2003). The method can also be a form of corruption. For example, a contractor making a second thinning in a cypress (*C. lusitanica*) stand in East Africa simply and illegally, though with the connivance of local forestry staff, was observed to take all the biggest and best trees.

A form of 'selection thinning' is common in woodlots managed in social and community forestry where trees are cut once they reach a certain size. In the Bilate project, Ethiopia, any eucalypt tree can be cut when it has reached a diameter (dbh) of 12 cm: the local community decided on this diameter threshold to supply poles, the commodity in most demand.

Influence of thinning method on stand development

It was seen earlier that thinning intensity influences both tree and stand development. However, because different thinning methods remove different kinds of trees, thinning method itself also has an immediate effect on a stand. Regardless of intensity, thinning method affects: (i) mean stand height; (ii) mean diameter and diameter distribution; and (iii) the ratio of average size of thinned trees (V_t) to the average size of trees remaining (V_t). Table 16.2 lists these effects.

In summary, low thinnings increase the average size of the remaining trees because smaller than average trees are mostly removed, while crown thinning has the opposite effect. Systematic thinning is neutral in its immediate effect. So, thinning intensity and method affect subsequent diameter distribution and tree size; the former by redistribution of future increment on varying numbers of trees, the latter by removing smaller, average, or larger than average trees compared with the stand mean.

Thinning and wood quality

Thinning, as well as affecting the quantity of usable timber from a stand also affects the quality. Removal of leaning, and misshapen trees, and those with basal sweep or crooked stems, reduces the amount of reaction wood and spiral grain remaining in the stand, and the trees left to grow on will have a higher percentage utilization. Also, because the remaining trees are encouraged to grow to large size their sawlog and veneering potential is improved since less of the log is wasted.

Table 16.2 Immediate effects of thinning method on stand parameters

Method	Description	Mean stand height (h)	Diameter mean (dg)	Distribution	V_t/V_r
Systematic	Systematic removal of trees favouring no kinds	Unchanged	Unchanged	Trees thinned	1.0
Selective low crown	Mostly removal of smaller trees	Increased	Increased		0.6–1.0
	Selected trees favoured by removing competing dominants + co-dominants	Decreased	Decreased		1.0–1.2 (1.2 is an extreme tending to a 'selection' thinning)

V_t = average volume of trees thinned out; V_r = average volume of trees remaining after thinning; n = number of trees; d = diameter.

Thinning may also have some negative effects.

1. *Increase in taper.* This tends to reduce slightly the percentage utilization of the log. Adlard and Richardson (1979) report this effect with *P. patula* in Tanzania.
2. *Increase in coarseness of branches and knot size.* For this reason high pruning is often done in conjunction with thinning.
3. *More rapid diameter growth.* This may lead to slightly less dense wood, slightly shorter fibre length, and the amount of juvenile wood may increase. Dominants in a stand tend to have lower wood specific gravity, and therefore slightly lower strength characteristics than other canopy classes, and over dominants a forester has least control. The important aim in growing timber is evenness in growth; this is produced by regular, moderate thinnings. Infrequent, very heavy thinnings lead to greater variation in wood properties.
4. *Damage to the stand and soil.* Extraction of thinnings may snap branches and rub bark of remaining trees increasing the risk of infection by pests or disease including wood decaying fungi. Also, along extraction tracks soil may be compacted and root damage may occur.

Thinning decisions

Whether to thin at all is the first of several questions to be decided. But, if thinning is intended, three decisions must be taken: (i) when to thin? (ii) what intensity to apply? (iii) which method to use? However, no questions about thinning should be decided until plantation objectives are established

and predictions made for the outcome of possible alternative regimes.

Management objectives and thinning policy

The reasons for thinning were listed at the beginning of the chapter. But it is the purpose of growing trees, which lies behind these reasons and which is the main consideration in what thinning policy to apply.

Broadly, there are five main uses for plantation wood. In order of increasing desirable log size these are: firewood; building poles; pulpwood, and reconstituted board products; sawn timber, low and high grade; and veneer, peeling for plywood and slicing for decorative quality. After species choice (Chapter 8), which determines the kind of wood produced and rate of growth, the thinning regime, and length of rotation are the main determinants of log size. In plantation management clear end-use objectives are usually specified, and thinning at the right time, in the right way, and at the right intensity is an important tool to achieve them.

An example which is still appropriate was in the 1976–80 Working Plan for araucaria plantations near Bulolo in PNG. This programme was begun in 1951 to produce veneer quality timber for plywood production. The objectives were specified as follows:

To manage the Bulolo-Wau forests on a continuous operations basis by plantation establishment designed to:

1. produce a final crop with a minimum knotty core, in 40 years, of 100 trees ha^{-1} with a diameter breast height of 60 cm or more, and a yield of 450 m³ ha⁻¹ of which 50% to be suitable for high grade veneer production; and

2. produce a series of intermediate thinnings suitable for economic exploitation. The annual rate of plantation establishment will be 500 ha.

It is clear that the first objective requires an average diameter growth of 1.5 cm per year for final crop trees to reach the target of 60 cm in 40 years. Of the thinning intensities compared earlier (Figs 16.6 and 16.7) only the two heaviest achieved this rate up to 25.5 years and may satisfy this size specification. And, it is evident that without any thinning few of the trees would reach the desired size in the intended rotation; so thinning is an essential part of management of these plantations. In reality, the objectives are not being met owing first to neglect of thinning (Skelton 1981) then to excessively heavy, very infrequent thinnings, combined with much logging damage. While the plantations can grow at the required rate, they are not realizing their potential due to poor thinning practices.

Such prescribed thinning regimes to deliver particular products can now be found in many countries, such as Kenya's revised technical orders (Anon. 1996). More sophisticated prescriptions may take into account regional variations in a country, species, as well as product grown. The spacing and thinning prescriptions for Mondi and SAPPI pine plantations in the eastern provinces of South Africa (Zwolinski and Hinze 2000) are an example. Within one forest different regimes will often be stipulated, for example, in Atherton State Forest, northern Queensland (Anon. 1993) the following prescriptions apply

Hoop pine plantations are managed for the production of quality saw and veneer logs. At present, 850 stems ha⁻¹ are planted followed by a pre-commercial thinning to 750 stems ha⁻¹ One thinning is planned for age 32 years leaving a final crop of 400 stems ha⁻¹ . . . Current average clear fall age is 60 years . . . but reducing to 50 years in the long term. Caribbean pine stands are now managed for sawlog production: 350–400 stems ha⁻¹ only are planted. There is no thinning before clearfall. The best 300 stems ha⁻¹ are pruned to a minimum of 5.2 m.

In woodlots and small plantations managed in farm forestry and social forestry formal thinning primarily for the improvement of the remaining stand is usually not done because the objectives are supply of a variety of products. Firewood may be regularly gathered by cutting small trees while larger trees are left to grow to pole size. Thinning to obtain the firewood may be as, or more, important than influencing future stand development.

Forecast of output

The Bulolo-Wau example shows what thinning regimes can, or are intended to achieve. In relating thinning to end-use requirements a forecast of the kinds of yields from thinned stands is necessary. Not only must the total volume yield be predicted but, equally important, the assortment of that yield by size of log, for example, the proportion of logs with a minimum diameter of more than 24 cm. This needs to be known for both thinnings and the final crop at rotation age; only in this way can end-use requirement be related to the management regime applied.

The information about how different thinnings affect stand assortment is obtained from yield plots. Diameter breast height distributions or mean diameter are mostly used in making stand assortment forecasts, and tables, models, or graphs are developed which show the output as a percentage (utilization percentage) of different log sizes in relation to a specific diameter distribution, for example, Table 16.3. This table shows, for example, that in a stand of *E. grandis*, with a mean diameter breast height of 20.0 cm, 82% of the volume is made up of logs larger than 12 cm in diameter and 1.2 m in length. Instead of tabulated utilization percentages, sometimes product models (equations relating proportion of product with mean DBH and perhaps minimum DBH that yields worthwhile volume) may be developed to predict yields in terms of specific product types and product mixes (Von Gadow and Bredenkamp 1992).

Once forecasts of mean diameter for different spacing/thinning/rotation length regimes are available, tables like 16.3 or product models indicate the out-turn, and thus the intended management regime, including thinning, can be compared with the management objectives. Once the effects of thinning are known, questions about method, intensity, and when to begin can be answered more accurately.

Table 16.3 Percentage out-turn by volume of logs of 1.2 m or longer of various tip diameters in *Eucalyptus grandis* plantations

Mean diameter (b.h.) of stand (cm)	Tip diameter class (cm)			
	8	12	16	20
25	98.6	91.1	74.4	45.5
20	96.5	82.1	42.8	
15	91.0	45.0		
10	54.9			

Source: partial extract of percentage utilization table in Appendix 31 of WRI (1972).

The no-thinning option

The increasing cost of thinning is now making the possibility of not doing it at all more widely considered. It was seen that there is usually no loss of volume production by not thinning, provided extreme competition does not cause significant mortality and this can be avoided by keeping rotations short. However, even without direct mortality, there is evidence, especially in the case of pines, that unthinned densely stocked stands can increase the risk of insect attack. Lack of thinning precipitates *Sirex* woodwasp attack of pines, including *P. patula* and *P. elliottii*, when trees are under stress, such as from severe competition (Kirsten *et al*. 2000). This effect has also been reported for bark beetle infestation in stands of *P. caribaea* and *P. oocarpa* in Honduras (Reyes and Groothousen 1990). Control is effected by thinning before stands reach 15 years of age and removing about 40% of basal area.

Where maximum wood production is the objective and quality is not so important, thinning may not be needed. It is rarely carried out in plantations grown for pulpwood, fuelwood, or rough poles in which high yield, moderate tree size, and short rotations are desirable. Most of the extensive *Acacia mangium* plantations grown for pulpwood in Asia are not thinned (Srivastava 1993) but sawlog prescriptions for this species developed in Peninsular Malaysia recommend thinning (combined with high pruning) to maximize diameter growth over a 15-year rotation (Mead and Speechly 1991).

In plantations established primarily to protect the soil and prevent erosion, often no thinning is done or it is very light to avoid damage to the site from logging and to prevent large breaks in the canopy. However, in such protection stands, gradual thinning over a long rotation may encourage natural regeneration. The no-thinning option may also be chosen in plantations considered susceptible to wind damage, since the operation of thinning may render trees more susceptible to both windthrow and stem-break. However, on long rotations the reverse may hold: Adlard (1980) reported most damage in densely stocked stands of *P. patula* in Tanzania.

With a policy of no thinning, low pruning is usually limited to some parts of the crop to reduce the fire hazard. It is cheaper to remove branches from the bottom 2 m of a tree after it is cut down than to low prune it while standing.

Timing of first and subsequent thinnings

The first thinning has been described as the most important management and silvicultural operation in a rotation because it largely defines the course and flexibility of subsequent operations and the log size assortments it will be possible to produce (Lewis *et al*. 1976). Wadsworth (1997) similarly stresses its importance for many kinds of tropical plantations. First thinning is also the occasion when vigorous, coarsely branched or multiple-leader trees must be removed (de-wolfing) before they dominate too much and irreparably harm stand quality: this is particularly important in stands of *Pinus kesiya* and other species exhibiting only moderate form notably broadleaves such as mahogany. Silviculturally, the timing of any thinning is best judged by examining the live crown ratio. Thinning should be undertaken before a crown becomes too small; for many broadleaved species before the live crown ratio is reduced to 30–40%, and for pines about 40–50% (cf. Fig. 15.9(a) and (b)). Other approaches adopt basal area as the indicator and Mayhew and Newton (1998) report 22–24 m^2 ha^{-1} as the level to aim at for mahogany stands and hence one way of deciding when thinning should start. However, with all thinnings, economic considerations tend to cause deferment until trees are larger and costs per cubic metre of harvesting timber are lower. This tendency can impair subsequent stand development (Chapter 15) as well as enhance the risk of pest attack as noted above, and the suppression of many trees in a stand just as Perera (1988) reported for unthinned *P. caribaea* and *P. patula* in Sri Lanka.

Where early thinnings may be unsaleable, respacing some time before canopy closure may be advantageous; see Chapter 13. However, sometimes thinning is carried out at the optimum time to maintain stand growth but the timber is not extracted. This is called a pre-commercial thinning or thinning to waste and, along with reduced initial stocking, is becoming the preferred management in Queensland conifer plantations, with less reliance on commercial thinning to achieve silvicultural objectives (Shea 1987). The number of thinnings has been reduced from seven to only three during the rotation. For *P. caribaea* in Queensland, Shea *et al*. (1984) showed that one heavy pre-commercial thinning, of a widely spaced (5 × 4.4 m) crop, reducing stocking to 300 stems ha^{-1}, was more attractive economically than carrying out conventional thinnings of more closely spaced crops.

First thinning in tropical plantations is often done 2–4 years after canopy closure. In general, faster growth brings forward the time of first thinning and the need to repeat thinnings while wider initial spacing and heavier thinning have the opposite effect. The exact timing of thinning is often laid down by prescription. In Zambia, stand height is used with *P. kesiya* (Allan and Endean 1966) and Keogh (1980*b*) recommends 8 m height for teak in central America and the Caribbean. At Jari (Brazil) *Gmelina arborea* stands were thinned when the live crown ratio fell to 33% or when the current annual increment in basal area fell below the mean annual increment (Briscoe 1979) though thinning is now little practised in the mainly pulpwood stands (Welker 1986) and, indeed, *Gmelina* is no longer a preferred species.

Thinning intensity

If it is essential to achieve maximum yield from a stand then thinning must be at or less than marginal intensity (limiting basal area) and tree size becomes primarily determined by rotation length. However, in some situations it is worthwhile to thin more heavily than marginal intensity, despite some loss in total volume production. In particular if large logs command a high price, for example, for veneer, the good increment on relatively few well-spaced trees may be more valuable than the volume shortfall. Where heavy thinning is carried out for this purpose the term 'value increment' is some-times used for the enhanced growth of the selected trees.

In most plantations thinning is done at or near marginal intensity though it may not be recognized as such. However, because thinning is costly, and the out-turn often only covers the cost of the operation, it is important to ensure it is done efficiently. One way is to thin heavily at infrequent intervals. It is cheaper to thin 60 $m^3 ha^{-1}$ from a stand every 5 years than to thin just 12 $m^3 ha^{-1}$ every year, though both regimes are the same intensity. Provided a stand is dynamic and can respond to occasional heavy thinnings, as well as light ones, the thinning cycle may be varied within limits, to suit the local market or other economic considerations. In such situations the only thinning decision is to stipulate intensity.

Choice of thinning method

Systematic thinning
In terms of cost per cubic metre systematic methods are the cheapest form of thinning as: (i) little specialist knowledge is required; (ii) the need for supervision is small; (iii) trees do not need marking before thinning; (iv) take-down and extraction of felled trees is easier.

The main disadvantage of systematic thinning is that the condition and quality of the trees felled or left are ignored. Deformed and diseased trees in unthinned rows are left while good phenotypes may be lost in felled rows. Another disadvantage, especially of line thinning, is increased susceptibility of the stand to windthrow owing to one sided exposure of a tree's crown and discontinuity of root systems.

Systematic thinning is most suited to first, or first and second thinnings as an inexpensive way to reduce stocking and increase growing space per tree. In later thinnings, the method is less suitable because the original orderliness of the stand has disappeared and most of the advantages which make it cheaper no longer apply. However, as a method of early thinning, the ease of application and cheapness of systematic methods are leading to their increasing use.

Selective low thinning
This is the most costly method in the short term because it involves removing many small trees. Marking trees for thinning is not difficult since it is usually easy to decide which are the poorest. Also, because mostly smaller trees are felled, difficulties with take-down are not great.

Low thinning removes the smaller, less vigorous trees in a stand, thus future increment is concentrated on the best, faster growing ones of good form, that is, low thinning has an element of positive selection. Comparisons between line thinning and selective low thinning with *P. patula* in Colombia (Ladrach 1980) and *P. caribaea* in Queensland (Queensland Department of Forestry 1985) have shown precisely this small but positive growth advantage selective thinning gives to stand increment. In financial terms low thinning is an investment into future capital. For this reason low thinnings tend to realize the highest site growth potential and, in the long term, are the most profitable.

Crown thinning
Crown thinning has two important advantages. First, a few selected trees are favoured and their growth encouraged. Second, by felling larger trees in earlier thinnings there is initially a better financial return and the cost of thinning per cubic metre is generally less, though take-down can be difficult. The most difficult part of crown thinning is in marking. No two people will wholly agree on which trees to fell and which to leave, and marking can become very time-consuming.

The effects of crown thinning on productivity were noted previously. In practice, stands are often given crown thinning initially and low thinning later in the rotation. Crown thinning and heavy low thinning are the most commonly practised methods in tropical plantations.

Table 16.4 Thinning prescriptions for *Araucaria cunninghamii* and *Pinus caribaea* var. *hondurensis* for sawlogs in Queensland, Australia (G.M. Shea, pers. comm.)

Thinning number	Age (years)	Stocking before thinning (no. ha^{-1})	Stocking after thinning (no. ha^{-1})	Thinning method
A. cunninghanii (hoop pine)				
Planting	0	833	750	90% survival
1	30	750	400	selective
	50	400		clearfell
P. caribaea var. *hondurensis* (Caribbean pine)				
Planting	0	746	700	94% survival
1	2–3	700	500	pre-commercial
2	22	500	300	selective
	30	300		clearfell

Note: Variations in these prescriptions exist throughout Queensland depending on local demand for thinnings, distance to markets, growth rates, accessibility to stands, etc. The prescription shown for *P. caribaea* applies to stands in tropical Queensland (north of 23°S).

Thinning practice and control

With systematic thinning, control is no problem, intensity is defined by removal of every third, fourth row, etc., and it is clear which trees are to be felled. In selective thinning the trees to be felled (or left) must be marked in advance. Control of thinning in this case has to ensure that both the right kinds of trees and in the right quantity are felled. This requires trained markers; instruction in the kinds of trees to remove is fairly straightforward, but controlling the intensity of thinning uniformly over a stand can be difficult. In the tropics two control systems are widely used, one based on basal area, the other on number of trees.

Basal area control

In this system the object is to thin to a certain after-thinning basal area, for example, limiting basal area (p. 266). This is achieved by estimating the before-thinning basal area, marking the trees to be thinned, and then estimating the after-thinning basal area. The method is slow if individual tree diameters are measured, but this is not necessary if a relascope is used to estimate stand basal area. In practice, basal areas are checked by establishing temporary plots as the stand is being marked. Tables are often available which indicate the basal area to be removed in a crop of a particular species, age, and site quality. Hussain and Cheema (1987) recommend the basal area control method for *Eucalyptus camaldulensis*, Seitz (1990) quotes percentage of basal area to remove each thinning in *P. caribaea* stands to prevent bark beetle attack, and

the examples cited by Mayhew and Newton (1998) for mahogany were noted above.

Stem number control

This method is simpler and more common than basal area control; thinning follows prescribed stocking levels for a stand of a particular species, age, and growth rate. The number of trees to be thinned per hectare at each thinning is stipulated. This system is the most widely used, particularly in Africa. For example, it is the method laid down for all plantation species in Kenya (pines, cypress, juniper, *Vitex* and *Bischofia*) in the country's technical orders (Anon. 1996). Similarly in Nigeria recommendations for thinning teak are simply to reduce stems per hectare to 800, 600, 400, and 300 at ages 5, 10, 15, and 20 years, respectively, to produce a good stocking of large-size timber by age 50–60 years (Abayomi *et al.* 1985). Table 16.4 illustrates two examples in use in Queensland in the 1990s.

Thinning in lowland tropical hardwoods

Most of the discussion on thinning in this chapter has focused on industrial species like pines and eucalypts because such plantations are extensive and research and trials have been undertaken. However, with increasing interest in plantations of tropical broadleaved species (Evans 2001c), especially high quality hardwoods, it is important to modify generally recommendations in this chapter which have tended to focus on maximizing growth and yield. As most hardwoods have indifferent stem form, every thinning should aim *to maximize quality over quantity*. This should begin in the crucial cleaning and climber cutting stages of early plantation maintenance to ensure that non-crop species do not overwhelm the plantation, but once a crop is established the following points should be considered. (i) Thin always to favour the best formed trees; (ii) use moderate crown thinning or heavy low thinning to provide adequate growing space for the favoured trees; (iii) consider high pruning favoured trees to further enhance quality; (iv) minimize frequency of thinning and reuse logging tracks to reduce soil erosion and compaction; (v) when thinning take every opportunity to develop patches of natural regeneration or other non-main crop species if polycyclic systems are pursued.

Thinning and regeneration

Thinning, especially later thinnings in a rotation, often leads to the appearance of natural regeneration.

Such regeneration may or may not be wanted, since at final felling opportunity may be taken to introduce a new species or use superior planting stock.

However, this phenomenon has recently come to the fore on many former rainforest sites with the potential of rehabilitation. Where a plantation has re-introduced forest conditions after a past clearance of the land, thinning can encourage natural regeneration of native species. Good examples occur in Ecuador where failed ranching lands have been planted to *Cordia alliodora* and as the plantation matures regeneration of native hardwoods such as *Virola* and *Brosimum* species is often abundant (see Fig. 22.3). This is further discussed in Chapter 22.

CHAPTER 17

High pruning

In tropical tree plantations pruning is mainly carried out in two very different situations: to improve stem and wood quality in some industrial crops and as a tool in agroforestry and social forestry plantings, both to manipulate the trees to favour food crops, for example, in alley cropping, and to provide fodder, leaf mulch, fuelwood, smooth, snag-free poles, etc. Note that pruning in arboriculture and horticulture is to improve crown shape, stimulate flowering and fruiting, train branches, and sometimes simply for aesthetic purposes. This chapter primarily concerns high pruning in industrial plantations; Chapters 18, 20–21, and 23 make further reference to pruning practices in agroforestry and rural development forestry including pollarding.

While branches remain on the stem of a tree, the wood laid down contains knots. In industrial forestry, high pruning is the removal of branches from the stem so that knot-free timber is produced. It is one of the few opportunities there is to improve wood quality. High pruning is distinguished from low pruning or brashing (Chapter 13), though obviously low pruning also results in knot-free timber at the base of the tree. Indeed, low pruning is often incorporated into a pruning schedule as the first pruning of a series carried out over several years.

Need to prune

There are two main considerations, one silvicultural and one technical.

Silvicultural considerations

How long dead branches remain on a stem varies much between species. With some species dead branches soon fall off (natural pruning), with others they persist for many years. Clearly, the amount of natural pruning markedly affects the need to high prune, and species can be loosely classified in this way. For example, both *Cupressus lusitanica* and *Pinus patula* have persistent branches, and must be pruned if knot-free timber is desired (Fig. 17.1), while such species as *Terminalia superba* (Fig. 17.2), *Cordia alliodora*, and most eucalypts (Fig. 1.1) are good

Figure 17.1 Persistent branches on 14-year-old *Pinus patula*.

Figure 17.2 Good natural pruning on *Terminalia superba*.

natural pruners and rarely need artificial pruning. *Eucalyptus grandis* and *P. caribaea* are intermediate and their branches persist 2–3 years after dying.

Natural pruning is, to some extent, influenced by stand density. Dense, unthinned stands, with trees close together, encourage earlier suppression and death of side branches while they are small. However, the effect on accelerating natural pruning is not great. Of greater importance is the increased persistence of branches low down the stem in open stands. Where wide spacing and heavy thinning are practised (Chapter 16), artificial pruning is essential for all but the most freely natural pruning species if high-quality timber is required.

With some species, for example, teak, pruning does not always achieve its aim, since adventitious branches or epicormic shoots are easily formed and often develop next to the pruning scar (Briscoe and Nobles 1966).

For some hardwoods pruning is required to improve stem form. Even mahogany that is damaged by shoot borer and consequently of poor form, can be much improved by judicious pruning. Mahogany responds well to pruning and scars heal quickly but pruning is expensive.

Technical considerations

The important technical consideration is whether clear, knot-free, and consequently high-grade timber is required. For example, wood of such quality is of little importance in plantations grown for: (i) firewood or fuel and domestic uses except poles; (ii) pulpwood and particle board; (iii) low-grade sawtimber for uses such as shuttering and packing cases; (iv) protection such as shelterbelts, erosion control, and dune stabilization—in fact, persistent lower branches are often an advantage.

In contrast, clear, knot-free timber is highly desirable or essential for: (i) veneer production for decorative use, match making and plywood, to ease peeling, improve appearance, and reduce blemishes; (ii) high-grade constructional timber where uniform strength, and good machining, finishing, and seasoning qualities are important; (iii) various types of poles, for example, transmission, which need to be smooth for handling, and free of surface snags or holes to prevent entry of fungi or termites into the heartwood which is often not penetrated by preservatives (WRI 1972).

If a plantation is grown for an end-use requiring high-grade, clear timber, some pruning must be carried out on all species where branches persist for more than a short time after suppression and death. In the example of managing araucaria plantations in Papua New Guinea (PNG) to produce plywood,

pruning is an essential operation to restrict the knotty core to a minimum and produce the quality of wood required (Chapter 16). In a species with less persistent branches, such as *E. grandis*, gains in knot-free timber from pruning are usually insignificant and not worth pursuing; for example, the trial reported by Bredenkamp *et al.* (1980).

It is emphasized however, that when the technical desirability of growing knot-free timber can only be achieved by pruning, it should be reflected in a higher price for the timber. Pruning should be viewed as an investment to improve wood quality which should command a higher market value, sufficient to cover the compounded pruning costs, compared with knotty timber from unpruned trees.

Production of knot-free timber

The process can be divided into three steps: removal of branches, healing of pruning scars, and the laying down of new wood over the scars.

Branch removal

Branches are cut off almost flush with the trunk usually using a specially curved saw or pruning knife. Above 2.5 m in height, branches are reached with long-handled pruners (Fig. 17.3), by using ladders or by climbing. Where branch removal tears the bark of the main trunk it is safer to prune a branch in two stages: first cut the branch about 50 cm from the bole and then, as a second operation, cut off the remaining stump. Research in temperate areas has shown that the final cut next to the trunk should not be so close as to cut into the branch bark ridge (the creases of bark in the crotch of the fork), but angled slightly outwards to help prevent decay entering the main trunk.

High pruning is usually carried out two, three, or four times at short intervals with each pruning going successively higher until the required length of pruned stem is reached. For example, Mayhew and Newton (1998) report three prunings of mahogany (*Swietenia macrophylla*) grown in the Solomon Islands to produce a merchantable log. First pruning to 2.5 m at 12–18 months, followed by pruning to 4.5 m and finally to 7 m at 24 months and 36–48 months of age, respectively. If pruning was delayed until the whole tree could be safely pruned to 7 m in one go, the basal diameter would be large, and a wasteful, large core of big knots the result; see also Fig. 17.4. This restriction of the knotty core is important to make pruning worthwhile and may be incorporated in management prescriptions. In Queensland's Atherton State

Figure 17.3 Final high pruning of selected *Pinus taeda* trees, Eastern Transvaal, South Africa. Length of pruned stem is being lifted from 3 to 5 m. Pruning from the ground is cheaper and safer than with ladders or climbing, but note essential safety precaution of a hard hat. Normally 200–220 trees per day can be pruned under these conditions with a long-handled pruner.

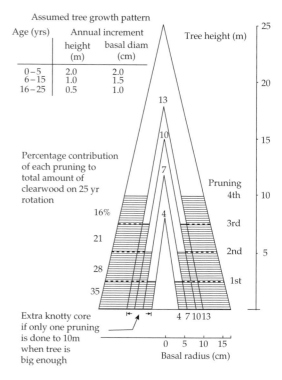

Figure 17.4 Schematic representation of a stem profile showing the effects of high pruning based on the schedule in Table 17.1. Shaded area shows zone of knot-free wood.

Forest, hoop pine (*A. cunninghamii*) is pruned in two stages to 5.2 m and timed to limit the knotty core to 15 cm diameter on average (Anon. 1993).

Healing of pruning scars

Occlusion of a pruning scar, by growth of callus tissue from the peripheral cambium, determines when the laying down of clear timber begins. Also, of course, it affects the period of time the uncovered wood surface is exposed to fungal and insect attack for species which do not exude a sealant of resin or gum.

The rate of occlusion differs between species and is also affected by tree health, vigour, size of scar, how flush the scar is to the trunk, and sometimes time of year or season of cutting. For example, on a healthy 5-year-old *Gmelina arborea* a scar 4 cm across healed over in 11 months (Fig. 17.5), whereas

on *P. caribaea* a scar of only 2 cm diameter took over a year to occlude. In a study of wound closure in *E. camaldulensis* pruning near the start of the growing season led to the most rapid occlusion (Perry and Hickman 1987).

In commercial forestry the use of wound sealants is much too costly and, moreover, is not very effective for aiding callus growth or preventing entry of disease.

Laying down knot-free wood

Clear knot-free timber is only laid down in the years following pruning (Fig. 17.6), and complete freedom from grain distortion is only achieved after the first 3–4 cm of new wood; pruning has no retrospective effect. Consequently, the sooner pruning commences, and the longer the rotation of the crop, the higher will be the proportion of knot-free timber. Pruning regimes are usually implemented early in the life of a stand to achieve the twin objectives of restricting the size of the knotty core as much as possible and allowing the maximum

(a)   (b)

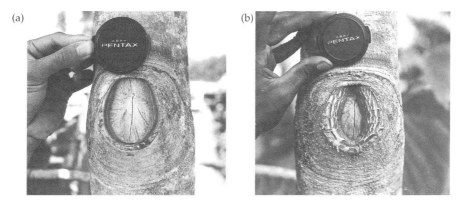

Figure 17.5 Healing of pruning scar on a *Gmelina arborea* tree (a) 4 weeks after pruning, note rim of callus; (b) 4 months after pruning.

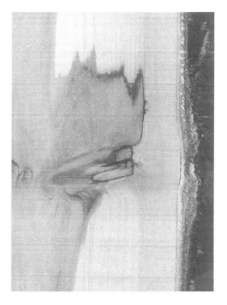

Figure 17.6 Layer of clear wood between pruned branch stub and outer bark exposed when sawing a thinning of *Araucaria cunninghamii* (note possible fungal staining).

amount of time to elapse for clear timber to be laid down (Fig. 17.4).

Timing and intensity of high pruning

It has been noted that one of the main aims of pruning is to keep the knotty core as small as possible. Clearly, the sooner pruning begins the smaller this will be, but obviously there is a limit to how early it is practicable to start.

In most young pine plantations, say under 5 years old, all branches, even near the ground, will be alive. The first pruning is often done in such situations, and live branches are cut off; this is called 'green' or 'live' pruning. As well as ensuring a small knotty core, the knots will be tight and the pruning scars will heal quickly, but to cut off live branches may increase resin infiltration into the wood (Bredenkamp and van Vuuren 1987) and, more importantly reduces the photosynthetic area of the tree's crown, which in turn reduces growth. This is unimportant if pruning is restricted to the oldest and most heavily shaded branches. However, for economic reasons, it is rarely practicable only to remove such branches which perhaps are growing on only the bottom 1 m of stem. Pruning only 1 m of the stem would be costly and time-consuming and usually 2–3 m is pruned in a single operation.

If a first pruning to 2.5 m is carried out on a young tree 5.0 m high, 50% of the live crown is cut off, assuming branches are live to ground level. On the other hand, if the pruning was delayed until the tree was 8.0 m high, and even if all branches were still alive crown reduction would only be about 30%. This is the concept of 'pruning intensity'. There is trade-off between the need to restrict the knotty core, the economically feasible length of stem to prune, and the reduction of the crown which slows growth.

The concept of pruning intensity applies to every pruning up to the desired height; Fig. 17.7.

While pruning 20% of the live crown may sometimes slightly depress diameter increment of *P. patula* compared with unpruned trees, generally only high pruning intensities over 50% lead to significant reduction in diameter growth and affect height growth (Marshall and Foot 1969; Karani 1978; Morris 1981). For mahogany (*S. macrophylla*) Vincent (1972) found that even pruning to half total height did not

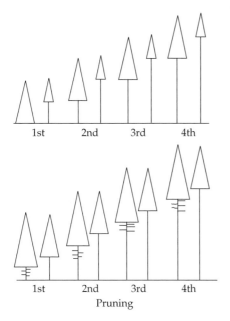

Figure 17.7 Schematic representation of different pruning intensities. Top: intense pruning with 50% live crown removed at each pruning. Bottom: moderately intense pruning with about 20% live crown removed.

affect growth, though Mayhew and Newton (1998) recommend that after pruning the live crown should be 35–40%. With eucalypts pruning more than 25% of the live crown is considered severe. For *Cupressus lusitanica* (e.g. Fig. 17.7) trials in Kenya and Tanzania indicate that height growth appears unaffected by pruning and diameter growth only significantly depressed at intensities exceeding removal of branches to more than two-thirds tree height (Kaumi 1980; Mwihomeke 1983). High pruning is usually only carried out in *Acacia mangium* plantations being grown for sawlogs and veneer. Removal of more than 30% of the live crown can result in a significant loss of increment and pruning scars greater than 2 cm diameter have a high risk of fungal infection (Srivastava 1993).

Two secondary effects of severe pruning, in addition to loss in volume, are reduction in stem taper, since greater diminution of diameter occurs near to the base of a tree, and some increase in juvenile core-wood density. In addition, the defect of stem bumps (nodal swellings), notable in *P. patula*, may be exacerbated by pruning (Schutz 1990). Further discussion of the effects of green pruning are found in Brown (1962), Adlard (1969), Plumptre (1979), and Wadsworth (1997).

Where all trees in a stand are green pruned this effect of depressing growth temporarily is not important. But it becomes extremely important if only selected trees are pruned.

Often because high pruning is costly, and because many trees in a stand are thinned out before rotation age, only some trees are selected for pruning. These are trees of good form, vigorous growth, and which are likely to constitute the final crop. If these trees are not free from competition with unpruned neighbours, the slowing in growth due to green pruning may result in their being overtopped and losing their dominance; the pruning then becomes a waste of money. Therefore, selective pruning must be done in association with thinning so that the pruned trees are favoured. Where selective pruning is commenced in young stands Plumptre (1979) suggests that with pines unpruned trees could even be topped to eliminate competition while still suppressing weeds. This might be impractical, but it makes the point!

Percentage of live crown depth has been used as the index of pruning intensity, but it may be expressed as a percentage of total tree height, or number of live branch whorls cut off, or simply pruning to a fixed height above ground. The last two are useful for instructing pruning gangs.

Table 17.1 gives a pruning schedule, applicable to most coniferous species in the tropics, designed to produce a 10 m clear stem in four operations.

Economics of pruning

Pruning is normally an expensive operation, though if the foliage or branches have a value, for example, for distillation of essential oils, pruning may pay for itself. This is the case with *C. lusitanica* and *E. citriodora* in Sri Lanka.

Important factors in pruning costs are: (i) the number of branches per unit length of stem (Fig. 17.8); (ii) branch thickness; (iii) height of pruning; and (iv) number of trees to be pruned per hectare.

Pruning height is particularly important because when branches cannot be reached from the ground, using ladders or climbing greatly increases the cost. For example, in Malawi third and fourth prunings to 6.8 m and 11.0 m cost 2.5 and 5 times respectively the cost of the first prunings to 2.1 m. Also, as Fig. 17.3 shows, the higher prunings yield the least amount of extra knot-free wood. Consequently, it becomes rapidly uneconomic to prune higher than about 10 m up the stem. Usually the final pruning height is determined by mill requirements for log length; trees are pruned to yield one or two butt logs with knot-free timber, for example, to 10 m for two 5 m logs. However, owing to increasing

Table 17.1 Typical pruning schedule

Pruning operation	Pruned height (m)	Approximate timing	
		Stand height (m)	Stand condition
First (low pruning)	2.5	6.0	Just after canopy closure
Second (high pruning[a])	5.0	9.0	Prior to first thinning
Third (high pruning[a])	7.5	12.0	At time of first thinning
Fourth (high pruning[a])	10.0	15.0	Prior to second thinning

[a] High pruning may be selective.

Figure 17.8 Second pruning of *Cupressus lusitanica* just completed (Malawi). Thirty scars are present on the upper stem.

labour costs even the practice of pruning to 10 m is less widespread than in the past, and pruning to about 5–7 m is now more common. In Queensland, Australia, for example, both *Araucaria cunninghamii* and *P. caribaea* are pruned only to 5.4 m in just two operations, a low pruning to about 2 m when predominant stand height is 9.5 m, using a hand-saw or light chain-saw, followed 2 years later by high pruning with an extendable pole-saw to 5.4 m. A similar two stage high pruning to 5.5 m is recommended for producing high quality sawlogs of

Acacia mangium in Malaysia (Mead and Speechly 1991).

There is no doubt that high pruning is expensive. Unfortunately, there is often much doubt whether the investment made is rewarded with a better market price. Individual timber users may not discriminate between wood grades or may simply be accustomed to using clear wood. It is often very difficult to relate the price of timber sold directly to the value added by pruning, since a specific identifiable premium for better quality is hard to determine.

In the case of plywood, plantations grown for veneer quality must have a pruning regime to produce satisfactory peeler logs which command high prices. If pruning is not done, apart from one or two naturally pruning species, the logs cannot be used for peeling. This clear distinction, and relatively simple identification of the extra value because of quality, is not common.

However, in considering the economics of pruning there is one other important aspect not directly related to the costs and returns of the operation itself. After weeding and cleaning there is usually little to be done in a plantation until first thinning. But it is in this period that pruning, both low pruning for protection and access and high pruning to improve wood quality, may be commenced. And, it will often be advantageous, in terms of overall management, to undertake pruning, even if direct economic returns are uncertain, since the operation can be viewed as a reserve of work to be carried out when other activities are slack. Since pruning can be done almost at any time it is a useful stop-gap operation.

Pruning is worthwhile if, as far as can be judged, the extra revenue for the timber exceeds the compounded costs of doing it. To achieve this in Queensland, it was suggested that stumpage price of pruned trees should be three times that of unpruned trees (Robinson 1968). The premium paid for pruned hoop pine suggests high pruning is economically worthwhile for 400 trees ha^{-1} (Hogg and Nester 1991). However, other considerations may complicate the picture and make the operation worth doing without any clear financial gain directly evident.

Pruning practice

In the tropics, pruning plantation-grown trees, notably conifers, is much more widely practised than in temperate countries. It is, or has been, for example, a key ingredient in plantation sawlog production in most eastern and southern Africa countries such as Kenya, Malawi, Tanzania, South Africa, Swaziland, and Zimbabwe. There are several reasons: (i) it is less expensive per tree because of lower wage rates and a less critical attitude to working conditions such as climbing trees; (ii) faster growth rates mean short rotations and earlier financial returns; and (iii) the wider spacings generally adopted make pruning an essential part of silviculture for growing sawlogs of reasonable quality.

For pine plantations grown for sawtimber, successive prunings may be done between ages 6 and 12 years and the trees felled at age 25 years. In temperate countries, these figures would be at least doubled and the compounded costs consequently much greater. In the southern Indian state of Kerala, it has been recommended that high pruning begins when dominant trees reach 10–15 cm diameter at breast height (Bhat 1990*a*). This size criterion, similar to the one used in hoop pine plantations in Queensland (Anon. 1993), is better than age and ensures that no stands are pruned too soon or too late.

Future of pruning as a silvicultural operation

Two contrary trends are apparent:

1. The relative cost of pruning, compared with some other silvicultural operations, will increase, because it is likely to remain highly labour intensive.
2. The overall supply of clear timber from natural stands will gradually diminish as stands become exhausted and timber supplies increasingly depend on younger second-growth forest and plantations; clear knot-free timber may increasingly command better prices.

Because of the relatively low wage levels still existing in most tropical countries, and the policy fairly widely adopted of maximizing employment at the expense of depressed wage rates, pruning may remain a practicable proposition. If, as seems likely, the demand and price for high-quality timber rises, this operation in tropical countries seems a particularly fruitful investment, with the advantage of fast growth rates. This makes pruning an attractive silvicultural operation in the tropics. It is a rare example of an investment that is a value-adding operation, easily and simply carried out in the rural locations of the forest plantation – the raising of timber quality from knotty to clear grade.

CHAPTER 18

Rotations, regeneration, and coppice

Introduction

At some stage three questions must be decided about any plantation.

1. For how long is the crop to grow, that is, what is the rotation length?
2. What is the best way to harvest the trees, dispose of the debris, and prepare for the next crop?
3. How is the next crop to be established?

These questions about regeneration cannot be decided in isolation. For example, if natural regeneration is planned then the rotation must be long enough for trees to produce seed abundantly. The main methods of regeneration used in tropical plantations, that is, clear-felling and replanting, and coppice systems, were briefly discussed in Chapter 4. This chapter considers them more fully, along with brief mention of natural regeneration, but, first, two important points are restated.

1. Regeneration by clear-felling and replanting allows the crop to be changed. Silvicultural improvements which may have arisen during the course of a rotation, such as better provenance selection, genetically superior seed, clonal stock, intercropping, fertilizers, improved establishment techniques, etc., can all be introduced. Rapid progress in tropical plantation silviculture, makes this an important opportunity for change.
2. Coppice, although it can only be applied with certainty to some broadleaved species, is an important, widespread, and successful method of regeneration in the tropics for many fuelwood, pole, and pulpwood crops. The latter part of this chapter considers coppice, and briefly the related practices of pollarding and shredding.

Rotation length

The rotation is the planned number of years between planting and felling. Put simply, it is the period of time a crop of trees is allowed to grow. Rotation length is an important tool for controlling

tree size: the longer the rotation the larger a tree can be grown. But rotation length also markedly influences yield, profitability, and regeneration methods. Any one of these factors can be the main determinant of length, thus foresters identify several different kinds of rotation and the outline below follows Fenton (1967).

Physical rotation

This rotation is determined by the site or other environmental factors which may prevent a stand reaching maturity.

Frequent and regular cyclones or fires can physically limit the size a stand might reach. Unexpected and severe droughts can cause extensive deaths especially where crops are on shallow soils. This was the experience with *Pinus patula* in some compartments of the Usutu Forest in Swaziland in the early 1990s (Morris 1993*a*). Even the build-up of a lethal pest or pathogen might also limit the life of a crop. This occurred with *Leucaena leucocephala* in the 1980s and 1990s in many parts of Asia and Africa owing to widespread defoliation by psyllids.

Because afforestation very often takes place on poor sites, physical limits to tree life may be commoner than usually supposed (Fenton 1967).

Silvicultural rotation

This rotation is not common in tropical plantations, but is the one in which natural regeneration is most easily obtained. Though most plantation species in the tropics set seed freely and relatively early (Chapter 9) and natural regeneration is possible (Fig. 13.9 and see p. 291 later in this chapter) it is not often encouraged for reasons already discussed.

Technical rotation

The technical rotation yields the most out-turn of a specified size and type to satisfy a particular end-use. A good example was quoted in Chapter 16 from the working plan for araucaria plantations in Papua New Guinea (PNG), where the rotation is 40 years, the stands appropriately thinned, so that '50% of the final crop will attain a diameter breast height of 60 cm or more'. The large diameter is required for the intended end-use of peeling the logs for plywood. Technical rotations may not only specify lower limits of size but sometimes upper limits as well. Many village woodlots and other social forestry projects are managed on short rotations simply so that the produce can be easily handled as large sticks and small poles and are easy to harvest, carry, and cut up. In Ethiopia, the highly prized pole products from eucalypt plantations are cut from densely stocked plantations/coppices on 3–6-year rotations; Fig. 12.11.

It is important to allow stands that are pruned to grow on for many years afterwards so that a worthwhile layer of clear, knot-free wood is laid down. This usually results in extending a technical rotation. Pruned hoop pine (*Araucaria cunninghamii*) in Queensland, Australia, is grown for 50 years (Anon. 1993) and high-quality teak in India often for 60 or 70 years.

A wood quality factor affecting technical rotation length in pulpwood crops is fibre length. Juvenile wood tends to have shorter wood fibres than adult wood and cutting trees on very short rotations can reduce pulp quality. Bhat (1990*b*) reported that *Eucalyptus tereticornis* at 3 years of age produced fibres 29% shorter than at 9 years, and advocated that rotations must be at least 9 years to maintain satisfactory pulp quality.

Financial rotation

The financial rotation yields the highest financial return under a particular set of circumstances. Economic analyses may be done in several ways see Uys (1990), von Gadow and Bredenkamp (1992), and a helpful chapter on this topic by Pearse (1990). Rotation lengths may be optimized to achieve the highest net discounted revenue per hectare, or per unit amount of capital invested (i.e. interest received on the capital outlay), or to optimize the internal rate of return (interest rate), or the discounted cashflow for the life of the plantation project and so on. For example, Rawat (1990) using maximum net present worth per hectare and interest rates of 12–15%, showed that 9–11 years is the optimum rotation for *Eucalyptus* hybrid woodlots in India to maximize returns to the farmer and that growing at 2 or 2.5 m spacing is superior economically to closer spacing on shorter rotations. However, different methods of analysis may give different optimum rotations. Economic analyses, especially using high interest rates, tend to favour short rotations (see Sedjo 1984), and explains the relatively high internal rates of return (>15%) claimed in the 1980s for many social forestry projects, for example, Spears (1984, 1987), World Bank (1986).

An economic generalization applicable to any crop nearing the end of the rotation is to fell it once current value increment (the increase in stand value due to each year's new growth) falls below the maximum average annual income of the stand, that is, the total accumulated value of the stand divided by its age. This is the financial equivalent of the rotation of maximum volume production and the relationships are similar to those in Fig. 18.1.

Rotation of maximum volume production

This rotation yields the greatest average annual out-turn of timber. In terms of wood yield it realizes the full growth potential of a species on a site.

As seen in Chapter 15, tree growth follows a definite pattern. This rotation is complete when the annual growth increment of the stand falls to the level of the overall mean annual increment (see Fig. 18.1). When the current annual increment (CAI) falls below the mean annual increment (MAI), the MAI itself will begin to fall gradually since each year less than the average increment is being put on. Therefore, the point of intersection of the CAI and MAI curves must also be the peak MAI will reach; this is the rotation of maximum volume production if the stand is felled at this point (R1).

However, where the next crop is likely to be more productive (shown in Fig. 18.1 by the broken line of MAI), due to improved planting stock, superior seed, better site preparation, etc., a shorter first

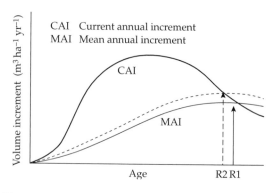

Figure 18.1 Generalized relationship between current annual increment and mean annual increment with age. Rotation of maximum volume production is indicated (R1 or R2).

rotation (R2) is justifiable if the aim is for maximum volume production from the site. This is because there is no gain from growing on a crop once its CAI falls below the forecast MAI of the next crop (broken line), since less than the best yield is being obtained when it is known that the replacement crop would achieve more (the CAI for the successor crop is not shown in Fig. 18.1).

This rotation is attractive because it realizes the maximum potential of a site. However, for slow growing crops it is rarely the most profitable, and only becomes so under the improbable combination of high-value product, low cost of production, and low interest rates (Fenton 1967).

In the tropics the high yields obtained are not only due to fast growth as such but to much earlier culmination of maximum MAI. In some temperate countries, and in many tropical areas, plantations with a maximum MAI of 20 m^3 ha^{-1} a^{-1} can be grown, but in the tropics this is achieved on 10–15-year rotations or sometimes even less whereas in temperate countries 40–80 years is usual. Consequently, many tropical plantations are grown for as long or longer than the rotation of maximum MAI usually for technical reasons to achieve the size of material required. Camino and Alfaro (1998) cite data from teak plantations in several Central American countries where MAI peaks at between 10 and 15 years, but planned rotation lengths are at least 20–25 years, unless very large-size trees are required when up to 50–70 years may be needed (Keogh 1980b).

The rotation of maximum volume production varies with species, site quality, spacing, and

thinning regime but few data are available for tropical species though a detailed study has been made in India by Ghosh and Kaul (1976). Table 18.1 lists some examples.

Other factors affecting rotation length

Frequently none of the above rotation types is exactly applied though one of them may have been intended. A number of examples from pine projects in southeast Africa show why, for sound reasons, rotation length may change.

1. *Opening of a new market.* In the Usutu forest, Swaziland, felling began in 1962–63 when the oldest *P. patula* were about 12 years of age because the new pulpmill had just been commissioned and required a steady supply of pulpwood. Good growth rates and the large size of the forest made it feasible to cut such crops even though the rotation of maximum volume production is about 15 years.

2. *Failure of a market to materialize.* The oldest stands in Viphya Pulpwood Project in Malawi were not felled at the age of maximum volume production because the intended wood pulp market did not develop.

3. *Rate of harvesting may not match the annual increment of a forest.* If the amount of wood cut is less than a forest's annual increment of new wood the rotation will inevitably lengthen since the whole forest will not be worked during the period of the intended rotation. The converse of overcutting also occurs. In the late 1990s large areas of th Usutu forest were burnt and this has led

Table 18.1 Rotations of maximum volume production for plantations

Species	Where grown	Rotation (years)	Notes
Pinus caribaea var. *hondurensis*	Queensland, Australia	25–33	Thinned to limiting basal area 30 m^2 ha^{-1}
Pinus patula	Swaziland	14–17	Unthinned, planted at 2.74 m spacing
Araucaria cunninghamii	PNG	18–22	From plot data used for Figs 16.5 and 16.6
Eucalyptus deglupta	PNG	2–8	Determined by initial spacing
Paraserianthes falcataria	Philippines	2–3	Revilla (1974)
Eucalyptus tereticornis	India	6–11	Sharma (1979), Kondas (1982) Culminates earlier for first coppice crop
Tectona grandis	Central America	10–22	Keogh (1980b)
Eucalyptus globulus	Peru	18–19	Rivadeneyra and Cabrejo (1980)
Gmelina arborea	Nigeria	7–8	Akachuku (1981)

to some premature felling of pine stands to make up supply.

4. *Unequal age-class distribution*; see Chapter 4. Again to cite an example from the Usutu forest, when the oldest crops of the second rotation were first felled in 1979 they were only 15 years old, since virtually no new land had been planted between 1959 and 1963 and hence there were no first rotation crops between 15 and 18 years of age. However, third rotation stands were felled close to the optimum age, since extensive new afforestation in the early 1970s has largely eliminated the abnormality in age classes. (Figure 4.3 shows these changes.)

5. *Amenity considerations*. Retention of large trees, to screen clear-felling operations or to enhance the environment of a village or town, may lead to greatly extended rotations for some stands. None of the pine stands in the middle of Mhlambanyati, the company town of the Usutu Pulp Co., Swaziland, were felled until they were over 35 years old. They were heavily thinned as a valued amenity and only felled (and then replanted) when trees became unsafe.

6. *Market conditions or availability of extraction equipment*. Felling may be delayed until demand for wood is strong, or in inaccessible stands until suitable extraction equipment is available.

7. *Profitability*. In times of poor profitability easily accessible stands may be cut to reduce immediate harvesting costs even though they may be some years away from the optimum rotation age. Both major pine plantation projects in Swaziland follow this expedient from time to time.

8. *Theft and excision of plantations from a forest estate*. Problems of this nature occurred in the 1990s in pine and cypress plantations in Kenya, and clearly led to sub-optimal rotation length.

First rotation harvesting and preparation for the next rotation: silvicultural considerations

Inevitably clear-felling is a sudden and major change to the ecosystem. But, silviculturally, harvesting a plantation should inflict minimal damage to the site and not prolong the period soil is exposed. Five aspects of harvesting are of concern and may affect silviculture.

Felling method

In industrial plantations powered chain-saws are now mostly used, though for felling small trees bow-saws and cross-cut saws still have a place and in uniform crops feller/buncher machines or

Figure 18.2 Feller/bunch harvesting *Eucalyptus grandis* trees for charcoal in Brazil.

processors may be used (Fig. 18.2). Stumps are cut as low as possible to minimize waste and allow machine access. Careful planning and control of felling direction can reduce machine movement needed to extract the trees and hence the amount of soil disturbance and compaction. Of silvicultural importance is the need to plan extraction so that damage to the site, particularly with heavy machines on clayey soils, is kept to a minimum.

In social (rural development) forestry hand-saws and axes are the most widely used tools. Poor axemanship can damage stumps and harm coppicing potential, but, in general, such manual working including carrying away the firewood or fodder does little direct damage to a site.

Size of cut

The French word 'coupe' is used to describe the area to be cut in one compartment or a whole forest. Harvesting method, economics, and landscaping considerations largely influence the scale of felling in a compartment but a large coupe is usually the aim. For a whole forest the size of the cut can be used to modify the age-class distribution (Fig. 4.3). Also, the annual replanting capability, for example, supply of seedlings, size of planting gang, equipment for site preparation, etc., must be commensurate with the areas being cut each year.

The quantity of trees cut in farm or village woodlots or in various agroforestry systems is largely determined by individual need, the desire to realize capital for some personal or community requirement, or to benefit an adjacent agricultural crop.

Damage to the site during extraction

Extracting trees from a site causes damage, though the severity and the form varies. Damage may arise from: (i) soil compaction; (ii) waterlogging; (iii) oil spillage; (iv) stream/drain blockage; and (v) scouring of soil surfaces and erosion.

The method of extraction is the most important factor affecting damage. The main systems are listed below in an increasing order of potential for damage:

(i) systems using animals such as cattle and mules;
(ii) aerial cable ways where logs are lifted over the ground, or only the tip touches the surface;
(iii) forwarder systems where a vehicle picks up and carries logs;
(iv) rubber-tyred vehicles for skidding or dragging logs; and
(v) tracked vehicles for skidding or dragging logs.

Studies show that skidding systems can damage up to 35% of the ground surface (Hatchwell *et al*. 1970). Fortunately, on steep slopes where the erosion risk is highest the most damaging extraction systems are unsuitable and animals or aerial cableways have to be used (Oberholzer 2000).

Soil compaction due to intensive machine traffic can damage soil structure resulting in soil compaction. The effect is to increase soil resistance to root penetration, reduce aeration, impede drainage, and hinder nutrient availability and hence slow tree growth (Greacen and Sands 1980). The long-term effects of extraction show up as patchy growth in the next crop. Studies of stem diameter distribution in second rotation stands have shown poor growth to be associated with extraction tracks (Fig. 18.3). The effects of forest operations, including harvesting, on soil compaction and the loss of soil, organic matter, and nutrients due to disturbance in general are discussed by Gonçalves *et al*. (1997) and in Brazilian plantations in particular by Gonçalves and Stape (2002).

How to overcome this new source of site variability is difficult to decide, but avoidance or at least minimizing damage is the best approach. It is one of the reasons why use of mule extraction in Swaziland has increased from 15% to about 40% of harvesting operations (Fig. 18.4). Careful planning

Figure 18.3 Aerial photograph of second rotation *Pinus patula* in the Usutu forest, Swaziland, showing regular depression in growth coinciding with the tractor extraction routes used when harvesting the previous crop, which caused localized compaction and disturbance of organic matter. (Photo S.E.G. Brook.)

Figure 18.4 Mule extraction of *Pinus patula* pulpwood in the Usutu forest, Swaziland.

of the logging operation and extraction routes will minimize soil damage. Ripping between rows of stumps may be possible to reduce compaction and improve soil physical characteristics (e.g. Gava 2002; Chapter 12, this volume). It is also one reason reduced-impact systems are now being explored and recommended (Elias 2000)—see below.

Management of harvesting residues

After harvesting, a large quantity of organic debris is left behind. A 29-year-old *Pinus elliottii* plantation in Queensland, Australia, had a total ground biomass of 284 t ha^{-1}, of which 28% (19.8 t ha^{-1} of litter, 26.2 t ha^{-1} of foliage, branches, and stem tips, 33.0 t ha^{-1} of stumps) remained after logging (Simpson *et al.* 1999). In Brazil, a 7-year-old stand of *Eucalyptus grandis* had a total above ground biomass of 164 t ha^{-1} of which 24% (23.7 t ha^{-1} of litter, 6.3 t ha^{-1} of foliage and branches, 8.9 t ha^{-1} of bark) remained (Gonçalves *et al.* 1999). In Nigeria, *Gmelina arborea* and *Pinus caribaea* plantations have branches, foliage, and fruits/cones typically comprising 20% and 22% respectively of total biomass (Kadeba and Aduayi 1982; Nwoboshi 1982). A 9-year-old *Acacia mangium* plantation in Indonesia had a total above ground biomass of 190 t ha^{-1} of which 65 t ha^{-1} (34%) was harvesting residues (foliage, branches, and bark) and in addition there was 17 t ha^{-1} of litter (Hardiyanto *et al.* 2000). These are sites where litter decomposition rates are high but for *P. patula* stands on sites where decomposition rates are low, the litter can accumulate on the forest floor at about 6 t ha^{-1} a^{-1}. A 17-year-old *P. patula* stand at high elevation (1450 m) in Swaziland can have a forest floor mass of 79 t ha^{-1} and at a lower elevation (1150 m) it can be 43 t ha^{-1} (Morris 1993*b*), so when harvesting residues are added the quantity of organic debris is very substantial. See also data for *P. patula* and *Cupressus lusitanica* in Table 14.1.

Harvesting residues hinder replanting, both in the ease of doing it and in achieving a regular planting pattern. Also, in dry weather a considerable fire hazard is created. There are several alternatives for handling harvest residues but each has advantages and disadvantages: (i) stack and burn—Fig. 18.5; (ii) mechanically pulverize; (iii) stack into windrows and leave; (iv) spread evenly over the site (Fig. 18.6).

In selecting a method, individual site conditions are important because between final felling and re-establishment, the soil surface is exposed and site degradation can take place. Soil erosion is more likely because the moderating effect of the canopy on the force of rain is removed, the insulating or protective effect of the humus and litter may be disturbed, and the underground network of roots dies. Also indiscriminate burning can destroy the litter/humus layer as well as the debris leading to nutrient losses through rapid leaching, washing away of surface ash during heavy rain, and volatilizing of nitrates. This loss is important where there has been considerable forest floor accumulation of organic matter and thus nutrient accretion.

In Indonesia the government has banned the burning of harvesting residues in plantations but in most countries it is the forest manager who must make the decision on whether or not to burn. Burning leaves a clean site (Chapter 12), the debris is no longer a fire hazard, and the ash, which is rich in base nutrients, may stimulate growth (ashbed effect), but burning does fully expose the soil. To examine these effects, and especially the significance of erosion and soil loss, experiments have been carried out comparing burning versus no burning site preparation for wattle (*Acacia mearnsii*) and eucalypt (*E. grandis*) plantations in South Africa. Sherry (1952, 1954, 1961, 1964) found that burning of harvesting debris of *A. mearnsii*, which always consumed the litter layer, increased water run-off greatly throughout the 9-year rotation of the succeeding wattle crop compared with sites where

Figure 18.5 Clear-felling debris piled in windrows and burnt: Usutu forest, Swaziland. This practice was abandoned in the 1970s.

Figure 18.6 Recently clear-felled site, temporarily free of weeds, where the debris has been 'opened' in readiness for planting.

the debris was not burnt but piled into windrows at right angles to the contour. Both the lower retention of water and greater soil loss in the burnt treatments probably lowered site quality since basal area of stands on burnt sites was poorer. In a 7-year-old *E. grandis* plantation there is about 80 t ha^{-1} of logging slash and sizable amounts of nutrient capital can be lost through burning and leaching. In particular, losses of nitrogen (about 300 kg ha^{-1}) after slash burning, and calcium and magnesium from leaching are significant (du Toit and Scholes 2002).

In Swaziland burning of debris in pine plantations has been found to stimulate the fungus *Rhizina undulata* which causes high planting losses. For this reason burning of debris ceased in 1972 (Germishuizen 1984). Conversely, Old *et al.* (2000) warned a no-burn policy in tropical acacia plantations may result in a build-up of root rot inoculum in harvesting residues that will affect subsequent rotations.

Silviculturally the ideal treatment is for the debris to be broken up into small pieces and left scattered as a mulch. In Brazil, where minimum cultivation techniques are used in eucalypt plantations, harvesting residues are retained on site and stacked along the stump lines, with trees planted in ripped lines between the stumps (Gava 2002). However, where large branches and the tops of trees are present few machines can cope satisfactorily. The present solution adopted in the Usutu forest, Swaziland, is to undertake salvage logging of small-size stem-wood after the main harvesting and either to leave debris scattered over the ground and to introduce a silvicultural operation called 'opening slash' (Fig. 18.6) or leave well-packed windrows and plant in-between. A planting spot is opened up in the debris with a mattock and is done

most easily immediately after harvesting while branches are still green and pliable. For opening 1332 spaces ha^{-1} for planting, 3–4 man-days are required. Debris, treated in this way, becomes a mulch and conserves soil moisture. Indeed, at Usutu in Swaziland this makes planting possible in the dry winter resulting in the advantages of extending the planting season, planting at a cooler time of the year, and reduction of the period between felling and replanting.

It is rarely necessary to remove stumps, though if they become a serious source of infection for diseases or pests this costly operation may be necessary.

Whole-tree utilization, where all parts of a tree are harvested and used, overcomes the problem of debris disposal, but brings other dangers of nutrient depletion, especially on infertile sites, which far outweigh the small gain in dry matter production.

Summarizing the effects of harvesting residue retention or removal in a coordinated series of experiments in eucalypts, pines, and acacia, Tiarks and Nambiar (2000) found no growth response in the subsequent rotation to retaining harvesting biomass on the site in five plantations, and seven sites on which the residues increased growth, with a trend for increased growth as more residues were retained. This suggests that the effects of harvesting residue management on subsequent rotations will depend very much on the soil fertility and other environmental factors at a particular site.

Reduced impact logging in plantations

Reduced impact logging (RIL), sometimes called reduced impact timber harvesting (RITH), is increasingly applied in natural forest to minimize damage to the site and to standing trees not being harvested (Dykstra and Heinrich 1996*b*). The same principles of care in directional felling, planning of extraction routes, use of light vehicles with low ground pressures, and timing of harvests to minimize soil damage apply equally to plantations. Few studies have been undertaken though Elias (2000) found many benefits in developing these more sensitive techniques in Indonesia.

Establishing the next crop

On many areas site conditions for the second and subsequent rotation are radically different from those when afforestation began.

1. Frequently the first crop will have suppressed and eradicated the typical weed species and the replanting site will be relatively weed free (Fig. 18.6) or invaded by different weeds (Fig. 18.7(a)).

(a)

(b)

Figure 18.7 Invasive species in restocking in Swaziland. (a) A pine species trial showing 'with' and 'without' weed control treatments and massive invasion of bugweed (*Solanum mauritanium*) in left-hand plot. (b) Natural regeneration of wattle (*Acacia mearnsii*) invading grassland at the edge of a compartment.

However, where the first crop was heavily thinned or if it was a light-crowned species this suppression of weeds may be incomplete and the advantage of less weed competition for the second crop not so important.

2. The site will be covered with stumps which usually precludes mechanical cultivation.

3. There may be prodigious regeneration of the former species (Fig. 13.9) or viable seeds on the ground awaiting the opportunity to germinate.

4. On old forest sites the large stumps, which would continue to re-shoot for many years during the first rotation of the planted crop, and the abundance of secondary species, will have mostly disappeared. Large logs and stumps will have rotted and no longer be an obstruction.

5. On infertile soil-less sites or man-made ground, one rotation of trees will have initiated the development of a soil structure and rendered it more amenable to tree growth.

6. Where soil cultivation was originally needed to obtain good growth of the first crop, for example,

eucalypts on grassland, the favourable influence of forest growth on soil conditions often renders further cultivation unnecessary.

All these factors may lead to different silvicultural practices compared with those employed for the first crop.

Site preparation
Vegetation clearance is usually not needed. Full mechanical cultivation may be impractical, though it is probably less important; see point 6 above, and subsoiling or ripping may be feasible and beneficial on some sites. Leaving debris scattered helps to suppress weeds and conserves moisture. Special treatments, such as localized cultivation and fertilizing, may be needed on compacted extraction routes.

Planting
This depends on whether natural regeneration is used, though usually replanting of seedlings or clonal stock is favoured for reasons given earlier.

Spacing the plants is easy using the line of old stumps, and stocking can be altered by changing the planting distance along the lines. Planting in the old stump line is usually preferred to facilitate future access.

Weeding

While, generally, weed control is less of a problem with the second and later crops, the changing weed dynamics needs careful monitoring. In the humid tropics in particular the growth of weeds can be a major problem and, for example, at Jari Celulose in the Brazilian Amazon vegetation competition has reduced productivity through all rotations (McNabb and Wadouski 1999).

Three different problems can occur on replanting sites. (i) Although the quantity of vegetation is often much less than when the site was originally planted, new weeds can invade rapidly, for example, bug-weed (*Solanum mauritianum*) on pine sites in southern Africa (Fig. 18.7(a)) or a quite different weed flora can develop owing to changed land-use practices nearby, for example, pasture development and the consequent enormous increase in grass seed. (ii) Unwanted natural regeneration can be a most serious 'weed'. Eradication is cheapest while seedlings are young. On third rotation sites in the Usutu forest, Swaziland, natural regeneration of *P. patula* is hand pulled when about 1-year-old. Invasion can also occur of other 'timber' species which are not wanted. According to Roux (1984), 9% of pine sites in South Africa are invaded by eucalypts and 5% by acacias (Fig. 18.7(b)). These genera are included in the 10 most important alien invaders in forestry! (Goodall and Klein 2000.) (iii) Another problem can be vigorous coppice regrowth of the previous crop. Low quality *Gmelina arborea* sites at Jari, Brazil, have been converted to *P. caribaea*, but killing the stumps proved difficult. Application of glyphosate to young coppice shoots (1–2 m high) was a successful means of control (Welker 1986). This herbicide can be used as a stump application to kill *E. grandis*, provided it is applied on the day of cutting (WRI 1982) and other treatments are described in Chapter 13.

Early protection

The changed site conditions can alter protection problems. In particular the debris and stumps, as well as being a fire hazard, can be important breeding sites for fungi and insects pests. In Swaziland dipping roots of pine plants in the insecticide BHC used to be standard practice to protect against the weevil *Hylastes angustatus* which breeds in stumps and logs (Bevan and Jones 1971). Recent research into the weevil's life-cycle and population dynamics has indicated silvicultural measures which minimize damage: (i) avoid felling in February and March when weevil numbers peak, (ii) plant immediately after felling, and (iii) use very healthy unstressed plants. Only high-risk plantings may need prophylactic treatment, for example, dipping in Permethrin. In Brazil, leaf-cutting ants (*Atta* spp.) were initially a serious problem in eucalypt plantations but with monitoring and an efficient low-cost control operation the damage is now insignificant (Campinhos 1999).

In previously unwooded areas the first rotation of forest provides new cover which may lead to increasing animal populations, especially deer and buck. Felling coupes at the end of the rotation often provide shelter and ideal grazing, and replants may need special protection.

Natural regeneration

Many plantation species exhibit profuse natural regeneration. This is mostly from seed but occasionally it also arises from suckering where shoots emerge from the roots of a tree some distance from the stem; in the case of *Faidherbia albida* this can be as much as 25 m away. As a regeneration system it is little used in industrial plantations, not because trees have not reached biological maturity to set seed, but to take the opportunity to introduce improved genetic stock or change the species on replanting. Nevertheless, it remains an option which can lead to cheaper re-establishment and should not be neglected; see Seitz and Corvello's (1983) discussion of *Pinus elliottii* regeneration on Parana grasslands in Brazil.

Cordia alliodora exhibits profuse natural regeneration throughout central America and, as noted in Chapter 2, plantations of the species in Ecuador are used to encourage natural regeneration of native rain forest species such as *Virola* and *Brosimum*.

In the drier tropics, natural regeneration is commonly effected by animals eating seeds and depositing them in their dung. Goats and camels often aid regeneration in this way, particularly of the many hard seeded *Acacia* and *Prosopis* species whether naturally occurring or in plantation (Ahmed 1986)—in the latter case perhaps betraying some illicit browsing!

Plantation species as invasives

The risk of planted trees themselves becoming weeds was mentioned in Chapter 8. Where good natural regeneration is taking place, that is, there is plentiful fertile seed produced and ground conditions are suitable, unwanted regeneration can soon become invasive and a threat to indigenous flora.

In general eucalypts are rarely invasive, though they are sometime seen to be so in South Africa, for example, pines are more so (*Pinus patula*) in the Eastern Highlands of Zimbabwe, and *Acacia* spp. often start spreading quickly beyond where they were originally planted (Fig. 18.7(b)). Goodall and Klein (2000) list 18 alien tree species in South Africa exhibiting invasive tendencies.

The risk of becoming invasive, or a tendency to weediness, can be illustrated by evaluating the widely planted lowland tropical acacia, *A. mangium*. First, acacias are hard-seeded legume species which produce seed at an early age and which stays viable for many years; two of the pre-requisites for the potential to become weeds (Fig. 18.7(b)). Second, to what extent this is a problem depends on local environmental conditions. *A. mangium*, like most acacias, is a pioneer which colonizes bare ground, especially after fire. There is usually a substantial seed bank in the soil under existing plantations and the seeds may be spread by birds far beyond the plantation. Although *A. mangium* has not yet become a major weed in any country, the threat of invasion into primary rain forest is a possibility, especially after a degree of disturbance by fire, etc. In natural conditions *A. mangium* usually forms a narrow strip along the edges of rain forest where fires are common but does not extend into the rain forests themselves.

Coppice

Definition

Coppice is a forest crop raised from shoots produced from the cut stumps (called stools) of the previous crop (Fig. 4.5(a)). It also describes the operation of felling and regenerating in this way and is sometimes called 'ratooning'. Where a few large trees grow amongst the coppice they are called standards and the system 'coppice with standards' (Fig. 18.8). Though a widespread European practice coppice with standards is rare in the tropics. It has been tried with *Eucalyptus saligna* and *E. maidenii* in Rwanda, and Ethiopian farmers have proved it is a practicable way for growing *E. globulus* for firewood, poles, or even sawlogs in the same stand. This system has also been used to manage teak forests with 25–50 trees ha^{-1} retained as standards on the basis of their large diameter while the rest of the trees are clear-felled to produce coppice (Pandey and Brown 2000).

In coppicing, the originally planted trees are felled and the next crop develops from vigorous shoots (coppice) which sprout from the stumps. If large-size material is needed, usually only one or two strong, well-attached shoots are grown on for the whole rotation and eventually become almost indistinguishable from a planted tree (Fig. 18.9). This process may be repeated several times, but in practice the number of coppice crops is often restricted to three or four, because each time a few stumps die.

Species

The ability to produce coppice is very variable and is determined mainly by species and condition of the stump. Few coniferous species coppice, though young stumps of some species will produce shoots, for example, *Pinus oocarpa*, *Cunninghamia lanceolata* (Fig. 18.8(a)), and some *Araucaria* spp. Many tropical broadleaved species show remarkable coppicing power, in particular some genera in the families of Leguminosae (*Acacia*, *Albizia*, *Cassia*, and *Leucaena*), Myrtaceae (*Eucalyptus*, *Eugenia*, and *Melaleuca*), and Verbenaceae (*Gmelina* and *Tectona*). However, even in these genera not all species coppice well. For example, the important plantation species *Acacia mangium*, *Casuarina equisetifolia*, and *Eucalyptus deglupta* have poor coppicing ability. Also, differences have been observed between provenances of the same species, for example, *Acacia auriculiformis* from PNG, compared with Queensland (Ryan and Bell 1989).

Physiology

Coppice shoots arise either from dormant buds on the side of the stool or from adventitious buds developing in the cambial layer around the edge of the cut surface (callus shoots). In a study of coppicing of Mysore gum (*E. camaldulensis/tereticornis*) in Tamil Nadu, India, 56% of surviving stools developed callus shoots in addition to those from dormant buds, though generally there were twice as many 'dormant' shoots arising from a stool than 'callus' ones (Kondas 1982). With many eucalypts there is an additional swelling of meristematic tissue near the base called a lignotuber (Fig. 10.8(b)). Many coppice shoots sprout from this tissue and its absence in a few species, for example, *E. deglupta* and *E. nitens*, partly explains their poor coppicing ability.

Usually, several shoots are produced per stool, the larger the stool in a stand, the more the shoots.

(a) (b)

Figure 18.8 (a) Young third rotation *Cunninghamia lanceolata* developing as multiple coppice shoots: southern China. (b) Seven-year-old *Eucalyptus grandis* coppice recently cut for firewood for tea curing, but with a few stems left to grow to large size (standards) (Waghi valley, PNG).

Figure 18.9 Six-year-old coppice of *Paraserianthes falcataria*. The coppice shoot has almost completely enclosed the old stump (indicated).

Yields

Coppice growth is sometimes exceptional. Troup (1952) cites data from coppice stands of *Eucalyptus globulus* in the Nilgiri Hills, India, showing a mean annual increment of 40 m^3 ha^{-1} a^{-1} over a 25-year rotation. However, though initial growth of coppice is often rapid, and some coppice crops do show an earlier culmination of maximum MAI, the overall performance of coppice stands is not usually very different from planted stands. A good example of this is provided by a comparison of yields of various coppicing regimes with *E. camaldulensis* in Morocco (Riedacker *et al.* 1985). Coppice does not raise site growth potential, apart from any effects of increased stocking, but only accelerates early growth. Also, as stumps die stocking per hectare gradually declines and there is some evidence that older stools produce coppice which becomes 'mature' (foliage and tree form) sooner. Yields per hectare may decline with each coppice crop after the first, for example, Kaumi's (1983) result over four rotations of fuel yields of *E. grandis* and *E. saligna* in Kenya cut on a 5-year cycle (Fig. 18.10).

Normally, the yield from the first coppice crop is higher than the seedling crop, as Fig. 18.10 shows, but thereafter declines with each coppicing. In India, where four coppice rotations of *E. globulus* are grown after the initial seedling crop, a fall in yield is expected of 9% in the third rotation and 20% in the fourth (Jacobs 1981). However, in Brazil the first coppice crop in eucalypts may have a 50% reduction in productivity on some sites due to soil nutrient depletion, especially of phosphorus (Barros and Novais 1996).

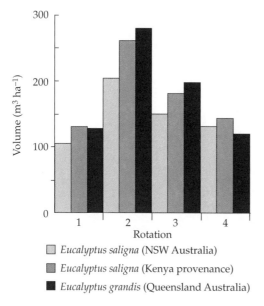

- ☐ *Eucalyptus saligna* (NSW Australia)
- ▩ *Eucalyptus saligna* (Kenya provenance)
- ■ *Eucalyptus grandis* (Queensland Australia)

Figure 18.10 Productivity of successive 5 year cuts (rotations) from eucalypt coppice in Kenya—yield of stacked billets per hectare. (Reproduced by permission from Kaumi 1983.)

Silviculture and coppice working

An excellent summary of good coppicing practice for tropical countries, and especially village or farmer scale social forestry projects, is provided in a short article by Pawlick (1989).

Tree-felling and factors affecting coppice growth
The season of felling affects coppice regeneration. Felling during the dry season delays sprouting and increases the risk of the stump drying out. Cutting at the end of the wet season is sometimes recommended when soil moisture levels and a tree's carbohydrate reserves are high. In Nigeria, *Gmelina arborea* benefits from cutting in the late wet season/early dry season period (Okorie 1990).

The method of felling can also be important. According to Schönau (1975), the practice in southern Africa of bark stripping eucalypts before cutting, to aid direction of felling and drying out, can be harmful. A low, cleanly cut stump without tearing bark from wood is ideal. And it is found that saws, rather than axes, yield better coppice since the rougher surface stimulates more rapid callus development on the stool. However, hacking at stems with blunt instruments, so typical of villager and farmer coppicing, is seriously damaging by depressing resprouting potential and increasing risk of disease infection.

Where brash is left scattered over a site it should not cover the stools to avoid distortion of the tender, emerging coppice shoots.

With eucalypts the height of the stool does not affect the number or growth of coppice but tall stools reduce the yield of the previous stand and make the shoots less wind-firm (WRI 1972). With many species a stool height of 30–50 cm is recommended to maximize sprouting potential (Pawlick 1989), but this height is clearly only acceptable for hand-worked coppices. For *Calliandra callothyrsus* a trial at PICOP in the Philippines showed that a high 50-cm stool sprouted more shoots and grew more vigorously than stools cut at 30 cm (Tomaneng 1990). In Thailand, *Acacia auriculiformis* often fails to coppice if cut at ground level, produces moderate coppice on 50-cm-high stumps, and vigorous coppice on tall 1 m stumps especially if one live branch is still attached. Also, the overall size of a stool affects coppice growth. Small stools arise from subdominant trees, which is evidence of their genetic inferiority, and produce poor coppice 'once a runt always a runt' (Stubbings and Schönau 1979). Very large stools may inhibit sprouting owing to thick bark. However, where large stools do sprout coppice it is likely to be vigorous (Venter 1972), since large stools arise from the most vigorous trees of the previous crop and they will also have a large root system. Kondas (1982) demonstrates this effect with *E. camaldulensis* of increasing stool productivity with increasing diameter except for the very largest 5%. One year after cutting, large stools (mean 19.5 cm) supported coppice with four times the basal area of small stools (mean 7.5 cm).

Thinning coppice shoots
A most important but time-consuming operation in coppice management for pole production is to reduce the shoots on a stool; sometimes called 'singling of coppice'. A stool produces many shoots and, if left untended, several will be suppressed naturally but often five or more will continue to grow. Such shoots are spindly, they lean away from the stool, have one-sided crowns, and have poor contact with the stool. On short rotations for firewood or production of large quantities of stick-size material, for example, for mud and wattle hut walls, such unthinned crops are often acceptable, even desirable.

Reduction of coppice shoots is usually done in two stages. In southern Swaziland *E. grandis* coppice is thinned to 3–4 shoots per stool at 9 months and to only 1–2 at 18 months (Fig. 18.11). It is important not to thin coppice too early since this stimulates development of additional unwanted shoots from epicormic buds. If above average stump mortality occurs commonly two shoots are

Figure 18.11 *Eucalyptus grandis* coppice, age 1.9 years, recently thinned to 1–2 shoots per stool.

left on all remaining ones to augment the stocking. At each reduction a stool is left with shoots of similar size well spaced around it. A single reduction to one or two shoots per stool over-exposes the remaining shoot(s) and increases the risk of breakage at the point of attachment. Coppice shoot growth is fast and time must be allowed to let callus develop around the stump to provide a secure base (Fig. 4.5(a)).

Coppice reduction is expensive. Howland (1969) found that with *E. saligna* coppice in Kenya, thinning was only worthwhile if the sticks produced could be sold. Volume and value yields were greatest when reduction was made at 18 months to 2–3 shoots per stool.

Thinning may also reduce total biomass production over short rotations, but growth of the remaining best quality shoots is enhanced. Also, this loss in yield is unimportant where thinning of coppice is part of agroforestry systems and the tool is used to maintain a favourable environment for the associated food crop.

Coppice rotations, stump mortality, and gapping up

The size of the roundwood required will largely determine the rotation length. The number of rotations that can be grown without replanting depends on stump mortality (i.e. the number that die and do not re-shoot when cut) and the poorer vigour from older stools. In Madagascar, *E. robusta*

was introduced a hundred years ago to provide fuel for the railways and some of the old stumps are still being coppiced on cycles or 3–5 years to provide domestic fuelwood and charcoal (Bertrand 1999). Similarly in India, *E. globulus* has been coppiced successfully in the Nilgiri Hills on 10-year rotations for more than 100 years (Jacobs 1981), and the same is true of *Markhamia lutea* in western Kenya (Pawlick 1989). As a general rule, three or four coppice rotations can be cut before full-scale replanting need be considered.

In Kenya, *E. saligna* coppiced for firewood is worked on a 12-year rotation and three to four crops are expected before replanting is necessary. In Brazil, *E. grandis* grown for industrial charcoal (Fig. 18.2) and pulpwood is worked on a 6–9-year rotation and replanted after two coppice crops. In the PNG highlands *E. grandis* and *E. robusta* are managed on a 6-year rotation for industrial fuelwood (Fig. 18.8). In Swaziland, a rotation of about 16 months was used in some areas of *E. smithii*, managed specifically for essential oils and coppice, was still being harvested 20 years or more after the first cut (Coppen and Hone 1992).

Teak coppices vigorously, sometimes even after attaining a large diameter. In the dry zone of India, where teak plantations grow slowly, they are managed on longer coppice rotations of about 40–50 years sometimes with standards. As more teak is planted, a market has developed for immature thinnings for posts, poles, furniture components, and small flooring boards (Pandey and Brown 2000). In Thailand, farmers plan to harvest half their crop at 10 years old

(22–25 cm diameter) and the remainder at 20 years (Mittelman 2000). Relatively short coppice rotations of teak are likely to become more common.

Stump mortality

Stumps die because of over stocking on infertile sites, poor felling practices, incipient decay, burning debris nearby, unhealthy parent trees, extraction damage, increasing age and decay of roots, etc. Stool mortality in *E. grandis* is in the range of 2–8% per rotation (Stubbings and Schönau 1979; Kaumi 1983) and annual stump mortality for other eucalypts is commonly in the range 5–10%.

Gapping up

A common European practice is to replace dead stumps at each coppicing by planting new seedlings or layering a shoot from an adjacent stool. This operation of 'gapping up' ensures full stocking and continual addition of new plant material to the site and negates the need to replace the whole coppice crop. It has proved feasible with *E. grandis* in Brazil (Freitas *et al.* 1979) and should be considered more widely except where, of course, complete introduction of improved genetic stock is desirable.

Pollarding and shredding

Pollarding is rather like coppicing at a higher level—see Figs 4.5(b) and 20.11. A tree is cut back at about 3 m above ground, to yield fodder or firewood, rather than to near ground level as in coppicing. The object is primarily to ensure that the regrowth that develops is out of reach from livestock browsing, especially by cattle, goats, sheep, and camels, but also it is a way of producing one large pole-length from the bole while continuing to obtain regular supplies of firewood, etc. Many savanna and semi-arid tree species are managed in this way by pastoralists, but it is equally applicable to planted trees around villages and on farms. It is a common practice with planted *Grevillea robusta* in East Africa and *Azadirachta indica* in West Africa when the trees are well-established. Pollarding

typically commences once the stem reaches about 10 cm in diameter.

Various modified forms of pollarding are used in agroforestry and in foliage production, for example, for essential oil production, where trees are decapitated or very severely pruned to maintain light levels or stimulate fresh growth at a convenient height for crop management. The distinction between coppicing and pollarding is not marked, but whenever growth is encouraged from the top of a stem rather than from a stump, it is called 'pollarding'.

The practice of shredding—see Fig. 4.5(c), where branches are stripped from the length of the stem is widespread in the more arid sub-tropics—is really another variant in tree management relying like coppicing and pollarding on the capacity of a tree to sprout new shoots. In natural forest the equivalent is the profusion of new shoots along the stem after fire has scorched and killed the crown but not the bark along the trunk. This is commonly seen with eucalypts as well as other species.

Coppicing and pollarding in plantation forestry

Coppicing is a widespread silvicultural practice to produce successive crops, cheaply and conveniently, of usually small roundwood material such as fuelwood, posts, poles, and pulpwood. Many 'plantations' of eucalypts and gmelina are managed in this way. Equally, it is a very important regeneration method in social forestry projects and, with suitable species, an excellent way of perpetuating woodlots and stands established for all the domestic wood products needed from stick to pole size.

Pollarding is mainly used as an individual tree management method which, alongside coppicing and pruning, manipulates the position and effect of the tree in its immediate environment. Beside roads and houses it prevents a tree getting too big (Fig. 20.11); in fields, as part of agroforestry, pollarding maintains light levels or provides crop mulch; in open rangeland it provides fodder for livestock without harming the tree itself. It is widely used and applicable to many trees planted in the tropics for social and domestic needs.

CHAPTER 19

Protection—an introduction

Introduction

A planted forest of any kind: commercial, amenity, or village woodlot, is a considerable investment and an accumulating asset of raw material. Its protection is an essential part of silviculture. The nature of a plantation, its uniformity in age, species, and genetic make-up, together with the environmental conditions of the site it occupies, renders it more susceptible to some forms of damage, notably fire and climatic hazards. Whether this also applies to diseases and pests is considered in Chapter 24. Successful plantation forestry is only possible provided there is adequate protection. Managers accept there are risks associated with plantations but are becoming increasingly risk adverse. They are seeking tools and techniques that enable them to assess risks and put in place management strategies to minimize them (Arbez *et al.* 2002).

Damage is unpredictable, irregular, not always easily detected at first, and varied in severity, all of which can lead to its importance being underestimated. Damage can be catastrophic, especially to small countries with limited forest and small-scale forest growers. In April 1973 one 20-min hail storm killed several hundred hectares of 20-year-old *Pinus patula* plantations in northern Swaziland because damaged tissue became infected by the fungus *Sphaeropsis sapinea* (*Diplodia pinea*). Similar damage in Australia, South Africa, and elsewhere has led to managers replacing pines such as *P. patula* and *P. radiata* with less susceptible species. In 1982 a typhoon damaged about half of *Paraserianthes falcataria* plantations of the Paper Industries Corporation of the Philippines (PICOP) in Mindanao (Poole 1989) and on Hainan Island, China more than half the trees in a young stand of *Acacia mangium* were seriously damaged during another typhoon and 9-h rain storm (Pan and Yang 1987). In 1988 a hurricane destroyed over 3000 ha of pines comprising a quarter of all Jamaica's pine plantations (Bunce and McLean 1990). A single unseasonal, severe frost in tropical climates can kill or severely damage acacias (Yang *et al.* 1994) and eucalypts (Rockwood 1991). Such examples can be multiplied many times, especially if destruction by fire is included. Indeed,

many plantations are threatened by fires every year in the dry season, though the amount burnt can vary enormously.

A proper perspective needs to be maintained when discussing problems of protection. Losses through reduced growth or death in intensively managed plantation forests due to pests, diseases, fire, etc., are generally less than in native forests (Gadgil and Bain 1999), and the great majority of plantations are successful because adequate protection is maintained and serious damage is rare.

Forest protection is a large subject. There are many kinds of damage and numerous species of fungi, insects, and other harmful agents. Tree species vary in their susceptibility to damage, and site conditions and forestry practice both influence protection problems. Moreover, weed competition and nutrient deficiencies are sometimes considered as protection problems, but are treated separately (Chapters 13 and 14). The purpose of this chapter is first to consider how to identify and assess protection problems and, second, to comment on approaches used to overcome problems.

It is convenient to divide protection problems into two groups:

1. Inorganic and man-made damage—through fire, climate extreme, air pollution, domestic animals, theft, etc. Both cause and damage are usually obvious to see and can occur suddenly.
2. Organic damage—through fungi, insects, mammals, or micro-organisms, etc. This type of damage is often difficult to detect and may not even be visible, for example, wood decay. The causal organism may also be difficult to identify and damage frequently develops slowly.

Inorganic and man-made damage

It is often possible to anticipate damage of this kind to some extent because the cause of the problem is usually obvious: fire burning trees, wind causing stem breakage, etc. Anticipation of problems is an integral part of plantation planning and essential if proper preventive measures are to be taken. A simple example is to ensure adequate firebreaks are maintained between compartments. Species' trials, site evaluation, and the preplanting survey are tools which provide much of the information around which a protection strategy can be based.

Climatic damage

This is considered only briefly since the influence of climate on management of plantations (silviculture) and the need for protection have been referred to already in several places, notably Chapters 1, 8, 12, and 13. However, for all types of climatic damage, whether drought, flooding, high winds, hail, frost, or lightning, not only must one know how to assess damage and what direct preventive measures to take if possible, but also be aware of contributory factors, and other effects the damage can have.

Indirect effects of climatic damage
Trees which are damaged or under stress, for example, from drought, are more susceptible to other forms of damage, notably insect pests and fungal diseases. If a tree is already damaged or weakened secondary pests can assume much greater importance, infections can take place through damaged tissue, and nutrient deficiencies may be induced (Chapter 14). Speight and Wainhouse (1989) provide an analysis of this host plant: stress interaction for damaging forest insects. The example of thinning to relieve between-tree competition as a means of reducing bark beetle attack in *Pinus caribaea* and *P. oocarpa* stands in Honduras (Reyes and Groothousen 1990) was noted in Chapter 16. The fungus *Botryosphaeria dothidea* is a common cause of disease in *Acacia mearnsii* plantations in South Africa where trees are wounded or stressed by frost, drought, hail, cold or hot winds, insect damage, etc. (Roux *et al.* 2000). Other examples are cited later in the chapter. To some extent the converse relationship also occurs, for example, where roots are damaged and the tree is then windthrown as occurred in Uttar Pradesh, India, where many eucalypts blew down owing to root decay by *Ganoderma incidium* (Annual Forest Research News Bulletin, India 1973–74). In 1985, *Phellinus noxious* root rot undoubtedly contributed to the severity of cyclone damage to one-third of all *Cordia alliodora* trees in 450 ha of plantations in Vanuatu during a 6-h period (Neil and Barrance 1987; Neil 1988).

However, serious and destructive climatic damage not only directly influences the growth of trees but, because of its social consequences, can affect forest growth and tree planting activities over whole regions. Drought will reduce growth and

even kill trees and, if prolonged, other destructive influences on forest cover begin to take place.

Such effects occurred in the Sahel droughts of 1970–76, 1983–86, and again in 1990–91. Though herds of domestic animals declined and many people became refugees and dependent on famine relief for food, the shortage of grass led to acute over-grazing and many trees, such as, *Acacia tortilis* and *Faidherbia albida*, were overcut to provide material for cattle stockades, hut construction, fodder, and fuelwood. This loss of slow-growing trees worsens the situation when another drought occurs. In good years, with plentiful grazing, domestic herds rapidly increase and intensify the pressure if drought conditions return. The decline in vegetative cover this induces hastens soil erosion and worsens the grazing value of the land still further, creating a permanent loss. Drought greatly increases pressure on the few trees in arid and semi-arid regions.

Assessment of climatic damage
Severe damage is mostly obvious (dead trees, windthrown stands, flooded compartments) and counts of mortality and estimates of areas affected can be made by field inspection or, in certain situations, aerial survey, or satellite imagery. Where damage is less severe, for example, growth is just depressed or some trees have stem-break or tops blown out, a more detailed survey is needed. This can be included as part of normal inventory work or as a special exercise to investigate the scale and extent of damage in question. For example, the effect of moderate hail damage in *Pinus patula* plantations in Swaziland of reducing or obliterating 1 year's height increment only became evident during stem analysis studies of growth (Evans 1978a).

Where the damaged stands have to be salvaged (see below) assessing the physical effects only indicates part of the loss in value. Often, the amount of saleable timber is reduced, harvesting may be more costly, timber degrade may have occurred, for example, blue stain, and the produce from damaged stands will often not be part of planned production and so a market must be found. Also, where trees have been uprooted, replanting is more expensive.

Approaches to protection
Limiting impact of climatic damage is the main approach, primarily through careful choice of species, as discussed in Chapter 8. Species differ markedly in their tolerance of drought, flooding, frost, and to some extent wind damage; see, for example, Tables 8.3 and 13.1. However, some silvicultural practices can reduce some kinds of damage,

for example, timing of planting and using moisture-conserving site preparation methods in arid zones (Chapter 23), and avoidance of thinning in stands prone to wind damage. Control or treatment of climatic damage is mostly impossible so prevention through careful planning is the best course of action.

Fire

Every year wild fires sweep through large areas of natural forests and grasslands in the tropics and extension of these fires into plantation areas is a constant threat. In the Congo, for example, uncontrolled fires of varying intensity affect 15% of the 40 000 ha of eucalypt plantations annually. Fire will often be the most important danger facing a newly established plantation.

1. Extensive burning of grassland and as a tool in shifting cultivation is widely practised and is a traditional part of much land husbandry (Fig. 19.1). It is difficult to persuade neighbouring landowners of the risk fire presents and restricting the right to burn is extremely difficult where there is little respect for the rights of other land-users.
2. Burning of debris from forest clearing to plant agricultural crops, such as oil palm, can be a hazard and was a major contributing factor to the extensive forest fires in Indonesia in 1997–98.
3. In most of the drier tropics there is a long dry season with high temperatures that rapidly dry off vegetation and make it highly flammable.
4. Much land used for afforestation is grass covered, and dry grass burns very easily.

The greatest danger from fire is while a plantation is young before the canopy has closed and suppressed the ground vegetation (Fig. 19.2), though in very dry conditions with strong winds, mature plantations can also be destroyed.

Forest fire planning
Fire management policies and standards need to be established by those with overall responsibility for plantations. The amount of funds required and their allocation should be based on a definition of the required forest fire protection level. This may be done on the basis of a fire damage assessment which values potential direct damage to the established crop and the cost of its replacement. It may also include damage to future site productivity, watersheds, and other environmental values (Caban 1998). Good planning involves collection, maintenance, and analysis of relevant environmental, logistical, and economic data much of which is best displayed on maps (Chandler *et al.* 1983).

All plantation enterprises should have a fire plan. It is a written document bringing together all relevant information about fire protection for a plantation. It lays down the fire protection strategy and contains details of: (i) how fire-danger rating is assessed; (ii) what is to be done; (iii) who is responsible for doing what; and (iv) where the main effort in protection is to be used. Effective plans can be prepared to protect a plantation against fire or at least limit the damage caused. Fire management planning is a continuous process requiring annual revision to assess risks as stands change in their susceptibility to fire, equipment gets old, and there are staff changes.

In protecting any plantation the aims are to: (i) prevent outside fires spreading in; (ii) prevent fires being ignited inside; and (iii) limit the spread of a fire. Fire protection strategy requires: provision of fixed defences, the means of firefighting and suppression, communications, and publicity and warnings.

Fixed defences
A firebreak slows the progress of a fire because there is little flammable vegetation or because it is planted with a species that does not burn easily,

Figure 19.1 Grass covered slopes near Bulolo, Papua New Guinea (PNG), which have just been burnt.

Figure 19.2 Burnt (foreground) *Pinus caribaea* plantation near Oomsis, PNG. Note effects of previous fire in background. Young pine plantations in grassland are highly susceptible to fire damage.

and provides an assembly place from which to tackle the fire. They are constructed along a boundary (Fig. 12.1) and between compartments (Figs 4.1 and 12.3). There are no fixed rules about the width of firebreaks, though if the unplanted ground is grazed by domestic animals, using a wide break becomes less wasteful. Firebreaks have a high maintenance cost and will not prevent the spread of a fire driven by high winds.

Another fixed defence is reduction of combustible material on the forest floor. In older plantations, branches, litter and undergrowth accumulate (the fuel load) and can be a serious fire hazard. Most of this material can be removed by burning it under controlled conditions, called compartmental or prescribed burning. A country or forest region generally lays down rules or prescriptions governing the use of controlled burning for each species, for example, Ross (1977) for plantations in Zambia, and Soares (1981) for tropical pines in Minas Gerais (Brazil). Prescribed burning is used in natural forests of eucalypts in Australia and pines in the southern United States to reduce the fuel load but is much less common in plantations (Luke and McArthur 1978; Wade and Lundsford 1990). A key ingredient in such prescriptions is favourable weather with very little wind, and prediction of periods of quiet weather is important in planning safe operations. However, any practice which reduces the amount of combustible material (grazing in plantations, clean weeding around eucalypts) lowers the fire hazard.

A third important fixed defence is the choice of species. Less susceptible species, both to the effects of fire and to burning, are planted at high risk locations such as next to roads and along boundaries. Generally broadleaved species are less flammable than conifers but some conifers, for example, *Pinus caribaea*, have great powers of recovery from fire damage. Fire resistance in eucalypts is mainly determined by bark thickness so young plantations, and even older trees of a few species, for example, *Eucalyptus deglupta*, are easily killed. Plantation hardwoods such as *Acacia mangium* and *Paraserianthes falcataria* are fire sensitive. Some other broadleaved species create flammable conditions at certain times of the year such as with the leaf litter which accumulates under *Tectona grandis* and *Gmelina arborea*. Planting fire-resistant species as firebreaks, once extensively used in southern Africa with eucalypts between compartments of pine, is now deprecated or at best used only in conjunction with other methods. The growing plant and ground fuel attributes of potential fire retardant plants for green firebreaks have been studied by Tran and Wild (1998).

Firefighting and suppression

A fire will only burn if oxygen, heat, and combustible material are all present: this is the 'fire triangle'. If one factor is reduced or eliminated the fire will go out. The three main ways of fighting forest fires each aim to reduce a different component of this triangle.

1. *Water*. Water suppresses fire by lowering temperature because much heat is needed to vaporize it. It is applied from back packs, portable pumps, direct from vehicles, or sometimes from fixed wing aircraft or helicopters. A supply must be available to replenish water-dispensing units and roadside dams or special tanks are located in a plantation. Additives may be used to increase 'wetness' and surface contact of water or to increase its bulk, for example, foaming agents, or its viscosity to reduce runoff. The term 'fire retardant' applies to all chemicals used in forest firefighting to enhance the putting out of fires.

2. *Beaters*. The fire is hit with a flat surface which temporarily excludes oxygen. There are numerous types of beater, but all are basically a handle with a flap at one end. In India green branches are still commonly used (Sidhu and Thakur 1998). Nevertheless, managers should avoid cost-cutting measures that reduce the quality of suppression equipment such that it is ineffective or fails on the fire line. Beaters are only suitable for surface fires which are safe to approach to within about a metre. Spades and shovels are sometimes used for beating out flames and for smothering the fire with earth.

3. *Constructing temporary breaks to contain a fire.* For low intensity, ground fires rake hoes or similar hand tools are commonly used to construct fire lines free of combustible material. Large plantation projects usually have bulldozers available which can be used to contain a fire by clearing a break around it. Backfiring (backburning) and advance felling also have this effect. The decision to backfire should not be taken lightly as it involves losing more area to the fire.

Sophisticated firefighting equipment, such as helicopters and fixed wing aircraft, has been developed to fight fires but in many countries in the tropics simple hand tools are still the only equipment used. The range of equipment used for fire suppression is described in specialized texts on firefighting, for example, Chandler *et al.* (1983) and Overton (1996). Whatever means are used to extinguish a fire two points are essential. First, the equipment must be kept serviceable and operated by people who know how to use it. Second, when the main fire is extinguished guards should be left to patrol the area to continue damping down to prevent the fire restarting.

Communications
Fires burn quickly. Rapid detection and movement of firefighting teams to the scene will greatly reduce damage.

For large plantations detection is usually by fire watchers posted at times of high fire danger in lookout towers (Fig. 19.3). In the Haldwani project, Uttar Pradesh, India, a network of 32 fire towers led to earlier detection and much smaller average size of fire as firefighters were sooner on the scene (Saigal 1990). Where there are no towers observers can be posted at good vantage points with views over the plantation. However, rapid detection is only the first step, the alarm must be raised, and information about a fire, its location and possible size, relayed to the firefighting teams. This is done by two-way radio or telephone, but if unavailable pre-arranged signals are used to give general warnings. At the very least a fire watcher should have some means of transport to get quickly to the forest depot. The US Forest Service began using aircraft to detect forest fires as early as 1915 and aerial spotting of fire is still used during periods of very high danger in some large industrial plantation projects, for example, Usutu forest, Swaziland. More sophisticated detection methods using remote sensing satellites or infrared equipment in aircraft are used in some circumstances. During the 1997–98 fires in Indonesia that burnt more than 3 million ha satellite imagery was used to detect hot spots and monitor progress of the fires (Mori 2000).

Figure 19.3 Old-style fire tower used in the plantations of Aracruz Florestal, Brazil, in the 1980s.

Rapid detection of fire and raising the alarm are the first two essentials of good communications. Appropriate dispatching procedures and conditions enabling firefighting teams to reach the scene quickly are also essential. Thus a plantation requires at least a rudimentary system of roads and tracks passable by vehicles. Roads leading to the plantation boundary and a boundary road are of special value so that rapid access is gained to an approaching fire.

Causes of forest fires
Though some forest fires in the tropics are caused by lightning, most are caused by people, both accidentally and deliberately, and are associated with land-use practices and changes. For example, the huge loss of over 12 000 ha of Fiji pine plantations due to arson in 1987, and the annual destruction of thousands of hectares of plantations in Indonesia, especially on *Imperata* grasslands. The causes of fire in a reforestation area in South Sumatra, Indonesia were reported as: (i) fire escaping from farmers' land clearing, (ii) hunters burning grass to create young shoots to attract wild animals, (iii) lack of care with cigarettes and campfires, and (iv) deliberate fire initiation linked to social protest (Sabaruddin 1988 in Wibowo *et al.* 1997).

Nevertheless, the chance of fires damaging tree plantations can be significantly reduced by: (i) educating people about the danger of fire and providing warnings both generally, for example, over the radio or mounting national fire-prevention weeks and other publicity, and specifically such as where public roads enter a forest; (ii) by restricting public access to a forest at times of high fire danger. Maintaining good relations with neighbouring land-users also helps to reduce the risk of fire encroaching from their land especially if fire is used in land management, such as burning sugar-cane fields after harvest or grassland to encourage new growth. A community-based management approach is increasingly being used to better integrate people and their use of fire into vegetation management and land-use (Goldammer *et al.* 2002).

A plantation can become the target of local displeasure with 'government' or plantation companies, and setting it alight is an all too common manifestation of some grievance. Sensitive interaction with the local community during the plantation planning phase can minimize grievances. In smaller-scale rural development forestry projects, although the fire hazard is still high, local participation and greater individual care greatly reduce the risk of someone starting a fire or of a fire getting out of control on neighbouring land.

Other damage caused by human activities

This can be categorized as: (i) theft of trees, wood, and other produce; (ii) squatting on land set aside for afforestation; see Chapter 5; (iii) grazing domestic animals in the plantation and over-zealous cutting of branches for fodder or firewood; and (iv) air pollution.

Damage of this nature generally reflects a prevailing social or industrial problem. Though sometimes damage is very serious, it is rarely deliberate with the intent of doing harm; it arises in the social context from other pressures on people such as shortage of food, no wood for burning, not enough land for building a house or cultivating a vegetable plot, nowhere to graze livestock, or too little fodder owing to drought, etc. Also, as discussed in Chapter 5, ownership and rights over trees, even attitudes to property, vary greatly and certainly cannot be assumed to be based on western patterns and practices.

Theft from the forest
Theft of firewood and building poles is common in the tropics. Where shortage of firewood for cooking and heating is acute, 'stealing' from a nearby plantation is almost inevitable since people must eat. In some cases the right to collect firewood is granted but often demand outstrips supply. In Pakistan both gathering wood from the ground and cutting off

Figure 19.4 *Pinus roxburghii* trees in the Murree hills, Pakistan, with stems stripped of almost all branches by firewood collectors.

lower branches is permitted, but shortage of wood leads to abuses (Fig. 19.4). In regions of extreme shortage where firewood prices are high even fence posts may be taken and newly planted trees uprooted.

There is no straightforward solution to theft when hunger, cold and lack of shelter are the real causes. Clearly, firewood can be collected and supplies made available in an orderly way, especially when all stakeholders are involved. However, to prevent theft, firewood may have to be sold at a very low price, or at least at no profit, especially where the right to cut wood is considered inalienable. In the case of firewood, many authors (e.g. Miller *et al.* 1986) stress the need for a multiple approach, that is, more economical use of forest produce including agroforestry with less waste, more efficient use of the wood itself (notably using simple stoves which conserve heat), promotion of social forestry including establishment of new fuelwood plantations and better management of indigenous woodland.

Grazing domestic animals
This has and continues to be a most destructive agent of both natural forest and plantations in much of the semi-arid tropics. Cattle, sheep, goats, camels, etc., eat tree foliage and must be excluded from a plantation while it is young. Young plantations often have dense ground vegetation between the trees which increases the attractiveness for grazing. Moreover, land acquired for afforestation may have

been casually grazed but when planting begins the sudden curtailment of this 'right' may be deeply resented by graziers used to pasturing their animals on the land. They may let their animals into the plantation to graze or cut branches for fodder.

Protection against unlawful grazing can be very expensive. In Sahel countries both fencing and shepherding (employing watchmen) are essential for establishment of plantations. If trees are unprotected they are soon devastated for firewood and by grazing. In Ethiopia, as noted in Chapter 12, a low boundary wall is erected to demarcate the plantation and shepherds are posted at intervals to keep out livestock. At one stage of the Bilate community forestry project more than 55 shepherds were employed along the 27 km boundary, not because the risk of browsing damage was excessive, but villagers agreed this was the best way to protect their new woodland from their own livestock (Evans 1989). Similarly, with plantations in Nigerian savanna fencing, mass propaganda, clean weeding, and watchmen are needed to combat damage from man and animals (Turabu 1977).

However, these solutions do not always solve the basic problem of people wishing to pasture their animals. Where a farmer's land has been reduced or displaced by tree-planting, perhaps as part of a food-for-work programme during famine, exclusion of grazing need only last while trees are small. Additionally, cutting grass and herbage from between the rows of trees for fodder is entirely feasible. This is practised in many community plantations in Africa and Asia and has the added benefit of reducing the fire hazard.

Further consideration of these problems is included in Chapters 21 and 23 because they often occur on sites where tree-planting is already difficult.

Air pollution

Much concern has been expressed in temperate countries about the possible harmful effects of ambient levels of air pollution on the health of trees and forests (Nilsson and Sallnäs 1990). Causal relationships have proved very elusive, as distinct from clear evidence of damage to trees downwind from industrial works such as smelters owing to noxious emissions, e.g. in the vicinity of Richards Bay, Natal, and several industrial complexes in Brazil. Pollution levels in the tropics are likely to be low compared with the world's more industrialized regions and measurably harmful effects are unlikely at the present time. The trees shown in Fig. 19.5 in Mexico City are dying but whether from pollution (smog) or, for example, from changes in soil drainage is hard to demonstrate. Norfolk Island pine (*Araucaria heterophylla*) in coastal plantings near Sydney, Australia has been damaged from detergent in windborne sea spray. The effects of sulfur dioxide and hydrogen fluoride on the growth of *Eucalyptus tereticornis* have been reported by Murray and Wilson (1988) and for a review of the response to environmental pollution of trees (mainly temperate species) see Kozlowski and Constantinidou (1986).

Organic damage

In the introduction it was pointed out that numerous organisms can harm trees: fungi, insects, animals, birds, micro-organisms, etc. No attempt is made here to consider these separately or even in groups; for general reference to pests and diseases see Browne (1968), Gibson (1975, 1979), and Speight (1997). Moreover, there are many national and regional accounts of pests and diseases of plantation trees.

Figure 19.5 Mature planted trees in the centre of Mexico City showing severe crown dieback: air pollution might be a cause here and of damage to subtropical pine forests in the mountains beyond the city boundary.

They include: Hodges and McFadden (1986) for tropical America; Hutacharern *et al.* (1990) and Day *et al.* (1994) for Asia-Pacific, Sharma *et al.* (1985) for southern India, Nair (2000) for Indonesia and Sang (1987) for eastern and central Africa. For diseases there are species/genus accounts such as Gibson (1979) for pines, Keane *et al.* (2000) for eucalypts, Boa and Lenné (1994) for tree legumes, Old *et al.* (2000) for tropical acacias, and Sharma and Sankaran (1987) for *Paraserianthes falcataria*.

The outstanding initial growth and productivity of exotic species, such as *Eucalyptus grandis*, in plantations have been attributed to separation from their natural enemies but gradually they have been attacked by new pests and diseases. Nair (2001) analysed experience with *Acacia mangium, Eucalyptus* spp., *Gmelina arborea, Hevea brasiliensis, Leucaena leucocephala, Paraserianthes falcataria, Pinus caribaea, Swietenia macrophylla,* and *Tectona grandis,* the most commonly planted species in the tropics. He concluded that monocultures are at greater risk of pest outbreaks than natural forests but that in plantations of exotic species are at no greater risk than indigenous species. Diversity of plantations, achieved by planting two or more species, also minimizes risk (Wingfield *et al.* 2001).

This part of the chapter outlines the general approach to dealing with organic damage through four stages: detection, identification, analysis, and action.

Detection

A serious problem, though not obviously catastrophic, can pass unnoticed for a long time. This is especially true if a plantation is not frequently visited and if forest staff is not trained to look for signs of damage. Detection of damage is the first main step in combating organic protection problems. Extensive surveys to identify existing pests and diseases and assess the incidence of risk are a prudent measure. Such surveys are being carried in the extensive plantations of *Acacia mangium* in Asia (e.g. Wylie *et al.* 1998; Old *et al.* 2000) and in many eucalypt plantations for example in Chile (Wingfield *et al.* 1995) and Vietnam (Old and Ivory 1999). Such surveys require local staff to be trained in detection methods and the networking in pest and disease management among countries.

However, it is important to note at the outset that usually it is damage to the tree which is first observed rather than the causal insect or a fungal fruiting body. The following may signal a protection problem: the forest floor, condition of the tree, and stand growth itself.

Forest floor
Examination of the forest floor not only can reveal, for example, sporophores of butt-rotting fungi, but may also indicate what is taking place in the canopy above, particularly when direct inspection of tree crowns is not easy.

1. If the forest floor under pine stands is covered with bits of green needles a defoliating insect is the likely cause, for example, *Lymantria* spp. and *Imbrasia cytherea* on *Pinus patula* (Fig. 19.6). Needle-cast fungi may cause similar symptoms but the needles are usually mottled in colour (and there is no sound of frass falling which can be heard during serious insect defoliation). Heavy defoliation allows more light to penetrate the canopy and this may be visible when looking through a stand. If higher light levels continue for long an influx of weeds follows. Indeed, unusual occurrence of weeds on the forest floor is also a useful sign of a prolonged increase in light conditions, perhaps due to poor needle or leaf retention or a group of dead trees from lightning damage.
2. The presence of damaged cones or fruits on the forest floor indicates predator damage by birds or arboreal animals. Squirrels in Cameroon and parrots in Ghana cause heavy damage to the fruit of *Triplochiton scleroxylon*. Cockatoos damage cones of *Araucaria hunsteinii* in PNG and pines in Australia. Debris from this kind of damage is

Figure 19.6 *Imbrasia cytherea* larva feeding on *Pinus patula*.

Figure 19.7 (a) *Cryphonectria cubensis* canker on stems of *Eucalyptus saligna* in Brazil. (b) Kino vein in *E. citriodora* in Thailand.

readily seen on the forest floor and is not always unwelcome if no economic loss is occurring and wildlife is benefitting.

3. Scattered, broken-off shoots lying on the floor indicate insect shoot borer activity, which is usually easy to confirm by examining the shoots themselves.

Condition of the tree
The following signs all indicate ill health.

1. Loose bark. In advanced stages of fungal attack bark often comes away from the stem, though many eucalypts do shed outer bark naturally.
2. Damaged bark. Scarring, ribbing, cankers may all indicate damage (Fig. 19.7), for example, the depressed grey-black stem cankers indicative of *Botryodiplodia* dieback of *Paraserianthes falcataria*.
3. Resin bleeding. Often occurs in response to pests and diseases, for example, *Armillaria mellea* infection of pines, also kino veins in *Eucalyptus citriodora*—Fig. 19.7(b), possibly indicating stem-shake.
4. Misshapen stems. Forking and multiple leaders can be caused by insect attack, though lightning

and nutrient deficiencies can also be responsible. Crookedness and twisted stems in Meliaceae species is widespread owing to the shoot boring of *Hypsipyla* spp. (Fig. 8.6).
5. Dead tops and branches in the crown. Bark stripping by monkeys in stands of *Pinus taeda* in Malawi usually causes the crown to die above where the damage is done. Elephants do similar damage to pines in Zambia (Loyttyniemi and Mikkola 1980). Also *Armillaria mellea* can cause similar symptoms but usually affects the entire tree. Dieback not associated with bark damage is often due to root rot diseases; *Phytophthora cinnamomi* damage, commonly occurring on damp sites, is often first identified in this way.
6. Presence of emergence holes. This is evidence of a fairly advanced stage of infestation. Observation of pin-hole borer damage on *Swietenia macrophylla* in Fiji was the first sign of a serious ambrosia beetle problem (Roberts 1977).
7. The presence of fungal brackets, toadstools, larvae, pupae, etc., are all useful indicators but their importance depends on the species and the number found, that is, population size—see below.

8. Foliage. Several signs may indicate organic damage but none is infallible.

(a) Colour. Mottling. Fungal or viral infections can easily be confused with nutritional disorders.

(b) Leaf size. Smaller than average leaves indicate ill health.

(c) Sparse foliage. Defoliation, see earlier.

(d) Browsed foliage. Eaten by animals.

(e) Patterning or skeletonizing of leaves indicate larval damage, for example, the teak skeletonizer *Hyblaea puera*.

(f) Other signs are misshapen leaves, for example, rolled up, sticky foliage covered in honeydew from aphid damage, etc.

Poor growth

A damaged stand frequently looks unhealthy, not only because leaves may be pale or sparse or the trees misshapen, but because growth increment is poor, and recent shoots are short. This appearance of stress is often most noticeable from a distance, and overall there may be a slight colour change or some patchiness.

Trees are exposed to many stresses both abiotic and biotic that reduce growth and survival and separation of these factors and assessment of the primary cause is a challenge. Atypical foliage and poor growth can indicate the incidence of pests or disease but are also signs of nutrient disorders, air pollution, and poor physical soil conditions. Nevertheless, what is important is that there are many signs to aid early detection of a problem provided there is careful monitoring and the observer is trained to understand their significance.

Identification

Once damage is detected, correct identification of the causal organism is needed to determine the likely scale of damage, if its biology is reasonably understood, and what control methods may be appropriate. In the tropics, identification, particularly of insect pests, is a major problem because many insects are not yet classified and some not even discovered. The number of species is vast (there are over 250 000 Coleoptera spp.) and there are few forest entomologists and pathologists. These difficulties can make identification a slow process. The mycoplasma-like organism causing spike disease in sandalwood (*Santalum album*) in India took many years to identify. In 1990 a disease called 'neem decline' affected neem (*Azadirachta indica*) throughout the Sahel region of West Africa. Several investigations have so far failed to identify the primary organism associated

with this condition. There is also much confusion over the identity and nomenclature of leaf diseases in eucalypts as many fungi have not yet been described and many of those that have are inadequately described and illustrated (Swart 1988). For many 'new' occurrences several months or years can elapse before the identity is definite if the species is not indigenous or if the adult stage of an insect is small or a fungal fruit-body not easily obtained. Walker (1987) analyses these problems for plant pathogens and the role of quarantine to minimize new introductions into Australia.

Where serious damage is detected and the cause uncertain it is necessary to consult a pathologist or entomologist. Providing this service of identification, and subsequent investigation if necessary, is an important function of many national forest research stations. Also, a large plantation project often has its own research staff, which may include a forest protection specialist. Wherever a timber species is intensively managed some pest and disease problems can be expected. A good safeguard against these dangers is to have specialists, backed up by staff and equipment, available to investigate reported outbreaks. Zimbabwe's forest pathologist reported a serious threat to *Eucalyptus grandis* plantations from a new canker disease and described a survey to assess the problem and steps taken to initiate identification (Masuka 1990). Papers presented at a workshop in Indonesia describe assessment procedures during surveys of diseases of tropical acacias in several countries and the results of identifications made at the Institute of Horticultural Development, Knoxville, Australia (Old *et al.* 1997).

Several international organizations provide an identification service, such as the excellent Forestry and Agricultural Biotechnology Institute (FABI) in Pretoria, South Africa, the Commonwealth Mycological Institute at Kew, United Kingdom, the Agricultural Research Institute in Kenya, CIRAD-Forêt in France and several African countries, and the Research and Training Center for Tropical Agriculture in Costa Rica (CATIE).

Samples should be collected for identification when damage is detected. If damage is serious a visit to the plantation by a specialist is essential. Samples for identification should include: (i) specimens showing all aspects of the damage, living and dead tissue, decayed wood, emergence holes, defoliated shoots, etc.; (ii) healthy, undamaged, specimens for comparison; (iii) specimens showing partial damage or transitional between (ii) and (i); (iv) a full description of the tree species, provenance, site, recent history, location and date of sample collection, and other relevant information which may aid identification. Once at a laboratory, identification of an organism, if it is unknown, is rarely straightforward. Identification of immature stages (eggs, larvae, spores, mycelia) may have to await further development under laboratory culture to produce an adult form or sporophore. The importance of the specialist receiving live material is readily apparent. Even when a mature form is obtained, and family or genus identity is made, species typing may still be very difficult.

Analysis of the organism/tree interaction

Two components of the problem must be examined together, the ecology and habits of the species itself, and the decision of whether to try to control the damage.

With a new pest or disease nothing or very little will be known about its mode of attack, apart from the damage seen when detection was first made, for example, as occurred in the 1980s with the damaging psyllid (*Heteropsylla cubana*) defoliation of *Leucaena leucocephala*. Study must therefore be initiated into its life history and habits (its autecology), for example, the study of van den Berg (1975) of the pine emperor moth *Imbrasia cytherea* which defoliates *Pinus patula* in southern Africa (Fig. 19.6). Such studies aim to identify conditions which predispose plants to damage and point to the best means of controlling or managing the pest. In the case of the widespread leucaena psyllid an advisory team was set up in Thailand to coordinate the many research programmes in Southeast Asia and the Pacific to investigate how to control this devastating pest.

In PNG, weevil damage (*Vanapa oberthueri*) of hoop pine (*Araucaria cunninghamii*) is found associated with damaged trees following pruning or thinning. Care with these operations and maintenance of hygienic conditions greatly reduces the amount of damage. Large areas of planting with, inevitably, some stands 'off-site' suffering stress and with low vigour, has made trees susceptible to pest damage. The cypress aphid (*Cinara cupressi*) was found in Malawi in on *Cupressus lusitanica* in 1986 and subsequently spread to several countries in eastern and southern Africa causing millions of dollars worth of damage. This aphid also attacks native *Juniperus* spp. A biological control programme was started in 1991 and after extensive surveys in *Cupressus*-growing areas two candidate parasitoids were identified and one is being field-tested. Complementary research on host tree resistance has also been undertaken (Murphy 1998). In Kenya in the 1970s introduction of the predator *Tetraphleps rao* from Pakistan helped to control an epidemic of woolly aphid (*Pineus pini*) attacking *P. patula*, but only after research identified the problem and investigated

means of biological control. During this latter work it was found that *T. rao* would prey on all stages of the aphid life-cycle and also on other aphid species in the same order.

Cryphonectria cubensis (Fig. 19.7) is the most important stem canker disease of eucalypt plantations in the tropics. Since it was first described on eucalypts in Cuba in 1917 it has spread to eucalypts in South America, Africa, Asia, and Australia causing significant economic loss. Detailed studies of the organism/tree interaction has shown conditions of high temperature and rainfall favour development of the disease and that some species and provenances within species are less susceptible than others (Old and Davison 2000). Such studies provide indications of how the risk of severe damage can be avoided by species and provenance selection and species-site matching. In Thailand, Vietnam, and other parts of Southeast Asia leaf pathogens have caused serious damage to large areas of eucalypt plantations (Old and Ivory 1999) and measures such as using bioclimatic modelling to predict areas of high risk, more rigorous species-site matching, and selection of disease resistant clones have begun (ACIAR 2001).

The need to control
Autecological studies often take many years, while a pest outbreak can develop into a major problem in a few weeks. Consequently a decision may have to be taken to attempt control measures while little is known about the organism. The decision largely depends on the scale of damage and this is evaluated using five criteria.

1. *Population density of the organism.* Proportion of trees damaged, concentration of damage.
2. *Extent.* Total area affected, perhaps classified into serious, moderate, and light damage.
3. *Rate of spread.* An indication is needed whether damage is confined, is spreading gradually, or rapidly. This normally reflects the build-up in population of the damaging organism which in insects can be very rapid.
4. *Severity.* The importance of any damage sustained depends on whether it is only temporary, perhaps with loss of increment, or whether trees are killed. For example, of the many teak defoliators only *Hyblaea puera* causes serious increment loss in India (Nair *et al.* 1996), and causes damage in most teak growing areas in Indonesia (Nair 2000).
5. *Value of trees at risk.* Occasionally trees are so valuable, for example, breeding stock in a seed orchard, that almost any attempt at control is worth trying. Precisely this problem occurred with *Armillaria* infection of clonal seed orchards of *Pinus oocarpa* in Zimbabwe (Masuka 1989).

Controlling organic damage

The decision to control
Several factors may have to be considered.

1. Scale of the problem.
2. Size of the plantation with the endangered species, that is, the possible extent of damage if an epidemic occurred.
3. Whether control is practicable. In general, direct control of fungal disease is difficult and a change of species or better forest hygiene is the simplest solution in the long term. With insect pests direct control of the population may be possible though usually only after study of the insect's life-history has confirmed that control can be effective.
4. Economics of control; cost versus likely damage.
5. Requirements for temporary control to allow time for alternative courses of action to be considered.
6. Longer-term effects of any action on population ecology and, more generally, the environment.

Means of control
Quarantine measures designed to reduce the risk of introducing damaging organisms are fundamental to plantation health management. However, once the pest or disease has arrived the manager usually has several choices.

1. *Stop the organism from spreading in the plantation area.* This is impossible with fungi and insects but animals can be fenced out, trapped, or poisoned. Full control is particularly important in the forest nursery.
2. *Avoid practices favourable to the organism.* Some examples were noted earlier, but this includes minimizing the amount of breeding material, preventing damaging infection by specific treatment of stumps or possibly wounds, maintaining a healthy stand, thinning or other operations which remove diseased trees and other sources of infection, and carrying out operations cleanly without causing damage to trees. In Brazil, harvesting plantations of *Eucalyptus cloeziana* is timed to coincide with seasonal climatic conditions that are least favourable to the rust *Puccinia psidii* developing on young coppice shoots.
3. *Change silvicultural practice.* For example, the observations of Sharma and Sankaran (1987) that *Botryodiplodia* diebacks of *Paraserianthes falcataria* is much reduced if fires and other bark injury in plantation are avoided and, when *P. falcararia* is established using taungya, a tapioca crop is not cultivated since the tubers provide a high level of inoculum, which provides the basis for changing silvicultural practice. The evidence that enrichment planting of Meliaceae species under an overstorey rather than in the open sometimes reduces damage from the shoot borer *Hypsipyla robusta* (e.g. from Indonesia, Sri Lanka, and Malaysia) is conflicting and to a large extent anecdotal but several silvicultural techniques show promise of improving plantation performance (Hauxwell *et al.* 2001).
4. *Reduce stress to trees.* The interaction between stress and increased susceptibility to pest and diseases has been noted earlier with the example of timely thinning, correct species choice, attention to nutrition, etc.
5. *Direct chemical or biological control.* The choice lies between direct control, by killing the damaging organism using an insecticide or fungicide, and biological control. An example of the latter, ecologically more desirable, approach was cited in the case of the cypress aphid (*Cinara cupressi*) in East Africa. The choice of approach depends on many factors, but mainly whether direct control is possible or biological control measures are available. Chemical control is often avoided due to the high cost relative to the growth losses or for environmental reasons.
6. *Genetic control.* Change to a less susceptible species, or select for resistance within the susceptible species. This does not solve the immediate problem of damage to an existing plantation. Several examples of species being unsuitable for planting because of pest/disease damage were cited in Chapter 8. In response to the stem canker, *Cryphonectria cubensis*, plantation managers in some parts of Brazil stopped planting *Eucalyptus saligna* and changed to more resistant species, others used moderately resistant provenances of *E. grandis* or selected, propagated, and planted resistant clones. Resistant eucalypt hybrid clones have also been used in Brazil and South Africa to reduce disease damage, and hybrids show promise in other genera (Wingfield *et al.* 2001).

It is clear from the above that no one approach to control is either likely to be wholly effective or most efficient. Managers should combine the advantages of the various control methods such as selecting species adapted to the site and applying silvicultural methods that maintain plantation hygiene and make conditions unfavourable for potentially damaging organisms. Integrated pest and disease management strategies that are preventive rather than reactive to maintain forest health should be the aim.

Preventing recurrence and monitoring
Where animal and insect populations fluctuate it is useful to carry out a regular check for evidence of a build-up to potentially epidemic proportions. Monitoring should be at regular intervals, monthly

or yearly, so that change can be noted. Both potentially damaging organisms and the condition of tree stands are monitored.

Monitoring forest health
Regular surveys are important to ensure early detection of new pests and diseases and monitoring of the state of existing populations. Inspection can form part of inventory work (Chapter 4) provided the team is trained to recognize potential problems. During harvesting the incidence of root and butt rot can be recorded. Aerial survey may also be helpful in locating damaged trees in the plantations (e.g. Old and Stone 2002) which can be further checked on the ground. A combination of aerial, roadside, and random surveys may be very cost-effective (e.g. Bulman *et al*. 1999) and development of a national forest contingency plan to facilitate a rapid and constructive response for any new pest and disease outbreaks is desirable (e.g. Hosking and Gadgil 1987).

A network of insect traps should readily indicate whether epidemic numbers of a species are developing (Fig. 19.8). Greig and Foster (1982) were able to survey *Heterobasidion* (*Fomes*) *annosus* infection in Jamaican *Pinus caribaea* plantations by direct inspection of stumps, exposure of placed pine discs, and muslin traps, and Bunce and McLean (1990) report use of insect pheromone traps to monitor bark beetle populations in Jamaica's hurricane devastated pine plantations. Sample counts of large animals such as deer, visually or by dung (faecal pellets) counts, provide an estimate of total population size. This, combined with the incidence of damage, indicates the need to cull. Whenever disease or pest organisms are observed and identified a record should be made. As data accumulate a general picture will emerge of the most prevalent organisms.

Salvaging a damaged crop

Where damage is serious, after fire, cyclone, an insect epidemic, or extensive dieback, the only course open may be to clear what is left of the crop and replant. In 1987 and 1988 the Fiji Pine Commission wrote off 3300 ha and 3700 ha of plantation respectively owing to the huge fires of 1987. In Queensland, Australia, the optimum treatment of cyclone damaged *Pinus caribaea* was to carry out a pre-commercial thinning (thinning to waste) and then leave to rotation age (Anderson and Harvey 1983 in Bell *et al*. 1983).

If some merchantable timber remains it should be salvaged as soon as possible after damage has occurred. Delay can worsen the situation in several ways.

1. Blue-stain fungus can invade dead trees of many species, notably pines, within 5 or 6 weeks and cause timber degrade.
2. Dead trees provide an abundance of breeding material for insects and may become a source of infection for nearby undamaged stands. In the tropics insect numbers can often build up extremely quickly in moist, warm conditions.
3. Damaged trees, or those seriously weakened, are more prone to attack from other organisms.
4. Misshapen trees, which have suffered damage but are still alive, for example, stem-snap or dead-top, if allowed to remain add increment that is of very low value.

There are several other reasons for not delaying salvaging a damaged crop. A damaged crop is occupying a site which could be replanted and support a new, more-productive, stand. Damaged stands also usually allow more light to reach the forest floor resulting in an influx of weeds, and, severely damaged plantations are unsightly. The account by Bunce and McLean (1990) of the impact of hurricane Gilbert in Jamaica is a good example of assessment, monitoring, and implementation of salvaging damaged plantations.

Figure 19.8 Water trap for insects. One of the devices to monitor insect populations in the Viphya plantations, Malawi in the 1980s.

Salvaging operations

Where damage is extensive, and because of the time constraints above, salvaging may be limited to taking the best trees from the damaged stand. It will usually be worthwhile doing this as quickly as possible, provided the produce is saleable or can be stored without further degrade, even if temporarily it involves employing extra labour.

Small-scale damage is best dealt with in the course of normal felling with yield from the damaged stand substituting for that of a healthy one. Sometimes salvaging and cleaning up after damage can be undertaken at no direct cost by using a contractor whose profit comes from the recovery of saleable material. However, there is not always a market for small-size timber and where severe damage is reasonably predictable, for example, in areas prone to cyclones, one should consider planning for regular salvage.

Regeneration

Regeneration following extensive damage may be difficult, not because trees will not grow but owing to many obstacles to achieving a uniform stand.

Previously it was urged that speed in salvaging is essential, but occasionally it may repay to delay deciding the future of a crop. The capacity of some species to recover from damage can be remarkable. Many of the young *Pinus caribaea* in Fig. 19.2 were severely scorched but recovered and sprouted a vigorous leader from the tip and there was no need to replant the stand. Many eucalypts produce vigorous epicormic and coppice shoots to replenish the burnt crown or dead stem. Similarly, *Gmelina arborea* is rarely killed by fire, burnt stems can be cut and the stumps will usually coppice. In thicket stage crops, where there is little to salvage, delay in clearing will give time to see if the stand will recover. Whether or not to replant will depend on consideration of several variables and the experience of the manager. The decision making may be assisted by a model prepared by Caulfield (1987) that provides an economically objective criterion against which the possible replanting decision in damaged, even-aged plantations can be assessed.

Where a stand is cleared because of wholesale death of trees profuse natural regeneration can occur. This is common following fire in pines since cones are opened by heat, and in eucalypts where seeds shed on to the burnt ground regenerate rapidly and add to shoots produced by underground lignotubers. Fire may also stimulate massive regeneration of acacias and other legumes from the soil seed bank. These

developments can complicate future management due to the very high stocking levels that often result. In the case of windthrown stands, the main obstacle to re-establishment is the debris left behind which makes tree-planting difficult and provides breeding material for pests.

Protection strategy

Apart from the national responsibility of enforcing international quarantine regulations to minimize spread of pests and diseases, the main concern is to minimize the impact of pests and diseases on tree survival and productivity. Where resources and expertise permit 'intergrated pest management' (IPM) should be practised. This basically involves using a combination of pest management tactics to reduce economic loss caused by pests to tolerable levels, with minimal environmental side-effects (Wylie 2000).

Management of pests and diseases initially involves trying to avoid the problem by taking great care in site selection and then in the selection and site-matching of resistant species followed by good silvicultural practices. It requires an understanding of applied ecology, identification of the causal organism, and some knowledge of its biology to develop strategies that predict and measure the effect of the organism and contain damage within acceptable limits. The 'economic threshold', the point beyond which it becomes economic to take active control measures and below which the presence of the pest or disease is tolerated, needs to be determined on a site-specific basis (Wylie 2001).

Useful summaries of IPM for fast-growing plantations are provided by van Rensburg (1984) and Speight and Wylie (2000) for insect pests and Wingfield (1984) for diseases. Simpson and Podger (2000) and Gadgil *et al.* (2000) give overviews of the options and constraints in managing eucalypt diseases that are also relevant for other tree species and damaging organisms. Wingfield *et al.* (2001) predict emerging technologies such as rapid screening techniques and genetically modified trees will improve our ability to deal with pests and diseases in the future. IPM in agroforestry systems in the tropics and future needs have been reviewed by Dix *et al.* (1999).

The manager or owner of plantations needs to tailor protection management strategies to the specific local conditions. Management objectives for a plantation need to be clearly defined so that pest and disease impacts can be judged in context. Intensive monitoring may be too expensive and lack of staff with protection expertise in many countries will hinder application of IPM. However, the following comments apply to most situations:

1. Efforts should be made to minimize the risk of damage from pests, diseases, and other damaging agents through good planning and appropriate silvicultural measures.
2. Diseases and pests should be managed, rather than simply reacted to. Integrated pest and disease management should be part of modern plantation technology with the role of research, at least for larger enterprises, self-evident.
3. Protection from only the most serious form of damage is a waste unless several lesser problems are also kept in check. For example, in a young newly established plantation where the combined control of weed competition, fire, and grazing damage enables unhindered growth, leads to earlier canopy closure. Protection against only one of these damaging agents will probably prolong the period of susceptibility to all three and is a false economy. Similarly, there is no point in providing adequate protection if growth is impaired due to nutrient deficiency or poor cultivation.

Finally, the need to protect planted forests is likely to become ever more important. Both large-scale and small-scale plantation growers aim to gradually improve productivity per unit area and reduce the risks to their investment. Minimizing or avoiding damage to the crop through good protection management strategies is a major contribution to this goal.

PART IV

Tree-planting and plantation forestry: in rural development, soil conservation, rehabilitation, environmental considerations and sustainability

CHAPTER 20

Tree-planting in rural development and agroforestry

Introduction

Trees, tree-planting, and woodlands are now recognized as an integral part of rural development. Fruit trees planted in home gardens and shelterbelt development are examples of the link between tree-planting and food security (e.g. Falconer and Arnold 1988; Hoskins 1990). Agroforestry can assist soil conservation (Young 1997) and help combat desertification (Baumer 1990). There has been greater recognition of the need for more sustainable agricultural practices and in many countries local people have had their access reduced to shrinking natural forest resources. Tree-planting has been the means many farmers have used to improve the sustainability of their cropping practices and to enable them to place greater reliance on on-farm resources to provide both timber and non-timber tree products. Trees are frequently used to generate cash for periodic needs in rural communities relying mainly on subsistence agriculture.

As emphasized in earlier chapters, social and community forestry projects frequently now fulfil the tree or forestry component in rural development. Community forestry is focused on the role of forest-dependent communities in managing resources and sharing the benefits flowing from these resources. It can include natural forest management, reforestation, agroforestry, and roadside plantations. In these developments it is commonly the non-timber benefits of trees, which are most important, and there has been an increasing role for trees and shrubs able to provide several benefits. Depending on the species, they can enrich the soil with nitrogen, control erosion, provide fuelwood and building materials, serve as a source of fodder for livestock, and produce fruits for human consumption. Such species have been referred to collectively as Multi-Purpose Tree Species (MPTS).

This chapter does not summarize rural development and agroforestry but examines only the tree and plantation components of the many different systems and approaches used. This large and important topic is therefore introduced from this single perspective.

Social forestry concepts and agroforestry practices have been covered in several textbooks, many publications of the International Centre for Research in Agroforestry (World Agroforestry Centre), FAO and the Rural Development Forestry Network of the Overseas Development Institute (ODI), and articles in the *Agroforestry Systems* journal. These provide more comprehensive and detailed accounts than can be given here. The following are examples of this literature: *Tree growing by rural people* (FAO 1985b), *People and trees: the role of social forestry in sustainable development* (Gregersen *et al*. 1989), *Keepers of the forest: land management alternatives in South-east Asia* (Poffenberger 1990), *Community forestry: ten years in review* (Arnold 1992), *Tree management in farmer strategies: responses to agricultural intensification* (Arnold and Dewees 1995), *The saga of participatory forest management in India* (Saxena 1997), *Agroforestry systems in the tropics* (Nair 1989); *An introduction to agroforestry* (Nair 1993b), *Patterns of farmer tree growing in Eastern Africa: a socioeconomic analysis* (Warner 1993), *Sustainable small-scale forestry: socio-economic analysis and policy.* (Harrison *et al*. 2001), *Agroforestry: classification and management* (MacDicken and Vergara 1990), *Agroforestry in sustainable agricultural systems* (Buck *et al*. 1999), *Agroforestry innovations for* Imperata *grassland rehabilitation* (Garrity 1997) and *Improving smallholder farming systems in* Imperata *areas of Southeast Asia: alternatives to shifting cultivation* (Menz *et al*. 1998). Several newsletters provide accounts of recent developments and initiatives in agroforestry and rural development e.g. World Agroforestry Centre's *Agroforestry Today*, FAO's *Forests, Trees and People Newsletter* and ODI's *Rural Development Forestry Network Newsletter*, and *Agroforestry Abstracts* enable rapid access to much of the literature.

The rapidly increasing literature on community forestry and agroforestry reflects a greater understanding of the principles and practices that underpin the successful incorporation of trees, tree-planting, and woodlots in farming systems and rural development. It is worth remembering that community forestry and agroforestry projects have often not always been successful. While inappropriate technologies have been a problem, more often it has been failure to address the complex matters of full stakeholder participation, land tenure, marketing, pricing and taxation policies, etc., which has resulted in enterprise failures. Some of these non-technical issues were discussed in Chapters 5–7.

In recent years there have been changes in how agroforestry is perceived, the term 'agroforestry' has been debated and various definitions have emerged (see Nair 1989; Underwood 1993; Leakey 1996). Confusion has arisen because of the use of agroforestry as an approach to land-use and as a set of integrated land-use practices. Sinclair (1999) suggests a two stage definition to cover both the practices and the umbrella concept. As a concept *'the approach is interdisciplinary and combines the consideration of woody perennials, herbaceous plants, livestock and people, and their interactions with one another in farming and forest systems. It embraces an ecosystem focus considering the stability, sustainability and equitability of land-use systems in addition to their productivity. Consideration of social as well as ecological and economic aspects is implied'*. As a set of land-use practices it involves *'the deliberate combination of trees (including shrubs, palms, and bamboos) and agricultural crops and/or animals on the same land management unit in some form of spatial arrangement or temporal sequence such that there are significant ecological and economic interactions between tree and agricultural components'*. Both agroforestry and community forestry practices imply tree-planting and management for sustainable rural development.

This chapter embraces all tree-planting practices associated with rural development and the classification of topics follows in part that of Sinclair (1999). Distinction is often made between the humid, semi-arid, and montane tropics, and certainly the species, food crops, and the protective values of trees do differ, but this important distinction will only be used in illustration; subsequent chapters address more directly tree-planting and desertification and soil erosion along with other environmental concerns.

Use of trees in rural development and agroforestry

Planted trees can have a variety of roles in agriculture. Many farming practices can be combined with tree growing, and trees can diversify the landscape modifying the environment in ways critical to sustainable agriculture. While agroforestry and tree-planting for rural development have had a high profile in recent years, it is important to acknowledge that trees have been an inherent component of traditional farming systems in many countries. Indigenous knowledge remains important as farmers pragmatically have used both native and exotic species to meet their economic needs and improve their livelihoods. Table 20.1 classifies these uses for

convenience, but they are not exclusive alternatives. A simple eucalypt woodlot can provide shade, shelter, soil protection, firewood and poles, and support bee keeping. *Erythrina poeppigiana,* and *Cordia alliodora* shade trees over coffee in Costa Rica provide firewood, mulch, fodder, and some timber as well as critically controlling the light environment, enhancing soil fertility and generally improving the coffee plantation microclimate.

Table 20.1 first classifies by the main components present, for example, trees and crops, and then by their arrangement: 'rotational' where mixing of trees and crops is primarily over time, and 'spatial' where mixing occurs at the same time on one site either intimately (spatial mixed) or systematically in strips or rows (spatial zoned). The right-hand column refers to examples and headings used in this chapter or where the topic is considered. No reference is made in Table 20.1 to nitrogen-fixing or multi-purpose trees because they are not an agroforestry practice as such but only a widely used component. The role of this important group of trees as soil improvers was discussed in Chapter 14.

Woodlots

A woodlot, called 'farm forests' in India or 'tree farms' in the Philippines, is any area of farmland with trees, the purpose of which is more than just providing shade and shelter. However, woodlots need not necessarily be restricted to farms and farmland, and any small plantation or group of trees, on a farm, around a village, on waste ground, or beside a road may fit the purpose of a woodlot (e.g. Fig. 6.3).

Woodlots, whether planted or natural, fulfil two important needs, in addition to providing protection or shelter. They are a source of domestic fuel and, if given rudimentary management, will produce small roundwood such as posts and poles (Fig. 20.1), and may provide food and/or medicines. These contribute to the five physical requirements for living: fuel, shelter, water, food, and clothing. In many parts of the tropics people, especially in rural areas, may not even have these five basic necessities. Frequently, such poverty is most acute in remote areas characterized by poor soil conditions, irregular and low rainfall, and declining or even absent natural forest cover.

In areas devoid of tree cover, establishment of woodlots can be effective in raising rural living standards. The fuel and wood needs of the rural poor were a motivating force for the development of woodlots in social forestry programmes beginning in the late 1970s, especially in Africa and Asia (Pandey 1995). Two of the intended benefits illustrate the intention. By planting a woodlot near a village less time is needed for collecting firewood and other

Table 20.1 A classification of uses of planted trees in rural development and agroforestry systems (modified from Sinclair 1999)

Tree component predominant with	*Example cited*
Some environmental benefits only	Woodlots
Reclamation forestry to rehabilitate poor ground leading to multiple use	(Chapter 23)
Trees and crops (agrosilviculture)	
Rotational	
Cropping before tree-planting	Pre-establishment
Cropping during tree establishment	Taungya
	Large-scale establishment
	Intercropping
Trees to improve soil after cropping	Improved tree fallow
Dispersed irregular	
Trees on cropland	Farm and shade trees
Plantation crop combinations	Full intercropping
Multistorey tree gardens	Home gardens
Dispersed zoned	
Hedgerow intercropping	Alley cropping
Tree intercropping	Line planting
Boundary planting	Border planting
Trees aiding erosion-control structures, for example, bund or terrace planting	(Chapter 23)
Windbreaks and shelterbelts	(Chapters 21 and 23)
Trees and livestock (silvopastoral)	
Dispersed irregular	
Tree plantations with pastures	Forest grazing
Dispersed zoned	
Planted trees as fence	Living fences
Planted trees as fodder	Fodderbanks
Trees for other purposes	
Bee-pasturage and honey production	Apiculture
Essential oils and other extractives	
Resin tapping	
Silk production	Sericulture
Tannins	Tanbark
Amenity planting	

domestic wood needs (Fig. 2.4) and more time is available for more productive activities such as soil cultivation, food preparation and cooking, spinning, tool making, etc. Second, a woodlot provides wood for fuel so that dung and other organic matter need not be burnt (Fig. 2.3) and can be used for fertilizer to improve crop yields. In addition, because produce from woodlots is usually marketable they are a source of income (a cash crop) for the owner or the village. Many of these economic and social benefits were evident in an assessment of the West Bengal Social Forestry Project in which eucalypt woodlots were established (Nesmith 1991). Wood produced in the woodlots was sold for poles and pulp, fuel needs

of the poorer families were met from the litter of leaves and twigs and small branches within the woodlots. Income generation is an important feature and reflects the role of trees as savings and security for people in rural areas (Arnold 1990; Chambers and Leach 1990). The importance of tree income from woodlots in meeting large livelihood expenses, such as marriages and education, was clearly demonstrated in the Kolar district of Karnataka, India (Ravindran and Thomas 2000). In India in community forestry schemes, the majority of farmers prefer to grow fruit trees, followed by timber and pole species, based on profitability and market demand. They may interplant the main species with other species to

Figure 20.1 Poles used for house construction collected from one of the many small *Eucalyptus camaldulensis* woodlots in Zimbabwe

produce fodder and/or fuelwood for local use (Hedge 1997).

The success of rural development programmes sponsored by development assistance agencies, development banks and non-government organizations (NGOs) that have promoted woodlot development has varied. Successful examples include the promotion of planting by hill-farmers in Haiti (USAID) and the Dominican Republic (ENDA-Caribe) (NGO) and widespread woodlot development in many parts of the tropical highlands of eastern Africa, for example, Kenya and Rwanda, in this instance often by NGOs. Less successful have been woodlots aimed solely at solving fuelwood problems. Shepherd (1990) concluded that a vaguely targeted woodlot approach will solve neither rural nor urban fuelwood supply problems. This has been apparent in many Sahelian countries of Africa where fast-growing exotic species such as eucalypts and neem (*Azadirachta indica*) have proved too expensive and too reliant on government inputs (Winterbottom and Hazlewood 1987). In Tanzania, communal woodlots promoted by the government were poorly supported by villagers due to lack of villager participation in planning, demands on labour, loss of private land, lack of control, and unclear benefits (Skutsch 1983; Warner 1993). Failure to understand

and address the complex social, political, and economic needs of rural people also resulted in many unsuccessful community forestry and woodlot developments in South Africa (Ham and Theron 1999). Too many of these early tree-planting projects were directed at a perceived energy supply problem rather than a response to local needs for trees and tree products (Dewees 1995).

Spontaneous tree growing in woodlots in which farmers use their own capital resources have been successful where a market for wood exists and it can be readily transported. Woodlots are planted on farms in Rwanda and Burundi when wood resources have to be met entirely by on-farm production. They are usually on marginal land, where there is a strong market for wood products, and/out migration of males as they make relatively little demand on household labour and do not require a large capital outlay (Warner 1993). Since teak (*Tectona grandis*) was introduced into villages in the 1930s in Côte d'Ivoire, farmers have planted it in 0.5–10 ha woodlots close to their houses mainly to produce poles (Maldonado and Louppe 1999). In the Philippines, farmers are growing *Gmelina arborea* in woodlots (Magcale-Macandog *et al.* 1998), in West Java they grow *Paraserianthes falcataria* (Gunawan *et al.* 1998) and in Vietnam they are establishing plantations of bamboo (*Dendrocalamus membranaceus*) (Woods 2003). In China, 90% of nearly 4 million ha of bamboo plantations are managed by village and farmer households (Zhong *et al.* 1997).

Outgrower schemes in which smallholders grow trees in woodlots to provide raw material for forest industries are also becoming more common in many parts of the world and various models are described in Mayers (2000). Outgrower schemes are discussed in more detail in Chapter 5.

Planning and planting of woodlots varies enormously. At one extreme are the establishment recommendations for plantations on the farms of wealthy farmers where, apart from the scale, the principles laid down are little different from those for an industrial plantation. At the other extreme is a woodlot to be established, often in harsh climatic conditions, around a village where poverty is so acute that there is little opportunity for self-help. For the latter it is often necessary to provide the seedlings, advice, or even help in planting and tending, and to undertake the whole establishment project as part of a supportive extension programme. As stated in earlier chapters, the success of a plantation, especially a small-scale plantation such as a private woodlot, depends as much on fully involving stakeholders, tree and land tenure arrangements, assured markets, infrastructure to transport products, government policies in relation to taxes and charges, availability of credit facilities, etc., as it does on having and implementing an appropriate technology i.e. silviculture.

Woodlots offer many opportunities to practise 'more complex plantation forestry' using more than one species, and several management practices such as a mixture of coppice with standards to provide a range of products for on-farm use or sale and environmental services (Kanowski and Savill 1992; Kanowski 1997). In Ethiopia, farmers with woodlots as small as 0.25 ha plant *Eucalyptus globulus* seedlings at very close spacing then progressively thin them out and manage most of them on a coppice system to produce leaves and twigs for fuel, vegetable stakes, and poles of various sizes for house construction. They also grow a few large trees to be sold for sawn timber when cash is required. Even this intensive management requires less labour than growing agricultural crops. Use of mixed-species plantings may provide an insurance against market changes and minimize the risk of losses from pest and diseases. It is critical to provide adequate protection of woodlots. In areas where the need for the forest produce is acute, whether firewood for burning or foliage for cattle fodder as in much of the Sahel, newly planted trees may need to be protected from theft or excessive 'pruning' of branches. This difficulty, noted in the previous chapter, often reflects more fundamental social problems and privation, but it has been a contributory cause to some plantation failures.

Most woodlots have several features in common:

1. Land used is often ill-suited to agriculture; the poorest land is allocated to tree growing along with 'waste' ground such as roadsides, rocky or poorly drained land and property boundaries. FAO (1985*b*) estimate that only 2–5% of available land should be planted with no net loss in agricultural production.
2. Species planted should be well adapted to local conditions, fast-growing, easily established, preferably able to coppice, and suitable to provide the desired products for on-farm use or for sale. Other important features of species' choice are discussed in Chapter 8.
3. Planting stock should be hardy and easily handled. Container-grown plants greatly aid survival in drier areas. Use of easily propagated material, for example, stumps of teak, (Chapter 10) or root cuttings of *Paulownia* spp. (Chapter 11), simplifies planting and reduces the chance of failure.
4. Spacing between plants should usually be closer than for industrial plantations, 1×1 m for fuelwood and 2×2 m for small pole products are commonly used to promote rapid ground cover, allow for higher failure rate, and maximize production of small roundwood in a short time.

5. Rotations for woodlots are dictated by intended end-use. For firewood, rotations are typically 3–6 years so that the produce is easily handled.

Woodlot size is important. For farm planting size is determined by how much land an owner decides to devote to trees, but it should be no more that what can be reasonably managed and should leave enough other land to produce food crops and other basic needs. In northern Mindanao, Philippines, the average farm size is 2.5 ha and the majority of farmers have planted half or less of their farms to *Gmelina arborea*. Farmers planted the trees to provide timber for future family houses but most harvested them to meet immediate financial needs such as school fees, to buy farm equipment, household appliances and animal feed (Magcale-Macandog 1999). For outgrower schemes growing pulpwood, a Filipino tree-farmer supplying PICOP has 5–10 ha of woodlot but in KwaZulu-Natal, South Africa woodlots averaged 1.2 ha (Arnold 1997*a*). In Kakamega district of Kenya about one-quarter of each of the tiny farms averaging 0.6 ha are planted to eucalypt woodlots. In Bangladesh, where landholdings per household average 0.8 ha, farmers have established 'micro-woodlots' of closely spaced trees for wood production and have grown selected arable crops under them for a few years (Hocking *et al.* 1997*a*). Where woodlot development is to provide fuel for a village a working assumption is that a family uses 2–5 m^3 of stacked firewood per year. For example, if, on a 5-year rotation, a woodlot produces 50 m^3 ha^{-1} a village of (say) 100 families (600 people) would require at least 20 ha to meet their fuel need. However, more often a woodlot is used to produce poles and other products in addition to firewood and so the area required must be adjusted accordingly.

Trees with crops

The main kinds of tree:crop combinations are indicated in Table 20.1 and all are a form of multiple cropping, which is so widely practised with farm crops in the tropics. It is useful to retain the distinction between growing crops one after the other on the same piece of land (sequential or rotational cropping) and growing them together at the same time (mixed or intercropping) though in practice they often merge into one another.

The choice of system used depends on many factors but underlying them is a general relationship between population density and intensity/complexity of land-use. For an individual farmer the use, density, and configuration of trees in the farming systems depends of complex interaction related

Rotational systems

Pre-establishment cropping is where land for tree-planting is first used for growing food. Sometimes land is devoted to food production for one or two seasons before planting commences. The practice ensures good site preparation for the plantation and often less weeding and early tending of trees will be needed. Another advantage is that any fertilizer applied to aid growth of the food crop may have some residual benefit for the trees. Of course, if satisfactory food crops are grown on land destined for tree-planting, the question may be asked whether the land should in fact be devoted to forestry? However, though land may yield food crops for 1 or 2 years, it does not necessarily mean it is suited for long-term sedentary farming. In the fragile, readily degraded soils of the humid tropics, food growing immediately after land clearance makes use of the short period when the ground is almost weed-free; tree-planting a few months later is in place of allowing profuse regrowth to take over the site.

Figure 20.2 shows this kind of land-use and is the first stage in the development of the Kuriva plantation in Papua New Guinea (PNG). Food crops are planted immediately after the remnant forest, following selective logging, was cleared and burnt. About 4–5 months later teak stumps are inserted to establish the plantation. The bananas are grown for 2–3 years after the teak is planted.

A second example of rotational cropping in Sind Province, Pakistan, is described in MacDicken (1994). *Acacia nilotica* is established with irrigation in the first year. The trees are then grown for 5–6 years to yield wood for home consumption and sale. When the trees are harvested the improved land is then often planted to cotton. *A. nilotica* trees are again planted after the cotton cropping phase. Wood production is a primary objective and soil improvement a secondary benefit.

A reversal of the sequence food crop-tree crop occasionally occurs. An example is underplanting of moribund coconut plantations in the Philippines with *Pinus kesiya*. The pines are introduced late in the life of the coconut plantation and eventually replace it.

Establishment intercropping and taungya

Taungya is a system of establishing forest plantations in which farmers plant the seeds or seedlings and tend the trees. As an incentive they are allowed to cultivate crops for the first few years between the seedlings of the plantation. Cropping is usually carried out for the first 2 or 3 years in the life of the tree crop before canopy closure and the ground becomes densely shaded. Taungya developed initially as a means of ensuring tree cover following shifting cultivation. The term 'taungya' originated in Burma (Myanmar) and its application in government forestry is usually attributed to Dietrich Brandis, a forest officer in-charge of Burmese teak forests from 1856 to 1862. The taungya regeneration method was introduced into Java in 1873 and from 1895 almost all teak forests have been regenerated by this system locally called 'tumpang sari'. In the original taungya in Java, food crops for home consumption were grown but in response to changing needs the agricultural component has been diversified to include

Figure 20.2 Bananas, sweet potatoes, and taro growing on a recently cleared site just prior to planting teak (Kuriva, Papua New Guinea).

horticultural species, fodder, and fuelwood crops (Stoney and Bratamihardja 1990). From Asia the taungya system was extended to Africa and more recently to tropical America (Wadsworth 1997; Beer *et al.* 2000). It is practised in many countries under various names, e.g. shamba (East Africa). 'Integral taungya' is a form of taungya offering a more integrated approach to rural development allowing longer term involvement of workers through planting perennial cash crops and having permanent agricultural plots (Raintree 1986). The historical reasons for taungya development were mentioned in Chapter 3 and the value of taungya as an inexpensive plantation establishment system remains.

The basic steps in taungya are similar wherever it is carried out.

1. Forest villages are established or shifting cultivators encouraged to settle in one place to allow some supervision of the people who will be employed in taungya work.
2. Each family or farmer is allocated about 0.5–1.5 ha of land to clear actually called a 'taungya' or 'shamba' (Fig. 20.3).
3. Debris from clearing operations is usually burnt to provide a clean weed-free site in which to plant the crops. In savanna some overhead cover may be left.
4. After preparation of the ground food crops are planted. Normally annual crops are cultivated but perennial crops like bananas, papaya, and cocoa may be grown.
5. Tree seedlings are planted at the same time as or shortly after food crops. Large plants are often used (40–70 cm tall) because they are sturdy and visible. Spacing between rows of trees may be wider than for a normal plantation to delay time of canopy closure and hence shading of the food crop.

6. The cultivator (farmer) tends and protects both the food crops and the planted trees.
7. Raising food crops among the trees in the plantation continues until overhead shade prevents satisfactory growth. This length of time may vary from several months to 4–5 years depending on tree growth, initial spacing, and the kind of food crops grown.
8. A taungya is abandoned by the cultivator when it is no longer suitable for growing food. Usually a new area nearby will already be being worked. Often, as with the shamba system in Kenya (Fig. 20.3), several small taungyas are worked at one time; each year the cultivator clears a new one, continues cultivation on one or two with only young trees, and abandons an old one.

Almost any tree species and any annual food crop may be grown under taungya. For example, in Nigeria alone taungya is practised with the following tree species: *Azadirachta indica, Gmelina arborea, Khaya grandifoliola, K. senegalensis, Mansonia altissima, Tectona grandis,* and *Triplochiton scleroxylon,* and the following food crops: yams, maize, ground-nuts, sweet potato, cotton, melon, tobacco, rice, beans, and cassava. The effect of growing a food crop on initial tree growth is bound to vary from site to site, with the tree and crop species and with the cultural conditions applied. Tree growth may be slightly reduced or it may be unaffected. In Tanzania, *Eucalyptus camaldulensis* intercropped with maize showed comparable growth, with clean weeding and superior growth to spot-weeding (Chingaipe 1985). Similarly, *Gmelina arborea* growth was unaffected by intercropping with maize and yams in Nigeria (Agbede 1985) and taungya systems with fast-growing trees are becoming ever more popular

Figure 20.3 Maize being grown in shambas in Kerita forest, Kenya. The *Cupressus lusitanica* trees are still small enough probably to allow one more maize crop to be cultivated before the tree canopy casts too much shade.

in that country (Agbede and Adedire 1989). Tree growth depressing food crop yields is common where spacing between trees is close, for example, 2 × 2 m or 3 × 3 m. The specific concern over the effect of eucalypts is discussed in Chapter 24. Also, as the soil remains exposed throughout the intercropping period, severe erosion may occur on some soil types, especially on steep slopes (Bruijnzeel 1997).

It is useful to list the main advantages and usefulness of taungya as it has been so widely used:

1. It does not impose a radical change in life-style on the many indigenous communities in the tropics used to shifting cultivation and related practices, since their practice of land clearing, food growing, and moving on are mimicked.

2. By ensuring trees are planted while food crops are cultivated, grass invasion is prevented, soil cover retained, and a forest stand left when the taungya is abandoned.

3. Where taungya is used for large-scale development, employment is provided for the whole family, food is produced locally rather than imported, and the land is seen to produce wood and food at the same time.

4. The forester benefits in several ways: cheaper establishment, better protected trees, often a more settled labour-force, and greater self-sufficiency in food within the plantation project. Indeed, in Mexico the sale of maize grown in taungya reduces the total plantation establishment cost by 27% (Haufe 1977).

5. The amount of food production can be significant. In Trinidad, Ramdial (1980a) cites average yields of food crops from teak taungyas as 5.0–6.6 t ha^{-1} and pine taungyas as 2.3–3.9 t ha^{-1}.

It would be surprising if there were no problems in the application of the taungya system. Sometimes cultivators may neglect the trees to favour their crops and it may be difficult to get them to move on to other areas. On the other hand where there is plenty of land close to the village there is little interest in the practice, for example, in East Kalimantan, Indonesia (Sutisna 2001).

Until 1987, most tree plantations of the Forestry Department in Ghana were established by taungya. It was discontinued due to many failures in application and management of the system rather than silvicultural problems. Farmers faced with the possibility of becoming landless when the trees were fully established often damaged or killed them. It was concluded that taungya requires socially acceptable arrangements to ensure security of land tenure and an effective land allocation system (Odoom 1999). Even in Ghana it is agreed that taungya is an appropriate system for farmers who practise shifting agriculture on soils unable to support continuous cropping.

The taungya system and its many variants are likely to continue to be important in plantation development in the tropics. While other agroforestry practices may become equally widespread, taungya remains an inexpensive system, which often does not require radical change in life-style by the cultivator. The taungya method of combining farming and forestry is only one kind of intercropping; a special one fulfilling a sociological role to provide employment for one family unit with their own area of land to work. It is usually of particular value in areas where there is heavy population pressure on land resources, although social problems as described above for Ghana may arise.

Large-scale establishment intercropping

Large-scale intercropping, sometimes called 'departmental taungya' when it is carried out as a standard establishment practice by a national forest service or plantation enterprise, can be adopted in almost any plantation development, where soil and site conditions are suitable for a food crop. There are many possible combinations of forest and food crops. Unlike the taungya system, small size is not a requirement for intercropping during establishment of the tree crop.

A common and widely applicable form of establishment intercropping, where land is not steep and can be cultivated, is to grow an annual food crop between rows of trees, just as in taungya, but on a large scale or even as a standard practice in plantation establishment (Fig. 20.4). In Brazil, Companhia Agricola e Florestais cultivated 3–5 rows of soybeans between rows of *Eucalyptus grandis*, which were 3 m apart. Yields up to 2.4 t ha^{-1} were achieved and there is evidence that the trees benefited from residual nitrogen left in the soil after harvesting this legume.

Many annual crops can be intercropped in this way for the first year or two after tree-planting: other examples include dryland rice, cassava, millet, peanuts, and sugarcane. Silviculturally, the only modification to normal practice is usually to plant the trees at wide spacing between rows (rectangular planting) to provide the food or cash crop adequate room and to allow tractor access if mechanical cultivation and tending is carried out. Any fertilizer applied to the food crop will usually benefit the trees. Similarly, the weeding necessary for the food crop may also be of benefit though the food crop itself may compete to some extent with the trees for available moisture and nutrients.

Figure 20.4 Maize grown between rows of *Eucalyptus grandis* (Shiselweni, Swaziland).

Improved tree fallow

The third main type of rotational cropping, which involves tree-planting (or planned regeneration), is where the tree crop is grown primarily for its soil improving role. The trees are the 'fallow' between food crops. The trees or shrubs used are usually nitrogen-fixing species with the ability to return atmospheric nitrogen to the soil through root decomposition and/or leaf fall. Binkley and Giardina (1997) state that few data are available on soil changes under nitrogen-fixing trees in the tropics but these probably include increases in total and available nitrogen and variable changes in other properties such as pH. They give data on species influence on a Vertisol soil in Costa Rica after 6 years where *Leucaena leucocephala* had the greatest N availability (about 30 mg kg^{-1}) followed by about 11 mg kg^{-1} by *Casuarina equisetifolia* and *Senna siamea* (a non-N-fixing legume) and lowest (<5 mg kg^{-1}) by *Eucalyptus robusta* (a non-N-fixing species).

Indigenous agroforestry fallows have recognized the benefits of using nitrogen-fixing species. For example, in the highlands of PNG fallows are planted in subsistence food gardens during the second or third cropping cycles. As yields of sweet potato decline the land is planted with *Casuarina oligodon* and allowed to remain under the casuarinas for 7–10 (perhaps 20 years when population pressure was less) before it is cleared for replanting with sweet potato (Thiagalingam 1983; Askin *et al.* 1990). In the Philippines, a fallow system using stump cuttings of *Leucaena leucocephala* on a 2–4 year rotation to improve the soil for maize and rice is described by MacDicken (1991). *Acacia senegal* in the Sudan may be planted with food crops then grown on for another 16 years for gum arabic and building up of soil fertility. In Ecuador 2 years of food cropping are followed by 8 years of 'fallow' with *Inga edulis* sometimes intercropped with bananas and forage legumes (Young 1997). All these examples include nitrogen-fixing trees as almost all improved tree fallows exploit their capability to enrich the soil.

In one sense this role for trees is the same as their use in rehabilitating degraded land where a pioneer species reverses the process of degradation. For example, use of *Acacia mangium* in degraded *Imperata* grasslands in Mindoro, Philippines, to create conditions suitable for eventual establishment of native hardwoods and cocoa crops. This acacia, as well as fixing nitrogen, is able to compete with and eventually suppress *Imperata* grass.

Dispersed irregular tree crop systems

Farm and shade trees

There are two broad categories of tree growing on cropland: (i) in semi-arid areas the retention or planting of trees at low densities of 20–50 ha^{-1}, creating a savanna appearance, with cultivation of cereal or other annual crops, and (ii) provision of light dappled shade from 50–200 trees ha^{-1} in management of perennial crops such as coffee, tea, and cocoa.

In many parts of the tropics it is common to see 'farm trees' dispersed throughout cultivated land. The trees are cared for to yield products such as fuelwood, poles, fruits and nuts and fodder, but are often grown to enrich the soil, particularly in drier sandy areas, and confer a more favourable microclimate. *Faidherbia albida* and *Azadirachta indica*, are commonly used in this way in fields in the West African Sahel and Ethiopia, *Grevillea robusta* and *Markhamia lutea* are favoured farm trees in the East African highlands, and mango (*Mangifera indica*) is frequently planted in both East and West Africa. *Melia azedarach* is a common farm tree in East Africa and in many parts of Asia. In the Vietnamese hills *M. azedarach* is planted with agricultural crops, then left to close canopy and dominate during the fallow period, thereafter it is thinned out to produce wood products and some trees retained to provide light shade over the crop (Woods 2003). On drier sites in

Kenya the native *Melia volkensii* is compatible with all crops grown (Stewart and Blomley 1994) and in the arid regions of South Asia and parts of the Middle East *Prosopis cineraria* is used as a farm tree to produce timber, fuel, and pods for both fodder and human consumption. Increased yields of crops such as millet, sorghum, and peanuts are reported from the vicinity of such trees, owing to greater soil organic matter and higher nitrogen levels; see Chapter 14 (plants as soil improvers). In some instances, for example, traditional small-scale farming in Mexico and Guatemala, leaf litter is brought from nearby forest and worked into the soil prior to cultivation (Wilken 1978).

Many perennial crops in the tropics, such as coffee, tea, cocoa, and some orchards, are grown under tree shade (Fig. 20.5). Some crops, such as cocoa, require shade when young. Others, notably coffee and tea, may not be at their most productive under shade (less light energy is available for growth) but the correspondingly smaller demand for nutrients allows very satisfactory crops to be grown with little or no fertilizing. This is an important consideration where farmers cannot afford inorganic fertilizers or fertilizer is unavailable. Coffee plantation yields are typically higher with shade, for example, coffee raised under *Erythrina poeppigiana* shade trees in Costa Rica have 40% greater yields without nitrogen fertilization (Ramirez *et al.* 1989). Where very intensive management, including fertilizing, is applied greater production is generally possible without shade. In Indonesia, tea gardens used to have a tree cover of species such as *Paraserianthes falcataria* as a source of nitrogen and organic matter but now the availability of cheaper inorganic fertilizer and new less shade-tolerant tea varieties have resulted in a change from agroforestry to a pure crop (van Noordwijk 1996).

Trees with light foliage, ability to fix atmospheric nitrogen and/or root systems that provide little competition with crop plants are favoured as shade trees. Light-crowned species such as *Moringa oleifera* and, *Melia azedarach* can be integrated easily with some crops. Legumes are often used including *Acacia decurrens, Albizia* spp., *Erythrina* spp. (Fig. 20.7), *Gliricidia sepium* and *Leucaena leucocephala*. Shade trees which root more deeply than the understorey crop, for example, *Azadirachta indica* and *Grevillea robusta*, help augment the surface soil through their leaf-fall, though some nutrients will be lost from the ecosystem when the tree is harvested. Trees that respond to pruning and pollarding, for example, *Erythrina poeppigiana* and *G. robusta*, can be managed to provide effective shade with minimum competition.

There are several advantages of growing crops under shade in the tropics:

- climatic damage is reduced: wind, hail, intense heat, sun-scorch, etc.,
- soil erosion is minimized and substantial quantities of organic matter and nutrients returned to the soil,
- undergrowth is less,
- pests and diseases on crops such as coffee and tea may be reduced,
- working conditions under shade are more pleasant.

Shade trees can play an important agricultural role and are particularly valuable in the tropics where all the inputs to obtain optimum yields of monoculture crops, especially expensive fertilizers, cannot be afforded. The important role of leaf-fall and the use of tree foliage mulches to augment soil nutrients were discussed in Chapter 14. However, shade trees also have a value themselves. The trees must be pruned and thinned to control growth rate and density of canopy; thus there is a regular supply of firewood, building poles, and sometimes even timber. In Costa Rica and other Central American countries, many coffee and cocoa plantations are shaded by two tree layers. *Erythrina poeppigiana* trees, which are pollarded twice a year to control shading levels are grown as a lower layer and are important for providing mulch and for nitrogen-fixation. The overstorey is usually *Cordia alliodora*, which is leafless for 5 months, and which produces a readily saleable timber.

Figure 20.5 *Leucaena leucocephala* shading cocoa (Markham valley, Papua New Guinea).

Trees giving dense shade throughout the year are needed in hot, arid areas to relieve stress on grazing animals in the hottest part of the day. They should be evergreen, have a wide, spreading crown with a dense canopy, long-lived, unpalatable to livestock, and able to tolerate soil compaction due to animals sheltering under them.

Full intercropping
This is where a food crop is cultivated during most or all of the life (rotation) of the associated tree crop. Trees are spaced sufficiently far apart and managed intensively, mainly thinning and pruning, so that food cropping is possible well beyond the establishment stage when trees are young. This form of intercropping as a system merges with shade trees grown with arable crops noted in the previous section.

Farmers recognize certain species as being compatible with crops and others as having a negative effect. In the Siaya area of the western highlands of Kenya trees are planted and retained in cropland with *Calliandra calothyrsus, Croton megalocarpus, Leucaena leucocephala, Markamia lutea, Sesbania sesban, Samanea saman,* and *Grevillea robusta* favoured while *Eucalyptus* spp., *Senna siamea, Lantana camara* and *Thevetia peruviana* are avoided (Warner 1993). *Grevillea robusta* intercropped with maize or maize and beans is a common sight in the African highlands (Fig. 20.6). Nitrogen-fixing and multipurpose tree species are a frequent but not essential component of intercropping systems. In Bangladesh farmers plant trees in their fields and recognize that depressed crop yields can be compensated for by valuable tree products (Hocking *et al.* 1997*b*).

Where trees are planted in rows, east–west orientation minimizes shading, but does not eliminate the need to manage the tree crop if continuous arable cropping is sought.

The intercropping system described is a formal way of growing trees and food together, but there are a variety of less formal systems, which logically fall into this category of agrosilviculture. For example, many pine and *Araucaria* plantations in PNG have passion fruit, papaya, pumpkins, and chillies growing in them through natural colonization and by local people planting them, and unplanted ground between widely spaced rows of trees for erosion control can yield grass for thatching or herbage cut for fodder. In Kerala, India, peppers and betel for home consumption are cultivated to climb up stems in *Bombax ceiba* (*B. malabaricum*) plantations grown for matchwood (Arnold 1990).

Multi-storey tree gardens
These are commonly known as home gardens, forest gardens, compound farms, or mixed gardens. They are usually located close to the homestead where they can be intensively cultivated and protected. Annual and perennial species are grown in association with crop rotations, ranging from just months (beans and maize) to many years (trees for poles and even timber) with many intermediate ones: cassava and pineapples 1–3 years, coffee and bananas 4–10 years, etc. This mix of rotations in time, though

Figure 20.6 Intercropping of maize, beans, and *Grevillea robusta* trees, Nyabisindu, Rwanda.

not of the actual examples which are cited only for illustration, is complemented by a layered vertical structure of low ground crops, for example, peanuts, tall ground crops (maize), low shrubs (cassava), tall shrubs (bananas), small trees (for firewood and fruit), and tall trees, along with cultivation of vines and climbers sometimes up 'trees' grown as stakes and regularly pollarded and/or shredded (side branches pruned from much of the stem) for the purpose. Fruit trees often are the most important woody species in the home garden. Many of the following species important in home gardens in the Philippines are also important elsewhere in the tropics: jackfruit (*Artocarpus heterophylla*), breadfruit (*Artocarpus altilis*), pomelo (*Citrus grandis*), guava (*Psidium guajava*), avocado (*Persea americana*), Banana (*Musa* sp.), cocoa (*Theobroma cacao*), santol (*Sandoricum koetjapi*), and coffee (*Coffea* spp.) (Guzman 1999). With careful planning home gardens can be managed to yield a continuous supply of food, fuel, and timber.

Home gardens are common where pressure on land is high and in this situation they combine numerous ecological, social, and economic functions. They often produce a significant proportion of farm income, make intensive use of land, require considerable labour input, and enable farmers to spread farm-work, output, and income throughout the year. This form of agroforestry has long been practised in the more densely populated regions of the moist tropics, especially in south and Southeast Asia (Fig. 20.7), West Africa, the highlands of East Africa and Central America. They are a feature of countries such as Indonesia (Java), Nigeria, Sri Lanka, and Tanzania. It is not possible to describe all the different crops and types of home gardens; but

examples will be found in Withington *et al*. (1988), Arnold (1990), Warner (1993), Michon and Mary (1994), and most general agroforestry textbooks.

The complexity and layered structure of home gardens ecologically closely resemble natural tropical ecosystems and this may lead to greater sustainability than simpler systems of forestry or agriculture (King 1979). To illustrate their complexity, a description of Kandyan home gardens in Sri Lanka is indicative (Perera and Rajapakse 1989). They found that 15% of the central hill land of Sri Lanka was occupied by home (forest) gardens typically averaging 0.2–0.4 ha in size. Between 500 and 1250 trees ha^{-1} were grown, consisting of 23 species on average, along with 17 herbaceous species, six climbers, and three endemic species. Coconut, jackfruit, and *Alstonia* spp. were typical overstorey trees, avocado, *Gliricidia* sp., mango, and fish-tail palm as midstorey, and papaya and coffee as lower storey crops. Pepper and cloves were also common.

Dispersed zoned tree crop systems

In many agroforestry systems the tree component is established in strips or zones specifically to obtain certain environmental benefits or to avoid occupying scarce farmland. The commonest examples are shelterbelts or windbreaks spaced at intervals to reduce wind-speed and air turbulence, and planting along terraces or other erosion control works to help stabilize the soil. These are considered in chapters 21 and 23 as part of the important topic of protective afforestation in the tropics. Here, three other examples only are noted: the use of trees primarily as side shelter and a source of nitrogen-rich mulch for the

Figure 20.7 Home garden in Indonesia. Maize and cassava as ground crop; cocoa, papaya, and bananas as a perennial lower-storey crop; coconuts and scattered trees of *Erythrina variegata*, often pollarded for firewood, as the upper-storey crop. Nearby there are planted crops of cloves, *Sebania seshan* for firewood, and *Paraserianthes falcataria* for shade, shelter, and timber.

Figure 20.8 Alley cropping of maize and groundnuts with hedges of *Leucaena leucocephala* (centre and far right). The *Leucaena* is ready for cutting and the foliage will be spread as a nutrient-rich mulch around the maize and where indicated (Cameroon).

arable crop (alley cropping), trees grown primarily for wood production but at a spacing wide enough to permit growing food crops (line planting), and planting trees to demarcate boundaries or to occupy land of least farming potential (border planting).

Alley cropping

The concept of alley cropping is simple. A fast-growing nitrogen-fixing tree or shrub species is grown in rows sufficiently far apart to allow cultivation of 4–6 rows of food crops and its foliage cut frequently to provide a nutrient-rich mulch around the food crop. A feature of the system is the capacity of the trees to produce a large quantity of biomass for green manure and the need for regular pruning to reduce competition with the associated crops. In the humid tropics alley cropping can significantly increase soil fertility in 2–3 years (Kang *et al*. 1990; Kang 1993). As well as soil fertility improvement, alley cropping in the form of contour hedgerows can be useful for soil conservation on sloping lands. Legumes and actinorrizhal species are almost exclusively used. A combination of *Leuceana leucocephala* and maize (Fig. 20.8) has commonly been successful, but many other crops and many other trees such as *Sesbania, Erythrina, Flemingia, Gliricidia, Guazuma, Prosopis,* and *Calliandra* species can be alley cropped. As well as nitrogen-fixing ability, the species must coppice or pollard well and have rapid early growth.

Because the tree component is kept low, typically less than 2 m, and regularly pruned, it is akin to a hedgerow hence its generic name (Table 20.1). The timing of pruning is critical and normally takes place at about one-quarter of the way through the life of the food crop to cut down immediately direct competition for light and moisture and to provide mulch at the time of maximum food crop growth. Height of pruning is also important; keeping it low reduces competitive effects, but obviously not so low that coppicing potential is impaired; (see Chapter 18). The optimum pruning schedules for particular combinations of food crops and woody legumes will often need to be researched for a particular site or locality, for example, the Duguma *et al*. (1988) study in Nigeria. The woody parts of stem and branch prunings are suitable for firewood; only the twigs and leaves are used for mulch or, if mulch is not needed, usually for fodder.

On sloping lands contour hedgerows can form a semi-permeable barrier to surface water movement, the mulch reduces the impact of raindrops on the soil and minimizes splash and sheet erosion. Although the technique has been strongly promoted and examples of successful application of this technique reported, for example, in Nigeria (Lal 1989), in Philippines (Maclean *et al*. 1992; Palmer 1999), in Indonesia (Rusastra *et al*. 1998), there is considerable doubt concerning the effects of the vegetative barriers and the sustainability of the system. While off-field erosion may be reduced, soil degradation may occur through soil redistribution within the alleys and in many cases annual crop yields within the alleys do not show any advantage compared with open field results suggesting more research is required to modify management practices to cope with the problem (Garrity 1996).

Successful alley cropping depends on intensively managing the nitrogen-fixing woody shrub to minimize its competitive effect, though this is hard to achieve. Often the adjacent rows of the food crop grow less well than those midway between the woody hedges. Research shows that frequently this is due to root competition for moisture rather than shading effects.

Alley cropping has been successful on relatively fertile soils but attempts to extend it to acid infertile soils in the humid tropics or to the semi-arid tropics have been disappointing. Progress has been made in selecting acid-tolerant, fast-growing trees but economical techniques for reducing tree-crop competition have not yet been developed (Ong 1996).

Line planting
This model has the trees planted in rows at a distance wide enough to enable farmers to grow cash crops between the lines and is like a farm-scale intercropping (p 325). The trees in the lines are grown primarily for wood production. This model is used by PICOP Resources Inc. in the Philippines to produce poles, sawlogs, and pulp-wood (Aggangan 2001). *Eucalyptus deglupta* or *Paraserianthes falcataria* are planted at a spacing of 1 × 10 m and the rotation about 8 years. The inter line agricultural crops are farmers' main source of income until the trees are harvested.

Border planting
This describes the use of trees to demarcate boundaries and/or locate them in places least in competition with agricultural crops. Tree-planting along field boundaries and around the edge of the compound offers an opportunity for supplementary production of fruits, fuel, fodder, etc., with little or no opportunity cost provided, the boundary is clear and arable crops least threatened. It is important to recognize that trees often take on legal meaning as boundary markers and can be the source of both tree tenure and land tenure disputes (Raintree 1986). Border planting is practised throughout the tropics, especially in densely populated areas. This practice was more formally adopted in China's famous 'four-around' planting, concentrating trees around houses, villages, along roads, and beside canals and watercourses. There are an estimated 50–100 million individual eucalypt trees in the four-around plantings (Zheng 1988; Wang and Zhou 1996).

Almost any species is suitable for border planting, although those which do not compete with adjacent crops for light, water, and nutrients are preferred. *Casuarina equisetifolia* is planted on the rice field bunds in India because its thin narrow crown casts little shade whereas eucalypts are often avoided because of their competitive root systems. Multipurpose species are mainly used as border trees as they are very much a part of the rural economy and farming systems.

Trees and livestock

As with trees and arable crops, there are different systems of combining tree growing with livestock management. Rotational practices, that is, the tree crop alternating with livestock, are rare, though occasionally land is intensively grazed prior to establishing a tree plantation as a way of reducing weeds and clearing scrub. Unfortunately it is often over-grazing of land that contributes to the need to plant trees for fuel or fodder to avert environmental disaster. However, a form of rotational management usually occurs where livestock graze in the plantations since such a practice is confined only to certain periods in the life (rotation) of the tree crops.

Table 20.1 shows three main kinds of tree and livestock systems: dispersed irregular systems where livestock graze/browse in tree plantations (forest grazing), and zoned systems where trees are used as living fences or for fodder production as part of livestock management.

Forest grazing

In a plantation the forage value of the large quantities of herbaceous matter (weeds) may be of feed value. Domestic animals can graze grass and browse herbs and shrubs growing beneath trees. Though this kind of grazing has been practised for hundreds of years in some parts of the world (for a review see Adams 1975) it has not generally been adopted in plantation forestry. This is mainly because grazing further complicates management and because grazing damage by domestic animals has always been a major protection problem in countries such as India. However, there is a renewed interest in forest grazing, especially in the Asia-Pacific region stimulated by the successful cattle grazing regimes in *Araucaria cunninghamii* plantations in Queensland, Australia, and, as Payne (1985) points out in his review of such systems in the tropics, it is now realized that livestock tree-crop integration can improve productivity per unit area of land.

Grazing animals in the forest benefits both plantation work and the livestock owner. In plantation forestry there are four main benefits.

1. The animals, by eating undergrowth, greatly reduce the fire hazard. This is probably the main reason for encouraging grazing in the tropics particularly where controlled burning has to be avoided. In young *Pinus caribaea* plantations in Fiji loadings of fine fuel are reduced by grazing (Fig. 20.9) from a dangerous 2500 kg ha^{-1} to a comparatively safe level of 800 kg ha^{-1} (Gregor 1972). Where under-growth is sparse, as in many more subtropical pine plantations such as the Viphya Pulpwood Project (Malawi) and the Usutu forest (Swaziland), domestic animals are

Figure 20.9 Cattle grazing under five-year-old *Pinus caribaea* in Fiji. Notice fencing, water trough, salt lick, and lack of ground vegetation.

used for the same purpose to graze open firebreaks between compartments.

2. A valuable supply of meat is obtained; dairy management may be done but it is usually less practicable. Such a supply of meat is important in large remote plantation projects and both the Jari scheme in Brazil and the Viphya Project in Malawi have, in the past, maintained herds of several hundred cattle. Typically, properly managed systems can yield 100–250 kg ha^{-1} year^{-1} of liveweight gain (Payne 1985).

3. Properly organized forest grazing may be profitable in its own right.

4. Grazing in plantations visibly demonstrates multiple land-use and shows that plantation forestry and farming need not be exclusive land-use activities. Where local villagers are livestock owners their traditional resistance to plantation development is often lessened if forest grazing is allowed, even to the point that they may actively encourage afforestation.

Management of grazing

Several factors need to be considered before commencing grazing since it increases the complexity of plantation management.

1. *Life of crop.* Domestic animals can seriously damage young trees and are usually excluded from stands until trees are 3–4 m tall. Not only does browsing cause damage, especially of leading shoots, but cattle will often lie against a tree which, if young, is pushed over and subsequently develops a basal sweep. Grazing can commence once trees are no longer likely to suffer damage; in the tropics this may be as early as 12–15 months after planting in eucalypt plantations and 4–5 years in slower growing pine

crops. Grazing continues while worthwhile forage lasts. In Fiji, where cattle are grazed in some *Pinus caribaea* (Fig. 20.9) stands, it is mostly done when stands are 4–9 years of age. However, earlier introduction of livestock has been found feasible if young trees are clean-weeded and visible and are at least 50 cm tall (Bell 1980). Such early grazing is found possible with hoop pine (*Araucaria cunninghamii*) but not true pines (*Pinus* spp.) in Queensland, but is practised with pine at Jari, Brazil (see below).

2. *Forage quality.* The forage quality of vegetation under plantations is mostly poor compared with range pastures. This may not be a serious disadvantage in the tropics because by being under trees the animals are protected from heat and heavy rain and will often graze continuously as in more temperate climates. On range pastures most feeding occurs in the mornings and evenings, the animals seeking shade, and resting, during the heat of the day. Continuous low-quality grazing may be no less nutritious than intermittent better quality grazing, and cattle and sheep are more adapted to the former. However, forage quality under a plantation can be improved. In particular legumes can be introduced such as *Phaseolus atropurpureous*, *Vigna luteola*, *Dolichos* spp., and *Stylosanthes guyanensis*. All, in fact, are introduced into PNG plantations where cattle are grazed and, as well as improving the forage value, their nitrogen-fixing ability helps to enrich the soil. At Jari in Brazil sweet grasses such as colonial grass, are sown between rows of *Pinus caribaea* planted at a rectangular spacing of 4.0 m × 2.25 m.

3. *Livestock management.* Because of the possibility of damage to young trees, grazing must always be controlled. Fencing grazing areas and gridding roads and tracks are normally essential. Suitably located pens will also be needed to collect animals

in one place for dipping and treating against disease and for sending to slaughter. Someone should be responsible for the animals at all times whether a full-time shepherd, herdsman, or a villager. Animal feed must always be adequate. During the dry season, when natural growth of forage is slow, feed supplement may be needed if animals are to remain healthy. Growing fodder trees, whose foliage is palatable, is one possibility (see later). Another kind of supplement needed is to augment the animals' mineral intake using a salt-lick (Fig. 20.9). A water supply or trough may also be needed. The optimum stocking density will depend on feed quality and growth, but typical are densities of 0.5–2.0 cows ha^{-1} of forest plantation. Higher rates can be achieved in more intensive systems, a lower tree density, more forage and rotational grazing. In addition, in countries with intensive pressure on land, for example, Rwanda and Indonesia, the dense patchwork of home gardens and tiny cultivated plots does not allow even controlled open-land grazing so most livestock is raised in stalls with fodder cut and carried to them i.e. zero grazing.

4. *Animals.* Cattle are the commonest domestic animal grazed in the hot tropics and it is usual to use crosses between pure breeds of European cattle and lower yielding but more heat-tolerant breeds. Where grazing forms a regular part of a forest management regime it is usually necessary to employ a full-time livestock specialist.

Fully integrating grazing and silviculture

The increased complexity of including domestic animals in plantation management generates extra work. Where grazing becomes part of multiple land-use practice and the forage value maximized by growing open stands (silvo-pastoral systems), major modifications to normal silvicultural practice may be needed.

1. As with intercropping, increased spacing between trees is important to maintain high light intensities at ground level. Trees may be planted at normal spacing in the row but 4–12 m between rows, or planted at wide spacing overall. In Ecuador, *Cordia alliodora* is planted at 5 × 5 m, and in Costa Rica *Alnus acuminata* from 7 × 9 m to 10 × 14 m to allow integrated cattle production.
2. Wide spacing produces rapid diameter growth and very thick branches. Pruning is essential to eliminate large knots if good quality saw-timber is to be produced.
3. Early and heavy thinnings are needed to maintain light levels under the trees. Thinning may be done to waste. In Ecuador, *Cordia alliodora* is thinned to 200 ha^{-1} at 3 years and 100 ha^{-1} at 5 years.

4. Grazing may depress growth on poorer sites. In Fiji, Vincent (1971) found that grazing under *Pinus caribaea* at a stocking of 1 cow ha^{-1} reduced basal area increment between 5 and 30%.
5. More attention must be given to forage and pasture management in terms of quality, productivity, and rotational use to ensure satisfactory yields and health of livestock.

Animals can benefit the tree crop through the nutrient cycling process but an integrated grass-legume system may be necessary to provide animal forage of an adequate quality throughout the year (Pell 1999).

Trees and livestock zoned

Living fences
Trees are commonly planted in the tropics as a stock-proof 'fence' because of the low establishment cost. Fast-growing, long-lived species with spines and stiff branches and unpalatable foliage are most suitable. Species easily propagated from large hardwood cuttings, referred to as 'stake' or 'pole' cuttings are rooted in position in the field, usually spaced 15–50 cm apart (Chapter 11).

This form of agroforestry has been developed to a high degree in Central America, especially with *Gliricidia sepium* (Fig. 20.10), where it is grown around cattle pastures and to mark boundaries (Stewart 1996). Budowski (1987) and Jolin and Torquebiau (1992) list many tree species used in Latin America as living fences. In East Africa *Euphorbia tirucalli* is widely used (Warner 1993); in the Sahel species such as *Acacia macrostachya*, *Euphorbia balsamifera*, *Parkinsonia aculeata*, *Prosopis julifera*, and *Ziziphus mauritiana* are recommended (Maydell 1986); and in the West African country of Cameroon farmers commonly use *Ficus* spp. and *Erythrina* spp. (Gautier 1995). Hocking (1993) also acknowledges the value of *Prosopis julifera* as a live fence but warns against its highly competitive root system if crops are planted near it.

Fodder trees and fodder banks
A common practice in the tropics is to use prunings from trees and shrubs, or even to grow trees specially, called fodder banks or intensive feed gardens, to provide fodder and browse for domestic animals. Tree fodders often contain high levels of crude protein and minerals, and many legumes and non-legumes have high levels of digestibility in ruminant livestock. Some species are very productive and maintain foliage well into the dry season. This source of feed is most important during a dry season when pasture is poor. In some countries, for

example, Sudan and northern Senegal, farmers actively protect their trees to conserve their supply of goat feed and in Nepal over 80% of fodder requirements of farmed livestock is supplied from trees. Antinutritive factors can be a problem as some tree species contain polyphenolics, toxic amino acids, cyanogenic glycosides, and alkaloids. Many species are suitable and species such as *Leucaena leucocephala* and *Calliandra calothyrsus* have been widely used in planted fodder banks. Comprehensive lists have been compiled by Singh (1982), Torres (1983), von Carlowitz (1986), Maydell (1986), Vercoe (1989), and Le Houérou (2001). Table 20.2 lists some more commonly planted species. There are many general accounts of fodder-tree management, nutritive value, and experience,

Figure 20.10 Living fence of *Gliricidia sepium* around cattle pasture in Costa Rica. The trees have been pollarded several times.

Table 20.2 Examples of important tree species grown in the tropics for timber or firewood, and which also are valuable for fodder

Species	Edible for cattle	Comments
Legumes		
Acacia arabica	Leaves and pods	Gum arabic; tannins from bark
A. catechu	Leaves from lower branches	Wood makes good charcoal
A. sepal	Foliage and bark is eaten	
Albizia lebbeck	Leaves	Important shade tree for plantations
Calliandra calothyrsus	Leaves	Good honey source
Dalbergia sissoo	Leaves, can be made into silage	
Faidherbia albida	Leaves and pods	Good for arid areas: in leaf in dry season
Gliricidia sepium	Leaves	Good firewood, leaves toxic to horses
Leucaena leucocephala	Leaves (toxic in excess)	Important multipurpose species
Parkia filicoidae	Leaves and pods	
Prosopis chilensis	Leaves and ripe pods	Very drought and cold resistant, a good shade tree
Samanea saman	Leaves and pods	Excellent and widely planted shade tree
Senna siamea	Leaves	Common firewood
S. fistula	Leaves and pods	
Non-legumes		
Artocarpus heterophyllus	Leaves and fruits	Also good human food
A. indicus	Leaves and fruits	Also good human food
Azadirachta indica	Edible fruit, can make oil seed cake	Drought resistant
Bombax malabaricum	Leaves and twigs	
Brosimum alicastrum	Leaves, seeds, and whole fruits	Fruit eaten by humans, latex for chewing gum
Shorea robusta	Leaves, seeds used for fat extraction	
Terminalia tomentosa	Branches lopped for fodder	

for example, Craswell and Tangendjaja (1985), Robinson (1985), Devendra (1990), and Paterson *et al.* (1998).

A tree grown for fodder also provides some shade, often firewood, and sometimes even human food. For example, dead leaves of breadfruit (*Artocarpus indicus*) are relished by cattle, breadfruit is a widely eaten food, and the wood is reasonable to work. *Faidherbia albida* comes into leaf at the start of the dry season when there is little other forage and during the critical forage supply period in the Sahel the pods containing highly nutritious seeds fall to the ground and are eaten by livestock, or in times of famine can even be eaten by humans (Maydell 1986). Tree growing to provide animal feed, especially in dry regions where grass is not always available, is an important additional form of integration between farming and forestry. But, as with other forms of social forestry, it is not only necessary to get the technology right but also all aspects of farmer involvement and support. Useful experience with introducing such fodder banks of *Gliricidia, Leucaena*, and other species in the tropics is reported by Francis and Atta-Krah (1989), Shelton (1994), and Stewart (1996). Low-growing fodder banks, typically of species of *Calliandra, Erythrina, Gliricidia, Leucaena, Markhamia, Parkinsonia, Prosopis, Senna,* and *Sesbania* are maintained as hedges and pruned on 6–18-month cycles for fodder, whereas taller trees are pollarded for fodder and fuelwood, perhaps every 4–8 years. Nitrogen-fixing species predominate, owing to high leaf protein levels, often in excess of 20%. Yields of up to 10 t ha^{-1} year^{-1} of dry fodder are quite possible and many experimental trials report much higher figures, for example, Bhatti *et al.* (1989). Livestock are either allowed to browse in controlled ways or the fodder is cut and carried to fields or stalls. The cut and carry system (zero grazing) is beneficial where animals find the wilted foliage more palatable than fresh material.

Trees planted for other purposes

Trees may provide directly or indirectly a range of 'other' products of which only a few are included here. These non-wood uses, and non-timber forests products (NTFPs) generally, often stimulate tree-planting by companies, communities, or farmers. This diversification in products and industry not only encourages greater land-use integration, but often becomes an important source of rural employment. Further, human requirements for recreation and for aesthetic values in the landscape (the pleasantness of having trees in the environment) leads to planting near or within villages, towns, and cities.

Apiculture

This is the keeping of bees for honey, and flowering of trees can be important for bee pasturage. Several *Acacia* species and many tropical eucalypts grown in plantations, including *E. deglupta, E. grandis,* and *E. camaldulensis*, flower profusely, often in most months of the year. Good quality honey is produced from eucalypts and apiculturists in many countries are taking advantage of eucalypts and acacias planted for timber and other products. Acacias produce little or no nectar but their pollen is a very important source of protein for bee-hive nutrition. In Ethiopia *E. camaldulensis* is highly regarded as it provides a source of nectar when there is little nectar production from indigenous species. In South Africa it is estimated that the honey potential of *E. grandis* is 1–4% of its timber value and in New South Wales, Australia, 30% of the economic value of a natural eucalypt forest can come from honey production. The role acacia and eucalypt plantations can play in honey production has been described by Moncur *et al.* (1991).

Apiculture may increase the risk of fire since it is a common practice to smoke out wild colonies of bees in the plantations. Nearly 60% of all fires in eucalypt plantations in southern Africa arise from this cause!

Essential oils and other extractives

An essential oil is a mixture of fragrant, volatile compounds commonly named after the plant from which it is derived. There are many trees that yield essential oils, those from Southeast Asia are described by Oyen and Dung (1999). Eucalypt oils are complex mixtures of organic compounds, which can be used for medicinal, industrial, perfumery, and flavouring purposes. Steam distillation of oils from eucalypt leaves is an important local industry in countries such as China, India, and South Africa. The composition and yield of the oils varies between and within species and according to the type and physiological age of the leaf harvested. Less than 20 species have been exploited commercially for oil production and only a few of these grow well in the tropics. Species yielding commercial quantities include *Eucalyptus citriodora, E. globulus,* and *E. smithii* (Boland *et al.* 1991; Doran 2002). Of particular interest is the relatively recent discovery that plantations established in drier parts of the tropics using the Petford provenance of *Eucalyptus camaldulensis* var. *simulata* offer immediate possibilities for medicinal grade essential oil production (Doran and Brophy 1990).

Essential oils are also extracted from the leaves *Melaleuca* spp. in Australia, Indonesia, New Caledonia, and Vietnam (Brophy and Doran 1996). Fragrant woods are also important sources of aromatic materials, some yield aromatic resins such as myrrh from *Commiphora* spp., and benzoin from *Styrax benzoin* while oils are distilled from sandalwood (*Santalum album*) and camphor (*Cinnamomum camphora*). Although these oils were originally extracted from natural forests, plantations are now being established for their production.

Many other products can be extracted from tropical tree species, but one which has aroused much interest is neem oil from the seeds and leaves of the widely planted *Azadirachta indica*. Large quantities of neem seed oil is used in India and other parts of South Asia for soap making with the residue 'neem cake' used for fertilizer and feeding livestock. The oil has medicinal properties and a neem oil-based contraceptive is marketed in India. Over 30 pesticides have been developed based on azadirachtin, one of many biologically active components extracted from leaves, seed, bark, or flowers of neem (Ahmed and Idris 1997).

Resin tapping
Southern pines (*Pinus elliottii* and *P. taeda*) are mostly used for resin tapping to obtain naval stores but experimental trials in Cuba and Malaysia show that *P. caribaea* can also be tapped productively (Low and Razak 1985; Alvarez and Stephan 1986). In tropical Asia *P. kesiya* and *P. merkusii* are both tapped for resin.

Sericulture
This is the culture of silk worms to produce silk thread. Silk worms are mainly raised on mulberry leaves (*Morus alba*), though other species are also suitable, for example, *Shorea robusta*, and can be a profitable rural industry. A ready supply of mulberry leaves is often obtained from trees planted around fields or along canal banks, which also yield canes for basket-making, firewood, timber, fodder, and edible fruit. Sericulture is an important industry in India, Pakistan, China, Vietnam, and Thailand.

Tan bark
Acacia bark was identified as an excellent source of tannin in Australia in 1814 and *A. mearnsii* is the species mainly cultivated for this product. The industry has long been carried out in the tropics and subtropics; see Chapter 3. There are many thousands of hectares of *A. mearnsii* in eastern and southern Africa, and substantial plantings in Brazil, India, and China (Brown and Ho 1997). Much of the bark produced comes from small plantations on farms (see 'woodlots' in this Chapter) and the wood is used for firewood, charcoal, poles, and pulpwood.

Amenity
The value of forests for recreation and aesthetic appeal is now widely appreciated and planting for amenity in the tropics can be important. A good example is in Hong Kong, where almost all of the original forest was removed and the hills became highly degraded. Since the late 1970s there has been the establishment of country parks to provide an extensive recreational facility for the large urban population. Now 14% of the land area has forest cover of which one third is planted and the remainder restored secondary forest (Zhuang 1997). *Pinus elliottii* and *P. massoniana* have been extensively used but there are also a few native species such as *Schima superba*, and increasingly, broadleaved species and mixed plantings are used, especially outside the country parks for amenity and landscape (Hong Kong 1989).

Management of forest for amenity need not exclude timber production but harvesting operations will usually be small-scale and the choice of which trees to plant will be little influenced by end-use of the wood.

CHAPTER 21

Protective afforestation

This chapter concerns the role trees and forests play in protecting the environment, which is an extremely important aspect of forestry. Forests are a key element in the landscape. They provide a natural buffer, which assists in maintaining ecological balance and supply raw materials to communities as well as providing protective services. Forest policies must reflect these multiple functions and forest management must respond accordingly. Forest practices need to ensure the health and vitality of forests if they are to be managed sustainably and fulfill their protective role especially the protection of water resources, soil protection, buffering local climate through control of wind and other conservation, and recreational functions (Göttle and Sene 1997).

Tree cover usually reduces soil erosion, slows wind speed, traps airborne sand and dust particles, moderates the force of rain, and slows water runoff after heavy rain. Also, planting trees on degraded land, mining sites, or sand-dunes, can be the first important step in soil rehabilitation and land reclamation. Though these benefits have long been recognized, and almost every national forest policy includes reference to them, the importance of the protective role of forests has been underestimated. In particular, the interdependence of one land-use on another has been insufficiently appreciated, a fact dramatically illustrated and becoming abundantly clear when increased flooding and silting of agricultural land is exacerbated by forest clearance in a mountain watershed, perhaps several 100 km away or even in another country. Several countries in the Asia–Pacific region have imposed logging bans in natural forests to protect watersheds and to reduce flooding in downstream areas (Durst et al. 2001). Devastating floods in downstream agricultural areas of the Yangtze River in 1998 causing 3650 deaths and $US30 billion of damage stimulated the Chinese government to protect natural forests in watersheds in 18 provinces by banning logging (Studley 1999; Yang 2001). Conversion of forest land to inappropriate agricultural systems, especially overgrazing, is often the chief cause of the worst flooding (Hamilton and Pearce 1985).

In protective afforestation, or any forest reserved and managed for protection, timber production is of secondary importance, a fact which markedly affects silviculture and management. The protective role of the trees and accompanying vegetation becomes the dominant consideration in all decisions such as what species to plant, whether to thin, how to regenerate the forest, whether to allow firewood collection and livestock grazing and so on. Chapter 23 deals specifically with some of these questions since the sites most urgently needing protective tree-planting are often those, which are inhospitable and where plantation establishment is difficult. However, it is stressed that tree-planting alone rarely, if ever, is a sufficient protective measure. This chapter concentrates on this aspect but, in the case of soil erosion for example, much myth and a lot of hope centre on what the planting supposedly will achieve when, in reality, it is the associated removal of grazing or the construction of terraces or the control of land-use, such as exclusion of fire, which accompanies tree-planting, which is so important (Thorne 1989). In the conservation and protection of mountain watersheds integrated planning and management is required to maintain optimum interrelations between forests, water, wildlife and soils, as well as a high level of participation of local communities and other stakeholders (Fernandez 1997).

Plantations have been little used in protection forestry. They have been limited mostly to shelterbelts on farmland and around towns or to stabilize sand-dunes. Maintenance of natural forest cover in upper watersheds and mountainous regions has been the main form of protection and should continue. Deforestation or degradation of many natural forests has already occurred so tree-planting and management of secondary forests for protection are the only options. Jong et al. (2001) have stressed the importance of managing secondary forests to meet environmental and local livelihood needs. Secondary forests can supply a wider range of goods and services, for examples, biodiversity conservation, than plantations but plantations may be the only possibility on bare, degraded lands. Plantation programmes have been used by some countries to compensate for loss of timber supply when logging bans were introduced in natural forests to protect watersheds (Durst *et al.* 2001).

Large afforestation programmes on deforested hills and on coastal sand-dunes hills of southern China have an important soil protection role as well as production of timber. Establishment of shelterbelts began in various parts of China in the 1950s (Song 1991). Sand–dunes along the coast of the South China Sea were stabilized with *Casuarina equisetifolia* plantations resulting in environmental and livelihood benefits (Turnbull 1983). In 1989 the Chinese government-sponsored the Yangtze Shelterbelt Program to conserve forests in the upper and middle reaches of the Yangtze River and in 1998 the Natural Forest Conservation Program's logging ban in natural forests was accompanied by a plan to establish 21 million ha of timber plantations in the upper reaches of the Yangtze River and the upper and middle reaches of the Yellow River (Yang 2001). The Government of India's Forest Conservation Act of 1980 was designed to reduce the loss of forest annual area and at the same time a Social Forestry Programme was initiated to reclaim degraded forests and village commons and to meet fuelwood demands. An estimated 100 million ha of degraded land (forest, village commons, and marginal farmland) has little vegetative cover and is subject to soil erosion (Ministry of Environment 1990). The mean area afforested annually from 1980 to 1998 was 1.4 million ha making it one of the largest afforestation programmes in the world (Ravindranath and Hall 1994). Although the benefits of this programme to landless rural population is debatable, it undoubtedly increased forest cover and assisted soil conservation. Protection from fire, grazing, and fuelwood collection in plantations resulting from the Social Forestry and Joint Forest Management programmes has encouraged natural regeneration of local species in these plantations providing an opportunity to develop secondary forests with greater conservation value (Bhat *et al.* 2001). Major tree-planting projects as part of soil protection measures have been made in many parts of Africa (e.g. Ethiopia) and Asia (e.g. Philippines). Also countries bordering the Sahara to the north and east, most notably Algeria, Libya, Morocco, Sudan, and Tunisia, have established plantations to combat desert encroachment while the numerous social forestry projects in the Sahel all help to control desert-forming processes.

Little mention is made here of whether extensive tree-planting influences the macroclimate; it is a subject of much uncertainty and there are few data. If tree belts do benefit the climate, any improvement is likely to be small and merely an added advantage to the main benefits of local soil stabilization, wind speed reduction, trapping blown sand, and providing fuel and fodder, so reducing pressure on natural vegetation.

Urgency of protective afforestation in the tropics

Ecosystems in the tropics are generally more fragile than those of temperate areas and more liable to rapid deterioration when disrupted. Large areas of land in the tropics have become degraded and the deleterious changes to the land resource include loss of topsoil, reduced soil fertility, increased salinity, and loss of biodiversity. In a global assessment of human-induced soil degradation, Oldeman *et al.* (1991) estimated over 2000 million ha of land had been degraded in the previous 50 years, mainly by water and wind erosion, and chemical soil deterioration. The main causes were removal of natural vegetation, overgrazing by livestock, improper management of agricultural land and over-exploitation of vegetation for domestic use (fuel-wood etc.). The importance of land degradation in the developing world and the implications for future food production, agriculture, and the environment have been reviewed by Scherr and Yadav (1996). While land degradation continues to occur there are examples of farmers improving their practices and trees being planted to effectively counter degradation and desertification, for example, in Africa (Carucci 2000; Pearce 2001).

Some reasons for land degradation and the continued destruction of forest due to pressures of population and land-use practices are discussed to demonstrate the importance and urgency of tree-planting programmes for protection as one component of sustainable land-use in the tropics.

Loss of vegetative cover

One reason for establishing plantations is that much land in the tropics has become denuded of vegetation (see Chapter 2). When it rains on a forest in the tropics a significant amount is intercepted by the canopy and evaporates back into the atmosphere, the remainder reaches the soil surface by crown drip, stem flow, or direct throughfall. The intensity of the rainfall, the nature of the vegetative cover, topography, and physical characteristics of soil determine the amount of infiltration and surface runoff. Erosivity of the soil, raindrop impact, and runoff all play a part in determining the extent of water erosion. The consequences of loss of vegetation vary greatly, but in mountainous areas and arid regions it is almost invariably highly damaging to the environment. They include soil erosion and accelerated water runoff from mountain slopes, irreversible development of hard iron pans in some tropical soils, and to desertification in arid areas.

Soil erosion

Soil erosion, defined simply as detachment and transport of soil, is inevitable in the tropics when vegetative cover is removed from the land. It is a normal geological phenomenon, which plays a predominant role in the formation of natural landscapes, but the rate at which it occurs varies greatly depending on climate, terrain, soil structure, and the amount of vegetative cover. Wind and water are the agents of erosion. It may occur over extensive areas as surface or sheet erosion, in rills, or as gullies (Fig. 21.1). Whatever type of erosion occurs much the commonest reason for serious damage is rapid loss of vegetative cover, especially forests. The less vegetation covering the soil, the more likely it is to erode, but not all soils erode in the same way or to the same degree. Soil structural quality is of particular importance to soil hydrological processes and storage of water and nutrients, and topsoil stability depends mainly on the amount of organic matter present (Cass *et al.* 1996). Soils low in organic matter, and with topsoil over an impermeable layer at less than 50 cm depth, are particularly prone to erosion.

The main effects of soil erosion are: (i) loss of topsoil for cultivation; (ii) loss of ground stability on steep slopes and development of landslides; (iii) siltation of rivers causing premature filling of dams and blockage of irrigation channels downstream, and development of deltas and deposition of sand banks, which impair river navigability.

Erosion is a universal problem. It has been estimated that rain-fed arable land in developing countries will shrink by 544 million ha (i.e. by over 30%) if conservation measures are not taken to prevent erosion and degradation (FAO 1984*a*). In Pakistan, an example of a very dry tropical/subtropical country, 36% of all land is affected to some degree by water erosion and 40% by wind erosion, and up to 50% of cultivated land in some parts suffers sheet erosion. Pakistan is an extreme example; its forests now cover less than 5% of the land surface, but similar damaging soil erosion on a more local scale occurs in many countries. In the mountainous and more humid conditions of the Philippines, an estimated 58% of the land area is susceptible to erosion and 30% of the estimated 17 million ha of forest land has some soil erosion (Cabrido 1985). Tree clearing for agriculture and soil disturbance associated with animal grazing and cultivation in Australia has produced massive soil erosion, for example, in the wheat-producing lands of Queensland annual soil loss may exceed 50 t ha^{-1} (Bird *et al.* 1992). The effects of erosion on agricultural production can be substantial. For example, in China, grain yield has been reduced by 19% (Rozelle and Huang 1995) and crop yields have declined by about 6% in Sub-Saharan Africa (Lal 1995).

Rate and quality of runoff from watersheds

The seriousness and frequency of damaging floods is increasing. In many cases, such as in the Indo-Gangetic plain of India and Pakistan, the Mekong

Figure 21.1 Severe gully erosion in Ethiopia. This gully and the one in the distance are very recent, only forming during the 10 years prior to taking the photograph. Poor land-use practices on slopes less than 1 km away allowed uncontrolled and erosive run-off to occur. Such loss of soil has contributed to Ethiopia's famines; a problem common to many countries.

River in Indochina, and the Yangtze and Yellow Rivers in China the most recent floods are also the worst and have affected millions of people in their catchments. During the Qing Dynasty (1644–1911) the Yangtze flooded every 10 years, from 1921 to 41 the frequency rose to every 6 years and since 1980 damaging floods have occurred every 2 years (Studley 1999). In these examples the proportion of forest cover in upper watersheds is at its lowest level in historical times. However, the apparent increase in flood damage in some countries can be attributed more to human activity in the flood plains rather than reduction of tree cover in catchments (Cossalter *et al*. 2003). The loss of forest itself need not directly be too damaging, but it is so often followed by poor farming and land-use practices, including overgrazing, lacking even rudimentary conservation measures (Fig. 24.5). Rainwater falling in these catchments, particularly during monsoonal downpours, rapidly drains off the more exposed land into streams and rivers, which reach peak levels higher and sooner than previously. Water quality can also deteriorate with increased turbidity, which can impact on aquatic life and also render the water less acceptable for drinking, livestock watering, or irrigation. The limited role tree-plantations can play in improving the situation is discussed later in this chapter.

Development of hard iron pans
A combination of geological, hydrological, and climatic factors cause iron-pan development, or the latent state of formation, which may be activated by exposure of the soil surface, on susceptible terrain features. A very hard 'ironstone' layer in the soil or as a surface crust develops because of the presence of sesquioxide-cemented material (concretions, nodules or slaggy, vesicular masses). This hardens irreversibly on exposure to repeated wetting and drying in a cycle, the general progress of which is towards desiccation, such as occurs when vegetative cover is removed. About 3% of tropical soils are affected, though in much of West Africa the proportion is greater, perhaps nearly 8% (approximations from FAO/UNESCO World Soil Map).

The much devalued old term 'laterite' and the somewhat confusing name 'plinthite' usually refer to iron-rich soil that may be termed 'soft iron pan' without severe induration and supports a tree or grass cover unless subsoil is exposed. Laterite and lateritic soils cover about 0.12 million km^2 in India. These areas are low in fertility, have a hard vesicular structure, and a deficient water supply making them difficult sites for tree growth (Prasad 1991). Even some apparently 'irreversible' hardening of iron pan will succumb to erosion processes gradually. Correct choice of species, fertilizer application, and soil water conservation are needed to reclaim these sites. Particularly strong-rooted, adaptable species such as *Acacia auriculiformis, Dalbergia sissoo, Eucalyptus camaldulensis, E. tereticornis,* and *Tectona grandis* have proved suitable. Such species can physically disrupt and penetrate an already breaking and crumbling hard iron pan, especially if it is thinner than 20–25 cm.

Desertification
Desertification is land degradation in arid, semi-arid, and dry subtropical areas due to adverse human impact. These areas are inhabited by about a billion people and an estimated 70% of these drylands are affected by desertification (FAO 1999*b*). Within historical times the area of desert and semi-desert, especially in the subtropics, has greatly increased. Desertification affects over 3 billion ha of land and is expanding at a rapid rate causing a decline in soil fertility and structure and impacting negatively on livelihoods and the environment (Nelson 1990; Middleton and Thomas 1997). Much of this is attributable to man's destruction of vegetation for fuel, fodder, grazing, and timber (Darling 1969), often exacerbated by climatic variations. The effects of climatic change on land are aggravated and even become irreversible if combined with overgrazing, burning, firewood gathering, or excessive soil cultivation. However, although clearance of natural vegetation and forest in the tropics and subtropics has been no greater and sometimes less than in temperate regions, the effects have mostly been much more damaging. In the 2000 years to 1900 Britain's forest cover diminished from about 45% of the land surface to 5% without extensive loss of topsoil and large areas becoming wholly unusable for farming. During the same period large areas of the Sahara changed from savanna to desert. Adoption of the UN Convention to Combat Desertification in Countries Experiencing Serious Drought and/or Desertification in 1994 has provided a framework to stimulate action to counter desertification affected regions, for example, in the Sudano-Sahelian zone of Africa (Carucci 2000) and Asia (Wang *et al*. 2000).

The severity of the tropical climate

The chief reasons why loss of vegetative cover in the tropics is so often damaging are that the climate is more erosive: intense rainstorms, prolonged periods of heavy monsoonal rain, and rapid drying of ground and desiccation of vegetation in the hot sun with relative humidities dropping to low levels in some parts.

Rainfall intensity and duration
The high energy load of individual rain storms in the tropics is a major factor causing soil erosion (Kelley 1983). Precipitation is usually heavy, of higher intensity, and often composed of raindrops of 3–5 mm diameter. Raindrop impact is highly erosive on exposed soils causing break-up of soil aggregates, compaction, sealing of the surface, reduced infiltration, and an increase in run-off, leading to erosion. Storms in northern Nigeria exceeding 20 mm of rainfall result in surface runoff and erosion (Kowal and Kassam 1976). In a year about 58% of rainfall was erosive and was an important cause of decline in soil fertility. In Zimbabwe it was found that generally intensities above 25 mm h^{-1} were erosive but that frequently highly erosive intensities, exceeding 100 mm h^{-1}, occurred for a time during many storms where the overall intensity per hour was low (Hudson 1963).

Added to the destructiveness of individual tropical rain storms is rainfall duration, especially in the monsoon. Not only are storms erosive but the strong seasonality of much rainfall results in long, almost continuous periods of heavy rain. When soils reach field capacity more rain results in flooding on flat sites, and surface runoff and erosion on slopes. Heavy rainfall is particularly damaging in dry areas where vegetation cover is sparse. An example of heavy monsoonal rains is illustrated by Bhopal in Madyha Pradesh, India, where 90% of annual rainfall (1330 mm) falls in just 4 months during the monsoon. Often, over 500 mm falls in only one month (July). For most of the year Bhopal is dry, semi-arid, and short of water, but for a short period it is often so wet that much of the annual rainfall total runs off saturated soils into watercourses. Even more extreme is Mt Kulal in northern Kenya where over 100 mm of rain fell in April 1977 causing much erosion. This amount of rain is not high by tropical standards, but during the previous 8 years Mt Kulal received an average of only 50 mm per year!

All inhabitants of the tropics know the muddy streams that develop within minutes of rain beginning to fall. Brown, silt-laden rivers are a feature of the tropics and evidence of the erosion occurring within their catchments.

Shifts in climatic zones
Temporary shifts in climatic zones occur throughout the world, but because of the damaging effects of many land-use and land-management practices their effects in the tropics are much greater. There was a southward shift of the rain belts in the Sahel in the late 1960s. For example, at Keita in Niger annual rainfall which had averaged 517 mm in the period 1956–66 fell to 317 mm during the period 1967–87 (Carucci 2000). Satellite imagery showed the Sahara increased in area by 15% between 1980 and 1984, although by 1993 it was smaller again but not back to the 1980 area (Tucker *et al.* 1994). This climatic shift led to one of the most acute periods of famine ever recorded in the Sahel. The great drought, only interspersed by a few more average years, caused widespread starvation and enormous losses of livestock. Droughts have always occurred, but where land-use practices are already contributing to environmental degrade, such climatic change can be catastrophic (Fig. 21.1). Most climatic modellers believe that the enhanced 'greenhouse effect' will cause climatic change although the exact consequences cannot be predicted with certainty. Both more and less precipitation has been predicted as a result of warming in tropical and subtropical regions depending on which global climate model is used (Andrasko 1990). Serious economic, ecological, and social impacts on forests by such climatic shifts due extended periods of drought, changes in forest fire regimes, pest and disease attacks, and the spread of weeds are possible (Solberg 1998).

Evapotranspiration
High solar radiation levels and high temperatures, which quickly lower relative humidity, lead to high rates of evapotranspiration in plants and rapid drying out of land. Tracks and roads in the tropics are usually dry within an hour of heavy rain. As soils become dry the cohesiveness of the surface declines sharply, and, on bare ground, the risk of wind erosion rises. Sand and dust storms during dry periods are the end result. This high rate of evapotranspiration means that the amount of 'dry-time' in the tropics per unit of precipitation is greater than elsewhere.

A destructive side-effect of high evapotranspiration levels is the ease with which fire can be used to burn off vegetation. Even in Lae, Papua New Guinea where annual rainfall exceeds 4500 mm and no month averages less than 200 mm, grass is burnt throughout the year whenever there are three or four dry days.

High evapotranspiration levels make rainfall less effective so annual rainfall in the range of 500–700 mm is considered as dry or semi-arid in the tropics but moderately moist in higher latitudes. This partly explains why loss of vegetation is so serious in the tropics and why the trend to desertification so widespread.

Fragility of tropical ecosystems

Evidence of past destruction shows that most tropical ecosystems are fragile. Compared with temperate ecosystems they are more prone to destruction and irreversible damage if disturbed. The main reasons are: (i) seasonality of rainfall and great year to year variability in more arid regions; (ii) exposure of the soil surface to the erosive forces removal of vegetation, as outlined above, quickly leads to loss of topsoil including the organic matter and seed bank it contains, thus both soil fertility and the chance of rapid recolonization by the pre-existing vegetation are lost; (iii) in forests (ii) is often made worse by rapid invasion of a cleared site by grasses, which are regularly burnt or grazed so preventing re-establishment of forest; (iv) not only is the soil more exposed to erosion and the possible development of hard iron pans as noted above, but most tropical soils are impoverished and contain only a small proportion of the total nutrient budget in an ecosystem (Chapters 1, 14, and 24), thus far greater loss occurs when vegetation is destroyed. One consequence of this fragility is that when the destructive forces are removed, for example, burning and grazing are controlled, recolonization is slow compared for example, with the regrowth of forest on abandoned fields in parts of northeast United States.

Population pressures

People must eat, cook food, and find shelter. Traditionally, the forest has provided many of these needs. Firewood is gathered for cooking and keeping warm, domestic livestock graze on green plants or wild animals are hunted, fertile land is found and cleared and cultivated for food crops, and timber is cut for house building. Cutting trees, shifting cultivation, grass burning, and gathering fodder are the land-use practices carried out to meet these needs.

As populations increase and people's aspirations rise, trees and forests come under ever greater pressure to satisfy these needs. Particularly damaging in many parts of the tropics and subtropics are the large herds of livestock kept partly for food but often for wealth, prestige, or in Islamic cultures, to provide animals for sacrifice (Stewart 1979). As the numbers of people increase so do the animal herds and pressure on vegetation is compounded. Increasing population is an important reason for continued loss of vegetative cover, with all its damaging consequences, in the tropics, but it is often not the main underlying cause. It is easy to blame high population growth when, in reality, bad laws, poor rural development policies, corruption, and nepotism are behind much deforestation.

Past deforestation and the great pressure on remaining resources combined with severe tropical climates, ecosystem fragility, and increasing populations make extensive plantation establishment essential. Plantations are not only able to provide wood products in the future but also can help mitigate soil erosion, flash floods, and the spread of desert. It is again stressed that this role for trees and tree-planting is primarily to relieve pressures from harmful practices and as one component of many conservation measures. Simply planting trees for protection without other measures is not adequate and is rarely successful.

Protective role of tree cover

Many of the influences of trees and forest on the environment have been understood for a long time (e.g. Kittredge 1948; FAO 1962). More recent work on some of the influences of tropical tree-plantations on soil, nutrients, and water has been reviewed by Nambiar *et al.* (1997). The purposeful retention of forest or the planting of trees to use these influences constitutes protection forestry. In the tropics this is carried out for several closely related purposes: soil stabilization and prevention of erosion, watershed management, provision of shelter and shade, reclamation of degraded sites, and to arrest desertification.

Soil stabilization and erosion control

Tree cover protects the soil and reduces erosion in many ways. High filtration rates in forests can reduce surface runoff and soil transport, and the binding action of tree roots helps soil stability on slopes and reduces erosion. Factors influencing effectiveness of planted forests to protect the soil are discussed below.

Rainfall interception
Tree crowns forming the forest canopy, together with the understorey and ground vegetation, are a barrier between falling rain and the soil surface, which confers several benefits. First, the force of rain is dissipated on the vegetative cover. This is not the case where large drips fall to the ground unimpeded from a high canopy as they may possess more kinetic energy and erosion potential than rain in the open. Hence the very great importance of low vegetation layers and soil surface covering as shown by in a study of *Acacia auriculiformis* plantations in Indonesia (Wiersum 1983), and his synthesis of data for a range of forest and tree crops (Wiersum 1985). Second, the water from rain falling on the crown takes longer to reach the ground as it

Table 21.1 Interception of rain by forest plantations

Species	Stand density (trees ha^{-1})	Throughfall (%)	Stemflow (%)	Interception (%)
Acacia mangium	1090	72	—	< 28
	1705	57	4	39
A. auriculiformis	1010	81	8	11
Araucaria cunninghamii	664	75	—	< 25
Eucalyptus saligna	1685	84	4	12
E. tereticornis	1658	81	8	11
Paraserianthes falcataria	600	82	—	< 18
Pinus caribaea	700	88	—	< 12
P. roxburghii	1156	74	4	22
Shorea robusta	668	55	7	38
	1678	67	8	25
Swietenia macrophylla	—	79	1	20
Tectona grandis	1742	73	6	21

Source: Based on Ghosh and Rao (1979), Bruijnzeel (1997).

trickles down the branches and trunk and drips off leaves. Third, the total quantity of water reaching the ground is reduced because some moisture evaporates from leaf surfaces (Table 21.1). Interception by the canopy of a vigorously growing plantation of *Acacia mangium* can be 25–40% of incident rainfall but is usually under 25% in pines (Bruijnzeel 1997). Reports of interception by eucalypts tend to be lower than for other species, as Table 21.1 illustrates, varying from about 10–23% but usually below 15% (Calder 2002). These effects can greatly moderate rainfall intensity at the ground surface and clearly some species provide better protection of the soil than others.

Wind speed reduction
Inside a forest and for a short distance to its leeward there is a reduction in wind speed, and consequently in wind erosion. This effect is important and is considered in detail later.

Soil covering and ground vegetation
If left undisturbed, the ground beneath trees becomes covered with a layer of debris called forest litter (dead leaves, twigs, branches, etc. Fig. 21.2). This litter layer, along with ground vegetation, is the most important protection for the soil surface as is shown by what happens when it is not present as, for example, sometimes occurs under teak plantations. Teak usually suppresses all ground vegetation and often the large dead, dry leaves are burnt, deliberately or accidentally, and the soil surface is exposed. This loss of an effective litter layer frequently leads to soil erosion. In a study of this problem in Trinidad's teak plantations, soil loss

Figure 21.2 Thick layer of litter (15 cm deep) accumulated under 26-year-old *Pinus patula*.

during 1 year was 9–21 times greater than under natural forest (Bell 1973). Ghosh (1978) found raindrop erosion increased from 10 to 90 times, depending on rainfall intensity, under *Shorea robusta* forest where the litter had been burnt. The common practice of raking up litter under plantations for fuel in countries such as China (Fig 14.1) increases the risk of soil erosion. Presence of litter and undergrowth alone virtually eliminated soil erosion in

A. auriculiformis plantations with or without a tree canopy but erosion occurred where local people removed it for fuel (Wiersum 1983) (Fig. 21.3). The dense needle mat under pine plantations in Jamaica was even better than rain forest in minimizing surface soil erosion (Richardson 1982).

Moisture retention
As well as protecting the soil from the direct impact of raindrops, the litter and humus layers absorb moisture. This further slows the movement of water into the soil (infiltration rate) and also a small amount will evaporate. The idea of litter and humus as a sponge is rather misleading since their water-holding capacity is often less than the soil below, but certainly infiltration into the soil is improved.

Binding action of roots
An examination of soil beneath a forest will reveal a mass of roots near the surface (Fig. 21.3). Under rain forest there is an almost uninterrupted mat of fine roots over every square metre of ground. The biomass of fine or small roots in tropical plantations is commonly up to $5 \, t \, ha^{-1}$ (Vogt *et al.* 1997), and Lugo (1992) reports the biomass of fine or small roots in secondary evergreen forest in Puerto Rico in the range $4.3–5.4 \, t \, ha^{-1}$. Comparing secondary forest and a plantation of mahogany (*Swietenia* spp.), the secondary forest had twice the root density in the top 20 cm of soil, both had a maximum small-root penetration depth of 70 cm and the root mass was four times greater in the secondary forest due partly to having a larger proportion of large diameter roots (Lugo 1992).

While trees are alive this root mat is continually renewed; if a root dies it is soon replaced. This living network of roots provides mechanical support on steep slopes and is the main contribution to slope strength and prevention of landslides (Rice 1978). But, if forest is cleared and stumps killed root regeneration stops, the old roots die and decay, and their binding effect on the soil soon disappears. Serious soil erosion and landslides can occur, especially in steep uplands in monsoonal areas, and a combination of physical structures such as check dams and revegetation may be required to control them. Exclusive use of vegetative measures for the control of severe erosion and major landslides is rarely successful. At Labok in Nepal a landslide area was successfully stabilized using engineering structures and planting *Alnus nepalensis*, broom grass (*Thysanolaene maxima*) and bamboos (Sthapit and Tennyson 1991).

The best evidence of these beneficial influences of forest cover is shown by studies of erosion under different land-uses with and without tree cover. Hamilton and King (1983) collated a large number of studies, which mostly show that erosion is least under undisturbed forest and dense grassland, tends to be a little higher under most tree-plantations and secondary forest, but is many times worse in coffee and tea plantations, heavily grazed land, cultivated ground, and recently prepared swiddens in shifting cultivation. For example, a 20-year study of watersheds in southern Kenya showed that sediment eroded annually from land used for agriculture (mostly grazing) up to 50 times faster ($1000 \, t \, km^{-2}$) than from forest land. Ambar (1986) reported huge variation in erosion rates under different land-uses in Indonesia (Table 21.2). In the Hazara catchment in Pakistan, tree-planting combined with closing slopes to grazing reduced runoff and sediment release from 30% to 1% (Abbas and Hanif 1987).

Nearly all studies confirm the favourable effect of forest cover on erosion, provided litter and undergrowth are present, and this is why ground cover is so important in water catchment areas to reduce the

Figure 21.3 Soil erosion under an *Acacia auriculiformis* plantation in Java, Indonesia after removal of litter for fuel by local people (photo K. F. Wiersum).

Table 21.2 Soil erosion rates under different land-uses in Indonesia

Land-use	Soil erosion (t ha^{-1} a^{-1})
Natural forest:	
undisturbed	0.0
with trees cut	0.4
with trees cut and litter removed	48.0
New agricultural fields:	
without terrace	11.0
with terrace	3.1
Old agricultural fields:	
without planted contours	136.1
with planted contours	43.5
Paraserianthes falcataria plantation:	
with litter and undergrowth	0.8
without litter and undergrowth	79.8
Teak plantation:	
first year with taungya	5.2
third and fourth years with taungya	1.0
Imperata grassland:	3.5
Lantana vegetation	5.1
Agricultural land (cultivated)	345.0

From Ambar (1986).

sediment loading of streams and rivers that flow from them. Intact natural forests are one of the most effective land-uses from the point of view of soil stabilization and streamflow stabilization. Is it necessary to replant denuded areas with trees to protect soil from erosion? Plantations differ from natural forest and it should be remembered that planted forests only protect the soil if they are properly managed for that purpose. Serious erosion can occur in dense plantations that shade out the understorey and in species such as teak that have a large drop size when rain drips from the canopy. There is also the potential for erosion during site preparation and harvesting.

Plantations and even annual agriculture can provide many of the same protective functions provided the land-use includes practices such as contour operations, maintenance of ground cover and buffer strips along watercourses (World Bank 1992). Trees may have a part to play in stabilizing soil in agricultural cropping systems (Young 1989). The role that planted trees can play in protecting soil from erosion needs to be assessed taking into account all the factors of the site under consideration and the capacity to provide appropriate management.

Watershed management

When water from a catchment is used for drinking, irrigation, or generation of hydroelectric power,

management of the watershed can affect both the quality and quantity of water supplied. The inorganic and organic constituents of the water reflect the mineralogy of the watershed, the character of the precipitation and the nature of the vegetative cover (Hewlett 1982). Changes in vegetative cover by deforestation or tree-planting therefore have significant effects on the hydrology of a watershed. What these effects are and their value can be contentious. Calder (2002) suggests that reports that forests increase runoff, regulate flows, reduce erosion, reduce floods, and improve water quality are 'seen to be either exaggerated or untenable' when examined critically. He has pointed out that foresters and hydrologists often have very different perceptions regarding the hydrological role of forests, with hydrologists concerned that trees intercept more rain during wet periods, and because of deeper root systems deplete groundwater by transpiring more water in dry periods (Calder 1996). He also suggests that it is unrealistic to attempt to generalize forest impacts as they affect extreme flows. Also, that some negative impacts related to floods and erosion may be more associated with forest management operations such as logging, site preparation and roading, rather than the presence or absence of the forests themselves. Foresters and the public may have a simplistic view of the impacts of forests in water catchments and so Calder's viewpoint is worth noting, especially when one attempts to make generalizations from reports of events and research that may be very site-specific. That said, the common observation across the tropics is that streams that once flowed from forested catchments often dry up following deforestation.

Useful reviews of information on water use by forests and forest plantations and its relevance to water catchment management have been produced by Bruijnzeel (1990, 1997), Calder *et al* (1992), O'Loughlin and Nambiar (2001), Calder (2002), and Vertessy *et al.* (2003). Some of the effects of forests on watershed management are discussed below.

Influence on water quality

Tree and ground cover can greatly reduce soil erosion. In watershed management this is not only important for preventing loss of fertile topsoil but means that water draining from a forested or well-grassed catchment will be largely free of sediment. The destructive effect of erosion is not only loss of soil from the eroding land itself but its transport and subsequent deposition elsewhere. It should be recognized that that the majority of soil that is eroded is redeposited elsewhere and that only a fraction of the sediment is deposited in major rivers and reservoirs. Finlayson (1998) has highlighted the difficulty of establishing any useful relationship between erosion and sedimentation but sediment deposition in reservoirs and irrigation channels is a widespread and costly consequence of deforestation in many tropical countries. A study of 17 major reservoirs in India showed annual sediment deposition rates were three times greater than was planned because 'very vast areas of forests have been deforested' (Tejwani 1977).

A second effect of forest cover in catchment areas is that alternative, often more polluting, land-uses are excluded. Water draining from a largely forested catchment is usually cleaner than from one where land is mostly used for grazing or arable farming, which may include the additional pollution hazards of fertilizers (especially nitrates and phosphates), pesticides, and human and animal wastes. For example, in Malaysia the quality of stream water was increasingly degraded as it passed from undisturbed forest, through a swamp area, logged-over forest and agricultural land (Yusoff *et al.* 2001). However, this advantage of forest cover may diminish somewhat where managed plantations are established on watersheds.

Many forest plantation operations cause disturbance and exposure of soil through ground preparation, planting, control burning, forest grazing, thinning, felling, and road construction. This is one of the reasons why, in forests planted for a protective function, timber production is of secondary importance to that of leaving tree and ground cover intact. It is also why chemicals such as herbicides and insecticides are avoided. The quality of water draining from a forested catchment is very much under the forester's control and increasingly codes of forest practice are being developed to ensure environmental protection. Foresters need to be aware of the physical, chemical, and biological indicators of water quality from catchments, such as turbidity, pH level, aquatic organisms, etc., and these are discussed in Walker and Reuter (1996).

In general, many undesirable effects of forests or forest management on water quality can be much reduced by leaving buffer zones of undisturbed vegetation next to all watercourses in a catchment.

Influence on water quantity—afforestation and streamflow

Forest cover, compared with open land, affects both the pattern and total quantity of water discharge from a catchment area. Both have important implications for flood control and land-use practices particularly downstream from the catchment. Trees planted for commercial or environmental purposes may use more water than the crops or grass they replace. The impacts will vary on different sites being influenced by the nature of water flows, landscape features, area and density of plantings, and management (O'Loughlin and Nambiar 2001).

Pattern of discharge

It was noted earlier that loss of vegetation leads to more severe flooding, but it does not follow that planting trees will automatically eliminate flooding problems.

It is claimed that forests in mountainous areas play a pivotal role in flood control (Meunier 1996). A forested watershed generally has lower peak flow rates after a storm than one recently denuded of forest or under arable farming or overgrazed, so forests exert some regulating influence on the flow of water. The mechanism is the same as that which reduces soil erosion, forest cover slows the movement of water through part of the hydrological cycle. This slowing, combined with the barrier and absorptive effects of the litter, lead to better infiltration of water into the soil, reduced surface runoff, and therefore slower drainage from the catchment into streams and rivers and reduced the velocity of flows.

Mathur *et al.* (1976), in studies of small catchments in India, found that afforestation of cut over *Shorea robusta* forest with *Eucalyptus grandis* and *E. camaldulensis*, compared with a natural covering of small shrubs and grasses, resulted in 28% less surface runoff and a 73% reduction in peak flow rates. Studies in the Ootacamund Hills (India) similarly yielded data showing that catchments under *Eucalyptus* and *Acacia* plantations had peak discharge rates only one-third of those under agriculture. The soil conservation programme in Kondoa, Tanzania, which incorporated tree-planting, terracing, and selective tree-cutting in woodland, regulated water flow and reduced flooding (peak flow rates) (Mugasha and Nshubenuki 1988). Further examples of the effects of clearing on catchment water yield and peak discharges are reported in Hamilton and King (1983) and Bruijnzeel (1997).

There are some observations that stream flow from forested land is prolonged during dry periods, for example, in the Ngoronit stream catchment in northern Kenya, but there is increasing evidence that planting fast-growing trees on grasslands will lead to greatly diminished streamflow, especially in the dry season (Bruijnzeel 1997; Calder 2002).

While various studies suggest plantations reduce run-off and peak flow rates and so prevent flooding, we need to be aware that their effects in normal rainfall events may be different from those in extreme rainfall events when large amounts of rain fall in a short time. Planting trees in a catchment will rarely guarantee that there will be no floods. Trees in catchments have a role to play, not least in preventing soil erosion and landslides, but their effectiveness in flood control should not be over-emphasized.

Total water yield

It is common practice to stand under a tree to keep dry (interception) and to plant trees, especially some species of eucalypts, to lower water tables. When forest is cut, it is frequently observed that the water table rises and the surface sometimes is wetter for a time, since interception and transpiration are both reduced. Because tree-planting affects the hydrology of a catchment the consequence of afforestation in a watershed must be considered especially in seasonally dry areas and where soils are shallow and are able to store very little water.

Trees, like nearly all plants, lose water through transpiration. As trees root more deeply than most grasses and herbs they are able to draw on reserves of moisture and continue transpiring at a high rate for longer during a dry period. This effect, combined with evaporation of water intercepted by the canopy, may result in an afforested watershed consuming more water than one that is not, and therefore a reduction in the total quantity of water draining from it. This water may be cleaner, less silty, and the flow more reliable and less prone to floods but if the total amount draining from a watershed is significantly reduced this could be a negative impact in arid areas where all water is at a premium.

Eucalypts have a reputation for high levels of water use but critical evaluation indicates very large differences between species in both stomatal responses and rooting patterns, which affect water use. When water becomes limiting most eucalypt species respond by closing their stomata and so transpire about as much as other tree species but a few species have less stomatal control and have greater transpiration rates (Calder 1992). Some species, such as *E. camaldulensis*, even show provenance variation in their stomatal response to water stress with the well-known Petford provenance responding to stress by stomatal closing but not shedding leaves whereas the Katherine provenance responded to stress by leaf shedding (Gibson and Bachelard 1994; Gibson *et al.* 1995). This suggests that *E. camaldulensis* seedlings from the dry tropics depend more on reduction of leaf area than control of transpiration to conserve water. Where eucalypts have access to the ground water table they may have very high transpiration rates, especially in

hot, dry conditions where atmospheric demand is high (Calder 1992). This accounts for the ability of some species to rapidly lower groundwater tables and dry out the ground. In a series of studies in Australia, Pakistan, and Thailand estimated annual water use by plantations of eucalypts, acacias, casuarinas, and *Prosopis pallida* ranged from 300 to 2100 mm. On a given site, water use differences between species largely reflected growth rates, in terms of sapwood area per hectare (Morris 1997; Mahmood *et al.* 2000).

It is important to recognize that measurements of water use by vegetation can only be properly interpreted with reference to the location and period in which the measurements are made. Availability of water and the evaporative demand greatly influence transpiration and may override differences in tree species, size of tree, water quality, and other factors. The water balance of agroforestry systems is even more complex than for tree monocultures and currently little is known about the way water is partitioned in them (Wallace 1996).

The canopy gradually increases from the time a plantation is established with a consequent increase in annual evapotranspiration as the stand ages. In a high rainfall area of New Zealand an indigenous mixed evergreen forest was converted to pine plantation in the mid 1970s. For the first 4 years water yields from catchments were 49–74% higher depending on harvesting method. Flood peaks also increased significantly after harvesting but within 15 years annual water yields had returned to pre-harvesting levels (Fahey and Jackson 1995). A similar effect was observed in French Guyana where run off increased 60% after primary rain forest was cleared and eucalypts planted. After about 5 years the run off was similar to pre-eucalypt establishment and after 6 years it was 10% less (Cossalter *et al.* 2003). The situation may be different in catchments where grassland or sparse vegetation is planted with pines or eucalypts. Results of long-term experiments in South Africa showing reductions in annual water yield from grassland and scrubland site planted with *Pinus patula*, *P. radiata*, and *Eucalyptus grandis* are summarized by Smith and Scott (1992) and cited in Bruijnzeel (1997). They concluded: (1) maximum reductions in annual water yields were about 400–500 mm and decreases followed a sigmoidal trend; (2) the effect was more rapid for the eucalypts than the pines. Typically the eucalypt produced a response in streamflow after about 3 years and pines after 5 years; (3) the proportion of catchment planted influenced both the rate and amount of change in water flow.

In South Africa it is recommended that where the water is needed and where alternative water sources are not economically available, sites outside the humid forestry zones or with long dry seasons are not planted. It is also South African practice not to plant trees close to streams and rivers to avoid copious water use by trees with a plentiful water supply and to encourage development of vegetation with relatively low transpiration rates, especially in dry winter months (Dye 1996). The 'Afforestation Permit System' now regulates tree-plantation development in South Africa, primarily to protect natural water resources (van der Zel 1995) and in 2003 a water levy on wood production is under consideration. Like South Africa, Australia is a dry country with an active forest plantation programme. Experience on the impact of plantations on river flows in Australia has been summarized by Vertessy *et al.* (2003). They note that the fraction of the catchment area planted, the plantation position within the catchment, and variations in stand age and site productivity all impact on water yields. They suggest careful planning can minimize negative impacts and that the beneficial environmental effects of plantations should be considered in the formulation of policy and legislation directed at plantation development.

Where there is a very high demand for water from certain catchments an alternative land cover which uses less water may be considered. Reviewing the options for planting in a Sri Lankan water catchment, Finlayson (1998) concluded that there is hardly ever a case for 'protection plantations' as they are unlikely to perform better, and may be much worse, than grass. However, reduction in water yields from afforested catchments must be kept in perspective as in much of the tropics the benefits of careful tree-planting to assist erosion control and flood prevention far outweigh the disadvantage of some reduction in total water yield. When establishing new plantations in areas where water availability is a critical issue, the planning of land-uses should consider overall ecosystem processes and socio-economic issues (O'Loughlin and Nambiar 2001).

Tree-planting in catchments—a summary of effects

1. While it is best to retain natural forest and vegetation, including grassland, undisturbed to minimize erosion and sediment release, where tree-planting is carried out it is important to minimize soil surface disturbance, and encourage and maintain an understorey, ground vegetation, and surface litter layers.
2. Planted forest for protection must be integrated with other land-uses and managed sustainably. Planting trees should be seen as one component of better land-use practices in catchment management that complements control of cultivation and grazing,

engineering structures, such as check dams and terraces, and buffer zones to moderate part of the water cycle.
3. When the trees become established and a canopy forms, afforestation of a catchment will reduce total water yield, peak and base flow rates, though where there is good infiltration base flows may be prolonged. Where there is a high demand for water an alternative vegetative cover such as grass should be considered.
4. Tree-planting probably does not increase rainfall, except in rare instances where the trees intercept fog or mist, known as 'occult precipitation' or 'cloud deposition'.

Provision of shelter

Trees provide shelter from driving rain, hot sun, and strong winds. Shelter from rain was considered earlier, and the value of shade for livestock and some food crops was noted in Chapter 20. Shelter from wind is the third major protective benefit of trees, especially belts of trees (shelterbelts, windbreaks). Planting of trees for this purpose is not new but the need for it, especially in more arid regions, is critical in developing sustainable land management systems and increasing agricultural production (Reifsnyder and Darnhofer 1989; Ben Salem 1991).

The primary purpose of shelterbelts is to reduce wind velocity, filter airborne particles of sand, dust, etc. so as to protect animals, agricultural crops, and human habitation. Design, establishment, and management of shelterbelts on farms are discussed by Reid and Bird (1990) and the range of species and establishment techniques used in arid and semi-arid areas are reviewed by Sheikh (1988) and Ritchie (1988). A shelterbelt will protect the largest area when it is aligned at right angles to the wind causing the problem. The effect of the shelterbelt depends on its height and its permeability. A denser shelterbelt can reduce the wind speed up to 80% but the effect is limited to close to the shelterbelt whereas with a permeability of 40–50% a much greater area will be partially sheltered. Species differ in crown characteristics and this can affect their porosity and effectiveness as a windbreak, for example, *Eucalyptus microcorys*, which has a deep crown has a more uniform porosity and makes a better windbreak than *E. grandis* which sheds its lower branches and is more open at its base (Sun and Dickinson 1996). The width of a windbreak has a major effect on permeability and its width should rarely exceed about twice the mature height of the trees:

1. An appropriately designed shelterbelt is permeable and will reduce wind speeds to leeward

up to a distance of at least 15 tree heights. For example, in Zimbabwe, a 6-m high belt of *Pinus patula* planted to protect fruit trees reduced wind speeds to leeward by 58, 33, 15, and 19% at distances of about 30, 60, 90, and 120 m, respectively (Payne 1968). In Nigeria, the moderating influence on wind speed of shelterbelts established to protect millet extends about 15 times tree height (Ujah and Adeoye 1984), and, in the Maggia valley, Niger, shelterbelts of *Azadirachta indica* (neem) interplanted with *Prosopis juliflora* and *Acacia nilotica* reduce wind speeds by up to 65%. In semi-arid areas of Australia it is claimed that planting 5% of the land to shelter can reduce wind speed by 30–50% and soil loss by 80% (Bird *et al.* 1992) and in more humid tropical conditions in North Queensland a shelterbelt decreased water stress and increased potato yields by about 7% providing considerable financial benefits (Sun and Dickinson 1994).

2. Temperature extremes and evapotranspiration stress are moderated within and to the leeward of shelterbelts (Delwaulle 1977c; Brenner 1996). This microclimatic influence has enabled cocoa to be grown in parts of Ghana and elsewhere in West Africa where otherwise it would have been a doubtful crop owing to the dry harmattan wind that blows off the Sahara desert at certain times of the year. In southern China the coastal windbreaks of *Casuarina equisetifolia* reduce moisture evaporation by 12% and wind speed by 60% resulting in substantially increased crop yields. In semi-arid regions, soil moisture storage is typically improved by 5–15% (Muthana *et al.* 1984; Ujah and Adeoye 1984). This improvement in soil moisture status is one of the main reasons for increased yields of crops grown between shelterbelts. Also, the moister soil is less liable to wind erosion.

3. The physical obstruction of a belt will trap airborne particles: sand, dust, topsoil, etc. One of the objectives of the 'green belt' of *Eucalyptus microtheca* and *E. camaldulensis* around parts of Khartoum (Sudan) was to provide shelter from dust and sand storms (Musa 1977).

4. The soil directly beneath a shelterbelt is stabilized, a fact employed when trees are used to help fix sand-dunes; see below.

5. The social and economic benefits from shelterbelt planting can be very substantial in terms of increased crop yields, fuel and timber production, fodder for livestock, shade and shelter, control of land degradation and other purposes. In Pakistan, financial analyses of shelterbelts of poplars protecting wheat and sugarcane in the Peshawar valley, eucalypts protecting wheat in the Sind, and *Dalbergia sissoo* protecting wheat in the Punjab indicated positive financial incentives

for establishing shelterbelt agroforestry in these regions (Subham 1990).

While windbreaks can have many beneficial environmental effects they may have management problems. There may be competition for moisture and between the trees in the windbreak and the species they are supposed to protect, there may be an increase in temperature on the lee side, which damages crops (Vandenbeldt 1992), and they may harbour pests such as grain-eating birds and rodents. Maximum benefits of shelter occur when water is not limiting.

Desertification

Decline in vegetative cover aggravates any tendency to desertification in an arid or semi-arid region due to a succession of unusually dry years. Logically, therefore, one of the goals in reclaiming land in danger of becoming desert is to replace lost trees and forest, and, in such a harsh environment, direct afforestation or introduction of agroforestry practices (Baumer 1990) is usually the only way. Recommendations 5, 7, and 14 of the United Nations conference on desertification in 1977 advocated planting programmes to establish shelterbelts, fuelwood plantations, woodlots, and to stabilize dunes. The UN Convention to Combat Desertification came into force in 1996 and promotes action to combat desertification and mitigate the effects of drought in countries seriously affected by these problems. Various forestry activities contribute to desertification control which involves integrated development and management of susceptible lands. In Mali, for example, these forestry activities include afforestation, natural plant management, silvo-pastoral systems, agroforestry systems, watershed management, and development of national parks (Berthe 1997).

To understand the role trees play it is useful to outline the main steps leading to desert conditions, a trend caused by poor rains but aggravated by over-grazing (Fig. 21.4), trampling and compaction of ground, savanna burning, and excessive gathering of firewood and fodder.

1. The first step is where too much pressure on land leads to a loss in the complete grass/vegetation cover and some patches of soil become exposed.
2. Step one results in even greater pressure on remaining grass and trees which in turn further exposes soil to wind erosion. Adjacent patches of degraded and exposed soil will start to coalesce.
3. Eventually a time comes when all trees disappear along with most of the vegetation; soil erosion

Figure 21.4 Desertification beginning owing to heavy grazing by cattle, goats, and sheep in Burkina Faso.

greatly increases and soil fertility declines rapidly. Degraded patches expand further forming desert-like areas that add to the natural desert.

4. Extensive wind erosion which takes place produces damaging dust and sand storms which blow up and move quickly over land now almost devoid of shelter. Any remaining vegetation will only grow slowly, is under great grazing pressure, and suffers serious abrasion in storms. Often, ditches and irrigation channels become blocked and even whole fields submerged by blown sand or movement of sand-dunes. Once fields are covered or vegetation is gone land becomes worthless, at the very least temporarily, and the cultivator/grazier moves on to new ground and so adds to the pressure there.

5. Desert is left.

Tree-planting can halt this process and reverse the trend of environmental degrade as an important component of better management of the entire renewable resource base (Ben Salem 1991; Berthe 1997). The object of such planting, for example, the 'green belt' projects and social/community forestry programmes of many countries bordering the Sahara, is not to create a physical barrier to desert encroachment (since this suggests incorrectly that most desert expansion is due to moving sand-dunes and land being swamped), but to bring about localized environmental improvement and provision of people's needs that relieve the desert-causing pressures. Belts of trees, along with introduction of agroforestry practices, as part of an overall integrated land-use strategy, can ameliorate the harsh environment and living conditions in the many ways already discussed; namely, reduced soil erosion, increased supplies of firewood, provision of shade and shelter, less extreme desiccation during strong winds, and trapping of sand and dust. The momentum generated in reversing the trend will itself help to stop the land-use practices that aggravate desertification. The several benefits of tree-planting combine together to effect improvement, just as the initial excessive pressure leads to rapid decline towards desert; unfortunately, the restoration process through tree-planting takes much longer.

Reclamation of special kinds of sites

This last role of tree-planting is to use trees as agents of rehabilitation of a site. The general case of reclaiming land from desert, or at least preventing it becoming desert could also be included here, but tree-planting to combat desertification is primarily to relieve the desert causing pressures rather than direct rehabilitation of land apart from where it includes sand-dune stabilization. Trees are used for several purposes in reclamation to:

1. Stabilize soil by preventing the soil surface from moving as root systems develop and ramify in the substrate;

2. Directly protect bare ground from erosive forces provided grazing is prevented and ground vegetation encouraged to develop;

3. Begin the gradual process of soil development through litter fall and the build-up of organic matter, soil organisms, and microflora. Nitrogen-fixing species are particularly useful since many tolerate harsh environments and are able to grow on nitrogen-deficient substrates typical of industrial wastes and sand-dunes. Also, their foliage, which is rich in nitrogen and base nutrients, is usually readily decayed and incorporated.

Sites for reclamation planting include former mining sites, unstable sand-dunes, salt-affected land, etc. On such sites, trees are planted because little else will grow readily and often because of the need to stabilize the ground surface, the presence of toxic substances, or extreme infertility. Examples include the stabilization of sand-dunes by *Casuarina equisetifolia* in China, India, Senegal, and Vietnam (Midgley *et al.* 1983; Lam 1998) and the use of a range of species, such as *Eucalyptus camaldulensis* and *Acacia ampliceps*, to reclaim saline and sodic sites (Marcar and Khanna 1997). Afforestation of industrial wastes include the planting of *A. auriculiformis* on copper tailings in the Philippines and tin tailings in Malaysia, *E. tereticornis* on copper tailings in PNG, *Pinus caribaea* var. *hondurensis* on restored opencast iron/nickel workings in the Dominican Republic, and eucalypts on waste land from opencast tin mining on the Jos plateau in Nigeria (Fig. 21.5). Many examples of the use of trees on mining wastes in India are given by Prasad (1991) who notes that acacias are generally better than eucalypts at improving the soil on such sites. Rehabilitation and restoration of degraded lands and forests are discussed in more detail in the next two chapters.

Implementing programmes of protective afforestation

The consequences of past deforestation, the urgent need for extensive tree-planting, and the reasons why planting is of such value for protecting land have been considered. Although these have long been recognized, there has been little protective planting and much it has failed. The reasons for failure are rarely technical, as emphasized in earlier chapters, since silvicultural methods are available for establishing trees on most sites usually encountered. Both the need for protective afforestation and how to grow trees on the sites are known, failure has occurred in initiating and implementing projects.

Failure to initiate programmes

1. *Scale.* Often the size of reclamation works needed is so large and the scale of planting so great that a poor country does not have the resources to tackle the problem effectively.

2. *Location.* Enthusiasm to carry out a programme is diminished when the main benefits from protective afforestation are received by another province or country, for example, downstream from an eroding catchment. Extensive afforestation, strict control of cutting, and prevention of overgrazing and burning and other harmful land-use practices in the foothills and lower slopes of the Himalayas would help to alleviate the danger of flooding in the lower Ganges and Indus basins, which are a 1000 km or more from the scene of environmental degrade where protective afforestation is needed. Furthermore, the upper catchments themselves are often inaccessible and the terrain difficult. Only recently is the link between forests and food

Figure 21.5 *Eucalyptus camaldulensis* planted on waste spoil from opencast tin mining at Jos, Nigeria (photo P. G. Batchelor).

security becoming widely recognized (Hoskins 1990).

3. *Lack of political commitment.* Areas needing protection are frequently geographically remote from centres of government and administration and often suffer from lack of political commitment and financial support. This has been one of the problems in combating desertification in Asia (Wang *et al.* 2000).

4. *Long duration.* Reclamation projects for erosion control or sand-dune fixation for example, tend to be long-term with few immediate benefits. This can lead to their downgrading in terms of political priorities.

5. *Lack of cash returns.* Few saleable products are derived, only general environmental improvements occur. This has made economic justification of programmes difficult in the past and has deterred commercial investment. However, the economic consequences of not investing in protective plantations may be substantial.

6. *Laws of land reservation.* Where the government seeks to obtain land for general environmental improvement works by direct reservation there is usually local resistance to the loss of land for traditional uses. Thus such an approach can be politically unpopular and often impossible to carry out except where strong governments enforce their will.

7. *Integration and coordination.* The complexity and interdisciplinary nature of all such schemes require coordination of several government ministries at national and regional level. Effective watershed management involves forestry, agriculture, water authorities, engineers, electricity boards, etc., and the task of coordinating these many disciplines and full involvement of stakeholders requires a comprehensive organization lacking in many developing countries. Programmes to control desertification effectively similarly require unity of purpose across many agencies and institutions. Lack of an integrated approach is bound to fail; see analyses of Chowdhry (1987), Grainger (1990), and Ben Salem (1991).

Failure to carry out specific projects

The requirements for successful implementation of projects were outlined in Chapters 5–7 and much is now known about the essentially sociological reasons behind many failures in the past (Fortmann 1988; Baumer 1990; Kerkhoff 1990; Tassin 1995). Several of the key points are as follows.

1. There is invariably much opposition to any scheme that reserves land from traditional uses such as the rights to graze and gather firewood when traditional users have not been consulted and

involved, for example, in the Phu Wiang watershed in Thailand (Ginneken and Thongmee 1991).

2. More generally there has often been a failure to gain the interest and involvement of local people and to utilize indigenous knowledge in all aspects of a project, for example, Michaelsen (1991) and Sharma and Sharma (2001). See also Chapter 6.

3. Often there has been a failure to come to terms with the socio-economic conditions of local people, particularly the habits of the shifting cultivator or pastoralist. Traditional land-users are often little aware of or concerned about national government and its ambitions.

4. Questions of land tenure, tree ownership, or rights to trees have often been neglected, leading to much uncertainty or sectional bias and empowering only the already powerful.

5. Many projects have been over-ambitious, expectations have been unrealistic and problems, such a failed plantings, have quickly led to disappointment, disillusionment, and loss of interest.

6. Many of the above difficulties stem from too few and often inadequately trained personnel to supervise project implementation, and lack of adequate pre-project appraisal and stakeholder participation.

The problems outlined above are not peculiar to certain countries. Even in China, where the greatest protective afforestation programme of any country in the world has been carried out, there are many examples of poor tree survival due to lack of post-planting maintenance, newly planted trees stripped of branches for firewood and where ambitious schemes fell far short of what was hoped. Nevertheless, China's achievements in protective afforestation have been substantial, particularly in recent years, and are regarded by many as models.

Pre-requisites for successful implementation of protection projects

1. There must be political will to plan, pass laws, and encourage the project. Hence the project must be politically acceptable, its benefits clearly stated and evident, and planners and decision makers directed to develop the work. International agreements, such as the UN Convention to Combat Desertification, and regional cooperation, such as the Thematic Programme Network on Agroforestry and Soil Conservation in Asia, help raise political awareness and commitment.

2. There must be adequate funding for the whole life of a project. Combating environmental degradation is usually expensive yielding few directly saleable products, and often the improvement works across many international frontiers, for example, northern Sahara Green Belt. Aid funding from external sources, with low interest rates and long repayment periods, are best suited to this kind of expenditure. In appraising such works the benefits to all sectors must be brought out (agriculture, water, engineering, forestry, etc.) and often these are best shown by examining the consequences of not carrying out a protection project. Protection projects should form part of an integrated and long-term land-use strategy.

3. At all levels the local people should be involved. This participation, and indeed with all stakeholders, should include full discussion of aims and benefits and airing of difficulties and objections, provision of employment, joint decision-making such as agreeing boundaries, visiting trial areas, and being granted concession rights on the restored land wherever this is possible, and so on. It is essential not to alienate the local inhabitants, particularly by authoritarian methods and to ensure genuinely equitable sharing of the benefits.

4. An effective silvicultural package that is straightforward, easily implemented, and with a high chance of success is needed at each stage of a project. Choice of species, planting methods, tending and protection needs, all need to be well established and known to succeed on the sites to be planted.

Management and silviculture in protection planting

In protection forestry all management and silviculture practised must maintain the protective function. Yield of forest products is secondary to the protective benefit the tree crop confers. The yields that are obtained will depend on a crop's protective importance. An example is Trinidad's Northern Range Reafforestation Project, which was designed primarily to reduce soil erosion, flash floods, and improve water quality (Ramdial 1980*b*). Land was zoned and above 700-m altitude denuded areas afforested with *Pinus caribaea* (Fig. 21.6), partially denuded areas encircled with broadleaved species, notably *Swietenia macrophylla*, *Cordia alliodora*, and *Cedrela mexicana*, and vegetative soil conservation barrriers, contour drains and mini-terraces established. All planting operations were carried out by cutting lines along the contour and there was no clear-felling of existing woodland to establish plantations. There is no intention to harvest the pine. Below 700 m, where the completeness of tree cover is less important, the land was planted with fruit trees and other perennial food crops and agroforestry encouraged on steeper slopes between 20 and 30 degrees (Dardaine 1989).

Choice of species

Desirable characteristics for a species for erosion control in descending order of importance (based on Kunkle 1978):

• Good survival and growth on impoverished sites.
• Ability to produce a large amount of litter.
• Strong and wide-spreading root system with numerous fibrous roots; in landslide zones deep roots are essential.
• Ease of establishment and need for minimal maintenance.

Figure 21.6 Protection planting of *Pinus caribaea* var. *hondurensis* on upper slopes in the Northern Range Reafforestation Project, Trinidad.

- Capacity to form a dense crown and to retain foliage year-round, or at least through the rainy season.
- Resistant to insects, disease, and animal browsing.
- Good capacity for soil improvement, such as through nitrogen-fixation.
- Provision of economic returns, particularly on a short-term basis such as fruit, nuts, fodder, or beverage products.

The tree species selected must be hardy enough not only to survive but to grow well in what is often a harsh environment. Also it is often an advantage for the species to regenerate naturally so that any gaps are colonized readily. For year-round wind protection, shelterbelts in arid regions need species to be evergreen and have persistent branches to ground level to prevent wind gusting and trap blown soil, since 90% of all blown soil never rises higher than 30 cm. Where fields and gardens only need protection during the rainy season the species choice is wider. Delwaulle (1977c) discusses species choice and other aspects of shelterbelts for savannas and the arid tropics, Maydell (1986) lists species suitable for soil protection/ soil improvement in the Sahel, and Young (1989) stresses the importance of high above-ground biomass, high rate of nitrogen-fixation, and satisfactory litter breakdown characteristics when choosing trees and shrubs for soil improvement.

Establishment and management

Only general considerations are noted here as the main afforestation methods are considered in Chapter 23.

Planting stock

Seedlings must be sturdy, healthy, and container-grown for high survival. Large size (50–100 cm), but with a high root:shoot ratio, is also desirable to improve seedling survival in weed competition since conventional weeding undermines the very protective role being sought from vegetative cover. Difficult access to many sites planted for protection purposes makes tending operations more expensive and this may result in suboptimal maintenance.

Protection

Full protection of planted trees against fire, grazing and man is essential. This is one of the most difficult and costly components of protective afforestation. It can be a particular problem where there is a high density of cattle or goats in the area. As noted in Chapter 19, either a physical barrier or employment of full-time shepherds may be essential. Tree protection costs in semi-arid Africa were responsible for 20–40% of the total establishment budget in projects financed by the Tearfund, a large British Christian relief and development charity. In the Khyber Pass (Pakistan), the vegetation is so sparse that roadside plantings of *Acacia modesta* are individually protected by a beehive-shaped wall to prevent browsing by goats (Fig. 21.7). In Africa a similarly shaped enclosure or ring fence made of branches is a common sight in the drier parts.

Controlled burning to reduce fuel levels as a fire protection measure is not carried out in protection plantations since the loss of litter facilitates soil erosion. To the contrary, every effort should be made to maintain ground vegetation and a permanent litter layer.

Figure 21.7 Extreme protection around newly planted trees beside the Khyber Pass, northern Pakistan, to prevent browsing by goats.

Pruning, thinning, and felling

These operations are much less important than in industrial plantations. Some shelterbelts may be thinned to maintain good individual tree growth and keep live branches to ground level. Extensive felling, especially clear-felling, is avoided so that the ground surface is not exposed. Pruning of shelterbelt trees should not be allowed to make them too open at the base, but selective or rotational pollarding/pruning or coppicing can be used to maintain aerodynamic structure, principally the need to maintain permeability to avoid wind eddying on the leeward side.

Rotations

On erodible slopes the objective is continuous ground cover, but where some cutting does take place it is usually done selectively on long silvicultural rotations with natural regeneration encouraged. Shelterbelts established in arid regions to reduce wind speed and trap dust and sand, often serve as a supply of firewood so working small coupes or strips on a rotational basis with a species that coppices well is an advantage.

Other operations

It may be necessary to construct roads in protection areas to facilitate management but as little forest produce is extracted from protection plantations the amount of road needed is generally small. On steep slopes it is important to align and construct roads, culverts, and related engineering carefully since these are so often responsible for inducing erosion in the first place. Sthapit and Tennyson (1991) describe the precautions taken to prevent road erosion in the Shivapuri watershed in Nepal.

Supervision

The remoteness of and poor access to many forests planted for protection makes their management difficult, yet proper management is necessary for success. The objectives of protection planting, the difficulty of sites encountered, and the multi-purpose approach often adopted makes supervision by trained personnel essential. A shortage of such staff continues to be one of the most serious limitations in carrying out protective afforestation in many tropical countries.

CHAPTER 22

Tree planting for ecosystem rehabilitation and restoration

Globally, natural forest has been exploited, cleared, and has gradually declined in extent (Chapter 2). In some countries the area of forests has been decreasing for centuries, in others it is a recent occurrence. The rate of resource depletion has increased sharply in recent years in most tropical countries. Deforestation and/or forest degradation has resulted in losses of forest productivity in terms wood and non-wood products and environmental services. Biodiversity has been greatly reduced, especially in the species-rich tropical rain forests, and few forests can recover unaided. While some deforestation may be justified, the global community is concerned about the environmental consequences. The focus of this chapter is using tree-planting, and sometimes forest plantations, to aid recovery of indigenous forest ecosystems, known generically as 'forest restoration'. It is a role that is increasing in importance.

Forest restoration is being addressed at the international level by the United Nations Forum on Forests and the Convention on Biological Diversity, and restoration of vegetation in arid and semi-arid areas is a basic aim of the Convention to Combat Desertification. While planted forests will rarely be able to restore all the products and services that natural forests provide or the needs of key interest groups, they can play a vital role in restoring forest benefits at the landscape level (Maginnis and Jackson 2003). At the local level, foresters will need to plan forest plantation development to provide diversity within the landscape (Chapter 4) and modify plantation practices when rehabilitation and/or restoration are their prime objectives. Forest restoration has been described as a major challenge for foresters in the twenty-first century (Sayer 2002).

Different approaches will be needed to address specific constraints at different sites as the degree of degradation may set limits on what restoration goals are achievable. This is explored in the context of rehabilitation and restoration of ecosystems and how plantations can be used to achieve objectives of biodiversity conservation for degraded forests and lands.

Definitions

Ecosystems can be characterized by their species composition, 'structure', that is, complexity, and 'function', that is, biomass and nutrient content. If any of these are reduced the ecosystem is degraded. So when a forest has been selectively logged and its complexity, and possibly species' richness, is reduced it is considered 'degraded' even though its biomass and productivity remain high. Three levels of increasing site disturbance have been identified by Aber (1987): (i) disruption or removal of the native plant community, without severe soil disturbance; (ii) damage to both vegetation and soil; and (iii) vegetation completely removed and the soil converted to a state outside natural conditions. Three approaches, 'restoration', 'rehabilitation', and 'reclamation', are commonly applied to reverse the degradation process and assist recovery of these different degrees of disturbance, see Fig. 22.1.

Restoration. This term generally means 'to bring back to the original state or condition'. So, the thrust of ecosystem restoration is to attempt to re-establish the presumed productivity and species diversity of the forest condition at a particular site before it was disturbed and degraded. It attempts to closely match the ecological processes and functions of the original forest (Lamb 2001). Generally much less emphasis has been placed on ensuring that ecological processes are reinstated because it is assumed that these will be restored once biodiversity is re-established.

One of the basic questions of restoration ecology is how to define what is meant by the 'original ecosystem'. Replacing the exact species composition of the original forest is impossible as generally this is not known with any certainty. The goals, and the relationship between the theory and practice of restoration ecology are discussed in Jordan *et al.* (1987). Those responsible for restoration will have to initiate successional development where natural recovery is impossible or accelerate it when it is feasible. How restoration is done will depend on

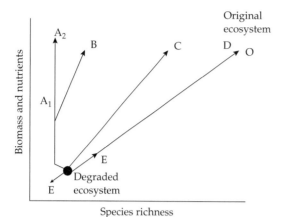

Figure 22.1 Relationship complexity and productivity of ecosystem development for reclamation, rehabilitation, and restoration. Species and complexity describe ecosystem structure, and biomass and nutrient content describe ecosystem function. Reclamation involves plantation monoculture of exotic species (A) which may have a biomass less than A1 or greater than A2, the original biomass depending on weedicide and fertilizer use. These may acquire a diverse understorey over time (B). Rehabilitation involves both native and exotic species (C), while restoration (D) leads to a new ecosystem approximating to the original ecosystem (O). If the degraded ecosystem is neglected it may recover or degrade further (E). Based on Lamb (1994) and Bradshaw (1987).

the extent of degradation and resources available and it will not necessarily follow the way the ecosystem developed under natural conditions. It should be sufficient to get the ecosystem close to natural conditions and then let nature take its course.

Rehabilitation. Returning any converted or damaged forest lands to a functioning forest is considered 'rehabilitation' by Brown and Lugo (1994). Lamb (2001) defines 'rehabilitation' as 'to re-establish the productivity and some, but not necessarily all, of the plant and animal species thought to be originally present at the site. For ecological or economic reasons, the new forest might also include species not originally present at the site. The protective function and many of the ecological services of the original forest may be re-established'. Rehabilitation is much easier to achieve than restoration and, by including commercial species, can often have a more favourable economic outcome. In fact, Lamb (1998) suggests pragmatically that using commercial timber plantations with modified designs is a promising approach to restore on a large scale a proportion of the former biodiversity of degraded forest lands. It is likely to be a more attractive option than restoration for developing countries in the tropics.

Reclamation. This term is widely used to refer to the revegetation of highly degraded site, such as mined or salt-affected lands. It aims 'to recover productivity of a degraded site mostly using exotic tree species. The original biodiversity is not recovered although the protective function and many of the ecological services may be re-established' (Lamb 2001). Harrington (1999) suggests that the term 'reclamation' generally reflects the imposition of a human value system rather than an ecological goal. There is no attempt to re-establish any of the original biodiversity at the site and one or more exotic species is used to achieve stability and productivity (Lamb 1994). Using species with exceptional physiological tolerances to rapidly improve site conditions means that species of *Acacia*, *Eucalyptus*, and *Pinus* and others used in traditional forest plantations are frequently employed for reclamation and rehabilitation of degraded lands.

Planting for restoration, rehabilitation, and reclamation can occur at an individual site but there is growing support for '*forest landscape restoration*'. This is a process aiming to regain the ecological integrity and enhance human well-being in deforested or degraded forest landscapes (Maginnis and Jackson 2003). It is more than just providing tree cover. It aims to provide at the landscape level a variety of forest products, services, and ecological processes that benefit both people and the environment. It therefore has both socio-economic and ecological dimensions (WWF 2002). There is clearly a role for planted forests to complement other strategies, such as ecological corridors, preservation of remnant native forests and woodlands, and agroforestry systems, in forest landscape restoration.

This chapter is restricted to practices to enhance and conserve biodiversity by restoration and rehabilitation of forests and lands. Plantations for reclamation of various inhospitable sites are described in Chapter 23.

Degradation and rehabilitation

While forest and land degradation can be caused by natural causes such as wild fires, often the most severe cases are the result of human actions. Forest and land degradation is a complex phenomenon with many socio-economic, political, and cultural causes (Chapter 2). Many pressures result in the exploitation and clearing of natural forests and several factors suggest that natural forests in the tropics, will continue to decline in area and that much remaining secondary forest and cleared forest lands will become degraded. Land degradation is also a complex process which changes soil conditions, microclimate, and availability of seed sources.

In the tropics much attention has been focused on the degradation of rain forests because they have the greatest biological diversity, are a major component of the global carbon cycle, and provide a wide range of goods and services at local and international levels. Properly executed selection fellings with reduced impact logging methods in tropical rain forest will result in minimal degradation and only a transient loss of structure and function. But short cutting cycles, illegal re-entry to further harvest trees in logged areas, and careless harvesting practices have resulted in the more serious tropical rain forest degradation (Appanah 1998). Forest degraded by harvesting becomes further degraded where fires occur annually and ultimately all woody vegetation may disappear as in the *Imperata*-dominated grasslands of Southeast Asia. The soils under degraded forests may or may not deteriorate, at least in the short-term (Ohta *et al.* 2000).

Degraded lands usually involve eroded or impoverished soils, hydrologic instability, reduced primary productivity, and biodiversity (Parrotta 1992). To reverse the degradation processes requires an understanding of ecosystem functioning, the motivation to manage ecological succession, and sufficient resources to undertake the task. Decisions have to be made on whether to allow natural succession to proceed unaided, or to provide some protection, such as the closure of mountain areas in China, to allow natural recovery, or to intervene actively by planting indigenous and/or exotic species. Much will depend on the degree of degradation and ecological skills of those undertaking the rehabilitation. Mining, clearing for agriculture, such as pasture development in the Amazon basin, and the actions of war, such as the spraying of defoliants on the forests of Vietnam, all arrest successional changes of vegetation and need positive management actions to re-establish the original ecosystem. The most severe cases will require reclamation efforts as the lack of vegetation and changed soil and microclimatic conditions will preclude ecosystem rehabilitation. In less degraded situations, the presence of exotic weeds and the lack of suitable seed sources on or near the site will prevent natural succession without planting and other interventions.

Plantations will influence the local environment in various ways, and under some circumstances, may allow biodiversity to increase over time if new species can reach and colonize the sites (Lugo *et al.* 1993). They may modify *microclimate*: (i) less fluctuation in temperature and humidity in the stand, and (ii) better light environment for seed germination and seedling growth; and *soil physical and chemical properties*: (i) reduction in erosion rates, (ii) increased soil organic matter and cation exchange capacity, and (iii) increased soil nutrient flow through litter decomposition and soil microbiological activities.

These changes initially stop or slow the degradation process and enable less hardy native plants species to regenerate in the understorey e.g. Lamb and Tomlinson 1994; Fimbel and Fimbel 1996; Huttel and Loumeto 2001, see Fig. 22.2. This is accompanied by changes in vertebrate fauna (birds, small mammals, etc.) and soil macrofauna (earthworms, termites, etc.) (e.g. Bernard-Reversat 2001).

Management strategies

The International Tropical Timber Organization (ITTO) in collaboration with FAO, the Center for International Forestry Research, the World Conservation Union, and the World Wide Fund for Nature has developed useful guidelines for planning and implementing restoration and rehabilitation of degraded tropical forests (ITTO 2002). They recommend a holistic approach of

Figure 22.2 Natural regeneration of local species under a clonal eucalypt plantation in the Congo.

forest landscape restoration, taking into account other landscape components in a particular locality. The landscape context helps planners and resource managers identify management strategies and options that meet the local communities' needs. Determining site-specific integrative ecosystem designs is a crucial part of management.

Participatory and adaptive management planning is required to determine what management strategies are needed to achieve restoration or rehabilitation. Objectives, methodologies, and monitoring must be addressed.

The primary objective for restoration planting will be ecological in nature and the aim to re-establish the original plant and animal communities to a site in a particular time frame and within the resources available. The first stage may be ecologically crude, such as the establishment of some sort of permanent cover as quickly and as cheaply as possible (Bradshaw 1987). But even this requires a good ecological knowledge and implementation of appropriate practices including discussions on socio-economic issues with stakeholders.

Where the objective is rehabilitation, the potential for making an economic return will need to be addressed in addition to increasing the structural complexity and biodiversity of the plantation. With both types of plantations it is important to address local social issues with stakeholders in the same way as for industrial plantations (Chapters 4 and 6). It is possible that closing areas for rehabilitation or restoration may have immediate impacts on the livelihoods of local people if they rely on the site for grazing animals, collection of fuel and non-wood forest products, or access to other areas. Such matters need to be discussed and resolved during the planning phase. Landholders may assist with rehabilitation if there is the possibility of an economic return from an otherwise unproductive area of land but may be quite indifferent or even antagonistic if they cannot perceive a direct benefit. This was the case in a dry part of Yunnan province in China, where villagers grazed their goats on highly eroded and virtually treeless slopes until they had the opportunity to plant neem, *Azadirachta indica*, to produce a saleable product (Liu 2003). Without the incentive of recouping their plantation costs and making a profit, they had little interest in rehabilitating the site.

During the planning phase it is necessary to determine what preplanting measures are needed to stabilize the site, improve soil conditions, remove exotic weeds, and provide protection, such as fencing. There is the very important task of deciding what species to plant, in what configuration, and whether it is necessary to inoculate the planting stock with mycorrhizas or nitrogen-fixing bacteria. Foresters need to be aware that the practices used for establishing industrial plantations may need to be modified significantly for restoration and rehabilitation plantations, although much of the operational planning will be similar.

It is also worth considering in the planning phase how to judge whether the plantation is achieving its restoration or rehabilitation objectives. It is rarely possible to predict accurately how the forest will develop so rigid management prescriptions should be avoided in favour of an adaptive management approach.

Species selection

Most of the principles of species selection described in Chapter 8 can be applied in rehabilitation and restoration plantations with species' choice strongly influenced by the degree of degradation and the objective of the planting. Proper attention to species-site relationships is especially important. It must be emphasized that the choice of species can have a major influence of the rate and direction of the rehabilitation processes.

In purely restoration plantings in areas where the soil is left largely intact, as in some recently logged rain forests, planting native species chosen from nearby undisturbed communities may be feasible (e.g. Mori *et al.* 2000). Then, provided the forest is not too fragmented, succession may occur naturally with seed from adjacent intact forest communities. In tropical rain forests natural succession can be problematical due to fragmentation. To overcome this and restore vegetation on abandoned pastures in Amazonia, planting 'tree islands' of species that grow rapidly in full sunlight and produce fleshy fruits soon after planting to attract seed-carrying birds and bats has been recommended (Nepstad *et al.* 1991). Similarly, ecological rehabilitation by planting 'framework species' can create, direct, and accelerate succession (Tucker and Murphy 1997). Perhaps 20–30 native species can be selected with characteristics that enable them to rapidly shade out weeds on the site and attract seed dispersing birds and animals. The mixture of framework trees should include pioneer species that are fast-growing and able to occupy the site quickly and slower-growing, large-seeded species which are less easily dispersed by birds and other wildlife (Forest Restoration Research Unit 2000). Even when suitable dispersers are present, regeneration of large-seeded tree species in clearings is a very slow process (Parrotta *et al.* 1997a; Wunderle 1997).

The use of indigenous species for industrial forestry in particular has been discussed in Chapter 8 and it was concluded that where a native species meets a need there is no need to use an exotic alternative (Fig. 22.3). For biodiversity conservation and restoration planting purposes native species are strongly preferred or essential and more attention needs to be paid to systematically determining their physiological tolerances, silvicultural characteristics, and propagation needs. Given the great diversity of native tree species, screening of a broad range of species, including those without known commercial value, is warranted. In recent years more attention has been paid to such studies, for example, in Brazil (Knowles and Parrotta 1995), Costa Rica (Butterfield 1995), and Thailand (Hardwick *et al.* 1997; Forest Restoration Research Unit 2000). From a practical point of view, species which seed regularly, can be grown easily and quickly in the nursery, and are tolerant of a range of site conditions, are the most useful.

Although exotic pines and eucalypts may not be very attractive to native wildlife (Wunderle 1997) they do modify site conditions and with appropriate management there is an increase of biodiversity in the understorey and soil (Fimbel and Fimbel 1996; Geldenhuys 1997; Bernard-Reversat 2001). Nitrogen-fixing hardwood species, such as *Acacia auriculiformis*, *Paraserianthes falcataria*, and *Alnus* spp., may improve soil nutrient levels more rapidly than non-nitrogen-fixing stands through higher rates of nutrient cycling and facilitate development of native understorey vegetation. Site conditions, such as soil nutrient status and drainage, temperature level, and quality of light may be more important than the particular canopy species (Geldenhuys 1997).

Rural populations in developing countries often rely heavily on wood and non-wood products from local forests to support their livelihoods and sometimes to provide a cash income. In densely populated areas, such as India and China, over-exploitation of forest resources results in forest degradation and ultimately in total loss of forest. Farmers and village communities are interested in restoring forest cover provided the species used yield products for their domestic needs or for sale. The species selected for planting will reflect these needs. For example, in northern India villagers value *Dalbergia sissoo* for timber, *Melia azedarach* for minor timber products, flower buds of *Bauhinia variegata* as a vegetable, *Ougeinia dalbergioides* for agricultural implements, *Prunus cerasoides* as a sacred species, and all of them for green fodder (Negi and Joshi 2002). On previously deforested hillsides in Guangxi province, China, some farmers grow *Cinnamomum cassia* and harvest its bark for Chinese medicine and distil its leaves and branches to produce oil (Fig. 22.4). They also grow *Illicium micranthum* to produce a spice, and bamboo for baskets, building materials, and for paper pulp making (Liu 2003). These economic plants provide ground cover and a diversity of habitat, may reduce pressure on harvesting in the natural forests, and generally contribute to the rehabilitation of forest lands.

Plantation design

Restoration and rehabilitation plantations

Industrial plantations are invariably planted with a uniform distribution of trees but in restoration and ecological rehabilitation plantations there is the possibility that planting in clumps or some other

Figure 22.3 A 17-year-old plantation of native *Cordia alliodora* in Ecuador with vigorous regeneration of *Virola*, *Brosimum*, and other native hardwoods.

Figure 22.4 Harvesting leaves of *Cinnamomum cassia* for oil production in Guangxi, China.

configuration will be more effective in producing the desired conditions for ecosystem development. Harrington (1999) suggests clump planting may result in a more natural pattern of variation in plant distribution and microclimate.

In forest plantations where the objectives of management have changed to convert them to native forests, for example, in zones along stream banks, commercial trees may be gradually thinned out to leave an early stage, biologically diverse secondary forest. Catalytic plantations of commercial trees may be managed in the same way. The value of the timber extracted may in some circumstances defray the costs of the conversion.

Industrial plantations

Few degraded forests will recover unaided and, as significant labour inputs and resources are usually required to fully restore original ecosystems, only small areas have been restored (Lamb 1998). Because of this limitation, there has been an increasing interest in using industrial plantations to increase biodiversity in the landscape (Lamb 1998, 2001) and within plantations (e.g. Lugo *et al.* 1993; Parrotta *et al.* 1997*b*). This may be achieved by underplanting with native species or, where the plantations have catalysed the colonization of the understorey by native species, adopting practices that encourage the growth and development of these species.

The catalytic effect of plantations is due to development of humus and litter layers, changes in microclimatic conditions, and increased structural complexity of the vegetation as already discussed—that is, it has reintroduced a simple forest habitat or environment. These changes enable increased germination of seed from neighbouring native forests by wildlife attracted to the plantation, suppression of weed competition, and improved seedling survival and growth. Providing silvicultural management does not interrupt the succession, a mixed forest of the plantation species and the early succession trees and other plants develop. For example, in parts of Sri Lanka planted pine forests have been invaded by a range of tree species and will gradually be converted to native forest by steadily removing the pines (Bandaratillake 1991).

There are several possibilities for managing the tree crop (Keenan *et al.* 1997; Lamb 1998):

1. The plantation may be harvested as originally planned. In this case the benefits of the increased biodiversity are transitory but they may have provided valuable habitat during the life of the plantation and contributed to landscape diversity.
2. The plantation may be selectively logged. If the planted trees are gradually removed a secondary forest can quickly develop. Economic returns will be derived from the plantation while a diverse secondary forest is developing.
3. The plantation may be abandoned as a production area and managed for biodiversity conservation. This may occur if there have been changed management objectives, or if the plantation has grown poorly, or if the market for the plantation produce has disappeared, or if the plantation is of indigenous species and by growing on to maturity and beyond it accelerates the recovery of

niches and habitats associated with large tree size, natural decay, dead wood, unplanned gaps, etc.

4. Both plantation and natural regeneration components may be selectively logged. This is a possibility where the understorey includes trees or shrubs with commercial potential. The plantation may then be managed as a mixed-species forest. Enrichment planting of commercial trees in the understorey may contribute to this type of plantation management.

There will be conflict between some of the management practices used to manage industrial plantations for maximum wood production and practices that improve biodiversity values in plantations. Nevertheless, modification of industrial plantation design may enhance biodiversity in the landscape with minimal loss of productivity and profitability. Small changes in design can sometimes contribute significantly to biodiversity conservation, provide habitat for some species, and connectivity of patches of tree cover. With careful planning, integration of natural features and access routes will assist plantation development and incorporate corridors and refuges for wildlife conservation. Embedding areas of industrial plantation in a matrix of natural vegetation is a relatively low cost option that may have benefits through watershed protection and reducing the risk of large areas of plantations being damaged in wildfires (Lamb 1998) or improved plantation stability (Campinhos 1999). Positive benefits for biodiversity conservation may also flow by careful scheduling in time and space of harvesting in plantation forests (Lindenmayer *et al.* 2003). Guidelines for plantation design and management for biodiversity conservation have been proposed by Lindenmayer (2002). They include:

1. Retaining at least 30% of remnant or re-established native vegetation within the plantation area in extensive developments (e.g. >1000 ha).
2. Restoration of native vegetation along gulley lines to link with existing patches of remnant vegetation.
3. Exclusion of domestic livestock from patches of native vegetation to allow natural regeneration, and control of weeds in the area.

There are clearly many options for industrial plantations to be managed to contribute to biodiversity conservation on a scale that ranges from modest to very significant.

Site management practices

Practices used for restoration and rehabilitation plantings depend very much on the nature of the degradation at a specific site. They are generally

Table 22.1 Site management practices for restoration and rehabilitation areas

Problem	Practice(s)
Unstable soil surface	Contour operations, mulching, vegetation strips
Compaction	Ripping, subsoiling
Toxicity	Irrigation, liming, plant tolerant species
Low organic matter	Mulching, establish vegetation
Low fertility	Fertilizing, plant nitrogen-fixing plants
Lack of symbionts	Inoculation
Exposed site	Add shade or shelter
Light level	Provide shade trees or thin out overstorey
Animal damage	Fencing, repellents, trapping, baiting
Fire	Fire breaks, control external vegetation
Weed competition	Weedicides, manual slashing

designed to stabilize and improve soil conditions, improve microclimate, and protect the vegetation from fire, and grazing and trampling by domestic animals. Table 22.1 lists some of the practices used.

Criteria for success

We need criteria and indicators to judge whether the plantation is achieving its objectives. For restoration planting the primary objective will be ecological in nature. Where the objective is rehabilitation, the potential for making an economic return will need to be addressed in addition to increasing the structural complexity and biodiversity of the plantation. It is not possible to have a universal set of criteria as forest type, environmental conditions, and time since the planting began will differ. Lamb (1993) has suggested a number of potential indicators of ecological success in restoration projects:

1. A litter layer is present and there is no evidence of soil erosion.
2. Tree crown cover conforms to that expected for the site conditions.
3. Most of the original trees and understorey plants are present and reproducing or evidence that colonization is continuing to add to species richness.
4. Tree health and vigour is satisfactory without further fertilization or weed control.
5. Weeds are absent or there is no evidence of spread.
6. Wildfire can be tolerated if site is fire prone.
7. Most abundant original fauna are present or evidence that colonization is continuing to add to species richness.
8. Exotic animals are absent or in low numbers.

9. Plant or animal indicator species for particular successional changes are present.
10. Target keystone or rare species are present.

These indicators will need to be expanded and modified where there are social and economic objectives to be met in rehabilitation projects.

Case studies

The following case studies have been selected to illustrate the range of restoration and ecological rehabilitation activities being undertaken in the tropics. The sites include light to moderately disturbed rain forest in Southeast Asia to highly degraded mine sites in New Caledonia and Brazil.

Restoration of rain forest in East Kalimantan, Indonesia

Human activities (commercial logging, and slash and burn agriculture) and drought-linked wildfires have converted much of the primary rain forests of dipterocarps in East Kalimantan to a mosaic of farm lands, *Imperata* grasslands, shrubby savanna, and secondary *Macaranga* forest. Following widespread wild fires heavily disturbed sites became dominated by a few pioneer species such as *Macaranga gigantea* and *M. triloba*. A range of research activities designed to facilitate restoration of the dipterocarp forest are described by Guhardja *et al.* (2000) and Mori (2001). Recommended treatments vary according to the degree of degradation.

1. Lightly degraded forest: only release cutting or canopy opening of the natural regeneration.
2. Moderately degraded forest: sparse natural regeneration is supplemented by enrichment planting with several dipterocarp species either as patch (gap) planting or underplanting the pioneer stands. Species that are relatively light-demanding are preferred for patch planting.
3. Heavily degraded forest: patch planting or line planting through the secondary pioneer species with dipterocarps carefully selected for the adaptation to the local soil and light conditions. The strips are one half to one fifth of the height of surrounding tree height depending on the light requirements of the species planted. Where the pioneer species has started to decline underplanting may be used.
4. Grassland or shrubland: fast-growing commercial species such as teak (*Tectona grandis*), mahogany (*Swietenia macrophylla*), and pine (*Pinus* spp.) are established followed by underplanting with dipterocarps. If *Acacia mangium* is used then several

species of dipterocarps are planted in cleared strips, 7–9 m wide. Planting with a single species of dipterocarp is not recommended and careful species selection is very important. Otsamo *et al.* (1996) have recommended species for planting in grassland. Inoculation of dipterocarp seedlings with selected ectomycorrhizas, mulching, and addition of charcoal to the soil is also advised (Suhardi 2000).

Rehabilitation of highly degraded forest land in Thailand

Large areas of tropical rain forest have been highly degraded in Thailand by shifting cultivation or commercial logging. These actions have deprived local people of sources of firewood, bamboo, edible fruits, honey, edible fungi, and medicinal plants. In the past decade there have been increased efforts to restore or rehabilitate these degraded forest ecosystems. It is government policy to increase forest cover from the current 17% to 40%.

In 1990 a project to rehabilitate denuded forest lands in the central highlands (Fig. 22.5) by reforestation was initiated by the Royal Forest Department and FAO (Margescu 2001). Failure to address the socio-economic problems of the local communities resulted in some early failures. Subsequently seven site types were identified based on soil moisture availability, degree of erosion, altitude, and degree of human influence and demand. This enabled species selection that matched site conditions and local economic and social needs. Preference was shown for indigenous species, nitrogen-fixing species, and fast-growing pioneer species that would suppress *Imperata* grass. These included the local indigenous species, *Choerospondias axillaris*, and 30 other species including *Acacia* spp., *Bambusa* spp., *Eucalyptus* spp., *Hopea odorata*, *Pinus* spp., *Tectona grandis*, and *Xylia kerrii*. Multipurpose species, such as *Azadirachta indica*, *Leucaena leucocephala*, *Mangifera caloneura*, and *Sesbania grandiflora* according to local preferences were planted on areas bordering agricultural fields and settlements. The trees were planted on the contour at a spacing of 6 m × 2 m to allow farmers to grow maize between the rows for at least 2 years. As well as establishing cover on the bare hillsides, the project helped reduce the incidence of fire and pressure on remnant patches of natural forest allowing natural regeneration and succession to increase the biodiversity.

In 1994 the Forest Restoration Research Unit (FORRU) was set up in a joint initiative between

Figure 22.5 Denuded forest lands in the central highlands of Thailand.

Chiang Mai University and Doi Suthep-Pui National Park under the Royal Thai Forest Department. FORRU began by gathering basic ecological data on tree species growing in northern Thailand including research on seed production, collection, and germination, nursery practices, and field establishment. It recognizes that restoring the exact species composition of the original tree community is impossible and so initially plants 20–30 'framework' species. These species are planted at different densities and given different cultural treatments. The framework species facilitate regeneration of other species with the result that after 6–10 years there may be up to 80 species on the site. Species commonly planted are *Erythrina subumbrans, Ficus altissima, Gmelina arborea, Hovenia dulcis, Melia toosendan, Prunus cerasoides Sapindus rarak*, and *Spondias axillaris*. FORRU also provides technical advice for local community tree-planting initiatives (Forest Restoration Research Unit 2000; Thomson 2001).

Mangrove restoration in the Philippines.

Mangrove forests are a feature of many coastal ecosystems throughout the tropics. Their complexity varies but they provide a habitat, breeding ground, and nursery for a great variety of fishes, crustaceans, amphibians, reptiles, birds, and mammals. They also serve as a buffer against typhoons and wave action. They are often cleared or degraded and the objective of restoration of mangrove ecosystems is usually for biological conservation and/or for sustainable utilization and protection of coastal areas (Saenger 2002).

In the Philippines about 80% of the mangrove forests have been destroyed or seriously degraded by human activities including urbanization, clearing to provide land for fish farms, conversion to salt pans, and cutting for a range of wood products. There are a number of initiatives to restore mangrove ecosystems but there have been many set backs due to lack of basic technical information, poor management decisions, and socio-economic problems (Fortes 1995; Walters 1997).

Regeneration by planting and direct seeding is undertaken but requires careful site selection and species-site matching (DENR 1994). Enrichment planting may be used where the mangroves are less degraded (e.g. Bravo 1995). Initial plantings should be on sites with a relatively stable substrate and sheltered from strong winds. Consideration must also be given to fresh water supply, degree of tidal inundation, soil type, and soil condition.

Species with economic value are recommended. *Rhizophora apiculata, R. mucronata*, and *R. stylosa* are suitable for planting on degraded areas but each has specific soil requirements. Species of *Avicennia, Bruguiera, Ceriops*, and *Nypa fruticans* are also matched to different sites and soil types.

Site preparation includes removal of stumps, branches, and other tidal debris that could damage young plants during tidal movement. Depth of planting depends on the length of the propagules, usually about one-third of total length. Spacing is 1 m × 1 m in areas exposed to wave action and about 1.5 m × 1.5 m in vegetated and more protected areas. Stakes or fencing may be erected around the planting area for protection. A number of pests and diseases attack mangrove species and need to be controlled if the attack is severe (DENR 1994).

Effects of eucalypts and other exotic tree plantations on biodiversity in the Congo

In the Democratic Republic of the Congo there are over 40 000 ha of eucalypts (mainly *E. grandis* × *E. urophylla* hybrids) and 2500 ha of *Pinus caribaea* planted for wood production. These have been established since 1978 on savanna sites of mainly infertile, sandy soils, and sparse vegetation. Studies reported by Loumeto and Huttel (1997) and Bernard-Reversat (2001) indicate there is a reduction in fire occurrence, and changes in microclimate and soil conditions in the plantations which favour development of understorey vegetation. The catalytic effect of the plantations on colonization and understorey development changes with age. An understorey rich in pioneer species developed (Fig. 22.1), but with greater diversity in pine and *Acacia auriculiformis* plantations. Thickets build-up in the plantations after about 10 years. Improved soil processes occurred as the eucalypt plantations aged due to changes in the soil-litter system which enhanced nutrient cycling. Increased soil organic matter appeared to be responsible for increasing density of earthworms, termites, and litter fauna leading to improved soil functioning.

While further research is required to determine the effects of silvicultural practices on soil organic matter and understorey development, the observations suggest that biodiversity has been increased in these grassland areas through commercial eucalypt and pine plantations.

Restoration forests on a bauxite-mined site in Brazil

In 1979, a Brazilian mining company, Mineração Rio do Norte SA decided to restore primary forest cover destroyed during bauxite mining in Pará State and initiate an annual reforestation programme of 100 ha. The activities have been described by Knowles and Parrotta (1995) and Parrotta *et al.* (1997a).

Site preparation involved levelling the overburden, replacing about 15 cm of topsoil, and deep ripping planting lines. An intensive study of fruiting phenology, dispersal, seed viability, and propagation methods of many tree species in adjacent primary forest was made and this enabled a diverse mixture of up to 90 species to be planted in a given year.

The programme has successfully re-established native forest cover on an operational scale at reasonable cost. It has provided an environment that favours regeneration of smaller-seeded woody species dispersed by mammals and birds from the local primary rain forest. However, additional interventions, such as enrichment planting, may be needed to accelerate regeneration of large-seeded species not adequately dispersed into the restoration areas.

Rehabilitation of nickel mining sites in New Caledonia

Revegetation of mining sites is difficult and costly. It can usually be described as 'reclamation' and some of the practices have been described in Chapter 23. However, greater public concern over environmental damage by mining companies, and legislation to enforce mine-site rehabilitation are making companies take a more serious approach to the problem (Le Roux 2002).

In New Caledonia there is a major effort for ecological rehabilitation of nickel mining sites by Le Nickel company (Sarrailh 2002). Previously two leguminous exotics, *Calliandra calothyrsus* and *Acacia ampliceps*, were used with some success. Then two local, fast-growing, nitrogen-fixing species, *Casuarina collina* and *A. spirobis*, grew well but did not encourage understorey natural regeneration. Now species found in scrubland where the mining occurs are showing promise. They include: *Grevillea exul*, *G. gilliwrayi* (Proteaceae), *Carpolepis laurifolia*, *Arillastrum gummiferum* (Myrtaceae), *Gymnostoma deplancheanum*, *G. chamaecyparis* (Casuarinaceae), and *Serianthes calyina* (Mimosaceae). Research is in progress on the introduction of symbiotic micro-organisms, fertilizers, organic matter, etc.

CHAPTER 23

Afforestation of inhospitable sites

It is very difficult to grow trees on some sites, but only in a few environments with extreme cold or no soil is it completely impossible. Apart from these environments, with careful treatment of the site and correct choice of species, trees can be established on many very unpromising sites. Although some species are remarkably tolerant of poor growing conditions, establishing trees on naturally inhospitable sites usually requires special techniques. This chapter outlines the reasons which make sites inhospitable and considers the silvicultural techniques for afforesting five examples: arid-zone areas; steep and eroding slopes; man-made, industrial waste ground; saline and sodic soils; and mangroves.

This topic is important because many sites needing protective afforestation (Chapter 21) are also inhospitable. In fact, this is sometimes the reason why they are denuded and lacking tree cover. Protective afforestation makes use of the tree's influence on the local environment and this only occurs if the trees are satisfactorily established in the first place. The practices described in this chapter aim at reclamation or rehabilitation of the site, they do not attempt to achieve ecological rehabilitation or restoration, a topic covered in Chapter 22.

What makes sites inhospitable?

The simple answer is because of 'extremes'—extremes of climate (low rainfall, continual high wind), extremes of soil condition (no topsoil, high mineral toxicity), or even extremes of damage (severe over-grazing, burning). However, the idea of extremes is relative. For example, *Eucalyptus grandis* grows well where annual rainfall is above 1000 mm, while *E. camaldulensis* or *E. microtheca* grow well with much smaller amounts. There is no absolute definition of 'extreme', but in this context it applies to those sites where special, and usually more costly, establishment methods are needed to grow trees.

Climatic extremes

In Chapters 1, 19, and 21 various aspects of the severity of tropical climates were considered. Some of the most biologically diverse and some of the most barren ecosystems are found in the tropics. Rainfall variation largely accounts for this, with low rainfall being the most important limiting climatic factor. In most semi-arid (annual rainfall 250–600 mm) and all arid areas (less than 250 mm year^{-1}) special techniques are needed to establish trees.

However, though aridity is important, the difficulty of establishing trees in such regions is increased by the desiccating effects of low relative humidity, for example, 10–20%, and by greater extremes of heat and cold owing to clear skies. Air and soil temperatures up to 50°C are common. Such temperatures are close to the 55–65°C range considered lethal for most plants, are well above the optimum for photosynthesis and lead to high rates of respiration. In the subtropics, extreme cold, for example, −15°C, rather than heat can be more severely limiting to tree growth in very dry areas (Bands and Britton 1977). Furthermore, arid areas with little vegetation suffer much from abrasive dust and sand storms.

Other climatic extremes are of more local importance and include excessively high rainfall leading to flooding and waterlogging, persistent wind causing extreme exposure and cyclones where their frequency limits forest development. Species such as *Acacia auriculiformis* and *E. camaldulensis* tolerate temporary flooding (Yantasath 1987) and some species are adapted to survive even severe cyclones (Tassin 1995). On cyclone-prone Hainan Island in China, farmers have planted the slow-growing, typhoon-resistant, local *Acacia confusa* for fuelwood but recently have used the fast-growing *A. mangium* for firewood and small poles on rotations as short as three years to minimize the risk of cyclone damage.

Extremes of soil condition

Frequently, poor soil conditions are associated with extremes of climate. There are several problems associated with the 'soil', or more correctly the 'substrate', which make tree establishment difficult. They are considered separately, but often two or three occur together on one site.

Very thin or no topsoil

Loss of topsoil is an early consequence of erosion, while on most man-made waste and large areas of sandy or stony desert it is completely absent. Lack of topsoil, with its important organic matter fraction, is a serious limitation since it is a much superior growing medium for plants than most types of subsoil. Not only are the physical and chemical properties of topsoil valuable (texture, moisture absorption and retention, and nutrient concentrations) but it is a living surface layer with an abundance of micro-organisms which play a critical part in decay and reincorporation of litter and organic matter, and recycling of the nutrients they contain. The great proliferation of roots in topsoil shows its importance. Trees planted in subsoil, or any non-soil substrate, almost always grow more slowly, suffer higher mortality, are sometimes unstable because of poor root development and often are nutrient deficient. Many natural, non-topsoil substrates, including clays, sands, gravels, and shales, are susceptible to physical damage such as erosion, compaction and puddling, and very poor moisture relations, when exposed.

Sometimes topsoil is present but is extremely shallow and there may be 30 cm or less to bedrock or an impermeable layer. This severe physical limitation to rooting leads to poor growth, instability and great susceptibility to drought in trees.

Nutritional problems

Some of these were discussed in Chapter 14, but severe nutritional problems can make sites very inhospitable. The following are all encountered: (i) extreme infertility (lack of many nutrients) owing to excessive leaching and long weathering, or sometimes a very sandy soil texture; (ii) serious deficiency in one or more nutrients; (iii) toxic levels of one or more nutrients or heavy metals; (iv) extremes of acidity or alkalinity which interfere with nutrient balance and nutrient availability; (v) a build-up of inorganic salts and development of salinity.

Many sands in arid areas have calcareous crusts and are alkaline. On salt-affected sites the salts accumulate having been carried upward by capillarity and remain when the water evaporates. Accumulation of soluble salts, mostly chloride and sulfate, of sodium, calcium, magnesium, and potassium occurs in the upper soil horizons and often a white crust appears on the surface. They result in a poor soil structure and an alkaline reaction. About 74 million ha of land is salt-affected due to clearing of deep-rooted vegetation and irrigation, 43 million ha of which is due to irrigation in arid and semiarid regions (Dregne *et al.* 1991). In Pakistan alone about 6.5 million ha of agricultural land is adversely affected by soil salinity and the related problem of waterlogging (Marcar *et al.* 1998).

The coastal sands of Queensland in Australia are very impoverished (Fig. 12.7) and were considered unplantable until introduction of intensive cultivation and fertilization. Similar impoverished soils occur in parts of Guyana and Surinam. Many industrial wastes are extremely deficient in nutrients, usually wholly lacking organic matter, and may have the added problem of high concentrations of metals.

Poor physical conditions

Several different problems can occur.

1. Compaction is particularly serious on restored industrial waste ground, where machines have been used to reshape the material, and at log landing areas and along logging trails (Fig. 18.3).
2. Surface hardness (exposure of some tropical soils leads to formation of iron pans which are extremely hard and very difficult to cultivate); see Chapter 21. A related problem is development of a thin impervious crust in arid regions which greatly hinders moisture infiltration when it rains. Breaking up this crust is often the key to successful rehabilitation.
3. Impeded horizon and lack of structure lead to poor moisture relations, bad or excessive drainage, a barrier to root growth, and other physical restrictions.

Some of these conditions are worsened by disturbance and exposure. This most commonly occurs when vegetation is removed, or because of logging operations, or from the eroding and trampling damage of livestock on trails and around water points.

Unstable and moving substrate

Movement of the soil surface due to wind or gravity creates a hostile environment for the establishment of almost all kinds of plants. Sand-dune stabilization and much erosion control work have to combat this problem. Movement of the soil or substrate surface is damaging in four main ways: (i) movement away from the site exposes roots and ultimately plants lose their anchorage; (ii) at some other place deposition must occur and plants are buried; (iii) movement of soil and sand particles is abrasive to any plants in their path; and (iv) retention of organic matter and the beginnings of topsoil development, with all its benefits, are impossible.

Land-use practices

Extremes of climate and soil make many soils inhospitable for tree-planting, but as is stressed in Chapter 21, such difficulties are made worse by some land-use and husbandry practices. All the pressures of over-grazing, burning, indiscriminate soil cultivation, tree cutting, and excessive firewood and fodder collection in arid areas or on steep slopes have detrimental effects on tree-planting: (i) protective benefits of existing vegetation is further reduced, and the soil rendered less stable; these trends are opposite to those protective afforestation seeks to achieve and they make successful tree-planting more difficult; (ii) planted trees are likely to suffer more damage and will need careful and rigorous protection.

Silvicultural challenges of inhospitable sites

To a varying extent the extremes of climate, soil, and damaging land-use practices, increase the difficulty of establishing trees. Tools are available to assist successful afforestation under such conditions: (i) using a species/provenances known to be adapted to the climate and on the soil concerned; (ii) modifying the site to improve surface stability, water relations, nutritional status, etc.; (iii) applying protection measures to minimize all forms of damage. Application of these general solutions to the problem of planting inhospitable sites is illustrated by considering afforestation in arid-zone areas, on steep and eroding slopes, on man-made, industrial waste ground, and on saline soils. Useful general references drawing on experience of wasteland afforestation in Asia are Khan (1987) and FAO (1989*d*).

Arid-zone afforestation

Trees are planted in arid regions to provide one or more functions.

1. Protection: to shade crops, reduce windspeed through shelterbelts and windbreaks, and stabilize moving sand.

2. Production: to provide wood and non-wood products including fodder, human food, and pharmaceutical products.
3. Rehabilitation: to use land unsuitable for agricultural crops, especially salt-affected and water-logged land, and to use municipal waste water.
4. Landscape value: to improve the aesthetics of harsh landscapes and to provide recreational opportunities.

Planting trees in arid areas has received considerable attention (Goor and Barney 1968; Kaul 1970; Felker 1986; FAO 1989*c*, 1999*b*; Baumer 1990; Grainger 1990). In addition to these general accounts, regional/country descriptions are provided for each of Latin America, Asia, and Africa in Felker (1986), for dry tropical Africa (Delwaulle 1978, 1979), southern Africa (Bands and Britton 1977; Esterhuyse 1989), Pakistan (Sheikh 1986) and Australia (Hall *et al*. 1972). Numerous reports deal with specific localities or technical issues and sociological experience. In some arid areas tree-planting on private land has been stimulated by the depletion of trees on common property areas and the potential for economic gain, for example, in the arid lands of Rajasthan, India (Jodha 1997).

Major problems of planting trees in arid areas are: (i) low and erratic supply of moisture and low relative humidities; (ii) desiccating wind, often with abrasive dust or sand; (iii) extremes of temperature; (iv) moving substrates, especially on sand-dunes; (v) maintaining adequate protection particularly in areas where supplies of animal fodder and firewood are limited.

Most arid lands have poor soils, lack of water, a harsh climate and population pressure. These inhospitable lands are commonly inhabited by nomadic or semi-nomadic people, have a poor economic base and little political commitment to improve them. Hence, it is critical to address the unique ecological, social, economic, and political constraints in any plan to afforest them (El-Lakany 1995). In most instances trees in arid lands will be planted in some form of agroforestry system.

Species for planting
If irrigation is not intended, trees must be drought resistant and either unpalatable as browse or have good powers of recovery. The vegetation already surviving in the region is an important source of potentially useful species, for example, *Tamarindus indica* in northern Kenya, *Erythrina senegalensis* in much of the Sahel, and *Prosopis cineraria* in the Middle East and Pakistan. Nevertheless, it is the genus *Acacia* that has proved the most widely

useful for planting in dry areas, and several species grow well in plantation including *A. arabica, A. nilotica, A. senegal, A. seyal,* and *A. tortilis. Faidherbia albida,* though not suited to the very driest areas, is especially valuable because it bears leaves during the long dry season and sheds them at the beginning of the wet, thus providing shade and fodder when most other vegetation is dried up or bare. Several species of *Prosopis* including *P. glandulosa, P. pallida,* and *P. juliflora* have been widely distributed (Pasiecnik 2001) and are important for planting in semi-arid areas in South America (Habit and Saavedra 1988), Africa (Johansson 1995), and the Indian subcontinent (Shams-ur-Rehman 1998; Tewari *et al.* 2000). Eucalypts have been tried extensively and several species (*E. camaldulensis, E. citriodora, E. microtheca,* and *E. tereticornis*) grow well in dry tropical conditions, but none thrives without irrigation where annual rainfall is less than about 400 mm. Eucalypts have the advantage of not being palatable to most livestock.

Other species which grow well in very dry conditions include *Azadirachta indica* (neem), much planted in West Africa, *Parkinsonia aculeata, Senna siamea,* and *Tamarix aphylla.* Lesser-known species with some potential include *Conocarpus lancifolius,* especially if there is water fairly near the surface, and several species of *Cassia, Mimosa,* and *Zizyphus.* Many Australian species, especially *Acacia* spp., have been used in the dry tropics and others have considerable potential (Thomson *et al.* 1994; Doran and Turnbull 1997). Other important references to species suitable for arid areas were noted in Chapter 8 and include Burley *et al.* (1986) and von Carlowitz (1986). *Leucaena leucocephala,* widely planted for firewood and fodder in agroforestry projects, has shown somewhat greater tolerance of dry conditions than expected and some seed origins and related species can be considered suitable for marginally semi-arid, for example, Brazil's semi-arid northeast (Lima 1986; Hughes 1989).

Container-grown plants are almost essential in arid areas since the ball of soil attached to the roots provides some moisture and nutrients which greatly aids initial survival (see also Chapters 10 and 12).

Tree growing can become feasible and the choice of species greater where irrigation augments moisture supply. Large-scale use of waste water for irrigating tree plantations is still relatively limited and is usually practised more to dispose of waste than for improving forestry production (Braatz and Kandiah 1996).

Large-scale trials of irrigated fuelwood plantations have been investigated for energy supply to Niamey, Niger, and pulp and paper mill effluent used successfully in India for irrigating plantations of *Eucalyptus camaldulensis, Acacia auriculiformis,* and *Leucaena leucocephala* (Neelay and Dhondiyal 1985). In Pakistan 230 000 ha of plantations are irrigated (Sheikh 1990). The oldest are at Changa Manga, where annual rainfall is about 250 mm and *Dalbergia sissoo* (Fig. 3.2). *Morus alba,* and poplar (*Populus* spp.) (Fig. 23.1) have been grown under irrigation. Brackish water is used successfully in some places in establishing trees on coastal sands but sea water is harmful. Ahmed *et al.* (1985) concluded that diluted underground saline water could be used for irrigating *Melia azederach* and *Azadirachta indica* in sandy deserts, though it was less satisfactory than freshwater. The chances of using brackish water for irrigation of tree crops will have a better chance of success on sandy/loamy textured soils because of the danger

Figure 23.1 *Populus deltoides* × *P. nigra,* clone I 214, grown using irrigation at Changa Manga, Pakistan.

of excessive salt accumulation in clayey soils. El-Lakany (1995) has warned that successful arid land tree-planting depends on serious consideration of species selection and that lists of so-called 'drought tolerant and resistant species' are often misleading unless the trees are tested in a specific land-use system.

Site preparation

Two objectives are uppermost in site preparation in arid areas: (i) to maximize the benefit of all available moisture—to trap all rain and dew that falls or condenses on the ground and to concentrate it around the planted trees; and (ii) where necessary to stabilize the soil surface, especially to fix wind-blown sand-dunes.

Water conservation

Water harvesting techniques have been used for runoff agriculture for thousands of years in dry areas of the Middle East and southwest United States (Shah 1992). The basic principle is to catch rainfall from a wide area and direct it to the planting position. Water losses from run-off are eliminated as far as possible and surplus water is led to or retained around planted trees. The extra water improves seedling survival and growth (Sheikh et al. 1984) but the effect of water conservation treatments can vary with species (Table 23.1).

Rain water is harvested in various ways, but three systems are common:

1. Rain can be 'harvested' from a hillside by collecting surface runoff, using cut-off drains across the lower slope and leading the water to the plantation site, usually located at the bottom. This is sometimes called 'water spreading' and, in Pakistan for example, runoff may be encouraged by covering the surface with mud and wheat straw (Sheikh 1986).

2. Sloping ground may be trenched along the contour (Fig. 23.2(d)) which conserves both water and soil, or prepared in a 'fishbone' pattern of ridges and troughs (Fig. 23.3). This pattern significantly increases soil moisture content around plants for up to 6 months after the end of the rains (Grewal et al. 1987). On sloping sites in Niger, 500–600 trenches ha^{-1} each 3 m long × 60 cm × 60 cm are prepared along the contour and the trees planted in them (Carucci 2000).

3. The 'micro-catchment' idea in Fig. 23.3 can be extended to flat sites by constructing a low ridge enclosing several square metres of ground with the surface sloping gently to one point, usually a corner, where water gathers. The tree is planted at the lowest point (Fig. 23.2(a)). Planting in the centre of a saucer-shaped pits or in trenches between mounds where water is directed to the base of the trees has proved effective in improving survival and early growth in arid areas of India (Gupta 1995; Gupta et al. 2000). As the trees grow larger the micro-catchments may provide insufficient moisture to maintain a high growth rate and in situ water harvesting may only be beneficial in the early years after establishment, for example, in the initial 4 years for Albizia lebbeck (Gupta et al. 2000).

In addition to these general approaches, practices which conserve moisture, such as deep or open pit sunken planting (Fig. 23.2(b)), where the tree is planted in depressions below the average ground surface and/or application of mulches have enhanced survival and initial growth (e.g. Gupta and Muthana 1986; Siddiqui and Noor 1993). In gullied areas, a series of small check-dams can be constructed in gullies to create planting sites behind the dams and to improve water soakage.

Table 23.1 Survival percentage (%) and height growth (m) of *Eucalyptus camaldulensis*, *Acacia nilotica* and *A. modesta* at 8 years old with and without water conservation at Kharian, Pakistan (modified from Raza-ul-Haq and Chaudhry 2001)

Water conservation technique	Eucalyptus camaldulensis		Acacia nilotica		Acacia modesta	
	%	m	%	m	%	m
Contour trench	92	8.3	73	3.6	58	1.5
Microcatchment	93	8.0	82	5.0	52	1.6
Gradoni	92	8.5	87	4.2	62	1.6
Simple pit (control)	72	4.6	80	3.3	53	1.6

Figure 23.2 Moisture conservation practices: (a) lower corner of a 15 × 15 m micro-catchment, and (b) sunken pit with recently planted *Prosopis juliflora* in south Turkana, Kenya (annual rainfall 200–300 mm); (c) 'Gradoni'—narrow bench terraces—near Suliemanya, northern Iraq (annual rainfall 350–700 mm); (d) contour ditching and walls (bunds) mainly for soil erosion control but also to conserve water next to trees planted along the bank or beside wall—Bilate project, Ethiopia (annual rainfall 400–650 mm).

Irrigation

Irrigated forestry plantations have been established since 1864 in the Indus plains of Pakistan and from the 1930s in the Gezira area of the Sudan in Africa (Armitage 1985). Irrigation can increase tree crop productivity considerably and there are major differences in response between species. In the Burra Irrigation and Settlement Project, Kenya, the mean annual increment (MAI) of irrigated *Prosopis* spp. was up to 50 times greater than surrounding natural bushland and greater than any other species tested (Johansson 1995). Watering by hand is usually impractical except for small plantings and drip irrigation from piped water is often uneconomic. Flood irrigation is more frequently used, especially in Asia.

Traditional flood irrigation, the method used in Figs 3.2 and 23.1, involves leading water on to the surface of the plantation at 2–3-week intervals during the dry season. Plantations must be flat with low ridges around the compartment boundary to contain the water. Generally, irrigation during a year raises the total water received by a site to about 800–1000 mm of rain equivalent.

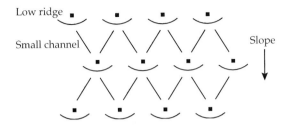

Figure 23.3 'Fishbone' pattern of shallow channels and low ridges to conserve water around planted trees.

At Changa Manga in the Punjab of Pakistan 12 irrigations are carried out during the six driest months and the plantation flooded to a depth of 6–15 cm each time. However a number of plantations, such as Changa Manga, have been adversely affected by salinity and waterlogging due to seepage of irrigation water from unlined canals (Siddiqui 1994). In India, firewood plantations of *Prosopis juliflora* are commonly flood irrigated to maximize shoot growth (Pasiecnik 2001). Irrigation water can be lost through evaporation, deep drainage, and runoff so practices must be developed to reduce these losses. Polglase *et al.* (2002) have emphasized that (i) broad-scale flood irrigation is inefficient for tree plantations, (ii) there is a high demand for water to maximize tree growth, (iii) guidelines are lacking for conservative use of water and expected growth responses per unit of water applied.

Trickle or drip irrigation uses much less water than flood irrigation and runs less risk of encouraging salinization. However, it requires supplies of clean water, expertise to install the delivery system, and a power source to deliver the water unless natural pressure sources are available. It is easily controlled and water is applied only to the trees. El-Lakany (1990) reports using drip irrigation to deliver 3 L per tree daily during the first year and 21 L weekly thereafter to raise casuarinas in a desert area in Egypt with an annual rainfall of 50–80 mm.

Instead of irrigating around the newly planted tree directly, a common practice in small-scale plantings is to insert a moderately permeable container (pitcher) in the root zone and fill this with water which then gradually seeps out. Water is used efficiently and very little evaporates or runs off. It is labour intensive, but one economy is to install a large pitcher mid-point among a group of four trees. Typically, the pitcher is filled with water weekly during establishment.

Irrigating tree plantations for wood production is an increasing practice to make productive use of waste water and the nutrients it contains, especially nitrogen and phosphorus. The waste water can come from rural industries, such as canneries and pig farms, urban heavy manufacturing industries, and municipal sewage. Most effluent requires some treatment before being applied to the land. Care must be exercised to avoid inducing salinity and sodicity, reduced water infiltration, and excessive alkalinity or acidity as they can lead to negative outcomes such as groundwater pollution, soil degradation, erosion, and low productivity. In India, *Eucalyptus tereticornis*, *Populus deltoides*, and *Leucaena leucocephala* have been tested (Braatz and Kandiah 1996) and in Australia the most popular species for irrigated plantations are *E. globulus*, *E. grandis*, *E. camaldulensis*, *E. saligna*, *E. maculata*, and *Pinus radiata*. In 1999 there were more than 2500 ha of plantations being irrigated with effluent in Australia. The design and management of sustainable, effluent-irrigated plantations based on this Australian experience are described by Myers *et al.* (1999).

Sand-dune stabilization
Moving sand-dunes occur in many arid areas and also on exposed coasts. If rainfall is adequate dunes can be stabilized directly by planting grasses, bushes, or trees and then encouraging natural revegetation once the area is fully protected. The grass, *Ipomea pes-caprae*, is used in Madagascar, but is not favoured in Senegal because it competes with the trees (Andeke-Lengui and Dommergues 1983). Bushes such as *Atriplex solanum* or trees such as species of *Casuarina*, *Acacia*, *Eucalyptus*, or *Prosopis* are commonly planted. Plant cover gives a small windbreak effect and shelters the surface, gradually enriches the soil, traps sand, and binds the surface of the dune.

However, if rainfall is low, the moving surface must be stabilized before trees can be planted. Common methods are: erecting physical barriers (short fences to trap the sand), and covering the dune surface with a mulch or thatch of brushwood or palm leaves, which is pegged down, or chemical mulching with a heavy oil or latex.

Fences or pallisades are erected along a dune projecting 0.5–1.0 m above the surface. They are established either in parallel lines, usually facing the prevailing wind or as a square lattice, but with either system, fences are spaced about 5–10 m apart. Local materials, such as palm leaves or grass sheaves, are used. *Commiphora* cuttings in Somalia (Zollner 1986) and in Mauritania fences made of the flexible branches of the indigenous shrub *Leptadenia pyrotechnica* (Jensen and Hajej 2001) have been successful. These fences serve as low windbreaks and sand traps and help to exclude livestock. They stabilize the surface long enough to allow fast-growing trees, such as species of *Acacia, Balanites, Casuarina, Eucalyptus, Prosopis, Tamarix,* and *Terminalia,* and perennial grasses to become established.

The use of chemical mulches, particularly oil-based sprays, to fix dunes has been widely publicized but are only used in a few Middle-East countries, for example, Iran, Kuwait, and Libya. In Iran the mulch oil is applied from a long mounted boom towed by a tractor. After it forms a 1–3-cm-thick crust, the dune is hydra-seeded with grasses and shrubs, followed by insertion of cuttings or planting with trees when the surface is stable. However, the grass may compete successfully for available moisture and tree growth can be poor. In Iran, more than 20 000 ha of dunes and blown sands have been stabilized this way and afforested with species of *Haloxylon, Tamarix, Zizyphus, Acacia, Prosopis,* and *Eucalyptus camaldulensis* (Skoupy 1982).

Protection

'Any planted stand in arid tropical Africa not protected during the first three years of life is doomed to failure' (Delwaulle 1977*b*). There is acute pressure on all vegetation in arid areas from grazing livestock and for collection of fodder, fuel, and building materials, thus newly planted trees must be protected until they are well established. In Nigeria, the minimum protection period is to the end of the second growing season. On sand-dunes exclusion of livestock is particularly important since they damage the newly protected surface and initiate erosion as well as threatening the trees. Newly planted seedlings also need protecting against the desiccating and abrasive effects of wind and blown sand. The short fences provide this where dunes are being stabilized, but elsewhere nursing of trees may be needed. Planting a living fence to keep out animals and give some shelter, is recommended. Species suitable for arid areas need to be resistant to browsing and so are often very spiny. The following have been found satisfactory: *Acacia ataxacantha, A. nilotica, A. seyal, Balanites aegyptica, Bauhinia rufescens, Commiphora* spp., *Euphorbia balsamifera, E. kamerunica, E. poissoni, Parkinsonia aculeata, Prosopis chilensis,* and *Ziziphus* spp.

Eroding slopes

Erosion is not determined by the presence or absence of trees but by the degree of human disturbance and its effect on vegetative cover, litter and organic matter that protect the soil surface. Maintaining tree and vegetation cover, or re-establishing it by planting and excluding livestock, on mountain slopes is a major part of watershed management to protect the soil from erosion (Chapter 21). The problems confronting afforestation of catchments particularly with steep slopes, mostly arise from soil instability and a tendency to erode. The particular difficulties are: (i) high rainfall amounts, in total or in certain seasons, lead to rapid surface runoff down the slopes if unchecked and sheet erosion, gullies, and landslides occur frequently; (ii) thin or no topsoil because of past erosion; (iii) moving substrate; and (iv) continuing pressure from the previous vegetation removal by firewood collecting, grazing, cultivating new land, etc.

It is again stressed that control of soil erosion on slopes is only solved by an integrated approach involving all aspects of land-use combining, for example, careful road construction, control of grazing and burning to preserve vegetative cover, introduction of good colonizing species, terracing of fields and other contour-aligned practices, careful farming to avoid over-use and overstocking of ground, maintenance of some perennial crops, and so on. The participation and support of the local population in the application of good land management practices is critical. This section concerns the tree-planting component and the works needed to aid establishment. For a general account of slope treatment measures and practices, particularly for watershed management, refer to Balci *et al.* (1988).

Direct control of erosion

Planting trees on eroding slopes without other erosion control measures is a slow or impossible task and unlikely to control erosion. Young trees, before they are properly established, confer little protection, suffer from movement of soil around their roots and remain prone to landslides and gully washouts. Some physical checks or barriers must be used to halt, at least temporarily, mass soil movement down a slope to allow trees to become established. Generally, effective erosion control requires both engineering and vegetation improvement works, such as tree-planting, together. An assessment of rehabilitation measures in the denuded catchment of the Swat river in Pakistan concluded that stone check dams and plantations of *Pinus roxburghii* mixed with broadleaved tree species and were more effective at reducing surface runoff and soil erosion than plantations without engineering structures

(Arifeen and Chaudhri 1997). Control of soil movement is aided by reducing runoff of surface water, especially torrents that occur frequently in mountain regions. Not only does water cause erosion directly but as soil becomes wet its cohesion is reduced and the chance of landslip increases.

Direct measures to control gully erosion include erection of barriers to slow water movement and trap soil and silt, and construction of channels and walls to divert water from areas most at risk. Barriers include check dams, made of stones, logs, or short fences, retaining walls around terraced fields, and silt traps.

Control of sheet erosion down a slope is traditionally done by terracing, contour trenches, or furrows, and low retaining walls (bunds; Fig. 23.4(a)), all of which slow downward movement of water. Figure 23.2(d) illustrates such work in Ethiopia. The tree-planting which followed was not overall but confined to along the trenches/walls to reinforce their erosion prevention role.

Tree-planting

The criteria influencing choice of species for erosion control were listed in Chapter 21. Trees are rarely planted over a whole catchment, even if this is desirable, since the needs of other land-users must be taken into account (Fig. 23.4(b)). Strategic tree-planting integrated with other land-uses is the best compromise, for example, Fig. 21.6. In Rwanda, for example, hilltops are planted with eucalypts and contour bands of *Cedrela* and *Grevillea* species are planted at intervals down the slopes (Fig. 23.5). Legumes and grasses are also established among the bands of trees to stop surface runoff. Between the bands the ground is farmed. In Mindanao, Philippines, land husbandry has been taken even further, and very steep slopes brought into productive use without risk of erosion by very careful attention to conservation measures incorporating contour hedgerows and a range of trees and shrubs (Sloping Agricultural Land Technologies—SALT) (Partap and Watson 1994; Palmer 1999). Figure 23.6 illustrates practices now widely used. SALT is a cropping system involving many component technologies which can be modified and adapted to meet individual farmers' conditions. Hence the generic SALT technologies have been modified to suit local conditions and needs, and illustrate the importance of a flexible and participatory approach in introducing new soil conservation technologies to farmers (Cramb 2000).

Maintaining tree and vegetative cover or undertaking new planting should be concentrated on areas of high erosion risk. Complete cover is desirable on hilltops, upper slopes, within and along the fringes of gullies and slopes down to watercourses. Any unused ground should also be planted and/or

(a)

(b)

Figure 23.4 Erosion control works. (a) Bund (wall) and ditch to stop surface erosion (Ethiopia)—note small cross-dams in ditch to prevent water flow in case contour levelling was not exact. (b) Valley in Pakistan with a riverbed full of erosion debris, but adjacent land now being well husbanded to control erosion by terracing, and tree-planting along field boundaries and in clumps next to the river.

Figure 23.5 Capping of a hilltop in Rwanda with eucalypts.

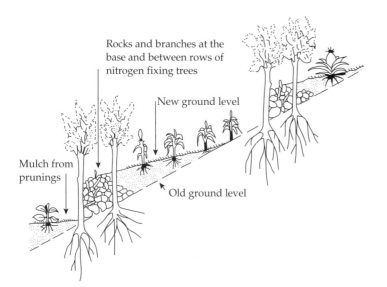

Figure 23.6 Construction of a 'green' terrace. Plant double rows of nitrogen-fixing trees along the contour and infill them at the base with 'waste' materials-rocks, branches, stumps, etc. not needed elsewhere. As crop cultivation continues over several years, the addition of mulch and other soil conservation measures gradually builds up a green terrace.

Source: MBRLC (1988).

simply protected from grazing to increase overall vegetation cover and provide additional supplies of firewood or fodder. Planting lines of trees along the contour or between individual holdings is a useful addition so that each land-user has some trees to care for.

On steep slopes planting sites are generally prepared by hand to avoid damage to the soil by vehicles and the need to construct roads, although in some situations a tractor and plough may be appropriate and less expensive (Sheikh *et al.* 1984). Planting follows along the contour, usually setting trees into small pits cut as a step into the slope. Sometimes this step is made continuous, like a very narrow (1.2 m wide) inward sloping bench terrace running along the contour, and is called a 'gradoni' (from Italian usage) (Fig. 23.2(c)). Where food crops are grown, the same effect is achieved by low walls, banks, or rows of trees against which soil accumulates, that is instead of cutting into the slope there is a building up of a step (Fig. 23.6). Large, healthy, container-grown plants help early survival and easy recognition of where trees are planted. However, in remote locations the burden of carrying many large bulky plants mostly prevents their use and smaller ones have to be used. Trees should be planted early in the wet season and kept well weeded but it is important to restrict weeding to around the tree and not to kill off all weeds on the site since the weed cover itself greatly helps to control erosion. Complete cultivation and weeding, as has often been employed in Chinese afforestation projects, leads to much erosion in the early years, especially around footpaths, because there is no surface ground cover. In Ethiopia, at the site in Fig. 23.3(d), so rapid was natural revegetation and woody regrowth when livestock grazing was excluded, that erosion control was largely effected before the planted trees, destined primarily for firewood, poles and fodder, were fully established; Fig. 7.1 shows the result.

In establishing trees the objective is to achieve rapid colonization and cover of the site by good growth in the first year. This is important for several reasons: (i) rapid cover is valuable since erosion is controlled sooner; (ii) larger trees are less susceptible to damage from fire and grazing; (iii) continual activity and attention to the crop to secure rapid establishment emphasizes the importance of the work to local people; and (iv) successful tree establishment demonstrates what can be achieved and the reluctance of land-users to undertake such work will be lessened if results are soon evident. Correct species choice, careful planting and tending, and full protection during the establishment phase, especially exclusion of grazing, are essential for successful protective afforestation.

Finally it should be noted that industrial tree plantations, not in any way associated with protective afforestation, can sometimes lead to erosion problems. Soil under plantations which are intensively managed with all understorey vegetation eliminated, or where the litter is harvested for fuel or organic fertilizer, may have erosion problems. The example of erosion under teak was mentioned in chapter 21 (see also Fig. 21.3). An example of the care that should be taken to protect watersheds, minimize soil losses, and maintain productivity when developing industrial plantations in the tropics has been described for *Pinus* plantations in Queensland (see Fig. 23.7). Field survey is used to describe site drainage, erodibility, and slope characteristics to determine site preparation classes and finally to design runoff control systems, such as diversion drains and cultivated strips (Foster and Costantini 1991*a,b,c*). In the establishment of *Araucaria cunninghamii* plantations grasses and cereals are sown in the inter-rows orientated across the slope to reduce soil erosion (Constantini 1989).

Man-made industrial wastes

Mined out land and associated overburden dumps, though small in total area in the tropics, can locally be an important rehabilitation problem since they are often close to highly populated areas. On the Jos plateau, one of the most densely populated parts of northern Nigeria, land for cultivation is at a premium, firewood and building poles are in much demand, and the pressure for grazing is very heavy, consequently revegetation of the 27 000 ha of land disturbed by opencast tin mining was a priority. In Malaysia, tin tailings comprised of slime or sand cover about 100 000 ha, most remain unproductive and have potential for conversion to timber production (Ang 2001). Natural colonization on most mining areas is much too slow to prevent erosion and silting of streams with all the attendant environmental damage, so seeding or planting must be carried out.

Figure 23.7 Strips of uncultivated ground (absorption strips) to help check surface erosion (Kuranda Forest, Queensland).

The silvicultural problems of afforesting industrial wasteland mostly arise from the substrate into which the trees are planted.

1. Usually there is no topsoil and subsoil, only coarse unweathered material of poor physical characteristics and very low in organic matter. It is often highly erodible, easily compacted, poorly or excessively drained, and may contain large boulders mixed with the finer material.
2. Substrate surface is often moving having been dumped or reshaped to slopes greater than the natural angle of repose, while its complete exposure to wind and rain renders it prone to sheet and gully erosion.
3. Substrate is usually extremely infertile. Organic matter and nitrogen are always deficient, phosphorus levels are often low, and pH is commonly very high or very low. Exposed substrates of open-cut nickel mines in New Caledonia are low in nitrogen, phosphorus, potassium, and calcium; may be rich in magnesium and nickel; and have a pH from 4 to 7.5 or higher (Jaffré et al. 1994).
4. Many substrates, especially washings and tailings, also have high and sometimes toxic levels of the ore being extracted.
5. Microbial activity is often poor or absent (Prasad 1991).

Afforestation
Because of the variety and specialized nature of the problems of industrial waste sites any sizeable afforestation project must be based on the results of research trials carried out on the ground to be restored. The general principles influencing the establishment of plantations on man-made ground were outlined by Knabe (1967). More detailed accounts include: afforestation of former tin-mining land in Nigeria (Wimbush 1963; Orode *et al.* 1977; Buckley 1988)—see Fig. 21.5, tree-planting trials on copper tailings in Papua New Guinea (Hartley 1977) and afforestation of mined areas and overburdens of dolomite, iron-ore, bauxite, coal, and lignite in India (Prasad 1991).

It should be noted that if topsoil has been retained or is available for spreading over the surface to a depth of at least 20 cm then revegetation becomes relatively easy. When industrial projects are planned removal and storage of soil from the land to be disturbed should always be provided for in the development, so that it is available later for the restoration or landscaping works. Direct seeding is a viable option on many sites where topsoil has been retained and spread. In the past topsoil was rarely kept and in

the discussion below mainly refers to situations where no topsoil is assumed available for the site restoration.

Species
The first step is to determine which species will grow on the substrate, and invariably pilot trials are needed. Pines, eucalypts, acacias, and casuarinas have been found suitable, and two generalizations can be made: (i) only species which are naturally pioneers of impoverished sites merit inclusion in a trial; and (ii) nitrogen-fixing species are particularly valuable. For example, Alexander (1986) found *Faidherbia albida* to be a better soil-improver on tin-mined land at Jos than either *Pinus caribaea* or *Eucalyptus torelliana* and in India *Acacia auriculiformis* was a better soil improver than *E. camaldulensis* and *Grevillea pteridifolia* on bauxite mining sites (Prasad 1991). Other examples include: *Casuarina equisetifolia* that effectively colonized tin tailings in Thailand (Thaiusta 1990) and has been effective when direct sown on bauxite mining sites in northern Australia (Dahl 1996). Prasad and Chadhar (1987) found *Acacia campylacantha* and *A. auriculiformis* superior to other species on dolomite mine overburden in Madhya Pradesh, India. In New Caledonia local nitrogen-fixing species, *Acacia spirobis*, and *Casuarina collina* grew well on a range of nickel mining sites (Jaffré *et al.* 1994) but direct sowing through hydroseeding seems to be the best method of revegetating the sites at a lower cost (Sarrailh and Ayrault 2001). Species that are not nitrogen-fixing but tolerate low levels of phosphorus, for example, *Grevillea pteridifolia*, have also grown well. Evaluation is also required for other plants, such as grasses and herbaceous legumes, for revegetation. Reconstitution of the original vegetation on mine sites is discussed in Chapter 22.

Site preparation
Because of compaction problems on many sites deep tine ploughing or ripping is essential. Heavy mechanical equipment and large ploughs are needed but they are normally already available on site as part of the machinery needed in mining operations.

Addition of nitrogenous or compound fertilizer and/or organic matter at the planting spot is usually necessary. Trees and other plants established on the site need to be checked regularly for signs of toxicity or deficiency. Weed growth is rarely a problem; if it is, the need to afforest is much less urgent.

Planting

Seedlings should not be too large else they tend to sway and loosen in the planting hole. Man-made waste sites are almost always bare and exposed, and top heavy plants are unsuitable. As with other plantings in protective afforestation survival of trees is helped by use of container grown stock. *Eucalyptus pilularis* in larger than average containers survived best on surface mined dunes in Queensland (Garthe 1983). In many instances the inoculation of seedlings of nitrogen-fixing legumes with *Rhizobium, Casuarina* spp. with *Frankia* and a range of species with selected mycorrhizal fungi will be beneficial for initial growth.

Protection

As with other kinds of inhospitable sites protection is important, but exclusion of grazing animals is essential because usually the planted trees are the only vegetation present and therefore will be browsed if animals gain access.

Saline and sodic soils

As noted previously in this chapter, salt-affected soils are a significant problem in many parts of the drier tropics. Estimates of the area of land affected by salinity were reviewed by Ghassemi *et al.* (1995) They concluded that the dynamic nature of the problem and inconsistencies in data from various sources make it difficult to estimate the area affected. One estimate that illustrates the scale of the problem is that about 74 million ha of land is salt-affected due to deforestation and irrigation, with irrigation responsible for 43 million ha in arid and semi-arid regions (Dregne *et al.* 1991).

Such soils often have a white, surface crust, are alkaline, and may be waterlogged. In Pakistan about 6.5 million ha of agricultural land is adversely affected by soil salinity and waterlogging (Marcar *et al.* 1998), in India at least 7 million ha are severely affected (Shrivastava *et al.* 1988), and in Australia there are about 4 million ha of salt-affected soils. These soils inhibit the growth of most plants due to the presence of soluble salts (saline soils), a high level of exchangeable sodium (sodic or alkaline soils) or both (saline-sodic soils). Several options are available to revegetate these soils depending on the how severely they are affected by salt and waterlogging. Growth of many tree species is poor or impossible, but such land need not be neglected. It provides opportunities for industrial plantations and for farm and community forestry, provided salt-tolerant species are used, appropriate silvicultural practices employed, and the land is only slightly to moderately salt-affected (Fig. 23.8). Apart from planting trees for wood or fodder-production they may be used to provide other benefits such as lowering saline water tables to improve agricultural productivity and using saline water that would otherwise have polluted rivers or seeped from irrigation channels on to agricultural land. A comprehensive annotated bibliography on the productive use and rehabilitation of salt-affected lands with trees, shrubs, and grasses has been prepared by Marcar *et al.* (1999).

Poor survival and growth of seedlings planted on saline sites is usually the result of one or more of

Figure 23.8 *Acacia ampliceps* planted on mounds on a moderately salt-affected site near Korat, northeast Thailand.

the following: (i) dehydration due to the high osmotic potential of the soil solution, (ii) nutritional imbalances resulting from high pH, and (iii) too much water (waterlogging) or water deficiency (drought).

Some species, known as 'halophytes', occurring on salt-affected sites have evolved salt tolerating mechanisms. Species of *Tamarix, Prosopis*, and *Atriplex* are examples of woody halophytes and their growth is stimulated by low salt concentrations but reduced at higher concentrations. Most trees of commercial interest are non-halophytes but some show a degree of salt tolerance. Some provenances of *Eucalyptus camaldulensis* have shown promise and in Pakistan trials suggest that this species can produce a MAI of 10–15 $m^3\,ha^{-1}\,yr^{-1}$ for a 5–7 year rotation on a saline-sodic soil of moderate salinity when irrigated with good quality water. Evaluations of species for tolerance to salinity, sodicity, and waterlogging have been summarized by Marcar and Khanna (1997).

Site preparation and cultural operations that can improve tree survival and growth include:

1. Manual pit digging or mechanical auguring of planting holes, and mechanical ripping to break up clay layers or calcium carbonate hardpans on sodic soils.
2. Mounding of areas subject to waterlogging, for example, using a double ridge mound or ridge-trench method (Tomar 1997).
3. Water harvesting or irrigation to relieve drought stress and to wash away accumulated salts. Micro-catchments and other techniques are described under 'water conservation' in this chapter.
4. Mulching to reduce evaporation and salt accumulation at or near the soil surface, for example, Ansari *et al.* (2001).
5. Fertilizing and adding soil amendments such as farmyard manure and gypsum, for example, Singh *et al.* (1989), Ahmed (1991), Marcar and Khanna (1997).

Guidelines for establishing tree on salt-affected land in Australia including ripping, mounding, mulching, weed control, and chemical soil amendments are given by Marcar *et al.* (1995). Tomar (1997) has described Indian experience of afforesting salt-affected soils and Ahmed (1991) provides details of an agroforestry system using *Prosopis juliflora* on alkaline wastelands on the Indo-Gangetic plain. When tree cover is established it ameliorates the inhospitable conditions by reducing soil pH and exchangeable sodium, and increasing organic matter; see an Indian case study by Khanduja *et al.* (1987).

Mangroves

Perhaps the most extreme salinity and water-logging conditions occur in some mangrove communities in the intertidal zone on the coast or in river estuaries. Saenger (2002) has provided an excellent summary of mangrove ecology, silvicultural and restoration techniques, and management requirements. Globally about 80 species in several genera are regarded as mangroves. They occur on about 65–75% of the tropical coastlines and have a total area of about 18 million ha (Aksornkoae 1995). Mangrove species are well-adapted to their saline, waterlogged environment and some produce root adaptations, pneumatophores, which facilitate gaseous exchange, especially oxygen. Many mangrove communities have been destroyed or severely degraded by human activities including urbanization, clearing to provide land for fish farms, conversion to salt pans, cutting for timber, poles, charcoal, and pulpwood for rayon production.

Mangrove forests are generally regenerated naturally (FAO 1994) but planting is the only option where they have been completely cleared. Examples of mangrove reforestation activities in Australia, Asia, the Middle East, and Central and South America are given in Field (1996) (Fig. 23.9(a)). In Bangladesh well over 100 000 ha have been reforested for coastal protection, land reclamation and wood production. Nursery and planting techniques have been developed for *Sonneratia apetala* and *Avicennia officinalis* and these two species dominate the monospecific plantations (Fig. 23.9(b)). The results in Bangladesh demonstrate the technical feasibility of large-scale mangrove afforestation and the potential for significant social and economic benefits (Siddiqui and Khan 1996). Many mangrove forests were destroyed by herbicides and defoliants during the Vietnam war and over 50 000 ha have been replanted in the Mekong Delta, mainly with *Rhizophora apiculata* (Hong 1995, 2001).

Principles for the design and execution of mangrove restoration projects are given by Stubbs and Saenger (2002). General guidelines for planting have been prepared by Field (1996) who suggests consideration of the following site characteristics before embarking on a plantation project:

• stability of the site
• rate of siltation
• nature of the soil
• salinity of the water
• exposure to waves and tidal currents
• depth of tidal inundation

(a)   (b)

Figure 23.9 Mangrove reforestation: (a) intertidal nursery beds of *Sonneratia apetala* seedlings for planting in Bangladesh, and (b) four-year old *Rhizophora apiculata* planted at Nusa Dua, Bali, Indonesia. (Photo P. Saenger)

- availability of fresh water (rain or runoff)
- insolation and wind exposure
- presence of pests
- availability of propagules or seeds
- cooperation from local community.

Recognition of the great value of mangroves to coastal aquatic life, especially as fish and crustacean breeding grounds, has resulted in more attention to their conservation and restoration. A case study for the Philippines is given in Chapter 22.

CHAPTER 24

Ecological factors and sustainability of plantations

In this final chapter three issues are addressed that affect planted forests to a greater or lesser extent. They are specifically included because debate and sometimes controversy surround them, namely, biodiversity in plantations compared with natural forest formations, the eucalypt controversy, and plantation sustainability.

Biodiversity in planted forests

Any casual examination of blocks of planted trees consisting of one species, and often of one age, compared with natural forests of many species and ages shows they are less rich in species. Obviously, and virtually by definition, this is true of tree species, but also applies to associated flora and fauna. This has led to claims that plantations are biological deserts.

Plantations and natural forest compared

Plantation ecology is usefully outlined by contrasting the differences between most natural ecosystems and most kinds of plantations. In order to emphasize the contrast, tropical rain forest is compared with industrial plantations, since savanna, dry woodland types, or arid-zone ecosystems mostly differ from plantations to a lesser extent. There are exceptions to all the generalizations given below, and some are noted.

Species composition
Plantations consist of few species, often only one or two (Fig. 24.1(a)). Most mature rain forest consists of many tree species and 70–100 ha^{-1} are common (Fig. 24.1(b)).

However, ecologically even more important than the total number of species is the frequency with which each species occurs. Floristic diversity (biodiversity) is much greater in a stand comprising

20 species with 10 trees of each than one with 181 of one species and only single representatives of the other 19. In most rain forest it is rare for one species of tree to make up more than 10% of the total, whereas in a planted forest all trees in a stand are usually only one species.

A second important difference is that a plantation may consist of exotic species while natural forest is obviously made up of native ones. See Zobel *et al.* (1987) for a detailed analysis of issues surrounding exotic forest.

However, though plantations are typically of one species this characteristic is not exclusively artificial and unnatural. Even within untouched rain forest, stands consisting of more than 80% of one species are found. Examples of this degree of dominance occur with *Anisoptera polyandra* in the lowlands of Papua New Guinea, and *Agathis borneensis* and *Shorea albida* in Sarawak (Brünig *et al.* 1978). Also in savannas and more arid regions one or two species tend to dominate, for example, *Acacia tortilis* in much of eastern Africa.

Age- and size-class distributions
In a plantation trees are of the same age and show relatively small variation in size. In most natural forest all ages and all sizes of trees are represented. However, as with species composition, uniformity in age and size is not exclusively artificial. Even-aged stands do occur naturally following catastrophic disturbances such as fire, flooding, hurricane damage, or volcanic eruption and as indicated by Mayhew and Newton (1998) are a common feature of forest dominated by mahogany (*Swietenia macrophylla*).

Stand profile and structure
A plantation consists essentially of one layer, the canopy of trees which make it up, though some ground flora may be present and several layers are often present in agroforestry (Fig. 20.7). Rain forest usually has many layers: emergent trees, upper canopy trees, lower canopy trees, tall shrubs,

(a)

(b)

Figure 24.1 (a) *Pinus patula* monoculture in Swaziland. The whole ecosystem is dominated by this one species, no other higher plant is present in this picture. (b) Logging clearing in primary tropical rain forest showing the variety of species, canopy layers, and other features of great biodiversity that is being lost (Sabah, Malaysia).

bushes, ground vegetation, and plants such as lianes and epiphytes which add further diversity to the structure. In addition, the occurrence of gaps and openings enormously increases the amount of 'edge' or boundary as opposed to interior, which is of such diversity ecologically, compared with a plantation. Within the soil, monoculture crops may use less of the available rooting space than multi-specific vegetation.

Primary productivity
Primary productivity is an ecological concept that describes the rate at which the sun's energy is stored by green plants in the ecosystem in a form which can be used as food materials. The concept illustrates an important difference between planted and natural forest.

In natural forest, though the total rate of production of organic matter (gross primary productivity) may be very high, relatively little of it is stored (as wood). Most is consumed in respiration and by heterotrophic organisms such as animals, insects, fungi: the forest is in a state of equilibrium. In plantations the rate of storage of organic matter (called net community productivity) is very high, and it is designed to be so. Clear-felling is done near the time when maximum mean annual increment (MAI) of stored organic matter (wood) culminates (Fig. 18.1) to maximize yield from the site. The relative productivities of stem wood shown in Table 2.2 illustrate this.

The plantation ecosystem is similar to early successional stages in natural forest, but unlike natural communities the plantation does not grow on to reach an equilibrium state where decay and breakdown balance new growth, but is cut to maximize wood yield from the site. Plantation management removes stored organic matter from the site at regular intervals, in natural forest the system is virtually closed—all that is produced is consumed within the ecosystem. It is for this reason that plantations have a role in ameliorating the 'greenhouse' effect by fixing carbon whereas the role of untouched forest is near neutral though there is a little evidence that overall biomass levels of the world's natural forest is rising. Of course, if rain forest is cleared (Fig. 22.1(b)) and burnt (deforestation) much carbon dioxide is released.

Organic matter and nutrient cycling
This is related to primary productivity, and natural forest and plantations differ in four main ways.

1. Nutrient uptake in natural forest is relatively constant from year to year, with nutrients being rapidly recycled in the ecosystem. In plantations nutrient uptake varies with the age of the stand (Chapter 14).
2. Plantations are less efficient at trapping released nutrients from decomposing litter than natural forest which has a mass of fine roots near the surface and fuller occupation of the whole rooting zone owing to the many different species present.
3. Relatively little organic matter accumulates beneath mature rain forest, typically 5–15 t ha^{-1} which decomposes in 4–12 months (Whitmore 1990); in plantations much litter may build-up (Fig. 21.2), for example, Morris (1993*b*) reported figures ranging from 42 to 78 t ha^{-1} under 17-year-old *Pinus patula* in Swaziland and very slow decay. Egunjobi and Onweluzo (1979) found *Pinus caribaea* litter in Nigeria took 3.6 years to mineralize, whereas angiosperm litter typically took 2–6 months.
4. Plantation management may lead to significant nutrient losses from the site where trees are harvested; see later.

The above outline comparing plantations and natural rain forest has shown the relative 'poorness ecologically of the plantation environment. However, this does not mean it is ecologically dead.'

Habitat diversity

The abundance and variety of plants and animals—many trophic levels of producers (green plants) and consumers (animals, insects, fungi, etc.)—are closely related to the variety of habitats present. One kind of habitat, a stand of one species of one age, supports fewer other species than where many habitats are found together. In plantations, stands may have such a dense canopy that virtually no other plants exist (Fig. 24.1). This is typical of many younger stands of pine, some eucalypts, and teak before thinning has occurred owing to the lack of light reaching the forest floor.

However, this extreme condition only occurs for part of the life of a plantation. Both during establishment and as a stand nears maturity, more open conditions lead to a greater preponderance of other plants and hence other kinds of habitats. Indeed, unlike some intensively cultivated farm crops, such as maize or pineapples, only for about one-third of the rotation is the cultivated tree species almost the only plant present. This leads to a second point, noted in Chapter 4, that most plantation forests consist of many age-classes ranging from newly planted to mature stands. Within one stand there is often little diversity but over a forest many different habitats exist (open ground, young trees and 'weed' shrubs, closed thickets, pole-stage stands, and mature open ones) there is spatial diversity rather than full diversity on one site. The feeding and grazing of many mammals and birds change from day to day and year to year in response to this suitability of particular stands; see, for example, Odendaal and Bigalke (1979). This is an important ecological argument for seeking to achieve a normal forest containing approximately equal areas of all age-classes.

Habitat diversity is further increased in plantations by the presence of roads and tracks and their verge vegetation (edge effect), by firebreaks, fire ponds, gullies and streams, rocky areas, the deliberate leaving of natural forest, and other features causing gaps where land is left unplanted (Fig. 12.3). In a fully stocked forest these unplanted areas constitute about one-fifth of the total area. For example, in the Usutu forest, Swaziland, surveys of flora in firebreaks have revealed them to be hugely important repositories of numerous flowers, shrubs, and native trees rarely found elsewhere either in forest plantations or on farmland, whether cultivated or grazed grassland (veldt), owing to human activity (Evans and Masson 2001).

In summary, though a plantation is a much less diverse habitat than most natural forest, it is not uniform: (i) several species may be planted, in particular along fire breaks a different species may be used; (ii) a stand grows through many different stages providing different kind of habitats; (iii) in most planted forests there will be stands in many different stages of development; (iv) of necessity much land is left unplanted so providing openings, clear strips, even ponds and small lakes, all of which allow other kinds of vegetation to develop; and (v) wind and fire damage often create gaps and openings.

However, though a planted forest may be more diverse than is readily apparent, there remains the influence of management imposed factors which usually reduce diversity. Clearance of pre-existing vegetation, intensive tending operations at regular intervals, thinning out dead and dying trees, and clear-felling long before trees reach physical maturity and naturally fall, all reduce diversity. The forester aims for fast growth, maintenance of crop health, and felling before physical maturity: these factors tend to diminish biodiversity.

Richness of flora and fauna

The lack of habitat diversity in plantations, compared with most natural forest, is important not only for wildlife conservation itself but because many people in the tropics depend on wildlife as a source of protein (bushmeat). For this reason plantations have often been criticized because such monocultures do not provide room and protection for complex wildlife systems.

The relationship between habitat diversity and richness of plant and animal life (biodiversity) largely derives from fewer niches, lack of food—diversity and abundance, and lack of cover and habitats for breeding. As vegetation type becomes simpler, biodiversity generally diminishes. For example, Senanayake (1987) cites research in Sri Lanka which shows from both bird species and soil fauna, diversity progressively increased from pine monoculture, to eucalypt monoculture, to home gardens, to natural forest. Interestingly, as home garden complexity increased with up to 40 or 50 species grown for fruit, fuel, foliage, etc., it supported almost as diverse fauna as natural forest.

Despite the above generalizations not all studies have shown simply that homogenous plantations are less rich in fauna than natural forest. In Brazil, four ecosystems were surveyed for small mammal populations and the highest relative density of small mammals occurred in 31-year-old *Araucaria angustifolia* plantations and the lowest in 10-year-old *Eucalyptus saligna* plantations; the two areas of heterogenous native forest were intermediate in mammal density (Dietz *et al.* 1975). Also, in a study in Sabah (east Malaysia), although wildlife was generally less abundant in plantations of *Eucalyptus deglupta*, *Gmelina arborea*, and *Paraserianthes falcataria* compared with adjacent natural forest, it was most abundant at the edge between 6-year-old eucalypt/Gmelina plantation and secondary forest

(Duff *et al.* 1985). One other point is that though plantations may support less diverse wildlife than natural forest, the total biomass of animal life under plantations is often no less than in natural forest.

So far comparison has been between natural forest and the plantation. However, where plantations replace another simple vegetation type, such as Imperata grassland, or poor degraded savanna, or are used to rehabilitate denuded industrial waste or eroded land, they can lead to increased diversity. The plantation adds habitat complexity, for example, from grassland only to grassland plus trees, with the associated edge zone. And, in time, the grassland is shaded out and other understorey species may establish (Fig. 24.2).There is no doubt that compared with most natural forest systems in the tropics plantations are poor wildlife habitats, but compared with the huge tracts of already denuded or degraded land they will usually add diversity. Indeed, there are several examples showing that new plantations in such environments provide a refuge for animals which has led directly to their increase. Following afforestation on the Nyika and Viphya plateaux in Malawi the leopard is again found in these regions. In Venezuela, afforestation

Figure 24.2 Plantation of native *Cordia alliodora* (age 15 years) at Silanche, Ecuador, established on abandoned ranch land and now showing a rich understorey, regeneration of native species, and diverse structure.

with *Pinus caribaea* of poor savannas with little fauna has led to a great increase in the deer population and the return of the jaguar.

In the Usutu forest (Swaziland), which the first author has been able to observe regularly since 1968, the numbers and variety of animals have increased as the forest has become older and more varied in structure and today greatly exceed that of the pre-existing grassland. Though the forest is mostly monoculture (Fig. 24.1) all of the following are now commonly seen: seven species of antelope and buck, monkeys, baboons, porcupines, antbears, bushbabies, warthogs, guinea fowl, lynx, spring hares, rock rabbits, three species of mongoose, honey badger, cape fox, civet, and cerval. These animals are rare or absent in adjacent grassland. This surprising diversity does not seem to be because the plantation habitat itself provides much extra food or breeding sites, but because the large plantations provide shelter and refuge from people. In the Usutu forest hunting is forbidden, people find the monoculture relatively unattractive for recreation, and the distance of sight, hearing, and smell are all reduced; thus animals are safer and more protected. And, as populations of a species increase its predators also increase and the food chain is strengthened and wildlife enriched. In densely populated parts of the tropics the relative exclusion of man from plantation forests can make them an important refuge for many birds and animals.

Finally, in the Bilate project in Ethiopia, which the first author has also observed since 1982, the effect of the necessary exclusion of livestock from the land being planted was first rapid recovery of natural vegetation followed by return of wildlife to the area. Within 4 years, animals at the top of the food chain such as hyaena were seen. Compared with what the land was like, the biodiversity of the Bilate community forestry project is inestimably greater (Fig. 7.1).

Management practices to encourage habitat diversity

Studies show that several practices can enhance the ecological value of plantations.

1. Benchmark reserves. Where land was previously under forest, islands of natural forest should be left. Roche (1978) suggests that for rain forest their minimum area should be 200 ha. If 10% of the pre-existing vegetation is retained it is likely that about half of the naturally occurring tree species will be conserved. Duff *et al.* (1986) showed in their study of Sabah plantations the benefit to gibbons,

bears, and clouded leopards of leaving islands of natural forest as refuges, but Struhsaker *et al.* (1989) found that exotic conifer plantations actually caused dieback in adjacent rain forest in Uganda.

2. Corridors of natural vegetation. Much diversity can be added if ground is left unplanted or uncleared along streams, in gullies and beside firebreaks. These natural breaks in a plantation have useful fire protection and hydrological roles as well as enhancing habitat diversity. As a general rule in the course of plantation development, natural woodland should be left wherever it occurs. In many of the grasslands, cerrados, and savannas of the tropics such woodland is now confined to inaccessible gullies or poorly drained ground of little planting potential anyway.

3. Use of native species. Where two tree species grow equally well but one is exotic, for ecological benefit it is preferable to plant the native one; see discussion in Chapter 8.

4. Use of several species. Few substantial plantations depend entirely on one tree species and for good reasons of optimum matching of species and site and differing management objectives for a stand, several species are usually employed. This adds diversity since flowering, fruiting, shade levels, rooting habit, effect on soil, foliage palatability, and so on, all vary. For example, both Odendaal and Bigalke (1979) and Duff *et al.* (1986) report that eucalypts are unpalatable to buck and deer compared with other plantation trees. Indeed, the latter authors suggest planting eucalypts along the edge of plantations to confer some protection from browsing.

5. Several management practices, as was noted earlier, aid biodiversity within a plantation forest—unplanted ground next to water-courses and rocky areas, conserving islands of natural vegetation, having many age-classes, restricting uncontrolled hunting and protecting against fire, livestock grazing, etc. In addition silvicultural treatments such as wide spacing, restriction of weed control to spot application next to a tree, and regular thinning, maintain natural vegetation and the cover and food source it provides. Extending rotation length to the time of prolific fruiting and leaving some trees unfelled and dead wood, both standing (snags) and fallen, on a site all increase food abundance and variety and ecological niches.

This short review of biodiversity in plantations adds to the assertion in Chapter 2 that clearance of natural forest simply to provide ground for tree plantations is not only unnecessary but plainly undesirable for wildlife conservation. Conversely, sensitive afforestation of some of the huge tracts of poor grassland, and denuded or eroding ground, not only brings unproductive land back into use, will increase carbon storage, but may enhance substantially its wildlife value.

Eucalypt environmental controversy

Of all widely used plantation species in the tropics, eucalypts have attracted by far the most criticism. As pointed out in Chapter 8 some countries have restricted or even banned their planting. There are three main concerns: (i) excessive water use and depression of food crops grown nearby; (ii) suppression of ground vegetation and resulting unsuitability for soil erosion control; and (iii) generally poor wildlife value, even by plantation standards. There is some substance to each of these which merits comment: Poore and Fries (1985) examine the subject in detail, ACIAR (1992) issued a leaflet entitled 'Eucalypts: curse or cure?' and Florence (1996) devotes 12 pages to the topic in his major work 'Ecology and silviculture of eucalypt forest', but to exclude such an immensely useful genus will deny countless peoples across the tropics a valuable pole and fuel tree quite apart from industrial roles.

Water use by eucalypts

In Chapter 21 it was noted that forest stands generally use more water than other land-uses such as grassland. The concern with eucalypts is that their use is excessive. Many eucalypts are well-adapted to grow in dry conditions, that is one of their great values, but it is becoming clear that they are not 'drought evaders', minimizing leaf water-loss by early stomatal closure, but drought-tolerant and capable of transpiring even under considerable moisture stress. This is enhanced by the capacity of many eucalypts to develop extensive and deep root systems. Undoubtedly, their survival in many arid areas is enabled by access to groundwater not available to other species. The converse is that many eucalypts with plentiful water supply grow exceedingly fast with growth rates generally well correlated with actual evapotranspiration. Indeed, there are several examples of eucalypts being used to dry out marshy ground, for example, E. robusta in papyrus swamp in Uganda, and E. camaldulensis in the Pontine marshes of Rome—see also Chapter 21 (p 344).

One of the most comprehensive studies of this question has been at the Karnataka project in India, for it is in India where most concern has been expressed. In particular, it is argued that food crops grown beside eucalypts suffer severe moisture competition and show poor yields. The sociological aspect was mentioned in Chapter 6. At Karnataka, where annual rainfall is about 800 mm, evidence is accumulating that plantations of E. tereticornis and E. camaldulensis do not use more water than adjacent degraded natural forest (Harding et al. 1991), but there is no recharge below the root zone. At the sites studied there is no excessive use of water by eucalypts through the direct abstraction of ground water. High transpiration rates are observed for eucalypts during and immediately following the monsoon but the rates decrease rapidly as soil water becomes limiting (Roberts et al. 1991). Under agricultural crops water use is low and annual recharge will occur. Because of the potential vigour of the species in question, put simply, they will use all the water available, whereas agricultural crops will not. Thus, when grown beside one another the annual recharge of soil moisture for the food crop may not occur with consequent loss of yield.

Unsuitability for erosion control

It was also noted in Chapter 21 that presence of ground vegetation and litter was the key to effective control of soil erosion. As with teak, some eucalypts appear to suppress ground vegetation very effectively and are therefore poor trees to use for protective afforestation or in tree pastures. This characteristic tends to be associated with species that cast a relatively heavy shade for eucalypts with foliage displayed more horizontally than vertically—the transversaria group, and are commonly found naturally on the fringe of Australian rain forest. Of the commoner species they include E. pellita, E. urophylla, and to a lesser extent E. grandis, E. citriodora, and E. tereticornis. Such species are mostly unsuitable for soil erosion control. In contrast, the arid zone species E. camaldulensis does not exhibit this tendency at all and, indeed, rarely suppresses ground vegetation which, along with its typically narrow fastigiate crown, makes it one of the few eucalypts that can be considered for agroforestry or in erosion control if natural vegetation is absent.

Use of E. camaldulensis in agroforestry is fairly common in Thailand (Fig. 24.3) and, not unexpectedly, was found superior to E. citriodora for tree pastures in Nigeria (Igboanugo et al. 1990).

Associated wildlife

Poore and Fries (1985) and ACIAR (1992) examine the contention that eucalypt plantations are particularly poor habitats. Certainly, their small hard fruits and very tiny seeds are poor food for birds and, as noted earlier, the foliage of many species is unpalatable to deer—and, in the case of E. globulus,

Figure 24.3 Spaced rows of *Eucalyptus camaldulensis* and pasture in Thailand. Note narrow crown form.

even to goats. The lack of ground vegetation further reduces biodiversity. But it is not all one-sided: many eucalypts flower profusely and are a rich nectar source (including of value for bee pasturage), and their habit of shedding often stringy bark provides nesting material.

Provided plantation design includes the points made earlier about management to aid biodiversity, as is now practised, for example, at Aracruz Florestal, the distinction between eucalypts and other exotic plantation species is unlikely to be of great wildlife significance. Overall, eucalypts are of enormous value in the tropics. But they are no more a panacea than any other kind of tree: there are sites and situations both suited and ill-suited to their use.

Sustainability of plantations

Forest plantations are an increasingly important resource worldwide, a trend that is expected to continue strongly. This chapter, indeed this book, concludes by examining the evidence concerning the 'narrow-sense' sustainability of forest plantations. It asks the question: is growing trees in plantations a technology that can work in the long term or are there inherent flaws biologically which will eventually lead to insuperable problems for such silviculture? It is important in the tropics where rotations are short, yields often high, and opportunities for remedial inputs more uncertain.

Types of sustainability

The question of sustainability in plantation forestry has two components. There are the general or broad issues of whether using land and devoting resources to tree plantations is a sustainable activity from the economic, the environmental, or from the social sense. They can be labelled 'broad sense' sustainability.

The second component, 'narrow-sense' sustainability, is largely a biological and silvicultural issue. The question raised is: can tree plantations be grown indefinitely for rotation after rotation on the same site without serious risk to their well-being? More specifically, can their long-term productivity be assured, or will it eventually decline over time? These questions are pertinent owing to the increasing reliance on planted forests but are also scientifically challenging since in previous centuries trees and woodlands were seen as 'soil improvers' and not 'impoverishers'. Are today's silvicultural practices more damaging because of greater intensity and the high timber yields achieved, especially in the tropics, typically two to eight times that of natural forest increment? And, of course, are resources such as genetic improvement, targeted fertilizer application, and sophisticated manipulation of stand density, along with rising atmospheric carbon dioxide, likely to lead to crop yield improvement, or could they disguise evidence of genuine site degrade or increasing risk of damaging pests and diseases?

This chapter looks at evidence, mainly drawn from the tropics and sub-tropics, to address four elements of narrow-sense sustainability: a fuller analysis is in Evans (1999*b*). (a) What changes to a site may plantation forestry induce and hence threaten future rotations? (b) What risks are tree plantations exposed to? (c) What factual evidence is there for and against productivity change over time? (d) What silvicultural interventions can help sustain yields?

Site change induced by plantation forestry—the biological, physical, and chemical changes plantations may bring about

Two important questions are: (i) do the silvicultural practices commonly applied, such as exotic species, monocultures, clear-felling systems, etc., cause site change, and (ii) are such changes more or less favourable to the next crop? Does growing one crop influence the potential of its successor?

This is a much-researched topic and only the main themes are summarized. Two recent books have presented the science: Dyck *et al.* (1994) *'Impacts of forest harvesting on long-term site productivity'*, and Nambiar and Brown (1997) *'Management of soil, nutrients and water in tropical plantation forests'*. More dated but still relevant is Chijioke's (1980) review of the impacts of fast-growing species on tropical soils. However, it is important to be cautious: tree rotations are long, even in the tropics, compared with most research projects!

Assessing changes in soil
Demonstrating that soil changes may be caused by forestry practices is usually difficult to establish conclusively both in fact and in scale. An absence of sound baseline data is common and, moreover, is the reported change actually induced by plantation silviculture?

The second question is whether the observed changes represent degradation or improvement. There are remarkably few examples of changes supposedly induced by *growing* trees that lead to less favourable conditions for that species. Equally, the irreversibility of changes has rarely been demonstrated, apart from obvious physical losses such as erosion of topsoil. A gradual trend, perhaps observed over several decades, can be quickly reversed as stand conditions change. As Nambiar (1996) points out 'the most striking impacts on soils and hence productivity of successive crops occur in response to harvesting operations, site preparation, and early silviculture from planting to canopy closure'.

Most reports of site change in plantation forestry derive from matched plots. Increasingly today long-term observational experiments are being specifically designed to investigate change, for example, CIFOR's tropics-wide study (Tiarks *et al.* 1998). Modelling is widely used but suffers in precision at site level because of assumptions made.

The observational approach suffers bias in that investigation is often carried out specifically because there is a problem which has already revealed itself in poor tree growth or health. It also suffers from soils being notoriously variable, a difficulty exacerbated on many sites by the kind of ground often used for plantations. A second, little known, source of variability is that measured values of many soil parameters can change radically during 1 year.

The above points underline the danger of drawing conclusions from limited investigations covering only a few years of a rotation. Short-term studies can be grossly misleading especially when extrapolating over whole rotations and successive rotations.

Soil chemical status
Plantations may have three impacts: nutrient removal from soil as trees grow and then are harvested; changes in the chemistry of the soil surface as the litter layer and organic matter are dominated by one species and hence result in uniform composition and decay characteristics; and site preparation practices such as ploughing, drainage, and fertilizing which directly affect soil physical parameters and in turn nutrient and moisture availability.

Soil as a mineral store Soils vary enormously in their role as a nutrient reservoir—see Chapter 14. Thinking has been conditioned by arable farming that treats soils as a medium in which to grow crops where nutrient supply is largely maintained by annual fertilizer inputs; and by the fact that in most temperate soils the store of plant nutrients far exceeds that in the above-ground biomass. In forestry, where fertilizer inputs are limited and trees perennial and generally deep rooting, the focus is less exclusively on soil reserves and more on where the dynamics of nutrient supply is mediated—that is, largely at the soil surface. Indeed, forests are highly efficient recyclers of nutrients and in the tropics, where recycling can be most efficient, nutrients in mineral soil often no longer represent the dominant proportion of the ecosystem. The soil often plays only a small part in the nutrient exchange and it is the surface organic, root-bearing zone, especially the annual turnover of fine roots, which is important in concentrating energy flow from decomposing organic matter back into living organic matter. The integrity of this layer and how it is handled in plantation silviculture is critical to sustainability.

Nutrient removal Nutrient removal in plantation forestry occurs when any product is gathered or harvested. Many studies have been made; Goncalves *et al.* (1997) alone list 12 tropical examples. Critical to plantation sustainability is what proportion the nutrients lost represent of the whole store. This ratio of nutrient export : nutrient store is advocated as a key measure of long-term ecosystem stability (though it rather begs what is the store and how it can be measured?). For example, Lundgren (1978) found that *Pinus patula* plantations in Tanzania led to annual removals of 40, 4, 23, 25, and 6 kg ha^{-1} of nitrogen (N), phosphorus (P), potassium (K), calcium (Ca), and magnesium (Mg) respectively. These rates of removal are about one-third of those of maize (Sanchez 1976) and in the Tanzania study represented less than 10% of soil store, that is, a stability ratio of <0.1. In contrast Fölster and Khanna (1997) report data for *Eucalyptus urophylla* × *grandis* hybrid stands with three very different site histories at Jari in northeast Amazonia suggesting imminent impoverishment: 'most of the previously grown *Gmelina*, *Pinus*, or *Eucalyptus* had already extracted their share of base cations from the soil and left it greatly impoverished.' with an unsustainable stability ratio of >1. However, caution is needed. Others (e.g. Rennie 1955; Binns 1962; Johnson and Todd 1990) have predicted from comparison of removals in harvested biomass with available quantities in soil that calcium nutrition will be a problem; yet trees continue to grow on soil where conventional soil analysis suggests there is virtually no calcium.

Understanding these dynamics helps identify at what points on the continuum of plantation growth throughout the world of sites, species and productivities the ratio becomes critical for long-term stability. There appear to be few examples of reaching such limits. It is worth remembering that nutrient removals by forest crops are typically only one-fifth to one-tenth that of arable farming, see Miller (1995).

Litter and Residues The influence of litter on soil chemical status may be important since leaves of different species decay at different rates. For example, in southern Africa substantial accumulations may develop under *P. patula* on certain sites (see Morris 1993*b*) while this is unusual beneath the more lightly canopied *P. elliottii*. In broadleaved stands accumulation of litter is uncommon though not unknown. Even under teak and *Gmelina*, which usually suppress all other vegetation, the large leaves readily decay. Similarly under the light crowns of eucalypts and the nitrogen rich foliage of leguminous trees such as *Acacia*, *Leucaena*, and *Prosopis* spp. and non-legume N-fixers such as casuarinas litter build up is rare owing to rapid decay of the rich organic matter.

Measured changes in soil chemistry The above processes indicate that plantation forestry practice could influence soil chemical status, but what has been observed? Most studies have either compared conditions in plantation sites with those before establishment or examined trends as a plantation develops. Few have examined changes over successive rotations. Few consistent trends emerge.

In the many tropical studies both *increases* and *decreases* in pH, carbon, nitrogen, and macro-nutrients under plantations compared with natural forest or pre-existing conditions have been reported—see references in Evans (1999*b*). Recent investigations have concerned acid rain impacts, though distinguishing these from direct tree effects on soil acidity is difficult. On the whole tree impacts are relatively small compared with the soil nutrient store.

Soil physical condition
Plantation forestry may impact soil physical conditions, and hence sustainabilty through (i) site preparation and establishment operations, (ii) the effects of tree growth itself, for example, on water uptake, and (iii) harvesting practices. They are discussed in Evans (1999*c*, 2001) and comment is only made here on vegetation suppression.

Indirect impact of vegetation suppression Plantations of teak and *Gmelina* and also many conifers in both tropical and temperate regions may suppress all ground vegetation. Where this exposes soil, perhaps because litter is burnt or gathered, erosion rates increase. Under teak, Bell (1973) found soil erosion 2.5–9 times higher than under natural forest. The protective function of tree cover derives more from the layer of organic matter that accumulates on the soil surface than from interception by the canopy. In India, raindrop erosion was nine times higher under *Shorea robusta* plantations where litter had been lost through burning (Ghosh 1978). Soil erosion beneath *Paraserianthes falcataria* stands was recorded as 0.8 t ha^{-1} year^{-1} where litter and undergrowth were kept intact but an astonishing 79.8 t ha^{-1} year^{-1} where it had been removed (Ambar 1986). Wiersum (1983) found virtually no soil erosion under *Acacia auriculiformis* plantations with litter and undergrowth intact, but serious where local people gathered the litter.

Organic matter dynamics

What happens to the litter and organic matter layer at the soil surface is critical to the question of sustainability for three reasons:

- the surface litter layer helps prevent soil erosion
- litter and organic matter represent a significant nutrient store, albeit a dynamic one
- the litter : organic matter : mineral soil interface is the seat of nutrient cycling and microbial activity.

Any activity that disturbs these roles in the ecosystem can have large effects of which perhaps most serious of all, and still practised in some countries, is regular and frequent litter raking (Fig. 14.1). In commercial plantation forestry the cost of managing debris and site preparation, when restocking plantations, is expensive and is a high proportion of the establishment costs. But as Nambiar (1996) points out 'one shoddy operation can leave behind lasting problems'.

Weed spectrum and intensity

Establishment of plantations greatly affects ground vegetation with many operations designed directly or indirectly to reduce weed competition to ensure that the planted tree has sufficient access to site resources. A neglected but critical phase is managing the weed problem through crop harvesting and restocking. In subsequent rotations the weed spectrum often changes. Owing to past weed suppression, exposure of mineral soil in harvesting, and the accumulation of organic matter, conditions for weed species change. Birds and animals may introduce new weed species, grass seed may be blown into plantations and accumulate over several years, and roads and rides in plantations can become sources of weed seeds. Weed management must be an holistic operation. As with a failure to handle organic matter carefully, where yield declines have been reported, often the significance of weeds has been insufficiently recognized on restocked sites in second or third rotations.

Risks exposed to plantations

Pest and disease incidence in monocultures

A serious threat to plantations can arise from a massive build-up of a pest or disease. It has been much disputed whether monoculture itself is more susceptible to devastation from these causes. The broadly accepted ecological principle of stability dates back to the 1950s and is that the stability of a community and its constituent species is positively

related to its diversity. Following this reasoning foresters have stressed that substitution of natural forest by even-aged monoculture plantations may remove many of the natural constraints on local tree pest and pathogens and thus increase risk of attack. Some evidence supports this, see Gibson and Jones (1977), though these authors point out that increased susceptibility mostly arises from conditions in plantations rather than because only one tree species is present.

The relative susceptibility of monocultures to organic damage is complex ecologically. The influence of diversity on stability of (say) insect populations depends on what population level is deemed acceptable. Often stable, equilibrium levels are too damaging and so artificially low populations sought through control. Speight and Wainhouse (1989) stress that artificially created diversity, that is, mixed crops, does not necessarily improve ecological stability and is certainly inferior to naturally occurring diversity, complexity of organization, and structure is as important (Bruenig 1986).

It is prudent, nevertheless, to spell out why plantations are perceived to be in danger.

1. Plantations of one or two species offer an enormous food source and ideal habitat to any pest and pathogen species adapted to them.
2. Uniformity of species and closeness of trees including branch contact above ground and root lesions in the soil, allow rapid colonization and spread of infection.
3. Narrow genetic base in plantations, for example, one provenance or no genetic variation (e.g. clones) reduces the inherent variability in resistance to attack.
4. Trees grow on a site for many years and permit pest or disease to build up over time.
5. Many plantations are of introduced species and without the insect pests and pathogens that occur in their native habitat. The many natural agencies controlling pests and diseases will also be missing. Thus many argue that exotic plantations experience a period of relative freedom from organic damage, perhaps for the first one or two rotations. Zobel's *et al.* (1987) analysis of the threat to exotics concluded that evidence does not confirm that stands are more at risk, other than clonal plantations, and that problems arise mainly when species are ill-suited to a site.

Examples of devastating outbreaks of fungal disease and insect pests are listed in many publications, see, for example, Ciesla and Donnaubauer (1994) and Evans (1999c), and they illustrate the scale and potential threat pest and diseases represent. They have prevented the planting of some species, impaired the productivity of others, but overall have not caused such widespread damage as to seriously question plantation silviculture as a practice.

There remain two serious concerns: (i) Environmental change—changing climate, increasing atmospheric pollutants of CO_2 and nitrogen compounds, will add stress to established plantations while higher nitrogen inputs may increase insect pest risk and diseases problems (Lonsdale and Gibbs, 1996). (ii) New pests and diseases will emerge: (a) from new hybrids or mutations; (b) from new introductions arising from increasing global trade, for example, *Cryphonectria* canker in eucalypts in South Africa; and (c) from native pests adapting to introduced trees.

Risks associated with plantation forestry practices
Many pest and disease problems in plantations arise from the nature of forest operations, and not directly from growing one species of tree in a uniform way (monoculture). They are summarized briefly.
Harvesting and other residues. Large amounts of wood residue from felling debris and the presence of stumps are favourable for colonization by insect pests and as sources of infection. Usually modification of silviculture or application of specific protection measures, for example, dipping the roots of seedlings in insecticide to prevent damage by *Hylobius* beetles and *Hylastes* weevils, can contain such problems.
Site and species selection. Extensive planting of one species, whether indigenous or exotic, inevitably results in some areas where trees are ill-suited to the site and suffer stress. This may occur where large monospecific blocks are planted or where exotics are used extensively before sufficient experience has been gained over a whole rotation, for example, *Acacia mangium* in Malaysia and Indonesia and the discovery of widespread heart rot.
Thinning and pruning damage. Thinning and pruning can damage trees and provide infection courts for disease. Neither practice seriously threatens plantation sustainability.

Evidence of productivity change

Problems with data
For forest stands (crops) hard evidence of productivity change over successive rotations is meagre with few reliable data. The long cycles in forestry make data collection difficult. Records are rarely maintained from one rotation to the next, funding

for long-term monitoring is often a low priority, detection of small changes is difficult, and often the exact location of sample plots is poorly recorded (Evans 1984). Also, few plantations are second rotation, and even fewer third or later rotation, thus the opportunity to collect data is limited.

The few comparisons of productivity between rotations have mostly arisen because of concern over yields, namely 'second rotation decline', or stand health. Thus the focus has been on problems: the vast extent of plantations where no records are available suggest no great concern and no obvious decline problems. Thus data in the older literature may be biased to problem areas while more recent studies may be less so, such as CIFOR's 'site management and productivity in tropical forest plantations' that incorporates systematic establishment of sample plots.

Review of evidence comparing yields in successive rotations
Apart from the early concerns about sustainability in Europe—see Evans (1999b) for details, three major studies have reported productivity in successive rotations along with some anecdotal evidence and occasional one-off investigations.

Pinus radiata in Australia and New Zealand
Significant yield decline in second rotation *Pinus radiata* appeared in South Australia in the early 1960s (Keeves 1966) with an average 30% drop in most forests in the state. In the Nelson area in New Zealand, on a few impoverished ridge sites there was transitory second rotation yield decline (Whyte 1973). These reports, particularly from South Australia, were alarming and generated much research. By 1990 it was clear for South Australia that harvesting and site preparation practices which failed to conserve organic matter and an influx of weeds, especially grasses, in the second rotation were the main culprits. By rectifying these problems and using genetically superior stock second and third rotation pine now grow substantially better than the first crop (Boardman 1988; Woods 1990; Nambiar 1996). This is now the case in other states of Australia where *P. radiata* is grown.

Of direct relevance to the tropics, a careful study in Queensland of first and second rotation *P. elliottii* of the same seed origin showed no evidence of yield decline, but a 17% increase in volume per hectare at 9 years where organic matter was left undisturbed (Bevege and Simpson 1980). Also, organic matter conservation is being applied to minimize soil erosion when felling and replanting *Araucaria* and pine plantations (Constantini *et al.* 1997).

In New Zealand the limited occurrence of yield decline was mostly overcome by cultivation and use of planted stock rather than natural regeneration (Whyte, personal communication). On most sites successive rotations gain in productivity. However, Dyck and Skinner (1988) do conclude that inherently low quality sites, if managed intensively, will be susceptible to productivity decline.

Pines in Swaziland

Long-term productivity research in the Usutu forest, Swaziland began in 1968 as a direct consequence of second rotation decline reports from South Australia. For 35 years measurements have been made over three successive rotations of *Pinus patula* plantations, grown for pulpwood, from a forest-wide network of long-term productivity plots. Plots have not received favoured treatment, but simply record tree growth during each rotation resulting from normal forest operations by South African Pulp and Paper Industries (SAPPI) Usutu.

The most recent analysis appear in Evans (1996, 1999*c*), Evans and Boswell (1998) and in Evans and Masson (2001). Tables 24.1 and 24.2 (simplified from Evans 1999*c*) show second and third rotation growth data obtained from plots on exactly the same sites. First rotation data were derived from stem analysis and paired plots and are less accurate: some are reported in Evans (1996).

Tables 24.1 and 24.2 summarize results from arguably the most accurate datasets available on narrow-sense sustainability. Over most of the forest on granite derived soils (Table 24.1) third rotation height growth is significantly superior to second and volume per hectare almost so. There had been little difference between first and second rotation (Evans 1978). In a small part of the forest (about 13% of area), on phosphate-poor soils derived from slow-weathering gabbro, decline occurred between first and second rotation, but this has not continued into the third rotation where there is no significant difference between rotations (Table 24.2). Figure 24.4 illustrates the development of mean height over all three rotations for these two generalized soils types. The importance of the Swaziland data, apart from the long run of measurements, is that no ameliorative treatment has ever been applied to any long-term productivity plot. According to Morris (1987) some third rotation *P. patula* is probably genetically superior to the second rotation. However, the 1980s and especially the period 1989–92 have been particularly dry, Swaziland suffering a severe drought along with the rest of southern Africa (Morris 1993*a*, Hulme 1996). This will have adversely impacted third rotation growth. These data are also of interest because plantation silviculture practised in the Usutu forest over some 72 000 ha is intensive with pine grown in monoculture, no thinning or fertilizing, and on a rotation of 15–17 years which is close to the age of maximum MAI. Large coupes are clearfelled and all timber suitable for pulpwood extracted. Slash is left scattered (i.e. organic matter

Table 24.1 Comparison of second and third rotation *Pinus patula* on granite and gneiss derived soils at 13/14 years of age (means of 38 plots)

Rotation	Stocking (S/ha)	Mean height (m)	Mean DBH(cm)	Mean tree vol. (m³)	Volume per hectare (m³ ha⁻¹)
Second	1386	17.5	20.1	0.205	294
Third	1248	18.7	21.2	0.233	326
% change		+7.1	+5.6		+11.0

Table 24.2 Comparison of second and third rotation *Pinus patula* on gabbro dominated soils at 13/14 years of age (means of 11 plots)

Rotation	Stocking (S/ha)	Mean height (m)	Mean DBH(cm)	Mean tree vol. (m³)	Volume per hectare (m³ ha⁻¹)
Second	1213	16.7	20.0	0.206	244
Third	1097	16.8	21.7	0.227	255
% change		+0.05	+8.3		+4.6

Source: Modified from Evans (1999*c*).

(a)
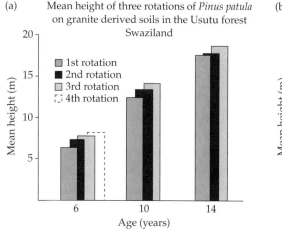

(b)
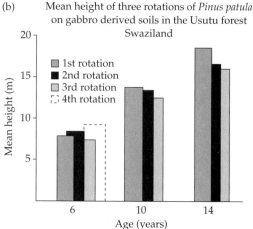

Figure 24.4 Mean height growth for three successive rotations of *Pinus patula* (a) on granite derived soils, (b) on gabbro derived soils. (Provisional data of young 4th rotation *P. Patula* shown for age 6 years)

Source: first author.

conserved) and replanting done through it at the start of the next wet season. These plantations are managed as intensively as anywhere and, so far, there is no evidence to point to declining yield. The limited genetic improvement of some of the third rotation could have disguised a small decline, but evidence is weak. Also, it can be strongly argued that without the severe and abnormal drought, growth would have been even better than it is. Overall, the evidence suggests no serious threat to narrow-sense sustainability.

A review of sustainability for wood production and wider environmental issues at the Usutu Forest in Swaziland is in Evans and Masson (2001). Assessment of fourth rotation productivity has commenced (Fig 24.4).

Chinese fir (Cunninghamia lanceolata) *in sub-tropical China*
There are about 6 million ha of Chinese fir plantations in subtropical China. Most are monocultures and are worked on short rotations to produce small poles, though foliage, bark and sometimes roots are harvested for local use. Reports of significant yield decline have a long history. Accounts by Li and Chen (1992) and Ding and Cheng (1995) report a drop in productivity between first and second rotation of about 10% and between second and third rotation up to a further 40%. Ying and Ying (1997) quote higher figures for yield decline. Chinese forest scientists attach much importance to the problem and pursue research into monoculture, allelopathy, and detailed study of soil changes. However, the widespread practices of whole tree harvesting, total removal of all organic matter from a site including

litter raking, and intensive soil cultivation that favours bamboo and grass invasion (Fig. 24.5) all contribute substantially to the problem (Evans 1999*b*). Ding and Cheng (1995) conclude that the problem is 'not Chinese fir itself, but nutrient losses and soil erosion after burning (of felling debris and slash) were primary factors responsible for the soil deterioration and yield decline . . . application of P fertilizer should be important for maintaining soil fertility, and the most important thing was to avoid slash burning . . . These (practices) . . . would even raise forest productivity of Chinese fir.'

Teak in India and Java
In the 1930s evidence emerged that replanted teak (*Tectona grandis*) crops (second rotation) were not growing well in India and Java (Griffith and Gupta 1948). Although soil erosion is widespread under teak and loss of organic matter through burning leaves is commonplace the research into the 'pure teak problem', as it was called in India, did not generally confirm a second rotation problem. However, Chacko (1995) describes site deterioration under teak as still occurring with yields from plantations below expectation and a decline of site quality with age. Four causes are adduced: poor supervision of establishment; over-intensive taungya (intercropping) cultivation; delayed planting; and poor after-care. Chundamannii (1998) similarly reports decline in site quality over time and blames poor management.

In Java, Indonesia, where there are about 600 000 ha of teak, site deterioration is a problem and 'is caused by repeated planting of teak on the same sites' (Perum Perhutani 1992).

Figure 24.5 Poor third rotation Chinese fir (foreground) owing to competition from grasses and bamboo, and damaging site preparation (far hillside) that removes all organic matter by burning and promotes soil erosion despite terracing.

Southern pines in the United States
Plantations of slash (*P. elliottii*) and loblolly (*P. taeda*) pines are extensive in the southern states. Significant plantings began in mid 1930s as natural stands were logged out (Schultz 1997) and with rotations usually 30 years or more, some restocking (second rotation) commenced in the 1970s. In general growth of the second crop is variable—see examples in Evans (1999*b*). A coordinated series of experiments in United States is assessing long-term impacts of management practices on site productivity (Powers *et al*. 1994).

Other evidence
Other evidence is limited or confounded. For example, Aracruz Florestal in Brazil has a long history of continually improving productivity of eucalypts owing to an imaginative and dedicated tree breeding programme so that regularly new clones are introduced and less productive ones discontinued (Campinhos and Ikemori 1988). The same is true of the eucalypt plantations at Pointe Noire, Congo (P. Vigneron personal communication). Thus recorded yields may reflect genetic improvement and disguise any site degrade.

In India one recent report (Das and Rao 1999) claims massive yield decline in second rotation clonal eucalypt plantations which the authors attribute to very poor silviculture.

At Jari in the Amazon basin of Brazil silvicultural practices have evolved with successive rotations since the first plantings between 1968 and 1982. A review of growth data from the early 1970s to present day suggest that productivity is increasing over successive rotations due to silvicultural inputs and genetic improvement (McNabb and Wadouski 1999).

In Venezuela, despite severe and damaging forest clearance practices, second rotation *Pinus caribaea* shows much better early growth than the first rotation (Longart and Gonzalez 1993).

Within-rotation yield class/site quality drift

Inaccuracy in predicted yield
For long rotation (>20 years) crops it is usual to estimate yield potential from an early assessment of growth rate to identify the site quality or yield class. A change from predicted to final yield can readily occur where a crop has suffered check in the establishment phase or fertilizer application corrects a specific deficiency. However, there is some evidence for very long rotation (>40 years) crops in temperate countries that initial prediction of yield or quality class underestimates final out-turn, that is, crops grow better in later life than expected. Either the yield models used are now inappropriate or growing conditions are 'improving'. Across Europe the latter appears to be the case (Spiecker *et al*. 1996; Cannell *et al*. 1998) and is attributed to rises in atmospheric CO_2 and nitrogen input in rainfall, better planting stock, and cessation of harmful practices such as litter raking.

However, as noted, the opposite is occurring with teak in India. High initial site quality estimates do not yield the expected out-turn and figures are revised downward as the crops get older.

Relation of quality (yield) class with time of planting
Closely related to the above is the observation that date of planting is often positively related to productivity, that is, more recent crops are more

productive than older ones regardless of inherent site fertility. This shift is measurable and can be dramatic, see example from Australia in Nambiar (1998). Several UK and European examples are cited in Evans (1999*b*).

The impact of these two related observations is that present forecasts of plantation yields are likely to be underestimates; yields generally appear to be increasing.

Interventions to sustain yield

The steady transition from exploitation and management of natural forest to increasing dependence on plantation forestry is following the path of agriculture. Many of the same biological means to enhance yield are available. They are outlined here, having been covered more fully in earlier chapters, to emphasize the opportunities available.

Genetic improvement

The forester only has one opportunity per rotation to change the chosen crop. Change in species, seed origin, use of new clones, use of genetically improved seed and, possibly in the future, genetically modified trees all offer the prospect of better yields in later rotations.

Species change
There are surprisingly few examples of wholesale species change from one rotation to the next which suggests that in most cases foresters have been good silviculturists. Examples of changes are cited in Evans (1999*b,c*, 2001).

Better seed origins, provenances, and land races
The impact of all these genetic improvements will affect yield and out-turn directly and indirectly through better survival and greater suitability to the site which may lead to increased vigour and perhaps greater pest and disease resistance. Countless studies affirm the benefits of careful investment in this phase of tree improvement.

Clonal plantations
Some of the world's most productive tree plantations use clonal material, including both eucalypts and poplars—see Chapter 11. It is clear that both the potential productivity and the uniformity of product make this silviculture attractive. Although clonal forestry has a narrow genetic base, careful management of clone numbers and the way they are interplanted can minimize pest and disease

problems. Roberds and Bishir (1997) suggest that use of 30–40 unrelated clones will generally provide security against catastrophic failure.

Tree breeding
Through an array of selection, crossing, and propagating techniques traits can be favoured that may improve vigour, stem and wood quality, pest and disease resistance, and other parameters such as frost tolerance. There are many examples of successful tree improvement strategies most of which are only beginning to bear fruit owing to long tree rotations and the slow process of tree breeding, particularly in orchard establishment and promotion of flowering, and in field testing of selections and progenies. Improvements in the order of 20–50% are considered relatively easy to achieve (Franklin 1989). From plus-tree selection alone, based on 24 published reports, Cornelius (1994) reported genetic gain values of 15% in height and 35% in volume. Genetic tree improvement offers by far the greatest assurance of sustained and improved yields from successive rotations in the medium and long term.

Genetically modified trees
There are no widely planted examples at present where genetic engineering has modified trees. The expectation is that these techniques will be used to develop disease resistance, modified wood properties, cold or drought tolerance rather than increase in vigour.

Role of different silvicultures

Silvicultural knowledge continues to increase through research and field trials and greater understanding of tree and stand physiology. While large yield improvements appear unlikely, incremental gains can be expected. Important examples include the following:

1. Manipulation of stocking levels to achieve greater output of fibre or a particular product, by fuller site occupancy, less mortality, and greater control of individual tree growth.
2. Matching rotation length to optimize yield—the rotation of maximum MAI—offers worthwhile yield gain in many cases.
3. In some localities prolonging the life of stands subject to windthrow by silvicultural means will increase yield over time.
4. Use of mixed crops on a site may aid tree stability, may lower pest and disease threats, but is unlikely to

raise productivity over growing the best suited species (FAO 1992).

5. Silvicultural systems that maintain forest cover at all times—continuous cover forestry—such as shelterwood and selection systems are likely to be neutral to slightly negative in production terms while benefiting tree quality, aesthetics, and probably biodiversity value.

6. Crop rotation, as practised in farming, appears unlikely. There are examples of forest plantations benefitting from a previous crop of nitrogen-fixing species, for example, *Acacia mearnsii* but industry is likely to require a similar not a widely differing species when replanting.

Fertilizing
Most forest use of fertilizer is to correct known deficiencies, for example, micronutrients such as boron in much of tropics, and macronutrients such as phosphorus on impoverished sites in many parts of both the tropical and the temperate world. In most instances fertilizer is only required once in a rotation. Fertilizer application is likely to be the principal means of compensating for nutrient losses on those sites where plantation forestry

practice does cause net nutrient export to detriment of plant growth.

Site preparation establishment practices
Ground preparation to establish the first plantation crop will normally introduce sufficient site modification for good tree growth in the long term. Substantial site manipulation is unlikely for second and subsequent rotations, unless there was failure first time round, except to alleviate soil compaction after harvesting or measures to reduce infections and pest problems.

Weed control strategies may change from one rotation to the next owing to differing weed spectrum and whether weeds are more or less competitive. The issue is crucial to sustainability since all the main examples of yield decline reflect worsening weed environments, especially competition from grasses and bamboos.

Organic matter conservation
It is clear from many investigations that treatment of organic matter both over the rotation and during felling and replanting is as critical to sustainability as coping with the weed environment. While

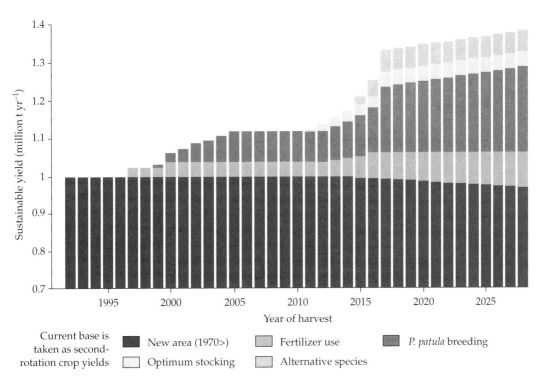

Figure 24.6 Estimated yield gains in the Usutu Forest, Swaziland, 1992–2028, arising from improvements in silviculture and genetic tree improvement (by permission Dr A. Morris, SAPPI (Usutu)).

avoidance of whole tree harvesting is probably desirable on nutrition grounds, it is now evident that both prevention of systematic litter raking or gathering during the rotation and conserving organic matter at harvesting are essential.

Holistic management
If all the above silvicultural features are brought together a rising trend in productivity can be expected. But if any one is neglected it is likely that the whole will suffer disproportionately. For example, operations should not exclusively minimize harvesting costs, but examine collectively harvesting, re-establishment and initial weeding, that is, as an holistic activity, so that future yield is not sacrificed for short term savings. Evidence of a rising trend reflecting the interplay of these gains is reported in Nambiar (1996) for Australia and reproduced in Evans (1999*b*) along with an example from Swaziland (Fig. 24.6).

Holistic management also embraces active monitoring of pest and disease levels, and researching pest and disease biology and impacts will aid appropriate responses such as altering practices, for example, delayed replanting to allow weevil numbers to fall, and careful reuse of extraction routes to minimize soil compaction and erosion.

Conclusions

Four main conclusions can be drawn about narrow-sense sustainability.

1. Planted forests and plantation forestry practices do affect sites and under certain conditions may cause deterioration, but are not inherently unsustainable. Care with harvesting, conservation of organic matter and management of the weed environment are critical features to minimize nutrient loss and damage to soil conditions.

2. Planted forests are at risk from pests and diseases. History of plantation forestry suggests that most risks are containable with vigilance and underpinning of sound biological research.

3. Measurements of yield in successive rotations of trees suggest that, so far, there is no widespread evidence that plantation forestry is unsustainable in the narrow sense. In the few cases where yield decline has been reported poor silvicultural practices appear largely responsible.

4. Several interventions in plantation silviculture point to increasing productivity in the future, providing management is holistic and good standards maintained. Genetic improvement especially offers the prospect of substantial gains over several rotations.

Bibliography

Aala-Capuli, F. (1993). Mass propagation of bamboo in the Philippines. In *Recent advances in mass clonal multiplication of forest trees for plantation programmes* (ed. J. Davidson), pp. 284–290. Proceedings of a Regional Symposium, Cisarua, Indonesia, 1–8 December 1992. RAS/91/004 Field Document 4. Food and Agriculture Organization of the United Nations, Los Banos, Philippines.

ABARE-Jaakko Pöyry (1999). *Global outlook for plantations.* ABARE Research Report 99.9. Australian Bureau of Agricultural and Resource Economics, Canberra.

Abayomi, J.O., Ekeke, B.A., and Nwaigbo, L.C. (1985). *Some preliminary results of thinning trials in Nigeria.* Proceedings of the 15th Annual Conference on Forestry Association, Nigeria, pp. 290–299.

Abbas, S.H. and Hanif, M. (1987). Effect of watershed management operations on run off and sediment release in Hazara. *Pakistan Journal of Forestry*, **37**, 89–98.

Abeli, W.S. (2000). Forestry work and its impact on human factors: the perspective of developing countries. In *Forests and society: the role of research*, Vol. 1 (ed. B. Krishnapillay *et al.*), pp. 535–544. XXI IUFRO World Congress, Kuala Lumpur, 7–12 August 2000, Sub-plenary Sessions, IUFRO, Vienna, and Forest Research Institute Malaysia, Kuala Lumpur.

Aber, J.D. (1987). Restored forests and the identification of critical factors in species-site interactions. In *Restoration ecology: a synthetic approach to ecological research* (ed. W.R. Jordan III, M.E. Gilpin, and J.D. Aber), pp. 241–250. Cambridge University Press, Cambridge, UK.

Abod, S.A. (1982). Plantation forestry practices in Malaysia. In *Tropical forests-source of energy through optimisation and diversification* (ed. P.B.L. Srivastava, A.M. Ahmad, K. Awang, A. Muktar, R.A. Kader, F. Che' Yom, and S.S. Lee), pp. 349–360. Proceedings of an International Forestry Seminar, Selangor, Malaysia, November 1980. Universiti Pertanian, Serdang, Malaysia.

ACIAR (1992). *Eucalypts: curse of cure.* Australian Centre for International Agricultural Research, Canberra, Australia.

ACIAR (2001). *Minimising impacts of eucalypt diseases in Southeast Asia.* Research Notes 25, 6/01. Australian Centre for International Agricultural Research, Canberra.

Acosta, R.T. (2000). Smallholders and communities in plantation development: lessons from two ITTO-supported projects. Proceedings of the International Conference on Timber Plantation Development, Manila, 7–9 November 2000, pp. 253–261. ITTO, FAO and Department of Environment and Natural Resources, Manila.

Adams, S.N. (1975). Sheep and cattle grazing in forests. A review. *Journal of Applied Ecology*, **12**, 143–152.

Adegbehin, J.O. (1988). Meeting the increasing wood demand from Nigerian forestry. *Journal of World Forest Resources Management*, **3**, 31–46.

Adeghehin, J.O., Abayomi, J.O., and Nwaigbo, L.C. (1988). *Gmelina arborea* in Nigeria. *Commonwealth Forestry Review*, **67**, 159–66.

Adendorff, M.W. and Schön, P.P. (1991). Root strike and root quality: the key to commercial success. In *Intensive forestry: the role of eucalypts*, Vol. 1 (ed. A.P.G. Schönau), pp. 30–38. South African Institute of Forestry, Pretoria, South Africa.

Adjers, G. and Srivastava, P.B.L. (1993). Nursery practices. In Acacia mangium: *growing and utilization* (ed. K. Awang and D. Taylor), pp. 75–100. Winrock International and the Food and Agriculture Organisation of the United Nations, Bangkok.

Adjers, G., Hadengganan, S., Kuusipalo, J., Nuryanto, K., and Vesa, L. (1995). Enrichment planting of dipterocarps in logged-over secondary forests: effect of width, direction and maintenance method of planting line on selected *Shorea* species. *Forest Ecology and Management*, **73**, 259–270.

Adlard, P.G. (1969). Quantitative effects of pruning *Pinus patula. Commonwealth Forestry Review*, **48**, 339–349.

Adlard, P.G. (1980). *Growing stock levels and productivity: conclusions from thinning and spacing trials in young* Pinus patula *stands in southern Tanzania.* Occasional Paper 8. Commonwealth Forestry Institute, Oxford.

Adlard, P.G. (1993). *Monitoring.* Study 11. Shell/WWF Tree Plantation Review, Shell International Petroleum Company and World Wide Fund for Nature, London.

Adlard, P.G. and Richardson, K.F. (1979). *Stand density and stem taper in* Pinus patula Schiede et Deppe. Occasional Paper 7. Commonwealth Forestry Institute, Oxford.

Agbede, O.O. (1985). Improving agroforestry in Nigeria: effect of plant density and interaction on crop production. *Forest Ecology and Management*, **11**, 231–239.

Agbede, O.O. and Adedire, M.O. (1989). *Taungya under short rotation fast growing trees: potentials and limitations of increasing forestry and agricultural yields in Nigeria.* Proceedings of a Conference on Fast Growing Trees and

Nitrogen Fixing Trees, October 1989. Philipps University, Marburg, Germany.

Aggangan, R.T. (2000). Tree farming in the Philippines: some issues and recommendations. In *Socio-economic evaluation of the potential for Australian tree species in the Philippines* (ed. S. Harrison and J. Herbohn), pp. 33–43. ACIAR Monograph 75, Australian Centre for International Agricultural Research, Canberra.

Ahlbäck, A.J. (1995). On forestry in Vietnam, the new reforestation strategy and UN assistance. *Commonwealth Forestry Review*, **74**, 224–229.

Ahmed, A. el Houri (1986). Some aspects of dry land afforestation in the Sudan with special reference to *Acacia tortilis* (Forsk.) Hayne, *A seyal* Willd and *Prosopis chilensis* (Molina) Stuntz. In *Tree plantings in semi-arid regions* (ed. P. Felker), pp. 209–221. Proceedings of a Symposium on Establishment and Productivity of Tree Plantings in Semi-arid Regions, Kingsville, Texas. Elsevier, Amsterdam, The Netherlands.

Ahmed, J. and Mahmood, F. (1998). Changing perspectives on forest policy. Pakistan country study. *Policy that works for forests and people, Series 1.* IUCN Pakistan, Islamabad, and International Institute for Environment and Development, London.

Ahmed, P. (1991). Agroforestry: a viable land use of alkali soils. *Agroforestry Systems*, **14**, 23–27.

Ahmed, R., Khan, D., and Ismail, S. (1985). Growth of *Azadirachta indica* and *Melia azederach* on coastal sand using highly saline water for irrigation. *Pakistan Journal of Botany*, **17**, 229–233.

Ahmed, S. and Idris, S. (1997). *Azadirachta indica* A.H.L. Juss. In *Auxiliary plants. Plant Resources of South-East Asia 11* (ed. I. Faridah Hanum and L.J.G. van der Maesen), pp. 71–76. Backhuys Publishers, Leiden, The Netherlands.

Akachuku, A. E. (1981). Estimation of volume and weight growth in Gmelina arborea with X-ray densitometry. In *Forest Site and Productivity* (ed. S. P. Gessel). pp. 153–59. Martinus Nijhoff, The Netherlands.

Aksornkoae, S. (1995). Ecology and biodiversity of mangroves. In *Ecology and management of mangrove restoration and regeneration in East and Southeast Asia* (ed. C. Khenmark), pp. 20–36. Proceedings of Ecotone IV, UNESCO-MAB, Bangkok.

Alder, D. and Montenegro, F. (1999). A yield model for *Cordia alliodora* plantations in Ecuador. *International Forestry Review*, **1**, 242–250.

Alexander, M.J. (1986). *Soil characteristics and the factors influencing their development on mine spoil of the Jos Plateau.* Report 11, Jos Plateau Environmental Resources Development Programme, Nigeria.

Allan, D.C. (1979). The cerrados of Brazil. A 'mid-west' of tomorrow? *Span*, **22**, 4–6.

Allan, T.G. (1977). Plantation planting and weeding in Savanna. In *Savanna afforestation in Africa*, FAO Forest Paper 11, pp. 139–148. Food and Agriculture Organization of the United Nations, Rome.

Allan, T.G. and Akwada, E.C.C. (1977). Land clearing and site preparation in the Nigerian savanna. In *Savanna afforestation in Africa.* FAO Forest Paper 11, pp. 123–138. Food and Agriculture Organization of the United Nations, Rome.

Allan, T.G. and Endean, F. (1966). *Manual of plantation techniques.* Department Instruction, Forest Department, Lusaka, Zambia.

Allen, H.L., Dougherty, P.M., and Campbell, R.G. (1990). Manipulation of water and nutrients—practice and opportunity in southern US pine forests. *Forest Ecology and Management*, **30**, 437–453.

Alvarez, A. and Stephan, G. (1986). Establishing highly productive stands of *Pinus caribaea* for resin tapping by making use of differences in the speed of regeneration of resin. *Beiträge für die Forstwirtschaft*, **20**, 138–140.

Amakiri, M.A. (1983). *Weeds and weed control in a forest plantation in the rainforest zone of Nigeria* (ed. M. Deat and P. Marnottel), pp. 272–281. Proceedings 2nd Biannual Conference West African Weed Science Society.

Ambar, S. (1986). Conversion of forest lands to annual crops and Indonesian perspective. In *Land use, watersheds, and planning in the Asia-Pacific region.* RAPA Report 1986/3, pp. 95–111. FAO, Bangkok.

Andeke-Lengui, M.A. and Dommergues, Y. (1983). Coastal sand dune stabilization in Senegal. In *Casuarina ecology, management and utilization* (ed. S.J. Midgley, J.W. Turnbull, and R.D. Johnston), pp. 158–166. CSIRO, Melbourne, Australia.

Anderson, J. and Farrington, J. (1996). Forestry extension: facing the challenges of today and tomorrow. *Unasylva*, **47**(184), 3–12.

Anderson, T.M. and Harvey, A.M. (1983). Managing cyclone-damaged *Pinus caribaea* stands. Report 8, Department of Forestry, Queensland, Australia. (Unpublished.)

Anderson, T.M., Bacon, G.J., and Shea, G.M. (1981). *Thinning strategies for Honduras Caribbean pine in plantations; an analysis of precommercial and commercial thinnings.* Technical Paper 25, Department of Forestry, Queensland, Australia.

Andrasko, K. (1990). Global warming and forests: an overview of current knowledge. *Unasylva*, **163**, 3–11.

Ang, L.H. (2001). *The suitability of tin tailings as a timber production area.* Paper to 16th Commonwealth Forestry Conference/19th Biennial Conference of the Institute of Foresters of Australia, 18–25 April 2001, Fremantle, Australia.

Anon. (1985). *The state of India's environment 1984–85.* Centre for Science and Environment, New Delhi.

Anon. (1989). *Country report: India.* Thirteenth Commonwealth Forestry Conference, New Zealand, September 1989.

Anon. (1993). *The plantation softwood resources of Queensland: a detailed analysis of the softwood processing opportunities that will flow from State Forest plantations through to the year 2020.* Queensland Forest Service, Department of Primary Industries, Brisbane, Australia.

Anon. (1996). *Treatment of Pinus patula plantations.* Technical Orders, Section 4.04.07 Forestry Department,

Ministry of Environment and Natural Resources, Government of Kenya, Nairobi.

Anon. (2000*a*). For forests and investors, big risks in tangled web of debt by pulp mills. *CIFOR Annual Report 2000*, p. 17. Center for International Forestry Research, Bogor, Indonesia.

Anon. (2000*b*). *Forest genetics for the next millennium.* Proceedings of an International Conference, Durban, South Africa, 8–13 October 2000. Institute of Commercial Forestry Research, Scottsville, South Africa.

Anon. (2001). *Bukidnon Forest Incorporated sustainable forestry in the Philippines manual.* Nimmo-Bell and Company, New Zealand.

Ansari, R., Marcar, N.E., Khanzada, A.N., Shirazi, M.U., and Crawford, D.F. (2001). Mulch application improves survival but not growth of *Acacia ampliceps* Maslin, *Acacia nilotica* (L.) Del. and *Conocarpus lancifolius* L. on a saline site in southern Pakistan. *International Forestry Review*, **3**, 158–163.

Antoine, J., Dent, F.J., Sims, D., Brinkman, R., and Bie, S. (1998). Resource management domains in relation to land-use planning. In *International workshop on resource management domains* (ed. J.K. Syers and J. Bouma), pp. 29–43. IBSRAM Proceedings 16. International Board for Soil Research and Management, Bangkok.

Appanah, S. (1998). Management of natural forests. In *A review of dipterocarps: taxonomy, ecology and silviculture* (ed. S. Appanah and J.M. Turnbull), pp. 133–149. Center for International Forestry Research, Bogor, Indonesia.

Appanah, S. (2001). Sustainable management of Malaysian rain forests. In *The forests handbook*, Vol. 2 (ed. J. Evans), pp. 341–35. Blackwell Science, Oxford, UK.

Appanah, S. and Turnbull, J.M. (ed.) (1998). *A review of dipterocarps: taxonomy, ecology and silviculture.* Center for International Forestry Research, Bogor, Indonesia.

Appanah, S. and Weinland, G. (1996). Experiences with planting dipterocarps in Peninsular Malaysia. In *Dipterocarp forest ecosystems—towards sustainable forest management* (ed. A. Schulte and D. Schöne), pp. 411–445. World Scientific, Singapore.

Applegate, G. (2002). Codes of practice and reduced impact logging in tropical forests: an overview. In *Pathways to sustainable management* (ed. A.G. Brown), pp. 41–47. Proceedings of the Second Hermon Slade Workshop, Ubud, Indonesia, 5–8 June 2001. The ATSE Crawford Fund, Melbourne, Australia.

Applegate, G.B. and Bragg, A.L. (1989). Improved growth rates for red cedar (*Toona australis* F. Muell. Harms.) seedlings in growtubes in north Queensland. *Australian Forestry*, **52**, 293–297.

Aracruz (2002). Aracruz social and environment report 2001. http://www.aracruz.com.br (15 September 2002).

Arbez, M. (2001). Ecological impacts of plantation forests on biodiversity and genetic diversity. In *Ecological and socio-economic impacts of close-to-nature forestry and plantation forestry: a comparative analysis* (ed. T. Green), pp. 7–20. EFI Proceedings 37. European Forest Institute, Joensuu, Finland.

Arbez, M., Birot, Y., and Carnus, J.-M. (ed.) (2002). *Risk management and sustainable forestry.* EFI Proceedings 45, European Forest Institute, Joensuu, Finland.

Arbuthnot, A. (2000). Clonal testing at Mondi Kraft, Richards Bay. In *Forest genetics for the next millennium.* Proceedings of an International Conference, Durban, South Africa, 8–13 October 2000, pp. 61–64. Institute of Commercial Forestry Research, Scottsville, South Africa.

Arifeen, S.Z. and Chaudhri, A.K. (1997). Effect of vegetation and engineering control structures on surface runoff and sediment yield at Fizagat, Swat. *Pakistan Journal of Forestry*, **47**, 29–35.

Armitage, F.B. (1985). *Irrigated forestry in arid and semi-arid lands: a synthesis.* IDRC 234e, International Research Development Centre, Ottawa.

Armitage, I. (1998). *Guidelines for the management of tropical forests. I. The production of wood.* FAO Forestry Paper 135. Food and Agriculture Organization of the United Nations, Rome.

Arnold, J.E.M. (1990). Tree components in farming systems. *Unasylva*, **160**, 35–42.

Arnold, J.E.M. (1991). *Tree products in agroecosystems: economic and policy issues.* IIED Gatekeeper Series 28. International Institute for Environment and Development, London.

Arnold, J.E.M. (1992). *Community forestry: ten years in review.* Community Forestry Note 7. Food and Agriculture Organization of the United Nations, Rome.

Arnold, J.E.M. (1997*a*). *Trees as out-grower crops for forest industries: experience from the Philippines and South Africa.* Rural Development Forestry Network Paper 22a. Overseas Development Institute, London.

Arnold, J.E.M. (1997*b*). Retrospect and prospect. In *Farms, trees and farmers. Responses to agricultural intensification* (ed. J.E.M. Arnold and P.A. Dewees). Earthscan Publications, London.

Arnold, J.E.M. (2001). *Forestry, poverty and aid.* Occasional Paper 33, Center for International Forestry Research, Bogor, Indonesia.

Arnold, J.E.M. and Dewees, P.A. (ed.) (1995). *Tree management in farmer strategies: responses to agricultural intensification.* Oxford University Press, Oxford, UK.

Arnold, J.E.M. and Dewees, P.A. (ed.) (1997). *Farms, trees and farmers. Responses to agricultural intensification.* Earthscan Publications, London.

Aronoff, S. (1989). *Geographic information systems—a management perspective.* WOL Publications, Ottawa.

Ashton, P.S., Givnish, T.J., and Appanah, S. (1988). Staggered flowering in the Dipterocarpaceae: new insights into floral induction and the evolution of mast fruiting in the aseasonal tropics. *American Naturalist*, **132**, 44–66.

Askin, D.C., Boland, D.J., and Pinyopusarerk, K. (1990). Use of *Casuarina oligodon* subsp. *abbreviata* in agroforestry in the North Baliem Valley, Irian Jaya, Indonesia. In *Advances in casuarina research and utilization* (ed. M.H. El-Lakany, J.W. Turnbull, and J.L. Brewbaker), pp. 213–219. Desert Development Center, American University in Cairo, Cairo.

Assman, E. (1970). *The principles of forest yield study.* Pergamon Press, Oxford, UK.

Aulerich, D.E. (1991). Harvest planning prior to plantation establishment. In *Recent developments in tree plantations in the humid/subhumid tropics of Asia* (ed. Sheikh Ali Abod *et al.*), pp. 308–320. Universiti Pertanian Malaysia, Serdang, Malaysia.

Awang, K. and Taylor, D. (ed.) (1993). Acacia mangium: *growing and utilization.* MPTS Monograph Series 3. Winrock International and FAO, Bangkok.

Ayaz, M. (1987). Performance of tools in tree felling and conversion in Changa Manga forest plantation. *Pakistan Journal of Forestry,* **37**, 141–50.

Babu, S.C. (1997). *Natural resource and sustainability monitoring. A conceptual framework, issues and challenges.* IFPRI Outreach Division Discussion Paper 17. International Food Policy Research Institute, Washington DC.

Bacon, G.J. and Bachelard, E.P. (1978). The influence of nursery conditioning treatments on some physiological responses of recently transplanted seedlings of *Pinus caribaea* Mar. var. *hondurensis* B & G. *Australian Forest Research,* **8**, 171–183.

Bacon, G.J. and Jermyn, D. (1977). Root dip evaluation in short-term storage trials with $1 + 0$ slash pine seedlings. *Australian Forestry,* **40**, 167–172.

Bahuguna, V.K. (2001). Production, protection and participation in forest management: an Indian perspective in achieving balance. In *Forests in a changing landscape,* pp. 1–16. Proceedings of 16th Commonwealth Forestry Conference/19th Biennial Conference of the Institute of Foresters of Australia. Promaco Conventions, Canning Bridge, Western Australia.

Bahungda, V.K., Lal, P., and Dhawan, V.K. (1987). Standardization of nursery techniques (seed sowing methods and watering schedules) of *Eucalyptus* FRI-4 under north Indian moist tropical climatic conditions. *Indian Forester,* **113**, 541–549.

Balaji, K., Ram Mohan, P., Saxena, M.R., Ravi Sankar G., Raghavswamy, V., and Gautam, N.C. (1994). Land use/land cover studies for environmental impact assessment—a study of Visakhapatnam–Vijayawada pipeline corridor using remote sensing techniques. In Proceedings of the ISPRS Commission VII Symposium, Rio de Janeiro, Brazil, 26–30 September 1994. National Institute for Space Research, São José dos Campos, São Paulo, Brazil. *Resource and Environmental Monitoring,* **30**(7a), 271–274.

Balasundaran, M., Sharma, J.K., Maria Florence, E.J., and Mohanan, C. (1995). Leaf spot diseases in teak and their impact on seedling production in nurseries. In *Caring for the Forest: research in a changing world* (ed. Korpilahti, E., Salonen, T., and S. Oja), p. 170. Congress Report Abstracts of Invited Papers, IUFRO XX World Congress, Tampere, Finland, 6–12 August 1995. The Finnish IUFRO World Congress Organising Committee, Tampere, Finland.

Balci, A.N., Sheng, T.C., and Dembner, S. (ed.) (1988). *Watershed management field manual: slope treatment measures and practices.* FAO Conservation Guide 13/3. Food and Agriculture Organization of the United Nations, Rome.

Ball, J.B. (2000). The field programme of FAO in forestry. In Aid to Forestry. Special Issue. *International Forestry Review,* **2**, 167–168.

Balocchi, C.E. (1996). Gain optimisation through vegetative multiplication of tropical and subtropical pines. In *Tree improvement for sustainable tropical forestry* (ed. M.J. Dieters, A.C. Matheson, D.G. Nikles, C.E. Harwood, and S.M. Walker), pp. 304–306. Proceedings of QFRI-IUFRO Conference, Caloundra, Australia, October/November 1996, Queensland Forestry Research Institute, Gympie, Australia.

Bandaratillake, H.M. (1991). Reforestation with pines in Sri Lanka. In *Recent developments in tree plantations in the humid/subhumid tropics of Asia* (ed. Sheikh Ali Abod *et al.*), pp. 122–133. Universiti Pertanian Malaysia, Serdang, Malaysia.

Bands, D.P. and Britton, P. (1977). *Guide to tree planting: arid zone.* Pamphlet 198. Department of Forestry, Pretoria, Republic of South Africa.

Barber, J. (1969). Control of genetic identity of forest reproductive material. Second World consultation on forest tree breeding, Vol. 2, 11/3. Washington DC.

Barber, K., Butler, R., Caird, D., and Kirby, M. (1996). Hierarchical approach for national forest planning and implementation. In *Hierarchical approaches to forest management in public and private organizations.* Proceedings of workshop held in Toronto, Canada, 25–29 May 1995. Information Report PI-X-124, pp. 36–44. Petawawa Forestry Institute, Canadian Forest Service.

Barnes, R.D. (1984). Genotype-environment interaction in the genetic improvement of fast-growing plantation trees. In *Symposium on site and productivity of fast growing plantations* (ed. D.C. Grey, A.P.G. Schönau, C.J. Schutz, and A. van Laar). IUFRO, Pretoria and Pietermaritzburg, South Africa.

Barnes, R.D. (1995). The breeding seedling orchard in the multiple population breeding strategy. *Silvae Genetica,* **44**, 81–88.

Barnes, R.D., Styles, B.T., Plumptre, R.A., and Ivory, M.H. (2001). Tropical pines. In *Tree crop ecosystems 19. Tree crop ecosystems* (ed. F.T. Last), pp. 163–192. Elsevier, Amsterdam, The Netherlands.

Barr, C. (2001). *Banking on sustainability: structural adjustment and forestry reform in post-Suharto Indonesia.* World Wide Fund for Nature, Washington DC and Center for International Forestry Research, Bogor, Indonesia.

Barrett, R. and Mullin, L.J. (1966). *A review of introduction of forest trees in Rhodesia.* Forest Research Bulletin 1. Salisbury, Rhodesia.

Barring, U. (1974). *Treatment of young stands—chemical weed control.* Proceedings of the Symposium on Stand Establishment, pp. 377–406. IUFRO, Wageningen, The Netherlands.

Barros, N.F. and Novais, R.F. (1996). Eucalypt nutrition and fertilizer regimes in Brazil. In *Nutrition of eucalypts* (ed. P.M. Attiwill and M.A. Adams), pp. 335–355. CSIRO Publishing, Melbourne, Australia.

Barros, S. and Campodonico, M.I. (ed.) (1999). *Assessment of sustainable forest management in planted forests—a first approach.* Proceedings of International Experts Meeting on the Role of Planted Forests in Sustainable Forest Management. Santiago, Chile, April 1999. Corporacion National Forestal de Chile, Chile.

Basden, S.C. (1960). Notes on deficiency symptoms in forest nurseries. *Papua New Guinea Agricultural Journal,* **13**(2).

Baskin, C.C. and Baskin, J.M. (2001). *Seeds: ecology, biogeography, and evolution of dormancy and germination.* Academic Press, London.

Bass, S. (1993). *Social environment.* Study 7. Shell/WWF Tree Plantation Review, Shell International Petroleum Company and World Wide Fund for Nature, London.

Bass, S. (2001). Working with forest stakeholders. In *The forests handbook,* Vol. 2 (ed. J. Evans), pp. 221–232. Blackwell Science, Oxford, UK.

Bass, S., Dubois, P., Moura-Costa, P., Pinard, M., Tipper, R. and Wilson, C. (2000). *Rural livelihoods and carbon management.* International Institute for Environment and Development (IIED), London.

Battaglia, M., Mummery, D., and Smith, T. (2001). Economic analysis of site survey and productivity modelling for selection of plantation areas. In *Site selection and productivity estimation: a CRC-SPF workshop* (ed. M. Cherry and C. Beadle), pp. 64–67. Technical Report 55. CRC for Sustainable Production Forestry, Hobart, Australia.

Baule, H. (1973). *Effect of fertilizers on resistance to adverse agencies.* Proceedings of the FAO/IUFRO Symposium on Forest Fertilization, Paris, pp. 181–214.

Baumer, M. (1990). *Agroforestry and desertification.* Technical Centre of Agricultural and Rural Cooperation, Wageningen, The Netherlands.

Bawa, K.S. and Hadley, M. (ed.) (1990). *Reproductive ecology of tropical forest plants.* Man and the Biosphere Series 7. UNESCO and Parthenon, Carnforth, UK.

Baxter, A.G. (2000). *Slash × Honduras pine hybrids: an overview of nursery production systems in south east Queensland, Australia.* Paper to Western Forest and Conservation Nursery Association Conference, Hawaii, August 2000.

Bayley, A.D. and Kietzka, J.W. (1997). Stock quality and field performance of *Pinus patula* seedlings produced under two nursery growing regimes and during seven different nursery production periods. *New Forests,* **13**, 341–356.

Beadle, C.L. (1997). Dynamics of leaf and canopy development. In *Management of soil, nutrients and water in tropical plantation forests* (ed. E.K.S. Nambiar and A.G. Brown), pp. 169–205. ACIAR Monograph 43. Australian Centre for International Agricultural Research, CSIRO Australia and Center for International Forestry Research, Indonesia.

Beaufils, E.R. (1973). *Diagnosis and recommendation integrated systems (DRIS). A general scheme for experimentation and calibration based on principles developed from research in plant nutrition.* Soil Science Bulletin 1, University of Natal, Pietermaritzburg, South Africa.

Beer, J., Ibrahim, M., and Schlonvoigt, A. (2000). Timber production in tropical agroforestry systems of Central America. In *Forests and Society: the Role of Research,* Vol. 1 (ed. B. Krishnapillay *et al.*), pp. 777–786. XXI IUFRO World Congress, Kuala Lumpur, 7–12 August 2000, Sub-plenary sessions IUFRO, Vienna and Forest Research Institute Malaysia, Kuala Lumpur.

Beetson, T.B., Taylor, D.W., and Nester, M.R. (1991). Effect of tree shelters on the early growth of four Australian tree species. *Australian Forestry,* **54**, 60–65.

Bell, D.T., McComb, J.A., van der Moezel, P.G., Bennett, I.J., and Kabay, E.D. (1994). Comparisons of selected and cloned plantlets against seedlings for rehabilitation of saline and waterlogged discharge zones in Australian agricultural catchments. *Australian Forestry,* **57**, 69–75.

Bell, T.I., Evo, T., and Sakumeni, A. (1983). *Cyclones and stability in Fiji's pine forests.* Fiji Pine Research Paper 14. Fiji Pine Commission/Fiji Forestry Department.

Bell, T.I.W. (1973). Erosion in the Trinidad teak plantations. *Commonwealth Forestry Review,* **52**, 223–233.

Bell, T.I.W. (1980). *Establishing* Pinus caribaea *plantations on red covered areas.* Fiji Pine Research Paper 6. Fiji Pine Commission/Fiji Forestry Department.

Bell, T.I.W. and Eva, T. (1982). *Field survival of bare-root container grown* Pinus caribaea *seedlings in the Fiji dry zone.* Fiji Pine Research Paper 11. Fiji Forestry Department/Fiji Pine Commission.

Ben Salem, B. (1991). Prevention and control of wind erosion in arid regions. *Unasylva,* **42**(164), 33–39.

Bene, J.G., Beall, H.W., and Coté, A. (1977). *Trees, food, and people: land management in the tropics.* International Development Research Centre, Ottawa, Canada.

Bennett, C. (1998). Outcome-based policies for sustainable logging in community forests: reducing forest bureaucracy. In *Incomes from the forest* (ed. E. Wollenberg and A. Ingles), pp. 203–220. Center for International Forestry Research, Bogor, Indonesia.

Bennett, F.D. (1980). *Advances in insect pest control. 2. Arthropods for biological control of forest pests and weeds.* Paper presented at the 11th Commonwealth Forestry Conference, Trinidad.

Bergen, K., Colwell, J., and Sapio, F. (2000). Remote sensing and forestry: collaborative implementation for a new century of forest information solutions. *Journal of Forestry,* **98**(6), 5–9.

Bernard-Reversat, F. (ed.) (2001). *Effect of exotic tree plantations on plant diversity and biological soil fertility in Congo savanna: with special reference to eucalypts.* Center for International Forestry Research, Bogor, Indonesia.

Bertault, J.-G. and Kadir, K. (ed.) (1998). *Silvicultural research in a lowland mixed dipterocarp forest of East Kalimantan.* CIRAD-forêt, Montpellier, France, and Forest Research and Development Agency and P.T. INHUTANI 1, Jakarta.

Bertault, J.-G., Dupuy, B., and Maitre, H.F. (1993). Silvicultural research for sustainable management of rain forest. In *The quest for sustainability: 100 years of silviculture and management in the tropics* (ed. P.J. Wood,

J.K. Vanclay, and W.M. Wan Razali), pp. 1–14. Forest Research Institute Malaysia, Kuala Lumpur.

Berthe, Y. (1997). The role of forestry in combating desertification. In *Protective and environmental functions of forests*, Vol. 2. Proceedings of the XI World Forestry Congress, Antalya, Turkey, 13–22 October 1997, pp. 307–311. Ministry of Forestry, Ankara, Turkey.

Bertrand, A. (1999). La dynamique seculaire des plantations paysannes d'*Eucalyptus* sur les Hautes Terres malgaches. *Le Flamboyant*, **49**, 45–48.

Bevege, D.I. (1967). *Thinning of slash pine in Queensland with special reference to basal area control*. Proceedings of the World Symposium on Man-made Forests and their Industrial Importance, pp. 1665–1682. Food and Agriculture Organization of the United Nations, Rome.

Bevege, D.I. and Simpson, J.A. (1980). Second rotation investigations in plantation conifers. Slash pine. *Australian Forest Research Newsletter*, **6**, 177–78.

Bewley, J.D. and Black, M. (1994). *Seeds, physiology of development and germination*. Plenum Press, New York and London.

Bhat, D.M., Murali, K.S., and Ravindranath, N.H. (2001). Formation and recovery of secondary forests in India: a particular reference to Western Ghats in South India. *Journal of Tropical Forest Science*, **13**, 601–620.

Bhat, K.M. (1990*a*). Wood quality improvement by pruning. *Evergreen. Newsletter of Kerala Forest Research Institute*, **24**, 4–5.

Bhat, K.M. (1990*b*). *Wood quality improvement of eucalypts in India*, Vol. 5. Proceedings of the 19th IUFRO World Congress, Montreal, Canada, p. 411.

Bhatti, M.B., Sultani, M.I., Aslam, M., and Syed, E. (1989). Forage production from *Leucena leucocephala* (Lam.) de Wit, as influenced by cutting intervals and plant spacings. *Pakistan Journal of Forestry*, **39**(2), 57–62.

Binkley, C.S. (1999). Ecosystem management and plantation forestry: new directions in British Columbia. *New Forests*, **18**, 75–88.

Binkley, D. and Giardina, C. (1997). Nitrogen fixation in tropical forest plantations. In *Management of soil, nutrients and water in tropical plantation forests*. (ed. E.K.S. Nambiar and A.G. Brown), pp. 297–337 ACIAR Monograph 43. Australian Centre for International Agricultural Research, Canberra.

Binns, W.O. (1962) Some aspects of peat as a substrate for tree growth. *Irish Forestry*, **19**, 32–55.

Bird, P.R., Bicknell, D., Bulman, P.A., Burke, J.A., Leys, J.F., Parker, J.N., Van Der Sommen, F.J., and Voller, P. (1992). The role of shelter in Australia for protecting soils, plants and livestock. *Agroforestry Systems*, **20**, 59–86.

Birks, J.S. and Barnes, R.D. (1990). *Provenance variation in* Pinus caribaea, P. oocarpa *and* P. patula *ssp.* tecunumanii. Tropical Forestry Paper 21. Oxford Forestry Institute, Oxford, UK.

Bishir, J. and Roberds J.H. (1999). On numbers of clones needed for managing risks in clonal forestry. *Forest Genetics*, **6**(3), 149–155.

Blair, G. and Lefroy, R. (ed.) (1991). *Technologies for sustainable agriculture on marginal uplands in Southeast Asia*. ACIAR Proceedings 33. Australian Centre for International Agricultural Research, Canberra.

Blake, J., Rosero, P., and Lojan, L. (1976). Interaction between phenology and rainfall in the growth of *Cordia alliodora*. *Commonwealth Forestry Review*, **55**, 37–40.

Blanchez, J.-L. and Dubé, Y.C. (1997). Funding forestry in Africa. *Unasylva*, **48**(188), 8–14.

Boa, E. and Lenné, J. (1994). *Diseases of nitrogen fixing trees in developing countries: an annotated list*. Natural Resources Institute, Overseas Development Administration, Chatham, UK.

Boardman, R. (1988). Living on the edge—the development of silviculture in South Australian pine plantations. *Australian Forestry*, **51**, 135–156.

Boden, D.I. (1991). Intensive site preparation on steep land: the effect on hydrological processes and growth of *Eucalyptus grandis* at 18 months. In *Intensive forestry: the role of eucalypts*, Vol. 1 (ed. A.P.G. Schönau), pp. 505–518. South African Institute of Forestry, Pretoria, South Africa.

Boden, D.I. and Herbert, M.A. (1986). Site preparation of Eucalyptus grandis at Glendale, Natal Midlands: provisional estimates of profitability. *ICFR Annual Research Report 1986*, pp. 119–122. Institute for Commercial Forest Research, Pietermaritzburg, South Africa.

Boland, D.J. (1997). Selection of species and provenances for planting. In *Australian trees and shrubs: species for land rehabilitation and farm planting in the tropics* (ed. J.C. Doran and J.W. Turnbull), pp. 39–58. ACIAR Monograph 24. Australian Centre for International Agricultural Research, Canberra.

Boland, D.J. (ed.) (1989). *Trees for the tropics: growing Australian multipurpose trees and shrubs in developing countries*. Australian Centre for International Agricultural Research, Canberra.

Boland, D.J., Brooker, M.I.H., and Turnbull, J.W. (1980). *Eucalyptus seed*. CSIRO, Australia.

Boland, D.J., Brophy, J.J. and House, A.P.N. (ed.) (1991). *Eucalyptus oils: use chemistry, distillation and marketing*. Inkata Press, Melbourne, Australia.

Bolstad, P.V., Kane, M., and Galindo, J. (1988). Height-growth gains 40 months after fertilization of young *Pinus caribaea* var. *hondurensis* in eastern Colombia. *Turrialba*, **38**, 233–241.

Bonner, F.T. (1992). Seed technology: a challenge for tropical forestry. *Tree Planters' Notes*, **43**, 142–145.

Bonner, F.T. (1996). Commercial seed supply of recalcitrant and intermediate seed: present solutions to the storage problem. In *Intermediate/recalcitrant tropical forest tree seeds* (ed. A.S. Ouedraogo, K Poulsen, and F. Stubsgaard), pp. 27–33. Proceedings of a Workshop, Humlebaek, Denmark, 8–10 June 1995, IPGRI, Rome and DANIDA Forest Seed Centre, Humlebaek, Denmark.

Bonner, F.T., Vozzo, J.A., Elam, W.W., and Land, S.B. Jr. (1994). *Tree seed technology training course. Instructor's manual*. General Technical Report SO-106, Southern Forest Experimental Station, US Department of Agriculture, Forest Service, New Orleans, USA.

Booth, T.H. (1990). Mapping regions climatically suitable for particular tree species at the global scale. *Forestry Ecology and Management*, **36**, 47–60.

Booth, T.H. (1998). Identifying climatically suitable areas for growing particular trees in Latin America. *Forest Ecology and Management*, **108**, 167–173.

Booth, T.H. (ed.) (1996). *Matching trees and sites*. ACIAR Proceedings 63. Australian Centre for International Agricultural Research, Canberra.

Booth, T.H., Jovanovic, T., and New, M. (2002). A new world climatic program to assist species selection. *Forest Ecology and Management*, **163**, 111–117.

Borota, J. and Proctor, J. (1967). *A review of softwood thinning practice and research in Tanzania*. Proceedings of the World Symposium on Man-made Forests and their Industrial Importance, pp. 1683–1694. Food and Agriculture Organization of the United Nations, Rome.

Boyle, T.J.B., Cossalter, C., and Griffin, A.R. (1997). Genetic resources for plantation forestry. In *Management of soil, nutrients and water in tropical plantation forests* (ed. E.K.S. Nambiar and A.G. Brown), pp. 25–63. ACIAR Monograph 43. Australian Centre for International Agricultural Research, Canberra.

Braatz, S. and Kandiah, A. (1996). The use of municipal waste water for forest and tree irrigation. *Unasylva*, **47**(185), 45–51.

Bradley, P.N. (1991). *Woodfuel, women and woodlots. Vol. l. Foundations of woodfuel and development strategy in E. Africa*. Macmillan.

Bradshaw, A.D. (1987). Restoration: an acid test for ecology. In *Restoration ecology: a synthetic approach to ecological research* (ed. W.R. Jordan III, M.E. Gilpin, and J.D. Aber), pp. 23–29. Cambridge University Press, Cambridge, UK.

Brand, D. (1998). Opportunities generated by the Kyoto Protocol in the forest sector. *Commonwealth Forestry Review*, **77**, 164–169.

Bravo, D.R. (1995). Restoration and management of the Pagbilao mangrove genetic resource area. In *Ecology and management of mangrove restoration and regeneration in East and Southeast Asia* (ed. C. Khenmark), pp. 190–213. Proceedings of Ecotone IV, UNESCO-MAB, Bangkok.

Braza, R.D. (1987). Resistance of four plantation tree species to white grubs, Leucopholis irrorata (Chevrolat) (Coleoptera: Scarabidae). *Sylvatrop. The Philippines Forest Research Journal*, **12**(1–2), 1–7.

Bredenkamp, B. (1984). *The Langepan C.C.T.* Directorate of Forestry, Republic of South Africa.

Bredenkamp, B.V. and Vuuren, N.J.J. van (1987). Pruning and resin infiltration of *Pinus caribaea* var. *caribaea* in Zululand. *South African Forestry Journal*, **140**, 29–34.

Bredenkamp, B.V., Malan, F.S., and Conradie, W.E. (1980). Some effects of pruning on growth and timber quality of *Eucalyptus grandis* in Zululand. *South African Journal of Forestry*, **114**, 29–34.

Brenner, A.J. (1996). Microclimatic modifications in agroforestry. In *Tree-crop interactions. A physiological approach*. (ed. C.K. Ong and P. Huxley), pp. 159–187. CAB International, Wallingford, UK and International Centre for Research in Agroforestry, Nairobi.

Briscoe, B.C. (1990). *Field trials manual for multipurpose tree species*. 2nd edn (ed. N. Adams), Manual 3. Multipurpose Tree Species Network Research Series, Winrock International Institute for Agricultural Development, Arlington, United States.

Briscoe, C.B. (1979). Improvement utilisation of tropical forests—silviculture in plantation development—Jari. *Y Coedwigwr*, **31**, 57–71.

Briscoe, C.B. and Nobles, R.W. (1966). *Effects of pruning teak*. Research Note ITF 11, Institute of Tropical Forestry, US Forest Service, Rio Pedras, Puerto Rico.

Brister, G.H. (1990). *Caribbean pine yield tables derived from temporary sample plot data*. Vol. 4. Proceedings of the 19th IUFRO World Congress, Montreal, p. 405.

Brophy, J.J. and Doran, J.C. (1996). *Essential oils of tropical Asteromyrtus, Callistemon and Melaleuca species*. ACIAR Monograph No. 40. Australian Centre for International Agriculture Research, Canberra.

Broun, A.F. (1912). *Silviculture in the tropics*. Macmillan, London.

Brown, A.G. and Ho, C.K. (ed.) (1997). *Black wattle and its utilisation*. Publication No. 97/72. Rural Industries Research and Development Corporation, Canberra.

Brown, A.G., Turnbull, J.W., and Booth, T.H. (1997). The Australian environment. In *Australian trees and shrubs: species for land rehabilitation and farm planting in the tropics*. (ed. J.C. Doran and J.W. Turnbull), ACIAR Monograph No. 24, pp. 1–18. Australian Centre for International Agricultural Research, Canberra.

Brown, B.N. (2000). Management of disease during eucalypt propagation. In *Diseases and pathogens of eucalypts* (ed. P.J. Keane, G.A. Kile, F.D. Podger, and B.N. Brown), pp. 119–151. CSIRO, Melbourne, Australia.

Brown, B.N. and Baxter, AG.M. (1991). *Nursery hygiene in concept and practice* (ed. J.R. Sutherland and S.G. Glover), pp. 133–140. Proceedings of the First Meeting of IUFRO Working Party S2.07-09, Victoria, British Columbia, Canada, 22–30 August 1990. Information Report BC-X-331. Pacific Forestry Centre, Pacific and Yukon Region, Forestry Canada, Victoria, British Columbia, Canada.

Brown, B.N. and Ferreira, F.A. (2000). Disease during propagation of eucalypts. In *Diseases and pathogens of eucalypts* (ed. P.J. Keane, G.A. Kile, F.D. Podger, and B.N. Brown), pp. 119–151. CSIRO, Melbourne, Australia Chakra.

Brown, C. (2000). *The global outlook for future wood supply from forest plantations*. Global Forest Products Outlook Study Working Paper Series, No. GFPOS/WP/03, Food and Agriculture Organization of the United Nations, Rome.

Brown, G.S. (1962). The importance of stand density in pruning prescriptions. *Commonwealth Forestry Review*, **41**, 246–257.

Brown, K. and Pearce, D. (ed.) (1994). *The causes of tropical deforestation*. UCL Press, London.

Brown, S. (1996). Present and potential roles of forests in the global climate change debate. *Unasylva*, **47**(185), 3–10.

Brown, S. and Lugo, A.E. (1994). Rehabilitation of tropical lands: a key to sustaining development. *Restoration Ecology*, **2**, 97–101.

Browne, F.G. (1968). *Pests and diseases of forest plantation trees*. Clarendon Press, Oxford, UK.

Bruce, J.W. (1989). *Community forestry: rapid appraisal of tree and land tenure*. Food and Agriculture Organization of the United Nations, Rome.

Bruce, J.W. and Fortmann, L. (1999). Contemporary uses of tree tenure. In *Agroforestry in sustainable agricultural systems* (ed. L.E. Buck, J.P. Lassoie, and E.C.M. Fernandes), pp. 237–244. Lewis Publishers, Boca Raton, Florida, USA.

Bruenig, E.F. (1986). *Forestry and agroforestry system designs for sustained production in tropical landscapes*, Vol. 2, pp. 217–228. Proceedings of the First Symposium on the humid tropics. EMBRAPA/CPATU, Belem, Brazil.

Bruijnzeel, L.A. (1990). *Hydrology of moist tropical forests and effects of conversion: a state of knowledge review*. UNESCO, Paris and Free University, Amsterdam.

Bruijnzeel, L.A. (1997). Hydrology of forest plantations in the tropics. In *Management of soil, nutrients and water in tropical plantation forests* (ed. E.K.S. Nambiar and A.G. Brown), pp. 125–167. ACIAR Monograph No. 43. Australian Centre for International Agriculture Research, CSIRO Australia and Center for International Forestry Research, Indonesia.

Brundrett, M., Bougher, N., Dell, B., Grove, T., and Malajczuk, N. (1996). *Working with mycorrhizas in forestry and agriculture*. ACIAR Monograph No. 32. Australian Centre for International Research, Canberra.

Brundtland, G. H. (1987). Our common future. Report of the world commission on environment and development. Oxford.

Brune, A. (1990). Reproductive biology and tropical plantation forestry. In *Reproductive ecology of tropical forest plants* (ed. K.S. Bawa and M. Hadley), pp. 349–354. UNESCO, Paris and Parthenon, Carnforth, UK.

Brünig *et al.* (1978). (from 2nd edn).

Buck, L.E., Lassoie, J.P., and Fernandes, E.C.M. (ed.) (1999). *Agroforestry in sustainable agricultural systems*. Lewis Publishers, Boca Raton, Florida, USA.

Buckley, G.P. (1988). Soil factors influencing yields of *Eucalyptus camaldulensis* on former tin-mining land in the Jos plateau region, Nigeria. *Forest Ecology and Management*, **23**, 1–17.

Budowski, G. (1987). Living fences in tropical America, a widespread agroforestry practice. In *Agroforestry: realities, possibilities and potential* (ed. G.L. Gholz), pp. 169–178. Martinus Nijhoff, The Netherlands.

Bulgannawar, G.N. and Math, B.B.M. (1991). The role of *Acacia auriculiformis* in afforestation in Karnataka, India. *Advances in tropical acacia research* (ed. J.W. Turnbull), pp. 110–115. ACIAR Monograph No. 35. Australian Centre for International Research, Canberra.

Bulman., M.S., Kimberley, M.O., and Gadgil, P.D. (1999). Estimation of the efficiency of pest detection surveys. *New Zealand Journal of Forest Science*, **29**, 102–115.

Bunce, H.W.F. and McLean, J.A. (1990). Hurricane Gilbert's impact on the natural forests and *Pinus caribaea* plantations in Jamaica. *Commonwealth Forestry Review*, **69**, 147–155.

Burdon, R.D. (1989). When is cloning on an operational scale appropriate? In *Breeding tropical trees: population structure and genetic improvement strategies in clonal and seedling forestry* (ed. G.I. Gibson, A.R. Griffin, and A.C. Matheson), pp. 9–27. Oxford Forestry Institute, Oxford, UK.

Burdon, R.D. (2003).Genetically modified forest trees. *International Forestry Review*, **5**, 58–64.

Burley, J. (1980). Selection of species for fuelwood plantations. *Commonwealth Forestry Review*, **59**, 133–147.

Burley, J. (1987). Problems of tree seed certification in developing countries. *Commonwealth Forestry Review* **66**(2), 151–9.

Burley, J. and Wood, P.J. (1976). *A manual on species and provenance research with particular reference to the tropics*. Tropical Forestry Paper No. 10. Commonwealth Forestry Institute, Oxford, UK.

Burley, J., Hughes, C.E., and Styles, B.T. (1986). Genetic systems of tree species for arid and semiarid lands. In *Tree plantings in semi-arid regions* (ed. P. Felker), pp. 317–343. Proceedings of a Symposium on Establishment and Productivity of Tree Plantings in Semi-arid Regions, Kingsville, Texas. Elsevier, Amsterdam, The Netherlands.

Burrough, P.A. (1986). *Principles of geographical information systems for land resources assessment*. Oxford University Press, Oxford, UK.

Busby, R.J.N. (1968). Reforestation in Fiji with large-leaf Mahogany. Paper presented at the 9th Commonwealth Forestry Conference, India.

Butterfield, R. (1990). Native species for reforestation and land restoration: a case study from Costa Rica. Proceedings 19th IRFRO World Congress, Montreal. Vol. 1, pp. 3–14.

Butterfield, R.P. (1995). Promoting biodiversity: advances in evaluating native species for reforestation. *Forest Ecology and Management*, **75**, 111–121.

Butterfield, R.P. and Fisher, R.F. (1994). Untapped potential. Native species for reforestation. *Journal of Forestry*, **92**(6), 37–40.

Byron, R.N. (2001). Keys to small-holder forestry. *Forests, Trees and Livelihoods*, **11**, 279–294.

Caban, A.G. (1998). Economics of fire damage assessment. In *Forest fire research*, Proceedings of the III International Conference on Forest Fire Research, Vol. 2 and the 14th Conference on Fire and Forest Meteorology, (ed. D.X. Viegas), pp. 2653–2665. Assoçiacao para o Dessenvolvimento da Aerodinamica Industrial, Coimbra, Portugal.

CABI (1997). *The forestry compendium—a silvicultural reference. Module 1. Asia*. CD Rom, CAB International, Wallingford, UK.

CABI (2000). *The forestry compendium—a silvicultural reference. Module 2. Rest of the World*. CD Rom, CAB International, Wallingford, UK.

Cabrido, C.A. Jr. (1985). An assessment of national soil erosion control management programs in the

Philippines. In *Soil erosion management* (ed. E.T. Craswell, J.V. Remenyi, and L.G. Nallana), pp. 13–20. ACIAR Proceedings No. 6. Australian Centre for International Agricultural Research, Canberra.

Cai, M., Liu, D. and Turnbull, J.W. (2003). Rehabilitation of degraded forests to improve livelihoods of poor farmers: a synthesis of four case studies in south China. In *Rehabilitation of degraded forests to improve livelihoods of poor farmers in south China* (ed. D. Liu). Center for International Forestry Research, Bogor, Indonesia.

Calder, I.R. (1992). Water use of eucalypts—a review. In *Growth and water use of forest plantations* (ed. I.R. Calder, R.L. Hall, and P.G. Adlard), pp. 167–179. Wiley and Sons, New York, USA.

Calder, I.R. (1996). Water use by forests at the plot and catchment scale. *Commonwealth Forestry Review*, **75**, 19–30.

Calder, I.R. (2002). Eucalyptus, water and the environment. In *Eucalyptus: the genus* Eucalyptus (ed. J.J.W. Coppen), pp. 36–51. Taylor and Francis, London.

Calder, I.R., Hall, R.L., and Adlard, P.G. (ed.) (1992). *Growth and water use of forest plantations*. Wiley and Sons, New York, USA.

Camino, R.V. de and Alfaro, M.M. (1998). *Teak in Central America*. Recursos Naturales Tropicales SA.

Campbell, C.A. (1990). Protection from grazing. In *Trees for rural Australia* (ed. K.W. Cremer), pp. 183–198. Inkata Press, Melbourne.

Campinhos, E. (1980). More wood of better quality through intensive silviculture with rapid-growth improved Brazilian *Eucalyptus*. *Tappi*, **63**(11), 145–147.

Campinhos, E. (1999). Sustainable plantations of high yield *Eucalyptus* trees for production of fiber: the Aracruz case. *New Forests*, **17**, 129–143.

Campinhos, E. and Ikemori, Y. (1988). Selection and Management of the basic population *Eucalyptus grandis* and *E. urophylla* established at Aracruz for the long term breeding progarmme. In *Breeding tropical trees: population structure and genetic improvement, strategies in clonal and seedling forestry* (ed. G.L. Gibson, A.R. Griffin, and A.C. Matheson), pp. 169–175. Proceedings of the IUFRO Conference, Pattaya, Thailand, November 1988, Oxford Forestry Institute.

Campinhos, E., La Torraca, S.M., Laranjeiro, A.J., and Penchel, R.M.F. (1993). The use/place of clonal multiplication in tree breeding and propagation programmes—prerequisites for success and reasons for failure. In *Recent advances in mass clonal multiplication of forest trees for plantation programmes* (ed. J. Davidson), pp. 9–30. Proceedings of a Regional Symposium, Cisarua, Indonesia, 1–8 December 1992. RAS/91/004 Field Document No. 4. Food and Agriculture Organization of the United Nations, Los Banos, Philippines.

Campinhos, E.N., Iannelli-Servin, C.M., Cardoso, N.Z., Almeida, M.A., and Rosa, A.C. (2000). Hidrojardim clonal Champion: uma otimização na produção de mudas de eucalipto. *Silvicultura*, **80**, 42–46.

Cannell, M.G. (1983). Plant management in agroforestry: manipulation of trees, population densities and mixtures of trees and herbaceous crops. In *Plant research and agroforestry* (ed. P.A. Huxley), pp. 455–487. International Council for Research in Agroforestry, Nairobi.

Cannell, M.G.R., Thornley, J.H.M., Mobbs, D.C., and Friend, A.D. (1998). UK conifer forests may be growing faster in response to N deposition, atmospheric CO_2 and temperature. *Forestry*, **71**, 277–296.

Cannon, P.G. (1982). *Site preparation trials on Andept soils with pine and eucalyptus species: results after three years*. Research Report, Investigacion Forestal, Carton de Colombia No. 87, 9 pp.

Carle, J. and Holgren, P. (2003). *Definitions related to planted forests*. Paper 22 to UNFF Intersessional Experts Meeting on the Role of Planted Forests in Sustainable Management, Wellington, New Zealand, 24–30 March 2003.

Carle, J., Vuorinen, P., and Del Lungo, A. (2002). Status and trends in global forest plantation development. *Forest Products Journal*, **52**(7), 3–13.

Carlson, C. (1998). Softwood nutrition research. *ICFR Annual Research Report 1998*, pp. 63–67. Institute for Commercial Forest Research, Scottsville, South Africa.

Carr, M. (1988). *Sustainable industrial development. Seven case studies*. Intermediate Technology Publications, London.

Carrere, R. and Lohmann, L. (1996). *Pulping the south: industrial tree plantations and the world paper economy*. World Rainforest Movement and Zed Books, London.

Carson, M.J., Carson, S.D., Richardson, T.E., Walter, C., Wilcox, P.L., Burdon, R.D., and Gardner, R.C. (1996). Molecular biology applications to forest trees—fact or fiction? In *Tree improvement for sustainable tropical forestry* (ed. M.J. Dieters, A.C. Matheson, D.G. Nikles, C.E. Harwood, and S.M. Walker), pp. 272–281. Proceedings of QFRI-IUFRO Conference, Caloundra, Australia. October/November 1996, Queensland Forestry Research Institute, Gympie, Australia.

Carter, E.J. (1987). *From seed to trial establishment*. DFR User Series No. 2. CSIRO Division of Forestry, Canberra.

Carucci, R. (2000). Trees outside forests: an essential tool for desertification control in the Sahel. *Unasylva*, **51**(200), 18–24.

Cass, A., McKenzie, N. and Cresswell, H. (1996). Physical indicators of soil health: a technical perspective. In *Indicators of catchment health* (ed. J. Walker and D.J. Reuter), pp 89–107. CSIRO, Melbourne, Australia.

Cassells, D.S. (2001). Processes for resolving conflict: managing land-use change. *International Forestry Review*, **3**, 206–213.

Catchpole, K.J. and Nester, M.R. (2002). Decision support software tools for pine plantations. *New Zealand Journal of Forestry*, **46**(4), 15–18.

Catesby, A. and Walker, S.W. (1998). An assessment of the relative amenability to vegetative propagation by leafy cuttings of 14 tropical and subtropical *Eucalyptus* and *Corymbia* species. In *Overcoming impediments to reforestation*, pp. 80–83. Proceedings of the 6th international workshop on tropical forest rehabilitation in the Asia Pacific region, BIO-REFOR, 2–5 December 1997. Brisbane, Australia.

Caulfield, J.P. (1987). Decision analysis in damaged forest plantations. *Forest Ecology and Management*, **22**, 155–165.

Cawse, J.C.L. (1979). Dry season planting and related establishment techniques. *South African Forestry Journal*, **111**, 34–38.

Caylor, J. (2000). Aerial photography in the next decade. *Journal of Forestry*, **98**(6), 17–19.

Cellier, G. (1999). Small-scale planted forests in Zululand, South Africa: an opportunity for appropriate development. *New Forests*, **18**, 45–57.

Chacko, K.C. (1995). *Silvicultural problems in management of teak plantations*. Proceedings of the 2nd Regional Seminar on Teak 'Teak for the Future', Yangon, Myanmar May 1995, pp. 91–98. FAO, Bangkok.

Chai, O.K.E., Dick, J. McP., Grace, J., Gillies, A.C.M., Wickneswari, R., Chai, P.K.P., and Andel, S. (1998). The aspects of tree improvement programmes for *Shorea macrophylla* in Sarawak, Malaysia. In *Recalcitrant seeds* (ed. M. Marzalina, K.C. Khoo, N. Jayanthi, F.Y. Tsan, and B. Krishnapillay), pp. 235–247. Proceedings of IUFRO Seed Symposium 1998, Kuala Lumpur, 12–15 October 1998. Forest Research Institute Malaysia, Kuala Lumpur.

Chakraborty, K. (1994). *Forestry nursery manual for West Bengal*. Forest Directorate, Government of West Bengal, Calcutta, India.

Chambers, R. (1987). Trees as savings and security for rural poor. Institute of Development Studies. University of Sussex. (Unpublished.)

Chambers, R. (1990). Interview on 'Farming Today', BBC, 20 October 1990.

Chambers, R. (1997). *Whose reality counts?* Intermediate Technology Publications, London.

Chambers, R. and Leach, M. (1990). Trees as savings and security for the rural poor. *Unasylva*, **161**, 39–52.

Chambers, R., Pacey, A., and Thrupp, I. (ed.) (1989). *Farmers first*. Bootstrap Press, New York, USA.

Champion, H.G. and Seth, S.K. (1968). *General silviculture for India*. Government of India, New Delhi.

Chamshama, S.A.O. and Hall, J.B. (1987). Effects of nursery treatments on *Eucalyptus camaldulensis* field establishment and early growth at Mafiga, Morogoro, Tanzania. *Forest Ecology and Management*, **21**, 91–108.

Chamshama, S.A.O., Mugasha, A.G., and Langerud, B.R. (1996). *Effect of top pruning on growth and survival of Prosopis chilensis seedlings* (ed. A.C. Yapa), pp. 194–198. Proceedings: International Symposium on Recent Advances in Tropical Forest Tree Seed Technology and Planting Stock Production. ASEAN Forest Tree Seed Centre Project, Mauk-Lek, Saraburi, Thailand.

Chandler, C., Cheney, P., Thomas, P., Trabaud, L., and Williams, D. (1983). *Fire in forestry: forest fire management and organization*, Vol. 2. Wiley, New York.

Chandrasekharan, C. (1997). International cooperation and resource mobilization for sustainable forestry development. In *Policies, institutions and means for sustainable forestry development*, Vol. 5. Proceedings of the XI World Forestry Congress, Antalya, Turkey, 13–22 October 1997, pp. 365–387. Ministry of Forestry, Ankara.

Chaplin, G.E. (1990). *Silvicultural manual*. Forest Record Number 6, Forestry Division, Ministry of Natural Resources, Honiara, Solomon Islands.

Chapman, G.W. (1973). *A manual on establishment techniques in man-made forests*. Food and Agriculture Organization of the United Nations, Rome.

Chapman, G.W. and Allan, T.G. (1978). *Establishment techniques for forest plantation*. FAO Forestry Paper No. 8. Food and Agriculture Organization of the United Nations, Rome.

Chijioke, E.O. (1980). *Impact on soils of fast-growing species in the lowland humid tropics*. FAO Forestry Paper 21, FAO, Rome.

Chijioke, E.O. (1988). Soil factors and growth of Gmelina arborea in Omo forest reserve. *Forest Ecology and Management*, **23**, 245–251.

Chin, H.F. and Diekmann, M. (1999). Ensuring seed quality in germplasm movement. In *Recalcitrant seeds* (ed. M. Marzalina, K.C. Khoo, N. Jayanthi, F.Y. Tsan, and B. Krishnapillay), pp. 81–86. IUFRO Seed Symposium 1998 Proceedings, Forest Research Institute Malaysia, Kepong, Malaysia.

Chinese Academy of Forestry. (1986). Paulownia *in China: cultivation and utilization*. Asian Network for Biological Sciences, Singapore and International Development Research Centre, Ottowa.

Chingaipe, T.M. (1985). Early growth of *Eucalyptus camaldulensis* under agroforestry conditions of Mafiga, Morogoro, Tanzania. *Forest Ecology and Management*, **11**, 241–244.

Chowdhry, K. (1987). Wastelands and the rural poor: essentials of a policy framework. *Tiger Paper*, **14**(2), 1–6. FAO, Bangkok.

Christian, C.S. and Stewart, G.A. (1953). *General report on survey of the Katherine-Darwin region, 1946*. Land Research Series I. CSIRO, Canberra.

Chudnoff, M. and Geany, T.E. (1973). Terminal shoot elongation and cambial growth rhythms in *Pinus caribaea*. *Commonwealth Forestry Review*, **52**, 317–324.

Chundamannii, M. (1998). *Teak plantations in Nilambur: an economic review*. KFRI Research Report No. 144, Kerala Forest Research Institute, India.

Chuong, H. and Dien, P.D. (1987). Studies on clinical and ecological variations of some tree species used in plantations for pulp and paper raw material production. *Vietnam forestry: summaries of forest research 1981–85*, Hanoi, pp. 19–22.

Chuong, H., Doran, J.C., Pinyopusarerk, K., and Harwood, C.E. (1996). Variation in the growth and survival of *Melaleuca* species in the Mekong Delta of Vietnam. In *Tree improvement for sustainable tropical forestry* (ed. M.J. Dieters, A.C. Matheson, D.G. Nikles, C.E. Harwood, and S.M. Walker), pp. 31–36. Proceedings of QFRI-IUFRO Conference, Caloundra, Australia, October/November 1996, Queensland Forestry Research Institute, Gympie, Australia.

Ciesla, W.M. and Donaubauer, E. (1994). *Decline and dieback of trees and forests*. FAO Forestry Paper 120, FAO, Rome.

Ciesla, W.M., Diekmann, M., and Putter, C.A.J. (ed.) (1996). Eucalyptus *spp. Technical guidelines for the safe movement of germplasm*, No. 17. International Plant Genetic Resources Institute, Rome.

Cockford, K.J., Dunsdon, A.J., Birks, J.S., and Barnes, R.D. (1990). *Provenance performance and genetic parameter estimates for* Pinus caribaea *var*. hondurensis. In Final Report of Evaluation of Tropical Pine Provenance and Progeny Tests, pp. 45–69. ODA Research Scheme R.4346, Oxford Forestry Institute, Oxford, UK.

Colchester, M. (2001). *Participatory investigation of the prospects for securing land rights and self-governance among the Masayakat adat in Indonesia.* Report of ICRAF/Aman collaborative project.

Constantini, A. (1986). Tolerance of *Pinus caribaea* var. *hondurensis* transplants to eleven herbicides. *Australian Forestry*, **49**, 241–245.

Constantini, A. (1989). Definition of a plant zone for weed management during the establishment of *Araucaria cunninghamii* plantations. *Forest Ecology and Management*, **29**, 15–27.

Constantini, A., Grimmett, J.L., and Dunn, G.M. (1997). Towards sustainable management of forest plantations in south-east Queensland. 1 Logging and understorey residue management between rotations in steep country *Araucaria cunninghamii* plantations. *Australian Forestry*, **60**, 221–225.

Conteras-Hermosilla, A. (1997). Country sector planning. In *Policies, institutions and means for sustainable forestry development*, Vol. 5. Proceedings of the XI World Forestry Congress, Antalya, Turkey, 13–22 October 1997, pp. 279–284. Ministry of Forestry, Ankara.

Contreras-Hermosilla, A. (2000). *The underlying causes of forest decline.* Occasional Paper No. 30. Center for International Forestry Research, Bogor, Indonesia.

Conway, F.J. (1988). Agroforestry outreach project, Haiti. In *The greening of aid* (ed. C. Conroy and M. Litvinoff), pp. 78–83. Earthscan, London.

Coppen, J.J.W. and Hone, G.A. (1992). *Eucalyptus oils: a review of production and markets*, NRI Bulletin 56, Natural Resources Institute, Chatham, UK.

Cornelius, J. (1994). The effectiveness of plus-tree selection for yield. *Forest Ecology and Management*, **67**, 1–3, 23–34.

Cossalter, C. (1986). Introduction of Australian acacias into dry tropical West Africa. In *Tree plantings in semi-arid regions* (ed. P. Felker), pp. 367–389. Proceedings of a Symposium on Establishment and Productivity of Tree Plantings in Semi-arid Regions, Kingsville, Texas. Elsevier, Amsterdam, The Netherlands.

Cossalter, C. (1994). *Forest seed orchards: basic principles and main strategic considerations.* Paper to training course on Seed Production Area/Seed Orchard Establishment, ASEAN-Canada Forest Tree Seed Centre Project, Palembang, Indonesia, 21–25 November 1994. CIFOR Working Paper November 1994. Center for International Forestry Research, Bogor, Indonesia.

Cossalter, C., Pye-Smith, C., and Kaimowitz, D. (2003). *Fast wood forestry—good or bad land use.* Center for International Forestry Research, Bogor, Indonesia.

Cotterill, P.P. and Dean, C.A. (1990). *Successful tree breeding with index selection.* CSIRO, Melbourne, Australia.

Craib, I.J. (1934). The place of thinning in wattle silviculture and its bearing on the management of exotic conifers. *Zeitschrift für die Welt forstwirtschaft*, **1**, 77–108.

Craib, I.J. (1939). *Thinning, pruning and management studies on the main exotic conifers grown in South Africa.* Science Bulletin 196. Department of Agriculture and Forestry, Pretoria,Union of South Africa.

Craib, I.J. (1947). *Silviculture of exotic conifers in South Africa.* Paper presented at the 5th British Empire Forestry Conference.

Cramb, R.A. (ed.) (2000). *Soil conservation technologies for smallholder farming systems in the Philippine uplands: a socioeconomic evaluation.* ACIAR Monograph No. 78, Australian Centre for International Agricultural Research, Canberra.

Crane, W.J.B. (1990). Planting of trees. In *Trees for rural Australia* (ed. K.W. Cremer), pp. 145–153. Inkata Press, Melbourne, Australia.

Craswell, E.T. and Tangendjaja, B. (ed.) (1985). *Shrub legumes in Indonesia and Australia.* ACIAR Proceedings No. 3. Australian Centre for International Agricultural Research, Canberra.

Cremer, K.W. (ed.) (1990). *Trees for rural Australia.* Inkata Press, Melbourne, Australia.

Cromer, R.N. and Williams, E.R. (1982). Biomass and nutrient accumulation in a planted *E. globulus* fertilizer trial. *Australian Journal of Botany*, **30**, 265–278.

Cromer, R.N., Cameron, D.M., Rance, S.J., Ryan, P.A., and Brown, M. (1993). Response to nutrients in *Eucalyptus grandis*. 1. Biomass accumulation. *Forest Ecology and Management*, **62**, 211–230.

Cuevas, E., Brown, S., and Lugo, A.E. (1991). Above- and below-ground organic matter storage and production in a tropical pine plantation and a paired broadleaf secondary forest. *Plant and Soil*, **135**, 257–268.

Curtin, R.A., Squire, R.H., and Mackowski, C.M. (1991). Management of native hardwood forests in state forests of North Coast New South Wales. In *Forest management in Australia* (ed. F.H. McKinnell, E.R. Hopkins, and J.E.D. Fox), pp. 77–106. Surrey Beatty, Chipping Norton, NSW, Australia.

Dabas, M. and Bhatia, S. (1996). Carbon sequestration through afforestation: role of tropical industrial plantations. *Ambio*, **25**, 327–330.

Dahl, N. (1996). *Casuarina equisetifolia*: its use and future in mine rehabilitation in northern Australia. In *Recent casuarina research and development* (ed. K. Pinyopusarerk, J.W. Turnbull, and S.J. Midgley), pp. 201–203. CSIRO Forestry and Forest Products, Canberra.

Dalmacio, M.V. and Banangan, F. (1976). Direct seeding of *Pinus kesiya* as affected by time of seeding, site preparation and seed coating. *Silvatrop. The Philippines Forest Research Journal*, **1**, 215–222.

Daniel, T.W., Helms, J.A., and Baker, E.S. (1979). *Principles of silviculture*, 2nd edn. McGraw-Hill, New York.

Daniels, F.W. and Van den Sijde, H.A. (1975). Cold stratification of *Pinus elliottii, Pinus taeda* and *Pinus patula*. *Bosbou in Suid-Afrika*, **16**, 63–68.

Dardaine, S. (1989). *Progress Report 1985–88: Forestry Division, Ministry of the Environment and National*

Service, Trinidad and Tobago. Thirteenth Commonwealth Forestry Conference, New Zealand.

Darling, F.F. (1969). *Wilderness and plenty*. (1969) Reith Lectures. Ballantine, London.

Dart, P., Umali-Garcia, M., and Almedras, A. (1991). Role of symbiotic associations in nutrition of tropical acacias. In *Advances in tropical acacia research* (ed. J.W. Turnbull), pp. 13–19. ACIAR Proceedings No. 35, Australian Centre for International Agricultural Research, Canberra.

Das, S.K. and Rao, Muralidhara, C. (1999). High-yield Eucalyptus clonal plantations of A.P. Forest Development Corporation Ltd—a success story? *Indian Forester*, **125**, 1073–1081.

Davey, S.M., Hoare, J., and Rumba, K. (2002). Science and its role in Australian regional forest agreements. *International Forestry Review*, **4**, 39–55.

Davidson, J. (1974). Reproduction of *Eucalyptus deglupta* by cuttings. *New Zealand Journal of Forestry Science* **4**, 204–210.

Davidson, J. (1976). *Three-dimensional forestry in Papua New Guinea*. Paper presented at the Third Meeting of the Papua New Guinea Botanical Society, May 1976.

Davidson, J. (1996). Developing Plantgro plant files for forest trees. In *Matching trees and sites* (ed. T.H. Booth), pp. 93–96. ACIAR Proceedings No. 63, Australian Centre for International Agricultural Research, Canberra.

Dawkins, H.C. (1958). *The management of natural tropical high-forest with special reference to Uganda*. Commonwealth Forestry Institute Paper No. 34. Oxford, UK.

Dawkins, H.C. (1963). Crown diameters: their relation to bole diameter in tropical forest trees. *Commonwealth Forestry Review*, **42**, 318–333.

Day, R.K., Rudgard, S.A., and Nair, K.S.S. (1994). *Asian tree pests: an overview*. FORSPA Publication 12. CAB International, Asian Development Bank, FAO of United Nations and United Nations Development Programme, Bangkok.

De Assis, T.F. (2000). *Production and use of Eucalyptus hybrids for industrial purposes*. In Symposium on Hybrid Breeding and Genetics, Noosa, Australia, 9–14 April 2000 (ed. H.S. Dungey, M.J. Dieters, and D.G. Nikles), pp. 63–74. Department of Primary Industries, Brisbane, Australia. (Compact Disk).

De Assis, T.F. (2001). *Evolution of technology for cloning Eucalyptus in large scale*. In Proceedings of IUFRO International Symposium on Developing the Eucalypt of the Future. Valdivia, Chile, 10–15 September 2001, Invited Paper, pp. 1–16. (Compact Disk).

De Freitas, M. (1996). Planted forests in Brazil. In *Caring for the forest: research in a changing world*, Vol. 2 (ed. E. Korphilati, H. Mikkelä and T. Salonen), pp. 147–154. IUFRO XX World Congress, Tampere, Finland. Congress Report, Finnish IUFRO World Congress Organising Committee, Jyväskalyä, Finland.

de la Cruz, R.E. (1996). *Mycorrhizae and beneficial soil microorganisms*. In Proceedings: International Symposium on Recent Advances in Tropical Forest Tree Seed Technology and Planting Stock Production. (ed.

A.C. Yapa), pp. 177–183. ASEAN Forest Tree Seed Centre Project, Mauk-Lek, Saraburi, Thailand.

de la Cruz, R.E. and Yantasath, K. (1993). Symbiotic associations. In Acacia mangium: *growing and utilization* (ed. K. Awang and D. Taylor), pp. 101–111. Winrock International and the Food and Agriculture Organisation of the United Nations, Bangkok.

DeBell, D.S., Cole, T.G., and Whitesell, C. (1997). Growth, development and yield in pure and mixed stands of *Eucalyptus* and *Albizia. Forest Science*, **43**, 286–298.

Dell, B. (1990). *Nutrition of horticultural tree crops—a research manual*. Murdoch University, Perth, Australia.

Dell, B. (1996). Diagnosis of nutrient deficiencies in eucalypts. In *Nutrition of eucalypts* (ed. P.M. Attiwill and M.A. Adams), pp. 417–440. CSIRO Publishing, Melbourne, Australia.

Dell, B. and Malajczuk, N. (1994). Boron deficiency in eucalypt plantations in China. *Canadian Journal of Forest Research*, **24**, 2409–2416.

Dell, B., Malajczuk, N., and Grove, T.S. (1995). *Nutrient disorders in plantation eucalypts* ACIAR Monograph No. 31. Australian Centre for International Agricultural Research, Canberra.

Dell, B., Malajczuk, N., Xu, D., and Grove, T.S. (2001). *Nutrient disorders in plantation eucalypts*, 2nd edn. ACIAR Monograph No. 31. Australian Centre for International Agricultural Research, Canberra.

Delwaulle, J.C. (1977a). Soil mixture, use of containers and other methods of plant raising. In *Savanna afforestation in Africa*, pp. 93–96. FAO Forest Paper 11. Food and Agriculture Organization of the United Nations, Rome.

Delwaulle, J.C. (1977b). Shelterbelts and environmental forestry. In *Savanna afforestation in Africa*, pp. 173—180. FAO Forest Paper 11. Food and Agriculture Organization of the United Nations, Rome.

Delwaulle, J.C. (1977c). Shelterbelts and environmental forestry. In Savanna afforestation in Africa, pp. 173–80. FAO, Rome.

Delwaulle, J.C. (1978). Forest plantations in dry tropical Africa: techniques and species to use. II. The nursery. *Bois et Fôrets des Tropiques*, **182**, 3–18.

Delwaulle, J.C. (1979). Forest plantations in dry tropical Africa. Techniques and species to use. III. *Bois et Fôrets des Tropiques*, **183**, 3–17.

Delwaulle, J.C. (1983). Creation et multiplication vegetative pan bouturage d'Eucalyptus hybrides en Republique Populaire de Congo. Silvicultura **8**, 775–8.

Denison, N.P. (1981). Recent developments in nursery and planting systems in SA. Forests Investments Limited. *South African Forestry Journal*, **118**, 26–30.

Denison, N.P. and Kietzka, J.E. (1993). The development and utilisation of vegetative propagation in Mondi for commercial afforestation programmes. *South African Forestry Journal*, **165**, 47–54.

DENR (1994). *Mangrove regeneration and management*. Fisheries Sector Program, Department of Environment and Natural Resources, Diliman, Quezon City, Philippines.

Desloges, C. and Gauthier, M. (1997). Community forestry and forest resource conflicts—an overview. In *Social dimensions of forestries contribution to sustainable development*, Vol. 5, pp. 109–121. Proceedings of the XI World Forestry Congress, Antalya, Turkey, 13–22 October 1997. Ministry of Forestry, Ankara, Turkey.

Desmond, D.F. (1989). *Forest tree nurseries in agricultural high schools: an analysis of Ecuadorean experiences.* Social Forestry Network Paper 9e. Overseas Development Institute, London.

Devendra, C. (ed.) (1990). *Shrubs and tree fodders for farm animals.* Proceedings of a workshop at Denpasar, Indonesia, 24–29 July 1989. International Development Research Centre, Ottawa, Canada.

Dewees, P.A. (1995). Farmer responses to tree scarcity: the case for woodfuel. In *Tree management in farmer strategies: responses to agricultural intensification* (ed. J.E.M. Arnold and P.A. Dewees), pp. 174–197. Oxford University Press, Oxford, UK.

Dewees, P.A. and Saxena, N.C. (1995). Wood product markets as incentives for farmer tree growing. In *Tree management in farmer strategies. Responses to agricultural intensification* (ed. J.E.M. Arnold and P.A. Dewees), pp. 198–241. Oxford University Press, Oxford, UK.

Di Stephano, J. and Mazzer, L. (2002). *The effectiveness of plastic fences and tree guards for reducing browsing damage: a case study from East Gippsland.* Research Report 383, Forest Science Centre, Department of Natural resources and Environment, Victoria, Australia.

Diatloff, G. (1977). Biological control of the weed *Lantana camara* by *Octotoma championi* and *Uroplata* sp. (near *bilineata*). *Agronomia Costarricense*, **1**, 165–167.

Dickinson, G.R., Nester, M.R., Lee, D.J., Lewty, M.J., and Raddatz, C.G. (2001). *Developing a hardwood pulpwood industry in Queensland: early results from genetics and silviculture trials in the Maryborough–Gladstone region.* Proceedings of 16th Commonwealth Forestry Conference/19th Biennial Conference of the Institute of Foresters of Australia, pp. 165–173. Promaco Conventions, Canning Bridge, Western Australia.

Dieters, M.J., Matheson, A.C., Nikles, D.G., Harwood, C.E., and Walker, S.M. (ed.) (1996). *Tree improvement for sustainable tropical forestry.* Vols 1 and 2. Proceedings of QFRI-IUFRO Conference, Caloundra, Australia, October/November 1996, Queensland Forestry Research Institute, Gympie, Australia.

Dietz, J.M., Couto, E.A., Alfenas, A.C., Faccini, A. and Silva, G.F. da (1975). Efeitos de duas plantacoes de florestas homogeneas sobre populacoes de mamiferos pequenos. Brasil Florestal **6**, 54–7.

Ding, Y.X. and Cheng, J.L. (1995). Effect of continuous plantation of Chinese fir on soil fertility. *Pedosphere*, **5**, 57–66.

Diouf, D., Neyra, M., Sougoufara, B., and Lesueur, D. (2001). The Senegalese Action Plan: assessment and prospects for reafforestation operations from 1993 to 1998. *Bois et Forêts des Tropiques*, **270**(4), 5–13.

Dix, M.E., Bishaw, B., Workman, S.W., Barnhart, M.R., Klopfenstein, N.B., and Dix, A.M. (1999). Pest management in energy- and labor-intensive agroforestry systems. In *Agroforestry in sustainable agricultural systems* (ed. L.E. Buck, J.P. Lassoie, and E.C.M. Fernandes), pp. 131–155. Lewis Publishers, Boca Raton, Florida, USA.

Dommergues, Y., Duhoux, E., and Diem, H.G. (1999). *Les arbres fixateurs d'azote.* CIRAD, and Editions Espaces, Montpellier; FAO, Rome; and Institut de Recherche pour le Developpment, Paris.

Donald, D.G.M. (1971*a*). Water requirements in the South African nursery. *Forestry in South Africa*, **12**, 25–33.

Donald, D.G.M. (1971*b*). Cleaning operations in South African forestry. *Forestry in South Africa*, **12**, 55–65.

Donald, D.G.M. (1979). Nursery and establishment techniques as factors in productivity of man-made forests in South Africa. *South African Forestry Journal*, **109**, 19–25.

Donald, D.G.M. (1984). A study in nursery efficiency and plant quality. *South African Forestry Journal*, **128**, 12–14.

Doran, J.C. (1997). Seed, nursery practice and establishment. In *Australian trees and shrubs: species for land rehabilitation and farm planting in the tropics.* (ed. J.C. Doran and J.W. Turnbull), pp. 59–87. ACIAR Monograph No. 24. Australian Centre for International Agricultural Research, Canberra.

Doran, J.C. (2002). Genetic improvement of eucalypts with special reference to oil-bearing species. In *Eucalyptus: the genus Eucalyptus* (ed. J.J.W. Coppen), pp. 75–101. Taylor and Francis, London.

Doran, J.C. and Brophy, J.J. (1990). Tropical red gums—a source of 1.8-cineole-rich *Eucalyptus* oil. *New Forests*, **4**, 157–178.

Doran, J.C. and Turnbull, J.W. (1997). *Australian trees and shrubs: species for land rehabilitation and farm planting in the tropics.* ACIAR Monograph No. 24, Australian Centre for International Agricultural Research, Canberra.

Doran, J.C., Boland, D.J., Turnbull, J.W., and Gunn, B.V. (1983). *Handbook on seeds of dry zone acacias: a guide for collection, extracting, cleaning and storing the seed and for treatment to promote germination.* Food and Agriculture Organization of the United Nations, Rome.

Doran, J.C., Turnbull, J.W., and Kariuki, E.M. (1987). *Effects of storage conditions on germination of five tropical tree species.* Proceedings of the International Symposium on Forest seed Problems in Africa (ed. S.S. Kamra and R.D. Ayling), pp. 84–94. Report 7, Swedish University of Agricultural Sciences, Umea, Sweden.

Dransfield, S. and Widjaja, E.A. (ed.) (1995). *Bamboos.* PROSEA No.7. Backhuys, Leiden, The Netherlands.

Drechsel, P. and Zech, W. (1991). Foliar nutrient levels of broad-leaved tropical trees: a tabular review. *Plant and Soil*, **131**, 29–46.

Drechsel, P. and Zech, W. (1994). DRIS evaluation of teak (*Tectona grandis*) mineral nutrition and effects of nutrition and site quality on teak growth in West Africa. *Forest Ecology and Management*, **70**, 121–133.

Dregne, H., Kassas, M., and Razanov, B. (1991). A new assessment of the world status of desertification. *Desertification Control Bulletin*, **20**, 6–18. United Nations Environment Programme, Nairobi.

du Toit, B. and Scholes, M.C. (2002). Nutritional sustainability of *Eucalyptus* plantations: a case study at Karkloof, South Africa. *Southern African Forestry Journal*, **195**, 63–72.

Duff, A.B., Hall, R.A., and Marsh, C.W. (1986). A survey of wildlife in and around a commercial tree plantation in Sabah. *Malaysian Forester* **47** (3/4), 197–213.

Duguma, B., Kang, B.T., and Okali, D.U.R. (1988). Effect of pruning intensities of three woody leguminous species grown in alley cropping with maize and cowpea on an alfisol. *Agroforestry Systems*, **6**, 19–35.

Dumanski, J. and Craswell, E. (1998). Resource management domains for evaluation and management of agroecological systems. In *International workshop on resource management domains* (ed. J.K. Syers and J. Bouma), pp. 1–13. IBSRAM Proceedings No. 16. International Board for Soil Research and Management, Bangkok.

Duncan, E.A., van Deventer, F., Kietzka, J.E., Lindley, R.C., and Denison, N.P. (2000). The applied subtropical *Eucalyptus* clonal programme in Mondi forests, Zululand coastal region. In *Forest genetics for the next millennium*, pp. 95–97. Proceedings of an international conference, Durban, South Africa, 8–13 October 2000. Institute of Commercial Forestry Research, Scottsville, South Africa.

Dunsworth, G.B. (1997). Plant quality assessment: an industrial perspective. *New Forests*, **13**, 439–448.

Durst, P. and Brown, C. (2000). *Current trends and development of plantation forestry in Asia-Pacific countries.* Proceedings of the International Conference on Timber Plantation Development, Manila, 7–9 November 2000, pp. 159–169. ITTO, FAO and Department of Environment and Natural Resources, Manila.

Durst, P.B., Waggener, T.R., Enters, T. and Tan, L.C. (ed.) (2001). *Forests out of bounds: impacts and effectiveness of logging bans in natural forests in Asia-Pacific.* RAP Publication 2001/08, Food and Agriculture Organization of the United Nations, Regional Office for Asia and the Pacific, Bangkok.

Duryea, M.L. and Landis, T.D. (ed.) (1984). *Forest nursery manual: production of bareroot seedlings.* Nijhoff/Junk, The Hague.

Dyck, B. (2003). *Benefits of planted forests: social, ecological and economic.* Paper to UNFF Intersessional Experts Meeting 'Role of Planted Forests in Sustainable Forest Management', Wellington, New Zealand, 24–27 March 2003.

Dyck, W.J. and Skinner, M.F. (1988). *Potential for productivity decline in New Zealand radiata pine forests.* Proceedings of the 7th North American Forest Soils Conference, pp. 318–332.

Dyck, W.J., Cole, D.W., and Comerford, N.B. (1994). *Impacts of Forest Harvesting on long-term site productivity.* Chapman and Hall, London, p. 371.

Dye, P.J. (1996). Climate, forest and streamflow relationships in South African afforested catchments. *Commonwealth Forestry Review*, **75**, 31–36.

Dykstra, D.P. (1997). Information systems in forestry. *Unasylva*, **48**(189), 10–15.

Dykstra, D.P. and Heinrich, R. (1996a). *FAO model code of harvesting practice.* Food and Agriculture Organization of the United Nations, Rome.

Dykstra, D.P. and Heinrich, R. (1996b). *Forest codes of practice contributing to environmentally sound forest operations.* FAO Forestry Paper 133. Food and Agriculture Organization of the United Nations, Rome.

Eamus, D., Lawson, G.J., Leakey, R.R.B., and Mason, P.A. (1990). *Enrichment planting in the Cameroon moist deciduous forest: microclimatic and physiological effects*, Vol. 1. Proceedings of the 19th IUFRO World Congress, Montreal, pp. 258–270.

Earl, D.E. (1975). *Forest energy and economic development.* Clarendon Press, Oxford, UK.

Edwardson, T.E. (1978). *Prospects for forestry cooperatives in small holdings and forest communities, with special reference to developing countries*, Vol. 2. Proceedings of the 8th World Forestry Congress, Jakarta, pp. 201–212.

Effendy, A. and Hardono, D.S. (2000). *The large scale private investment of timber plantation development in Indonesia.* Proceedings of the International Conference on Timber Plantation Development, Manila 7–9 November 2000, ITTO, FAO and Department of Environment and Natural Resources, Manila.

Eggeling, J. (1942). *Elementary forestry.* Forest Department, Uganda.

Egunjobi and Onweluzo (1979). Litter fall, mineral turnover and litter accumulation in *pinus caribaea* stands at Ibadan, Nigeria. *Biotropica* **11**(4), 251–255.

Egunjobi, J.K. (1975). Dry matter production of immature stands of *Pinus caribaea* in Nigeria. *Oikos* **26**, 80–85.

El-Lakany, M.H. (1990). Provenance trials of *Casuarina glauca* and *C. cunninghamiana* in Egypt. In *Advances in casuarina research and utilization* (ed. M.H. El-Lakany, J.W. Turnbull, and J.L. Brewbaker), pp. 12–22. Desert Development Center, American University in Cairo, Cairo.

El-Lakany, M.H. (1995). Afforestation techniques for desert development. In *Caring for the forest: research in a changing world*, Vol. 2, (ed. E. Korpilahti, H. Mikkelä, and T. Salonen), pp. 369–376. Congress Report, IUFRO XX World Congress, Tampere, Finland, 6–12 August 1995. The Finnish IUFRO World Congress Organising Committee, Tampere, Finland.

Eldridge, K.G., Davidson, J., Harwood, C. and van Wyk, G. (1993). *Eucalypt domestication and breeding.* Clarendon Press, Oxford, UK.

Elias (2000). *Reduced impact tree harvesting in the tropical forest plantations.* Proceedings of the International Conference on Timber Plantation Development, Manila, 7–9 November 2000, pp. 171–181. ITTO, FAO and Department of Environment and Natural Resources, Manila.

Ellefson, P.V. (1991). Integration of forest sector plans with national plans for development: a strategic necessity. *Journal of World Forest Resource Management*, **5**, 73–83.

Ellis, R.H. and Hong, T.D. (1996). *Seed quality: seed development and storage.* In Proceedings of the International Symposium on Recent Advances in Tropical Forest Tree Seed Technology and Planting Stock Production (ed. A.C.Yapa), pp. 80–92. ASEAN Forest Tree Seed Centre Project, Mauk-Lek, Saraburi, Thailand.

Ellis, R.H., Hong, T.D., and Roberts, E.H. (1990). An intermediate category of seed storage behaviour. *Journal of Experimental Botany*, **41**, 1167–1174.

Enters, T. and Hagmann, J. (1996). One-way, two-way, which way? Extension workers: from messengers to facilitators. *Unasylva*, **47**(184), 13–20.

Enters, T., Durst, P.B., and Brown, C. (2003). *What does it take? The role of incentives in forest plantation development in the Asia-Pacific region.* Paper 20 to UNFF Intersessional Experts Meeting on the Role of Planted Forests in Sustainable Management, Wellington, New Zealand, 24–30 March 2003.

Esterhuyse, C.J. (1989). Agroforestry in South Africa. *South African Forestry Journal*, **149**, 62–68.

Eswaran, H., Kimble, J., Cook, T., and Beinroth, F.H. (1992). Soil diversity in the tropics: implications for agricultural development. In *Myths and science of soils of the tropics* (ed. R. Lal and P.A. Sanchez), pp. 1–16. SSSA Special Publication No. 29, Soil Science Society of America, Madison, Wisconsin, USA.

Evans, J. (1974). Some aspects of the growth of Pinus patula in Swaziland. *Commonwealth Forestry Review* **53**, 57–62.

Evans, J. (1976a). Spacing in a pulpwood plantation. *South African Forestry Journal*, **96**, 23–26.

Evans, J. (1976b). Plantations: productivity and prospects. *Australian Forestry*, **39**, 150–163.

Evans, J. (1978a). A further report on second rotation productivity in the Usutu Forest, Swaziland—results of the 1977 reassessment. *Commonwealth Forestry Review*, **57**, 253–261.

Evans, J. (1978b). Some growth effects of low rainfall and hail damage in *Pinus patula* plantations. *South African Forestry Journal*, **105**, 8–12.

Evans, J. (1979). The effects of leaf position and leaf age in foliar analysis of *Gmelina arborea*. *Plant and Soil*, **52**, 547–552.

Evans, J. (1980). Preliminary data on foliar nutrient levels in klinkii pine (*Araucaria hunsteinii* K. Schum). *Malaysian Forester*, **43**, 212–218.

Evans, J. (1982) *Plantation forestry in the tropics*, 1st edn, Oxford University Press, UK.

Evans, J. (1984) Measurement and prediction of changes in site productivity. In IUFRO Symposium on Site and Productivity of Fast Growing Plantations (ed. D.C. Grey, A.P.G. Schonau, C.J. Schutz, and van Laar), Pretoria and Pietermaritzberg, South Africa April 1994, pp. 970–920.

Evans, J. (1987a). Site and species selection-changing perspectives. *Forest Ecology and Management* **21**, 299–310.

Evans, J. (1989). Community forestry in Ethiopia—the Bilate project. *Rural Development in Practice*, **1**(4), 7–8, 25.

Evans, J. (1992). *Plantation forestry in the tropics*. 2nd edn. Clarendon Press, Oxford, UK.

Evans, J. (1996). The sustainability of wood production from plantations: evidence over three successive rotations in the Usutu Forest, Swaziland. *Commonw. Forestry Review*, **75**, 234–239.

Evans, J. (1999a). Planted forests of the wet and dry tropics: their variety, nature and significance. *New Forests*, **17**, 25–36.

Evans, J. (1999b) *Sustainability of forest plantations—the evidence.* Issues Paper, UK Department for International Development, London, p. 64.

Evans, J. (1999c) Sustainability of plantation forestry: impact of species change and successive rotations of pine in the Usutu Forest, Swaziland. *Southern Africa Forestry Journal*, 63–70.

Evans, J. (2001a). *Biological sustainability of productivity in successive rotations.* Forest Plantations Thematic Working Paper FP/2 (ed. D.J. Mead) Forest Department, FAO, Rome, p. 24.

Evans, J. (2001b). How to be successful in plantation development. *Tropical Forest Update*, **11**(3), 3–5.

Evans, J. (2001c). Tropical hardwood plantations. In *Ecosystems of the World. 19. Tree Crop Ecosystems* (ed. F.T. Last), pp. 217–227. Elsevier, Amsterdam, The Netherlands.

Evans, J. and Boswell, R.C. (1998). Research on sustainability of plantation forestry: volume estimation of *Pinus patula* trees in two different rotations. *Commonwealth Forestry Review*, **77**, 113–118.

Evans, J. and Hibberd, B.G. (1993). *Operations.* Shell/ WWF Tree Plantation Review. Study No. 9. Shell International Petroleum and World Wide Fund for Nature, Godalming, UK.

Evans, J. and Masson, P. (2001). Sustainable plantation forestry: a case study of wood production and environmental management strategies in the Usutu Forest, Swaziland. In *The forests handbook*, Vol. 2 (ed. J. Evans), ch. 18, pp. 357–372. Blackwell Science, Oxford.

Evans, J. and Wright, D. (1988). The Usutu Pulp Company—development of an integrated forestry project. In *Sustainable industrial development. seven case studies* (ed. M. Carr), pp. 151–171. Intermediate Technology Publications, London.

Fagg, P. (2001). *Eucalypt sowing and seedfall.* Native Forest Silviculture Guideline No. 8. Department of Natural Resources and Environment, Melbourne, Australia.

Fagg, P.C. and Cremer, K.W. (1990). Weed control and water conservation. In *Trees for rural Australia* (ed. K.W. Cremer), pp. 161–182. Inkata Press, Melbourne.

Fahey, B. and Jackson, R. (1995). Hydrological impacts of converting native forests and grasslands to pine plantation, South Island, New Zealand. In *Caring for the forest: research in a changing world* (ed. E. Korpilahti, T. Salonen, and S. Oja), p. 31. Congress Report Abstracts of Invited Papers, IUFRO XX World Congress, Tampere, Finland, 6–12 August 1995, The Finnish IUFRO World Congress Organising Committee, Tampere, Finland.

Falconer, J. and Arnold, J.E.M. (1988). *Forests, trees and household food security.* Social Forestry Network Paper 7a, Overseas Development Institute, London.

FAO (1962). *Forest influences.* Food and Agriculture Organization of the United Nations, Rome.

FAO (1967a). *Actual and potential role of man-made forests in the changing world pattern of wood consumption.* Secretariat Note. Proceedings of the World Symposium on Man-made Forests and their Industrial Importance, pp. 1–50. Food and Agriculture Organization of the United Nations, Rome.

FAO (1967*b*). *Taungya in Kenya: the shamba system*. Proceedings of the World Symposium on Man-made Forests and their Industrial Importance, pp. 1057–1068. Food and Agriculture Organization of the United Nations, Rome.

FAO (1974). *An introduction to planning forestry development*. Food and Agriculture Organization of the United Nations, Rome.

FAO (1975). *Pulping and papermaking properties of fast-growing plantation wood species*. Food and Agriculture Organization of the United Nations, Rome.

FAO (1978). *Forestry for rural communities*. Food and Agriculture Organization of the United Nations, Rome.

FAO (1981). *Forest resources assessment project 1980*. Food and Agriculture Organization of the United Nations, Rome.

FAO (1982). *Forestry in China*. FAO Forestry Paper 35. Food and Agriculture Organization of the United Nations, Rome. 307 pp.

FAO (1984*a*). *Land, food and people*. Food and Agriculture Organization of the United Nations, Rome.

FAO (1984*b*). *Land evaluation forforestry*. FAO Forestry Paper No. 48, 123 pp. Food and Agriculture Organization of the United Nations, Rome.

FAO (1985*a*). *Tropical forestry action plan*. Committee on Forest Development in the Tropics. Food and Agriculture Organization of the United Nations, Rome.

FAO (1985*b*). *Tree growing by rural people*. FAO Forestry Paper 64, 130 pp. Food and Agriculture Organization of the United Nations, Rome.

FAO (1986*a*). *Forestry extension organization*. FAO Forestry Paper 66. Food and Agriculture Organization of the United Nations, Rome.

FAO (1986*b*). *Guide for training in the formulation of agricultural and rural investment projects*. Food and Agriculture Organization of the United Nations, Rome. (Introduction and 5 parts.)

FAO (1987). *Appropriate wood harvesting techniques in plantation forests*. Food and Agriculture Organization of the United Nations, Rome.

FAO (1988*a*). *FAO/UNESCO soil map of the world. 1 : 5 000 000*. Revised legend. World Soil Resources Report 60. Food and Agriculture Organization of the United Nations, Rome. (Digitized 1991.)

FAO (1988*b*). *An interim report on the state of forest resources in the developing countries*. FO; MISC/88/7 Food and Agriculture Organization of the United Nations, Rome.

FAO (1989*a*). *Review of forest management systems of tropical Asia*. FAO Forestry Paper 89. Food and Agriculture Organization of the United Nations, Rome.

FAO (1989*b*). *Management of tropical moist forests in Africa*. FAO Forestry Paper 88. Food and Agriculture Organization of the United Nations, Rome.

FAO (1989*c*). *The role of forestry in combating desertification*. Conservation Guide No. 21. Food and Agriculture Organization of the United Nations, Rome.

FAO (1989*d*). *Wasteland development for fuelwood and other rural needs*. Regional Wood Energy Development Programme in Asia, GCP/RAS/111/NET. Field Document 19. Food and Agriculture Organization of the United Nations, Bangkok.

FAO (1992). *Mixed and pure forest plantations in the tropics and subtropics*. FAO Forestry Paper 103. Food and Agriculture Organization of the United Nations, Rome.

FAO (1994). *Mangrove forest management guidelines*. FAO Forestry Paper 117. Food and Agriculture Organization of the United Nations, Rome.

FAO (1995). *Planning for sustainable use of land resources: towards a new approach*. Land and Water Bulletin 2. Food and Agriculture Organization of the United Nations, Rome.

FAO (1998). *Global forest products consumption, production, trade and prices: global forest products model projections to 2010*. Working Paper GFPOS/WP/01 Food and Agriculture Organization of the United Nations, Rome.

FAO (1999*a*). *State of the world's forests 1999*. Food and Agriculture Organization of the United Nations, Rome.

FAO (1999*b*). *Forestry information notes on combating desertification*. Food and Agriculture Organization of the United Nations, Rome.

FAO (2001*a*). *Global forest resources assessment 2000. Main report*. FAO Forestry Paper 140. Food and Agriculture Organization of the United Nations, Rome.

FAO (2001*b*). *State of the world's forests 2001*. Food and Agriculture Organization of the United Nations, Rome.

FAO (2003). *Harmonizing forest-related definitions for use of various stakeholders*. Second Expert Meeting, Rome, 11–13 September 2002, 323 pp. Food and Agriculture Organization of the United Nations, Rome.

FAO/UNEP/ISRIC (1999). *Soil and terrain databases for Latin America and the Caribbean*. Land and Water Digital Media Series 5. Food and Agriculture Organization of the United Nations, Rome.

Fatimson, T. (1989). *Drylands agroforestry. Homestead trees and nurseries to support them. NGO experience in Tamil Nadu, south India*. Social Forestry Network Paper 9b. Overseas Development Institute, London.

Felker, P. (ed.) (1986). *Tree plantings in semi-arid regions*. Proceedings of a Symposium on Establishment and Productivity of Tree Plantings in Semi-arid Regions, Kingsville, Texas. Elsevier, Amsterdam, The Netherlands.

Felker, P., Smith, D., and Wiesmann, C. (1986). Influence of mechanical and chemical weed control on growth and survival of tree plantings in semi-arid regions. In *Tree plantings in semi-arid regions* (ed. P. Felker), pp. 259–267. Proceedings of a Symposium on Establishment and Productivity of Tree Plantings in Semi-arid Regions, Kingsville, Texas. Elsevier, Amsterdam, The Netherlands.

Fenton, R., Roper, R.E., and Watt, G.R. (1977). *Lowland tropical hardwoods. An annotated bibliography of selected species with plantation potential*. Ministry of Foreign Affairs, Wellington, New Zealand.

Fenton, R.T. (1967). *Rotations in man-made forests*. Proceedings of the World Symposium on Man-made Forests and their Industrial Importance, pp. 600–614. Food and Agriculture Organization of the United Nations, Rome.

Ferguson, I.S. (1996). *Sustainable forest management*. Oxford University Press, Oxford, UK.

Fernandez, E.B. (1997). Strategies for strengthening watershed management in tropical mountain areas. In *Protective and environmental functions of forests*, Vol. 2. Proceedings of the XI World Forestry Congress, Antalya, Turkey, 13–22 October 1997, pp. 247–255. Ministry of Forestry, Ankara, Turkey.

Fernandes, V. and Arrieche, R. (1982). Ensayos de herbicidas de uso pre-emergente para control de malezas en viveros de Pinus caribaea var. caribaea. Venezuela Forestal **6**, 38–64.

Fernandez, J.Q.P., Dias, L.E., Barros, N.F., Novais, R.F., and Moraes, E.J. (2000). Productivity of *Eucalyptus camaldulensis* affected by rate and placement of two phosphorus fertilizers to a Brazilian Oxisol. *Forest Ecology and Management*, **127**, 93–102.

Ferreira, F.A. and Muchovej, J.J. (1991). Diseases of forest nurseries in Brazil. In *Diseases and insects in forest nurseries* (ed. J.R. Sutherland and S.G. Glover), pp. 17–23. Proceedings of the First Meeting of IUFRO Working Party S2.07–09, Victoria, British Columbia, Canada, 22–30 August 1990. Information Report BC-X-331. Pacific Forestry Centre, Pacific and Yukon Region, Forestry Canada, Victoria, British Columbia.

Field, C. (1996). General guidelines for the restoration of mangrove ecosystems. In *Restoration of mangrove ecosystems* (ed. C.D. Field), pp. 233–250. International Society for Mangrove Ecosystems, Okinawa, Japan.

Field, C.D. (ed.) (1996). *Restoration of mangrove ecosystems*. International Society for Mangrove Ecosystems, Okinawa, Japan.

Fiji Pine Commission (1988). *Annual Report 1988*. Fiji Pine Commission, Lautoka, Fiji.

Fiji Pine Commission (1989). *Annual Report 1989*. Fiji Pine Commission, Lautoka, Fiji.

Fiji Pine Limited (2001). *Annual Report 2000*. Fiji Pine Ltd, Lautoka, Fiji.

Fimbel, R.A. and Fimbel, C.C. (1996). The role of exotic conifer plantations in rehabilitating degraded forest lands: a case study from Kibale Forest in Uganda. *Forest Ecology and Management*, **81**, 215–226.

Finch, A.C. (1977). Examples of Zambian plantation costs. In *Savanna afforestation in Africa*, FAO Forest Paper 11, pp. 254–258. Food and Agriculture Organization of the United Nations, Rome.

Finlayson, W. (1998). *Effects of deforestation and of tree planting on the hydrology of the Upper Mahaweli catchment: a review of published evidence*. Environment and Forest Conservation Division, Mahaweli Authority of Sri Lanka.

Finn, M. (2002). Modeling with Woodstock: an Australian case study. *New Zealand Journal of Forestry*, **46**(4), 15–18.

Fladung, M. (1999). *Transgenic trees for a better world* (ed. E.S. Ritter), pp. 339–345. Proceedings of Application of Biotechnology to Forest Genetics. Biofor-99, Vitoria-Gasteiz, Spain, 22–25 September 1999.

Florence, R.G. (1996). *Ecology and silviculture of eucalypt forests*. CSIRO, Melbourne, Australia.

Floyd, R.B. and Hauxwell, C. (ed.) (2001). Hypsipyla *shoot borers in Meliaceae*. ACIAR Proceedings No. 97. Australian Centre for International Agricultural Research, Canberra.

Follas, C. (2002). Kyoto Protocol—carbon sequestration rights. *New Zealand Journal of Forestry*, **46**(4), 24–28.

Fölster, H. and Khanna, P.K. (1997). Dynamics of nutrient supply in plantation soils. In *Management of soil, nutrients and water in tropical plantation forests* (ed. E.K.S. Nambiar and A.G. Brown), pp. 339–378. ACIAR Monograph No. 43. Australian Centre for International Agriculture Research, CSIRO Australia and Center for International Forestry Research, Indonesia.

Ford-Robertson, F.C. (ed.) (1971). *Terminology of forest science, technology, practice and products*. IUFRO/Society of American Foresters, Washington DC.

Forest Restoration Research Unit (2000). *Tree seeds and seedlings for restoring forests in northern Thailand*. Biology Department, Chiang Mai University, Chiang Mai, Thailand.

Foresta, H. de and Michon, G. (1993). Creation and management of rural agroforests in Indonesia: potential applications in Africa. In *Tropical forests, people and food. Biocultural interactions and applications for development* (ed. C.M. Hladik, A. Hladik, O.F. Linares, H. Pagezy, A. Semple, and M. Hadley), Man and the Biosphere Series, Vol. 13. UNESCO, Paris and Parthenon Publishing Group, Lancaster, UK.

Forge, K. and Black, A. (1998). Developing effective extension and advisory strategies for farm forestry in Australia. In *Plantation and regrowth forestry a diversity of opportunity*. (ed. R. Dyason, L. Dyason, and R. Garsden), pp. 436–437. Proceedings of Australian Forest Growers Biennial Conference Southern Cross University, Lismore, NSW, Australia.

Fortes, M. (1995). Causes of failure (and success?) of mangrove restoration in the Philippines. In *Ecology and management of mangrove restoration and regeneration in East and Southeast Asia* (ed. C. Khenmark), pp. 129–141. Proceedings of Ecotone IV, UNESCO-MAB, Bangkok.

Fortmann, L. (1988). Great planting disasters: pitfalls in technical assistance to forestry. *Agriculture and Human Values*, **5**(1/2), 49–60.

Fortmann, L. and Bruce, J.W. (ed.) (1988). *Whose trees? Proprietary dimensions of forestry*. Westview Press, Boulder, USA.

Foster, G.S. and Bertolucci, F.L.G. (1994). Clonal development and deployment: strategies to enhance gain while minimizing risk. In *Tropical trees: potential for domestication and the rebuilding of forest resources* (ed. R.R.B. Leakey and A.C. Newton), pp. 103–110. Her Majesty's Stationery Office, London.

Foster, P.G. and Costantini, A. (1991*a*) *Pinus* plantation establishment in Queensland: I. Field surveys for site preparation planning and site design. *Australian Forestry*, **54**, 75–82.

Foster, P.G. and Costantini, A. (1991*b*) *Pinus* plantation establishment in Queensland: II. Site preparation classes. *Australian Forestry*, **54**, 83–89.

Foster, P.G. and Costantini, A. (1991c) *Pinus* plantation establishment in Queensland: III. Site preparation design. *Australian Forestry*, **54**, 90–94.

Fox, J.E.D. (1976). *Environmental constraints on the possibility of natural regeneration after logging in tropical moist forest.* Proceedings of the XVI IUFRO World Congress, Div. 1, pp. 512–538.

Foy, T. (2001). Incentives for sustainable private sector management of forests in the public and private interest: privatisation and the leasing of state forests in South Africa. *International Forestry Review*, **3**, 223–230.

Foy, T.J. and Pitcher, M.J. (1999). Private sector/community partnerships for commercial forestry development: recent experiences from South Africa. *International Forestry Review*, **1**, 222–227.

Francis, G.K. (1986). Performance of exotic tree species in Puerto Rico. In *Management of the forests of tropical America: prospects and technologies* (ed. J.C.F. Colon, F.H. Wadsworth and S. Branham), pp. 377–388. Conference at Institute of Tropical Forestry, Rio Pedras, Puerto Rico, September 1986.

Francis, P.J. (1984). The role of cultivation in plantation establishment in subtropical eastern Australia. In Grey, D. C. et al. (ed.) (1984), pp. 579–87.

Francis, P.A. and Attah-Krah, A.N. (1989). Sociological and ecological factors in technology adoption: fodder trees in southeast Nigeria. *Experimental Agriculture*, **25**, 1–10.

Francis, P.J. and Shea, G.M. (1991). Management of tropical and subtropical pines in Queensland. In *Forest management in Australia* (ed. F.H. McKinnell, E.R. Hopkins, and J.E.D. Fox), pp. 203–213. Surrey Beatty, Chipping Norton, NSW, Australia.

Franklin, E.C. (1989). Selection strategies for eucalypt tree improvement—four generations of selection in Eucalyptus grandis demonstrate valuable methodology. In *Breeding Tropical Trees: population structure and genetic improvement, strategies in clonal and seedling forestry* (ed. G.L. Gibson, A.R. Griffin, and A.C. Matheson), pp. 197–209. Proceedings of the IUFRO Conference, Pattaya, Thailand, November 1988, Oxford Forestry Institute.

Franzel, S., Jaenicke, H., and Janssen, W. (1996). *Choosing the right trees. Setting priorities for multipurpose tree improvement.* ISNAR Research Report 8, International Service for National Agricultural Research, The Hague.

Fraser, A. (2000). *The role of financial/banking institutions in timber plantation development.* Proceedings of the International Conference on Timber Plantation Development, Manila, 7–9 November 2000, pp. 309–319. ITTO, FAO and Department of Environment and Natural Resources, Manila.

Fraser, A.I. (1973). *A manual on the planning of man-made forest.* Food and Agriculture Organization of the United Nations, Rome.

Fraser, A.I. (1976). *A manual on the management of plantation forests.* International Forest Science Consultancy, Edinburgh.

Freezaillah, C.Y. and Sandrasegaran, K. (1969). *Preliminary observations on the rooting characteristics of* Pinus caribaea *Mor. grown in Malaya.* Research Pamphlet No. 58, Forestry Research Institute, Kepong, Malaya.

Freitas, M., de Silva, A.P., da Gutierrez Netom F., and Caneva, R.A. (1979). Interplanting as an alternative in successive rotations of *Eucalyptus* stands. *Instituto de Pesquisas et Estudos Florestais (Publicacao Semestral)*, **19**, 1–6.

Freitas, M., Silva, A.P., Kageyama, P.Y., and Ferreira, M. (1983). The research programme with *Eucalyptus grandis* Hill ex Maiden at Champion Papel e Celulose S.A. *Silvicultura Sao Paulo*, **31**, 537–539.

Fryer, J.H. (1996). Site sampling and performance prediction for *Eucalyptus camaldulensis* in Central America. In *Matching trees and sites* (ed. T.H. Booth), pp. 112–117. ACIAR Proceedings No. 63, Australian Centre for International Agricultural Research, Canberra.

Fujisaka, S. (1991). The role of socioeconomic and policy research in effecting technology adoption. In *Technologies for sustainable agriculture on marginal uplands in Southeast Asia* (ed. G. Blair and R. Lefroy), pp. 77–82. ACIAR Proceedings No. 33, Australian Centre for International Agricultural Research, Canberra.

Gadgil, P.D. and Bain, J. (1999). Vunerability of planted forests to biotic and abiotic disturbances. *New Forests*, **17**, 227–238.

Gadgil, P.D., Wardlaw, T.J., Ferreira, F.A., Sharma, J.K., Dick, M.A., Wingfield, M.J., and Crous, P.W. (2000). Management of disease in eucalypt plantations. In *Diseases and pathogens of eucalypts* (ed. P.J. Keane, G.A. Kile, F.D. Podger, and B.N. Brown), pp. 519–529. CSIRO, Melbourne.

Galiana, A. and Prin, Y. (1996). The benefits of rhizobial inoculation in tropical forestry: response of *Acacia mangium* to the application of selected strains of *Bradyrhizobium* spp. In *Tree improvement for sustainable tropical forestry.* (ed. M.J. Dieters, A.C. Matheson, D.G. Nikles, C.E. Harwood, and S.M. Walker), pp. 449–454. Proceedings of the QFRI-IUFRO Conference, Caloundra, Australia, October/November 1996, Queensland Forestry Research Institute, Gympie, Australia.

Gallegos, C.M. (1981). Flowering and seed production of *Pinus caribaea* var. *hondurensis* (preliminary results of a worldwide survey). FAO Forest Genetic Resources Information No. 10, 17–22

Garrity, D.P. (1996). Tree-soil-crop interactions on slopes. In *Tree-crop interactions: a physiological approach* (ed. C.K. Ong and P. Huxley), pp. 299–318. CAB International, Wallingford, UK.

Garrity, D.P. (ed.) (1997). Agroforestry innovations for *Imperata* grassland rehabilitation. *Agroforestry Systems*, **36**, 1–284.

Garthe, R.J. (1983). *Establishment* of E. pilularis *on the sandmined areas of Fraser Island.* Report No. 11. Queensland Department of Forestry, Brisbane, Australia. (Unpublished.)

Gautier, D. (1995). The pole-cutting practice in the Bamileke country (Western Cameroon). *Agroforestry Systems*, **31**, 21–37.

Gava, J.L. (2002). Cultivo minimo de solos com textura arenosa e média em áreas planas e suave-onduladas.

In *Conservação e cultivo de solos para plantações florestais* (ed. J.L.M. Gonçalves and J.L. Stape), pp. 221–243. Instituto de Pesquisas e Estudos Florestais, Piracicaba (SP), Brazil.

Gaviria, D. (1997). Economic and financial instruments for sustainable forestry in Colombia. *Unasylva*, **48**(188), 32–35.

Geldenhuys, C.J. (1997). Native forest regeneration in pine and eucalypt plantations in Northern Province, South Africa. *Forest Ecology and Management*, **99**, 101–115.

Gelens, H.F. (1984). Land evaluation for forestation. In *Strategies and designs for afforestation, reforestation and tree planting* (ed. K.F. Wiersum), pp. 219–230. Pudoc, Wageningen, The Netherlands.

Gentle, S.W. and Humphreys, F.R. (1968). *Experience with phosphatic fertilizer in man-made forests of* Pinus radiata *in New South Wales*. Proceedings of the 9th Commonwealth Forestry Conference, Delhi.

Germishuizen, P.J. (1984). *Rhizina undulata as a pine seedling pathogen in southern Africa*, Vol. 2 (ed. D.C. Grey, A.P.G. Schönau, and C.J. Schultz), pp. 753–765. Proceedings of IUFRO Symposium on Site Productivity of Fast Growing Plantations. South Africa Forestry Research Institute, Pretoria and Pietermaritzburg.

Ghassemi, F., Jakeman, A.J., and Nix, H.A. (1995). Salinisation of land and water resources. Human causes, extent, management and case studies. University of New South Wales Press, Sydney, Australia.

Gholz, H.L. and Lima, W.P. (1997). The ecophysiological basis for productivity in the tropics. In *Management of soil, nutrients and water in tropical plantation forests* (ed. E.K.S. Nambiar and A.G. Brown), pp. 213–246. ACIAR Monograph No. 43. Australian Centre for International Agricultural Research, CSIRO Australia and Center for International Forestry Research, Indonesia.

Ghosh, R.C. (1978). *Evaluating and analysing environmental impacts of forests in India*, Vol. 7A. Proceedings of the 8th World Forestry Congress, Jakarta, pp. 475–484.

Ghosh, R.C. and Rao, B.K. (1979). Forests and floods. *Indian Forester*, **105**, 249–259.

Ghosh, R.G. and Kaul, G.N. (1976). *Effect of standard silvicultural systems on major uses and forest types in India*. Proceedings of the 16th IUFRO World Congress, Div. 1, pp. 447–462.

Gibson, A. and Bachelard, E.P. (1994). Relationships between site characteristics and survival strategies of *Eucalyptus camaldulensis* seedlings. In *Australian tree species research in China* (ed. A.G. Brown), pp. 91–95. ACIAR Proceedings No. 48 Australian Centre for International Agricultural Research, Canberra.

Gibson, G.L. and Barnes, R.D. (1985). Availability of Central American pines for ex situ conservation stands, provenances resource stands, breeding populations and provenance testing. FAO Forest Genetic Resources Information No. **13**, 37–41.

Gibson, A., Bachelard, E.P., and Hubick, K.T. (1995). Relationship between climate and provenance variation in *Eucalyptus camaldulensis* Dehnh. *Australian Journal of Plant Physiology*, **22**, 453–460.

Gibson, G.L., Barnes, R.D., and Berrington, J. (1988). Provenance productivity in Pinus caribaea and its interaction with environment. *Commonwealth Forestry Review* **62** (2), 93–106.

Gibson, G.L., Barnes, and Berrinton (1988). (Ref in 2nd edn.)

Gibson, G.L., Griffin, A.R., and Matheson, A.C. (ed.) (1989). *Breeding tropical trees: population structure and genetic improvement strategies in clonal and seedling forestry*. Oxford Forestry Institute, Oxford and Winrock International, Arlington, United States.

Gibson, I.A.S. (1975). *Diseases of forest trees widely planted as exotics in the tropics and southern hemisphere. Part I: Important members of the Myrtaceae, Leguminoseae, Verbenaceae, Meliaceae*. Commonwealth Forestry Institute, Oxford.

Gibson, I.A.S. (1979). *Diseases of forest trees widely planted as exotics in the tropics and Southern Hemisphere. Part II. The genus* Pinus. Commonwealth Mycological Institute, London.

Gibson, I.A.S. and Jones, T. (1977). Monoculture as the origin of major forest pests and diseases. In *Origins, of pest, parasite, disease and weed problems* (ed. J.M. Cherrett and G.R. Sagar), pp. 139–161. Blackwell, Oxford.

Gill, C.J. (1970). Flooding tolerance of woody species—a review. *Forestry Abstracts*, **31**, 671–688.

Gill, R.I.S. and Gosal, S.S. (1996). Micropropagation of economically important tropical forest trees. In *Tree improvement for sustainable tropical forestry* (ed. M.J. Dieters, A.C. Matheson, D.G. Nikles, C.E. Harwood, and S.M. Walker), pp. 230–233. Proceedings of QFRI-IUFRO Conference, Caloundra, Australia, October/November 1996, Queensland Forestry Research Institute, Gympie, Australia.

Ginneken, P. van and Thongmee, U. (1991). Attempting integrated watershed development in Phu Wiang, Thailand. *Unasylva*, **42**(164), 8–15.

Giraldo, L.G., del Valle, J.I., and Escobar, M. (1981). Growth of salmwood, *Cordia alliodora* (Ruix & Pavon) Oken, in relation to some climatic, edaphic, and physiographic factors in southwestern Antioquia, Colombia. In *Wood production in the neotropics via plantations* (ed. J.L. Whitmore), pp. 30–41. Proceedings of the IUFRO/MAB/Forest Service Symposium, Rio Pedras, Puerto Rico, September 1980.

Gleed, J.A. (1993). Development of plantlings and stecklings of radiata pine. In *Clonal forestry II: conservation and application* (ed. W.J. Libby and M.R. Ahuja), pp. 149–157. Springer-Verlag, Berlin Heidelberg, Germany.

Glover, N. (1987). Establishment and maintenance of NFT plantings. In *Nitrogen fixing trees—a training guide*, pp. 73–95. RAPA Publication, 1987/15. FAO, Bangkok.

Goldammer, J.G., Frost, P.G.H., Jurvélius, M., Kamminga, E.M., Kruger, T., Moody, S.I., and Pogeyed, M.L. (2002). Community participation in intergrated forest fire management: experiences from Africa, Asia and Europe. In *Communities in flames* (ed. P. Moore, D. Ganz, L.C. Tan, T. Enters, and P.B. Durst), pp. 32–52. Proceedings of an International Conference on Community Involvement in Fire Management. FAO Regional Office for Asia and the Pacific, Bangkok.

Golfari, L. and Caser, R.L. (1977). *Zoneamento ecologico da Regiao Nordeste para Eperimentacao Florestal.* PRODEPEF, PNUD/FAO/IBDF/BRA-45. Serie Technica No. 10. Brasilia, Brazil.

Golfari, L., Caser, R.L., and Moura, V.P.G. (1978). *Zoning for reforestation in Brazil and trials with tropical* Eucalyptus *and pines in central region.* UNDP/FAO Technical Report 12, Brazilian Institute for Forest Development Project BRA/76/027. Brasilia, Brazil.

Gonçalves, J.L.M. and Benedetti, V. (ed.) (2000). *Nutrição e fertilização florestal.* Instituto de Pesquisas e Estudos Florestais, Piracicaba (SP), Brazil.

Gonçalves, J.L.M. and Stape, J.L. (ed.) (2002). *Conservação e cultivo de solos para plantações florestais.* Instituto de Pesquisas e Estudos Florestais, Piracicaba (SP), Brazil.

Gonçalves, J.L.M. *et al.* (1999). Eucalypt plantations in the humid tropics: São Paulo, Brazil. In *Site management and productivity in tropical plantation forests* (ed. E.K.S. Nambiar, C. Cossalter, and A. Tiarks), pp. 5–12. Center for International Forestry Research, Bogor, Indonesia.

Gonçalves, J.L.M., Barros, N.F., Nambiar, E.K.S., and Novais, R.F. (1997). Soil and stand management for short rotation plantations. In *Management of soil, nutrients and water in tropical plantation forests* (ed. E.K.S. Nambiar and A.G. Brown), pp. 379–417. ACIAR Monograph No. 43. Australian Centre for International Agricultural Research, Canberra.

Gonçalves, J.L.M., Dematte, J.L.I., and Couto, H.T.Z. (1990). Relacões entre a productividade de sitios florestais de *Eucalyptus grandis* e *E. saligna* com as propriedades de alguns solos de textura arenosa e media no Estado de São Paulo. *IPEF,* **43/44**, 24–39.

Gonçalves, J.L.M., Serrano, M.I.P., Mendes, K.C.F.S., and Gava, J.L. (2000). Effects of site management in a *Eucalyptus grandis* plantation in the humid tropics: São Paulo, Brazil. In *Site management and productivity in tropical plantation forests: a progress report* (ed. E.K.S. Nambiar, A. Tiarks, C. Cossalter, and J. Ranger), pp. 3–9. Center for International Forestry Research, Bogor, Indonesia.

Gonslaves, J. (1990). Agroforestry seeds. *Sustainable Agriculture Newsletter,* **2**(3). CUSO, Bangkok.

Gonzalez, J.E. and Fisher, R.F. (1994). Growth of native forest species planted on abandoned pasture land in Costa Rica. *Forest Ecology and Management,* **70**, 159–167.

Goodall, J. and Klein, H. (2000). Invader plant control in forestry. In *South African. Forestry Handbook* (ed. D.L. Owen), pp. 253–264. South African Institute of Forestry, Pretoria, South Africa.

Goor, A.Y. and Barney, C.W. (1968). *Forest tree planting in arid zones.* The Ronald Press, New York, USA.

Gordon, A.G., Gosling, P.G., and Wang, B.S.P. (ed.) (1991). *Tree and shrub handbook.* International Seed Testing Association, Zurich, Switzerland.

Gosling, P.G. (1996). International standards and the testing of tropical tree seed. In *Innovations in tropical tree seed technology.* IUFRO Symposium of the Project Group P.2.04.00 'Seed Problems' (ed. K. Olesen), pp. 71–81. Danish Forest Tree Seed Centre, Humlebaek, Denmark.

Göttle, A. and Sene, E.M. (1997). Protective and environmental functions of forests. In *Protective and environmental functions of forests,* Vol. 2, Proceedings of the XI World Forestry Congress, Antalya, Turkey, 13–22 October 1997, pp. 233–243. Ministry of Forestry, Ankara, Turkey.

Grace, K. (2000). Certification of plantations and plantation timber. In *Proceedings of the International Conference on Timber Plantation Development, Manila, 7–9 November 2000,* pp. 297–308. ITTO, FAO and Department of Environment and Natural Resources, Manila.

Grainger, A. (1988*a*). Future supplies of high-grade tropical hardwoods from intensive plantations. *Journal of World Forest Resources Management,* **3**, 15–29.

Grainger, A. (1988*b*). Estimating areas of degraded tropical lands requiring replenishment of forest cover. *International Tree Crops Journal,* **5**(1/2), 31–62.

Grainger, A. (1990). *The threatening desert.* Earthscan, London.

Granhof, J. (1984). Extraction of pine seed by means of sun drying on elevated trays followed by tumbling. Technical Note No. 17, DANIDA Forest Seed Centre, Denmark.

Grayson, A.J. (2000). Introduction. In *Aid to Forestry.* Special Issue. *International Forestry Review,* **2**, 150–153.

Grayson, A.J. and Maynard, W.B. (1997). *The World's forests—Rio + 5: international initiatives towards sustainable management.* Commonwealth Forestry Association, Oxford, UK.

Greacen, E.L. and Sands, R. (1980). Compaction of forest soils: a review. *Australian Journal of Soil Research,* **18**, 163–189.

Greaves, A. (1980). *Review of* Pinus caribaea *Mor. and* Pinus oocarpa *Schiede international provenance trials.* Commonwealth Forestry Institute, University of Oxford, Oxford.

Greaves, A. (1981). Progress in *Pinus caribaea* Morelet and *Pinus oocarpa* Scheide international provenance trials. *Commonwealth Forestry Review,* **60**, 35–43.

Greaves, A. and McCarter, P.S. (1990). Cordia alliodora. *A promising tree for tropical agroforestry.* Tropical Forestry Paper No. 22. Oxford Forestry Institute, Oxford.

Greenwood, D.E. (1976). Nursery design and irrigation. In *Savanna afforestation in Africa.* FAO Forest Paper 11, pp. 86–92. Food and Agriculture Organization of the United Nations, Rome.

Gregersen, H., Arnold, J.E.M., Lundgren, A., and Contreras, H.A. (1995). *Valuing forest: context, issues and guidelines.* FAO Forestry Paper 127. Food and Agriculture Organization of the United Nations, Rome.

Gregersen, H., Arnold, J.E.M., Lundgren, A., Contreras, H.A., Montalembert, M.R. de., and Gow, D. (1993). *Assessing forestry project impacts: issues and strategies.* FAO Forestry Paper 114. Food and Agriculture Organization of the United Nations, Rome.

Gregersen, H., Draper, S., and Elz, D. (1989). *People and trees: the role of social forestry in sustainable development.* Economic Development Institute, World Bank, Washington DC.

Gregersen, H.M. (1984). Incentives for forestation: a comparative assessment. In *Strategies and designs for afforestation, reforestation and tree planting* (ed. K.F. Wiersum), pp. 301–311. Pudoc, Wageningen, The Netherlands.

Gregor, E.W. (1972). *Integration of grazing in tropical forestry—an experiment in combining cattle raising with pine plantations in Fiji*, Vol. 3. Proceedings of the 7th World Forestry Congress, Buenos Aires, pp. 3551–3556.

Greig, B.J.W. and Foster, L.E.P. (1982). *Fomes annosus* in pine plantations of Jamaica. *Commonwealth Forestry Review*, **61**, 269–275.

Grewal, S.S., Abrol, I.P., and Singh, O.P. (1987). Rainwater management for establishing agroforestry on alkali soils. *Indian Journal of Agriculture*, **57**(1), 30–37.

Grey, D.C. (1979). Site quality prediction for *Pinus patula* in the Glengarry area, Transkei. *South African Forestry Journal*, **111**, 44–48.

Griffin, A.R. (1990). Effects of inbreeding on the growth of forest trees and implications for management of seed supplies for plantation programs. In *Reproductive ecology of tropical forest plants* (ed. K.S. Bawa and M. Hadley), pp. 363–382. UNESCO, Paris and Parthenon, Carnforth, UK.

Griffin, A.R. (1996). Genetically modified trees—the plantations of the future or an expensive distraction? *Commonwealth Forestry Review*, **75**, 169–175.

Griffin, A.R. (2001). *Deployment decisions—capturing the benefits of tree improvement with clones and seedlings*. Proceednngs of IUFRO International Symposium on developing the eucalypt of the future. Valdivia, Chile, 10–15 September 2001, Invited Paper, pp. 1–16. (Compact Disk.)

Griffin, R., Harbard, J., Centurion, C., and Santini, P. (2000). Breeding *Eucalyptus grandis* × *globulus* and other interspecific hybrids with high inviability—problem analysis and experience at Shell Forestry projects in Uruguay and Chile. In *Symposium on hybrid breeding and genetics, Noosa, Australia, 9–14 April 2000* (ed. H.S. Dungey, M.J. Dieters, and D.G. Nikles), pp. 1–13. Department of Primary Industries, Brisbane, Australia. (Compact Disk).

Griffith, A.L. (1942). *Teak plantation technique*. Indian Forest Records, **5**, 123–219.

Griffith, A.L. and Gupta, R.S. (1948). Soils in relation to teak with special reference to laterisation. *Indian Forestry Bulletin*, **141**.

Grove, T.S. and Le Tacon, F. (1993). Mycorrhiza in plantation forestry. *Advances in Plant Pathology*, **9**, 191–227.

Guggenberger, C., Ndulu, P., and Shepherd, G. (1989). *After Ujamaa: farmer needs, nurseries and project sustainability in Mwanza, Tanzania*. Social Forestry Network Paper 9c. Overseas Development Institute, London.

Guhardja, E., Fatawi, M., Sutisna, M., Mori, T., and Ohta, S. (ed.) (2000). *Rainforest ecosystems of East Kalimantan*. Ecological Studies 140. Springer-Verlag, Tokyo.

Guizol, P. and Cossalter, C. (2000). *Assessing stakeholder agreements: a new research focus for CIFOR's Plantation Programme in Southeast Asia*. Proceedings of the International Conference on Timber Plantation Development, 7–9 November 2000, Manila, pp. 345–352.

ITTO, FAO and Department of Environment and Natural Resources, Manila.

Gunawan, M., Susilawati, S.H., Budhi, G.S., and Rusastra, I.W. (1998). *Paraserianthes falcataria* growing in Sumdang, West Java. In *Improving smallholder farming systems in Imperata areas of Southeast Asia: alternatives to shifting cultivation* (ed. K. Menz, D. Magcale-Macandog, and I.W. Rusastra), pp. 45–54. ACIAR Monograph No. 52. Australian Centre for International Agricultural Research, Canberra.

Gunn, B. (2001). *Australian Tree Seed Centre operations manual*. CSIRO Forestry and Forest Products, Canberra.

Guofang, S. (1990). *Choice of species in China's plantation forestry*. Beijing Forestry University, Beijing.

Gupta, G.N. (1995). Rain water management for tree planting in Indian desert. *Journal of Arid Environments*, **31**, 219–235.

Gupta, B.N. and Kumar, A. (1976). Estimation of potential germinability of teak (Tectona grandis Linn. f) fruits from twenty-three Indian sources by cutting test. Indian *Forester* **102**, 808–13.

Gupta, G.N., Sarita Mutha, and Limba, N.K. (2000). Growth of *Albizia lebbeck* on micro-catchments in the Indian arid zone. *International Tree Crops Journal*, **10**, 193–202.

Gupta, J.P. and Muthana, K.D. (1986). Effect of integrated moisture conservation technology on early growth and establishment of *Acacia tortilis* in the Indian desert. *Indian Forester*, **111**, 477–485.

Guzman, R.S. (1999). Multipurpose tree technologies for the Philippine uplands. In *Domestication of agroforestry trees in Southeast Asia* (ed. J.M. Roshetko and D.O. Evans), pp. 27–31. Forest, Farm and Community Tree Research Reports. Special Issue 1999. Taiwan Forestry Research Institute; Winrock International, Morrilton, Arkansas, USA; ICRAF, Nairobi.

Habit, M.A. and Saavedra, J.C. (ed.) (1988). *The current state of knowledge on* Prosopis juliflora. Plant Production and Protection Division, Food and Agriculture Organization of the United Nations, Rome.

Habiyambere, T. and Musabimana, F. (1992). Effect of spacing on the growth and production of *Grevillea robusta* in the semi-arid region of Bugesera in Rwanda. In Grevillea robusta *in Agroforestry and Forestry* (ed. C.E. Harwood), pp. 99–102. International Centre for Research in Agroforestry, Nairobi.

Hackett, C. (1991). *Plantgro, a software package for coarse prediction of plant growth*. CSIRO, Melbourne.

Hackett, C. (1996). A study of forest scientists' perceptions of tree environmental relationships: implications for predicting growth. In *Matching trees and sites*. (ed. T.H. Booth), pp. 77–84. ACIAR Proceedings No. 63. Australian Centre for International Agricultural Research, Canberra.

Hadi, S. (1978). *Problems associated with production of* Pinus merkusii *nursery stock*, Vol. 5. Proceedings of the Eighth World Forestry Congress, Jakarta, pp. 1331–1340.

Haines, R.J. (1994). *Biotechnology in forest tree improvement*. FAO Forestry Paper 118. Food and Agriculture Organization of the United Nations, Rome.

Haines, R.J. and Griffin, A.R. (1992). Propagation options for *Acacia mangium, Acacia auriculiformis* and their hybrid. In *Breeding techniques for tropical acacias* (ed. L.T. Carron and K.M. Aken), pp. 122–127. ACIAR Proceedings No. 37. Australian Centre for International Agricultural Research, Canberra.

Haines, R.J. and Nikles, D.G. (1987). Seed production in *Araucaria cunninghamii*—the influence of biological features of the species. *Australian Forestry*, **50**, 224–230.

Haishui, Z. (1988). The role of *Eucalyptus* plantations in southern China. In *Multipurpose tree species for small-farm use* (ed. D. Withington, K.G. MacDicken, C.B. Sastry, and N.R. Adams), pp. 79–85. Proceedings of a Symposium, Pattaya, Thailand, November, 1987. Winrock International and IDRC, Bangkok.

Hall, M. (1985). Tolerance of *Eucalyptus, Acacia* and *Casuarina* seedlings to pre-emergent herbicides. *Australian Forestry*, **48**, 264–266.

Hall, N. *et al.* (1972). *The use of trees and shrubs in the dry country of Australia.* Australian Government Publishing Service, Canberra.

Halos, S.C. and Go, N.R. (1990). *Micropropagation of two tropical pines*: Pinus caribaea *and* Pinus kesiya, Vol. 1. Proceedings of the 19th IUFRO World Congress. Montreal, Canada, pp. 28–39.

Ham, C. and Theron, J.M. (1999). Community forestry and woodlot development in South Africa: the past, present and the future. *South African Forestry Journal*, **184**, 71–79.

Hamilton, L.S. and King, P.N. (1983). *Tropical forested watersheds—hydrologic and soils response to major uses or conversions.* Westview Press, Colorado, USA.

Hamilton, L.S. and Pearce, A.J. (1985). *What are the soil and water benefits of planting trees in developing country watersheds?* Working Paper, East-West Center, Hawaii, USA.

Hamilton, P.C., Chandler, L.R., Brodie, A.W., and Cornelius, J.P. (1998). A financial analysis of small-scale *Gmelina arborea* Roxb. improvement program in Costa Rica. *New Forests*, **16**, 89–99.

Hance, R.J. and Holly, K. (ed.) (1990). *Weed control handbook: principles*, 8th edn. Blackwell Scientific Publications.

Hanum, F.I. and Maesen, L.J.G. van der (ed.) (1997). *Auxiliary plants.* PROSEA No. 11. Backhuys, Leiden, The Netherlands.

Harahap, R.M. and Soerianegara, I. (1977). *Heritability of some characteristics in teak.* Paper to Third World Consultation Forest Tree Breeding, Canberra.

Harcharik, D.A. (1997). The future of world forestry: sustainable forest management. In *Main report*, Vol. 7. Proceedings of the XI World Forestry Congress, Antalya, Turkey, 13–22 October 1997, pp. 148–152. Ministry of Forestry, Ankara, Turkey.

Harding, R.J., Hall, R.L., Swaminath, M.H., and Srinavasa Murthy, K.V. (1991). The soil moisture regimes beneath forest and agricultural crops in southern India – measurements and modelling. Proceedings of the Bangalore Seminar on Growth and Water Use of Forest Plantations. February 1991.

Hardiyanto, E.B. (1998). Approaches to breeding acacias for growth and form: the experience of PT. Musi Hutan Persada (Barito Pacific Group). In *Recent developments in acacia planting* (ed. J.W. Turnbull, H.R. Crompton, and K. Pinyopusarerk), pp. 178–183. ACIAR Proceedings No. 82. Australian Centre for International Agricultural Research, Canberra.

Hardiyanto, E.B., Ryantoko, A., and Anshori, S. (2000). Effects of site management in *Acacia mangiun* plantations at PT. Musi Hutan Persada, South Sumatra, Indonesia. In *Site management and productivity in tropical plantation forests: a progress report* (ed. E.K.S. Nambiar, A. Tiarks, C. Cossalter, and J. Ranger), pp. 41–49. Center for International Forestry Research, Bogor, Indonesia.

Hardwick, K., Healey, J., Elliott, S., Garwood, N., and Anusarnsunthorn, V. (1997). Understanding and assisting natural regeneration processes in degraded seasonal evergreen forests in northern Thailand. *Forest Ecology and Management*, **99**, 203–214.

Harrington, C. (1999). Forests planted for ecosystem restoration or conservation. *New Forests*, **17**, 175–190.

Harrison, S.R. and Herbohn, J. (ed.) (2000). *Socio-economic evaluation of the potential for Australian tree species in the Philippines.* ACIAR Monograph No. 75, Australian Centre for International Agricultural Research, Canberra.

Harrison, S.R., Herbohn, J.L., and Herbohn, K.F. (ed.) (2001). *Sustainable small-scale forestry: socio-economic analysis and policy.* Edward Elgar, Cheltenham, UK.

Hartley, A. (1977). The establishment of *Eucalyptus tereticornis* on tailings of the Bougainville copper mine, Papua New Guinea. *Commonwealth Forestry Review*, **56**, 239–245.

Hartmann, H.T., Kester, D.E., Davies, F.T. and Geneve, R.L. (1997). Plant propagation, principles and practice. 6th ed. Prentice Hall International, London.

Harvett, C. (2000). A manager's view of tree improvement—towards designer fibres: maximizing value from applied tree improvement. In *Forest genetics for the next millennium.* Proceedings of an international conference, Durban, South Africa, 8–13 October 2000, pp. 21–23. Institute of Commercial Forestry Research, Scottsville, South Africa.

Harwood, C. (1985). *Guidelines for large-scale fuel wood plantations in the Pacific Islands.* UN Pacific Energy Development Programme Report REG 84–10 rev 9/85, Suva, Fiji. 37 pp.

Harwood, C.E. (1989). Grevillea robusta: *an annotated bibliography.* CSIRO, Canberra, and International Council for Research in Agroforestry, Nairobi, Kenya.

Harwood, C.E. (1992). Natural distribution and ecology of *Grevillea robusta.* In Grevillea robusta *in agroforestry and forestry* (ed. C.E. Harwood), pp. 21–28. Proceedings of an International Workshop. International Centre for Research in Agroforestry, Nairobi, Kenya.

Harwood, C.E. (1998). Eucalyptus pellita—*an annotated bibliography.* CSIRO Forestry and Forest Products, Canberra.

Harwood, C.E. and Williams, E.R. (1992). A review of provenance variation in growth of *Acacia mangium.* In *Breeding technologies for tropical acacias* (ed. L.T. Carron and K.M. Aken), pp. 22–30. ACIAR Proceedings No. 37.

Australian Centre for International Agricultural Research, Canberra.

Hatchwell, G.E., Ralston, C.B., and Foil, R.R. (1970). Soil disturbances in logging. *Journal of Forestry*, **68**, 772–775.

Haufe, H. (1977). *Agrisilvicultural techniques as an alternative to shifting cultivation in Latin America.* Paper to FAO/SIDA Seminar on Conservation and Land Use in Latin America, Peru.

Hauxwell, C., Mayhew, J., and Newton, A. (2001). Silvicultural management of *Hypsipyla* species. In *Hypsipyla shoot borers in Meliaceae* (ed. R.B. Floyd and C. Hauxwell), pp. 151–163. ACIAR Proceedings No. 97. Australian Centre for International Agricultural Research, Canberra.

Hawkins, B.J. (1996). *Planting stock quality assessment* (ed. A.C. Yapa), pp. 107–111. Proceedings: International Symposium on Recent Advances in Tropical Forest Tree Seed Technology and Planting stock Production. ASEAN Forest Tree Seed Centre Project, Mauk-Lek, Saraburi, Thailand.

Haywood, J.D., Tiarks, A.E., and Sword, M.A. (1997). Fertilization, weed control, and pine litter influence loblolly pine stem productivity and root development. *New Forests*, **14**, 233–249.

He, F. and Hu, F. (1991). Tree crop forestry in China. In *Development of forestry science and technology in China* (ed. J. Deng, D. Yunyi, W. Li, and Z. Chen), pp. 71–78. China Science and Technology Press, Beijing.

Headley, M. (2001). National forest policies and strategies: fundamental or academic fad. In *Forests in a changing landscape*. Proceedings of 16th Commonwealth Forestry Conference/19th Biennial Conference of the Institute of Foresters of Australia, pp. 263–270. Promaco Conventions, Canning Bridge, Western Australia.

Hedegart, T. (1976). Breeding systems, variation and genetic improvement in teak (*Tectona grandis* L.f.). In *Tropical trees* (ed. J. Burley and B.T. Styles), pp. 109–123. Academic Press, London.

Hedge, N.G. (1997). Community forestry for sustainable development. In *Social dimensions of forestry's contribution to sustainable development*, Vol. 5. Proceedings of the XI World Forestry Congress, Antalya, Turkey, 13–22 October 1997, p. 82. Ministry of Forestry, Ankara, Turkey.

Heinrich, R. (1987). Appropriate wood harvesting in plantation forest in developing countries. In *Appropriate wood harvesting techniques in plantation forests*, pp. 79–94. Food and Agriculture Organization of the United Nations, Rome.

Helin, W.H. (1989). Species elimination trial Luuq, Gedo region, Somalia. *Commonwealth Forestry Review*, **68**, 273–279.

Hendreck, K.A. (1985). *Potting mixes and care of plants growing in them.* Discovering soils. No. 9. CSIRO, Melbourne, Australia.

Herbert, M.A. (1990*a*). Fertilizer/site interactions on the growth and foliar nutrient levels of *Eucalyptus grandis* in Natal. *Forest Ecology and Management*, **30**, 247–257.

Herbert, M.A. (1990*b*). Fertilization of trees at planting. *ICFR Annual Research Report 1990*, pp. 86–98. Institute for Commercial Forest Research, Scottsville, South Africa.

Herbert, M.A. (1991). The influence of site factors on the foliar nutrient content of *Eucalyptus grandis* in Natal. *South African Forestry Journal*, **156**, 28–34.

Herbert, M.A. (1996). Fertilizers and eucalypt plantations in South Africa. In *Nutrition of eucalypts* (ed. P.M. Attiwill and M.A. Adams), pp. 303–325. CSIRO Publishing, Melbourne, Australia.

Herbohn, J. (2001). Prospects for small-scale forestry in Australia. In *Economic sustainability of small-scale forestry* (ed. A. Niskanen and J. Väyrynen), pp. 9–20. EFI Proceedings No. 36. European Forest Institute, Joensuu, Finland.

Herbohn, J.L., Harrison, S.R., Herbohn, K.F., and Smorfitt, D.B. (ed.) (2001). *Developing policies to encourage small-scale forestry.* Proceedings from an IUFRO International Symposium, Kuranda, Australia, 9–13 January 2000. The University of Queensland, Brisbane, Australia.

Herrero, C., Blanco, J., Garcia, A., Geigel, E.B., and Romero, F. (1988). Effect of dose and type of phosphate on the development of *Pinus caribaea* var. *caribaea*. I. quartizite ferrallitic soil. *Agrotecnia de Cuba*, **20**, 7–16.

Hewlett, J.D. (1982). *Principles of forest hydrology*. University of Georgia Press, Athens, USA.

Heywood, J.D., Tiarks, A.E., and Sword, M.A. (1997). Fertilization, weed control, and pine litter influence loblolly pine stem productivity and root development. *New Forests*, **14**, 233–249.

Higa, A.R. and Resende, M.D.V. (1994). Breeding *Acacia mearnsii* in southern Brazil. In *Australian tree species research in China* (ed. A.G. Brown), pp. 158–160. ACIAR Proceedings No. 48. Australian Centre for International Agricultural Research, Canberra.

Higashi, E.N., Silveira, R.L.V.A., and Goncalves, A.N. (2000). Propagação vegetative de *Eucalyptus*; principios básicos e a sua evolucao no Brasil. Circular Tecnica 192, IPEF-ESALQ-USP, Sao Paulo, Brazil.

Higman, S., Bass, S., Judd, N., Mayers, J., and Nussbaum, R. (1999). *The sustainable forestry handbook*. Earthscan, London. 289 pp.

Hillis, W.E. and Brown, A.G. (ed.) (1984). *Eucalypts for wood production*, 2nd edn. CSIRO and Academic Press, Melbourne.

Hippi, I. and Rissanen, H. (1996). Modern technology in forest management planning: Ensofris and Ensovideo. In *Reforestation: meeting the future industrial wood demand* (ed. A. Otsamo, J. Kuusialpo, and H. Jaskari), pp. 42–54. Enso Forest Development Oy Ltd, Jakarta.

Hocking, D. (ed.) (1993). *Trees for dry lands*. International Science, New York.

Hocking, D., Hocking, A., and Ray, I. (1997*a*). Trees in Bangladesh paddy fields and homesteads: the future of forestry in crowded countries. *Commonwealth Forestry Review*, **76**, 255–260.

Hocking, D., Sarwar, M.G., and Youseff, M.F.A. (1997*b*). Trees on farms in Bangladesh. 4. Crop yields under traditionally managed mature trees. *Agroforestry Systems*, **35**, 1–13.

Hodges, C.S. and McFadden, M.W. (1986). Insects and diseases affecting forest plantations in tropical America. In *Management of the forests of tropical America: prospects and technologies* (ed. J.C.F. Colon, F.H. Wadsworth, and S. Branham), pp. 365–376. Conference at Institute of Tropical Forestry, Rio Pedras, Puerto Rico, September 1986.

Hogg, B. and Nester, M. (1991). Productivity of direct thinning regimes in south-east Queensland hoop pine plantations. *Commonwealth Forestry Review*, **70**, 37–45.

Holdridge, L.R. (1947). Determination of world plant formations from simple climatic data. *Science*, **105**(2727), 367–368.

Hollingsworth, I.D., Boardman, R., and Fitzpatrick, R.W. (1996). A soil-site evaluation index of productivity in intensively managed *Pinus radiata* (D. Don) plantations in South Australia. *Environmental Monitoring and Assessment*, **39**, 531–541.

Homer, F.M. (1997). A case for reforestation with native species in Trinidad, West Indies. In *Productive functions of forest*, Vol. 3. Proceedings of the XI World Forestry Congress, Antalya, Turkey, p. 61. Ministry of Forestry, Ankara, Turkey.

Hong Kong (1989). *Progress report 1985–88, Agriculture and Fisheries Department, Hong Kong*. 13th Commonwealth Forestry Conference, New Zealand.

Hong, P.N. (1995). Problems of mangrove degradation in Vietnam: results of restoration and management. In *Ecology and management of mangrove restoration and regeneration in East and Southeast Asia* (ed. C. Khenmark), pp. 142–153. Proceedings of Ecotone IV, UNESCO-MAB, Bangkok.

Hong, P.N. (2001). Reforestation of mangroves after severe impacts of herbicides during the Vietnam war: the case of Can Gio. *Unasylva*, **52**(207), 57–60.

Hosaka, R.T. and Schneider, P.R. (1984). Forest production planning in southern Brazil. *Allgemeine Forst-und Jagdzeitung*, **155**(1), 5–8.

Hosking, G.P. and Gadgil, P.D. (1987). Development of contingency plans for use against exotic pests and diseases of trees and timber. 4. Forest insect and disease protection in New Zealand: an integrated approach. *Australian Forestry*, **50**, 37–39.

Hoskins, M. (1990). The contribution of forestry to food security. *Unasylva*, **160**, 3–13.

Howland, P. (1969). Effects of singling coppice in *Eucalyptus saligna* wood fuel crop at Mugugu, Kenya. *East African Agricultural and Forestry Journal*, **35**, 66–67.

Howlett, D. (1993). *Environmental management*. Study No. 10. Shell/WWF Tree Plantation Review, Shell International Petroleum Company and World Wide Fund for Nature, London.

Hudson, N.W. (1963). Raindrop size distribution in high intensity storms. *Rhodesian Journal of Agricultural Research*, **1**, 6–11.

Hughes, C.E. (1989). New opportunities in Leucaena genetic improvement. In *Breeding tropical trees: population structure and genetic improvement strategies in clonal and seedling forestry* (ed. G.I. Gibson, A.R. Griffin, and A.C. Matheson), pp. 218–226. Proceedings of IUFRO Conference, Pattaya, Thailand. Oxford Forestry Institute/Winrock, Oxford.

Hughes, C.E. (1994). Risks of species introduction in tropical forestry. *Commonwealth Forestry Review* **73**, 243–252.

Hulme, M. (1996). *Climate change and Southern Africa: an exploration of some potential impacts and implications in the SADC region*. Climate Research Unit, University of East Anglia, UK and WWF International, 104 pp.

Hulugalle, N.R., Lal, R., and Ter Kuile, C.H.H. (1984). Soil physical changes and crop root growth following different methods of land clearing in western Nigeria. *Soil Science*, **138**, 172–179.

Hunt, M.A. (2001). Nutrient loading or nutrient starvation? Effects on survival and growth of planting stock. In *Site selection and productivity estimation* (ed. M. Cherry and C. Beadle), pp. 48–50. Technical Report No. 55. Cooperative Research Centre for Sustainable Production Forestry, Hobart, Australia.

Hurtig, E. (1987). Vocational training, safety, health and ergonomics in forestry work. In *Appropriate wood harvesting techniques in plantation forests*, pp. 167–174. Food and Agriculture Organization of the United Nations, Rome.

Husch, B. (1987). *Guidelines for forest policy formulation*. FAO Forestry Paper 81, Food and Agriculture Organization of the United Nations, Rome.

Hussain, A. and Ibrahim, M. (1987). Evaluation of *Sesbania spinosa* leaves applied as a green manure supplement to inorganic fertilisers. *Nitrogen Fixing Tree Research Reports*, **5**, 63–64.

Hussain, R.W. and Cheema, M.A. (1987). Possibility of application of thinning grades in terms of basal area in *Eucalyptus camaldulensis* plantations. *Pakistan Journal of Forestry*, **37**(1), 21–25.

Hussin, Y.A. and Bijker, W. (2000). Remote sensing applications for sustainable management of forests. Abstract of paper to *Forests and Society: the Role of Research*. XXI IUFRO World Congress, Kuala Lumpur, 7–12 August 2000. A full version of the findings in the abstract is available at http://www.itc.nl/forestry/URS/

Hutacharern, C., MacDicken, K.G., Ivory, M.H., and Nair, K.S.S. (ed.) (1990). *Pests and diseases of forest plantations in Asia-Pacific*. FAO Regional Office for Asia-Pacific, Bangkok.

Huttel, C. and Loumeto, J.J. (2001). Effect of exotic plantations and site management on plant diversity. In *Effect of exotic tree plantations on plant diversity and biological soil fertility in Congo savanna: with special reference to eucalypts* (ed. F. Bernard-Reversat), pp. 9–18. Center for International Forestry Research, Bogor, Indonesia.

IDB (1995). Proceedings of a Workshop on the use of Financial Incentives for Industrial Forest Plantations. 19 January 1995, Paper ENV-4. Environment Division, Inter-American Development Bank, Washington DC.

Igboanugo, A.E.I., Omijeh, J.B., and Adegebehin, J.0. (1990). Pasture floristic composition in different Eucalyptus species plantations in some parts of the northern Guinea savanna zone of Nigeria. *Agroforestry Systems* **12**, 257–68.

Ikemori, Y.K. (1987). Epicormic shoots from the branches of *Eucalyptus grandis* as an explant source for in vitro culture. *Commonwealth Forestry Review*, **66**, 351–356.

ILO (1979). *Appropriate technology for the establishment and maintenance of forest plantations.* Paper to Regional Seminar on the Application of Appropriate Technology in Forestry and Forest-based Industries, Manila.

ILRI (1977). *A framework for land evaluation.* International Institute for Land Reclamation and Improvement. Publication No. 22. (Also FAO Soils Bulletin 32.)

International Seed Testing Association (1993). International rules for seed testing. *Seed Science and Technology*, **21**, 1–288.

IPF (1997). *Final report. Intergovernmental Panel on Forests.* United Nations, New York, USA.

Ismariah, A. and Aminah, H. (1994). The economics of large scale propagation of tropical timber species by stem cuttings in Malaysia. *Commonwealth Forestry Review*, **73**, 253–260.

ITTO (1993). *ITTO guidelines for the establishment and sustainable management of planted tropical forests.* ITTO Policy Development Series No. 4. International Tropical Timber Organization, Yokohama, Japan.

ITTO (2002). *ITTO guidelines for the restoration, management and rehabilitation of degraded and secondary tropical forests.* ITTO Policy Development Series No. 13. International Tropical Timber Organization, Yokohama, Japan.

Ivory, M.H. (1975*a*). Pathology of tree species in West Malaysia. *Commonwealth Forestry Review*, **54**, 64–68.

Ivory, M.H. (1975*b*). Pathology of *Pinus* spp. in West Malaysia. *Commonwealth Forestry Review*, **54**, 154–165.

Iyppu, A.J. and Chandrasekharan, C. (1961). *Thinnings in teak.* Proceedings of the 10th Silvicultural Conference, Dehra Dun, India, pp. 725–730.

Jackson, I.J. (1977). *Climate, water and agriculture in the tropics.* Longman, London.

Jackson, J.E. (1983). Light climate and crop-tree mixtures. In *Plant research and agroforestry* (ed. P.A. Huxley), pp. 365–378. International Council for Research in Agroforestry, Nairobi.

Jackson, J.K. (1977*a*). Results of nursery research. In *Savanna afforestation in Africa.* FAO Forest Paper 11, pp. 106–111. Food and Agriculture Organization of the United Nations, Rome.

Jackson, J.K. (1977*b*). Irrigated plantations. In *Savanna afforestation in Africa*, FAO Forest Paper 11, pp. 168–172. Food and Agriculture Organization of the United Nations, Rome.

Jackson, J.K. (1984). Why do plantations fail? In *Strategies and designs for afforestation, reforestation and tree planting* (ed. K.F. Wiersum), pp. 277–285. Pudoc, Wageningen, The Netherlands.

Jackson, N.A., Wallace, J.S., and Ong, C.K. (2000). Tree pruning as a means of controlling water use in an agroforestry system in Kenya. *Forest Ecology and Management*, **126**, 133–148.

Jacobs, M.R. (1979). *Eucalypts for planting.* FAO Forestry Series No. 11, U.N. Food and Agriculture Organization, Rome.

Jacobs, M.R. (1981). *Eucalypts for planting.* FAO Forestry Series No. 11, Food and Agriculture Organization of the United Nations, Rome.

Jaffré, T., Rigault, F., and Sarrailh, J.-M. (1994). La végétalisation des anciens sites miniers. *Bois et Forêts des Tropiques*, **242**(4), 45–57.

Jagawat, H. and Verma, D.P.S. (1989). *Nurseries in Gujarat, North India: two views.* Social Forestry Network Paper 9d, Overseas Development Institute, London.

Jenkinson, A. (2000). APRIL takes a leaf out of the green book. *Pulp and Paper International*, **42**(8), 19–21.

Jensen, A.M. and Hajej, M.S. (2001). The Road of Hope: control of moving sand dunes in Mauritania. *Unasylva*, **52**(207), 31–36.

Jodha, N.S. (1997). Trends in tree management in arid land use in western Rajasthan. In *Farms, trees and farmers: responses to agricultural intensification* (ed. J.E.M. Arnold and P.A. Dewees), pp. 43–64. Earthscan Publications, London.

Johansson, S. (1995). Forestry in irrigated agricultural schemes with special reference to the Burra Irrigation and Settlement Project, Kenya. *Tropical Forestry Reports* 10. Department of Forest Ecology, University of Helsinki, Finland.

Johnson, D.W. and Todd, D.E. (1990). Nutrient cycles in forests of Walker Beach Watershed, Tennessee: roles of uptake and leaching causing soil change. *Journal of Environmental Quality*, **19**, 97–104.

Johnston, D.R., Grayson, A.J., and Bradley, R.T. (1967). *Forest planning.* Faber and Faber, London.

Johri, R.B. (1978). Man-made forest in India. *Forest News for Asia and the Pacific*, Vol. 2. FAO, Bangkok, pp. 12–13.

Jolin, D. and Torquebiau, E. (1992). Large cuttings: a jump start for tree planting. *Agroforestry Today*, **4**(4), 15–16.

Jones, N. (1974). Records and comments regarding flowering of *Triplochiton scleroxylon*. *Commonwealth Forestry Review*, **53**, 52–56.

Jones, N. (1994). *Field experiences concerning forest seed quality problems.* Proceedings: International Symposium on Genetic Conservation and Production of Tropical Forest Tree Seed (ed. R.M. Drysdale, S.E.T. John, and A.C. Yapa), pp. 277–282. ASEAN-Canada Forest Tree Seed Center Project, Mauk-Lek, Saraburi, Thailand.

Jones, P.G. (1998). Geographic information systems in natural resource management domains. In *International workshop on resource management domains* (ed. J.K. Syers and J. Bouma), pp. 79–90. IBSRAM Proceedings No. 16. International Board for Soil Research and Management, Bangkok.

Jong, W. de, Chokkalingam, U., Smith, J., and Sabogal, C. (2001). Tropical secondary forests in Asia: introduction and synthesis. *Journal of Tropical Forest Science*, **13**, 563–576.

Jordan III, W.R., Gilpin, M.E., and Aber, J.D. (ed.) (1987). *Restoration ecology: a synthetic approach to ecological research.* Cambridge University Press, Cambridge, UK.

Josiah, S.J. (1990). *Containerized nursery production in Haiti: lessons learned and research needs*, Vol. 3. Proceedings of the 19th World IUFRO Congress, Montreal, pp. 102–108.

Josiah, S.J. and Jones, N. (1992). *Root trainers in seedling production systems for tropical forestry and agroforestry.* Land Resources Series No. 4. Asia Technical Department, Agriculture Division, The World Bank, Washington DC.

Judd, T.S., Attiwill, P.M., and Adams, M.A. (1996). Nutrient concentrations in *Eucalyptus*: a synthesis in relation to differences between taxa, sites and components. In *Nutrition of eucalypts* (ed. P.M. Attiwill and M.A. Adams), pp. 123–153. CSIRO Publishing, Melbourne, Australia.

Kadeba, O. (1978). Nutrition aspects of afforestation with exotic tree species in the savanna regions of Nigeria. *Commonwealth Forestry Review*, **57**, 191–199.

Kadeba, O. and Aduayi, E.A. (1982). Biomass production in *Pinus caribaea* of different ages in the Savanna zone of Nigeria. In *IUFRO Symposium on forest site and continuous productivity* (ed. R. Ballard and S.P. Gessel), pp. 53–57. USDA General Technical Report PNW-163, 1983. Seattle, USA.

Kadir, W.R. and van Cleemput, O. (1995). *Nutrient translocation during the early growth of two exotic plantation species* (ed. A. Schulte and D. Ruhiyat), pp. 96–100. Proceedings of the International Congress on Soils of Tropical Forest Ecosystems 3rd Conference on Forest Soils, Vol. 6, Nutrient cycling/ecosystem studies. Mulawarman University Press, Samarinda, Indonesia.

Kaimowitz, D. (2000). Forestry assistance and tropical deforestation: why the public doesn't get what it pays for. *International Forestry Review*, **2**, 225–231.

Kaimowitz, D. and Angelsen, A. (1998). *Economic models of tropical deforestation: a review.* Center for International Forestry Research, Bogor, Indonesia.

Kaimowitz, D., Byron, R.N., and Sunderlin, W. (1998). Public policies to reduce inappropriate tropical deforestation. In *Agriculture and the environment: perspectives on sustainable rural development* (ed. E. Lutz, H. Binswanger, P. Hazell, and A. McCalla), pp. 302–322. World Bank, Washington DC.

Kandya, A.K. (1987). Forest seed in India: research and problems. In Kamra and Ayling (ed.) (1987), pp. 124–42.

Kane, M. (1989). *La supervivencia y el crecimiento inicial son buenos para Bombacopsis quinata plantado antes de la estacion de lluvias.* Informacion Investigaciones No. 7. Monterey Forestal Ltd, Cartagena, Colombia.

Kang, B.T. (1993). Alley cropping: past achievements and future directions. *Agroforestry Systems*, **23**, 141–156.

Kang, B.T., Reynolds, I., and Atta-Krah, A.N. (1990). Alley farming. *Advances in Agronomy*, **43**, 315–359.

Kang, B.T., Salako, F.K., and Hulugalle, N.R. (1998). Effect of tillage and woody hedgerows in alley cropping on the productivity of an Alfisol. In *Soils of tropical forest ecosystems* (ed. A. Schulte and D. Ruhiyat), pp. 144–149. Springer, Berlin and Heidelberg, Germany.

Kanowski, P., Savill, P.S., Adlard, P.G., Burley, J., Evans, J., Palmer, J.R., and Wood, P.J. (1991). *Plantation forestry.* World Bank Forestry Policy Issues Paper, Oxford Forestry Institute, Oxford.

Kanowski, P.J. (1997). Afforestation and plantation forestry for the 21st century. In *Productive functions of forests*, Vol. 3. Proceedings of the XI World Forestry Congress, Antalya, Turkey, 13–22 October 1997, pp. 23–34. Ministry of Forestry, Ankara, Turkey.

Kanowski, P.J. and Savill, P.S. (1992). Forest plantations: towards sustainable practice. In *Plantation politics* (ed. C. Sargent and S. Bass), pp. 121–155. Earthscan Publications, London.

Kaosa-ard, A. (1982). Storage technique of teak (*Tectona grandis* Linn. f.) planting stumps. In *Tropical forests: source of energy through optimisation and diversification* (ed. P.B.L. Srivastava, A.M. Ahmad, K. Awang, A. Muktar, R.A. Kader, F. Che'Yom, and S.S. Lee), pp. 327–333. Universiti Pertanian, Serdang, Malaysia.

Kaplan, E.D. (1996). *Understanding GPS: principles and applications.* Artech House Publishers, Boston, USA.

Karani, P.K. (1978). Pruning and thinning in a *Pinus patula* stand at Lendu plantation, Uganda. *Commonwealth Forestry Review*, **54**, 269–278.

Kartodihardjo, H. and Supriono, A. (2000). *The impact of sectoral development on natural forest conversion and degradation: the case of timber and tree crop plantations in Indonesia.* Occasional Paper No. 26(E). Center for International Forestry Research, Bogor, Indonesia.

Kaul, R.N. (ed.) (1970). *Afforestation in arid zones.* Junk, The Hague.

Kaumi, S.Y.S. (1983). Four rotations of a Eucalyptus fuel yield trial. *Commonwealth Forestry Review* **62** (1), 19–24.

Kaumi, S.Y.S. (1980). *The effects of height of pruning on the growth of* Cupressus benthamii *and* Pinus radiata. Forestry Technical Note No. 40. Kenya Agricultural Research Institute, Nairobi.

Kazmierczak and Shimabukuru (1994). *Geoprocessing as a support for forest production planning.* Proceedings of the ISPRS Commission VII Symposium, Rio de Janeiro, Brazil, 26–30 September 1994. National Institute for Space Research, São José dos Campos, São Paulo, Brazil. *Resource and Environmental Monitoring*, **30**(7a), 264–270.

Keane, P.J., Kile, G.A., Podger, F.D., and Brown, B.N. (ed.) (2000). *Diseases and pathogens of eucalypts.* CSIRO, Melbourne.

Keenan, R. and Grant, A. (2000). *Implications of the Kyoto Protocol for plantation development.* Proceedings of the International Conference on Timber Plantation Development, Manila, 7–9 November 2000, pp. 103–119. ITTO, FAO and Department of Environment and Natural Resources, Manila.

Keenan, R., Lamb, D., Woldring, O., Irvine, T., and Jensen, R. (1997). Restoration of plant biodiversity beneath tropical tree plantations in northern Australia. *Forest Ecology and Management*, **99**, 117–131.

Keenan, R., Sexton, G., and Lamb, D. (1999). Thinning studies in plantation-grown Queensland maple (*Flindersia brayleyana*). *International Forestry Review*, **1**, 71–78.

Keeves, A. (1966). Some evidence of loss of productivity with successive rotations of *Pinus radiata* in the south east of S. Australia. *Australian Forestry*, **30**, 51–63.

Keiding, H. (1993). Tectona grandis *Linn. F.* Seed Leaflet No. 4. Danida Forest Seed Centre, Humlebaek, Denmark.

Keipi, K. (1997). Financing forest plantations in Latin America: government incentives. *Unasylva*, **48**(188), 50–56.

Kelley, H.W. (1983). *Keeping the land alive: soil erosion-its causes and cures*. FAO Soils Bulletin 50, 84 pp. Food and Agriculture Organization of the United Nations, Rome.

Kellogg, G.E. and Orvedal, A.C. (1969). Potentially arable soils of the world and critical measures for their use. *Advances in Agronomy* **21**, 109–70.

Kengan, S. (1997). *Forest valuation for decision-making. Lessons of experience and proposals for improvement*. Food and Agriculture Organization of the United Nations, Rome.

Keogh, R.M. (1980*a*). Teak (*Tectona grandis* Linn. F.) provenances of the Caribbean, Central America, Venezuela and Colombia. In *Wood production in the neotropics via plantations* (ed. J.L. Whitmore), pp. 30–41. Proceedings of the IUFRO/MAB/Forest Service Symposium, Rio Pedras, Puerto Rico, September 1980.

Keogh, R.M. (1980*b*). Teak (*Tectona grandis* Linn. F.): volume growth and thinning practice in the Caribbean, Central America, Venezuela and Colombia. In *Wood production in the neotropics via plantations* (ed. J.L. Whitmore), pp. 58–71. Proceedings of the IUFRO/MAB/Forest Service Symposium, Rio Pedras, Puerto Rico, September 1980.

Keogh, R.M. (1987). *The care and management of teak* (Tectona grandis *L. f.*) *plantations*. Universidad Nacional, Heredia, Costa Rica.

Kerkhoff, P. (1990). *Agroforestry in Africa: a survey of project experience*. Panos, London.

Kermani, W.A. (1974). *Aerial seed sowing in riverain forest areas of Sind*. Paper presented at a Forestry Conference. Pakistan Forest Institute, Peshawar, Pakistan.

Keys, M.G., Dieters, M., Walker, S.M., and Huth, J.R. (1996). *Forest productivity gains from use of top quality seed and nursery plants: two case studies from Queensland, Australia* (ed. A.C. Yapa), pp. 14–20. Proceedings: International Symposium on Recent Advances in Tropical Forest Tree Seed Technology and Planting Stock Production. ASEAN Forest Tree Seed Centre Project, Mauk-Lek, Saraburi, Thailand.

Kha, L.D. (1996). Studies on natural hybrids of *Acacia mangium* and *A. auriculiformis* in Vietnam. In *Tree improvement for sustainable tropical forestry*. (ed. M.J. Dieters, A.C. Matheson, D.G. Nikles, C.E. Harwood, and S.M. Walker), pp. 328–332. Proceedings of QFRI-IUFRO Conference, Caloundra, Australia, October/November 1996, Queensland Forestry Research Institute, Gympie, Australia.

Kha, L.D. (2001). *Studies on the use of natural hybrids between* Acacia mangium *and* Acacia auriculiformis *in Vietnam*. Agricultural Publishing House, Hanoi.

Khaleque, K. (1988). Growing multipurpose fruit trees in Bangladesh—farmers perspectives of opportunities and obstacles. In *Multipurpose tree species for small-farm use* (ed. D. Withington, K.G. MacDicken, C.B. Sastry, and N.R. Adams), Proceedings of a Symposium, Pattaya, Thailand, November, 1987. Winrock International and IDRC, Bangkok.

Khan, I. (1987). *Wastelands afforestation. Techniques and systems*. Oxford and IBH Publishing, New Delhi.

Khan, M.A.W. and Chaudhary, N.R. (1961). *Eucalypt trials in India*. Proceedings of the 10th All-India Silvicultural Conference, Dehra Dun, pp. 535–545.

Khanduja, S.D., Chandra, V., Srivastava, G.S., Jan, R.K., Misra, P.N., and Garg, V.K. (1987). Utilization of alkali soils on the plans of northern India—a case study. In *Amelioration of soil by trees* (ed. R.T. Prinsley and M.J. Swift), pp. 54–61. Commonwealth Science Council, London.

Kijkar, S. (1991). *Producing rooted cuttings of* Eucalyptus camaldulensis. ASEAN-Canada Forest Tree Seed Centre, Mauk-Lek, Saraburi, Thailand.

Kilander, K. (1985). *Forest industries in developing countries—possibilities and constraints*. Paper E-II.2.A. Ninth World Forestry Congress, Mexico.

King, K.F.S. (1975). It's time to make paper in the tropics. *Unasylva*, **109**, 2–5.

King, K.F.S. (1979). Agroforestry and the utilisation of fragile ecosystems. *Forest Ecology and Management*, **2**, 161–168.

Kirsten, J.F., Tribe, G.D., van Rensburg, N.J., and Atkinson, P.R. (2000). Insect pests in South African forest plantations. In *South African forestry handbook* (ed. D.L. Owen), pp. 221–239. South African Institute of Forestry, Pretoria, South Africa.

Kittredge, J. (1948). *Forest influences*. McGraw-Hill, New York.

Knabe, G. (1967). *Man-made forest on man-made ground*. World Symposium on Man-made Forests and their Industrial Importance, pp. 1165–1173. Food and Agriculture Organization of the United Nations, Rome.

Knowles, O.H. and Parrotta, J.A. (1995). Amazonian forest restoration: an innovative system for native species selection based on phonological data and field performance studies. *Commonwealth Forestry Review*, **74**, 230–243.

Knudsen, O.K. (2000). The World Bank's forest policy and strategy. In Aid to Forestry. Special Issue. *International Forestry Review*, **2**, 169–170.

Kondas, S. (1982). Mysore gum coppice growth: vigour, productivity and regulation of cutting. In *Tropical forests-source of energy through optimisation and diversification* (ed. P.B.L. Srivastava, A.M. Ahmad, K. Awang, A. Muktar, R. A. Kader, F. Che' Yom, and S.S. Lee), pp. 317–325. Proceedings of an International Forestry Seminar, Selangor, Malaysia, November 1980. Universiti Pertanian, Serdang, Malaysia.

Koppen, W. (1923). *Die klimate der erde*. Walter de Gruyter, Berlin.

Kowal, K.M. and Kassam, A.H. (1976). Energy load and instantaneous intensity of rain storms at Samara, Northern Nigeria. *Tropical Agriculturist* (Trinidad), **53**, 185–198.

Kozlowski, T.T. and Constantinidou, H.A. (1986). Responses of woody plants to environmental pollution. Part 2. Factors affecting responses to pollution and alleviation of pollution effects. *Forestry Abstracts*, **47**, 105–132.

Kretzschmar, D.E. (1991). Site preparation for the growth of *Eucalyptus* species in South Africa. In *Intensive forestry: the role of eucalypts*, Vol. 1, (ed. A.P.G. Schönau), pp. 532–537. South African Institute of Forestry, Pretoria, South Africa.

Kriedemann, P.E. and Cromer, R.N. (1996). The nutritional physiology of the eucalypts—nutrition and growth. In *Nutrition of eucalypts* (ed. P.M. Attiwill and M.A. Adams), pp. 109–121. CSIRO Publishing, Melbourne, Australia.

Krishna-Murthy, A.V.G.R. (1976). *Bibliography on teak*: Tectona grandis *Linn. F.* Jugal Kishore, India.

Krishnapillay, B. and Tompsett, P.B. (1998). Seed handling. In *A review of dipterocarps: taxonomy, ecology and silviculture* (ed. S. Appanah and J.M. Turnbull.), pp. 73–88. Center for International Forestry Research, Bogor, Indonesia.

Kumar, A. (1979). Effect of fruit size and source on germination of teak (*Tectona grandis Linn f.*) seeds. *Sri Lanka Forester* 14, 58–63.

Kunkle, S.H. (1978). *Forestry support for agriculture through watershed management, windbreaks, and other conservation measures*, Vol. 3, pp. 113–146. Proceedings of the 8th World Forestry Congress, Jakarta.

Laarman, J.G. and Contreras, A. (1991). Benefits from development assistance projects in forestry: does the available evidence paint a true picture? *Unasylva*, 164, 45–49.

Ladipo, D.O. (1985). *Comparative study of the growth of seedlings, vegetative propagules, ortets and the effects of topophysis on the growth of* Triplochiton scleroxylon *cuttings*. Research Paper (Forest Series) No. 51, Forest Research Institute of Nigeria, Ibadan, Nigeria.

Ladipo, D.O., Britwum, S.P.K., Tchoundjeu, Z., Oni, O., and Leakey, R.R.B. (1994). Genetic improvement in West African tree species: past and present. In *Tropical trees: potential for domestication and the rebuilding of forest resources* (ed. R.R.B. Leakey and A.C. Newton), pp. 239–248. Her Majesty's Stationery Office, London.

Ladrach, W.E. (1980). Thinning of *Pinus patula* by the mechanical and selective method: results at 10 years. In *Wood production in the neotropics via plantations* (ed. J.L. Whitmore), pp. 155–164. Proceedings of the IUFRO/MAB/Forest Service Symposium, September 1980. Forest Service, Rio Pedras, Puerto Rico.

Ladrach, W.E. (1983). Genetic gains of *Cupressus lusitanica* through six years of tree improvement in Colombia. *Silvicultura*, 30, 343–346.

Ladrach, W.E. (1992). Plantation establishment techniques in tropical America. *Tree Planters' Notes*, 43, 125–132.

Lal, P. (2001). Private sector forestry research: a success story from India. *Bois et Forêts des Tropiques*, 267(1), 33–48.

Lal, R. (1989). Agroforestry systems and soil surface management of a tropical alfisol: (1) soil moisture and crop yields. *Agroforestry Systems*, 8, 7–29.

Lal, R. (1995). Erosion-crop productivity relationships for soils of Africa. *American Journal of Soil Science Society*, 59, 661–667.

Lal, R. (1997). Soils of the tropics and their management for plantation forestry. In *Management of soil, nutrients and water in tropical plantation forests* (ed. E.K.S. Nambiar and A.G. Brown), pp. 95–123. ACIAR Monograph No. 43. Australian Centre for International Agricultural Research, CSIRO Australia and Center for International Forestry Research, Indonesia.

Lam, D.C. (1998). Fixation des dunes vives par *Casuarina equisetifolia* au Vietnam. *Bois et Forêts des Tropiques*, 256(2), 35–41.

Lamb, A.F.A. (1955). Trinidad's teak forests. *Journal of the Agricultural Society of Trinidad and Tobago* (Society Paper No. 800), 3–10.

Lamb, A.F.A. (1969). Artificial regeneration within the humid lowland tropical forest. *Commonwealth Forestry Review*, 48, 41–53.

Lamb, A.F.A. (1973). *Pinus caribaea*. Fast growing timber trees in the lowland tropics No. 6. Department of Forestry, Oxford.

Lamb, D. (1975). *Kunjingini Plantations, 1965–1975*. Tropical Forestry Research Note SR 24. Office of Forests, Papua New Guinea.

Lamb, D. (1976). *Reforestation following chipwood logging. A review of silvicultural research at the Gogol valley*. Tropical Forestry Research Note SR 34. Office of Forests, Papua New Guinea.

Lamb, D. (1977). Relationships between growth and foliar nutrient concentrations in *Eucalyptus deglupta*. *Plant and Soil*, 47, 495–508.

Lamb, D. (1993). Restoration of degraded forest ecosystems for nature conservation. In *Conservation biology in Australia and Oceania* (ed. C. Moritz and J. Kikkawa), pp. 101–114. Surrey Beatty, Chipping Norton, UK.

Lamb, D. (1994). Reforestation of degraded tropical forest lands in the Asia-Pacific region. *Journal of Tropical Forest Science*, 7(1), 1–7.

Lamb, D. (1998). Large-scale ecological restoration of degraded tropical forest lands: the potential role of timber plantations. *Restoration Ecology*, 6, 271–279.

Lamb, D. (2001). Reforestation. In *Encyclopedia of biodiversity* (ed. S.A. Levin), pp. 97–108. Academic Press, San Diego, USA.

Lamb, D. and Tomlinson, M. (1994). Forest rehabilitation in the Asia-Pacific region: past lessons and present uncertainties. *Journal of Tropical Forest Science*, 7, 157–170.

Lambert, M.J. (1984). The use of foliar analysis in fertilizer research. *Proceedings of IUFRO Symposium on Site Productivity of Fast Growing Plantations*, Vol. 1 (ed. D.C. Grey, A.P.G. Schönau, and C.J. Schultz), pp. 269–291. South Africa Forestry Research Institute, Pretoria and Pietermaritzburg.

Landis, T.D., Tinus, R.W., McDonald, S.E., and Barnett, J.P. (1989). *Seedling nutrition and irrigation. Vol. 4. The container tree nursery manual*. Agricultural Handbook No. 674. US Department of Agriculture, Forest Service, Washington, DC.

Landis, T.D., Tinus, R.W., McDonald, S.E., and Bennett, J.P. (1990). *Nursery pests and mycorrhizae. Vol. 5. The container tree nursery manual*. Agricultural Handbook No. 674. US Department of Agriculture, Forest Service, Washington, DC.

Landsberg, H.E. (1961). Solar radiation at the earth's surface. *Solar Energy*, 5, 95–98.

Lanly, J.-P. (1982). Tropical forest resources. FAO Forestry Paper No. 30. Food and Agriculture Organization of the United Nations, Rome.

Lapis, E.B. and Genil, Z.N. (1979). Biology of *Ozola minor* (Moore), a defoliator of yemane (*Gmelina arborea*). *Sylvatrop. The Philippine Forest Research Journal*, **4**, 31–37.

Larsen, C.S. (1956). *Genetics in silviculture*. Oliver and Boyd, London.

Last, I. (2001). *Pinus* plantations on the coastal lowlands of south-east Queensland. In *Site selection and productivity estimation: a CRC-SPF Workshop* (ed. M. Cherry and C. Beadle), pp. 9–12. Technical Report No. 55. CRC for Sustainable Production Forestry, Hobart, Australia.

Lauridsen, E.B. (1986). *Gmelina arborea*. Seed Leaflet No. 6. Danida Forest Seed Centre, Humlebaek, Denmark.

Lauridsen, E.B. (1990). *Seed biology*. Lecture Note C-2. Danida Forest Seed Centre, Humlebaek, Denmark.

Lauridsen, E.B. (1996). Seed processing—effect on seed quality. In *Innovations in tropical tree seed technology* (ed. K. Olesen), pp. 113–130. IUFRO Symposium of the Project Group P.2.04.00 'Seed Problems'. Danish Forest Tree Seed Centre, Humlebaek, Denmark.

Lauridsen, E.B., Olesen, K., and Scholer, E. (1992). *Packaging materials for tropical tree fruits and seeds*. Technical Note No. 41. Danida Forest Seed Centre, Humlebaek, Denmark.

Laurie, M.V. (1974). *Tree planting practices in African savannas*. Forestry Development Paper No. 19. Food and Agriculture Organization of the United Nations, Rome.

Le Hourerou, H.N. (2001). Utilization of fodder trees and shrubs in arid and semiarid zones of West Africa and North Africa. *Arid Soil Research and Rehabilitation*, **14**, 101–135.

Le Roux, C. (2002). Rehabilitating open-cast mines and quarries. *Bois et Forêts des Tropiques*, **272**(2), 5–19.

Leakey, R. (1996). Definition of agroforestry revisited. *Agroforestry Today*, **8**(1), 5–7.

Leaky, R.R.B. and Newton, A.C. (ed.) (1994). *Domestication of tropical trees for timber and non-timber products*. MAB Digest 17. United Nations Educational, Scientific and Cultural Organization, Paris.

Leaky, R.R.B., Mesén, J.F., Tchoundjeu, Z., Longman, K.A., Dick, J.McP., Newton, A., Matin, A., Grace, J., Munro, R.C., and Muthoka, P.N. (1990). Low technology techniques for vegetative propagation of tropical trees. *Commonwealth Forestry Review*, **69**, 247–257.

Leaky, R.R.B., Newton, A.C., and Dick, J.McP. (1994). Capture of genetic variation by vegetative propagation: processes determining success. In *Tropical trees: potential for domestication and the rebuilding of forest resources* (ed. R.R.B. Leakey and A.C. Newton), pp. 72–83. Her Majesty's Stationery Office, London.

LeBude, A.V., Blazich, F.A., and Goldfarb, B. (1999). *Effects of jiffy forestry peat pellets on rooted stem cuttings of loblolly pine* (ed. M. Bowen and M. Stine), p. 120. Proceedings of the 25th Biennial Southern Forest Tree Improvement Conference, New Orleans, Louisiana, 11–14 July 1999. National Technical Information Service, US Department of Commerce, Washington DC.

Lechoncito, I.L. (1987). Contract reforestation–prospects and problems. Philippine Lunibennan **33** (1), 18–21.

Lemmens, R.H.M.J., Soerianegara, I., and Wong, W.C. (ed.) (1995). *Timber trees: minor commercial species*. PROSEA No. 5(2). Backhuys, Leiden, The Netherlands.

Leng, L.K. (1994). *The use of computer model in the evaluation of forest plantation operations* (ed. W.M. Wan Razali, S. Ibrahim, S. Appanah, and M. Farid), pp. 53–64. Proceedings of a symposium on Harvesting and Silviculture for Sustainable Forestry in the Tropics, 5–9 October 1992, Kuala Lumpur, Malaysia. Forest Research Institute Malaysia, Kuala Lumpur.

Leslie, A.J. (1987). A second look at the economics of natural management systems in tropical mixed forests. *Unasylva*, **39**, 46–58.

Letourneau, L.R. (1987). *Appropriate forest based industries*. In FAO Forestry Paper No. 69, pp. 9–22. Food and Agriculture Organization of the United Nations, Rome.

Leuning, R., Cromer, R.N., and Rance, S. (1991). Spatial distribution of foliar nitrogen and phosphorus in crowns of *Eucalyptus grandis*. *Oecologia*, **88**, 504–510.

Lewis, N.B., Keeves, A., and Leech, J.W. (1976). *Yield regulation in South Australian* Pinus radiata *plantations*. Technical report No. 23. South Australian Woods and Forests Department, Adelaide, Australia.

Lewty, M.I. and Frodsham, T.M. (1983). *Post-emergence weed control efficacy of three herbicides in a* Pinus *nursery*. Technical Note No. 13, Department of Forestry, Queensland, Brisbane, Australia.

Li, Y. and Chen, D. (1992). *Fertility degradation and growth responses in Chinese fir plantations*. Proceeding of the 2nd International Symposium on Forest Soils, Ciudad, Venezuela, pp. 22–29.

Libby, W.J. (1973). Domestication strategies for forest trees. *Canadian Journal of Forest Research*, **3**, 265–276.

Libby, W.J. and Ahuja, M.R. (1993). Clonal forestry. In *Clonal forestry II: conservation and application* (ed. W.J. Libby and M.R. Ahuja), pp. 1–8. Springer-Verlag, Berlin Heidelberg, Germany.

Liegel L.H. and Venator C.R. (1987) *A technical guide to forest nursery management in the Caribbean and Latin America*. General Technical Report SO-67, Institute of Tropical Forestry, Southern Forest experiment Station, US Department of Agriculture, Forest Service, Rio Pedras, Puerto Rico.

Liegel, L.H. (1983). Susceptibility of *Pinus caribaea* and *Pinus oocarpa* provenances to rain/wind damage from cyclonic storms in Puerto Rico. In *Planted forests in the neotropics: a source of energy* (ed. N.F. Barros) IUFRO/MAB/UFV Symposium. Vicosa, Brazil. (Published 1985.)

Lieth (1977). (Ref in 2nd edn.)

Lieth, H. (1977). Biological productivity of tropical lands. *Unasylva*, **114**, 24–31.

Lima, P.C.F. (1986). Tree productivity in the semi-arid zone of Brazil. In *Tree plantings in semi-arid regions* (ed. P. Felker), pp. 5–13. Proceedings of a Symposium on Establishment and Productivity of Tree Plantings in Semi-arid Regions, Kingsville, Texas. Elsevier, Amsterdam, The Netherlands.

Lind, L.L. and Martel, F. (1986). *Pine extension research study. Private plantation development in the proximity of Fiji*

Pine Commission forests. USAID/Fiji Pine Commission report. (Unpublished.)

Lindenmayer, D.B. (2002). *Plantation design and biodiversity conservation.* RIRDC Publication No. 02/019. Rural Industries Research and Development Corporation, Canberra.

Lindenmayer, D.B., Hobbs, R.J., and Salt, D. (2003). Plantation forests and biodiversity conservation. *Australian Forestry,* **66**, 62–66.

Little, K.M. and Gous, S.F. (2000). Vegetation management. In *South African. Forestry Handbook* (ed. D.L. Owen), pp. 106–108. South African Institute of Forestry, Pretoria, South Africa.

Liu, D. (2001). Tenure and management of non-state forests in China since 1950: a historical review. *Environmental History,* **6**, 239–263.

Liu, D. (ed.) (2003). *Rehabilitation of degraded forests to improve livelihoods of poor farmers in south China.* Center for International Forestry Research, Bogor, Indonesia.

Lohmann, L. (1990). Commercial tree plantations in Thailand: deforestation by any other name. *The Ecologist,* **20**(1), 9–17.

Longart, J.J. and Gonzalez, L. (1993). Methods of site preparation for second rotation plantations and their influence on the growth of *Pinus caribaea* var. *honduren-sis. Boletin Tecnico CVG—PROFORCA,* **5**, 18–30.

Lonsdale, D. and Gibbs, J.N. (1996). Effects of climate change on fungal diseases of trees. In *Fungi and Environmental change* (ed. J.C. Frankland, N. Magan, and G.M. Gadd), Symposium of the British Mycological Society, Cranfield University, UK, March 1994.

Loumeto, J.J. and Huttel, C. (1997). Understory vegetation in fast-growing tree plantations on savanna soils in Congo. *Forest Ecology and Management,* **99**, 65–81.

Low, C.K. and Abdul Razak, M.A. (1985). Experimental tapping of pine oleoresin. *Malaysian Forester,* **48**, 248–253.

Lowe, R.G. (1976). Teak (*Tectona grandis* Linn. f.) thinning experiment in Nigeria. *Commonwealth Forestry Review,* **55**, 189–202.

Lowery, R.F., Lamberth, C.C., Endo, M., and Kane, M. (1993). Vegetation management in tropical forest plantations. *Canadian Journal of Forest Research,* **23**, 2006–2014.

Loyttyniemi, K. and Mikkola, L. (1980). Elephant as a pest of pines in Zambia. *Tropical Pest Management,* **26**, 167–169.

Luangviriyasaeng, V. and Pinyopusarerk, K. (2002). Genetic variation in a second-generation progeny trial of *Acacia auriculiformis* in Thailand. *Journal of Tropical Forest Science,* **14**, 131–144.

Lugo, A.E. (1992). Comparison of tropical tree plantations with secondary forests of the same age. *Ecological Monographs,* **62**, 1–41.

Lugo, A.E., Parrotta, J.A., and Brown, S. (1993). Loss of species caused by tropical deforestation and their recovery through management. *Ambio,* **22**, 106–109.

Lugo, A.E., Wang, D., and Bormann, F.H. (1990). A comparative analysis of biomass production in five tropical tree species. *Forestry Ecology and Management,* **31**, 153–166.

Luke, R.H. and McArthur, A.G. (1978). *Bushfires in Australia.* Australian Government Publishing Service, Canberra.

Lund, G.H. (2000). *Definitions of forest, deforestation, afforestation and reforestation* [On-line] Manassas, VA: Forest Information Services, Available from World Wide Web. http://home.att.net/~gklund/DEFpaper.html (1 June 2002).

Lundgren, B. (1978). *Soil conditions and nutrient cycling under natural and plantation forests in Tanzanian highlands.* Reports in Forest Ecology and Forests Soils No. 31. Department of Forest Soils, Swedish University of Agricultural Science, Uppsala, Sweden.

Lynch, O. and Talbot, K. (1995). *Balancing acts: community-based forest management and national law in Asia and the Pacific.* World Resources Institute, Washington DC.

MacDicken, K.G. (1991). Impacts of *Leucaena leucocephala* as a fallow improvement crop in shifting cultivation on the Island of Mindoro, Philippines. *Forest Ecology and Management,* **45**, 185–192.

MacDicken, K.G. (1994). *Selection and management of nitrogen-fixing trees.* Winrock International, Morrilton, Arkansas, USA and FAO, Bangkok.

MacDicken, K.G. and Vergara, N.T. (1990). *Agroforestry: classification and management.* John Wiley, New York, USA.

MacDicken, K.G., Hairiah, K., Otsamo, A., Duguma, B., and Majid, N.M. (1997). Shade-based control of *Imperata cylindrica*: tree fallows and cover crops. *Agroforestry Systems,* **36**, 131–149.

MacGillivray, A.W. (1990). *Forest use and conflict in Burma: 1750–1990.* M.Sc Thesis, University of London. (Unpublished.)

Mackinnon, A., Meidinger, D., and Klinka, K. (1992). Use of bioclimatic ecosystem classification system in British Columbia. *Forestry Chronicle,* **68**, 100–120.

Maclean, R.H., Litsinger, J.A., Moody, K., and Watson, A.K. (1992). The impact of alley cropping *Gliricidia sepium* and *Cassia spectabilis* on upland rice and maize production. *Agroforestry Systems,* **20**, 213–218.

MacRae, S. and Cotterill, P. (2000*a*). Breeding strategy for the next generation of clonal forests. In *Symposium on hybrid breeding and genetics, Noosa, Australia, 9–14 April 2000* (ed. H.S. Dungey, M.J. Dieters, and D.G. Nikles), pp. 44–47. Department of Primary Industries, Brisbane, Australia. (Compact Disk.)

MacRae, S. and Cotterill, P. (2000*b*). *Application of biotechnology in hybrid breeding.* Symposium on Hybrid Breeding and Genetics, Noosa, Australia, 9–14 April 2000. (ed. H.S. Dungey, M.J. Dieters, and D.G. Nikles), pp. 48–52. Department of Primary Industries, Brisbane, Australia. (Compact Disk.)

MacRae, S. and van Staden, J. (1999). Transgenic *Eucalyptus.* In *Transgenic trees,* Vol. 44 (ed. Y.P.S. Bajaj), pp. 88–114. Biotechnology in Agriculture and Forestry Springer-Verlag, Berlin and Heidelberg.

MacRae, S.P. (2003). *Eucalypt domestication and breeding—past triumphs and future horizons.* Paper to Eucalypts in Asia conference, Zhanjiang, China, 7–11 April 2003.

Madoffe, S.S. and Chamshama, A.S.O. (1989). Tree improvement activities in Tanzania. *Commonwealth Forestry Review*, **68**, 101–107.

Magcale-Macandog, D.B. (1999). Smallholder timber production and marketing: the case of *Gmelina arborea* in Claveria, northern Mindanao, Philippines. *International Tree Crops Journal*, **10**(1), 61–78.

Magcale-Macandog, D.B., Menz, K., Rocamora, P.M., and Predo, C.D. (1998). *Gmelina* timber production and marketing in Claveria, Philippines. In *Improving smallholder farming systems in Imperata areas of Southeast Asia: alternatives to shifting cultivation* (ed. K. Menz, D. Magcale-Macandog, and I.W. Rusastra), pp. 77–91. ACIAR Monograph No. 52. Australian Centre for International Agricultural Research, Canberra.

Maginnis, S. and Jackson, W. (2003). *The role of planted forests in forest landscape restoration*. Paper to UNFF Intersessional Experts Meeting 'Role of Planted Forests in Sustainable Forest Management'. Wellington, New Zealand, 24–27 March 2003, 13 pp.

Maguire, D.A., Schreuder, G.F., and Sheikh, M. (1990). A biomass/yield model for high-density *Acacia nilotica* plantahons in Sind, Pakistan. *Forest Ecology and Management*, **37**, 285–302.

Mahmood, K., Morris, J., Collopy, J., and Slavich, P. (2000). Groundwater uptake and sustainability of farm plantations on saline sites in the Punjab province of Pakistan. *Agricultural Water Management*, **48**, 1–20.

Malajczuk, N., Grove, T.S., Bougher, N.L., Dell, B., and Gong Mingqin (1994). Ectomycorrhizas and nutrients: their importance to eucalypts in China. In *Australian tree species research in China* (ed. A.G. Brown), pp. 132–139. ACIAR Proceedings No. 48. Australian Centre for International Agricultural Research, Canberra.

Maldonado, G. and Louppe, D. (1999). Plantations villageoises de teck en Côte d'Ivoire. *Bois et Forêts des Tropiques*, **262**(4), 19–30.

Malmivaara, E. (1996). Contemporary seedling production technology. In *Reforestation: meeting the future industrial wood demand* (ed. A. Otsamo, J. Kuusialpo and H. Jaskari), pp. 118–121. Enso Forest Development Oy Ltd, Jakarta.

Mantayla, J. (1993). *Nursery manual*. Technical Report No. 1, Central Nurseries Establishment Project. Enso Forest Development Oy Ltd, Jakarta.

Marcar, N. and Khanna, P.K. (1997). Reforestation of salt-affected and acid soils. In *Management of soil, nutrients and water in tropical plantation forests* (ed. E.K.S. Nambiar and A.G. Brown), pp. 481–525. ACIAR Monograph No. 43. Australian Centre for International Agricultural Research, Canberra.

Marcar, N., Crawford, D., Leppert, P., Jovanovic, T., Floyd, R., and Farrow, R. (1995). *Trees for saltland: a guide to selecting native species for Australia*. CSIRO, Division of Forestry, Canberra.

Marcar, N., Ishmail, S., Hossain, A., and Ahmad, R. (1999). *Trees, shrubs and grasses for saltlands: an annotated bibliography*. ACIAR Monograph No. 56. Australian Centre for International Agricultural Research, Canberra.

Marcar, N., Naqvi, M., Crawford, D., Arnold, R., Mahmood, K., and Hossain, A. (1998). Results from an *Acacia ampliceps* provenance-family trial on saltland in Pakistan. In *Recent developments in acacia planting* (ed. J.W. Turnbull, H.R. Crompton, and K. Pinyopusarerk), pp. 161–166. ACIAR Proceedings No. 82. Australian Centre for International Agricultural Research, Canberra.

Marchi, S.R., Pitelli, R.A., Bezutte, A.J., Corradine, L., and Alvarenga, S.F. (1995). Efeito de periodos de convivencia do controle das plantas daninhas na cultura de *Eucalyptus grandis*. In *Seminario sobre cultivo minimo do solo em florestas*. CNP Floresta, Curitiba, Brazil.

Margescu, T. (2001). Restoration of degraded forest land in Thailand: the case of Khao Kho. *Unasylva*, **52**(207), 52–55.

Margolis, H.A. and Brand, D.G. (1990). An ecophysiological basis for understanding plantation establishment. *Canadian Journal of Forest Research*, **20**, 375–390.

Marsh, E.K. (1978). *The cultivation and management of commercial pine plantations in South Africa*. Department of Forestry Bulletin No. 56, Pretoria, South Africa, 146 pp.

Marsh, E.K. and Burgers, T.F. (1973). The response of even-aged pine stands to thinning. *Forestry in South Africa*, **14**, 103–111.

Marshall, H.G.W. and Foot, D.L. (1969). *Silviculture of Pinus patula in Malawi*. Forestry Research Institute, Zomba, Malawi.

Marx, D.H., Cordell, C.E., Maul, S.B., and Ruelhe, J.L. (1989). Ectomycorrhizal development on pine by *Pisolithus tinctorius* in bare-root and container seedling nurseries: 1. Efficacy of various vegetative inoculation formulations. *New Forests*, **3**, 45–56.

Marx, D.H., Hedin, A., and Toe, S.F.P. (1985). Field performance of *Pinus caribaea* var. *hondurensis* seedlings with specific ectomycorrhizae and fertilizer after three years on a savanna site in Liberia. *Forest Ecology and Management*, **13**, 1–25.

Marx, D.H., Jarl, K., Ruelhe, J.L., Cordell, C.E., Riffle, J.W., Molina, R.J., Pawuk, W.H., Navratil, S., Tinus, R.W., and Goodwin, R.C. (1982). Commercial vegetative inoculation of *Pisolithus tinctorius* and inoculation techniques for development of ectomycorrhiza on container-grown seedlings. *Forest Science*, **28**, 373–400.

Marzalina, M., Khoo, K.C., Jayanthi, N., Tsan F.Y., and Krishnapilly, B. (ed.) (1999). *Recalcitrant seeds*. IUFRO Seed Symposium 1998 Proceedings. Forest Research Institute Malaysia, Kepong, Malaysia.

Mascarenhas, A.F. and Muralidharan, E.M. (1993). Clonal forestry with tropical hardwoods. In *Clonal forestry II: conservation and application* (ed. W.J. Libby and M.R. Ahuja), pp. 169–187. Springer-Verlag, Berlin Heidelberg, Germany.

Massey, P. (1976). *Natural resource information and site classification (for Viphya forest industry trials)*. Forestry Research Institute. Zomba, Malawi.

Masuka, A. (1989). The incidence of *Armillaria* root rot and butt diseases in *Pinus* spp. in plantation in Zimbabwe. *Commonwealth Forestry Review*, **68**, 121–124.

Masuka, A. (1990). A new canker of *Eucalytus grandis* Hill ex Maid, in Zimbabwe. *Commonwealth Forestry Review*, **69**, 195–200.

Matheson, A.C. (1990). Breeding strategies for MPTs. In *Tree improvement of multipurpose species*, Vol. 2 (ed. N. Glover and N. Adams), pp. 67–99. Multipurpose Tree Species Network Technical Series, Winrock International Institute for Agricultural Development, Arlington, United States.

Matheson, A.C. and Raymond, C.A. (1986). A review of provenance × environment interaction: its practical importance and use with particular reference to the tropics. *Commonwealth Forestry Review*, **65**, 283–302.

Mathur, H.N., Babu, R., Joshie, P., and Singh, B. (1976). Effect of clearfelling and reforestation on runoff and peak rates in small watersheds. *Indian Forester*, **102**, 219–226.

Matthews, J.D. (1963). Factors affecting the production of seed by forest trees. *Forestry Abstracts*, **24**, i–xiii.

Matthews, J.D. (1989). *Silvicultural systems*. Oxford University Press, Oxford.

Matthews, J.D. and Wareing, P.F. (1971). *Physiological and genetical factors determining productivity of species*. Proceedings of the 15th IUFRO Congress, Div. 1, pp. 136–143.

May, P.H. and Pastuk, M. (1996). Tropical forest management options, social diversity and extension in eastern Amazonia. *Unasylva*, **47**(184), 21–26.

Maydell, H. von and Gregersen, H.M. (1976). *Investment policy as a precondition for development of forestry and forest industries*. Proceedings of the 16th IUFRO World Congress, Div. 3, pp. 364–375.

Maydell, H.-J. von (1986). *Trees and shrubs of the Sahel. Their characteristics and uses*. Deutsche Gesellschaft fur Technische Zusammenarbeit (GTZ) GmbH, Eschborn, Germany.

Mayer, J. (1996). *Industrial timber plantation in Indonesia: a critical review with reference to the P.T. Finnantara Intiga project in Sanggau and Sintang, West Kalimantan*. Paper to Finland and Forests of the World Seminar III: Social and Environmental Impacts of Industrial Forestry in Asia, Helsinki 30–31 October 1996.

Mayers, J. (2000). Company–community forestry partnerships: a growing phenomenon. *Unasylva*, **51**(200), 33–41.

Mayers, J. and Bass, S. (2000). Overview report. *Policy that works for forests and people Series No. 7*. International Institute for Environment and Development, London.

Mayhew, J.E. and Newton, A.C. (1998). *The silviculture of mahogany* (Swietenia macrophylla). CAB International, Wallingford, UK.

MBRLC (1988). *A manual on how to farm your hilly land without losing your soil*. Mindanao Baptist Rural Life Centre, Davao del Sur, Philippines.

McGauhey, S.E. (1986). International financing for forestry. *Unasylva*, **151**, 2–11.

McGauhey, S.E. and Gregersen, H.M. (1988). *Investment policies and financing mechanisms for sustainable forestry development*. Inter-American Development Bank, Washington DC.

McKay, H.M. (1997). A review of the effect of stresses between lifting and planting on nursery stock quality and performance. *New Forests*, **13**, 369–399.

McNabb, K., Borges, J., and Welker, J. (1994). Jari at 25: an investment in the Amazon. *Journal of Forestry*, **92**(2), 21–26.

McNabb, K.L. and Wadouski, L.H. (1999). Multiple rotation yields in intensively managed plantations in the Amazon basin. *New Forests*, **18**, 5–15.

Mead, D.J. and Speechly, H.T. (1991). Growing *Acacia mangium* for high quality sawlogs in Peninsular Malaysia. In *Recent developments in tree plantations in the humid/subhumid tropics of Asia* (ed. Sheikh Ali Abod et al.), pp. 54–71. Universiti Pertanian Malaysia, Serdang, Malaysia.

Menz, K., Magcale-Macandog, D., and Rusastra, I.W. (ed.) (1998). *Improving smallholder farming systems in Imperata areas of Southeast Asia: alternatives to shifting cultivation*. ACIAR Monograph No. 52. Australian Centre for International Agricultural Research, Canberra.

Mesén, J.F., Boshier, D.H., and Cornelius, J.P. (1994). Genetic improvement of trees in central America, with particular reference to Costa Rica. In *Tropical trees: potential for domestication and the rebuilding of forest resources* (ed. R.R.B. Leakey and A.C. Newton), pp. 249–255. Her Majesty's Stationery Office, London.

Meunier, M. (1996). Forest cover and floodwater in small mountain watersheds. *Unasylva*, **47**(185), 29–37.

Michaelsen, T. (1991). Participatory approaches in watershed management planning. *Unasylva*, **42**(164), 3–7.

Michon, G. and Mary, F. (1994). Conversion of traditional village gardens and new economic strategies of rural households in the area of Bogor, Indonesia. *Agroforestry Systems*, **25**, 31–58.

Middleton, N.J. and Thomas, D.S.G. (ed.) (1997). *World atlas of desertification*, 2nd edn. Edward Arnold, London.

Midgley, S.J. (1990). Tree seed in forest development. In *Tropical tree seed research* (ed. J.W.Turnbull), pp. 10–12. ACIAR Proceedings No. 28. Australian Centre for International Agricultural Research, Canberra.

Midgley, S.J. (1996). Seed collection strategies in a changing world. In *Innovations in tropical tree seed technology* (ed. K. Olesen), pp. 175–187. IUFRO Symposium of the Project Group P.2.04.00 'Seed Problems' Danish Forest Tree Seed Centre, Humlebaek, Denmark.

Midgley, S.J. Turnbull, J.W., and Johnston, R.D. (ed.) (1983). *Casuarina ecology, management and utilization*. CSIRO, Melbourne, Australia.

Milimo, P.B., Dick, J. McP., and Munro, R.C. (1994). Domestication of trees in semi-arid East Africa. In *Tropical trees: potential for domestication and the rebuilding of forest resources* (ed. R.R.B. Leakey and A.C. Newton), pp. 210–219. Her Majesty's Stationery Office, London.

Miller, A.S., Mintzer, I.M., and Hoagland, S.H. (1986). *Growing power: bioenergy for development and industry*. Study No. 5, World Resources Institute, Washington DC.

Miller, H.G. (1981). Forest fertilization: some guiding concepts. *Forestry*, **54**, 157–167.

Miller, H.G. (1984). Dynamics of nutrient cycling in plantation ecosystems. In *Nutrition of plantation forests* (ed. G.D. Bowen and E.K.S. Nambiar), pp. 53–78. Academic Press, London.

Miller, H.G. (1989). Internal and external cycling of nutrients in forest stands. In *Biomass production of fast-growing trees* (ed. H.C. Periera and H.E. Landsberg), pp. 73–78.

Miller, H.G. (1995). The influence of stand development on nutrient demand, growth and allocation. *Plant and Soil*, **168/169**, 225–232.

Ministry of Environment (1990). *Developing India's wastelands*. Ministry of Environment and Forests, New Delhi.

Ministry of Forestry (1985). *Statistics of China's forests*. Ministry of Forestry, People's Republic of China.

Misra, R.M. (1985). A note on *Leptobyrsa decora* (Hemiptera: Tingitidae) a bio-control agent of *Lantana camara* (Verbenaceae). *Indian Forester*, **111**, 641–644.

Mitchell, M.R. (1989). Susceptibility to termite attack of various species planted in Zimbabwe. In *Trees for the tropics* (ed. D.J. Boland), pp. 215–227. ACIAR Monograph No. 10. Australian Centre for International Agriculture Research, Canberra.

Mittelman, A. (2000). Teak planting by smallholders in Nakhon Sawan, Thailand. *Unasylva*, **51**(201), 62–65.

Moeliono, M.M.M. (2002). Property rights and participatory forest management: an overview. In *Pathways to sustainable management* (ed. A.G. Brown), pp. 58–63. Proceedings of the Second Hermon Slade Workshop, Ubud, Indonesia, 5–8 June 2001, The ATSE Crawford Fund, Melbourne, Australia.

Mok, C.K., Cheah, L.C., and Chan, Y.K. (2000). Site management and productivity of *Acacia mangium* in humid tropical Sumatra, Indonesia. In *Site management and productivity in tropical plantation forests: a progress report* (ed. E.K.S. Nambiar, A. Tiarks, C. Cossalter, and J. Ranger), pp. 87–94. Center for International Forestry Research, Bogor, Indonesia.

Moncur, M.W., Kleinschmidt, G., and Somerville, D. (1991). The role of acacia and eucalypt plantations for honey production. In *Advances in tropical acacia research*. (ed. J.W. Turnbull), pp. 123–127. ACIAR Proceedings No. 35. Australian Centre for International Agricultural Research, Canberra.

Mondi (2002). Company website. http://www.mondi-forests.co.za (15 September 2002) and personal communication (A. Kenny).

Monteuuis, O. and Goh, D.K.S. (1999). About the use of clones for teak. *Bois et Forêts des Tropiques*, **261**(3), 28–37.

Morell, M. (1997). Financing community forestry activities. *Unasylva*, **48**(188), 36–43.

Mori, T. (2000). Effects of droughts and forest fires on dipterocarp forest in East Kalimantan. In *Rainforest ecosystems of East Kalimantan* (ed. E. Guhardja *et al.*), pp. 29–45. Ecological Studies 140. Springer-Verlag, Tokyo.

Mori, T. (2001). Rehabilitation of degraded forests in lowland Kutai, East Kalimantan, Indonesia. In *Rehabilitation of degraded forest ecosystems in the tropics* (ed. S. Kobayashi, J.W. Turnbull, T. Toma, T. Mori, and N.M.N.A. Majid), pp. 17–26. Center for International Forestry Research, Bogor, Indonesia.

Mori, T., Ohta, S., Ishida, A., Toma, T., and Oka, T. (2000). Overview of the changing forest ecosystems in East Kalimantan. In *Rainforest ecosystems of East Kalimantan* (ed. E. Guhardja *et al.*), pp. 309–317. Ecological Studies 140, Springer-Verlag, Tokyo.

Moro, L. and Gonçalves, J.L.M. (1995). Effect of forest biomass "ash" on the productivity of *Eucalyptus grandis* plantation and financial evaluation. *IPEF*, **48/49**, 18–24.

Morris, A.R. (1981). *Report of pruning trial C67*. Forest Research Report No. 14, Usutu Pulp Company, Swaziland. (Unpublished.)

Morris, A.R. (1987). *A review of* Pinus patula *seed sources in the Usutu Forest, 1950–86*. Forest Research document 8/87. Usutu Pulp Company. (Unpublished.)

Morris, A.R. (1993*a*). *Observations of the impact of the 1991/2 drought on the Usutu Forest*. Forest Research Document 6/9, Usutu Pulp Company, Swaziland. (Unpublished.)

Morris, A.R. (1993*b*). Forest floor accumulation under *Pinus patula* in the Usutu Forest, Swaziland. *Commonwealth Forestry Review*, **72**, 114–117.

Morris, J. (1997). Water use by eucalypt plantations. *ACIAR Forestry Newsletter* **22**, 1, 4.

Morris, J.M. (1989). *Earth roads: a practical manual for the provision of access for agricultural and forestry projects in developing countries*. Cranfield Press.

Morrison, E. and Bass, S.M.J. (1992). What about the people? In *Plantation politics. Forest plantations in development* (ed. C. Sargent and S. Bass), pp. 92–120. Earthscan, London.

Moura-Costa, P. and Stuart, M.C. (1998). Forestry-based greenhouse gas mitigation: a story of market evolution. *Commonwealth Forestry Review*, **77**, 191–202.

Mudge, K.W. and Brennan, E.B. (1999). Clonal propagation of multipurpose and fruit trees used in agroforestry. In *Agroforestry in sustainable ecosystems* (ed. L.E. Buck, P. Lassoie, and E.C.M. Fernandes), pp. 157–190. Lewis Publishers, Boca Raton, Florida, USA.

Mugasha, A.G. and Nshubemuki, L. (1988). Soil conservation in Kondoa, Tanzania: the case of the Rangi people in the Hado area. *Forest Ecology and Management*, **25**, 159–180.

Mulawarman, Roshetko, J.M., Sasongko, S.M., and Iranto, D. (2003). *Tree seed management—seed sources, seed collection and seed handling: a field manual for field workers and farmers*. International Centre for Research in agro forestry and Winrock International, Bogor, Indonesia.

Munro, D. and Holdgate, M.W. (ed.) (1991). *Caring for the Earth. A strategy for sustainable living*. Earthscan Publications, London.

Murphy, S.T. (1998). Protecting Africa's trees. *Unasylva*, **41**(192), 57–61.

Murray, F. and Wilson, S. (1988). Joint action of sulphur dioxide and hydrogen fluoride on growth of *Eucalyptus tereticornis*. *Environmental and Experimental Botany*, **28**, 343–349.

Musa, O.S. (1977). Der Grüngürtel von Khartoum. [The green belt of Khartoum.] *Beiträge sür de Forstwirtschaft*, **11**, 95–97.

Muthana, K.D., Madhander, S., Mertia, R.S., and Arora, G.D. (1984). Shelterbelt plantations in arid regions. *Indian Farming*, **33**(11), 19–20, 40.

Mwihomeke, S.T. (1983). *Effects of pruning on diameter, height growth of* Cupressus lusitanica *var. benthamii Carr. at Rongai, North Kimilanjaro Forest Project, Tanzania: results of experiment No. 125.* Tanzania Silviculture Technical Note No. 56.

Myers, B.J., Bond, W.J., Benyon, R.G., Falkiner, R.A., Polglase, P.J., Smith, C.J., Snow, V.O., and Theiveyanathan, S. (1999). *Sustainable effluent-irrigated plantations. An Australian guideline.* CSIRO, Melbourne, Australia.

Nair, K.K.N. (1988). *Mikania micrantha* H. B. K.—A noxious weed in the forests of Kerala. *Evergreen. Newsletter of Kerala Forest Research Institute,* **20**, 13–14.

Nair, K.S.S. (2001). *Pest outbreaks in tropical plantations: is there a greater risk for exotic tree species?* Center for International Forestry Research, Bogor, Indonesia.

Nair, K.S.S. (ed.) (2000). *Insect pests and diseases in Indonesian forests.* Center for International Forestry Research, Bogor, Indonesia.

Nair, K.S.S., Sudheendrakumar, V.V., Varma, R.V., Chacko, K.C., and Jayaraman, K. (1996). Effect of defoliation by *Hyblaea puera* and *Eutectona machaeralis* (Lepidoptera) on volume increment of teak. In *Impact of diseases and insect pests in tropical forests* (ed. K.S.S. Nair, J.K. Sharma, and R.V. Varma), pp. 257–273. Proceedings of IUFRO Symposium, Kerala, India, 23–26 November 1993 Kerala Forest Research Institute, Peechi, India.

Nair, P.K.R. (1989). The role of trees in soil productivity and protection. In *Agroforestry systems in the tropics* (ed. P.K.R. Nair), pp. 567–589. Kluwer Academic Publishers, Dortrecht, Netherlands.

Nair, P.K.R. (1993*a*). The diagnosis and design (D & D) methodology. In *An introduction to agroforestry* (ed. P.K.R. Nair), pp. 347–356. Kluwer Academic Publishers, Dordrecht, The Netherlands.

Nair, P.K.R. (ed.) (1993*b*). *An introduction to agroforestry.* Kluwer Academic Publishers, Dordrecht, The Netherlands.

Nambiar, E.K.S. (1996). Sustained productivity of forests is a continuing challenge to soil science. *Journal of Soil Science Society of America,* **60**, 1629–1642.

Nambiar, E.K.S. (1998). *Productivity and sustainability of plantation forests.* Proceedings of the Silvotecna X IUFRO Conference 'Site productivity improvement', Concepcion, Chile, June 1998.

Nambiar, E.K.S. and Brown, A.G. (ed.) (1997). *Management of soil, nutrients and water in tropical plantation forests.* ACIAR Monograph No. 43. Australian Centre for International Agricultural Research, Canberra; CSIRO Australia; and Center for International Forestry Research, Indonesia.

Nambiar, E.K.S. and Sands, R. (1993). Competition for water and nutrients in forests. *Canadian Journal of Forest Research,* **23**, 1955–1968.

Nambiar, E.K.S., Tiarks, A., Cossalter, C., and Ranger, J. (ed.) (2000). *Site management and productivity in tropical plantation forests: a progress report.* Center for International Forestry Research, Bogor, Indonesia.

Namkoong, G., Kang, H.C., and Brouard, J.S. (1988). *Tree breeding: principles and strategies,* Monographs in Applied Genetics, Vol. 11, Springer-Verlag, New York.

Napier, l.A. (1980). Seedling quality: a crucial factor in the establishment of bare-root pine plantations in the tropics. In Whitmore, J. L. (ed.) (1981), pp. 165–81.

Napier, I.A. (1981). Seedling quality: a crucial factor in the establishment of bare-root plantations in the tropics. In *Wood production in the neotropics via plantations* (ed. J.L. Whitmore), pp. 165–181. Proceedings of the IUFRO/MAB/Forest Service Symposium, September 1980 Forest Service, Rio Pedras, Puerto Rico.

Napier, I.A. (1983). Nursery techniques for the production of conifers in the tropics. In *Planted forests in the neotropics: a source of energy* (ed. N. F. Barros). IUFRO/MAB/UFV Symposium, Vicosa, Brazil, February 1983.

National Academy of Sciences (1975). *Under-exploited tropical plants with promising economic value.* Report No. 16, US National Academy of Sciences, Washington DC.

National Academy of Sciences (1977). Leucaena. *Promising forage and tree crop for the tropics.* Report No. 26, US National Academy of Sciences, Washington DC.

National Academy of Sciences (1980). *Firewood crops- shrub and tree species for energy production,* Vol. 1. National Academy of Sciences, Washington DC.

National Academy of Sciences (1983). *Firewood crops- shrub and tree species for energy production,* Vol. 2. National Academy of Sciences, Washington DC.

Neelay, V.R. and Dhondiyal, L.P. (1985). Observations on the possibility of using industrial effluent water for raising forest plantations. *Journal of Tropical Forestry,* **1**(2), 132–139.

Negi, G.C.S. and Joshi, V. (2002). Agroforestry trees restore degraded land in the Himalayas. *Agroforestry Today,* **13**(1–2), 19–21.

Neil, P.E. (1988). Root disease (*Phellinus noxius* (Corner) G. H. Cunn.) of *Cordia alliodora* in Vanuatu. *Commonwealth Forestry Review,* **67**, 363–372.

Neil, P.E. and Barrance, A.J. (1987). Cyclone damage in Vanuatu. *Commonwealth Forestry Review,* **66**, 255–264.

Nelson, R. (1990). *The 'desertification' problem.* World Bank Technical Paper 116. The World Bank, Washington DC.

Nepstad, D.C., Uhi, C., and Serrão, A.S. (1991). Recuperation of a degraded Amazonian landscape: forest recovery and agricultural restoration. *Ambio,* **20**, 248–255.

Nesmith, C. (1991). Gender, trees, and fuel: social forestry in West Bengal, India. *Human Organization,* **50**, 337–348.

Newman, E.I. and Reddell, P. (1987). The distribution of mycorrhizas amongst families of vascular plants. *New Phytologist,* **106**, 745–751.

Ng, F.S.P. (1981). Vegetative and reproductive phenology of dipterocarps. *Malaysian Forester,* **44**, 197–221.

Nghia, N.H. (2000). *Successful development of Acacia species in Vietnam.* APAFRI Publication Series No. 5. APAFRI, Kuala Lumpur, Malaysia.

Nghia, N.H. and Kha, L.D. (1998). Selection of *Acacia* species and provenances for planting Vietnam. In *Recent developments in acacia planting* (ed. J.W Turnbull,

H.R. Crompton, and K. Pinyopusarerk), pp. 130–135. ACIAR Proceedings No. 82 Australian Centre for International Agricultural Research, Canberra.

Nguyen-The, N. and Sist, P. (1998). Phenology of some dipterocarps. In *Silvicultural research in a lowland mixed dipterocarp forest in East Kalimantan: the contributions of the STREK project* (ed. J-G. Bertault and K. Kadir), pp. 95–110. CIRAD-foret, Montpellier, France.

Nichols, D. (1994). *Terminalia amazonica* (Gmel.) Exell: development of a native species for reforestation and agroforestry. *Commonwealth Forestry Review*, **73**, 9–13.

Nieuwenhuis, M. and O'Connor, N. (2000). Challenges and opportunities for small-scale tree nurseries in the East African highlands. *Unasylva*, **51**, (203), 56–60.

Nieuwolt, S. (1977). *Tropical climatology*. John Wiley, London.

Nik, M.M. and Paudyal, P.K. (1992). Pruning trial for *Acacia mangium* Willd. plantation in Peninsular Malaysia. *Forest Ecology and Management*, **47**, 285–293.

Nikles, D.G. (1996). The first 50 years of the evolution of forest tree improvement in Queensland. In *Tree improvement for sustainable tropical forestry* (ed. M.J. Dieters, A.C. Matheson, D.G. Nikles, C.E. Harwood, and S.M. Walker), pp. 51–64. Proceedings of QFRI-IUFRO Conference, Caloundra, Australia, October/November 1996, Queensland Forestry Research Institute, Gympie, Australia.

Nikles, D.G., Spidy, T., Rider, E.J., Eisemann, R.L., Newton, R.S., and Matthews-Fredrick, D. (1983). Genetic variation and windfirmness among provenances of *Pinus caribaea* Mor. var. *hondurensis* Barr. and Golf. in Queensland. *Silvicultura*, **29**, 125–130.

Nilsson, S. and Sallnäs, O. (1990). Air pollution and European forests: policy implications based on simulation models. *Unasylva*, **41**(163), 34–41

Nilsson, S. and Schopfhauser, W. (1995). The carbon sequestration potential of a global afforestation programme? *Climate Change*, **30**, 267–293.

Niskanen, A. and Väyrynen, J. (ed.) (2001). *Economic sustainability of small-scale forestry*. EFI Proceedings No. 36. European Forest Institute, Joensuu, Finland.

Noble, B.F. (1985). Comparative costs and survival rates of direct seeding, germinant and seedling plantings of denuded Benguet pine (*Pinus kesiya* Royle ex Gordon) areas. *Sylvatrop. The Philippines Forest Research Journal*, **10**(4), 259–270.

Noble, I. *et al.* (2000). Implication of different definitions and generic issues. In *Land use, land-use change and forestry* (ed. R.T. Watson, I.R. Noble, R. Bolin, N.H. Ravindranath, D.J. Ravado, and D.J. Dokken), pp. 53–126. Intergovernmental Panel on Climate Change Special Report. Cambridge University Press, Cambridge, UK.

Nwoboshi, L.C. (1982). *Potential impacts of some harvesting options on nutrient budgets of a Gmelina pulpwood plantation ecosystem in Nigeria* (ed. R. Ballard and S.P. Gessel), pp. 212–217. IUFRO Symposium on Forest site and Continuous Productivity USDA General Technical Report PNW-163, 1983. Seattle, USA.

Nwoboshi, L.C. (1984). Growth and nutrient requirements in a teak plantation age series in Nigeria. II. Nutrient accumulation and minimum annual requirements. *Forest Science*, **30**, 35–40.

Nwonwu, F.O.C. and Obliaga, P.C. (1988). Economic criteria in the choice of weed control methods for young pine (*Pinus caribaea* var. *hondurensis* Barr and Golf) plantations. *Weed Research, UK*, **28**, 181–184.

NZFRI (1992). Satellite navigation in forestry. *What's new in forest research*, No. 222. New Zealand Forest Research Institute, Rotorua, New Zealand.

O'Loughlin, E. and Nambiar, E.K.S. (2001). *Plantations, farm forestry and water*. Water and Salinity Issues in Agroforestry No. 8. RIRDC Publication No. 01/137. Rural Industries Research and Development Corporation, Canberra.

Oberholzer, F. (2000). Cable yarding. In *South African Forestry Handbook* (ed. D. L. Owen), pp. 323–328. S African Institute of Forestry, Pretoria, South Africa.

Odendaal, P.B. and Bigalke, R.C. (1979). Habitat selection by bushbuck in a disturbed environment. *South African Forestry Journal* **108**, 39–41.

Odoom, F.K. (1999). Securing land for forest plantations in Ghana. *International Forestry Review*, **1**, 182–188.

OECD (1976). OECD scheme for the control of forest reproductive material moving in international trade. OECD, Paris.

Ohba, K. (1993). Clonal forestry with sugi (*Cryptomeria japonica*). In *Clonal forestry II: conservation and application* (ed. W.J. Libby and M.R. Ahuja), pp. 66–90. Springer-Verlag, Berlin Heidelberg, Germany.

Ohta, S., Morisada, K., Tanaka, N., Kiyono, Y., and Effendi, S. (2000). Are soils in degraded dipterocarp forest ecosystems deteriorated. A comparison of Imperata grasslands, degraded secondary forests and primary forests. In *Rainforest ecosystems of East Kalimantan* (ed. E. Guhardja *et al.*), pp. 49–57. Ecological Studies 140, Springer-Verlag, Tokyo.

Okojie, J.A., Bailey, R.L., and Borders, B.E. (1988). Spacing effects in an unthinned 11-year-old *Terminalia superba* plantation in the dry lowland rainforest area of Nigeria. *Forest Ecology and Management*, **23**, 253–260.

Okorie, P.E. (1990). Age and seasonality and their implications in management of *Gmelina arborea* Roxb. coppice stands in Nigeria. Proceedings of the 19th IUFRO World Congress, Montreal, August 1990, Vol. 1 (2), p. 468.

Ola-Adams, B.A. and Charter, J.R. (1980). Some aspects of girth increment in *Khaya grandifolia*. *Indian Forester*, **106**, 604–607.

Old, K.M. and Davison, E.M. (2000). Canker diseases of eucalypts. In *Diseases and pathogens of eucalypts* (ed. P.J. Keane, G.A. Kile, F.D. Podger, and B.N. Brown), pp. 241–257. CSIRO, Melbourne

Old, K.M. and Ivory, M.H. (1999). *Pathogen threats to short rotation forest plantations in South East Asia and options for management* (ed. A. Sivapragasam *et al.*), pp. 153–157. Proceedings of 5th International Conference on Plant Protection in the Tropics, Kuala Lumpur, Malaysia,

15–18 March 1999. Malaysian Plant Protection Society, Kuala Lumpur.

Old, K.M. and Stone, C. (2002). *New approach to aerial assessment of forest health*. Onwood No. 36, 5. CSIRO Forestry and Forest Products, Canberra.

Old, K.M., Lee, S.S., and Sharma, J.K. (ed.) (1997). *Diseases of tropical acacias*. Proceedings of an international workshop, Subanjeriji (South Sumatra). Special Publication. Center for International Forestry Research, Bogor, Indonesia.

Old, K.M., Lee, S.S., Sharma, J.K., and Zi, Q.Y. (2000). *A manual of diseases of tropical acacias in Australia, Southeast Asia and India*. Center for International Forestry Research, Bogor, Indonesia.

Oldeman, L.R., Engelen, V.W.P., and Pulles, J.H.M. (1991). The extent of human-induced soil degradation. In *World map of human-induced soil degradation: an explanatory note* (ed. L.R. Oldeman, R.T.A. Hakkeling, and W.G. Sombroek), pp. 27–33. International Soil Reference and Information Centre, Wageningen, Netherlands.

Oliver, W.W. (1992). *Plantation forestry in the South Pacific: a compilation and assessment of practices*. Field Document 8. RAS/86/036, FAO and UNDP, South Pacific Forestry Development programme, Port Vila, Vanuatu.

Ong, C.K. (1996). A framework for quantifying the various effects of tree-crop interactions. In *Tree-crop interactions: a physiological approach* (ed. C.K. Ong and P. Huxley), pp. 1–23. CAB International, Wallingford, UK.

Orode, M.O., Adeka, B., and Allan, T.G. (1977). Mine reclamation areas. In *Savanna afforestation in Africa*, pp. 190–196. FAO Forest Paper 11. Food and Agriculture Organization of the United Nations, Rome.

Oseni, A.M. (1976). Forest plantation planning at the national level. In *Savanna afforestation in Africa*. FAO Forest Paper 11, pp. 214–219. Food and Agriculture Organization of the United Nations, Rome.

OTA (1984). *Technologies to sustain tropical forest resources*. Office of Technology Assessment, United States Congress, US Government Printing Office, Washington DC.

Otsamo, A. (1996). Technical solutions for plantation forestry on grasslands. In *Reforestation: meeting the future industrial wood demand* (ed. A. Otsamo, J. Kuusialpo, and H. Jaskari), pp. 55–67. Enso Forest Development Oy Ltd, Jakarta.

Otsamo, A. and Kurniati, L. (1999). Early performance of ten timber species planted under *Acacia mangium* plantation on an *Imperata* grassland site in South Kalimantan, Indonesia. *International Tree Crops Journal*, **10**, 131–144.

Otsamo, A., Adjers, G., Hadi, T.S., Kuusipalo, J., and Vuokko, R. (1997). Evaluation of reforestation potential of 83 tree species planted on *Imperata cylindrica* dominated grassland. *New Forests*, **14**, 127–143.

Otsamo, A., Hadi, T.S., Adjers, G., Kuusipalo, J., and Vuokko, R. (1995). Performance and yield of 14 eucalypt species on *Imperata cylindrica* (L.) Beauv. grassland 3 years after planting. *New Forests*, **10**, 257–265.

Otsamo, A.O., Nikles, D.G., and Vuokko, R.H.O. (1996). Species and provenance variation of candidate acacias for afforestation of *Imperata cylindrica* grasslands in South Kalimantan, Indonesia. In *Tree improvement for sustainable tropical forestry* (ed. M.J. Dieters, A.C. Matheson, D.G. Nikles, C.E. Harwood, and S.M. Walker), pp. 46–50. Proceedings of QFRI-IUFRO Conference, Caloundra, Australia, October/November 1996, Queensland Forestry Research Institute, Gympie, Australia.

Otsamo, R., Otsamo, A., and Ådjers, G. (1996). Reforestation experiences with dipterocarp species on grassland. In *Dipterocarp forest ecosystems—towards sustainable forest management* (ed. A. Schulte and D. Schöne), pp. 464–477. World Scientific Publishing, Singapore.

Ouedraogo, A.S., Poulsen K., and Stubsgaard F. (ed.) (1996). *Intermediate/recalcitrant tropical forest tree seeds*. Proceedings of a Workshop, Humlebaek, Denmark, 8–10 June 1995. IPGRI, Rome and DANIDA Forest Seed Centre, Humlebaek, Denmark.

Overton, F. (1996). *Fire fighting. Management and techniques*. Inkata Press, Melbourne, Australia.

Owen, D.L. (ed.) (2000). *South African Forestry Handbook*. SA African Institute of Forestry, Pretoria, South Africa.

Owens, J.N. (1994). Biological constraints to seed production in tropical forest trees. In *Proceedings: international symposium on genetic conservation and production of tropical forest tree seed*. (ed. R.M. Drysdale, S.E.T. John, and A.C. Yapa), pp. 40–51. ASEAN-Canada Forest Tree Seed Center Project, Mauk-Lek, Saraburi, Thailand.

Owusu, R.A. (1999). GM technology in the forest sector: a scoping stuffy for WWF. http://www.wwf.org.uk/ filelibrary/pdf/gmsummary.pdf.asp (11 December 2002).

Oyen, L.P.A. and Dung, N.X. (ed.) (1999). *Essential-oil plants*. PROSEA No. 19. Backhuys, Leiden, The Netherlands.

Päivinen, R., Gillespie, A.J.R., Davis, R., and Holmgren, P. (2000). Assessing state and change in global forest cover: 2000 and beyond. In *Forests and society: the role of research*, Vol. 1 (ed. B. Krishnapillay *et al.*), pp. 251–257. XXI IUFRO World Congress, Kuala Lumpur, 7–12 August 2000, Sub-plenary sessions, IUFRO, Vienna and Forest Research Institute Malaysia, Kuala Lumpur.

Palmer, J. (1999). Domestication of agroforestry trees: the Mindanao Baptist Rural Life Center experience in Bansalan, Davao del Sur, Philippines. In *Domestication of agroforestry trees in Southeast Asia* (ed. J.M. Roshetko and D.O. Evans), pp. 155–163. Forest, Farm and Community Tree Research Reports, Special issue. Taiwan Forestry Research Institute and Council of Agriculture, Taiwan, Republic of China; Winrock International, Morrilton, Arkansas, USA; and International Centre for Research in Agroforestry, Nairobi.

Pan, Z. and Yang, M. (1987). Australian acacias in the People's Republic of China. In *Australian acacias in developing countries* (ed. J.W. Turnbull), pp. 136–138. ACIAR

Proceedings No. 16. Australian Centre for International Agricultural Research, Canberra.

Pancel, L. and Wiebecke, C. (1984). Diagnostic methods for large-scale industrial forestation projects. In *Strategies and designs for afforestation, reforestation and tree planting* (ed. K.F. Wiersum), pp. 231–247. Pudoc, Wageningen, The Netherlands.

Pande, M.C., Tandon, V.N., and Shanker, P.P. (1987). Distribution of nutrients in age-series of *Eucalyptus* and *Acacia auriculiformis* plantations in Bihar. *Indian Forester*, **113**, 418–426.

Pandey, D. (1995). *Forest resources assessment project 1990. Tropical forest plantation resources*. FAO Forestry Paper No. 128. Food and Agriculture Organization of the United Nations, Rome.

Pandey, D. (2000). *Development of plantation forestry and joint forest management in India*. Proceedings of the International Conference on Timber Plantation Development, Manila, 7–9 November 2000, pp. 193–202. ITTO, FAO and Department of Environment and Natural Resources, Manila.

Pandey, D. and Ball, J. (1998). The role of industrial plantations in future global fibre supplies. *Unasylva*, **49**(193), 37–43. Also www.fao.org/docrep/w7990E/w7990E00.htm. (25 September 2002).

Pandey, D. and Brown, C. (2000). Teak: a global overview. *Unasylva*, **51**(201), 3–13.

Parrotta, J.A. (1992). The role of plantations in rehabilitating degraded tropical ecosystems. *Agriculture, Ecosystems and the Environment*, **41**, 115–133.

Parrotta, J.A. and Turnbull, J.W. (ed.) (1997). Catalyzing native forest regeneration on degraded tropical lands. *Forest Ecology and Management*, **99**(1,2), 1–290.

Parrotta, J.A., Baker, D.D., and Fried, M. (1994). Application of ^{15}N-enrichment methodologies to estimate nitrogen fixation in *Casuarina equisetifolia*. *Canadian Journal of Forest Research*, **24**, 201–207.

Parrotta, J.A., Knowles, O.H., and Wunderle, J.M. Jr. (1997a). Development of floristic diversity in 10-year-old restoration forests on a bauxite mined site in Amazonia. *Forest Ecology and Management*, **99**, 21–42.

Parrotta, J.A., Turnbull, J.W., and Jones, N. (1997b). Catalyzing native forest regeneration on degraded tropical lands. *Forest Ecology and Management*, **99**, 1–7.

Parry, M.S. (1956). *Tree planting practices in tropical Africa*. Forestry Development Paper No. 8, Food and Agriculture Organization of the United Nations, Rome, Italy.

Partap, T. and Watson, H.R. (1994). *Sloping agricultural land technology (SALT): a regenerative option for sustainable mountain farming*. ICIMOD Occasional Paper No. 23. International Centre for Integrated Mountain Development, Kathmandu.

Pasicolan, P.N. (1996). *Tree growing on different grounds: an analysis of local participation in contract reforestation in the Philippines*. Centre of Environmental Science, Leiden University, Leiden, The Netherlands.

Pasiecnik, N.M. (ed.) (2001). *The* Prosopis juliflora–Prosopis pallida *complex: a monograph*. HDRA Publishing, Coventry, UK.

Paterson, R.T., Karanja, J.M., Roothaert, R.L., Nyaata, O.Z., and Kariuki, I.W. (1998). A review of tree fodder production and utilization within smallholder agroforestry systems in Kenya. *Agroforestry Systems*, **41**, 181–199.

Paton, D.M., Willing, R.R., and Pryor, L.D. (1981). Root-shoot gradients in *Eucalyptus* ontogeny. *Annals of Botany*, **47**, 835–838.

Pawitan, H. (1996). The use of Plantgro in forest plantation planning in Indonesia. In *Matching trees and sites* (ed. T.H. Booth), pp. 97–100. ACIAR Proceedings No. 63. Australian Centre for International Agricultural Research, Canberra.

Pawlick, T. (1989). Coppice with care. *Agroforestry Today*, **1**(3), 15–17.

Payn, T.W., Schutz, C.J., and Clough, M.E. (1989). Determination of the most stable period for sampling patula pine foliage in the summer rainfall region of South Africa. *Communications in Soil Science and Plant Analysis*, **20**, 403–420.

Payne, C.B. (1968). Wind and fruit trees. *Hortus Rhodesia*, **9**, 24–27.

Payne, W.J.A. (1985). A review of the possibilities for the integrating of cattle and tree crop production systems in the tropics. *Forest Ecology and Management*, **12**, 1–36.

Pearce, F. (2001). Desert harvest. *New Scientist*, No. **2314**, 44–47.

Pearse, P.H. (1990). *Forestry economics*. University of British Columbia, Vancouver, Canada.

Pell, A.N. (1999). Animals and agroforestry in the tropics. In *Agroforestry in sustainable agricultural systems* (ed. L.E. Buck, J.P. Lassoie, and E.C.M. Fernandes), pp. 33–45. Lewis Publishers, Boca Raton, Florida, USA.

Peltier, R., Lawali, E.H., and Montagne, P. (1994). Aménagement villageois des brousses tachetées au Niger. *Bois et Forêts des Tropiques*, **242**(4), 59–76.

Pereira, J.S., TomÈ, M., Madeira, M., Oliveira, A.C. TomÈ, J. and Almeida, M.H. (1996). Eucalypt plantations in Portugal. In Nutrition of eucalypts (ed. P.M. Attiwill and M.A. Adams), pp. 371–387. CSIRO Publishing, Melbourne, Australia.

Perera, A.H. (1988). Growth and performance of pines in Sri Lanka. In *Reforestation with* Pinus *in Sri Lanka* (ed. H.P.M. Gunasena, S. Gnnatilleke, and A.H. Perera), pp. 29–37. Proceedings of a Symposium, University of Peradeniya, Sri Lanka, July 1988.

Perera, A.H. and Rajapakse, R.M.N. (1989). *A baseline study of Kandyan forest gardens in Sri Lanka: structure, composition and productivity*. Conference on Agroforestry Principles and Practice, July 1989, University of Edinburgh. Edinburgh, UK.

Perera, W.R.R. (1962). The development of forest plantations in Ceylon since the seventeenth century. *Ceylon Forester*, **5**, 142–147.

Perrin, R. and Sutherland, J.R. (ed.) (1993). *Diseases and insects in forest nurseries*. Institut National de la Recherche Agronomique, Paris.

Perry, E. and Hickman, G. (1987). Wound closure in *Eucalyptus*. *Journal of Arboriculture*, **13**(8), 201–202.

Perum Perhutani (1992). Teak in Indonesia. In H. Wood (ed.) *Teak in Asia* FORSPA publication 4. Proceeding of the Regional Seminar, March 1991, Guangshou, China. FAO (Bangkok).

Peter, A. von (1989). Use of fertilizers in developing countries. *Plant Research and Development*, **29**, 38–64.

Pham Van Mach (1995). Management of damping-off of *Pinus merkusii* and *P. caribaea* in northern Vietnam. In *Caring for the forest: research in a changing world* (ed. E., Korpilahti, T., Salonen, and S. Oja), p. 167. Congress Report Abstracts of Invited Papers, IUFRO XX World Congress, Tampere, Finland, 6–12 August 1995. The Finnish IUFRO World Congress Organising Committee, Tampere, Finland.

Pijl, L. van der (1982). *Principles of dispersal in higher plants.* Springer Verlag, Berlin, Heidelberg, and New York.

Pinto, P., Laroze, A., and Muñoz, F. (2001). Optimization models for scheduling silvicultural operations. In *Models for the sustainable management of temperate plantation forests* (ed. J-M. Carnus, R. Dewar, D. Loustau, M. Tomé, and C. Orazio), pp. 95–103. EFI Proceedings No. 41. European Forest Institute, Joensuu, Finland.

Pinyopusarerk, K. (1990). Acacia auriculiformis: *an annotated bibliography.* Winrock International Institute of Agricultural Development and ACIAR, Bangkok.

Pinyopusarerk, K. and House, A.P.N. (1993). Casuarina: *an annotated bibliography of* C. equisetifolia, C. junghuhniana *and* C. oligodon. International Centre for Research in Agroforestry, Nairobi.

Plantation 2020 Vision Implementation Committee (1997). *Plantations for Australia: the 2020 vision.* Ministerial Council on Forestry, Fisheries and Aquaculture/Standing Committee on Forestry, Canberra.

Plumptre, R.A. (1979). Pruning fast-growing pines for wood uniformity: can you have your cake and eat it? *Commonwealth Forestry Review*, **58**, 181–189.

Plumptre, R.A. (1984). Pinus caribaea, *Vol. 2—Wood properties.* Tropical Forestry Paper No. 17. Oxford Forestry Institute, Oxford.

Poffenberger, M. (ed.) (1990). *Keepers of the forest: land management alternatives in South-east Asia.* Kumarian Press, West Hartford, USA.

Poggiani, F. and Suiter Filho, W. (1974). Importância da nebulização intermitente e efeito do tratamento hormonal na formação de raízes em estacas de eucalipto. *IPEF Piracicaba*, **9**, 119–129.

Polglase, P.J., Theiveyanathan, S., Benyon, R.G., and Falkiner, R.A. (2002). *Irrigation management and groundwater uptake in young tree plantations growing over high watertables.* RIRDC Publication No. 02/146. Rural Industries Research and Development Corporation, Canberra.

Poole, B. (1989). Forest health issues in South-east Asian countries. *New Zealand Journal of Forestry Sciences*, **19**, 159–162.

Poore, M.E.D. (1989). *No timber without trees.* Earthscan, London.

Poore, M.E.D. and Fries, C. (1985). The ecological effects of eucalyptus. FAO Forestry Paper 59, FAO/SIDA, Rome.

Poschen, P. (1997). Forests and employment—much more than meets the eye. In *The economic contribution of forestry to sustainable development*, Vol. 4. Proceedings of the XI World Forestry Congress, Antalya, Turkey, 13–22 October 1997. pp. 61–77. Ministry of Forestry, Ankara, Turkey.

Potter, L. and Lee, J. (1998). *Tree planting in Indonesia: trends, impacts and directions.* CIFOR Occasional Paper No. 19. Center for International Forestry Research, Bogor, Indonesia.

Pottinger, A., Chamberlain, J., and MacQueen, D. (1996). Linking international evaluation of agroforestry tree species with farmers' objectives. *Agroforestry Forum*, **7**(4), 11–13.

Pouli, T., Tuilaepa, E., and Woods, P.V. (1995). *Promising indigenous tree species for use in plantations in Western Samoa.* Proceedings of Second International Conference on Forest vegetation Management, Rotorua, New Zealand, 20–24 March 1995, Forest Research Institute Bulletin 192, pp. 289–291, Forest Research Institute, Rotorua, New Zealand.

Poulsen, K. (1993). *Seed quality—concept, measurement and measurement to increase quality.* Lecture Note C-14, Danida Forest Tree Centre, Humlebaek, Denmark.

Poulsen, K. (1996). Ways and recommendations on how to test seed quality of tropical tree seeds. In *Innovations in tropical tree seed technology* (ed. K.Olesen), pp. 233–241. IUFRO Symposium of the Project Group P.2.04.00 'Seed Problems' Danish Forest Tree Seed Centre, Humlebaek, Denmark.

Poulsen, K. and Stubsgaard, F. (1995). *Three methods for mechanical scarification of hardcoated seed.* Technical Note No. 27. Danida Forest Seed Centre, Humlebaek, Denmark.

Powell, M.B. and Nikles, D.G. (1996). Performance of *Pinus elliottii* var. *elliottii* and *P. caribaea* var. *hondurensis*, and their F_1, F_2 and backcross hybrids across a range of sites in Queensland. In *Tree improvement for sustainable tropical forestry* (ed. M.J. Dieters, A.C. Matheson, D.G. Nikles, C.E. Harwood, and S.M. Walker), pp. 382–383. Proceedings of QFRI-IUFRO Conference, Caloundra, Australia, October/November 1996, Queensland Forestry Research Institute, Gympie, Australia.

Powers, R.F., Mead, D.J., Burger, J A., and Ritchie, M.W. (1994). Designing long-term site productivity experiments. In *Impacts of forest harvesting on long-term site productivity* (ed. W.J. Dyck, D.W. Cole, and N.B. Comerford), pp. 247–286. Chapman Hall, London.

Poynton, R.J. (1971). Silvicultural map of southern Africa. *South African Journal of Sciences*, **67**, 58–60.

Poynton, R.J. (1979). *Tree planting in southern Africa.Vol. 2. The eucalypts.* Department of Forestry, Johannesburg.

Poynton, S. (1996). *Producing high quality* Eucalyptus *seedlings using a best practices approach in Vietnam's Mekong Delta.* (ed. A.C. Yapa), pp. 112–118. Proceedings: International Symposium on Recent Advances in Tropical Forest Tree Seed Technology and Planting Stock Production. ASEAN Forest Tree Seed Project, Mauk-Lek, Saraburi, Thailand.

Prabhu, R., Colfer, C.J.P., and Dudley, R.G. (1999). *Guidelines for developing, testing and selecting criteria and indicators for sustainable forest management*. The Criteria and Indicators Tool Box Series No. 1. Center for International Forestry Research, Bogor, Indonesia.

Prado, J.A. and Toro, J.A. (1996). Silviculture of eucalypt plantations in Chile. In Nutrition of eucalypts (ed. P.M. Attiwill and M.A. Adams), pp. 357–369. CSIRO Publishing, Melbourne, Australia.

Prasad, R. (1987). Technological planning vis-a-vis plantations: a case study of Kesla project, Hoshangabad. *Journal of Tropical Forestry*, **3**, 198–206.

Prasad, R. (1991). Use of acacias in wastelands reforestation. In *Advances in tropical acacia research* (ed. J.W. Turnbull), pp. 96–102. ACIAR Proceedings No. 35. Australian Centre for International Agricultural Research, Canberra.

Prasad, R. and Chadhar, S.K. (1987). Afforestation of dolomite mine overburdens in Madhya Pradesh. *Journal of Tropical Forestry*, **3**, 124–131.

Prasad, Ram (1988). Effectiveness of aerial seeding in reclamation of Chambal ravines in Madhya Pradesh. *Indian Forester*, **114**, 1–18.

Pritchard, M.A. (ed.) (1989). *A systems approach to forest operations planning and control*. Forestry Commission Bulletin 82. Her Majesty's Stationery Office, London.

Pryor, L.D. (1978). *Eucalypts as exotics*. International Training Course on Forest Tree Breeding, pp. 219–221. Australian Development Assistance Agency, Canberra..

Puri, G.S. (1960). *Indian forest ecology*, Vols 1 and 2. Oxford Book and Stationery, India.

Purse, J.G. (1989). Development of techniques for vegetative propagation in pines—a review of progress and prospects. In *Breeding tropical trees: population structure and genetic improvement strategies in clonal and seedling forestry* (ed. G.L. Gibson, A.R. Griffin, and A.C. Matheson), pp. 298–310. Proceedings of an IUFRO Conference, Pattaya, Thailand, December 1988. Oxford Forestry Institute and Winrock International, Oxford.

Puttonen, P. (1995). Looking for the "silver bullet"—can one test do it all. *New Forests*, **13**, 9–27.

Quaile, D.R. and Mullin, L.J. (1984). Provenance and progeny testing in *Eucalyptus grandis* in Zimbabwe. In *Provenances and genetic improvement strategies in tropical forest trees* (ed. R.D. Barnes and G.L. Gibson), pp. 438–450. IUFRO Conference, Mutare, Zimbabwe.

Quayle, S. and Gunn, B. (1998). *Tree nursery manual for Namibia*. CSIRO Forestry and Forest Products, Canberra.

Quayle, S., Arnold, R., Gunn, B., and Mohns, B. (2001). *Tree nursery manual for the Sri Lankan plantation industry*. Estate Forest and Water Resources Development Project, Kandy, Sri Lanka.

Queensland Department of Forestry (1987). *Research Report 1985*, pp. 26, 30.

Rai, S.N. (1999). *Nursery and planting techniques of forest trees in tropical South-Asia*. Punarvasu Publications, Dharwad, Karnataka, India.

Raintree, J.B. (1986). Agroforestry pathways: land tenure, shifting cultivation and sustainable agriculture. *Unasylva*, **154**, 2–15.

Raintree, J.B. (1991). *Socio-economic attributes of trees and tree planting practices*. Community Forestry Note No. 9. Food and Agriculture Organization of the United Nations, Rome.

Raintree, J.B. (ed.) (1987). *Land, tree and tenure*. Land Tenure Center, Madison, USA and International Council for Research in Agroforestry, Nairobi.

Raintree, J.B. and Lantican, C. (1993). Forestry economics research and MPTS development. In *Forestry Economics Research in Asia* (ed. Songkram Thammincha, Ladawan Puangchit, and H. Wood), pp. 15–31. Faculty of Forestry, Kasetsart University, Thailand and IDRC, Canada.

Rajan, M.S. (1991). *Remote sensing and geographic information system for natural resource management*. ADB Environment Paper No. 9. Asian Development Bank, Manila.

Raman, S.S. (1976). Biological productivity of *Shorea* plantations. *Indian Forester*, **102**, 174–184.

Ramdial, B.S. (1980*a*). *The taungya as practised in Trinidad with emphasis on teak* (Tectona grandis L.). Field guide prepared for the 11th Commonwealth Forestry Conference, Port of Spain, Trinidad and Tobago.

Ramdial, B.S. (1980*b*). *Forestry in Trinidad and Tobago*. Paper presented at the 11th Commonwealth Forestry Conference, Trinidad and Tobago.

Ramirez, C., Sanchez, G.A., Kass, D.C.L. Sanchez, J.F., and Viquez, E. (1989). *Advances in* Erythrina *research*. Paper to CATIE. Conference on Fast Growing Trees and Nitrogen Fixing Trees, October 1989, Philipps University, Marburg, Germany.

Rampanana, L., Rakotomanana, J.L., Louppe, D., and Brunck, F. (1988). Dying of *Pinus kesiya* crowns in Madagascar. *Bois et Fôret des Tropiques*, **214**, 23–47.

Rance, S.J., Cameron, D.M., and Williams, E.R. (1982). Correction of crown disorders of *Pinus caribaea* var. *hondurensis* by application of zinc. *Plant and Soil*, **65**, 293–296.

Rao, Y.S. (1986). Some socioeconomic and institutional aspects of forest land uses. In *Land use, watersheds and planning in the Asia-Pacific region*. RAPA Report 1986/3, pp. 3–10. Food and Agriculture Organization of the United Nations, Bangkok.

Ravindran, D.S. and Thomas, T.H. (2000). Trees on farms, stores of wealth and rural livelihoods—insights and evidence from Karnataka, India. *International Forestry Review*, **2**, 182–190.

Ravindranath, N.H. and Hall, D.O. (1994). Indian forest conservation and tropical deforestation. *Ambio*, **23**, 521–523.

Rawat, J.K. (1990). Economic spacing and rotation decisions in farm forestry. *Indian Forester*, **116**, 341–347.

Raza-ul-Haq and Chaudhry, A.K. (2001). Effect of water conservation techniques on the establishment and growth of forest tree species in scrub zone. *Pakistan Journal of Forestry*, **51**, 41–45.

Reddell, P., Rosbrook, P.A., Zierl, A., Yang, Y., and Kang, L. (1996*a*). Frankia culture and inoculation technologies for *Casuarina* species. In *Recent casuarina research and development* (ed. K. Pinyopusarerk, J.W. Turnbull, and S.J. Midgley), pp. 63–67. CSIRO Forestry and Forest Products, Canberra.

Reddell, P., Webb, M.J., Poa, D., and Aihuna, D. (1999). Incorporation of slow-release fertilisers into nursery media: a highly effective technique for supplying nutrients during early field establishment of plantation trees in the humid tropics. *New Forests*, **18**, 277–288.

Reddell, P., Zierl, A., and Rosbrook, P.A. (1996*b*). Comparative symbiotic effectiveness of 24 *Frankia* isolates on seedlings of three *Casuarina* species. In *Recent casuarina research and development* (ed. K. Pinyopusarerk, J.W. Turnbull, and S.J. Midgley), pp. 68–73. CSIRO Forestry and Forest Products, Canberra.

Reid, R. and Bird, P.R. (1990). Shade and shelter. In *Trees for rural Australia* (ed. K.W. Cremer), pp. 319–335. Inkata Press, Melbourne.

Reifsnyder, W.S. and Darnhofer, T.O. (1989). *Meteorology and agroforestry*. International Centre for Research in Agroforestry, Nairobi.

Reis, M.G.F., Barros, N.F., and Kimmins, J.P. (1987). Acumulo de nutrients em uma sequencia de idade de *Eucalyptus grandis* W. Hill (ex Maiden) plantado no cerrado, em duas āreas com diferentes produtivades, em Minas Gerais. *Revista Avore*, **11**, 1–15.

Rennie, P.J. (1955). The uptake of nutrients by mature forest growth. *Plant and Soil*, **7**, 49–55.

Repetto, R. (1988). *The forest for the tree? Government policies and the misuse of forest resources*. World Development Institute, Washington DC.

Repetto, R. (1990). Deforestation in the Tropics. *Scientific American*, **262**(4), 36–42.

Retief, E.C.L. and Clarke, C.R.E. (2000). The effect of site potential on eucalypt clonal performance in coastal Zululand, South Africa. In *Forest genetics for the next millennium*. Proceedings of an international conference, Durban, South Africa, 8–13 October 2000, pp. 192–196. Institute of Commercial Forestry Research, Scottsville, South Africa.

Revilla, A.V. (1974). Yield prediction in forest plantations. Proceedings of a Forestry Research Symposium on Industrial Forest plantations, pp. 32–43. Philippine Forest Research Society.

Reyes, J.Ch. and Groothousen, C. (1990). *Preventive silviculture to control bark beetles in the pine forest of Honduras*, Vol. 1 (2). Proceedings of the 19th IUFRO World Congress, Montreal, Canada, p. 418.

Rice, R.M. (1978). *The effects of forest management on erosion and sedimentation due to landslides*. Vol. 3. Proceedings of the 8th World Forestry Congress, Jakarta, pp. 319–332.

Richardson, J.H. (1982). Some implications of tropical forest replacement in Jamaica. *Zeitschrift für Geomorphologie*, **44**(Suppl.), 107–118.

Richardson, K.F. (1989). Foresters 'farm' trees to meet growing timber needs. *Shell Agriculture*, **5**, 26–29.

Riedacker, A., Knockeart, C., and Zaidi, A. (1985). Wood production of eucalypt coppice; effect of rotation length and spacing. *Annales des Sciences Forestieres* (Paris), **42**(1), 39–52.

Rimando, E.F. and Dalmacio, M.V. (1978). Direct seeding of ipil-ipil (*Leucaena leucocephala*). *Sylvatrop. The Philippines Forest Research Journal*, **3**, 171–175.

Ritchie, G.A. (1984). Assessing seedling quality. In *Forest nursery manual: production of bareroot seedlings* (ed. M.L. Duryea and T.D. Landis), pp. 243–259. Martinus Nijhoff Dr W. Junk, The Hague, Netherlands.

Ritchie, K.A. (1988). Shelterbelt planting in semi-arid areas. *Agriculture, Ecosystems and Environment*, **22/23**, 425–440.

Ritter, E.S. (ed.) (1999). Proceedings of Application of Biotechnology to Forest Genetics. Biofor-99. Vitoria-Gasteiz, Spain, 22–25 September 1999.

Rivadeneyra, M.G. and Cabrejo, R.L. (1980). The growth of Eucalyptus globulus in Puno, Peru. In Whitmore, J.L. (ed.) (1981), pp. 96-103.

Robbins, A.M.J. (1985). A versatile, low-cost drying kiln for opening pine cones. Occasional Paper No. 26, Oxford Forestry Institute, UK.

Roberds, J.H. and Bishir, K.W. (1997). Risk analyses in clonal forestry. *Canadian Journal of Forest Research*, **27**, 425–432.

Roberts, B. (2001). *Eucalypt seed coating*. Native Forest Silviculture Guideline No. 5. Department of Natural Resources and Environment, Melbourne, Australia.

Roberts, E.H. (1973). Predicting storage life of seeds. *Seed Science and Technology*, **1**, 499–514.

Roberts, H. (1977). When ambrosia beetles attack mahogany trees in Fiji. *Unasylva*, **117**, 25–28.

Roberts, S. and Dubois, O. (1996). The role of social/farm forestry schemes in supplying fibre to the pulp and paper industry. *Towards a sustainable paper cycle*. Sub-study Series No. 4. International Institute for Environment and Development/WBCSD, London.

Roberts, J.M., Rosier, P.T.W., and Srinavasa Murthy K.V. (1991). Physiological studies in ground Eucalyptus plants in southern India and their use in estimating forest transpiration. Proceedings of the Bangalore Seminar on Growth and Water Use of Forest Plantations. February 1991 (in press).

Robinson, P.J. (1985). Trees as fodder crops. In *Attributes of trees as crop plants* (ed. M.G.R. Cannell, J.E. Jackson, and J.C. Gordon). Institute of Terrestrial Ecology, Edinburgh, UK.

Robinson, W.M. (1968). *Pruning and thinning practice in Queensland plantations*. Proceedings of the 9th Commonwealth Forestry Conference, India.

Roche, L. (1978). Community forestry and the conservation of plants and animals. Proceedings of the 8th World Forestry Congress, Jakarta, Vol. 7A, pp. 877–94.

Roche, L. (1986). Forestry and famine: arguments against growth without development. *Commonwealth Forestry Review*, **65**, 99–108.

Rocheleau, D. (1999). Confronting complexity, dealing with difference: social context, content, and practice in agroforestry. In *Agroforestry in sustainable agricultural systems* (ed. L.E. Buck, J.P. Lassoie, and E.C.M. Fernandes), pp. 191–235. Lewis Publishers, Boca Raton, Florida, USA.

Rockwood, D.L. (1991). Frost resilient E. grandis clones for Florida, USA. In *Intensive forestry: the role of eucalypts* Vol. 1 (ed. A.P.G. Schonau), pp. 455–466. South African Institute of Forestry, Pretoria, South Africa.

Rodríguez-Pedraza, C.D., Walker, R., and Oliveira, P.M. de. (1994). *Use of GPS with Landsat images to identify main crops in farms along the Transamazônica highway, Brazil.* Proceedings of the ISPRS Commission VII Symposium, Rio de Janeiro, Brazil, 26–30 September 1994. National Institute for Space Research, São José dos Campos, São Paulo, Brazil. *Resource and Environmental Monitoring.* **30**(7a), 35–37.

Rojas, J.C., Wright, J.A., and Allen, H.L. (1999). *Genotype × environment interaction of the urograndis hybrid of eucalypts in western Venezuela* (ed. M. Bowen and M. Stine), pp. 145–151. Proceedings of the 25th Biennial Southern Forest Tree Improvement Conference, New Orleans, Louisiana, 11–14 July 1999. National Technical Information Service, US Department of Commerce, Washington DC.

Romero, A.E., Ryder, J., Fisher, J.T., and Mexal, J.G. (1986). Root system modification of container stock for arid land plantings. In *Tree plantings in semi-arid regions* (ed. P. Felker), pp. 281–290. Proceedings of a Symposium on Establishment and Productivity of Tree Plantings in Semi-arid Regions, Kingsville, Texas. Elsevier, Amsterdam, The Netherlands.

Romero, G. (1997). *Relationship of seed-borne pathogens to nursery and plantation diseases of eucalypts and pines in Uruguay* (ed. Z. Prochazakova and J.R. Sutherland.), pp. 82–85. Proceedings of the ISTA seed pathology meeting, Opocno, Czech Republic, 9–11 October 1996. International Seed Testing association, Zurich, Switzerland.

Rose, R., Carlson, W.C., and Morgan, P. (1990). The target seedling concept. In *Target seedling symposium* (ed. R.Rose *et al.*), pp. 1–8. General Technical Report RM-200. US Department of Agriculture, Forest Service, Rocky Mountain Forest and Range Experiment Station, Fort Collins, Colorado.

Roshetko, J.M. and Evans, D.O. (ed.) (1999). Domestication of agroforestry trees in southeast Asia. *Forest, Farm and Community Tree Research Reports.* Special issue. Winrock International, Morrilton, Arkansas, USA.

Ross, M.S. and Donovan, D. (1986). *Land clearing in the humid tropics,* IUCN/IIED Tropical Forest Policy Paper No. 1. International Institute for Environment and Development, London.

Ross, W. (1977). Fire protection in industrial plantations of Zambia. In *Savanna afforestation in Africa.* FAO Forest Paper 11, pp. 196–212. Food and Agriculture Organization of the United Nations, Rome.

Roux, J., Dunlop, R., and Wingfield, M.J. (2000). Development of disease tolerant *Acacia mearnsii*. In *Forest genetics for the next millenium*, pp. 200–202. Institute for Commercial Forest Research, Scottsville, South Africa.

Roux, P.J.le (1984). *Plant invader species in pine plantations in South Africa* (ed. D.C. Grey, A.P.G. Schönau, C.J. Schutz, and A. van Laar), pp. 767–776. Symposium on site and Productivity of Fast Growing Plantations IUFRO, Pretoria and Pietermaritzburg, South Africa.

Rozelle, S. and Huang, J. (1995). Environmental stress and grain yields in China. *American Journal of Agricultural Economics,* **77**, 853–864.

Ruis, B.G.M.S. (2001). No forest convention but ten tree treaties. *Unasylva*, **52**, 3–13.

Rungu, C. (1996). Pretreatment with water and sulphuric acid of different seed sources of *Acacia nilotica.* In *Innovations in tropical tree seed technology.* IUFRO Symposium of the Project Group P.2.04.00 'Seed Problems' (ed. K. Olesen), pp. 242–245, Danish Forest Tree Seed Centre, Humlebaek, Denmark.

Rusastra, I.W., Susilawati, S.H., Budhi, G., and Grist, P. (1998). *Flemingia* hedgerows. In *Improving smallholder farming systems in Imperata areas of Southeast Asia: alternatives to shifting cultivation* (ed. K. Menz, D. Magcale-Macandog, and I.W. Rusastra), pp. 149–159. ACIAR Monograph No. 52. Australian Centre for International Agricultural Research, Canberra.

Ryan, P.A. and Bell, R.E. (1989). Growth, coppicing and flowering of Australian tree species in trials in southeast Queensland, Australia. In *Trees for the tropics* (ed. D. J. Boland), pp. 49–68. Australian Centre for International Agricultural Research, Canberra.

Ryan, P.A., Podberscek, Raddatz, C.G., and Taylor, D.W. (1987). *Acacia* species trials in southeast Queensland, Australia. In *Australian acacias in developing countries.* (ed. J.W.Turnbull), pp. 81–85. ACIAR Proceedings No. 16. Australian Centre for International Agricultural Research, Canberra.

Ryan, T. and Shea, G. (1977). *Exotic pine plantations in Queensland and the role of the Beerwah-Beerburrum forests.* Post consultation tour notes No. 8. Third World Consultation of Forest Tree Breeding, Canberra.

Sabaruddin, K.M. (1988). *Fire control of forest plantation in Benakat.* Reforestation Technology Institute, Benekat, South Sumatra, Indonesia.

Saenger, P. (2002). *Mangrove ecology, silviculture and conservation.* Kluwer Academic Publishers, Dordrecht, The Netherlands.

Saenger, P. and Siddiqui, N.A. (1993). Land from the sea: the mangrove afforestation program of Bangladesh. *Ocean and Coastal Management*, **20**, 23–39.

Saharjo, B.H. (1997). Fire protection and industrial plantation management in the tropics. *Commonwealth Forestry Review*, **76**, 203–206.

Saigal, R. (1990). Modern forest fire control: the Indian experience. *Unasylva*, **162**, 21–27.

Sakai, C., Subiakto, A., Heriansyah, I., and Nuroniah, H.S. (2001). Rehabilitation of degraded forest with *Shorea leprosula* and *S. selanica* cuttings. In *Rehabilitation of degraded forest ecosystems in the tropics* (ed. S. Kobayashi, J.W. Turnbull, T. Toma, T. Mori, and N.M.N.A. Majid), pp. 191–195. Center for International Forestry Research, Bogor, Indonesia.

Sanchez, P.A. (1976). *Properties and management of soils in the tropics.* Wiley Interscience, New York.

Sanchez, P.A. (1995). Science in agroforestry. *Agroforestry Systems*, **30**, 5–55.

Sanchez, P.A. and Palm, C.A. (1996). Nutrient cycling and agroforestry in Africa. *Unasylva*, **47**(185), 24–28.

Sandrasegaran, K. (1966). *Optimum planting distances and crop densities of the ten exotic species in Malaya utilising triangular spacing based on a consideration of crown diameter to stem diameter relationship*. Research Pamphlet 51, Forestry Research Institute, Kepong, Malaya.

Sang, F.K.A. (1987). Forest diseases and pests with special reference to eastern Africa experience and assess potential in Rwandan forestry practice. In *Compte-rendu du premier seminaire national sur la sylviculture des plantations forestieres au Rwanda* (ed. V. Pleines), pp. 455–516. Butare, September 1987. Department de Forestiere, Institut des Sciences Agronomiques du Rwanda.

Sargent, C. and Bass, S. (ed.) (1992). *Plantation politics*. Earthscan Publications, London.

Sarraihlh, J-M. (2002). La revégétalisation des exploitations minières: l'exemple de la Nouvelle-Caledonie. *Bois et Forêts des Tropiques*, **272**(2), 21–31.

Sarrailh, J.M. and Ayrault, N. (2001). Rehabilitation of nickel mining sites in New Caledonia. *Unasylva*, **52**(207), 16–20.

Sary, H., Yameogo, C.S., and Stubsgaard, F. (1993). *The CO_2 method to control insect infestation in tree seed*. Technical Note 42, Danida Forest Seed Centre, Humlebaek, Denmark.

Saxena, N.C. (1991). Marketing constraints for *Eucalyptus* from farm lands in India. *Agroforestry Systems*, **13**, 73–85.

Saxena, N.C. (1997). *The saga of participatory forest management in India*. CIFOR Special Publication. Center for International Forestry Research, Bogor, Indonesia.

Saxena, N.C. (2001). The new forest policy and joint forest management in India. In *The forests handbook*, Vol. 2 (ed. J. Evans), pp. 233–259. Blackwell Science, Oxford, UK.

Saxena, N.C. and Vishwa Ballabh (1995). Farm forestry and the context of farming systems in South Asia. In *Farm Forestry in South Asia* (ed. N.C. Saxena and Vishwa Ballabh), pp. 24–50. Sage, New Delhi.

Sayer, J. (2002). Restoration: the challenge for 21st century foresters. *Bois et Forêts des Tropiques*, **272**(2), 3–4.

Scande, M. and Hoekstra, F.A. (1999). Improving the storage longevity of neem (*Azadirachta indica*) seeds. *In Recalcitrant seeds* (ed. M. Marzalina, K.C. Khoo, N. Jayanthi, F.Y. Tsan, and B. Krishnapilly), pp. 64–73. IUFRO Seed Symposium 1998 Proceedings. Forest Research Institute Malaysia, Kepong, Malaysia.

Scherr, S.J. and Yadav, S. (1996). *Land degradation in the developing world: implications for food, agriculture and the environment to 2020*. Food, Agriculture and the Environment Discussion Paper 14. International Food Policy Research Institute, Washington DC.

Schirmer, J., Kanowski, P., and Race, D. (2000). Factors affecting adoption of plantation forestry on farms: implications for farm forestry development in Australia. *Australian Forestry*, **63**, 44–51.

Schlamadinger, B. and Karjalainen, T. (2000). Afforestation, reforestatation and deforestation activities. In *Land use, land-use change and forestry* (ed. R.T. Watson, I.R. Noble, R. Bolin, N.H. Ravindranath, D.J. Ravado, and D.J. Dokken), pp. 127–179. Intergovernmental Panel on Climate Change Special Report. Cambridge University Press, Cambridge, UK.

Schmidt, L. (2000). *Guide to handling of tropical and subtropical forest seed*. Danish Forest Seed Centre, Humlebaek, Denmark.

Schonau, A.P.G. (1984c). A factorial thinning experiment in *Eucalyptus grandis*. *Commonwealth Forestry Review*, **63**, 285–295.

Schönau, A.P.G. (1975). Ensure satisfactory eucalypt coppice regeneration by applying correct felling and bark stripping practices. *Newsletter No. 52*. South African Timber Growers Association.

Schönau, A.P.G. (1977). Initial responses to fertilizing *Eucalyptus grandis* at planting are sustained until harvesting. *Commonwealth Forestry Review*, **56**, 57–59.

Schönau, A.P.G. (1981). Seasonal changes in foliar nutrient content of *E. grandis*. *South African Forestry Journal*, **119**, 1–4.

Schönau, A.P.G. (1982). Additional effects of fertilizing on several foliar nutrient concentrations and ratios in *Eucalyptus grandis*. *Fertilizer Research*, **3**, 385–397.

Schönau, A.P.G. (1984a). Fertilization of fast-growing broadleaved species. In *Symposium on site and productivity of fast growing plantations* (ed. D.C. Grey, A.P.G. Schönau, C.J. Schutz, and A. van Laar), pp. 253–268. IUFRO, Pretoria and Pietermaritzburg, South Africa.

Schönau, A.P.G. (1984b). Silvicultural considerations for high productivity of *Eucalyptus grandis*. *Forest Ecology and Management*, **9**, 295–314.

Schönau, A.P.G. (1985). Basic silviculture for the establishment of eucalypt plantations with special reference to *Eucalyptus grandis*. *South African Forestry Journal*, **143**, 4–9.

Schönau, A.P.G. and Aldworth, W.J.K. (1990). *Site evaluation in black wattle with special reference to soil factors*, Vol. 1. Proceedings of the 19th IUFRO World Congress, Montreal, p. 423.

Schönau, A.P.G. and Boden, D.I. (1982). Preliminary biomass studies in young eucalypts. *South African Forestry Journal*, **120**, 24–28.

Schönau, A.P.G. and Herbert, M.A. (1982). Relationship between growth rate and foliar concentrations of nitrogen, phosphorus and potassium for *Eucalyptus grandis*. *South African Forestry Journal*, **120**, 19–23.

Schönau, A.P.G. and Herbert, M.A. (1989). Fertilizing eucalypts at plantation establishment. *Forest Ecology and Management*, **29**, 221–244.

Schönau, A.P.G. and Pennefeather, M. (1975). A first account of profits at harvesting as a result of fertilising *Eucalyptus grandis* at time of planting in South Africa. *South African Forestry Journal*, **94**, 29–35.

Schultz, R.P. (1997). *Loblolly pine: the ecology and culture of loblolly pine (Pinus taeda L.)* USDA Forest Service Agricultural Handbook 713.

Schultz, R.P. (1999). Loblolly—the pine for the twenty-first century. *New Forests*, **17**, 71–88.

Schumacher, E.F. (1973). *Small is beautiful. A study of economics as if people mattered*. Blond and Briggs, London.

Schutz, C.J. (1976). *A review of fertiliser research on some of the more important conifers and eucalypts planted in subtropical and tropical countries, with special reference to*

South Africa. Bulletin 53, Department of Forestry, Republic of South Africa.

Schutz, C.J. (1990). *Site relationships for some wood properties of pine species in planatation forests of southern Africa*, Vol. 1. Proceedings of the 19th IUFRO World Congress, Montreal, Canada, pp. 221–231.

Sedjo, R. and Lyon, K. (1996). *Timber Supply Model 96: a global timber supply model with a pulpwood component*. Discussion Paper 96–15. Resources for the Future, Washington DC.

Sedjo, R.A. (1983). *The comparative economics of plantation forestry: a global assessment*. Resources for the Future, Washington DC.

Sedjo, R.A. (1984). Industrial forest plantations: an economic assessment. In *Strategies and designs for afforestation, reforestation and tree planting* (ed. K.F. Wiersum), pp. 286–300. Pudoc, Wageningen, The Netherlands.

Sedjo, R.A. (1999). The potential of high-yield plantation forestry for meeting timber needs. *New Forests*, **17**, 339–359.

Sedjo, R.A. (2001). From foraging to cropping: the transition to plantation forestry, and implications for wood supply and demand. *Unasylva*, **52**(204), 24–32.

Seitz, R.A. (1990). *Tree architecture and increment analysis as orientation for silviculture in the tropics*, Vol. 1(2). Proceedings of the 19th IUFRO World Congress, Montreal, Canada, pp. 69–78.

Seitz, R.A. and Corvello, W.V. (1983). Natural regeneration of *Pinus elliottii* in the Parana grasslands. In *Planted forests in the neotropics: a source of energy* (ed. N.F. Barros). IUFRO/MAB/UFV Symposium Vicosa, Brazil.

Senanayake, F.R. (1987). Analog forestry as a conservation tool. Tiger Paper (FAO, Bankok Forest Newsletter) **14** (2), 25–30.

Serajuddoula, M.D., Khan, M.A.S., and Islam, M.R. (1996). Rooted cutting—an alternative source of improved planting material of Sonneratia apetala for coastal afforestation in Bangladesh. In *Tree improvement for sustainable tropical forestry* (ed. M.J. Dieters, A.C. Matheson, D.G. Nikles, C.E. Harwood, and S.M. Walker), pp. 535–536. Proceedings of QFRI-IUFRO Conference, Caloundra, Australia, October/November 1996, Queensland Forestry Research Institute, Gympie, Australia.

Sessions, J. and Bettinger, P. (2001). Hierarchial planning: pathway to the future. In *Precision Forestry*. Proceedings of the First International Precision Forestry Cooperative Symposium, Seattle, Washington. 17–20 June 2001, pp. 185–190. University of Washington, Seattle, USA.

Shah, B.H. (1992). Development of agroforestry model using water harvesting system in semiarid and arid zones. *Pakistan Journal of Forestry*, **42**, 190–199.

Shah, T. (1988). *Gains from social forestry: lessons from West Bengal*. Discussion paper No. 243, Institute of Development Studies, University of Sussex, UK.

Shams-ur-Rehman (1998). Growth assessment of some arid zone species and seed sources in Pakistan. *Pakistan Journal of Forestry*, **48**, 13–18.

Shanks, E. and Carter, J. (1994). *The organization of small-scale tree nurseries*. Rural Development Forestry Study Guide 1, Overseas Development Institute, London.

Sharma, R.P. (1979). Production potential and other crop characteristics of the first generation coppice of Eucalyptus hybrid. *Indian Forester* **105**, 89–99.

Sharma, D.D. and Sharma, A. (2001). Watershed management in rainfed areas: learning from the people. *The Indian Forester*, **127**, 845–854.

Sharma, J.K. and Sankaran, K.V. (1987). *Diseases of Albizzia falcataria in Kerala and their possible control measures*. KEFRI Research Report No. 47, Kerala Forest Research Institute, India.

Sharma, J.K., Mohannan, C., and Maria Florence, E.J. (1985). *Disease survey in nurseries and plantations of forest tree species grown in Kerala*. KFRI, Research Report 36, Kerala Forest Research Institute, India.

Shea, G.M. (1987). *Aspects of management of plantations of tropical and subtropical Queensland*. Revised edition of paper prepared by P.J. Hawkins and J.D. Muir for the Ninth Commonwealth Forestry Conference, 1968. Queensland Department of Forestry, Brisbane, Australia.

Shea, G.M., Harvey, A.M., and Anderson, T.M. (1984). Holistic evaluation of research findings in softwood plantations of Queensland (ed. D.C. Grey, A.P.G. Schönau, C.J. Schutz, and A. van Laar), pp. 787–798. Symposium on Site and Productivity of Fast Growing Plantations. IUFRO, Pretoria and Pietermaritzburg, South Africa.

Sheikh, M.I. (1986). *Afforestation of arid and semi-arid areas in Pakistan*. FAO and Pakistan Forest Institute, Peshawar, Pakistan.

Sheikh, M.I. (1988). Planting and establishment of windbreaks in arid areas. *Agriculture, Ecosystems and Environment*, **22/23**, 405–423.

Sheikh, M.I. (1990). Water requirements for optimum growth of *Eucalyptus camaldulensis, Salmalia malabarica, Morus alba* and poplar clones. Vol. 1. Proceedings of the 19th IUFRO World Congress, Montreal, Canada, pp. 564–584.

Sheikh, M.I., Shah, B.H., and Aleem, A. (1984). Effect of rainwater harvesting methods on the establishment of tree species. *Forest Ecology and Management*, **8**, 257–263.

Shelton, H.M. (1994). Establishment of forage tree legumes. In *Forage tree legumes in tropical agriculture* (ed. R.C. Gutteridge and H.M. Shelton), pp. 132–142. CAB International, Wallingford, UK.

Sheng, W. (1991). The management of *Cunninghamia lanceolata* plantations in China. In *Development of forestry science and technology in China* (ed. J. Deng, D. Yunyi, W. Li, and Z. Chen), pp. 108–114. China Science and Technology Press, Beijing.

Shepherd, G. (1988). *Putting trees in the farming system: land adjudication and agroforestry on the lower slopes of Mount Kenya*. Social Forestry Network Paper 8a, Overseas Development Institute, London.

Shepherd, G. (1990). *Forestry, social forestry, fuel wood and the environment: tour of the horizon*. Social Forestry Network Paper 11a, Overseas Development Institute, London.

Shepherd, G. (1997). Trees on farm and people in the forest: social science perspectives in tropical forestry. *Commonwealth Forestry Review*, **76**, 47–52.

Sherry, S.P. (1952). The effect of different methods of brush wood disposal upon site conditions in wattle plantations. *Annual Report of the Wattle Research Institute*, **51**/2, 33–41; **53**/4, 27–36 (1954); **60**/1, 32–40 (1961); **63**/4, 41–50 (1964).

Shrivastava, M.B., Tewari, K.N., and Minakshi, S. (1988). Afforestation of salt affected soils in India. *Indian Journal of Forestry*, **11**, 1–12.

Siddiqui, K.M. (1994). Tree planting for sustainable use of soil and water with special reference to the problem of salinity. *Pakistan Journal of Forestry*, **44**, 97–102.

Siddiqui, K.M. and Noor, M. (1993). Pakistan experience in dryland afforestation. *Pakistan Journal of Forestry*, **43**, 54–64.

Siddiqui, N.A. and Khan, M.A.S. (1996). Planting techniques for mangroves on new accretions in the coastal areas of Bangladesh. In *Restoration of mangrove ecosystems*. (ed. C.D. Field), pp.143–159. International Society for Mangrove Ecosystems, Okinawa, Japan.

Sidhu, D.S. and Thakur, A. (1998). Forest fires and their control in India. In *Forest Fire Research*, Vol. 1. (ed. D.X. Viegas), pp. 191–201. Proceeedings of the III International Conference on Forest Fire Research and the 14th Conference on Fire and Forest Meteorology. Associacao para o Dessenvolvimento da Aerodinamica Industrial, Coimbra, Portugal.

Sim, B.L. (1987). Research on *Acacia mangium* in Sabah: a review. In *Australian acacias in developing countries* (ed. J.W.Turnbull), pp. 164–166. ACIAR Proceedings No. 16. Australian Centre for International Agricultural Research, Canberra.

Sim, B.L. and Gan, E. (1991). Performance of Acacia species on four sites of Sabah Forest Industries. *Advances in tropical acacia research* (ed. J.W. Turnbull), pp. 159–165. ACIAR Monograph No. 35. Australian Centre for International Research, Canberra.

Sim, B.L. and Nykvist, N. (1991). Impact of forest harvesting and replanting. *Journal of Tropical Forest Science*, **3**, 251–284.

Simons, A.J. (1996). Delivery of improvement for agroforestry trees. In *Tree improvement for sustainable tropical forestry* (ed. M.J. Dieters, A.C. Matheson, D.G. Nikles, C.E. Harwood, and S.M. Walker), pp. 391–400. Proceedings of QFRI-IUFRO Conference, Caloundra, Australia, October/November 1996, Queensland Forestry Research Institute, Gympie, Australia.

Simons, A.J., MacQueen, D.J., and Stewart, J.L. (1994). Strategic concepts in the breeding of non-industrial trees. In *Tropical trees: potential for domestication and the rebuilding of forest resources* (ed. R.R.B. Leakey and A.C. Newton), pp. 91–102. Her Majesty's Stationery Office, London.

Simpson, J. (1998). Site specific fertilizer requirements of tropical pine plantations. In *Soils of tropical forest ecosystems* (ed. A. Schulte and D. Ruhiyat), pp. 115–124. Springer, Berlin and Heidelberg, Germany.

Simpson, J.A. and Grant, M.J. (1991). *Exotic pine fertilizer practice and its development in Queensland*. Queensland Forest Service Technical Paper No. 49, pp. 1–17.

Simpson, J.A. and Podger, F.D. (2000). Management of eucalypt diseases—options and constraints. In *Diseases and pathogens of eucalypts* (ed. P.J. Keane, G.A. Kile, F.D. Podger, and B.N. Brown), pp. 427–444. CSIRO, Melbourne.

Simpson, J.A., Osborne, D.O., and Xu, Z.H. (1999). Pine plantations on coastal lowlands of subtropical Queensland, Australia. In *Site management and productivity in tropical plantation forests* (ed. E.K.S. Nambiar, C. Cossalter, and A. Tiarks), pp. 61–67. Center for International Forestry Research, Bogor, Indonesia.

Simula, M. (1998). The economic contribution of forestry to sustainable development. *Commonwealth Forestry Review*, **77**, 4–10.

Sinclair, F.L. (1999). A general classification of agroforestry practice. *Agroforestry Systems*, **46**, 161–180.

Singh, G., Abrol, I.P., and Cheema, S.S. (1989). Effects of gypsum application on mesquite (*Prosopis juliflora*) and soil properties in an abandoned sodic soil. *Forest Ecology and Management*, **29**, 1–14.

Singh, R.V. (1982). *Fodder trees of India*. Oxford University Press, New Delhi.

Skelton, D. (1981). *Reforestation in Papua New Guinea*. Office of Forests, Papua New Guinea.

Skoupy, J. (1982). Afforestation of extreme sites. *Silvaecultura Tropica et Subtropica (Prague)*, **9**, 3–20.

Skutsch, M. M. (1983). Why people don't plant trees: socioeconomic impacts of existing woodfuel programmes-village case studies, Tanzania. Discussion paper D-73P, Centre for Energy Policy Research, Resources for the Future, Washington DC. (Unpublished).

Skutsch, M.M. (1985). Forestry by the people for the people: some major problems in Tanzania's village afforestation programme. *International Tree Crops Journal*, **3**, 147–170.

Smith, C.W. (1998). Harvesting impacts and soil tillage. In *Annual Research Report 1998* (ed. L. MacLennan), pp. 99–107. Institute for Commercial Forestry Research, Scottsville, South Africa.

Smith, C.W., Little, K.M., and Norris, C.H. (2001). The effect of land preparation at re-establishment on the productivity of hardwoods in South Africa. *Australian Forestry*, **64**, 165–174.

Smith, D.M. (1986). *The practice of silviculture*, pp. 527. John Wiley, New York.

Smith, D.R. (1999). *Successful rejuvenation of radiata pine* (ed. M. Bowen and M. Stine), pp. 158–167. Proceedings of the 25th Biennial Southern Forest Tree Improvement Conference, New Orleans, Louisiana, 11–14 July 1999. National Technical Information Service, US Department of Commerce, Washington DC.

Smith, F.W. (1986). Interpretation of plant analysis: concepts and principles. In *Plant analysis— an interpretation manual* (ed. D.J. Reuter and J.B. Robinson), pp. 1–12. Inkata Press, Melbourne.

Smith, J. and Scherr, S.J. (2002). Forest carbon and local livelihoods: assessment of opportunities and policy recommendations. CIFOR Occasional Paper No. 37. Center for International Forestry Research, Bogor, Indonesia.

Smith, J.E.N. (1976). *The silviculture of Pinus in Papua New Guinea*. Office of Forests, Papua New Guinea.

Smith, R.E. and Scott, D.F. (1992). The effects of afforestation on low flows in various regions of South Africa. *Water SA*, **18**, 185–194.

Smits. W.T.M., de Fraiture, A.C., and Yasman, I. (1994). Production of dipterocarp planting stock by cuttings in Indonesia. In *Tropical trees: potential for domestication and the rebuilding of forest resources* (ed. R.R.B. Leakey and A.C. Newton), pp. 267–272. Her Majesty's Stationery Office, London.

Smorfitt, D.B., Herbohn, J.L., and Harrison, S.R. (2001). In *Sustainable farm forestry in the tropics: social and economic analysis and policy* (ed. S.R. Harrison and J.L. Herbohn), pp. 77–88. Edward Elgar, Cheltenham, UK.

Snowdon, P. (2000). Nutritional disorders and other abiotic stresses in eucalypts. In *Diseases and pathogens of eucalypts* (ed. P.J. Keane, G.A. Kile, F.D. Podger, and B.N. Brown), pp. 385–410. CSIRO, Melbourne, Australia.

Soares, R.V. (1981). Controlled burning in pine plantations in Sacremento region (Minas Gerais). *Floresta*, **10**(2), 33–40.

Söderlund, M. and Pottinger, A. (2001). *The World's forests—Rio +8: policy, practice and progress towards sustainable management.* Commonwealth Forestry Association, Oxford, UK.

Soerianegara, I. and Lemmens, R.H.M.J. (ed.) (1993). *Timber trees: major commercial species.* PROSEA No. 5(1). Pudoc Scientific Publishers, Wageningen, The Netherlands.

Sohngen, B., Mendelsohn, R., Sedjo, R., and Lyon, K. (1997). *An analysis of global timber markets.* Discussion Paper 97–37. Resources for the Future, Washington DC.

Soil Survey Staff (1999). *Soil taxonomy*, 2nd edn. USDA/NRCS Soil Survey Staff, Agricultural Handbook 436, Washington DC.

Solberg, B. (1998). Economic aspects of forestry and climate change. *Commonwealth Forestry Review*, **77**, 229–233.

Solberg, B. (ed.) (1996). *Long-term trends and prospects in world supply and demand for wood and implications for sustainable forest management.* EFI Report No. 6. European Forest Institute, Joensuu, Finland.

Song, Z. (1991). A review of the development of shelterbelt system in China. In *Development of forestry science and technology in China* (ed. K. Shi), pp. 64–70. China Science and Technology Press, Beijing.

Sosef, M.S.M., Hong, L.T., and Prawirohatmodjo, S. (ed.) (1995). *Timber trees: lesser-known timbers.* PROSEA No. 5(3). Backhuys, Leiden, The Netherlands.

South, D.B., Zwolinski, J.B., and Allen, H.L. (1995). Economic returns from enhancing loblolly establishment on two upland sites: effects of seedling grade, fertilization, hexazinone and intensive soil cultivation. *New Forests*, **10**, 239–256.

Southgate, D. (1995). *Subsidized tree plantations in Ecuador: some issues.* Proceedings of a workshop on the use of financial incentives for industrial forest plantations. 19 January 1995. Paper ENV-4, Environment Division, Inter-American Development Bank, Washington DC.

Spears, J.S. (1980). *Overcoming constraints to increased investment in forestry.* Paper 11th Commonwealth Forestry Conference, Trinidad.

Spears, J.S. (1982). Preserving watershed environments. *Unasylva*, **137**, 10–14.

Spears, J.S. (1984). Role of forestation as a sustainable land use and strategy option for tropical forest management and conservation and a source of supply for developing country wood needs. In *Strategies and designs for afforestation, reforestation and tree planting* (ed. K.F. Wiersum), pp. 29–47. Pudoc, Wageningen, The Netherlands.

Spears, J.S. (1987). *Tropical deforestation: a suggested policy research and development agenda for the bank in the 1990s.* Discussion document, World Bank, May 1987. (Unpublished).

Speight, M.R. (1997). Forest pests in the tropics: current status and future threats. In *Forests and insects* (ed. A.D. Watt, N.E. Stork, and M.D. Hunter), pp. 207–227. Chapman and Hall, London.

Speight, M.R. and Wainhouse, D. (1989). *Ecology arid management of forest insects.* Oxford.

Speight, M.R. and Wylie, R. (2000). *Insect pests in tropical forestry.* CAB International, Wallingford, UK.

Spiecker, H, Mielikainen, K, Kohl, M, and Skovsgaard, J (Eds) (1996) Growth Trends in European Forests. European Forest Institute Research Report No. 5. Springer-Verlag, Berlin.

Srivastava, P.B.L. (1993). Silvicultural practices. In Acacia mangium *growing and utilization* (ed. K. Awang and D.Taylor), pp. 113–147. Winrock International and FAO, Bangkok.

Srivastava, P.B.L. and Elias, A.B.H. (1982). Phenology of shoot development of *Pinus caribaea* var. *hondurensis* under Malaysian conditions. In *Tropical forests-source of energy through optimisation and diversification* (ed. P. B. L. Srivastava, A. M. Ahmad, K. Awang, A. Muktar, R. A. Kader, F. Che' Yom, and S. S. Lee), pp. 335–348. Proceedings of an International Forestry Seminar, Selangor, Malaysia, November 1980. Universiti Pertanian, Serdang, Malaysia.

Stape, J.L., GonÁalves, J.L.M. and GonÁalves, A.N. (2001). Relationships between nursery practices and field performance for Eucalyptus in Brazil. *New Forests*, **22**, 19–41.

State Forestry Administration (2001). *China forestry development report 2000.* State Forestry Administration, Beijing.

Stewart, J.L. (1996). Utilization. In Gliricidia sepium: *genetic resources for farmers* (ed. J.L. Stewart, G.E. Allison, and A.J. Simons), pp. 33–48. Tropical Forestry Papers 33, Oxford Foresty Institute, Oxford, UK.

Stewart, J.L., Allison, G.E., and Simons, A.J. (ed.) (1996). Gliricidia sepium: *genetic resources for farmers.* Tropical Foresty Paper 33, Oxford Forestry Institute, Oxford, UK.

Stewart, M. and Blomley, T. (1994). Use of *Melia volkensii* in a semiarid agroforestry system in Kenya. *Commonwealth Forestry Review*, **73**, 128–131.

Stewart, P.J. (1979). Islamic law as a factor in grazing management: the pilgrimage sacrifice. *Commonwealth Forestry Review*, **58**, 27–31.

Stewart, R. and Gibson, D. (1995). *Environmental and economic development consequences of forest and agricultural sector policies in Latin America: a synthesis of case studies of Costa Rica, Ecuador and Bolivia.* (ed. H. Cortes-Salas, R. de Camino, and A. Contreras). Readings of the

Workshop on Government Policy Reform for Forestry Conservation and Development in Latin America. IICA, San Jose, Costa Rica.

Sthapit, K.M. and Tennyson, L.C. (1991). Bio-engineering erosion control in Nepal. *Unasylva*, **42**(164), 16–23.

Stone, E. L. (1990). Boron deficiency and excess in forest trees: a review. *Forest Ecology and Management*, **37**, 49–75.

Stone, E.C. (1955). Poor survival and physiological condition of planting stock. *Forest Science*, **1**, 89–94.

Stone, E.L. and Kalisz, P.J. (1991). On the maximum extent of tree roots. *Forest Ecology and Management*, **46**, 59–102.

Stoney, C. and Bratamihardja, M. (1990). Identifying appropriate agroforestry technologies in Java. In *Keepers of the forest: land management alternatives in South-east Asia* (ed. M. Poffenberger), pp. 145–160. Kumarian Press, West Hartford, USA.

Strandgard, M, Wild, I., and Chong, D. (2002). PLYRS plantation management system. *New Zealand Journal of Forestry*, **46**(4), 12–15.

Streets, R.J. (1962). *Exotic forest trees in the British Commonwealth*. Clarendon Press, Oxford.

Struhsaker, T.T., Kasenene, J.M., Gaither, J.C., jun., Larsen, N., Musango, S., and Bancroft, R. (1989). Tree mortality in Kibale forest, Uganda: a case study of dieback in a tropical rain forest adjacent to exotic conifer plantations. *Forest Ecology and Management* **29**, 165–85.

Stubbings, J.A. and Schönau, A.P.G. (1979). *Management of short rotation coppice crops of* Eucalyptus grandis *Hill ex Maiden*. Paper to Technical Consultation on Fast-growing Broadleaved Trees for Mediterranean and Temperate Zones. FAO, Lisbon.

Stubbs, B.J. and Saenger, P. (2002). The application of forestry principles to the design, execution and evaluation of mangrove restoration projects. *Bois et Forêts des Tropiques*, **273**(3), 5–21.

Stubsgaard, F. (1992). *Seed storage*. Lecture Note C-9. Danida Forest Seed Centre, Humlebaek, Denmark.

Stubsgaard, F. (1993). *Seed handling prior to processing*. Lecture Note C-6 (Revised). Danida Forest Seed Centre, Humlebaek, Denmark.

Stubsgaard, F. and Baadsgaard, J. (1989). *Planning seed collections*. Lecture Note C-3. Danida Forest Seed Centre, Humlebaek, Denmark.

Stubsgaard, F. and Moestrup, S. (1991). *Seed processing*. Lecture Note C-7. Danida Forest Seed Centre, Humlebaek, Denmark.

Stubsgaard, F. and Poulsen, K.M. (1995). *Seed moisture and drying principles*. Lecture Note C-5. Danida Forest Seed Centre, Humlebaek, Denmark.

Studley, J. (1999). Forests and environmental degradation in SW China. *International Forestry Review*, **1**, 260–265.

Subedi, B.P., Das, C.L., and Messerschmidt, D.A. (1993). *Tree and land tenure in the Eastern Terai, Nepal*. Community Forestry Case Study Series 9. Food and Agriculture Organization of the United Nations, Rome.

Subham, F. (1990). Financial analysis of selected shelterbelts systems in Pakistan. *Pakistan Journal of Forestry*, **40**, 247–252.

Suhardi (2000). Treatment to develop mycorrhiza formation on dipterocarp seedlings. In *Rainforest ecosystems of East Kalimantan* (ed. E. Guhardja *et al.*), pp. 245–250. Ecological Studies 140, Springer-Verlag, Tokyo.

Sulaiman, R. (1987). Survival rates of direct seeding and containerised planting of *Acacia mangium*. In *Australian acacias in developing countries* (ed. J.W. Turnbull), pp. 173–175. ACIAR Proceedings No. 16. Australian Centre for International Agricultural Research, Canberra.

Sun, D. and Dickinson, G.R. (1994). A case study on shelterbelt effect on potato (*Solanum tuberosum*) yield on the Atherton Tableland in tropical north Australia. *Agroforestry Systems*, **25**, 141–151.

Sun, D. and Dickinson, G.R. (1996). Suitability of *Eucalyptus grandis* and *E. microcorys* as windbreak species in tropical northern Australia. *Journal of Tropical Forest Science*, **8**, 532–541.

Sundberg, U. (1978). *Heat stress: a scourge impeding development*. FEP/12–5. Proceedings of the 8th World Forestry Congress, Jakarta.

Surendran, C., Ravichandran, V.K., and Parthiban, K.T. (1996). Macro- and micropropagation of *Casuarina junghuhniana*. In *Recent casuarina research and development* (ed. K. Pinyopusarerk, J.W. Turnbull, and S.J. Midgley), pp. 109–112. CSIRO Forestry and Forest Products, Canberra.

Sutherland, J.R. and Glover, S.G. (ed.) (1991). *Diseases and insects in forest nurseries*. Proceedings of the First Meeting of IUFRO Working Party S2.07-09, Victoria, British Columbia, Canada, 22–30 August 1990. Information Report BC-X-331. Pacific Forestry Centre, Pacific and Yukon Region, Forestry Canada, Victoria, British Columbia.

Sutisna, M. (2001). Taungya experiment for rehabilitation of burnt-over forest in East Kalimantan. In *Rehabilitation of degraded forest ecosystems in the tropics* (ed. S. Kobayashi, J.W. Turnbull, T. Toma, T. Mori, and N.M.N.A. Majid), pp. 115–121. Center for International Forestry Research, Bogor, Indonesia.

Swart, H.J. (1988). Australian leaf-inhabiting fungi. XXVI. Some noteworthy coelomycetes on *Eucalyptus*. *Transactions of the British Mycological Society*, **90**, 279–291.

Tagurdar, E.T. (1978). *Forest regeneration practices in the Paper Industries Corporation of the Philippines*. Forestry Handbook No. 3, Paper Industries Corporation of the Philippines, Manila.

Tampubolon, A.P. and Hamzah, Z. (1988). *Effect of water conservation measures on the growth of teak* (Tectoria grandis) *seedlings in a low rainfall zone*. Buletin Penelitian Hutan, Pusat Penelitiandan Pengem-bangan Hutan, Indonesia. No. 496, pp. 1–15.

Tassin, J. (1995). La protection des basins versants a Madagascar. *Bois et Forêts des Tropiques*, **246**(4), 7–22.

Tassin, J. and Hermet, M. (1994). Les dégats du cyclone Hollanda à la Réunion. *Bois et Forêts des Tropiques*, **240**(2), 29–36.

Taylor, S. and Perrin, M.R. (1999). The efficacy of stem guards in protecting *Pinus patula* seedlings in KwaZulu-Natal Midlands. *South African Forestry Journal*, **185**, 35–37.

Tejwani, K.G. (1977). Trees reduce floods. *Indian Farmer*, **26**, 57.

Tekatay, D. (1996). Germination ecology of twelve indigenous and eight exotic multipurpose species from Ethiopia. *Forest Ecology and Management*, **80**, 209–223.

Terry, P.J., Adjers, G., Akobundu, I.O., Anoka, A.U., Drilling, M.E., Tjitrosemito, S., and Utomo, M. (1997). Herbicides and mechanical control of *Imperata cylindrica* as a first step in grassland rehabilitation. *Agroforestry Systems*, **36**, 151–179.

Tewari, D.N. (1992). *A monograph on teak* (Tectona grandis Linn.F.). International Book Distributors, Dehra Dun, India.

Tewari, J.C., Harris, P.J.C., Harsh, L.N., Cadoret, K., and Pasiecznik, N.M. (2000). *Managing* Prosopis julifera (*vilayati babul*): *a technical manual*. CAZRI, Jodhpur, India, and HDRA Publishing, Coventry, UK.

Thai See Kiam (2000). *Forest plantation development in Malaysia and the potential of rubber wood as an important source of timber in the future*. Proceedings of the International Conference on Timber Plantation Development, Manila, 7–9 November 2000, pp. 227–237. ITTO, FAO and Department of Environment and Natural Resources, Manila.

Thaiusta, B. (1990). Estimating productivity of *Casuarina equisetifolia* grown on tin-mine lands. In *Advances in casuarina research and utilization* (ed. M.H. El-Lakany, J.W. Turnbull, and J.L. Brewbaker), pp. 94–101. Desert Development Center, American University in Cairo, Cairo.

Thiagalingam, K. (1983). Role of casuarinas in agroforestry. In Casuarina *ecology, management and utilization* (ed. S.J. Midgley, J.W. Turnbull, and R.D. Johnston), pp. 175–179. CSIRO, Melbourne, Australia.

Thomson, L.A.J. (1994). Acacia aulacocarpa, A. cincinnata, A. crassicarpa *and* A. wetarensis: *an annotated bibliography*. CSIRO Division of Forestry, Canberra.

Thomson, L.A.J. (2001). Management of natural forests for conservation of forest genetic resources. In *Forest genetic resources conservation and management: in managed natural forests and protected areas* (in situ), Vol 2, pp. 13–44. International Plant Genetic Resources Institute, Rome.

Thomson, L.A.J., Turnbull, J.W., and Maslin, B.R. (1994). The utilisation of Australian species of *Acacia*, with particular reference to those of the subtropical dry zone. *Journal of Arid Environment*, **27**, 279–295.

Thorne, J. (1989). Solutions to soil erosion. *New Scientist*, **3 June**, 45–49.

Thornthwaite, C.W. (1948). An approach towards a rational classification of climate. *Geographical Review*, **38**, 55–94.

Thornthwaite, C.W. and Hare, H.F. (1955). Climatic classification in forestry. *Unasylva*, **9**, 50–59.

Thornthwaite, C.W. and Mather, J.R. (1955). *The water balance*. Publications in Climatology 8(1). Centerton, New Jersey, USA.

Tiarks, A. and Nambiar, E.K.S. (2000). Summary of the third workshop in Kerala, India. In *Site management and productivity in tropical plantation forests: a progress report* (ed. E.K.S. Nambiar, A. Tiarks, C. Cossalter, and J. Ranger), pp. 105–108. Center for International Forestry Research, Bogor, Indonesia.

Tiarks, A.E., Nambiar, E.K.S., and Cossalter, C. (1998). *Site management and productivity in tropical forest plantations*. CIFOR Occasional Paper No. 17. Center for International Forestry Research, Bogor, Indonesia. 11 pp.

Timyan, J. (1996). *Bwa Yo: important trees of Haiti*. South-East Consortium for International Development, Washington DC.

Tomaneng, A.A. (1990). *Calliandra callothryus*: observations on coppicing characteristics. *Agroforestry Today*, **2**(2), 14.

Tomar, O.S. (1997). Technologies of afforestation of salt-affected soils. *International Tree Crops Journal*, **9**, 131–158.

Tompsett, P.B. (1992). A review of literature on storage of dipterocarp seeds. *Seed Science and Technology*, **20**, 251–267.

Tompsett, P.B. (1994). Capture of genetic resources by collection and storage of seed: a physiological approach. In *Tropical trees: the potential for domestication and rebuilding of forest resources*. ITE symposium No. 29 (ed. R.R.B. Leakey and A.C. Newton), pp. 61–71. Her Majesty's Stationery Office, London.

Torres, F. (1983). Role of woody perennials in animal agroforestry. *Agroforestry Systems*, **1**, 131–163.

Tran, C. and Wild, C.H. (1998). Green firebreaks-fire retardant plants. In *Forest Fire Research*, Vol. 2. Proceeedings of the III International Conference on Forest Fire Research and the 14th Conference on Fire and Forest Meteorology (ed. D.X. Viegas), pp. 2631–2639. Assoçiacao para o Dessenvolvimento da Aerodinamica Industrial, Coimbra, Portugal.

Trenkel, M.E. (1997). *Improving fertilizer use efficiency: controlled-release and stabilized fertilizers in agriculture*. International Fertilizer Industry Association, Paris.

Trexler, M.C. and Associates (1993). *Mitigating global warming through forestry: a partial literature review*. Report to GTZ for Enquette Commission, German Bundestag, Bonn.

Trexler, M.C., Faeth, P.E., and Kramer, J.M. (1989). *Forestry as a response to global warming. An analysis of the Guatemala agroforestry and carbon sequestration project*. World Resources Institute, Washington.

Troup, R.S. (1952). *Silvicultural systems*, 2nd edn (ed. E.W. Jones), Clarendon Press, Oxford.

Tsewana, A., Banasiak, M., Watt, M.P., and Blakeway, F. (2000). Developments in the propagation of mature *Eucalyptus grandis* via somatic embryogenesis. In *Forest Genetics for the Next Millenium*, p. 256. Institute for Commercial Forest Research, Scottsville, South Africa.

Tucker, C.J., Newcomb, W.W., and Dregne, H.E. (1994). Desertification on the south side of the Sahara: did the desert expand from 1980 to 1993. In *Resource and Environmental Monitoring*, Vol. **30**(7a), pp. 164–167.

Proceedings of the ISPRS Commission VII Symposium, Rio de Janeiro, Brazil, 26–30 September 1994. National Institute for Space Research, São José dos Campos, São Paulo, Brazil.

Tucker, N.I.J. and Murphy, T.M. (1997). The effects of ecological rehabilitation on vegetation recruitment: some observations from the wet tropics of North Queensland. *Forest Ecology and Management*, **99**, 133–152.

Tuomela, K., Otsamo, A., Kuusipalo, J., Vuokko, R., and Nikles, G. (1996). Effect of provenance variation and singling and pruning on early growth of *Acacia mangium* Willd. plantation on *Imperata cylindrica* (L.) Beauv. dominated grassland. *Forest Ecology and Management*, **84**, 241–249.

Turabu, A.H. (1977). Protection of plantations against animals and man. In *Savanna afforestation in Africa*, pp. 209–213. FAO Forest Paper 11. Food and Agriculture Organization of the United Nations, Rome.

Turnbull, J. W. (1983). The use of *Casuarina equisetifolia* for protection forests in China. In *Casuarina ecology, management and utilization* (ed. S.J. Midgley, J.W. Turnbull, and R.D. Johnston), pp. 55–57. CSIRO, Melbourne, Australia.

Turnbull, J. W. (1997). Australian vegetation. In *Australian trees and shrubs: species for land rehabilitation and farm planting in the tropics* (ed. J.C. Doran and J.W. Turnbull), pp. 19–37. ACIAR Monograph No. 24. Australian Centre for International Agricultural Research, Canberra.

Turnbull, J.W. (1981). Eucalypts in China. *Australian Forestry*, **44**, 222–234.

Turnbull, J.W. (1984). Tree seed supply—a critical factor for the success of agroforestry projects. In *Multipurpose tree germplasm* (ed. J. Burley and P. von Carlowitz), pp. 289–298. ICRAF, Nairobi.

Turnbull, J.W. (1996). *Influence of collection activities on forest tree seed quality*. (ed. A.C.Yapa), pp. 29–35. Proceedings: International Symposium on Recent Advances in Tropical Forest Tree Seed Technology and Planting Stock Production ASEAN Forest Tree Seed Centre Project, Mauk-Lek, Saraburi, Thailand.

Turnbull, J.W. (2000) Economic and social importance of eucalypts. In *Diseases and pathogens of eucalypts* (ed. P.J. Keane, G.A. Kile, F.D. Podger, and B.N. Brown), pp. 1–9. CSIRO, Melbourne, Australia.

Turnbull, J.W. (2002). Tree domestication and the history of plantations. In *The role of food, agriculture, forestry and fisheries and the use of natural resources* (ed. V.R. Squires), Encyclopedia of Life Support Systems developed under auspices of UNESCO. Eolss Publishers, Oxford, UK. http://www.eolss.net.

Turnbull, J.W. and Doran, J.C. (1987). Seed development and germination in the Myrtaceae. In *Germination of Australian native plant seed* (ed. P. Langkamp), pp. 46–57 and 186–197. Inkata Press, Melbourne and Sydney, Australia.

Turnbull, J.W. and Eldridge, K.G. (1983). *The natural environment of* Eucalyptus *as the basis for selecting frost resistant species*. Colloque International sur les *Eucalyptus* resistants au froid, pp. 43–62. Association Forêt-Cellulose (AFOCEL), Nangis, France.

Turnbull, J.W. and Griffin, A.R. (1986). The concept of provenance and its relationship to intraspecific classification in forest trees. In *Infraspecific classification in wild and cultivated plants* (ed. B.T. Styles), pp. 157–189. Clarendon Press, Oxford, UK.

Turnbull, J.W. and Vanclay, J.K. (1999). Codes of forest practice and related research needs. In *Sustainable Forest Management*, pp. 21–28. Proceedings of the Hermon Slade International Workshop, Melbourne, 30 November–4 December 1998. The Crawford Fund, Parkville, Melbourne, Australia.

Turnbull, J.W., Crompton, H.R., and Pinyopusarerk, K. (ed.) (1998a). *Recent developments in acacia planting*. ACIAR Proceedings No. 82. Australian Centre for International Agricultural Research, Canberra.

Turnbull, J.W., Midgley, S.J., and Cossalter, C. (1998b). Tropical acacias planted in Asia: an overview. In *Recent developments in acacia planting* (ed. J.W Turnbull, H.R. Crompton, and K. Pinyopusarerk), pp. 14–28. ACIAR Proceedings No. 82. Australian Centre for International Agricultural Research, Canberra.

Turner, J., Lambert, M.J., and Kelly, J. (1989). Nutrient cycling in a New South Wales subtropical rainforest: organic matter and phosphorus. *Annals of Botany*, **63**, 635–642.

Turvey, N. (1995). *Afforestation of Imperata grasslands in Indonesia. Results of industrial tree plantation research trials at Teluk Sirih on Palau Laut, Kalimantan Selatan*. ACIAR Technical Report No. 33, Australian Centre for International Agricultural Research, Canberra.

Turvey, N.D. (1996). Growth at age 30 months of *Acacia* and *Eucalyptus* species planted in *Imperata* grasslands in Kalimantan Selatan. *Forest Ecology and Management*, **82**, 185–195.

Ugalde, L.A. (1980). Yield and utilization of two selective thinning intensities of *Eucalyptus deglupta* Bl., in Turrialba, Costa Rica. In *Wood production in the neotropics via plantations* (ed. J.L. Whitmore), pp. 104–126. Proceeding of the IUFRO/MAB/Forest Service Symposium, Rio Pedras, Puerto Rico, September 1980.

Ujah, J.E. and Adeoye, K.B. (1984). Effects of shelterbelts in the Sudan savanna zone of Nigeria on microclimate and yield of millet. *Agricultural and Forest Meteorology*, **33**, 99–107.

Underwood, M. (1993). Agroforestry defined—its applicability to the forest industry. *South African Forestry Journal*, **167**, 73–79.

Uys, H.J.E. (1990). Determining optimal financial rotations under inflationary conditions. *South African Forestry Journal*, **154**, 18–23.

Valder, P. (1999). *The garden plants of China*. Florigeum, Sydney, Australia.

Valeri, S.V. and Corrandini, L. (2000). Nursery fertilization for the production of *Eucalyptus* and pine seedlings. In *Nutrição e fertilização florestal* (ed. J.L.M. Gonçalves and V. Benedetti), pp. 167–190. Instituto de Pesquisas e Estudos Florestais, Piracicaba (SP), Brazil. [In Portuguese].

Valli, I. (1996). *Production of high quality seedlings in central nurseries in Indonesia*. Proceedings: International

Symposium on Recent Advances in Tropical Forest Tree Seed Technology and Planting Stock Production (ed. A.C.Yapa), pp. 130–135. ASEAN Forest Tree Seed Centre Project, Mauk-Lek, Saraburi, Thailand.

Van den Berg, M.A. (1975). Bio-ecological studies on forests tests. 3. *Nudaurelia cytherea* Clerkii geertsema (Lepidoptera, Saturniidae). *Forestry in South Africa*, **17**, 1–15.

Van der Zel D.W. (1995). Accomplishments and dynamics of the South African afforestation permit systems. *South African Forestry Journal*, **172**, 49–57.

Van Laar, A. (1976). Thinning research in South Africa. In Aspects of thinning (ed. G. J. Hamilton), pp. 62–71. Forestry Commission Bulletin No. 55. HMSO, London.

Van Noordwijk, M. (1996). Mulch and shade model for optimum alley-cropping design depending on soil fertility. In *Tree-crop interactions: a physiological approach* (ed. C.K. Ong and P. Huxley), pp. 51–72. CAB International, Wallingford, UK.

Van Noordwijk, M., Lawson, G., Soumare, A., Groot, J.J.R., and Hairiah, K. (1996). Root distribution of trees and crops: competition and/or complementarity. In *Tree-crop interactions. A physiological approach* (ed. C.K. Ong and P.Huxley), pp. 319–364. CAB International, Wallingford, UK and ICRAF, Nairobi.

Van Rensburg, N. J. (1984). Forest insect pest management with reference to insect populations in fast-growing plantations. In *Symposium on site and productivity of fast growing plantations* (ed. D.C. Grey, A.P.G. Schönau, C.J. Schutz, and A. van Laar), pp. 375–385. IUFRO, Pretoria and Pietermaritzburg, South Africa.

Van Wambeke, A. (1992). *Soils of the tropics: properties and appraisal.* McGraw Hill Inc., New York, USA.

Van Wyk, G. (1983). Breeding result from two young *Eucalyptus grandis* progeny tests. *Silvicultura*, **31**, 572–575.

Van Wyk, G., Pierce, B.T., and Verryn, S.D. (1991). Two year results from a site by clone interaction trial series of *Eucalyptus grandis*. In *Intensive forestry: the role of eucalypts*, Vol. 1 (ed. A.P.G. Schönau), pp. 334–344. South African Institute of Forestry, Pretoria, South Africa.

Van Zuidam, R.A. (1977). Terrain classification for the Colombian Amazon region using SLAR imagery: a geomorphological approach. *ITC Enschede*, **4**, 705–706.

Vanclay, J. K. (1994). *Modelling forest growth and yield applications to mixed tropical forests.* Centre for Agriculture and Biosciences International, Wallingford, UK.

Vandenbeldt, R.J. (1992). Agroforestry in the semi-arid tropics. *Unasylva*, **43**(168), 41–47.

Varghese, M., Nicodemus, A., Nagarajan, B., Sasidharan, K.R., Siddappa, Bennet, S.S.R., and Subramanian, K. (2000). *Seedling seed orchards for breeding tropical trees.* Institute of Forest Genetics and Tree Breeding, Indian Council of Forestry Research and Education, Coimbatore, India.

Venator, C.R., Munoz, J.C., and De Barros, N.F. (1977). Root immersion in water; a promising method for successful bare-root planting of Honduras pine. *Turrialba*, **27**, 287–291.

Venter, A. (1972). The effect of stump size on vigour of coppice growth in *Eucalyptus grandis*. Forestry in South Africa, **13**, 51–52.

Vercoe, T.K. (1989). Fodder value of selected Australian tree and shrub species. In *Trees for the tropics: growing Australian multipurpose trees and shrubs in developing countries* (ed. Boland, D.J.), pp. 187–192. Australian Centre for International Agricultural Research, Canberra.

Vertessy, R.A., Zhang, L., and Dawes, W.R. (2003). Plantations, river flows and river salinity. *Australian Forestry*, **66**, 55–61.

Victor, M.A.M. (1977). O reflorestamento incentivado 10 anos depois. *Silvicultura*, **1**, 17–46.

Vigneron, P., Bouvet, J-M., Gouma, R., Saya, A., Gion, J-M., and Verhaegen, D. (2000). *Eucalypt hybrid breeding in Congo* (ed. H.S. Dungey, M.J. Dieters, and D.G. Nikles), pp. 14–26. Symposium on Hybrid Breeding and Genetics, Noosa, Australia, 9–14 April 2000. Department of Primary Industries, Brisbane, Australia. (Compact Disk).

Vincent, A. J. (1971). *An experiment to determine the effect on tree growth from grazing under* Pinus caribaea *var. hondurensis plantations.* Experimental Report No. 2. Silviculture. Department of Forestry, Suva, Fiji.

Vincent, A.J. (1972). *Development of a thinning and pruning schedule for both regular planted and line-planted mahogany* (Swietenia macrophylla) *in private woodlands.* Experimental Report No. 7. Silvicultural Research Division, Department of Forestry, Suva, Fiji.

Vincent, L. W. (1986). Site classification and prediction in young Caribbean pine plantations in grasslands of Venezuela. In *Forest site and productivity* (ed. S.P. Gessel), pp. 51–67. Martinus Nijhoff, The Netherlands.

Vise, S. (1990). *Land use for wood production in Fiji.* Department of Forestry, University of Melbourne, Melbourne, Australia.

Vogt, K., Asbjornsen, H., Ercelawn, A., Montagnini, F., and Valdes, M. (1997). Roots and mycorrhizas in plantation ecosystems. In *Management of soil, nutrients and water in tropical plantation forests* (ed. E.K.S. Nambiar and A.G. Brown), pp. 247–296. ACIAR Monograph No. 43. Australian Centre for International Agricultural Research, Canberra; CSIRO Australia and Center for International Forestry Research, Bogor, Indonesia.

Von Carlowitz, P.G. (1986). *Multipurpose tree and shrub seed directory.* International Council for Research in Agroforestry, Nairobi.

Von Carlowitz, P.G. (1990). *Techniques for tree establishment in dry zones—experiments and preliminary results*, Vol. 3, Proceedings of the 19th IUFRO World Congress, Montreal, Canada, p. 117.

Von Carlowitz, P.G., Wolf, G.V., and Kemperman, R.E.M. (1991). *Multipurpose tree and shrub database. Version 1.0.* International Centre for Research in Agroforestry, Nairobi.

Von Gadow, K. and Bredenkamp, B. (1992). *Forest Management.* Academica, Pretoria, South Africa.

Vries, P.G. de, Hildebrand, J.W., and Graaf, N.R.de (1978). *Analysis of 11 years' growth of Caribbean pine in a replicated*

Graeco-Latin square spacing-thinning experiment in Surinam. Mededelingen Landbouwhogeschool, 78–17, Wageningen, The Netherlands.

Vuokko, R. and Otsamo, A. (1996). Species and provenance selection for plantation forestry on grasslands. In *Reforestation: meeting the future industrial wood demand* (ed. A. Otsamo, J. Kuusialpo, and H. Jaskari), pp. 68–79. Enso Forest Development Oy Ltd, Jakarta.

Wade, D.D. and Lundsford, J. (1990). Fire as a forest management tool: prescribed burning in the southern United States. *Unasylva*, 41(162), 28–38.

Wadsworth, F.H. (1997). *Forest production for tropical America.* Agriculture Handbook 710, Forest Service, US Department of Agriculture, Washington DC.

Walker, J. (1987). Development of contingency plans for use against exotic pests and diseases of trees and timber. 1 Problems with the detection and identification of exotic plant pathogens of forest trees. *Australian Forestry*, 50, 5–15.

Walker, J. and Reuter, D.J. (ed.) (1996). *Indicators of catchment health: a technical perspective.* CSIRO, Melbourne, Australia.

Walker, S. and Haines, R. (1998). Evaluation of clonal strategies for tropical acacias. In *Recent developments in acacia planting* (ed. J.W Turnbull, H.R. Crompton, and K. Pinyopusarerk), pp. 197–202. ACIAR Proceedings No.82. Australian Centre for International Agricultural Research, Canberra.

Walker, S., Haines, R., and Dieters, M. (1996). Beyond 2000: clonal forestry in Queensland. In *Tree improvement for sustainable tropical forestry* (ed. M.J. Dieters, A.C. Matheson, D.G. Nikles, C.E. Harwood, and S.M. Walker), pp. 351–354. Proceedings of QFRI-IUFRO Conference, Caloundra, Australia, October/November 1996, Queensland Forestry Research Institute, Gympie, Australia.

Wallace, J.S. (1996). The water balance of mixed tree-crop systems. In *Tree-crop interactions: a physiological approach* (ed. C.K. Ong and P. Huxley), pp. 189–233. CAB International, Wallingford, UK.

Walters, B.B. (1997). Human ecological questions for tropical restoration: experiences from planting native upland trees and mangroves in the Philippines. *Forest Ecology and Management*, 99, 275–290.

Walters, G.A. (1980). Eucalyptus saligna *growth in a 15-year-old spacing study in Hawaii.* Research Paper PSW-151. Pacific SW Forest and Range Experimental Station. USDA Forest Service, Hawaii, USA, 6 pp.

Wang, B.S.P. (1977). Procurement, handling and storage of tree seed for genetic research. FAO/IUFRO Third World Consultation on Forest Tree Breeding, Canberra.

Wang, D. (1988). The history of the ornamental plants of China. *Camellia News*, 107, 14–16.

Wang, H. and Zhou, W. (1996). Fertilizer and eucalypt plantations in China. In *Nutrition of eucalypts* (ed. P.M. Attiwill and M.A. Adams), pp. 389–397. CSIRO Publishing, Melbourne, Australia.

Wang, H., Jiang, Z., and Yan, H. (1994). Australian trees grown in China. In *Australian tree species research in China* (ed. A.G. Brown), pp. 19–25. ACIAR Proceedings No. 48. Australian Centre for International Agricultural Research, Canberra.

Wang, S. (2001). Towards an international convention on forests: building blocks versus stumbling blocks. *International Forestry Review*, 3, 251–264.

Wang, S., Zheng, R., and Yang, Y. (2000). Combating desertification: the Asian experience. *International Forestry Review*, 2, 112–117.

Wardell, D.A. (1987). Control of termites in nurseries and young plantations in Africa: established practices and alternative courses of action. *Commonwealth Forestry Review*, 66, 77–89.

Wardell, D.A. (1990). The African termite: peaceful coexistence or total war? *Agroforestry Today*, 2(3), 4–6.

Warner, K. (1993). *Patterns of farmer tree growing in Eastern Africa: a socio-economic analysis.* OFI Tropical Forestry Papers No. 27. Oxford Forestry Institute, Oxford and International Centre for Research in Agroforestry, Nairobi.

Warner, K. (1997). Patterns of tree growing in eastern Africa. In *Farms, trees and farmers. Responses to agricultural intensification* (ed. J.E.M. Arnold and P.A. Dewees), pp. 90–137. Earthscan Publications, London.

Watson R.T., Noble I.R., Bolin R., Ravindranath N.H., Ravado D.J., and Dokken D.J. (ed.) (2000). *Land use, land-use change and forestry.* Intergovernmental Panel on Climate Change Special Report. Cambridge University Press, Cambridge, UK.

Watson, V., Cervantes, S., Castro, C., Mora, L., Solis, M., Porras, I.T., and Cornejo, B. (2000). Making space for better forestry. Costa Rica country study *Policy that works for forests and people Series No. 6.* Centro Cientifico Tropical, San José, and International Institute for Environment and Development, London.

Watt, G.R. (1973). *The planning and evaluation of forestry projects.* Paper 45. Commonwealth Forestry Institute, Oxford, UK.

Watt, M.P., Blakeway, F.C., Mokotedi, M.E.O., and Jain, S.M. (2003). Micropropagation of *Eucalyptus.* In *Micropropagation of woody trees and fruits* (ed. S. Mohan Jain and K. Ishii). Kluwer Academic Publishers, Dordrecht, The Netherlands.

Watt, M.P., Duncan, E.A., Ing, M., Blakeway, F.C., and Herman, B. (1995). Field performance of micropropagated and macropropagated *Eucalyptus* hybrids. *South African Forestry Journal*, 173, 17–21.

WCFSD (1997). *The causes of forest decline.* Background Paper 8, World Commission on Forests and Sustainable Development, Geneva, Switzerland.

Weaver, P.L. (1986). Enrichment planting in tropical America. In *Management of the forests of tropical America: prospects and technologies* (ed. J.C.F. Colon, F.H. Wadsworth, and S. Branham), pp. 259–278. Conference at Institute of Tropical Forestry, Rio Pedras, Puerto Rico, September 1986.

Webb, D.B., Wood, P.J., and Smith, J.P. (1980). *A guide to species selection in tropical and sub-tropical plantations.* Tropical Forestry Paper No. 15. Commonwealth Forestry Institute, Oxford, UK.

Webb, D.B., Wood, P.J., Smith, J.P., and Henman, G.S. (1984). *A guide to species selection in tropical and sub-tropical plantations*. Tropical Forestry Paper No. 15, 2nd edn, revised. Commonwealth Forestry Institute, Oxford, UK.

Webb, M.J., Poa, D., Hambleton, A., and Reddell, P.W. (1995). *Identifying and solving nutritional problems in establishing plantation of high-value tropical cabinet timbers: a case study from Kolombangara in the Solomon Islands*. (ed. A. Schulte and D. Ruhiyat), pp. 148–165. Proceedings of the International Congress on Soils of Tropical Forest Ecosystems 3rd Conference on Forest Soils, Vol. 4, Soils and plantation forestry. Mulawarman University Press, Samarinda, Indonesia.

Webb, M.J., Reddell, P., and Grundon, N.J. (2001). *A visual guide to nutritional disorders of tropical timber species*: Swietenia macrophylla and Cedrela odorata. ACIAR Monograph No. 61. Australian Centre for International Agricultural Research, Canberra.

Weed Science Society of America (1994). *Herbicide handbook*, 7th edn. Champaign, Illinois, USA.

Weetman, G.F. and Wells, C.G. (1990). Plant analyses as an aid in fertilizing forests. In *Soil testing and plant analysis*, 3rd edn. (ed. R.L. Westerman), pp. 659–690. Soil Science Society of America, Madison.

Weinland, G. (1998). Plantations. In *A review of diptercarps: taxonomy, ecology and silviculture* (ed. S. Appanah and J.M. Turnbull), pp. 151–185. Center for International Forestry Research, Bogor, Indonesia.

Welker, J. C. (1986). Site preparation and regeneration in the lowland humid tropics: plantation experience in northern Brazil. In *Management of the forests of tropical America: prospects and technologies* (ed. J.C.F. Colon, F.H. Wadsworth, and S. Branham), pp. 297–333. Conference at Institute of Tropical Forestry, Rio Pedras, Puerto Rico, September 1986.

Westley, S.B. and Powell, M.H. (ed.) (1993). Erythrina *in the old and new worlds*. Nitrogen Fixing Tree Research Reports. Special issue. Nitrogen Fixing Tree Association, Paia, Hawaii, USA.

Westoby, J. (1989). *Introduction to world forestry: people and their trees*. Basil Blackwell, Oxford.

Westoby, J.C. (1978). *Forest industries for socio-economic development*, Vol. 5. Proceedings of the 8th World Forestry Congress, Jakarta, pp. 19–32.

Westoby, J.C. (1985). *Foresters and politics*. Paper III.1.6. Ninth World Forestry Congress, Mexico.

White, K and Maclean, A. (2001). People, participation and public native forests. In *Forests in a changing landscape*. Proceedings of 16th Commonwealth Forestry Conference/19th Biennial Conference of the Institute of Foresters of Australia, pp. 153–163. Promaco Conventions, Canning Bridge, Western Australia.

White, K.J. (1990). Dalbergia sissoo: *an annotated bibliography*. Winrock International- F/FRED, Bangkok.

White, K.J. (1993). Small scale vegetative multiplication of *Eucalyptus* and its use in clonal plantations. In *Recent advances in mass clonal multiplication of forest trees for plantation programmes* (ed. J. Davidson), pp. 60–75. Proceedings of a Regional Symposium, Cisarua, Indonesia, 1–8 December 1992. RAS/91/004 Field

Document No. 4. Food and Agriculture Organization of the United Nations, Los Banos, Philippines.

Whiteman, A. (2002). The use of computer models for forestry investment and policy appraisal. *New Zealand Journal of Forestry*, **46**(4), 4–7.

Whitmore, J.L. (1972). *Pinus merkusii* unsuitable for plantations in Puerto Rico. *Turrialba*, **22**, 351–353.

Whitmore, T.C. (1990). An introduction to tropical rain forests. Clarendon Press, Oxford.

Whyte, A.G.D. (1973). Productivity of first and second crops of *Pinus radiata* on the Moutere gravel soils of Nelson. *New Zealand Journal of Forestry*, **18**, 87–103.

Wibowo, A., Suharti, M., Sagala, A.P.S., Hibani, H., and van Noordwijk. M. (1997). Fire management on *Imperata* grasslands as part of agroforestry development in Indonesia. In *Agroforestry innovations for Imperata grassland rehabilitation* (ed. D.P. Garrity). *Agroforestry Systems*, **36**, 151–179.

Wiersum (1985). Effects of various vegetation layers in an *Acacia auriculiformis* forest plantation on surface erosion in Java, Indonesia. In *Soil erosion and conservation* (ed. S.A. El-Swaify, W.C. Moldenhauer, and A. Lo), pp. 79–89. Soil Conservation Society of America, Ankeny, Iowa, USA.

Wiersum, K.F. (1983). *Effects of various vegetation layers of an* Acacia auriculiformis *forest plantation on surface erosion at Java, Indonesia*. Paper presented at Malama Aina, International Conference of Soil Erosion and Conservation, Hawaii, January 1983.

Wightman, K.E. (1999). *Good tree nursery practices: practical guidelines for community nurseries*. International Centre for Agroforestry Research. Nairobi. http://www.icraf.cgiar.org/res-dev/prog-2/Manuals/Community.htm (15 January 2001)

Wilken, G.C. (1978). Integrating forest and small-scale farm systems in middle America. *Forest Ecology and Management*, **1**, 223–234.

Willan, R.L. (1985). *A guide to forest seed handling, with special reference to the tropics*, 379 pp. FAO Forestry Paper 20/2, FAO/DANIDA. Food and Agriculture Organization of the United Nations, Rome.

William, G. (1990). *Forestry for rural development: the issue of incentives in Uganda*, Vol. 4. Proceedings of the 19th IUFRO World Congress, Montreal, Canada, pp. 180–187.

Williams, J. (2000). *Financial and other incentives for plantation establishment*. Proceedings of the International Conference on Timber Plantation Development, 7–9 November 2000, Manila, pp. 87–101. ITTO, FAO and Department of Environment and Natural Resources, Manila.

Williamson, M. (1993). *Forest management in industrial forest plantations in the Philippines*. Forestry Technical Services (FORTECH), Canberra.

Wilson, P.H. (1986). *Nursery job instruction sheets*, Forestry Division, Solomon Islands. (Mimeo.).

Wilson, P.J. (1998). Developing clones from *Eucalyptus globulus* and hybrid seedlings by stem cuttings propagation. *New Zealand Journal of Forestry Science*, **28**, 293–303.

Wimbush, S.A. (1963). Afforestation of restored tin mining land in Nigeria. *Commonwealth Forestry Review*, **42**, 255–262.

Wingfield, M.J. (1984). Diseases and their management in fast-growing plantations. In *Symposium on site and productivity of fast growing plantations* (ed. D.C. Grey, A.P.G. Schönau, C.J. Schutz, and A. van Laar), pp. 345–358. IUFRO, Pretoria and Pietermaritzburg, South Africa.

Wingfield, M.J., Crous, P.W., and Peredo, H.L. (1995). A preliminary list of foliar pathogens of *Eucalyptus* spp. in Chile. *South African Forestry Journal*, **173**, 53–57.

Wingfield, M.J., Roux, J., Coutinho, T., Govender, P., and Wingfield, B.D. (2001). Plantation disease and pest management in the next century. *Southern African Forestry Journal*, **190**, 67–71.

Winjum, J.K., Dixon, R.C., and Schroeder, P.E. (1997). Carbon storage in forest plantations and their wood products. *Journal of Forest Resource Management*, **8**, 1–19.

Winter, R.K. (1974). *The forest and man*, 393 pp. Vantage, New York, USA.

Winterbottom, R. (1990). *Taking stock: the tropical forestry action plan after five years*. World Resources Institute, Washington DC.

Winterbottom, R.T. and Hazlewood, P.T. (1987). Agroforestry and sustainable development: making the connection. *Ambio*, **16**, 100–110.

Winters, R.K. (1977). Edited addendum No. 1. *Terminology of forest science, technology, practice and products* (ed. F. C. Ford-Robertson). IUFRO/Society of American Foresters, Washington DC.

Withington, D., MacDicken, K.G., Sastry, C.B., and Adams, N. (ed.) (1988). *Multipurpose tree species for small-farm use*. Proceedings of a Symposium, November 1987, Pattaya, Thailand, Winrock International (USA) and IDRC (Canada), Bangkok.

Woessner, R.A. (1980). *Gmelina arborea* Roxb. Genetic improvement program at Jari, Para, Brazil. Jari Florestal, Brazil. In Wadsworth, *Forest production for tropical America* (ed. F. H. Wadsworth) Agriculture Handbook 710, 1997. Forest Service, US Department of Agriculture, Washington DC.

Woessner, R.A. and McNabb, K.L. (1979). Large scale production of *Gmelina arborea* Roxb. Seed—a case study. *Commonwealth Forestry Review*, **58**, 117–121.

Wong, C.Y. and Jones, N. (1986). Improving tree form through vegetative propagation of *Gmelina arborea*. *Commnonwealth Forestry Review*, **65**, 321–324.

Wood, H. (ed.) (1993). *Teak in Asia*. FORSPA Publication No.4, Bangkok.

Woods, P.V. (2003). *Spontaneous agroforestry: regreening barren hills in Vietnam*. PhD Thesis, University of Melbourne, Australia.

Woods, R. V. (1990). Second rotation decline in *P. radiata* plantations in South Australia has been corrected. *Water, Air and Soil Pollution*, **54**, 607–619.

World Bank (1986). *World Bank financed forestry activity in the decade 1977–86. A review of key policy issues and implications of past experience to future project design.*

World Bank, Washington DC, December 1986. (Unpublished).

World Bank (1992). *Strategy for forest sector development in Asia*. World Bank Technical Paper 182, Asia Technical Department Series. The World Bank, Washington DC.

World Bank (1998). *Assessing aid: what works, what doesn't and why*. Oxford University Press, New York, USA.

World Resources Institute (1998). Aracruz Celulose S.A. and Riocell S.A.: efficiency and sustainability of Brazilian pulp plantations. In *The business of sustainable forestry: case studies*, pp. 5(1)–5(31). The John D. and Catherine T. MacArthur Foundation, Chicago, USA.

Wormald, T.J. (1975). *Pinus patula*. Tropical Forestry Paper No. 7. Commonwealth Forestry Institute, Oxford.

WRB (1998). *World reference base for soil resources*. International Society of Soil Science/ International Soil Referenceand Information Centre, Food and Agriculture Organization of the United Nations, Rome.

WRI (1972). *Handbook of eucalypt growing*. Wattle Research Institute, Pietermaritzburg, South Africa.

WRI (1982). Killing stools of Eucalyptus grandis. *Annual Report 1982*, pp. 91–92. Wattle Research Institute, Natal, South Africa.

Wright, J.A. (1995) Operational gains and constraints with clonal *Eucalyptus grandis* in Colombia. In *Eucalypt Plantations: Improving Fibre Yield and Quality* (ed. B.M. Potts, N.M.G. Borralho, J.B. Reid, R.N. Cromer, W.N. Tibbits, and C.A. Raymond), pp. 308–310. Proceeding of the CRCTHF-IUFRO Conference, Hobart, Australia, February 1995 CRC for Temperate Hardwood Forestry, Hobart, Australia.

Wright, J.A. (1997). Realized operational gains from clonal eucalypt forestry in Colombia and current methods to increase them, pp. 203–205. Tappi Biological Sciences Symposium, October 1997, San Francisco, USA. Tappi Press, Atlanta, Georgia, USA.

Wright, J.A., Osorio, L.F., and Dvorak, W.S. (1996). Realised and predicted genetic gain in the *Pinus patula* breeding program of Smurfit Carton de Colombia. *South African Forestry Journal*, **175**, 19–22.

Wright, J.W. (1976). *Introduction to forest genetics*. Academic Press, London.

Wright, S.J. and Cornejo, F.H. (1990). Seasonal drought and the timing of flowering and leaf fall in neotropical forest. In *Reproductive ecology of tropical forest plants* (ed. K.S. Bawa and M. Hadley), pp. 49–61. Man and the Biosphere Series 7. UNESCO and The Parthenon Publishing Group, Carnforth, UK.

Wunderle, J.M. Jr. (1997). The role of animal seed dispersal in accelerating native forest regeneration on degraded tropical lands. *Forest Ecology and Management*, **99**, 223–235.

WWF (2002). *Forest landscape restoration*. WWF Position Paper, January 2002. WWF International, Gland, Switzerland.

Wyatt-Smith, J. (1987). *The management of tropical moist forest for the sustained production of timber: some issues*. IUCN/IIED Tropical Forest Policy Paper No. 4. International Institute for Environment and Development, London.

Wylie, R. (2000). *Integrated pest management in tropical forestry.* Proceedings of the International Conference on Timber Plantation Development, Manila, 7–9 November 2000, pp. 159–169. ITTO, FAO and Department of Environment and Natural Resources, Manila.

Wylie, R. (2001). Plague, pestilence and plantations. *Tropical Forest Update*, **11**(3), 6–7.

Wylie, R. *et al.* (1998). Insect pests of tropical acacias: a new project in southeast Asia and northern Australia. In *Recent developments in* Acacia *planting* (ed. J.W. Turnbull, H.R. Crompton, and K. Pinyopusarerk), pp. 234–239. ACIAR Proceedings No. 82. Australian Centre for International Agricultural Research, Canberra.

Xu, C. (2000). *Timber plantation in China.* Proceedings of the International Conference on Timber Plantation Development, Manila, 7–9 November 2000, pp. 185–192. ITTO, FAO and Department of Environment and Natural Resources, Manila.

Yadav, J.P. (1992). Pretreatment of teak to enhance germination. *Indian Forester*, **118**, 260–264.

Yan, H. (2001). *Frost prediction for Australian tree species in China.* PhD Thesis, Australian National University, Canberra.

Yanchuk, A.D. (2001). The role and implications of biotechnological tools in forestry. *Unasylva*, **52**(204), 53–61.

Yang, M., Zeng, Y., Zhang, X., and Zhang, X. (1994). Effect of low temperatures on *Acacia.* In *Australian tree species research in China* (ed. A.G. Brown), pp. 176–179. ACIAR Proceedings No. 48. Australian Centre for International Agricultural Research, Canberra.

Yang, X. (2001). Impacts and effectiveness of logging bans in natural forests: People's Republic of China. In *Forests out of bounds: impacts and effectiveness of logging bans in natural forests in Asia-Pacific* (ed. P.B. Durst, T.R. Waggener, T. Enters, and L.C. Tan), pp. 81–102, RAP Publication 2001/08, Food and Agriculture Organization of the United Nations, Regional Office for Asia and the Pacific, Bangkok.

Yantasath, K. (1987). Field trials of fast-growing, nitrogen-fixing trees in Thailand. In *Australian acacias in developing countries* (ed. J.W. Turnbull), pp. 176–179. ACIAR Proceedings No. 16. Australian Centre for International Agricultural Research, Canberra.

Yelu, W. (1998). Silviculture of Acacia mangium in Papua New Guinea. In Recent developments in acacia planting. (ed. J.W. Turnbull, H.R. Crompton and K. Pinyopusarerk), pp. 326–331. ACIAR Proceedings No. 82. Australian Centre for International Agricultural Research, Canberra.

Ying, JiHua, and Ying, J. H. (1997) Comparative study on growth and soil properties under different successive rotations of Chinese fir. *Journal of Jiangsu Forestry Science and Technology*, **24**, 31–34.

Young, A. (1989). *Agroforestry for soil conservation.* Science and Practice of Agroforestry No. 4, CAB International/ICRAF, Nairobi.

Young, A. (1997). *Agroforestry for soil management*, 2nd edn. CAB International, Wallingford, UK.

Yusoff, M.K., Heng, S.S., Majid, N.M., Mokhtaruddin, A.M., Hanum, I.F., Alias, M.A., and Kobayashi, S. (2001). Effects of different land use patterns on stream quality water in Pasoh, Negri Sembilan, Malaysia. In *Rehabilitation of degraded forest ecosystems in the tropics* (ed. S. Kobayashi, J.W. Turnbull, T. Toma, T. Mori, and N.M.N.A. Majid), pp. 87–97. Center for International Forestry Research, Bogor, Indonesia.

Zakaria, I. (1993). Reproductive biology. In Acacia mangium *growing and utilization* (ed. K. Awang and D. Taylor), pp. 21–34. Winrock International and FAO, Bangkok.

Zani, J. and Kageyama, P.Y. (1984). A producao de sementes melhoradas de especies florestais, com enfase en *Eucalyptus. IPEF Piracicaba*, **27**, 49–52.

Zech, W. (1984a). Investigations on the occurrence of potassium and zinc deficiencies in plantations of *Gmelina arborea, Azadiracta indica* and *Anacardium occidentale* in semi-arid areas of West Africa. *Potash Review* Sub. **22/31**(1), 1–5.

Zech, W. (1984b). Leaf analysis—a method to detect mineral deficiencies in fast growing plantations in West Africa. In *Symposium on site and productivity of fast growing plantations* (ed. D.C. Grey, A.P.G. Schönau, C.J. Schutz, and A. van Laar), pp. 691–699. IUFRO, Pretoria and Pietermaritzburg, South Africa.

Zech, W. and Drechsel, P. (1998). Nutrient disorders and nutrient management in fast growing plantations. In *Soils of tropical forest ecosystems* (ed. A. Schulte and D. Ruhiyat), pp. 99–106. Springer, Berlin Heidelberg, Germany.

Zhang, H. (2000). Paulownia *breeding, cultivation, utilization and extension in China.* APAFRI Publication Series No. 6. Asia Pacific Association of Forestry research Institutions, UPM, Serdang, Selangor, Malaysia.

Zheng, H. (1988). The role of *Eucalyptus* plantations in southern China. In *Multipurpose tree species for small farm use* (ed. D. Withington *et al.*), pp. 79–85. Winrock International and International Development Research Centre, Ottawa, Canada.

Zhong, A.L. and Hsiung, W.Y. (1993). Evaluation and diagnosis of tree nutrition status in Chinese fir (*Cunninghamia lanceolata* (Lamb) Hook.) plantations, Jiangxi, China. *Forest Ecology and Management*, **62**, 245–270.

Zhong, M., Xie, C., Fu, M., and Xie, J. (1997). Bamboo and rattan socio-economic development in China. *Forestry Economics*, **2**(1), 36–46.

Zhou, T., Zhou, J., and Shelbourne, C.J.A. (1998). Clonal selection, propagation, and maintenance of juvenility of Chinese fir and afforestation with monoclonal blocks. *New Zealand Journal of Forestry Science*, **28**, 275–292.

Zhuang, X. (1997). Rehabilitation and development of forest on degraded hills of Hong Kong. *Forest Ecology and Management*, **99**, 197–201.

Zobel, B. (1992). Vegetative propagation in production forestry. *Journal of Forestry*, **90**(4), 29–33.

Zobel, B.J. and Talbert, J. (1984). *Applied tree improvement.* John Wiley, New York.

Zobel, B.J., Campinhos, E., and Ikemori, Y. (1983). Selecting and breeding for wood uniformity. *Tappi*, **66**(1), 70–74.

Zobel, B.J., van Wyk, G., and Stahl, P. (1987). *Growing exotic forests*. John Wiley, New York.

Zollner, D. (1986). Sand dune stabilization in central Somalia. In *Tree plantings in semi-arid regions* (ed. P. Felker), pp. 223–232. Proceedings of a Symposium on Establishment and Productivity of Tree Plantings in Semi-arid Regions, Kingsville, Texas. Elsevier, Amsterdam, The Netherlands.

Zsuffa, L., Sennerby-Forsse, L., Weisgerber, H., and Hall, R.B. (1993). Strategies for clonal forestry with poplars, aspens and willows. In *Clonal forestry II: conservation and application* (ed. W.J. Libby and M.R. Ahuja), pp. 91–119. Springer-Verlag, Berlin Heidelberg, Germany.

Zwolinksi, J.B. and Hinze, W.H.F. (2000). Pines. In *South African forestry handbook*. (ed. D.L. Owen), pp. 116–120. S. African Institute of Forestry, Pretoria, South Africa.

Zwolinski, J. and Bayley, A.D. (2001). Research on planting stock and forest regeneration in South Africa. *New Forests*, **22**, 59–74.

Index to tree species, plants, and animals

General index

integrated land-use 60–2, 102, 316
 ideal features of 61–2
integrated pest management 308, 311
intercropping 47, 220, 283, 317, 319,
 320–3
Inter-American Development Bank,
 loans for plantation
 development 72
International Centre for Research in
 Agroforestry (ICRAF) 31,
 118, 129
International Seed Testing
 Association (ISTA) 134
International Tropical Timber
 Organization (ITTO) 31, 92, 355
International Union of Forest Research
 Organizations (IUFRO) 30, 129
Inter-Tropical Convergence Zone 5
introduced species see exotic species
invasive species 43, 127–8, 290, 291–2
inventory 47–8
investment in plantation projects
 77–80
 aid 21, 25, 28, 38, 72, 77, 80, 102, 108
 development bank loans 72
 funding in rural development
  in 79
 protective afforestation 334–52
 need for long term finance 79
 tax concessions, example of
  Brazil 78
Iran 370
iron pans in soil 337, 339, 364
irrigation
 newly planted trees 192, 320
 nurseries in 153, 154, 156, 170, 171,
  172, 175
 plantations, in 26, 27, 320, 347, 359,
  368–9
Ivory Coast 184

Jamaica 60, 61, 126, 297, 309, 341
 area of plantations 34
Jari Florestal e Agropecuaria, Brazil
 19, 30, 40, 61, 70, 75, 78, 181,
 207, 214, 215, 220, 386, 391
 agroforestry 329
 forest villages 81
 ground preparation 207
 many land-uses 59
 retention of trained labour 86
 thinning practice 263, 273
Java, Indonesia 26, 169, 230, 318, 326,
 341, 390–1
 protective afforestation 107, 320,
  341, 390–1
Jos plateau, Nigeria 348, 373, 374

Kalimantan, Indonesia 84, 155, 211,
 232, 322, 360
Kandyan home garden (Sri Lanka)
 326
Karnataka project (India) 383
Kenya 10, 21, 26, 27, 37, 44, 56, 58,
 65, 76, 82, 97, 109, 112, 129,
 176, 232, 261, 262, 268, 271,
 274, 280, 282, 286, 306, 307,
 318, 319, 324, 325, 365, 368
 area of plantations in 33
 coppice working 292, 293, 294, 295
 extension programmes in 66, 70,
  152
 shamba(taungya) 27, 321
 soil erosion 341
 Turbo project 29
 water catchments 338, 341, 343
 see also Green Belt Movement
Kerala, India 26, 261, 282, 325
Khartoum, Sudan, planting around
 23, 346
kino veins (in eucalypts) 305
KwaZulu-Natal 3, 4, 141, 187, 319

labour
 nurseries 153, 157–8
 planning and organization in
 plantation development 80–2
 planting 216–17
 productivity in forest operations
  99
 pruning 235, 267, 281, 282
 recruitment of 96
 weeding 173, 228, 229
land
 availability for plantations 18–19,
  45, 51, 80
 capability 51, 52–6
 evaluation 53, 54, 56, 62–4
 Landsat, imagery used in
  evaluation 63, 64
 legal title 80, 200–1
 obtaining for plantations 65–71
 race 392
 reservation 65, 349
 suitability see land capability
 tenure 7, 16, 31, 51, 56–9, 94
 use 7–8, 51–65, 94, 319, 342, 365
 see also integrated land use
Laos 72, 80, 125
 area of plantations in 34
laterite see iron pans
laws to encourage tree planting 68
layout and subdivision of plantation
 45
leaf area index (LAI) 257

leaf-cutting ants (*Atta* sp.) 175, 291
leasing land for planting 57, 65–6
Liberia 30, 180
 area of plantations in 33
Libya 335, 370
lifting plants in the nursery 153, 154,
 157, 169, 174, 181
lightning damage 298, 301, 304, 305
lignotuber of eucalypts 127, 163,
 292, 310
limiting basal area 262–5, 273, 274, 285
line-planting see enrichment planting
litter 6, 28, 207, 212, 238, 245, 246,
 262, 288
 collection 239, 324, 340–2
 effects on soil development 237,
  238, 248, 250, 251, 288, 300,
  340–3, 345
live crown ratio 262, 272
 effects of pruning 280
local inhabitants, need to involve 83,
 350
long-term productivity 384, 389
low pruning 47, 98, 234–6
 see also brashing

Madagascar 38, 75, 236, 237, 239,
 262, 295, 369
 area of plantations in 33
Madhya Pradesh, India 220, 374
Maggia valley project (Niger) 346
Maharashtra state, India 175
maintenance of plantations 223–36
 post-planting problems 223–4
Malawi 9, 20, 23, 29, 38, 45, 55, 61,
 65, 75, 81, 109, 164, 171, 172,
 177, 206, 305, 307, 309, 328,
 329, 381
 area of plantations in 33
 forest act permitting land
  reservation 61, 65
 fuelwood planting 59
 high pruning in 280–2
 line thinning of pine 267
 see also Viphya Pulpwood Project
Malaysia 8, 14, 15, 18, 19, 20, 27, 32,
 37, 41, 69, 70, 73, 95, 96, 115,
 116, 117, 137, 147, 165, 174,
 184, 202, 212, 214, 227, 231,
 239, 252, 256, 272, 281, 308,
 333, 343, 348, 373, 379, 381, 388
 area of plantations in 34
 reforestation after logging 70
Mali 109, 168, 214, 225, 346
 area of plantation in 33
management in plantation projects
 99–101